# Electron Correlations and Materials Properties

# Electron Correlations and Materials Properties

Edited by

**A. Gonis**
*Lawrence Livermore National Laboratory*
*Livermore, California*

**N. Kioussis**
*California State University—Northridge*
*Northridge, California*

and

**M. Ciftan**
*U.S. Army Research Office*
*Research Triangle Park, North Carolina*

Kluwer Academic / Plenum Publishers
New York, Boston, Dordrecht, London, Moscow

**Library of Congress Cataloging-in-Publication Data**

Electron correlations and materials properties / edited by A. Gonis,
N. Kioussis, and M. Ciftan.
p. cm.
"Proceedings of the First International Workshop on Electron
Correlations and Materials Properties, held June 28-July 3, 1998, in
Crete, Greece"--T.p. verso.
Includes bibliographical references and index.
ISBN 0-306-46282-6
1. Electron configuration Congresses. 2. Solid state physics
Congresses. I. Gonis, Antonios, 1945- . II. Kioussis, Nicholis.
III. Ciftan, Mikael. IV. International Workshop on Electron
Correlations and Materials Properties (1st : 1998 : Crete, Greece)
QC176.8.E4E34 1999
530.4'11--dc21 99-39537
CIP

Proceedings of the First International Workshop on Electron Correlations and Materials Properties, held June 28–July 3, 1998, in Crete, Greece

ISBN: 0-306-46282-6


233 Spring Street, New York, N.Y. 10013

A C.I.P. record for this book is available from the Library of Congress.

10 9 8 7 6 5 4 3 2 1



Printed in the United States of America

## Preface

Over the last thirty years or so, the attempts to identify the electronic origins of materials properties have proceeded along two distinct and apparently divergent methodologies. On the one-hand, so-called single-particle methods are based on the study of a single electron moving in an effective field formed by the other electrons and the nuclei in the system. Band theory, as this approach is referred to, has had impressive successes in determining the equilibrium properties, such as structural stability, volume, and charge densities, of specific materials, notably metals. Today, even coherent phase diagrams (based on a single underlying lattice) for binary metallic alloys can be studied with considerable accuracy. In spite of its serious and well-understood limitations regarding the handling of correlations, band theory has been embraced by the materials scientist. Its single-particle nature endows the method with an economy of concepts which leads to a clear identification of mechanisms driving physical behavior at the electronic level. This perceived clarity often tends to override legitimate concerns regarding the validity of the method or its ability to correctly identify the mechanisms in the first place.

The alternative methodology pursued in the study of quantum systems consists of what can be referred to as conventional many-body theory. This methodology is based on attempts to study explicitly the effects of interparticle correlations using a number of different formal approaches, including but not limited to, perturbation methods, Green-function equation of motion methods, configuration interactions, quantum Monte Carlo, and others. Undeniably, this approach has been able to shed much light in our understanding of correlation effects and has been successful in predicting spectra and other physical properties in atoms and molecules, as well as the width of the gap in the single-particle spectrum in some semiconductors. However, the system-specific signature of band theory, so useful to the materials scientist, tends to be largely wiped out in this methodology.

Yet, as technology marches toward the development of smaller, more sophisticated devices, and as interest in the properties of the elements of the rare earth group and the actinide series intensifies, there is increased urgency to understand the effects that electron-electron correlations have on determining the observable properties of materials. For example, let us consider some long-standing and some more recently obtained experimental indications of the significance of these effects:

1. Photoemission, and optical and magneto-optical spectra from solids have long indicated the presence of correlation effects and the inability of band theory to account for them.

2. Elemental solids such as Ce (rare earths) and Pu (actinides) exhibit phase transformations associated with large volume changes. These transformations are taken as a clear indication of correlation effects, yet their understanding from a first-principles point of view has still not been achieved.

3. The presence of large, normal persistent currents in Cu wires can be attributed directly to the localization properties of two interacting particles (states) in the system. These states cannot be understood within a single-particle picture, nor can they be distinguished in a calculation in which very many electrons and their interactions are taken into account.

4. The presence of so-called gate currents in insulating interfaces in semiconductor devices, such as that between Si and $SiO_2$, becomes increasingly important as the size of the device decreases below a certain limit. For sufficiently large sizes and operating voltages of about 10.0V, these currents, which give rise to voltages of about 0.3V, can be safely ignored. As the operating voltages decrease to about 1.5V with decreasing size of the device, these currents can become the determining factors in device operation and reliability.

In spite of the importance of correlation effects on materials behavior and the great efforts expanded in their theoretical and computational treatment, there is very little cross-communication among disciplines addressing different aspects of the problem. Band theorists tend to regard many-body theory with suspicion, as being rigorous but without much practical application. Many-body theorists on the other hand, turn away from band theory as not well founded in spite of its wide practical applicability. Traditional theoretical/computational studies of materials properties, based mostly on electronic structure methods, provide at best a very approximate treatment of correlation effects. It is therefore necessary to identify new and novel methods that would augment the realm of validity of electronic structure calculations so as to include the effects of correlations.

The aim of the *International Workshop on Electron Correlations and Materials Properties* is to provide a forum for the exposition of the experimental evidence of the effects of correlation on the physical, chemical, and mechanical properties of materials, as well as the theoretical/computational methodology that has been developed for their study. One important novelty of the workshop is the focus on understanding in detail the quantum nature of the electronic states in solids pertaining to correlation effects and their impact on observable behavior as provided by *both* band-theoretical and many-body approaches. The workshop brings together experimentalists and theorists involved in determining and assessing materials behavior in order to promote the interdisciplinary exchange of ideas and methods that have been developed in the study of correlation effects on materials properties. *The workshop acquaints experimentalists and theorists alike with the wide ranging importance of correlations in materials behavior, assesses the state of the field, provides a platform for integrating the conceptual effort expended in the study of correlations, and identifies new and promising directions for research.* Thus the workshop goes considerably beyond the aims of a scientific conference on correlations. In fact, it provides a forum for establishing research directions into materials and their properties, in which the effects of correlations are addressed as directly as possible.

In order to effect the desired cross-fertilization of various disciplines, as well as illustrate the underlying unifying features of many different approaches used in the study of correlation effects, the workshop includes experimental studies, phenomenological treatments, and ab initio methods of the effect of correlation effects in materials properties. The relatively equal weight given to experiment, phenomenological and *ab initio* theory is intended to promote the study of the effect of correlations within a combined experimental-theoretical approach. The workshop emphasizes the connection between theoretical/computational developments and their effect on the accuracy of calculated materials properties. In addition, it is hoped that it will provide motivation and encouragement for further work on correlation-related materials properties and developing innovative methods for their study.

This volume contains the proceedings of the first workshop and is intended to serve as a reference on the state of the art in the study of electron-electron correlations in materials. As the co-organizers of the workshop and the editors of the volume we wish to thank the participants for the generally high level of presentations, of insightful discussions and comments, and for submitting camera-ready manuscripts for publication. We are also grateful to our sponsors, the California State University Northridge, the US Army Research Office, the Glenn T. Seaborg Institute at Los Alamos National Laboratory, the

Glenn T. Seaborg Institute and the Materials Research Institute at Lawrence Livermore National Laboratory, without whose generosity the workshop would not have taken place. Last but not least, we thank the editorial staff at Plenum for guiding us in the publication for the proceedings and the patience during the time that the book was being assembled. We hope that the material found in this volume will prove useful to both seasoned scientists and those who wish to enter a career in the study of materials properties.

March 1999

A. Gonis
N. Kioussis
M. Ciftan

# Acknowledgments

We are grateful for the sponsorship of the following organizations and institutions.

**California State University Northridge**

**US Army Research Office**

**Glenn T. Seaborg Institute, Los Alamos National Laboratory**

**Glenn T. Seaborg Institute, Lawrence Livermore National Laboratory**

**Materials Research Institute, Lawrence Livermore National Laboratory**

CONTENTS

## ELECTRON CORRELATIONS AND MATERIALS PROPERTIES

Speakers are identified by an underline.

Invited speakers are indicated by an asterisk.

# INTERNATIONAL WORKSHOP ON ELECTRON CORRELATIONS AND MATERIALS PROPERTIES

Walter Kohn

Physics Department
University of California, Santa Barbara
Santa Barbara, CA 93106

## OPENING REMARKS

I am very happy, as I imagine we all are, to have been invited to such an interesting conference on this beautiful, fabled island where, even if we are not Greek, we can admire and feel proud of some of the great achievements of mankind near the beginning of its history.

To make the opening remarks at a meeting is a rather dangerous undertaking, since it gives everybody else the opportunity to disagree. On the other hand, I don't face what could be Norman March's problem at the end, that everything that could be, already has been said. I have run down the titles of the main papers and marked them m or h for moderately or highly correlated, by what criterion I will explain later. By my count the "highly correlated" papers carry the day, 13÷10. Since I consider myself to be a rather moderate person, I look forward to this meeting primarily as a learning experience.

My acquaintance with the subject of electron correlations in solids goes back almost exactly 50 years when, a fresh Ph.D. in nuclear physics, I was assigned to teach a graduate course in solid state physics. Thus I came across the famous papers on the alkali metals by Wigner and Seitz, in the middle thirties, as well as Wigner's 1938 paper on correlation energy, the birth place of what was later called the Wigner crystal. That was a highly correlated system if there ever was one. I remember admiring Wigner's combination of brilliance and chuzpah.

In the first paper, exchange-correlation effects were simply taken care of by the off-hand remark that it was very unlikely that 2 electrons would be simultaneously in the same unit cell. Fairly good agreement with experiment was obtained. In the 1938 paper Wigner calculated approximately, but very accurately, the energy of the Wigner *crystal* and then, without blinking, used the result to obtain the correlation energy of a dilute, that is highly correlated, electron *gas*.

*Electron Correlations and Materials Properties*, edited by Gonis *et al.*
Kluwer Academic/Plenum Publishers, 1999. 

About a decade later, Mott, always keeping an eye on the latest experiments, noticed that NiO, which by simple electron count, should have a partially filled valence band and thus be a metal, in fact was an insulator. This led him to consider a cubic lattice of H-atoms and to conclude that beyond a critical lattice parameter, $a > a_c$, which he estimated with a precision of a few per cent, the system would become an insulator, the so-called Mott insulator, in later usage; this, although nominally the valence band of this model was only half full. Spin was an important part of Mott's thinking, allowing double occupancy of the same unit cell at the cost of a positive energy U. I was persuaded but not convinced by Mott's reasoning and looked for a more basic characterization of the Mott – (or any other) insulator. About 15 years later I proposed, and Mott agreed with me, that an insulator's astonishing *total* absence of DC conductivity at zero temperature is due to a gauge invariance of the insulating ground state, which is not shared by metallic conductors.

Since the 1950's the so-called Hubbard model, which built on Mott's ideas, has become an extremely important tool for studying correlated electron systems. It has the merits of allowing the introduction of semi-phenomenological constants which reflect physical insights, and - especially for 2D-systems - of permitting useful numerical calculations.

The 1950's and 60's also saw the beginning of the Landau Fermi Liquids Theory and of Density Functional Theory (DFT). I propose, at least for myself, to use the designation "moderately correlated" for a system for which either generalized Landau theory or DFT, in the Kohn Sham form, constitute a satisfactory description. This omits a great many systems which are on the program of this conference, e.g. many with partially filled inner atomic shells. Combination of Hubbard model and DFT concepts are subject of ongoing current theoretical work.

Finally in preparing these remarks I asked myself the question whether Mott transitions necessarily involve spin. I believe the answer is no.

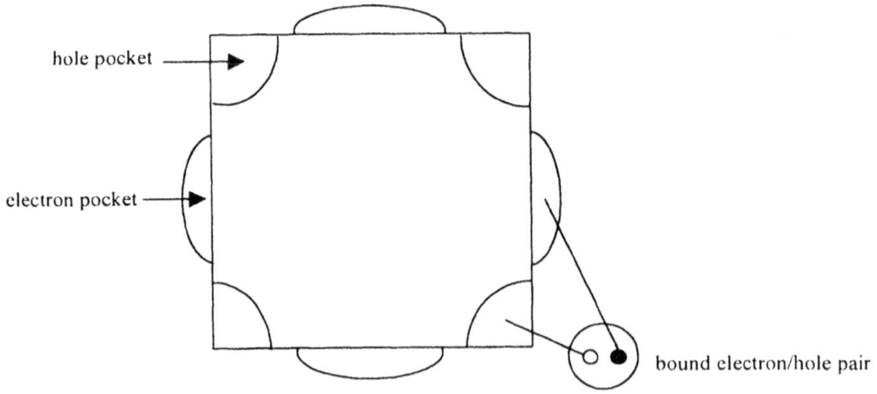


One spinless fermion per unit cell, with small electron and hole pockets.

For spinless, non-interacting, electrons, one per unit cell, and sufficiently small electron and hole pockets ($a > a_c$), the electrons and holes would appear to bind, leading to an insulating ground state. (See the figure). What is the nature of this state? I have some preliminary ideas and hope we can fit this question into our discussions this week.

# Part 1

# Experimental Indications of Correlation Effects in Materials

# EXPERIMENTAL STUDIES OF ELECTRON CORRELATION EFFECTS IN SOLIDS

G.A. Sawatzky
Laboratory of Solid State Physics
Materials Science Center, University of Groningen
Nijenborgh 4, 9747 AG Groningen, the Netherlands

## INTRODUCTION

In this paper I discuss the influence of electron correlation effects in solids of current interest in solid state physics and chemistry and some experimental techniques which have been used to determine the magnitude of the electron-electron interactions. Of course the things we are really interested in solids are their physical properties such as the electrical, magnetic, thermal etc. properties and the effects of electron correlation on these. The measurement of these properties in themselves seldom contain enough detailed information to draw conclusions as to the origin of a discrepancy with conventional one particle like theory. For example we want to know why $La_2CuO_4$ is an antiferromagnetic insulator with a 2 eV band gap. The measurements of physical properties are usually macroscopic in character and in many cases boil down to the determination of a few quantities like the linear temperature coefficient of the specific heat or the power law dependence of the electrical conductivity at low temperatures or the temperature dependence of the specific heat, or the temperature dependence of the magnetic susceptibility.

In discussing the physical properties of solids in relation to electron correlation effects we mean the response of solids to external perturbations which inadvertently involve the excited states of the solid and not only the ground state. This immediately brings us to a discussion which goes beyond the exact statements which can be made in density functional theory concerning the exact ground state density and energy [1]. In fact it is a study of the physical properties of solids which give us in my opinion a critical test of the physical usefulness of the single Slater determinant of one electron Bloch states used in the Kohn-Sham [2] procedure as a mathematical basis in which the charge density and total energy of the ground state can be expressed. The question asked in solid state physics is to what extent do the Bloch states which correctly describe the ground state density and energy in the Kohn Sham procedure also form a useful basis for predicting the physical properties of solids? In this regard it is important to realize that a single Slater determinant of one electron Bloch states contains in it no sign of correlation between the electrons where as in solids it is exactly this correlated motion of electrons which causes things like a strong enhancement of the magnetic susceptibility, a very high effective mass visible in the linear coefficient of the temperature dependent specific heat (the heavy fermion systems), superconductivity, the formation of local magnetic moments as

*Electron Correlations and Materials Properties*, edited by Gonis *et al.*
Kluwer Academic/Plenum Publishers, 1999. 

seen in the Currie-Weiss like temperature dependence of the magnetic susceptibility, the semiconductor metal transitions in many transition metal oxides like $V_2O_3$, or even the fact that $La_2CuO_4$ is a large band gap antiferromagnetically ordered insulator rather than a non magnetic metal as predicted if one takes the single Slater determinant of one electron Bloch states obtained in a band structure calculation as a reasonable representation of the true ground state eigenfunction.

The above are examples of systems in which it is the correlated motion of electrons which dominate in describing the physical properties of solids. The most obvious examples are found in materials involving 3d transition elements, rare earth elements and or actinides. For example compounds of the 3d transition metals exhibit an astonishing variety of physical properties both electrical and magnetic. The electric properties range from large band gap insulators ($V_2O_5$, $Cr_2O_3$, $\alpha$-$Fe_2O_3$) and ferroelectrics ($BaTiO_3$) to semiconductors ($Cu_2O$, $FeS_2$, MnS) and via semiconductor metal transitions ($VO_2$, $V_2O_3$, $Fe_3O_4$, $Ti_2O_3$) to metals ($CrO_2$, NiS, CuS) and even superconductors ($La_{2-x}Ba_xCuO_4$, $YBa_2Cu_3O_7$). In magnetism we find antiferromagnets ($\alpha$-$Fe_2O_3$, NiO, CoO, MnO) ferrimagnets ($\gamma$-$Fe_2O_3$, ferrites and garnets) and ferromagnets ($CrO_2$). Many of these materials have been and are of great importance in various applications because their properties can be tailored and fine tuned with chemical substitutions to, for example, optimize the magnetic anisotropy or the compensation or Néel temperature. Perhaps the most astonishing example of how extremely sensitive the properties can be to the details of the chemical composition is provided by the high $T_c$ superconductors. Here a change of x in $La_{2-x}Ba_xCuO_4$ or the stoichiometry in $YBa_2Cu_3O_{7-\delta}$ turns an antiferromagnetic insulator into a high temperature superconductor. The reason for this wide range of properties must lie in the sensitivity of the electronic structure to the details of the crystal structure, the chemical composition and the stoichiometry. This sensitivity is on the other hand a direct result of the extremely correlated motion of the 3d electrons in these systems not describable in terms of a single Slater determinant of one electron Bloch states.

The electronic structure of these compounds has been a controversial issue ever since de Boer and Verwey [3] in 1937 pointed out that compounds like NiO and CoO do not exhibit metallic behaviour as predicted by the Wilson and Bloch [6] theory of solids. In that theory a system with an odd number of electrons per unit cell must be a metal since there must be a band which is neither full nor empty and therefore must cross the Fermi energy. As a comment on the paper by de Boer and Verwey, Peierls was the first to point out that the repulsive on-site Ni d-d Coulomb interaction could be larger than the kinetic energy gained by delocalization and therefore each Ni ion would have 8 3d electrons, all polarity fluctuations would be frozen out, and only spin degrees of freedom would remain in the low energy scale. These are the basic ideas behind the Mott-Hubbard theory [4,5,7] in which an insulating state with a correlation gap is obtained if the d-d Coulomb interaction (U) is larger than the d band dispersion width (w). Also Anderson, in his theory of superexchange [8], recognized the importance of the d-d Coulomb interaction and used, as a starting point, the criterion $U >> w$ starting with localized 3d electrons rather than band like Bloch states. Perhaps I should remark at this point that the use of localized states for which the translational symmetry has been broken rather than one electron Bloch states can still in the end give us exactly the same total charge density and therefore is not contradictory to density functional theory. It simply is a different representation of the ground state which may in the end be closer to the true ground state eigenstate and which may yield a better physical picture of the possible excitations and their energy scales in the solid.

Another quantity of great importance in Anderson's theory of superexchange is the charge transfer energy ($\Delta$) which involves the transfer of an electron from the closed shell anion to the cation. These "virtual" charge fluctuations form the intermediate states for an interaction between the cation spins. In the Mott-Hubbard picture and the Anderson theory of superexchange it was (often implicitly) assumed that $\Delta >> U$. In this limit the anion states can be projected out leaving us with a correlated d-band system with effective hopping integrals and renormalised Coulomb interactions.

The basic questions which one would like to answer are how large are the local atomic Coulomb interactions which we represent here by U and other quantities like the interatomic charge transfer energies ($\Delta$), as well as one electron band widths (w) and the hybridization (T) (covalency) of the transition metal d and anion p orbitals and what is the nature of the low energy scale excited states. In this regard high energy spectroscopies have played an important role. For example the first evidence that the above assumption ($\Delta >> U$) was not applicable in divalent Cu compounds came from a photoelectron-Auger study of the Cu dihalides [9]. In this study it was shown that $U > \Delta$ or in other words the energy to remove a d electron is larger than that to remove an anion p electron in spite of the fact that the Cu 3d band is not full and the anion p band is full. It was not till later, however, that Fujimori and Minami [10]. and Sawatzky and Allen [11] pointed out the importance of this in terms of the understanding of the electronic structure and the nature of band gaps and superexchange in transition metal compounds (for recent review see [12-14]).

Another area of intense research in which localization caused by local Coulomb interactions like U plays a dominating role is that of magnetic impurities in metals and the extension to a lattice of such impurities leading to Kondo [15] and heavy fermion [16] behaviour. Friedel [17] suggested that 3d transition metal impurities would retain their local magnetic moment even when dissolved in a metal although the d states would broaden somewhat and shift due to hybridization with conduction band states. Anderson [18] introduced a much used model Hamiltonian which forms the basis for a discussion of magnetic impurities in metals. Here again the Coulomb interaction U between d electrons in transition metal impurities or f electrons in rare earth impurities plays an important role and determines whether or not a local moment exists. In the limit that U is large the Anderson Hamiltonian can in certain cases be transformed [19] to a Kondo Hamiltonian in which only the spin degrees of freedom of the impurity are retained and U enters in the size of the exchange interaction with the conduction electrons. In this latter case the low energy scale excitations are dominated by the very high density of states structure due to the Kondo or Abrikosov-Suhl resonance which develops at the Fermi energy due to many body effects.

In the above, I have tried to indicate very briefly some of the broader areas of research in which electron correlation effects and a knowledge of the local Coulomb interaction is of great importance in understanding the physical properties. In general one could say that the 3d and 4d transition metals, rare earth and actinide metals, their compounds and alloys exhibit properties like localized magnetic moments, Mott-Hubbard transitions, mixed valency and valence fluctuations, heavy fermion behaviour, high $T_c$ superconductivity, Kondo behaviour etc. all of which are most likely due to the importance of the atomic d-d and f-f electron-electron interactions and the resulting correlations between these electrons. In other wide band insulators semiconductors and metals not containing such elements a description of the properties in terms of band theory often gives good agreement with experiment unless one is dealing with very low electron density systems or low dimensional systems.

In an over simplified picture of the electronic structure of solids we can divide the valence orbitals in solids into two extreme classes, namely those in which the radial extent of the atomic orbitals is comparable or even larger than the interatomic distances in the solids. Such states will lose their atomic identity and form broad bands in the solid like the s-p bands of alkali metals or the transition metal compounds as well as the O2p bands of the oxides. In the other extreme we have valence orbitals which have a small radial extent comparable to the interatomic distance in the solid. Such orbitals will retain their atomic signature like the core orbitals of atoms but for us also the open shell valence orbitals (4f of rare earths, 3d in TM compounds) in the solid. These form narrow bands at best often well described by tight binding like procedures and electrons in such orbitals will experience strong local Coulomb and exchange interactions leading to multiplet splittings, Hund's rule stabilization of a local magnetic moment etc. if the shell is 'open' i.e. not empty and not full and strong suppression of polarity fluctuations which can lead to insulating behaviour. For such systems the low energy scale excitations are mainly spin fluctuations rather than charge fluctuations. Some of these characteristics are summarized in the table below.

Much of the most interesting solid state physics and chemistry has to do with systems in which these two types of states coexist and hybridize. Examples are the Kondo systems, heavy fermions, mixed valent and valence fluctuation compounds, magnetic systems, spin glasses, high $T_c$ superconductors, etc. We will see below that actually the 3d transition metal oxides form a class of systems in which localized highly correlated atomic states (3d orbitals) coexist and hybridize with rather broad itinerant valence (O 2p) and conduction band (TM 4s, 4p) states.

For all of these areas of research it is of utmost importance to have methods with which the effective Coulomb interactions in the solid can be quite directly measured. By quite directly I mean without too much interpretation. In this paper I will discuss examples for which various forms of high energy spectroscopy can provide such information although when one gets to the details concerning the exact size of the interactions one eventually does require a detailed often complicated theoretical analysis.

**Table 1.** A summary of some relevant characteristics of solids with two extreme valence orbitals. $R >> D$ radial extent of valence orbitals is comparable to or larger than the interatomic distance and $R << D$ the other extreme

| $R >> D$ | $R << D$ |
|---|---|
| Valence electrons loss atomic identity | Valence electrons remain atomic |
| Broad bands | Narrow bands |
| Small e-e interaction | Large e-e interaction |
| Low energy scale - charge fluctuations | Low energy scale - spin fluctuations |
| Non or weakly magnetic | magnetic |
| examples Al, Mg, Zn | Gd, CuO, $SmCo_5$ |
| Many interesting systems have coexisting $R >> D$ and $R << D$ valence orbitals (Kondo, High $T_c$, Heavy Fermions ......) | |

## SPECTROSCOPIC METHODS TO CHECK FOR CORRELATION EFFECTS

As mentioned above there are many examples of physical properties which contradict the results one would expect from a band structure study in which we take the one electron Bloch waves as representing true states of the system. The excited states of the systems are then assumed to be described by simply changing the occupation numbers of the one electron Bloch states or by using these states in perturbation theory to describe the response to some external perturbation. In many cases there are also direct spectroscopic methods that can be used to check for the validity of single particle pictures. In the next section we describe some of these with some examples taken from the classes of materials listed above.

We can quite directly check the predictions of band theory with angular resolved photoelectron spectroscopy (ARUPS) [20]. With ARUPS both the energy of the electrons and the corresponding momentum (k) can be determined providing us with a powerful technique for band mapping. In Fig. 1 the result of a study on Cu is shown [21]. We see nicely the strongly dispersing 4s band crossing the Fermi level and a manifold of weakly dispersing 3d bands between 2 and 5 eV below $E_F$. The band theory (solid lines) does an excellent job of describing these data bringing us to the conclusion that one particle theory works well for the first electron removal states which are the states below $E_F$. Similarly the inverse photoelectron spectra IPS yield the states above $E_F$ [22] which again are in reasonable agreement with band theory for Cu [23] for not too high energies [24].

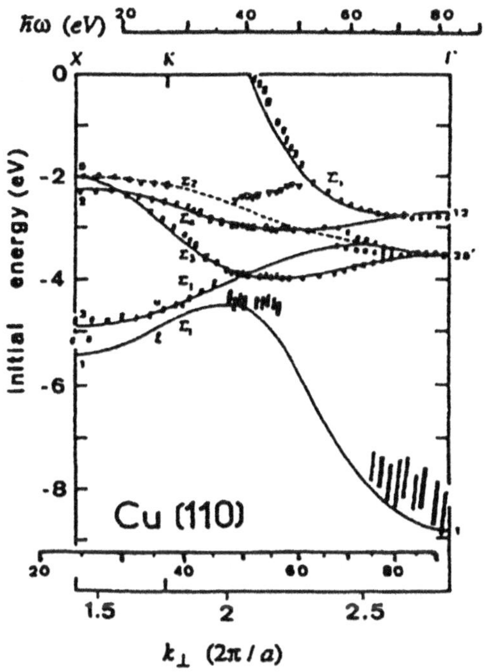


**Figure 1.** Experimentally determined valence band for copper along the QKX line. The full curves correspond to a band-structure calculation. The dashed curve (level $\Sigma_2$) is forbidden in both polarizations (s and p) for normal emission. The few points indicated along this line are obtained by moving the analyzer by a few degrees. The height of the rectangles gives the experimental uncertainty. Taken from ref. 21

We however get quite a shock upon studying the $L_{23}VV$ Auger spectrum of Cu. In Auger spectroscopy the final state has two electrons missing in the valence band and since the core hole (2p) in the intermediate state is strongly localized and the Auger matrix elements are atomic like we expected to see the local two hole density of states [25]. In a one particle theory the local two hole density of states is just the self-convolution of the one hole density of states which in turn is closely approximated by the self-convolution of the angular integrated X-ray photoelectron spectrum (XPS). The resulting broad (5 eV) structure together with the Auger spectrum [26] is shown in Fig. 2a.

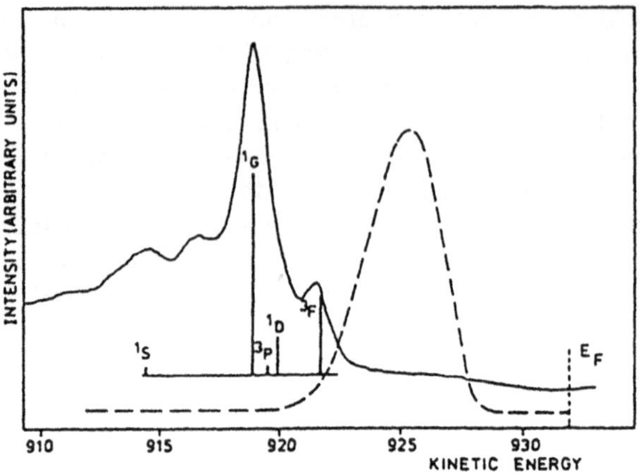


**Figure 2a.** The $L_3M_{45}M_{45}$ Auger spectrum of Cu (solid line). The dashed line represents the self convolution of the 3d band, $E_F$ is the Fermi energy. The energy scale is in electron volts.

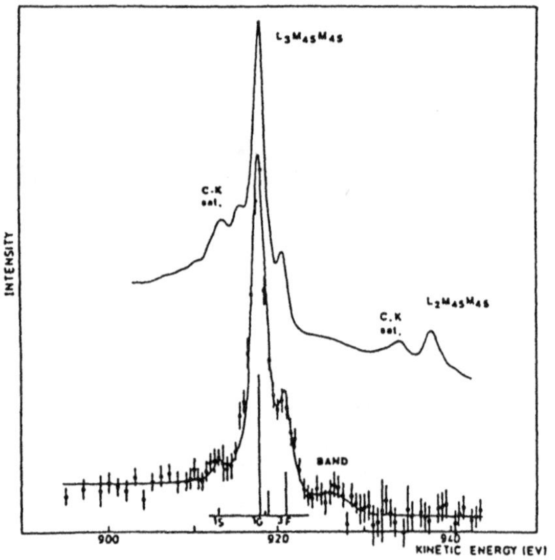


**Figure 2b.** The upper solid curve is the Auger spectrum as in Fig. 2a. The $L_{23}M_{45}M_{45}$ Auger spectrum in coincidence (APECS) with the $L_3$ ($2p_{3/2}$) line is shown as dots. The solid line fit to the experimental APECS spectrum shows both the atomic like multiplet split part and the band like part separated in energy by $U_{dd}$. The theoretical curve is obtained ussing the two particle Greens function theory described in references 28-30 and the atomic Slater integrals describing the multiplet splitting.

In Fig. 2b is shown the Auger spectrum [26] and the Auger photoelectron coincidence spectrum (APECS) [27]. (In APECS we can eliminate the Coster-Kronig preceded Auger structures and therefore get a more direct picture of the local two hole density of states.) We see two regions, one is marked "band" and the other consists of the high intensity sharp structures. The low intensity band region corresponds closely to the self-convolution of the d density of states shown in Fig. 2a. The higher energy atomic part which contains most of the intensity is sharply structured and shows the multiplet structure of an atomic $d^8$ state as indicated by the vertical bars. This behaviour is quite nicely explained by the Cini-Sawatzky [28,29] theory in which proper account is taken of the intra-atomic d-d Coulomb interaction $U_{dd}$. These two different sets of sets of two hole states can using the above theory be identified as states in which the two holes are primaraly of different Cu atoms forming the band like states and the states in which the two holes are on the same atom resulting in a $d^8$ like state which will be U higher in energy and will dispay the multiplet structure of a free Cu 2+ atom. In fact the separation of the "$d^8$" part and the band like part where the 2 holes are on different sites is a direct measure of $U_{dd}$ although there are small shifts due to hybridization [30]. These bound two hole states will still propagate through the solid but their dispersion will be strongly reduced because they have to move together much like a Frenkel exciton as described by the exact theory for the two hole problem. Here I will suffice by stating that if U is larger than the one hole bandwidth the $d^8$ states are discrete bound states which look very atomic like [31]. For Cu the average U is about 7 eV which is considerably smaller than the free atom value of 18eV but the multiplet structure which is determined by the higher order Coulmb integrals is not at all reduced from the free atom values. This proves a very important fact namely that it is only the monopole Coulomb integrals which are strongly screened in the solid and the multipole Coulomb integrals retain their atomic values.

The above brings us in direct contact with what Peierls said about NiO. For the case of the bound two hole states in Cu we can better start by forgetting about the translational symmetry and treat the on-site Coulomb interaction and the multiplet structure as if we were dealing with an isolated atom first and then look at what happens if we switch on the translational symmetry afterwards. In fact an exact solution to the Auger line shape for a model in which only the on-site Coulomb interactions and the band structure are taken into account shows the strong difference in the two hole local density of states for $U>W$ (atomic like) and $U<W$ (broad band like) line shapes [28-31].

Obviously the one particle band theory breaks down in describing the Auger spectra of Cu. We argued that this is a result of the strong repulsion of the two final state holes if they are on the same atom. In a single particle approach we can not distinguish between these two sets of states since the particles are assumed to be uncorrelated. The reason we do not notice a problem in the photoelectron spectrum is because the Cu d band is initially full so that removing one d electron results in only one d hole and therefore is a one particle problem if we can neglect the correlation with other 4s,4p etc. holes. The conclusion we reach is that for the electrons in atomic like wave functions with radial extents small compared to the interatomic distance we must explicitly take electron correlation into account in the wave function if the states we study contain more than one hole or electron per atom. This is what we tried to preclude in table 1.

It is interesting then to look at the photoelectron spectrum of Ni metal for which the d band is not full in the ground state. In perfect analogy to the Auger spectrum of Cu metal we expect to see a breakdown of the one electron picture for describing the photoelectron spectrum of Ni metal. Upon removing one d electron we can reach states with two holes ($d^8$) or one hole ($d^9$) on the same atom if we consider the ground state to have on the average 9.4 3d electrons per atom corresponding to a magnetic moment of 0.6 Bohr magnetons. This results in the much discussed "6 eV satellite" in the XPS spectrum of Ni metal [32] shown in Fig. 3. Also shown in Fig. 3 is the Ni $L_3$ VV Auger spectrum which corresponds to primarily $d^8$ like final states since it originates from mainly the $2p^5 3d^{10}$ (fully screened) [33] core line. The correspondence of the Auger spectrum and the XPS satellites suggests that our interpretation of the origin of the satellite is correct. Even more conclusive are the resonant photoemission results [34] which clearly show the strong resonance of the satellite at the 3p absorption threshold photon energy. For a detailed description I refer to the review of Davis [33]. Resonant photoemission is another very important spectroscopic tehnique in the study of strongly correlated systems and we will come back to it below. At this point I would like to remark that the so-called satellite

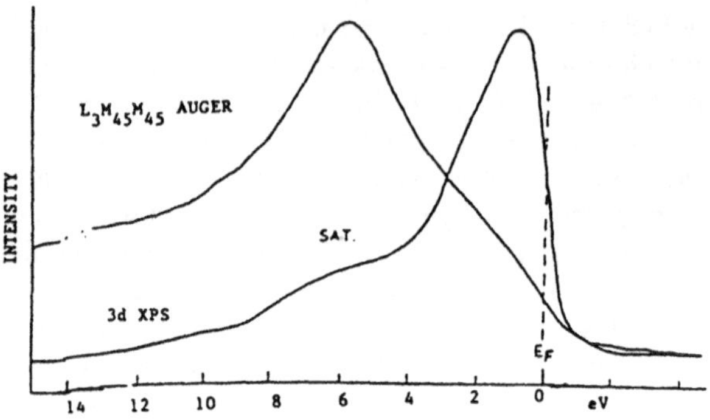


**Figure 3.** The $L_3M_{45}M_{45}$ Auger spectrum and the 3d valence band XPS spectrum of Ni, the Auger spectrum is shifted by the $2p_{3/2}$ core level binding energy so that $E_F$ coincides with that of the XPS spectrum.

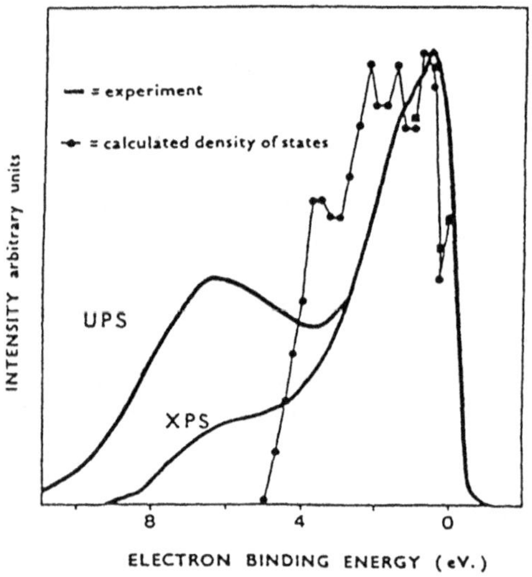


**Figure 4.** The heavy solid lines are the XPS and on resonance UPS spectra of Ni metal showing the "6 eV" satellite and its resonant behaviour. Also shown with dots is the calculated density of states. This figure is taken from Falicov [35].

structure is not the result of some many body excitation since in photoemission we only see the eigenstates of the same with one electron removed. In fact the little picture given above shows that the satellite structure is a result of the strongly correlated nature of the ground state. The satellite structure shows that the ground state should not be regarded as a single Slater determinant of one electron Bloch states but rather as a lnear combination of many such determinants.

By comparing the energy of the satellite ($d^8$) to that of the main line ($d^9$) we can get a rough estimate of U ≈ 3 eV which is considerably smaller than that for Cu metal (≈ 7 eV) and in fact comparable to the Ni 3d band width of also about 3 - 3.5 eV. Nonetheless U has a large influence on the electronic structure. This can be seen from Fig. 4 showing again the XPS and resonant photoemission spectra together with a band structure calculation. As expected the band structure calculation does not predict the satellite but in addition the band width is quite a bit larger than that of the $d^9$ band like part of the spectrum. This band narrowing is a direct result of U and can be qualitatively understood as follows. A d hole corresponding to a $d^9$state can hop to a neighboring site provided that site is in a $d^{10}$ configuration, i.e. $d^9_i d^{10}_{i+a} \rightarrow d^{10}_i d^9_{i+a}$ without an on site energy change. But if the neighboring site is also $d^9$ a hop would result in a $d^8$ state i.e. $d^9_i d^9_{i+a} \rightarrow d^{10}_i d^8_{i+a}$ which costs an energy U. For U large these hopping processes are not allowed on the low energy scale so that the number of degrees of freedom for hopping is reduced resulting in a band narrowing. For $U \rightarrow \infty$ the expected band width would be about 0.4 of that obtained from band theory. Also the exchange splitting of the 3d band in the ferromagnetic ground state is found to be considerably less than the spin polarized band structure calculation. Also this is a direct result of the configuration interaction between the $d^9$ like part and $d^8$ like part of the n-1 electron states and the fact that the 3F part of the $d^8$ like structure is much lower in energy than the singlet states because of Hunds rule.

## MOTT-HUBBARD-ANDERSON

Mott [4] and Hubbard [5] proposed that for materials containing atoms with open shell orbitals with a radial extent small compared to the interatomic spacing we could better start with localized electrons rather than putting the electrons in Bloch states. In this way we would quite naturally get the atomic limit in our previous example for a large interatomic spacing . As an ansatz we consider the solid to consist of a lattice of atoms each with exactly n d or f electrons neglecting for the moment the possible presence of the more free electron like outer shell s and p electrons. This we recall is quite different from a band theory approach for which the probability that an atom has n electrons is given by the binomial distribution [5]

$$P(n) = \frac{z!}{(z-n)!n!} c^n(1-c)^{z-n} \tag{1}$$

in which z is the degeneracy i.e. 10 for d or 14 for f and zc is the electron concentration per atom. The atomic ansatz therefore assumes that polarity fluctuations are completely suppressed. To determine the stability of our ansatz we consider the energy required to move an electron from atom i to atom j. This will turn out to be our definition of the electron-electron interaction U.

$$U = E(d^{n+1}) - E(d^n) + E(d^{n-1}) - E(d^n) \tag{2}$$

We see that U involves the energy difference between $d^{n-1}$ and $d^{n+1}$ which states differ by 2 electrons. Note that is a different definition from that used in Auger spectroscopy as described above for Cu and Ni.

$$U = (E(d^{n-2}) - E(d^n)) - 2(E(d^{n-1}) - E(d^n)). \tag{3}$$

Since all the one particle potential energy terms cancel in eq. (2) any theory including density functional which replaces exchange and correlation by a one particle potential in effect assumes that U=0. In this sense U is truly a measure of correlation effects and involves the difference in the ionization potential and the electron affinity.

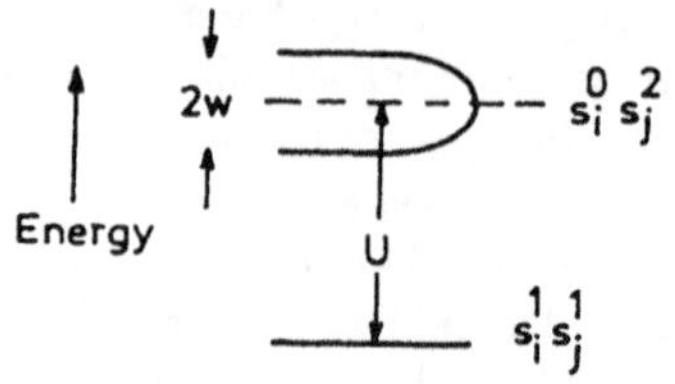


**Figure 5.** The total energy level diagram for a solid of hydrogen atoms with U >> W showing the stability of the ground state with strongly reduced polarity fluctuations. In fact the states $S^1{}_iS^1{}_j$ and $S^0{}_iS^2{}_j$ will mix for i and j nearest neighbors by an amount given approximately by $2(W/4)^2$ 1/U which mixes in some higher polarity states in the ground state and also causes an antiferromagnetic exchange between the unpaired spins on nearest neighbor atoms.

The state with one extra electron on site j and a hole on site i will in fact form a band of width 2W because both the hole and the electron can now freely propagate. The total energy level diagram is shown in Fig. 5 from which we see that if U>W our ansatz is stable and we will have a Mott-Hubbard insulator with close to n electrons on each atom and in the case of the model H lattice a local spin of ½ per atom. As expected the explicit inclusion of correlation in the wave function reduces polarity fluctuations. More generally each atom has n, d or f electrons which will interact just as in the free atom yielding a lowest energy electronic configuration given by Hunds rule. Since Hunds rule states that the lowest energy state is that with a maximum spin and a maximum orbital angular momentum the materials with U>>W will have local magnetic moments. Hunds rule and more generally the multiplet splitting of a $d^n(f^n)$ electronic configuration strongly complicates Fig. 5 and also raises the question as to which states should we consider in determining U. Herring [36] has taught us that we should for each case in eq. (2) take the lowest (Hunds rule) state to define U.

We still had to deal with a number of complicating aspects like the importance of the multiplet structure and the problem of which states should be treated as atomic like or which band like. It is clear that if the radius of an atomic orbital is small compared to the interatomic spacing in the solid we can to good approximation use atomic orbitals to describe these states. Although we will not be able to define sharp boundaries for states which can be treated as atomic like and states for which the wave function is strongly k dependent it is possible to at least discuss some extremes. In Fig. 6 is shown the radial extent of the charge distribution for the 3d, 4d, 5d and rare earth and actinide f states as compared to the Wigner Seitz radius of the elemental solid [37]. For the rare earths the f radial extent is so small compared to the interatomic distance that atomic behaviour is expected. Also the late actinides fall in this category. The 3d transition metals even for example Ni lie well above the rare earths as do the 4d, 5d and early actinides. Obviously the smaller the radial extent of the wave function, the smaller the band width and the larger is U. Harrison [38] has shown that for d bands $W \approx \mathbf{a}^{-5}$.

Let's look at the extreme case of the rare earths. First from a straight forward application of a band theory approach we expect to see a broad s,p,d band and for the non-magnetic case a narrow 4f band (<1 eV wide [38]). Since the 4f band is not full the Fermi level must lie in it so we get the qualitative structure as shown in Fig. 7. What we really mean by the states above and below $E_F$ drawn in Fig. 7 are the eigenstates of the system with one electron added (states above $E_F$) and those with one electron removed (states below $E_F$). We see that this is exactly how we also defined U in eq. (5). Also we see that these states can be directly measured by a combination of photoelectron spectroscopy PS and inverse photoelectron spectroscopy (IPS). A classic and beautiful demonstration is the measurement by Lang and Baer [39] on the rare earth elemental metals shown in Fig. 8. The sharp structures seen result from the 4f states. We see energy spreads in the 4f states of up to 20 eV and usually a large gap in the PS-IPS 4f states - all quite different from our band theory prediction.

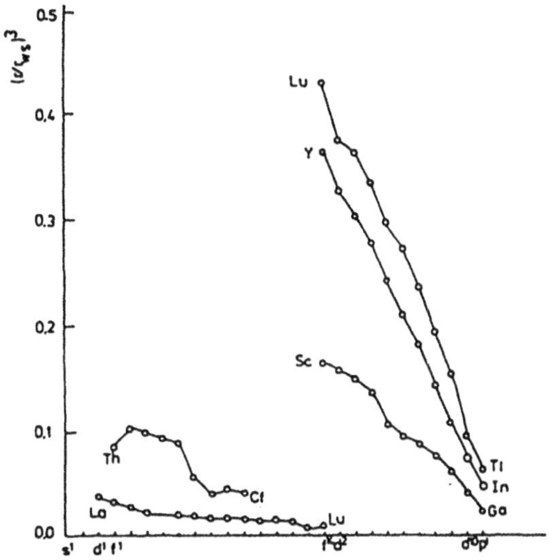


**Figure 6.** Ratio of l shell "volume" to Wigner-Seitz volume of the 3d, 4sd and 5d elements (l=2) and 5f elements (l=3) as a function of their position in the periodic system.

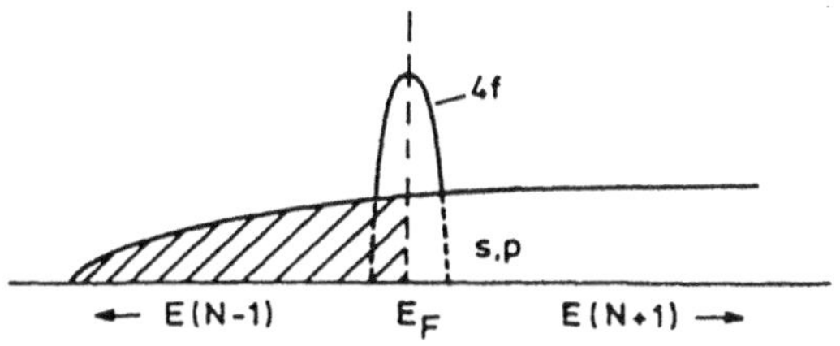


**Figure 7.** An artist's concept of the density of states from a one particle theory point of view for a rare earth metal with a narrow partly full 4f band and a wide partly full s,p,d band. Note that the states below $E_F$ are the eigenstates of the system with one electron removed and those above $E_F$ are the eigenstates with one electron added.

The structure seen in the spectra is the multiplet structure of the $4f^{n-1}$(PS) and $4f^{n+1}$ (IPS) electronic configurations as demonstrated by the correspondence with the free atom calculations shown with the solid bars. The height of the solid bars are the free atom fractional parentages assuming the ground state to be that determined by Hund's rule for a $4f^n$ configuration. This really is a textbook example demonstrating:

a) The importance of electron correlation
b) The stay at home principle for the ground state electronic configurations.
c) The power of PS-IPS to determine directly U
d) The importance of the multiplet structure which is apparently *not screened* by solid state effects.

In Fig. 9 are shown the energies of the lowest (closest to the Fermi level) $4f^{n-1}$ and $4f^{n+1}$ states as determined by PS-IPS. From this we can directly determine U which is the largest for

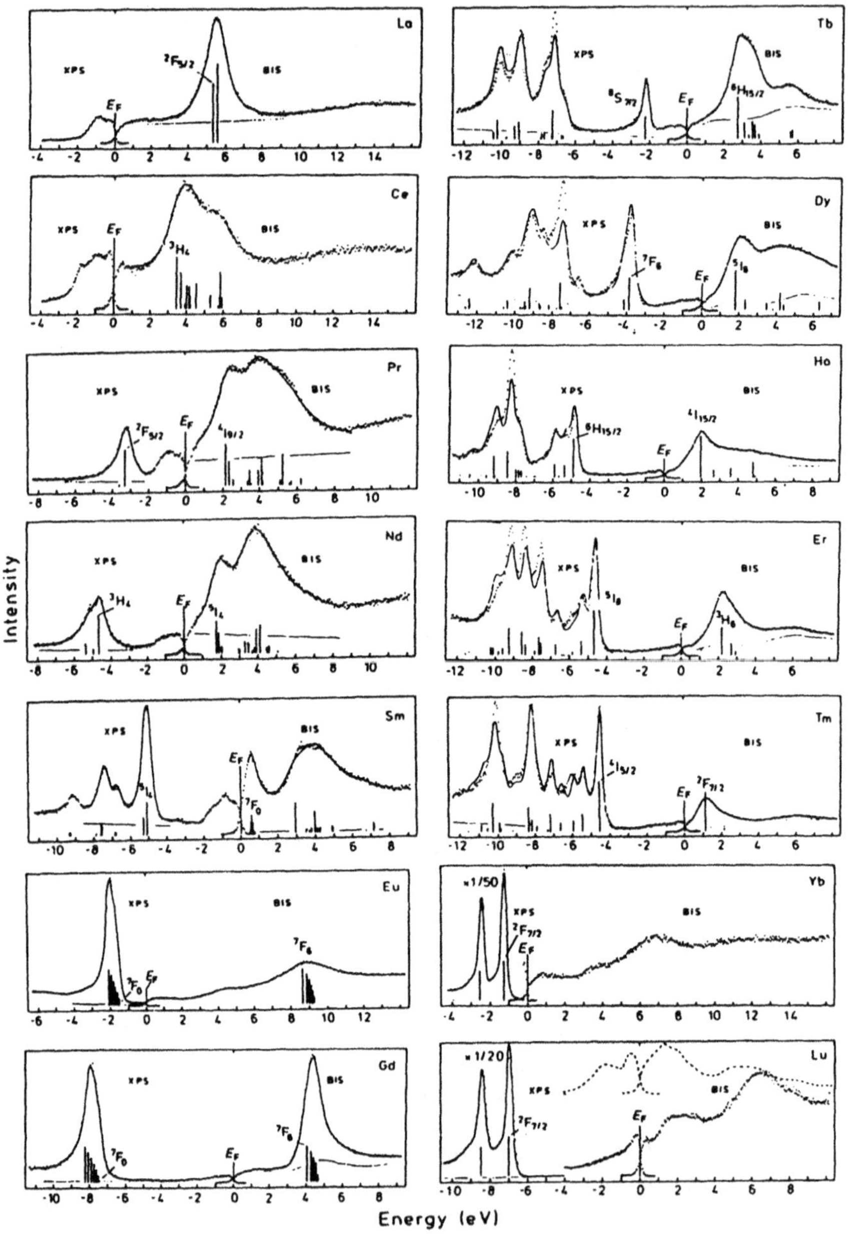


**Figure 8.** The photoemission and inverse photoemission spectra of the rare earths from Lang and Baer [39] showing the sharp contrast with the one particle picture of Fig. 7. The vertical bars are the calculated *atomic* multiplets with heights given by the coefficients of fractional parentage for electron removal ($E<E_F$) and electron addition ($E>E_F$) starting with the *Hund's* rule ground state .

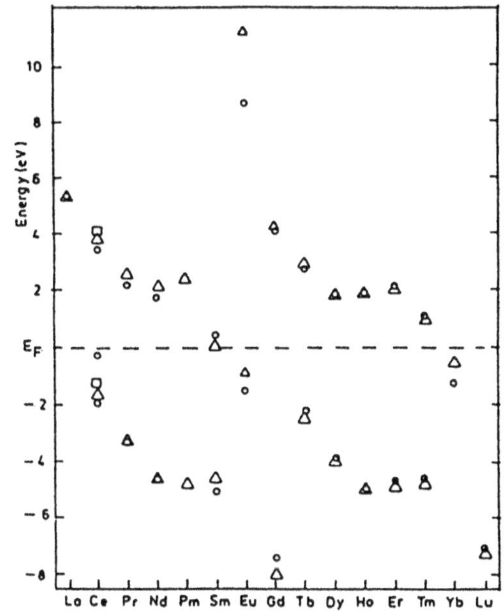


**Figure 9.** The lowest energy 4f electron removal and electron addition states for the rare earth metals taken from ref. 37. The open circles are from the data of Lang and Baer [39]. The triangles are from atomic theory with a screened monopole Coulomb interaction

a $4f^7$ ground state. We also see that for Ce, Eu and Yb the $4f^{n-1}$ states and for Sm and Tm the $4f^{n+1}$ states are quite close to $E_F$. By alloying or compound formation we can move the chemical potential ($E_F$) relative to the 4f states by about -2 to +1 eV relative to the elemental rare earths and we should for these elements be able to obtain other valence states in the ground state. The elements mentioned are indeed known to form mixed valent and valence fluctuation compounds.[40]

What about the multiplet structure and the apparent absence of screening? The Coulomb interaction between d or f electrons is given by the Slater $F^0$, $F^2$, $F^4$, $F^6$ integrals which describe the monopole ($F^0$) dipole ($F^2$) etc. like contributions. The multipole contributions result in the multiplet structure and they are due to the dependence of the Coulomb integrals on the relative spin (exchange) as well as orbital occupations. The monopole part ($F^0$) is strongly screened in the solid since it is the direct interaction between two charges. Say it costs us $\epsilon$-$E_{pol}$ to produce a local charge where $E_{pol}$ is the reduction in the ionization potential due to polarization and screening. To create two such charges well spatially separated would cost $2\epsilon$-$2E_{pol}$. But to create two charges on the same atom would cost $2\epsilon$-$4E_{pol}$ because the polarization or screening energy goes as $Z^2$ resulting in a screening of $F^0$ of $2E_{pol}$. However, the system really does not care much what the relative spin direction of the two holes (or electrons) or what their orbital ($m_l$) quantum number is if the radial extent is small compared to the interatomic distance. So for the states in materials which have radial extents much smaller than the interatomic spacing we have atomic wave functions with atomic multiplet structure but a strongly screened $F^0$ Coulomb integral. This shows that we expect $F^0$ to be strongly screened but the $F^2$, $F^4$, $F^6$ are expected to be atomic like as found for the rare earths. This is a quite general and for magnetism an extremely important result which holds not only for the rare earths but we saw it above also for the Auger spectrum of Cu and has also been demonstrated for numerous 3d, 4d and 5d transition metals. If the screening of $F^2$ and $F^4$ were as large as that for $F^0$ we would have few, if any, magnetic 3d transition metal systems. It is interesting to look at this screening problem in the light of what we generally learn about screening ussing the random phase approximation. In this approximation the screening of the Coulomb interaction is basically described by 1/dielectric constant and since the multiplet structure and exchange as well as U all involve the Coulomb interaction 1/ (r-r') they should all be screened by the same factor. The reason why this does not work here is mainly because of the breakdown of RPA for such local interactions. In fact to describe the screening of very local interactions in a polarizable medium the results are dominated by local field effects which are not included in RPA.(see for example the discussion by Nozieres and Pines [41]). The simple physics described above for the difference in screening of the monopole and multipole Coulomb integrals then turns out to be the correct way to tackle the problem. The experimental results clearly show how worng it really is to treat the screening of the monopole Coulomb interaction U in the same way as the mulitipole interactions as for example the Hund's rule coupling.

The multiplet structure is determined by $F^2$ and $F^4$ for the d electrons and $F^2$, $F^4$ and $F^6$ for the f electrons [42]. The multiplet spread is large for the rare earths as demonstrated in Fig. 9 but also very significant for the 3d elements. For a review of the magnitudes of the various interactions I refer to a paper by van der Marel et al [44].

For the rare earths we obviously are in the limit U >> W so the f electrons behave atomic like and in a Mott-Hubbard way. For the elemental transition metals, however, things are not so clear. PS and IPS measurements do not show the by correlation split band behaviour [45]. In fact we had just found above that for Ni U ($^1$G) Z 3 eV but using the lowest Hund's rule state we should look at U($^3$F) = 1.6 eV which in fact is less than W but as pointed out above not negligible [46]. So for the 3d transition metals we are more in the band like regime but for at least Mn, Fe, Co, Ni the effects of correlation are certainly important. For the 4d and 5d transition metals we expect increasingly good agreement with band theory and a decreasing importance of correlation as evidenced also by the predominantly non-magnetic behaviour. For the actinides we expect a behaviour intermediate between the 3d transition metals and the rare earths and perhaps even a change over as we progress from Pu to Am. If we dilute the 3d transition metal atoms or actinides so that the band width decreases the importance of electron correlation again emerges which brings us to the Anderson impurity, Kondo and heavy fermion systems.

## ANDERSON-KONDO HEAVY FERMION

Going back to the rare earths we found that the 4f states behaved atomically and we could just as well have discussed them as impurity states in a metal where the metallic character is determined by the broad sp(d) bands. Also for impurities or diluted systems we define U by eq. (2) realizing that what is important is the energy required to remove or add one electron to system. In Fig. 10 we have sketched the electronic structure of an impurity with n-d electrons in a metallic host. The states below $E_F$ ($d^{n-1}$) will have a finite lifetime because they are degenerate with a continuum of states (the sp band) with which they hybridize. This hybridization causes a broadening resulting in so-called virtual bound states [17,1]. If both $d^{n-1}$ and $d^{n+1}$ are far removed from $E_F$ and if $U>>\Gamma$ where $\Gamma$ is the virtual bound state width, the lowest Hund's rule state of $d^n$ will be the ground state and the impurity will have a local moment. This is a simplified description of Anderson's criterion for local moment formation showing again the competition of electron-electron interactions with the one electron terms leading to delocalization.

We could describe the rare earth elemental metals and alloys in this way but also now we can understand that transition metal impurities in non-magnetic metals can have large local

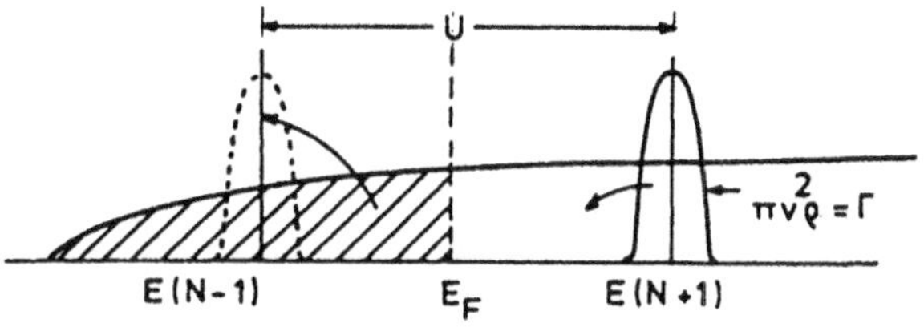


**Figure 10.** An artist's concept of correlation split impurity states hybridizing with a broad band. This results in a gap in the impurity states given by U and broadenings into virtual bound states of widths $\pi v^2 p$ where v is the impurity-host hybridization interaction and p is the host density of states.

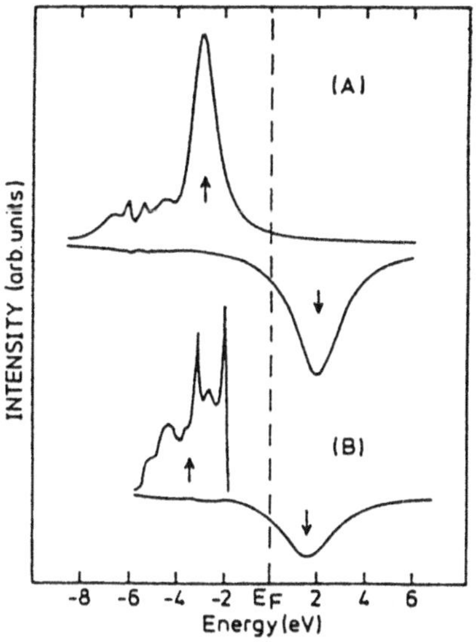


**Figure 11.** Calculated spin-projected local d density of states at a Mn site in a (A) Ag and (B) Cu surrounding.

moments. In Fig. 11 we show the local d spectral weight as determined from PS-IPS measurements of Mn impurities in Ag [43]. Clearly U>Γ explaining the local moment (s=5/2)($d^5$), for this system. From the $d^{n-1}$ - $d^{n+1}$ splitting we find

$$U = F^0 + 4J = 4.6 \text{ eV}. \quad (J = 1/14(F^2 + F^4)) \qquad (4)$$

Combining this with an Auger measurement leads to $F^0$ = 1 eV showing the strong screening from the atomic value of about 16 eV. [47] In a way Mn diluted in Ag looks similar to Gd metal showing nicely how dilution increases the relative importance of correlation.

There is another interesting phenomena which can occur for magnetic impurities in metals. The $d^{n-1}$ state of Fig. 11 can be filled by an electron hopping into it from the s, p band states yielding the finite width. However this possibility abruptly stops when we reach the Fermi level since at T=0 K the occupation abruptly cuts off at $E_F$. (An analogous thing happens for $d^{n+1}$.) This leads to the build up of a very sharp resonance at, or very close to, the Fermi level resulting in the Kondo effect [15]. This resonance is so narrow for Mn in Ag that we could not observe it in PS-IPS but it is evident from the minimum in the low temperature resistivity of

Mn Ag alloys [48]. Such a sharp resonance has been observed by Allen et al. [49] for the high Kondo temperature materials like Ce $Ir_2$ and Ce $Ru_2$ with PS-IPS and also by Baer et al. in a high resolution low temperature PS measurement of $CeSi_2$ [50].

A lot of attention has been placed on the experiments and theory of various Ce, Pr alloys with the result that we now have available a unified theory describing both the high energy spectroscopies, including core level spectroscopies, and the low energy properties like the Kondo effect and magnetic susceptibility with the same Hamiltonian and the same so-called 1/N many body theory developed by Gunnarsson and Schönhammer [51]. This is a really remarkable bit of physics in which we can so clearly see the use of high energy spectroscopies to determine model Hamiltonian parameters like U and the subsequent detailed predictions of the same theory for the physical properties.

## TRANSITION METAL COMPOUNDS

There are two reasons why a discussion of transition metal (TM) compounds follows logically. First of all the anions in the compound act as a dilutent so just as above we expect the influence of electron correlation effects for 3d transition metal compounds to be considerably larger than in the pure metals. Secondly since the 3d band widths often turn out to be very small (<0.5 eV) the problem could be akin to the rare earth metals where we found that the 4f states could be treated in an impurity model to good approximation. The difference is that for TM compounds we will be dealing with 3d impurities in a semiconductor rather than a metallic host density of states. To see this let's take a very simplified ansatz for a TM compound like a halide or an oxide. At stoichiometry as for example in $NiCl_2$ or NiO, or CuO the anions are assumed to be closed shell (i.e. $Cl^-$ or $O^{2-}$) and the TM cations have empty 4s, p states and n-d electrons where n is determined by charge neutrality. For the examples given Ni or Cu are 2+ and have n=8 for Ni and n=9 for Cu. In this ionic ansatz the Cl or O valence p band is full and the TM, 4s,p band is empty and the d "band" is partially full. In a band structure approach this d band would be in the gap formed by the O2p and TM 4s,p bands and at least for the non-magnetic states band theory would predict all of these to be d band metals whereas they all have substantial gaps (for example 4 eV for NiO). In Fig. 12 a band structure calculated density of states is shown for (non magnetic) $NiCl_2$ [52] displaying the large O2p - TM 4s,p gap, the very narrow (impurity like) d bands in this gap, and the Fermi level in one of the crystal field split d sub-bands.

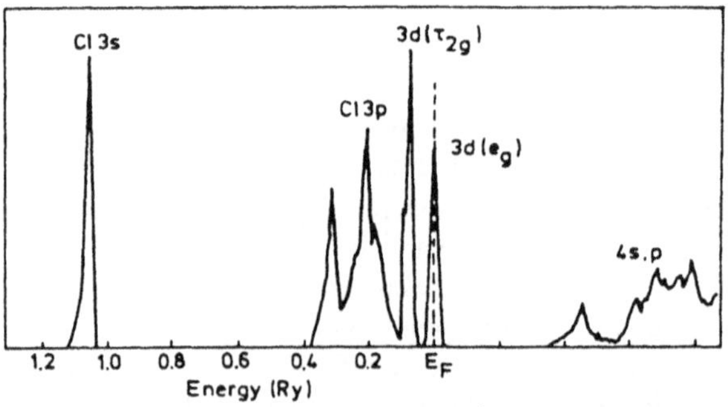


**Figure 12.** The calculated density of states of $NiCl_2$ showing the full Cl- 3p band, the empty Ni 4s,4p band and in this gap the very narrow crystal field split Ni 3d bands

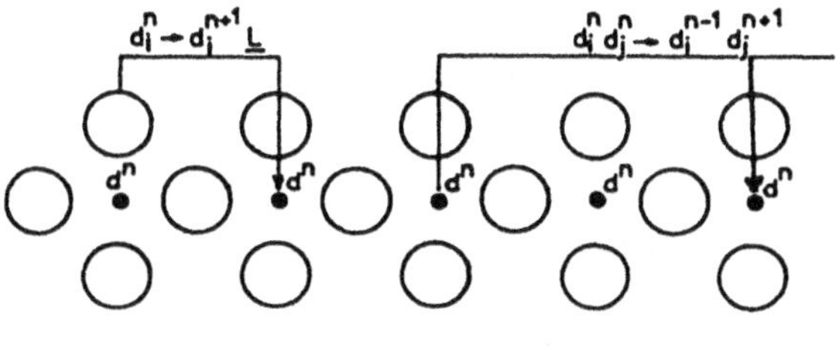


**Figure 13.** Representation of an ionic lattice consisting of TM ions ($d^n$) and closed shell anions. The most important charge fluctuation excitations are indicated.

Let us see what we would get from the ionic ansatz. The basic picture is shown in Fig. 13 with an ansatz for which each TM has a configuration $d^n$ and the anions are in an inert gas configuration (i.e. $2p^6$ for $O^{2-}$).
Similar to the discussion around Fig. 5 we consider the stability of this ansatz by looking at the possible charge fluctuations. These are:

$d^n_i d^n_j \rightarrow d^n_i{}^{-1} d^n_j{}^{+1}$ involving U as in eq. 2,

$d^n_i d^n_j \rightarrow d^n_i{}^{+1} d^n_j \underline{L}$ where $\underline{L}$ is a "ligand" O 2p hole and this involves the charge transfer energy W,

$d^n_i d^n_j \rightarrow d^n_i{}^{-1} d^n_j 4s_j$ which involves an energy $W_{d,s}$.

Let us for the moment neglect the last possibility and look at the total energy level diagram in analogy to Fig. 5. This is shown in Fig. 14 in which also the band width W of the 2p hole is included. We see that for U << W the charge transfer state can be neglected and we have a Mott-Hubbard insulator or a d band metal for U > w or U < w respectively as for Fig. 5. However for U >> W we have a charge transfer gap insulator for W > (W+w)/2 or a hole like metal (holes in the 2p band) for W < (W+w)/2. This leads us to a phase diagram shown in Fig. 15 in which the various possibilities are represented which is a simplified version of that presented recently by Zaanen et al. [53]. As shown first by Fujimori [10], Sawatzky and Allen [11] the late 3d TM compounds are in the charge transfer gap regime which has very important consequences also for the high $T_c$ materials. This conclusion was basically reached by a detailed analysis of the PS, IPS spectra of NiO as shown in Fig. 16. This figure shows that the first ionization states are of primarily O 2p character rather than Ni 3d character. This in itself shows that U > W and we are in the charge transfer gap regime of Fig. 15.

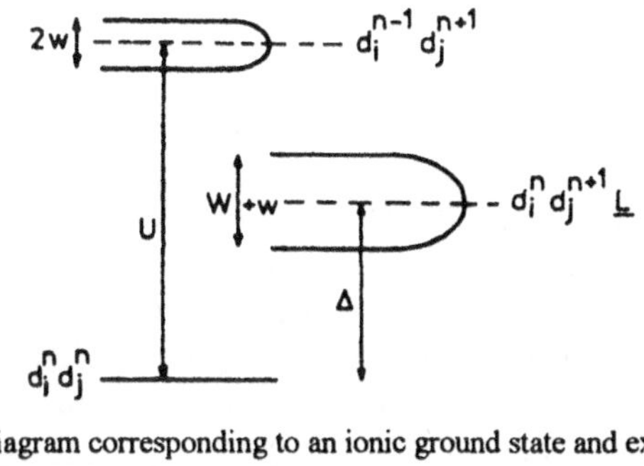


**Figure 14.** Total energy local diagram corresponding to an ionic ground state and excitations as shown in Fig. 13.

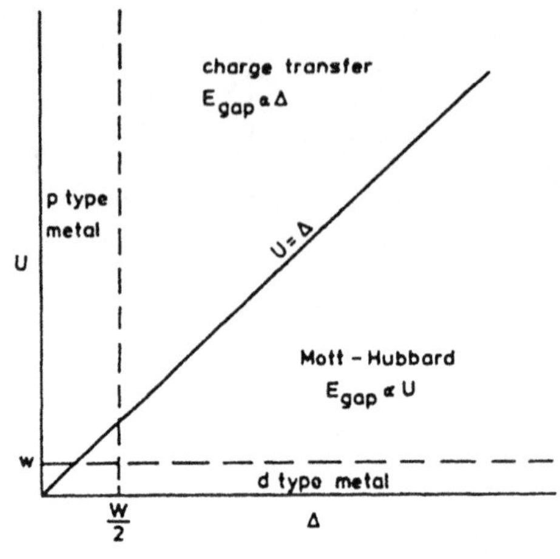


**Figure 15.** Simple phase diagram showing the various types of insulating and metallic states in transition metal compounds.

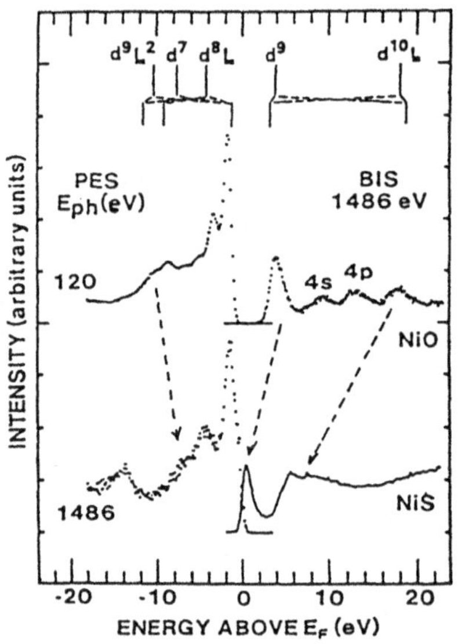


**Figure 16.** Top is the photoelectron - inverse photoelectron spectrum of NiO with the states labeled according to the ionic ansatz. The gap closes as we go from NiO to NiS as shown in the lower part.

Within the impurity limit we can describe the various forms of valence band spectroscopies such as optical, photoemission and inverse photoemission using a basis set of configurations as described in fig 17. The central diagram gives the states for the neutral system which gives us the ground state and the local excitations as for example determined by the optical spectrum. The eigenstates of the system with one electron removed are obtained from a calculation using the configurations on the left. This when combined with transition matrix elements yields the photoelectron spectrum. From the right figure we obtain the eigenstates of the system with one electron added which, together with matrix element effects, yields the inverse photoelectron or Bremstrahlung isochromat (BIS) spectrum. The band gap is determined from the minimum energy in each case as shown.

Shown in fig. 17 are only the states corresponding to one of the many multiplets in each of the three cases. To get the total spectral distribution the calculation will have to be done for each of the multiplets or for each of the irreducible representation (IR) of the appropriate point group. For example, to determine the ground state and optical spectrum we should calculate the eigenstates of the N electron system for each of the IR's as done in [54] for NiO.

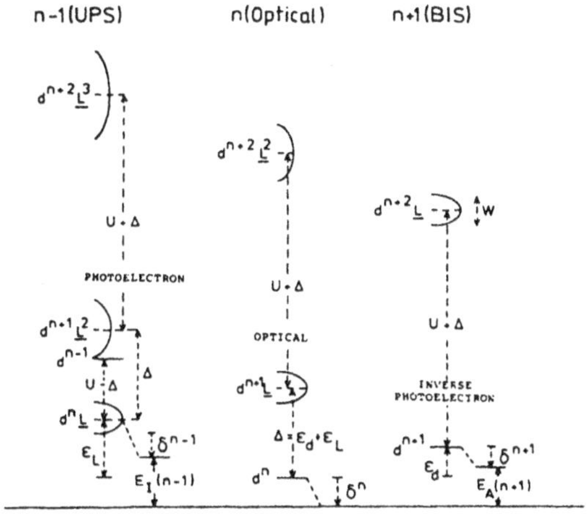


**Figure 17.** Total energy level diagram for the N -1, N, and N+1 particle states indicating the bases states for both the discrete and continuum states in the impurity approximation. Shown to the left in each case is the energy of the ground states for each case and in the center the spectroscopy which can be used to study the excitations in each case.

Since these experiments and theoretical interpretations numerous other investigations have been done which support these conclusions. These ideas have rather interesting consequences for the CuO layers in high $T_c$ materials. It is generally believed that the substitution of $Sr^{2+}$ for $La^{3+}$ in $La_{2-x}Sr_xCuO_4$ is charge compensated by holes in the CuO planes. For U >Δ these holes would be in the O 2p band as has in fact been confirmed by $O_{1s}$ electron energy loss [55] as well as $O_{1s}$ x-ray absorption spectroscopy [56]. At first glance this sounds very strange because it is after all the Cu 3d band which is open and the O2p band which is full according to band theory. We therefore would have expected the first electron removal state in CuO or in $La_2CuO_4$ to be from the open Cu 3d band. It is however the very strong correlation effects which cause the electron removal energies from a Cu $3d^9$ configuration in CuO to be considerably higher in energy than that for the O2p band. So the hole doped materials would have holes mainly in the O2p bands with the Cu remaining in a $d^9$ configuration. This picture has also been confirmed by resonant photoelectron spectroscopy.( For details see ref [34]). Here one looks at the photoelectron spectrum at photon energies passing through a core to 3d resonant absorption energy. For example in Cu we start from a say $3d^9$ state and with electron removal we would go to $3d^8$ or $3d^9L$ (L here stands for a hole in the O2p band or a ligand hole) if the electron is removed from the O2p states in photoemission. At the Cu 3p absorption threshold we can reach intermediate states of $3p^5 3d^{10}$ like states which will Auger decay into a final state which is again $3d^8$ and a free electron. At this resonant energy the two chanels ending up with $3d^8$ final states will interfere and we will be able to see directly where these $3d^8$ states are in energy. An example is given in Fig 18 showing the spectrum of CuO at and just above the resonance.

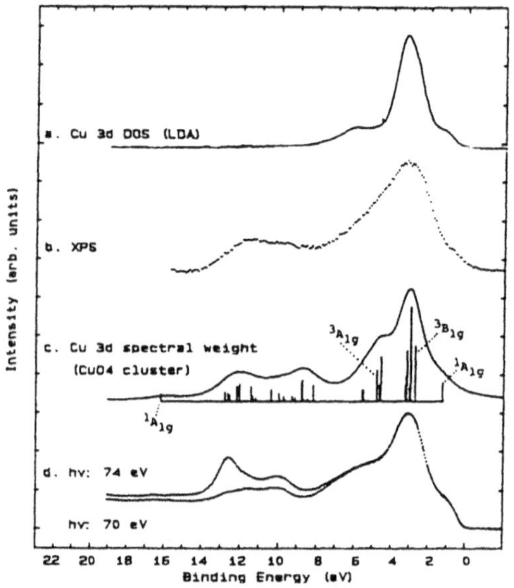


**Figure 18.** The photoelectron spectrum of CuO at the 3p-3d resonance photon energy of 74eV and just off resonance at 70eV. The strongly resonating structure has the expected multiplet structure of a d8 configuration in a D4H point group symmetry ( see ref [57]) . Also shown are the3d partial density of states obtained from LDA , the X ray photoelectron spectrum (XPS) and the Cu 3d spectral weight calculated for a CuO4 Cluster exhibiting the d8 multiplet structure.

There is no doubt from this measurement that the $3d^8$ states which are the 3d electron removal states are at very high energy and that the first electron removal states are of mainly O 2p character. Of course in general these states are mixed so that a real theoretical simulation is needed to extract the various parameters mentioned above which has been done [57]. In the case of CuO and the cuprates in general however it is clear that first electron removal state is of mainly O2p character. Realizing then that the first electron addition state is to a Cu 3d state ending up in a $3d^{10}$ configuration we are left with the strange conclusion that the band gap involves charge transfer states in spite of the fact that the 3d band is partially full in band theory.

There are many aspects in addition to the gap of transition metal compounds for which detailed knowledge of the electron-electron interaction is of extreme importance like the superexchange [57] and the O 2p - TM 3d exchange [58] just to mention a few. A detailed discussion however would take too much space so I refer to the quoted literature.

It is rather interesting to put all of the above into one picture and note the similarities, differences and the classes of phenomena. This is shown in Fig. 19. We start at the top 19(a) which describes a magnetic impurity for $U \gg \pi V2\rho$, a Kondo system for $E(d^{n-1})$ - $Ed(^{n})$ and $E(d^{n+1})$ - $E(d^{n}) \gg \pi V2\rho$, a mixed valence or valence fluctuation system for $E(d^{n-1})$ - $Ed(^{n})$ or $E(d^{n+1})$ - $E(d^{n}) < \pi V2\rho$ and $\pi V2\rho \leq kT$. In Fig. 19(b) is shown a 3d impurity in a semiconducting host for $U_{eff} < W$ and both the $d^{n-1}$ and $d^{n+1}$ states in the gap. This could be $Fe^{2+}$ in $CaF_2$ for example. For a lattice of such impurities and $U > W$ where W is the d bandwidth this would represent a Mott-Hubbard gap like for example in $V_2O_3$ below the semiconductor metal transition. Fig. 19c shows a large U case with only one of the d states ($d^{n-1}$) in the gap and the other ($d^{n+1}$) in the conduction band. Such a material would have quite heavy localized d holes ($d^{n-1}$) and light free electron like electrons. Fig. 19d shows the case for U >> W so that the gap is of a charge transfer type as in $NiCl_2$ for this case the electrons would be heavy and d like and the holes relatively light and anion p like.

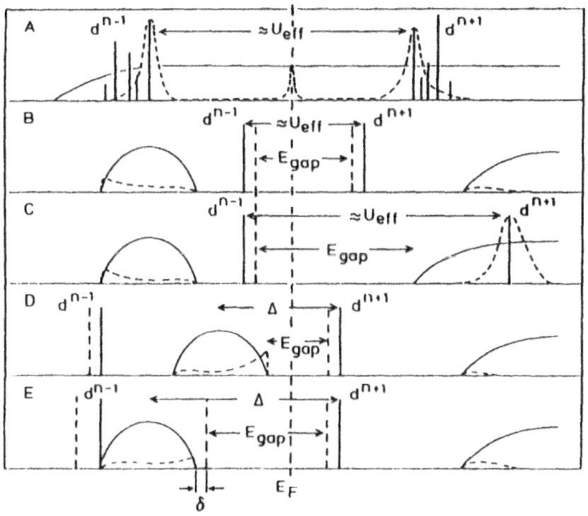


**Figure 19.** An artist's concept of the possible situations encountered for strongly correlated impurities in solids. Dashed lines indicate the effect of hybridization.

Fig. 19e is rather special. Although basically the same as 19d the hybridization between the $d^{n-1}$ state and the state with a hole in the valence band i.e. $d^9\underline{L}$ is strong enough to push a bound state out of the top of the valence band (dashed line) which now forms a state in the gap. This state has primarily ligand hole character but has the spin and symmetry of the $d^{n-1}$ states. This is very similar to the Haldane-Anderson [59] description of multiple charged impurity states in semiconductors. This state also has similarities to the Kondo state of Fig. 19a which is also a result of hybridization. In fact it has been proposed that this state in CuO is a local singlet because of a large O2p-Cu3d exchange and is a likely candidate for the quasi particle state providing the conductivity in the high $T_c$ materials.[60]. This so called Zhang-Rice singlet state is easily visualized in a small cluster or impurity like approach. Of course in the end we will have to take into account the translational symmetry. This has been done for the Zhang-Rice singlet like state resulting in a quasi particle with an effective band like dispersion given not by a hopping integral as in band theory but rather by the antiferromagnetic superexchange interaction. The model Hamiltonian much used to describe the electronic structure of the doped cuprates derived using the above is the so-called t,J model describing the hoping of the holes (bound into Zhang-Rice singlets) created now by chemical doping in an antiferromagnet spin background with a nearest neighbor (super) exchange interaction of J. A very complicated experimental verification of the idea that the lowest energy electron removal states in CuO are such Zhang-Rice singlets was recently made using spin polarized circular light polarized resonant photoelectron spectroscopy [61].

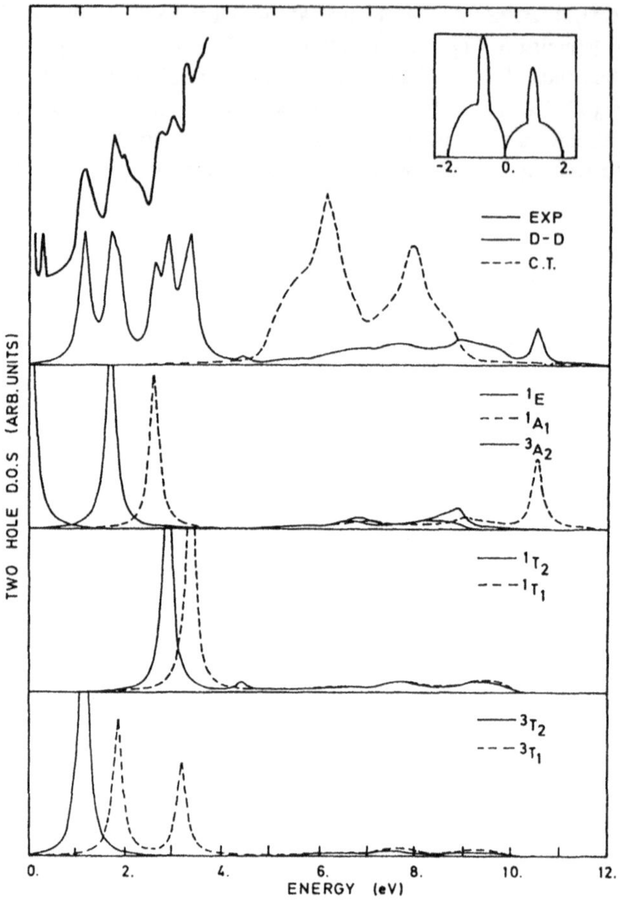


**Figure 20.** The spectral distribution of Ni d8 character for different two hole irreducible representations as found from an impurity two particle Greens function calculation. In the top figure the sum of these is compared to the experimental optical spectrum of ref 62. Also shown are the single impurity charge transfer transitions corresponding to the conductivity gap and in the inset the model O 2p density of states used in the calculation.

In addition to looking at the electron addition, removal and Auger spectroscopies the influence of electron correlation effects is also very evident from optical and electron energy loss spectroscopy. In these experiments we conserve the number of electrons in the solid and simple look at the excited states of the n electron system as given by the middle panel of fig 17. These excitations could involve local charge neutral excitations such as excitons which could very well have energies below the band gap excitations. To distinguish such charge neutral excitations from the real band gap excitations one can look at the photoconductivity onset. This onset will occur at photon energies corresponding to excited states in which the electron and hole are not bound. Any structure below this onset must be due to locally charge neutral excitations such as excitons. In the 3d transition metal compounds such excitonic states can be separated into two classes. First we have the local excitations to excited d electron states involving transitions between the various multiplet (crystal field states) states of the transition metal ion with a given number of d electrons. These excitations involve the higher order d-d Coulomb integrals and the crystal field splittings. The energy range can range from about 1 eV starting from the Hund's rule ground state to about 5 eV. For example the photoconductivity gap in NiO is 4.0 eV which is also the gap as determined from a combination of photoelectron and inverse photoelectron spectroscopy. Inside this energy scale we see in the optical conductivity sharp excitations starting at about 1 eV and extending all the way to the gap itself. These states are only weakly seen because they are optically allowed transitions but always get a bit of oscillator strength via spin orbit or electron phonon coupling. This low energy part of the spectrum is shown in Fig 20 [62] together with the assignemnts of the peaks using an impurity like two particle greens function calculation with parameters determined from the photoelectron spectra [63]. These excited states have the symmetries spanned by two d holes in Oh symmetry displaying again the importance of the local interactions in contrast to the translational symmetry. Of course translational symmetry is important as it was for the Auger spectra and this will cause the excitonic states to delocalize but now as a tightly bound electron hole pair. These states therefore behave like Frenkel excitons with a small dispersion just like the two hole bound states in the Auger problem.

In addition to these tightly bound Frenkel exciton states there may also be more loosely bound electron hole states in which the hole and the electron are separated by large distances but yet have an energy below that of the band gap. Such excitonic states are clearly seen in semiconductors like Si and GaAs but are also present as so called charge transfer excitons in transition metal oxides. In addition to optical spectroscopy to study excitonic states one can also use electron energy loss spectroscopy [63] also including spin polarization [64]. More recently synchrotron based x ray Raman or photon energy loss spectroscopy have also provided very useful information on the d-d excitonic states of the cuprates [65]. These latter methods are proving to be very powerful techniques which have the advantage of not being limited to q=0 excitations as in standard optics.

## CONCLUSIONS

In this paper I have attempted to give a simple physical picture of why electron correlation effects are important for describing the physical properties of materials and how we can measure the electron-electron interactions using examples out of some of the "hottest" topics in solid state physics and chemistry of the last few decades. I have especially emphasized Auger spectroscopy to directly see the importance of electron correlation by comparing two particle excitations to one particle excitations, and combinations of photoelectron and inverse photoelectron spectroscopies to determine the band gaps and the nature of these gaps. In these studies I have shown that the first ionisation states in strongly correlated magnetic systems are quite different from those obtained from band theory and that the energy spread in the single electron removal spectrum is often dominated by many body effects to a much larger extent than by band structure or translational symmetry effects. I have also used examples from optical spectroscopy and electron energy loss spectroscopy to identify low energy scale excitations in correlated insulators which are bound two particle excitonic states and therefore not present in band structure calculations. The nature of these states can give direct insight into the importance of electron correlation effects. Also for these states which, often are the lowest energy electronic excitations, translational symmetry is often of minor importance compared to the multiplet structures due to local Coulomb and exchange interactions.

All this leads us to the conclusion that in many cases such as in the 3d transition metal compounds and the rare earths among others local Coulomb interactions are much more important than the translational symmetry and that a better starting point to the electronic structure and excitations may be one using broken symmetry such as in effective impurity of cluster models to get the high energy scale multiplet structure upon which we might switch on the translation symmetry to see how these states propagate and eventually interact. Of course all this was done with the aim of understanding the physical properties of these systems and their dependence on chemical composition and external parameters.

## ACKNOWLEDGEMENTS

This investigation was supported by the Foundation for Fundamental Research on Matter (FOM) with financial support from the Netherlands Organization for the Advancement of Pure Research (NWO).

## REFERENCES

1. P. Hohenberg and W. Kohn, Phys. Rev. 136, B864 (1964);
2. W. Kohn and L.J. Sham, Phys. Rev. 140, A1133 (1964).
3. H.J. de Boer and E.J.W. Verweij, Proc. Phys. Soc. 48, 59 (1937).
4. N.F. Mott, Proc. Phys. Soc. Sect. A 62, 416 (1949).
5. J. Hubbard, Proc. Phys. Soc. Ser A, 277, 237 (1964).
6. F. Bloch, Z. Phys. 57, 545 (1929).
   A.H. Wilson, Proc. Roy. Soc. A133, 458 (1931).
7. N.F. Mott, in Metal Insulator Transitions, (Taylor & Francis, London 1974).
8. P.W. Anderson, Phys. Rev. 115, 2 (1959);
   Solid State Physics 14, 99 (1963).
9. G. van der Laan, C. Westra, C. Haas and G.A. Sawatzky, Phys. Rev. B23, 4369 (1981).
10. A. Fujimori and F. Minami, Phys. Rev. B30, 957 (1984).
11. G.A. Sawatzky and J.W. Allen, Phys. Rev. Lett. 53, 2339 (1984).
12. Spectroscopy of Mott Insulators and correlated Metals ( Springer series in solid state science 119, edited by A.Fujimori and Y. Tokura 1995)
13. M. Imada, A. Fujimori, and Y. Tokura Reviews of Moders Physics 70, 1039 (1998)
14. S. Hufner, Photoelectron Spectroscopy Principles and Applications (Springer Series in Solid State Science 82 1995)
15. J. Kondo, Prog. Theoret. Phys. (Kyoto) 28, 846 (1962);
    32, 37 (1964); 34, 204 (1965) see also the book by Abrikosov Fundamentals of the Physics of Metals ( North Holland Amsterdam 1988).
16. M.B. Brodsky in Valence Instabilities and Related Narrow Band Phenenomena, edited by R.D. Parks (Plenum, New York, 1977), p. 351.
17. J. Friedel, J. Phys. Radium 19, 573 (1958).
18. P.W. Anderson, Phys. Rev. 124, 41 (1961).
19. J.R. Schrieffer and P.A. Wolff, Phys. Rev. 149, 491 (1966).
20. F.J. Himpsel, Adv. Phys. 32, 1 (1983).
21. P.Thiry, O. Chandesris, J. Lecant, C. Guillet, R. Pinchaux and Y. Petroff, Phys. Rev. Lett. 43, 82 (1979).
22. N.V. Smith, Reports in Progress of Physics, 51, 1227-1294 (1988).
23. D.P. Woodruff and N.V. Smith, Phys. Rev. Lett. 48, 283 (1982);
    W. Jacob, V. Dose and A. Goldman, Appl. Phys. A41, 145 (1986).
24. For energies larger than about 5 eV above $E_F$ systematic deviations are found which are due to the exchange correlation hole (as used in DF theory) not being able to follow the added electron.
    W. Speier, R. Zeller and J.C. Fuggle, Phys. Rev. B32, 3597 (1985).
25. J.J. Lander, Phys. Rev. 91, 1382 (1953).
26. J.C. Riviére, AERE Harwell rep. no. AERE-R 10384 (1982).
27. H. Haak, G.A. Sawatzky and T.D. Thomas, Phys. Rev. Lett. 41, 1825 (1978).
28. M. Cini, Solid State Commun. 24, 681 (1977).
29. G.A. Sawatzky, Phys. Rev. Lett. 55, 418 (1985).
30. G.A. Sawatzky and A. Lenselink, Phys. Rev. B21, 1790 (1980).

31. J.C. Fuggle "High Resolution Auger Spectroscopy" in "Electron Spectroscopy IV", ed. C.R. Brundle and A.D. Baker, p.85 (Academic, London) 1981.
P. Weightman, Rep. Prog. Phys. 45, 753 (1982);
D.E. Ramaker "Auger Spectroscopy as a Probe of Valence Bonds and Bands", Springer series in Chemical Physics, 1982.
32. L.C. Davis, J. Appl. Phys. 59, R25 (1986).
33. S. Hüfner and G.K. Wertheim, Phys. Lett. A51, 299 (1975).
34. C. Guillot, Y. Ballu, J. Paigné, J. Lecante, K.P. Jain, P. Thiry, R. Pinchaux, Y. Petroff and L.M. Falicov, Phys. Rev. Lett. 39, 1632 (1977).
35. L.M. Falicov in "X-Ray and Inner Shell Processes", edited by P. Lagarde, F.J. Wuilleumier and J.R. Briand (Les editions de Physique, France) p. C9-891 (1987).
36. C. Herring, in Magnetism, Vol. IV, edited by T. Rado and H. Suhl (Academic, New York, 1964).
37. D. van der Marel and G.A. Sawatzky, Phys. Rev. 37, 10674 (1988).
38. W. Harrison in Electronic Structure and the Properties of Solids (Freeman, San Francisco, 1980).
39. J.K. Lang, Y. Baer and P.A. Cox, J. Phys. F11, 121 (1981).
40. Valence Instabilities and Related Narrow-Band Phenomena, edited by R.D. Parks (Plenum, New York, 1977);
Valence Fluctuations in Solids, edited by L.M. Falicov, W. Hanke, and M.B. Maple, (North-Holland, Amsterdam, 1981);
Proceedings of the 4th International Conference on Valence Fluctuations, Cologne, 1984, edited by E. Müller-Hartmann, B. Roden and D. Wohlleben, (North-Holland, Amsterdam, 1985);
Proceedings of the 5th International Conference on Crystalline Field and Anomalous Mixing Effects in f-electron Systems, Sendai 1985, edited by T. Kasuya, (North-Holland, Amsterdam, 1985).
41. P. Nozieres and D. Pines Phys. Rev. 109, 762 (1958)
42. J.S. Griffith in The Theory of Transition Metal Ions, (Cambridge University Press, 1961).
43. D. van der Marel, G.A. Sawatzky and F.U. Hillebrecht, Phys. Rev. Lett. 53, 206 (1984).
44. D. van der Marel and G.A. Sawatzky, Solid State Commun. 55, 937 (1985).
45. W. Speier, J.C. Fuggle, R. Zeller, B. Ackermann, K. Szat, F.U. Hillebrecht and M. Campagna, Phys. Rev. B30, 6921 (1984).
46. J.C. Fuggle, P. Bennet, F.U. Hillebrecht, A. Lenselink and G.A. Sawatzky, Phys. Rev. Lett. 49, 1787 (1982).
47. The atomic values can be obtained from atomic spectroscopic data.
C.E. Moore, in Atomic Energy Levels, Natl. Bur. Stand. (USA), Circ. No. 467.
48. A.N. Gerritsen and J.O. Linde, Physica, 17, 573 (1951).
49. J.W. Allen, S.-J. Oh, O. Gunnarsson, K. Schönhammer, M.B. Maple, M.S. Torikachvilli and I. Lindau, Adv. in Phys. 35, 275.
50. F. Patthey, J.-M. Imer, W.-D. Schneider, Y. Baer, B. Delley and F. Hulliger, Phys. Rev. B36, 7697 (1987).
51. O. Gunnarsson and K.Schönhammer, Phys. Rev. Lett. 50, 604 (1983);
Phys. Rev. B28, 4315 (1983). See also ref. 48.
52. R. Coehoorn, Thesis, University of Groningen, the Netherlands (1985).
53. J. Zaanen, G.A. Sawatzky and J.W. Allen, Phys. Rev. Lett. 55, 418 (1985).
54. J. Zaanen and G.A. Sawatzky, Progress in Theoretical Physics 101, 231 (1990).
55. J.A. Yarmoff, D.R. Clarke, W. Drube, U.O. Karlsson, A. Talab-Ibrahimi and F.J. Himpsel, Phys. Rev. B36, 3967 (1987).
56. N. Nücker, J. Fink, J.C. Fuggle, P.J. Durham, W.M. Timmerman, Phys. Rev. B37, 5158 (1988).

57. J.Ghijsen et al.Phys. Rev.B 38, 11322 (1988); 42, 2268 (1990): and L.H.Tjeng et al. Phys.Rev. Lett. 67, 501 (1991)
58. J. Zaanen and G.A. Sawatzky, Can. J. Phys. 65, 1262 (1987).
59. F.D.M. Haldane and P.W. Anderson, Phys. Rev. B13, 2553 (1976).
60. H. Eskes and G.A. Sawatzky, Phys. Rev. Lett. 61, 1415 (1988); F.C. Zhang and T.M. Rice, Phys. Rev. B37, 3759 (1988).
61. L.H.Tjeng et al.Phys.Rev. Lett. 78, 1126 (1997)
62. R.Newman and R.M. Cherenko Phys.Rev. 114, 1507 (1959)
63. H.P. Geserich, G. Scheiber, J. Geerk, H.C. Li, G. Linker, W. Assmus and W. Weber, Europhys. Lett. 6, 277 (1988).
64. B. Fromme, U. Brunokowski and E. Kisker Phys. Rev. B 58, 9783 (1998)
65. P. Kuiper et al.Phys. Rev. Lett. 80, 5204 (1998)

## Discussion

**Kohn:** Can you explain the physics behind the experimental observation that the on-site coulomb repulsion, that is the monopole part of the coulomb interaction, is strongly screened in the solid whereas the exchange and higher multipole components describing the atomic multiplet structure in the solid are not?

**Sawatzky:** I repeated the experimental evidence for this using the Auger spectrum of Cu metal as an example in which the on-site monopole part of the coulomb interaction is reduced from about 18eV in the free atom to about 6 eV in the solid, while the multiplet splitting visible in the Auger spectrum is very close to the free atom value showing that the higher order coulomb and exchange integrals are not or at most weakly screened in the solid. The reason I gave for this is that monopole interaction involves differences in charge, i.e., in the case of Cu Auger the final state is d8 whereas the initial state was d10 and the monopole coulomb interactions in the energy difference between a d8 state and 2 times a d9 state, i.e., the extra energy to place two holes on the same atom. The strong reduction in the solid is caused by the response of the polarizable surroundings to the introduction of such a charge. Recall that the reduction in the energy to create a local charge in polarizable medium is proportional to the charge squared so that the reduction in energy to produce a d8 state involving a local charge of 2 is 2 times as that to produce two separated charges as in two d9 states.

On the other hand, if the local orbitals in which the holes reside are rather atomic-like and have radial extents much smaller than the interatomic distance, then the surroundings does not really reach differently to whether the two holes have spin parallel or spin antiparallel or if they are in differently oriented spatial d orbitals. Therefore, the higher order coulomb and exchange interactions are not or hardly reduced from the free ion values.

In a private discussion with Kohn we came to the conclusion that another way to see this is to look at the nature of a coulomb vs. an exchange integral. The former involves density, density-like terms in the integral whereas the exchange involves overlap, overlap-like terms. The latter, therefore, will involve the shorter wavelength parts of the coulomb interaction, which are not screened in the convention sense. For more information on this, I can refer to work by Antonides and by van der Marel.

**Kohn:** Could you give a physical picture supporting your statement that the short-range electron-electron interaction on a C60 molecule and in finite size systems in general, are strongly screened whereas the longer range interactions are weakly screened or even antiscreened?

**Sawatzky:** To demonstrate this I drew a picture of a hollow sphere on the flip over and considered the spherical shell to represent the C60 molecule and also to represent the polarizable medium. If we now place a test charge (our electron) on one side of the sphere just outside it, then it is easy to see that there will be a net charge flow in the spherical shell resulting in a positive charge close to the negative test charge, and since the sphere itself is charge neutral a negative charge on the other side. It is then obvious that the short-range potential of the test charge will be reduced from the free 1/r behavior but the long range part, i.e., the potential on the other side of the sphere will, in fact, be larger than the bare 1/r potential. This argument can be extended to other shapes and in general finite size systems. For more information on this I refer to Gunnarson, and the thesis by van de Brink Thesis.

# PHOTOEMISSION IN STRONGLY CORRELATED CRYSTALLINE f-ELECTRON SYSTEMS: A NEED FOR A NEW APPROACH

A.J. Arko[1], J.J. Joyce[1], J. Sarrao[1], Z. Fisk[1], J.L. Smith[1], J.D. Thompson[1], M. Hundley[1], A. Menovsky[2], A. Tahvildar-Zadeh[3], and M.A. Jarrell[3]

[1]Los Alamos National Laboratory
Los Alamos, Nm 87545

[2]Natuurkundig Laboratorium
University of Amsterdam
The Netherlands

[3]Univ. of Cincinnati
Cincinnati, Ohio 45221-0011

## INTRODUCTION

The unusual properties of heavy fermion (or heavy electron) materials have sparked an avalanche of research over the last two decades in order to understand the basic phenomena responsible for these properties. The reader is referred to numerous review articles, such as Fisk et al., (1995), Grewe and Steglich (1991), Hess et al., (1993), Lawrence and Mills (1991), Ott and Fisk (1987), Stewart (1984), Thompson and Lawrence (1994). The term heavy fermion refers to materials (thus far only compounds with elements having an unfilled 4f or 5f shells) whose electronic properties suggest that the conduction electrons have a very heavy effective mass. This is inferred from the linear term in the low specific heat ($Cp = \gamma T + \sigma T^2$) which usually has an upturn at low temperatures and suggests [some of the first discoveries were by Andres et al. (1975), followed by Steglich (1979), Stewart et al (1984), Ott et al. (1983)] γ-values as large as 1500 mJ/mole-K$^2$. deHaas-van Alphen measurements have confirmed the existence of heavy masses [Springford and Reinders (1988), Lonzarich (1988), Onuki (1993)]. The quadratic term in the resistivity and the Pauli susceptibility also indicate the existence of a mass enhancement. It is well established that indeed it is the f-electrons that are responsible for the unusual properties [Grewe and Steglich (1991)].

Magnetic susceptibility measurements, χ, generally yield a Curie-Weiss behavior at high temperatures with a well developed moment [see, for example, Thompson and Lawrence (1994)], which would be consistent with a localized behavior of the f-electrons. Indeed, the f-electrons appear to behave as non-interacting single impurities at elevated temperature, while below some characteristic temperature, usually referred to as the Kondo Temperature or $T_K$, the susceptibility levels off or even decreases. This phenomenon is interpreted as a compensation of the magnetic moment by the ligand conduction electrons that are believed to align anti-parallel to the f-electrons to form a singlet state [see, for example, Grewe and Steglich (1991)]. At still lower temperatures a dramatic drop in the electrical resistivity, ρ, is associated with the formation of a coherent periodic lattice of these compensated f-electrons so that the low temperature electronic structure can be consistently interpreted as being due to a band of very heavy mobile fermions. This is often referred to as the Kondo lattice or the Anderson lattice. The concept of a Kondo or Anderson lattice appears to date back to Lawrence et al. (1981), but the reader is referred also to an excellent review of theoretical approaches by Lee et al. (1986).

*Electron Correlations and Materials Properties*, edited by Gonis *et al.*
Kluwer Academic/Plenum Publishers, 1999.

There is, however, another school of thought that claims that the f-electrons form well-defined Bloch states and very narrow bands at all temperatures [Zwicknagl (1992), Liu (1993, 1997), Sheng and Cooper (1995)]. While there is little quarrel with this view at low temperatures, conventional band theory (using the Local Density Approximation, or LDA) is clearly unable to explain the high-temperature properties as well as the very heavy mass. Renormalized band theory [Strange and Newns (1986), Zwicknagl (1992)] may yet prove useful, especially those that incorporate physics similar to the Kondo effect and its temperature dependence, perturbatively [Steiner (1994), Norman (1985), Norman et. al. (1985)]. However, the apparent success of the Single Impurity Model (SIM) in explaining macroscopic bulk phenomena suggests a correctness, at least at some level, of the model, though recent results (C. Booth, private communication) indicate that attempts to extract bandwidths from bulk data yield totally unphysical results. This suggests a need to re-evaluate the SIM even in connection with bulk data. The heart of the issue clearly then is the exact role of the f-electrons in the formation of the heavy fermion state. In this respect the effect of the periodic lattice is non-negligible. The periodic Anderson model (PAM) has made headway in treating this problem and will be discussed in detail though other periodic models are not yet ruled out.

As stated, the localized approach utilizes (SIM) as the starting point. The assumption here is that even in periodic systems (i.e., ordered compounds), where an f-electron exists within each unit cell, the f-f overlap is so small that at high temperatures the f-electrons can be treated as completely localized impurities in a sea of ligand conduction electrons having a density of states N(E). It is the very slight residual hybridization with N(E) which yields the low temperature behavior as the hybridization energy exceeds $k_BT$. The Gunnarsson and Schonhammer (1985, 1986) approximate solution of SIM calculates the particle-hole excitation spectrum (we will refer to it as GS) and has been hailed as a breakthrough in the understanding of heavy fermion systems. The non-crossing approximation, or NCA, has in addition been able to introduce [Bickers (1987), Bickers et. al. (1985 and 1987)] the effects of temperature, and has recently included the effect of finite f-f Coulomb repulsion, $U_{ff}$, [Pruschke (1989)] thus including the possibility of double occupancy of f-levels. Without a doubt the GS and NCA approximations are the most comprehensive, as well as the most widely accepted approximations of the particle-hole excitation spectrum in heavy fermions, in the Kondo limit. More recently numerical renormaliztion group (NRG) [Wilson (1975)] and Quantum Monte-Carlo (QMC) [Hirsch and Fye (1986)] approaches have been extended to treat dynamics [Oliveira and Wilkins (1981), Silver et. al. (1990), Jarrell and Gubernatis (1996)]. A universal behavior for the f-electron DOS comes out of these approximate solutions, the core of which is the prediction of scaling with $T_K$.

Photoelectron spectroscopy (often referred to as PES in the following sections), the most direct measurement of the electronic structure of a material, should in principle be able to shed considerable light on this matter. In general the distinction between a localized and a band-like state is trivially observed in band dispersion. However, in the case of heavy fermions such distinctions pose a serious problem since any dispersion of an f-band within the coherent state is expected to be of the order of $T_K$ which is generally far below the resolution of present day apparatii. Nonetheless, PES measurements on heavy fermions have been abundant [For a review of some of the early work on poly-crystals see Allen et al. (1986 and 1992)], owing to the very specific predictions of the GS and NCA approximations regarding the width, position, spectral weight, and temperature dependence of the f-electron DOS, which lends itself readily to a photoemission measurement.

Much of the past work was performed on poly-crystalline samples, scraped *in-situ* to expose a clean surface for PES. Further, despite the fact that the features studied were at least an order of magnitude narrower than the then-available experimental resolution (or perhaps because of it), good agreement with the GS approximation was often reported [Patthey et al. (1985, 1986, 1987a and b, 1990), Malterre et al. (1992a and b), Tjeng et al. (1993), Kang et al. (1990), Allen et al. (1986), Liu et al. (1992), Weibel et al. (1993)]. There have since been considerable advances both in the quality of specimens as well as experimental resolution, which raise questions regarding these conclusions. Much of the past work on poly-crystalline samples has been reported in several review articles, most notably Allen et al (1986, 1992), and it is not necessary here to review those efforts again, with the exception of subsequent work performed at high resolution. The primary focus of the present review will be on new measurements obtained on single crystals, cleaved or prepared *in situ* and measured at high resolution, which seem to suggest [Joyce et al. (1992a,b,c), (1993a,b),

(1994a,b), (1996), Arko et al. (1993), (1994), Andrews et al. (1994), (1995a,b), (1996), Blyth et al. (1992), (1993)] that agreement with the GS and NCA approximations is less than perfect, and that perhaps the starting models need to be modified, or that even an entirely new approach is called for. Of the promising new models the Periodic Anderson Model is most closely related to the SIM. Indeed, at high temperatures it reverts to the SIM. However, the charge polaron model of Liu (1997) as well as the two-electron band model of Sheng and Cooper (1995) cannot yet be ruled out.

Inasmuch as the bulk of the single crystal work was performed by our own Los Alamos group [Andrews et al. (1995a,b), (1996), Joyce et al, (1996)], this review will draw heavily on those results. Moreover, since the GS and NCA approximations represent the most comprehensive and widely accepted treatment of heavy fermion PES, it is only natural that we primarily concern ourselves with analysis of PES data in terms of these models, in order to thoroughly test their validity in light of the new data.

The evolution of theoretical models has been driven by experiment. To date, the f component of the one electron addition/removal spectra of alloys containing Ce ions, as well as ordered Ce compounds, are qualitatively similar. Due to the difficulty of separating out the f component of the spectrum, the only concentration dependence that has been identified is the linear dependence of the overall intensity. This unsatisfactory state of affairs has caused theorists to focus their attentions primarily on single impurity models that have the same qualitative features as observed in experiments. Thereby, effects that are present in ordered compounds have been, hitherto, largely ignored. In this review our main focus will be on ordered compounds, indeed, single crystals.

## THEORETICAL CONSIDERATIONS

### Kondo Models

Most models are based on the single impurity Anderson (1961) model, which describes an f shell that is bathed in a Fermi-sea of conduction electrons. The Hamiltonian, H, can be written as the sum of three terms,

$$H = H_f + H_d + H_{fd}$$

where $H_f$ describes the f electron system, $H_d$ describes the conduction electron systems and $H_{fd}$ describes the coupling between these two systems. The Hamiltonian $H_f$ can be written as

$$H_f = \sum E_f f^+_\alpha f_\alpha + \sum U_{ff}/2\, f^+_\alpha f^+_\beta f_\beta f_\alpha$$

where the operators $f^+_\alpha$ and $f_\alpha$ respectively create and annihilate an electron of spin $\alpha$ in the f shell. The first term, proportional to $E_f$, represents the binding energy of an electron in the f level, and the second term represents the coulomb repulsion between pairs of electrons in the f shell. The Pauli exclusion principle makes this term, proportional to $U_{ff}$, inactive when $\alpha = \beta$. The conduction electron states are described by $H_d$, as non-interacting conduction electrons in band states,

$$H_d = \sum e(\underline{k})\, d^+_{\underline{k},\alpha} d_{\underline{k},\alpha}$$

where the operators $d^+_{\underline{k},\alpha}$ and $d_{\underline{k},\alpha}$, respectively, create and annihilate electrons of spin $\alpha$ in the conduction band state labeled by the Bloch wave-vector $\underline{k}$. The quantity $e(\underline{k})$ is the dispersion relation of the conduction band states. The f electron and d electron systems are coupled via the hybridization process, $H_{fd}$, which can be written as

$$H_{fd} = \sum (V(\underline{k})\, d^+_{\underline{k},\alpha} f_\alpha + V^*(\underline{k})\, f^+_\alpha d_{\underline{k},\alpha})$$

The first term represents a process in which an electron of spin $\alpha$ is taken out of the f shell and placed in a conduction band state, the Hermitean conjugate represents the reverse process. The hybridization term conserves the spin index $\alpha$ . This term, when treated in second order perturbation theory yields a lifetime, or effective width of the f level given by $\Delta = \pi\, |V|^2 \rho(\mu)$, where $\rho(\mu)$ is the conduction band density of states at the fermi-level.

The solutions to the above Hamiltonians are lengthy and laborious. Gunnarsson and Schonhammer (1985, 1986) after a series of approximations succeeded in calculating the particle-hole excitation spectrum for the case of T = 0. Subsequently, Bickers, Cox and Wilkins (1985, 1987), using the non-crossing approximation (NCA) succeeded in calculating the spectrum at finite temperatures, but at infinite $U_{ff}$. A finite $U_{ff}$ calculation was eventually obtained by Pruschke (1989). It is beyond the scope of this paper, however, to review in detail the theoretical approaches. Rather it is the comparison of theory to experiment, and the possible sorting out of the models which attracts our attention. For this reason we will rather show the excitation spectrum predicted by NCA since the comparison to NCA will form the bulk of this paper. The inconsistencies encountered between experiment and NCA predictions will force us to consider other models, especially the periodic Anderson Model.

Nozieres (1985) has pointed out that the formation of the polarization clouds may be expected to only involve the conduction electrons within $k_BT_K$ of the fermi-energy. Thus the number of electrons forming the compensating polarization cloud may be expected to be of the order of $k_BT_K\rho(\mu)$, where $\rho(\mu)$ is the conduction band density of states evaluated at the fermi-level. Since $k_BT_K$ is much less than the depth of the occupied portion of the conduction band density of states, this yields an estimate of much less than one conduction electron per f ion. Thus, the estimated number of electrons in the polarization cloud is less than the number of f moments to be screened so an alternative description, other than the single impurity description, may have to be utilized for concentrated compounds.The periodic Anderson Model (PAM) appears to fit this description since in this picture the compensationg conduction electrons are allowed to hop from f-site to f-site thus effectively screening in spite of insufficient numbers, and forming a narrow hybridized f-band - *a magnetic polaron.*.

But before considering this approach, let us look at the NCA predictions. A graphic representation of the predictions of NCA is perhaps most useful to the reader. An unbroadened calculated spectrum is shown in Fig. 1a (code provided by D.L. Cox) for a typical heavy fermion material having a $T_K$ of 35K, a model (flat) conduction band 5eV wide, crystal field levels at -25 and -48 meV, $U_{ff}$ of 7 eV, and a bare f-level positioned at $\varepsilon_f \approx -2.3$ eV. These parameters approximately correspond to those of the heavy fermion compound $CeSi_2$ which has been heavily reported in literature [Patthey et al (1986), Joyce et al. (1992c), Takahashi et al. (1994), Malterre et al. (1992a,b), Garnier et al. (1997)]. It should be

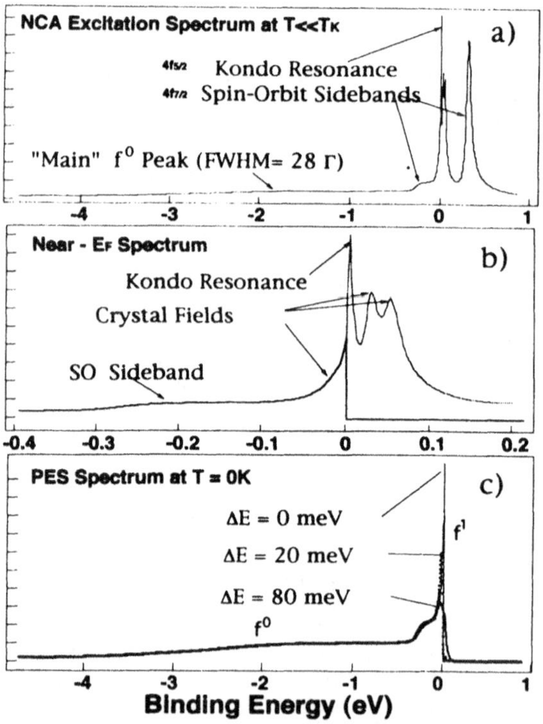


Fig. 1 (a) Particle-hole excitation spectrum calculated for T = 11 K using NCA code provided by D.L. Cox. Parameters approximately correspond to those expected for $CeSi_2$, namely: $T_K$ = 37 K, W = 5 eV (Lorentzian), $\Gamma$ = 0.1439, $\Delta_{SO}$ = 280 meV, $e_f$ = -2.4 eV, $\Delta_{CF1}$ = -25 meV, $\Delta_{CF2}$ = -48 eV, $U_{ff} = \infty$. (b) Blow-up of near-$E_F$ region. Dark line corresponds to convolution of spectrum with a T = 0 K Fermi function. (c) Calculated PES spectrum at T $\approx$ 0 K using three different instrument resolutions.

stressed that within the SIM the 4f DOS in Fig. 1a is obtained by averaging the hybridization over all conduction electron states, so that dispersion is precluded. To be more precise, we are not actually dealing with a 4f DOS at the Fermi energy but rather only the response of the system to a core hole. We will nevertheless refer to the intensity at the Fermi energy as the 4f DOS. This was nicely explained for the experimentalist by Huffner (1986) and Huffner and Schlapbach (1986). In any case it is argued that NCA yields an excellent representation of the 4f DOS at high temperatures, while at low temperatures (far below $T_K$ where a Kondo Lattice model is more appropriate) any dispersion in the coherent state is in any case smaller than $k_BT_K$ and thus negligible. The essential predictions of SIM (i.e., the scaling of 4f spectral features with $T_K$) are of necessity retained at all temperatures.

Although Ce contains only one f-electron, the GS theory [Gunnarsson and Schonhammer (1985, 1986)] as well as NCA [Bickers et. al. (1985, 1987)] predicts at least three features in a PES spectrum (occupied states) and at least two features in a BIS spectrum (remaining 13 empty states). In Fig. 1a the occupied features are labeled as the $f^0$ or "main" peak which represents the primary $f^1$ to $f^0$ excitation, while the $f^1$ portion of the spectrum consists of the narrow $4f_{5/2}$ feature (called the Kondo Resonance, KR) located at $k_BT_K$ above $E_F$ with only a small occupied tail extending below $E_F$ (see Fig. 1b), as well as a spin-orbit split sideband (the $4f_{7/2}$ feature) located at $\Delta_{SO}$-$T_{SO}$ below the KR (Here $\Delta_{SO}$ is the spin-orbit splitting while $T_{SO}$ is an effective Kondo temperature for the sideband, usually about an order of magnitude larger than $T_K$). Above the Fermi energy we have the bulk of the KR at $\approx k_BT_K$, an equally intense spin-orbit sideband at $+\Delta_{SO}$, as well as a broad $f^2$ feature (not shown) corresponding to double f-occupancy at finite $U_{ff}$, and situated at $U_{ff}$ above the occupied $f^0$ peak. The KR may additionally be split into crystal field levels situated at $\Delta_{CF}$ relative to $T_K$, which have their own effective Kondo temperature, $T_{CF}$, and which may actually carry more spectral weight than the KR. These are shown in the expanded view in Fig. 1b.

Because of particle-hole symmetry, the PES spectrum for Yb heavy fermions, where one has only one f-hole ($f^{13}$ vs $f^{14}$) rather than one f-electron, can be obtained from Fig. 1a simply by applying mirror symmetry about the Fermi energy [Bickers et al. (1987)]. Thus in Yb heavy fermions the bulk of the KR is predicted to be occupied which makes them ideal candidates for studying spectral weights, widths, and temperature dependencies. The ability to determine $n_f$, the f-hole occupancy (f-electron for Ce compounds), is likewise made possible in Yb compounds since the f-spectral weight is fully occupied. The $n_f$ more or less corresponds to the percentage of the f-hole in the $f^{13}$ state, which should be close to unity in the Kondo limit. The $f^{14}$ spectral weight in the Kondo limit is then roughly ($1$-$n_f$) with small adjustments for the differences in the degeneracy

From Bickers et al. (1987) we note several important points regarding Fig. 1a. (1) The width of the $f^0$ peak varies with the hybridization $\Gamma$ (width = $28\Delta$ when using a model DOS), and has a spectral weight of about one electron (i.e., $n_f \approx 1$). The width of the KR $\approx k_BT_K$, while its total spectral weight (including the unoccupied portion) scales with $T_K$ approximately as $\pi T_K/\Gamma$. The position of the SO sideband varies as $\Delta_{SO}$- $T_{SO}$ so that for increasing $T_K$ it shifts closer to $E_F$ as $T_{SO}$ also increases along with $T_K$. Thus most of the spectral features at $E_F$ increase with $T_K$ (the sidebands, of course, scale directly with $T_{SO}$ and $T_{CF}$) and readily lend themselves to measurement in PES,.

The existence of crystal field levels yields additional features near the Fermi energy. For the spectrum of Fig. 1a two crystal field levels have been introduced (at -24 meV and -48 meV) to approximately correspond to those reported for $CeSi_2$ in neutron diffraction measurements [Galera et al., (1989)]. A blowup of the near-$E_F$ region is shown in Fig. 1b with the crystal field levels delineated. Note that on the occupied portion of the spectrum these levels are relatively weak and in fact do not account for the bulk of the occupied spectral weight. Although crystal field levels have never been unambiguously observed in a PES measurement, their theoretical spectral weight cannot always be assumed to be small (it depends on the parameters), particularly in Yb compounds.

In an actual PES measurement the spectrum in Fig. 1a would be convoluted with a Fermi function to yield only the occupied states, as well as broadened by instrument resolution. The dark lined spectrum in Fig. 1b corresponds to a convolution with a T=0 K Fermi function. In Fig. 1c this T = 0K spectrum has been further broadened with a 20 meV Gaussian to match the stated resolution of Patthey et al. (1987a) as well as with an 80 meV Gaussian to match the best resolution of a measurement at resonance (hv = 120 eV). This then is the expected shape of a typical Ce heavy fermion PES spectrum, while a typical Yb spectrum would mimic the unoccupied states of Fig. 1a broadened by instrument resolution.

Besides the scaling with $T_K$, a major prediction of SIM and its extensions (in particular, NCA) is the temperature dependence of the KR as shown by Bickers et al. (1987). They showed that a substantial decrease in the KR intensity already exists at $0.23T_K$, while at $29T_K$ all the $f^1$ spectral features have been renormalized to nearly zero. A real spectrum at temperatures comparable to $T_K$ may be complicated somewhat by the existence of possible crystal field (or CF) levels where $T_K<T_{CF}<T_{SO}$. Nonetheless, at $29T_K$ all $f^1$ states should renormalize to near zero.

Effects of CF levels and finite $U_{ff}$ have been singled out as complicating the PES spectra of heavy fermions [Allen et al., 1986, 1992]. In particular, because of their larger effective Kondo temperature, CF levels may result in a lack of scaling with $T_K$. (Huffner (1992) views these as single particle states, although that argument cannot not hold for large-$T_K$ materials such as $CeBe_{13}$ where the width of the KR is theoretically broader than any CF states). Likewise, it is possible that spectral weight from the very broad $f^2$ states at finite $U_{ff}$ may spill over into the occupied portion of the spectrum and strongly affect the scaling. Joyce and Arko (1993a) have shown however that while CF states indeed add extra weight (though not much), the scaling with $T_K$ remains dominant within GS theory.

The effect (or lack thereof) of the finite $U_{ff}$ was also demonstrated by Joyce and Arko (1993a). Space limitations prevent extensive discussion Its net effect is merely to increase all intensities without changing the overall shape of the spectra, thus keeping all relative weights unaffected. The lack of an effect is not surprising since spillover of the very broad $f^2$ states into the occupied portion of the spectrum should resemble the extreme end of the tail of a Gaussian which cannot affect the $f^1$ features much more than the $f^0$ peak.

In summary, the important predictions of GS and NCA are those of scaling of 4f features with hybridization (and hence$T_K$ ) together with a strong temperature dependence of these features as the hybridization decreases with increasing temperature. The scaling is complicated somewhat by the exact choice of parameters, but the increase in the $f^1$ spectral weight with $T_K$ is at the very least monotonic.

## Periodic Models

We will see that PES results from single crystals fail to reproduce the NCA predicted spectra, differing qualitatively and not just quantitatively. In general there is much less temperature than predicted, while $n_f$ values, especially in Yb compounds, are dramatically too low (i.e., too much spectral weight in the $f^{14}$ peaks). Most importantly, dispersion is often observed in f-electron features at very high temperatures relative to $T_K$ (i.e., at temperatures where the susceptibility displays Curie-Weiss behavior), belying the concept of a single impurity.

While several theoretical approaches [Tahvildar-Zadeh, 1997,1998, Zwicknagel, 1992, Steiner et al., 1994, Liu, 1993, 1997, Shenh and Cooper, 1995] hold promise of accounting for dispersive behavior, we focus below on the Periodic Anderson Model [Tahvildar-Zadeh, 1997], PAM, which in its preliminary stages appears to capture the essential physics. Furthermore, it represents merely an extension of the SIM to periodic systems and should thus be most pallatable to proponents of the SIM. Quite possibly these different approaches, mostly based on the Anderson Hamiltonian, may eventually converge on the same final result, namely some form of renormalized bands [Zwicknagel, 1992, Steiner et al., 1994] displaying minimal temperature dependence. A necessary flattening of these bands at $E_F$ with temperature, due to the correlations, yields the heavy electron masses well as a mimicing, in parts of the Brillouin zone, of the feature called the Kondo resonance.

The PAM is believed to most correctly describe the strong correlation of electrons in Kondo lattice systems; i.e., stoichiometric compounds with 4f or 5f electrons in the valence shell. While for more than a decade the SIM has been the paradigm for comparison with PES, it cannot account for the coherent nature of electrons (i.e., periodic Bloch states) now observed both above and below $T_K$. The PAM accounts for these effects. Unfortunately, the difficult nature of PAM calculations necessitates the use of simple generic models rather than real systems. Indeed, the calculation is done in infinite dimensions. Nevertheless the predicted trends hold the promise of much better agreement with PES data.

The PAM Hamiltonian on a D-dimensional hypercubic lattice is,

$$H = \frac{-t^*}{2\sqrt{D}} \sum_{\langle i,j \rangle \sigma} (d_{i\sigma}^\dagger d_{j\sigma} + h.c.)$$

$$+ \sum_{i\sigma} (\varepsilon_d d_{i\sigma}^\dagger d_{i\sigma} + \varepsilon_f f_{i\sigma}^\dagger f_{i\sigma}) + V \sum_{i\sigma} (d_{i\sigma}^\dagger f_{i\sigma} + h.c.)$$

$$+ \sum_i U(n_{fi\uparrow} - 1/2)(n_{fi\downarrow} - 1/2).$$

In the above equation, $d(f)_{i\sigma}^{(\dagger)}$ destroys (creates) a $d(f)$-electron with spin s on site i. The hopping is restricted to the nearest neighbors and scaled as $t = t^*/2\sqrt{D}$. Here t* = 1, the width of the Gaussian density of states, as the energy scale. The unhybridized d-bandwidth is 2√D, and this broadens further with hybridization. U is the screened on-site Coulomb repulsion for the localized f-states and V is the hybridization between d-and f-states. This model, then, retains the screening and moment formation of the impurity problem, but is further complicated by the lattice effects and the interaction between the moments.

Metzner and Vollhardt [1989] observed that the irreducible self-energy and the vertex functions become purely local as the coordination number of the lattice increases. As a consequence, the solution of this interacting lattice model may be mapped onto the solution of a local correlated impurity coupled to an effective bath which is self-consistently determined [Pruschke et al., 1996]. The Quantum Monte Carlo (QMC) algorithm of Hirsch and Fye [1989] is employed to solve the remaining impurity problem. The basic outputs of this procedure are the f and d Green's functions of the model in Matsubara frequency. The maximum entropy method (MEM) is then employed to analytically continue these functions to real frequency [Jarrell et al., 1996].

Initial results indicate a substantial differences between the PAM and the SIM approaches, especially in the much slower temperature dependence of the PAM. While both yield a sharp peaking of the DOS at $E_F$ that can be interpreted as a Kondo resonance, thePAM calculations find that these are narrow bands both above and below $T_K$, as shown in Fig. 2. Here we present the spectral functions for various k-vectors in a generic simple

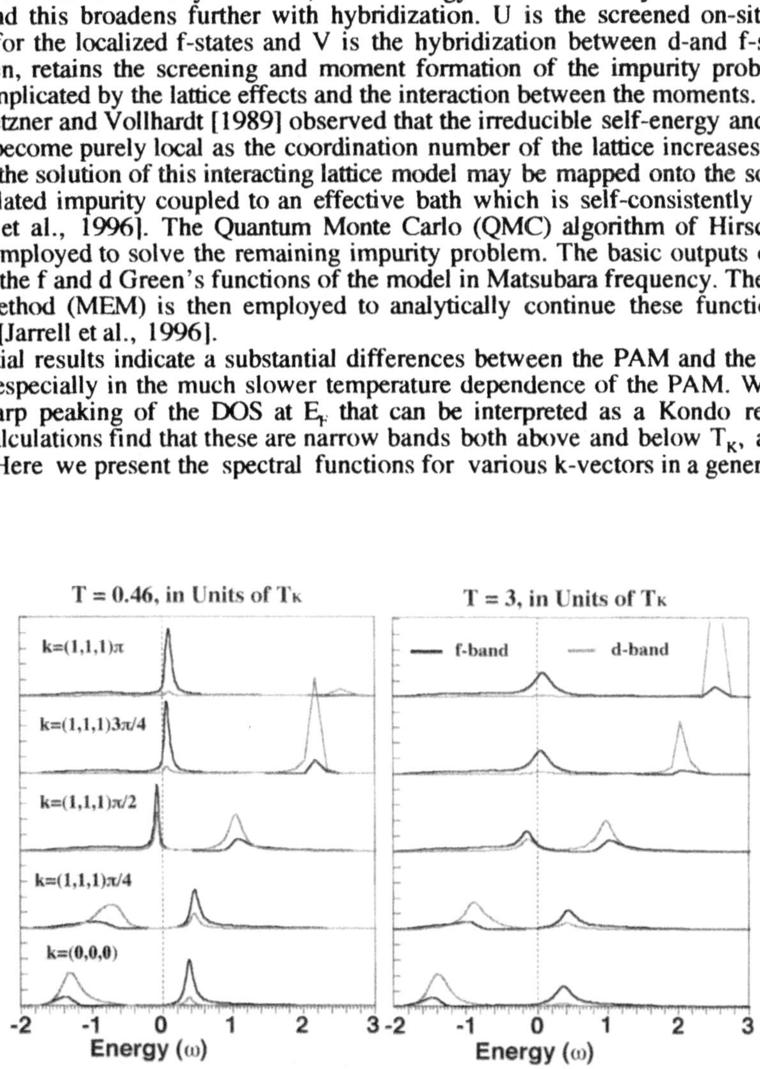


Fig. 2 PAM-derived spectral functions at the indicated points in the simple cubic Brillouin zone, for two temperatures relative to $T_K$. Dark lines indicate partial f-DOS while gray lines indicate partial d-DOS. Note that the narrowest features, both f and d, are at the Fermi energy.

cubic zone along the cube diagonal, at two different temperatures ($0.46T_K$ and $3T_K$). A d-band at 0.6 filling is allowed to hybridize with a band of singly occupied f-states, resulting in the two f-d mixed renormalized bands, with a clear hybridization gap. For this system, U = 1.5, hybridization V = 0.6, while TK = 0.02, in units of t*. Clearly dispersion is much larger than $T_K$. There is no f-intensity at $E_F$ for k = (0,0,0), the latter developing only at k = (1,1,1)p/2. As the temperature increases to $3T_K$ there is only a minimal transfer of spectral weight from the quasiparticle peaks to the "main" or $f^0$ peak (still slightly visible at $w \approx -1$) in contrast to the SIM. Instead one finds primarily a broadening of the quasiparticle peaks, very much in accordance with experimental results reported previously [see papers by Arko, Joyce and Andrews]. Calculations for $T = 10T_K$ yield similar results. As in the SIM, the $f^0$ peak, despite much reduced intensity, is found to increase with decreasing hybridization, and hence $T_K$. The single impurity physics is generally recovered at sufficiently high temperatures ($T > 10T_K$).

## EXPERIMENTAL RESULTS

### Single Crystal vs Poly Crystal PES

As indicated in the introduction, the emphasis in this chapter will be primarily on single crystal photoemission spectra and their comparison to the various models. It may be useful, however, to show from the start that indeed major differences exist between single crystal and poly-crystal heavy fermion PES spectra, in order to demonstrate the need for a new approach.

In Fig. 3 we show single- and poly-crystal spectra for $YbCu_2Si_2$ where the KR is identified with the $4f_{7/2}$ state and is fully occupied [Arko et. al., unpublished]. The samples were cleaved (not scraped) *in-situ* at 20 K, and measured at 20K. The $T_K$ for this material is approximately 35K. The spectra have been normalized so that the trivalent portion of the spectrum (presumed to be entirely due to the bulk) has equal intensities in both curves when measured peak to valley. While the Cu-3d feature at hv = -4 eV normalizes in this way to equal intensity in both materials (the slight shifts are due to angle resolved effects), the divalent portion of the Yb-4f spectrum within 2 eV of the Fermi energy is dramatically reduced in the poly-crystalline specimen relative to the trivalent portion, both for surface and bulk features. In terms of the 4f hole occupancy discussed below, the $n_f$ is apparently increased from 0.65 to about 0.8 or 0.9 in the poly-crystal. Stoichiometry does not seem to be the problem since the Cu-3d peak has the correct intensity. If the SIM truly applied in Yb heavy fermions, then all the $4f_{7/2}$ intensity at $E_F$ is of necessity due to the Kondo interaction. The SIM by definition implies that the 4f element is an impurity with no interaction with its neighbors at high temperatures. Why the existence of long range atomic order (or lack thereof) then so dramatically affects the 4f spectrum is at this point a mystery. Some suggestions will be presented later in the section 13 where we consider the possibility that these are band states even at high temperatures, and long range order may be impaired in poor poly-crystals.

It is natural to suspect poor sample quality in poly-crystals for the above result, and this is certainly expected to be a contribution. However, although the quenching of the $4f_{5/2}$ feature in poly-crystals is not totally universal, it occurs often enough that an effect other than poor sample stoichiometry is a contributing factor. On the other hand, inasmuch as it is single crystal data that is more out of line with bulk properties than poly-crystal data, it is occasionally suggested that this could be due to a surface effect in single crystals. Unfortunately this runs counter to conventional wisdom which states that best coupling to the bulk is obtained in single crystals at normal emission. Indeed, poly-crystals, especially scraped poly-crystals, are far more likely to exhibit strong surface effects. Further, only the KR and its sidebands are affected by the poly-crystal and scraping, the trivalent portions of the spectra remaining unaffected. Clearly we are hinting that the 4f states exist in very narrow bands whose phase coherence is readily destroyed by an imperfect surface.

To sum up, it is apparently crucial for a correct measurement of the near-$E_F$ intensities in heavy fermions that the surface is a cleaved single crystal, with a mirror surface especially desirable. Scraping of the surface should be particularly avoided since it was found that in $YbAgCu_4$, a material having a $T_K$ of about 35 K, the scraping of a coarse-grained surface totally eliminated all divalent 4f features [Arko et al., unpublished], much more consistent with thermodynamic data. Perhaps it is so that the destruction of long range order via scraping indeed puts us into the single impurity regime, while a good single crystal surface with long range order requires a treatment within the PAM.

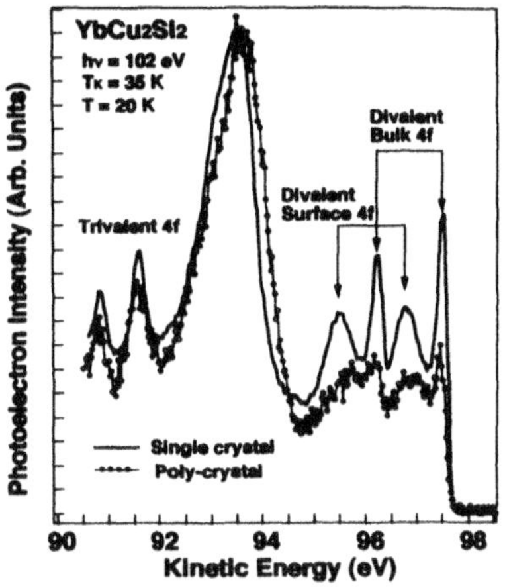


Fig. 3 Single crystal and poly-crystal $YbCu_2Si_2$ PESspectra taken under identical conditions Note the large reduction in poly-crystal divalent intensity even though the trivalent intensities are equal.

**Lack of Scaling with $T_K$ of the Near-$E_F$ Features**

Perhaps the single most important prediction of the GS and NCA models is that of scaling of the various spectral properties with hybridization and hence $T_K$ , the universal parameter. In particular, the spectral weight, position, and width of the KR is predicted to increase with $T_K$. While instrument resolution has been insufficient in the past to deal properly with the position and width of the $4f_{5/2}$ feature, researchers have concentrated on the spectral weight. Allen et. al. [1986] were first to find a relationship between $T_K$ and spectral weight of the near-$E_F$ features, although the limited resolution of the early research did not allow a quantitative comparison to theory.

Patthey et. al. [1990] published a monumental paper on the application of the SIM to spectral properties of Ce and Yb heavy fermions. For the first time they performed high resolution ($\Delta E \approx 20$ meV) systematic experiments on a number of materials with $T_K$'s ranging from less than 10 K for $CeCu_6$ to about 200K for α-Ce, observing the trends in the near-$E_F$ spectral weight. They obtained their data using HeI and HeII radiation, using the difference curves obtained by subtracting the spectra obtained at two different photon energies. More importantly, they also performed model calculations using NCA at the appropriate temperatures to compare to the results and found excellent agreement.

Joyce et al. [1992c], by contrast, pointed out that by using high quality single crystals there obtains an amazing similarity between all Ce heavy fermion spectra, regardless of $T_K$. This can be seen in Fig. 4 [Arko et al., 1997] where spectra for 7 Ce heavy fermions (including two directions for $CeBe_{13}$) are plotted. Indeed, one observes the interesting effect that spectral differences with momentum within the same crystal ($CeBe_{13}$) are in fact larger than differences between materials whose $T_K$'s differ by as much as 2 orders of magnitude (more on that later). All the spectra in Fig. 4 were taken at a photon energy of 120 eV which roughly corresponds to the 4d absorption edge and thus resonantly enhances the 4f emission. The energy resolution in each case is about 90 meV. All spectra represent raw data except in the case of $CeCu_2Si_2$ where the spectrum at anti-resonance (hv = 112 eV) has been subtracted from the resonance spectrum in order to eliminate the very strong Cu-3d emission. In Fig. 4 one should further note that $CeAl_3$ and $CeSi_2$ are cleaved poly-crystal specimens so that the poly-crystal effect discussed above is not universal (but common). The normalization of the spectra in Fig.4 is to the $f^0$ peak maxima at -2 to -3 eV, except for $CeBe_{13}$ (110) where the $f^0$ peak is clearly much smaller than that for $CeBe_{13}$ (100), and it just is not obvious which maximum is the true bulk $f^0$ (this is further discussed below). While nearly all the spectra in Fig. 4 represent 4f and 5d Ce emission only (due to the nearly negligible ligand emission), in the case of $CePt_{2+x}(111)$ and $Ce_3Pt_4Sb_4$ there is also substantial Pt-5d emission so that the $f^0$ peak intensities should be ignored for these. To the list of spectra in Fig. 4 one should also

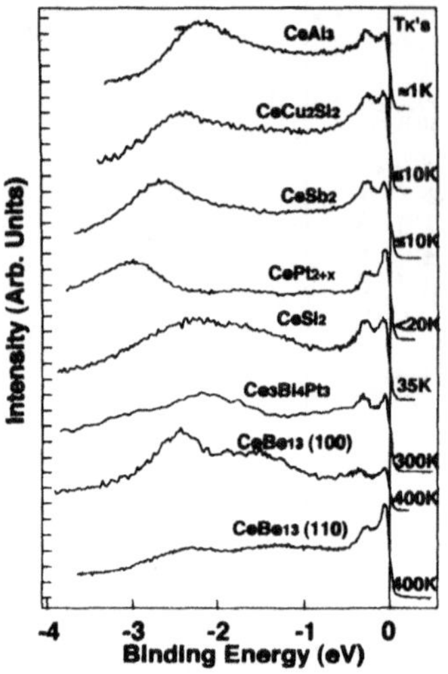


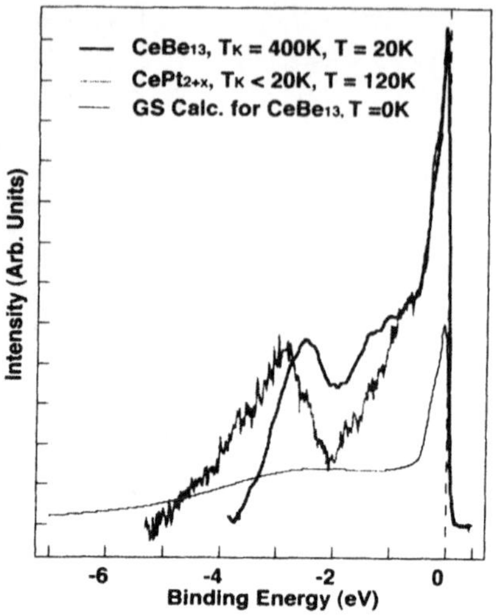


Fig. 4 PES spectra for several Ce heavy fermions with $T_K$'s ranging fron $\approx 1$ K to $\approx 400$ K. Note the lack of scaling of the $4f_{5/2}$ with $T_K$. Two different directions are shown for $CeBe_{13}$.

Fig. 5 Overlay of $CeBe_{13}$ (110) spectrum with $CePt_{2+x}$ (001) spectrum after subtracting backgrounds (from Andrews et al. [1995a]. The calculated GS spectrum approximates the parameters for $CeBe_{13}$.

add α-Ce ($T_K \approx 800$K, Weschke et al., [1991]) and $CeB_6$ ($T_K \approx 10$K, Kakizaki, [1995]) which likewise exhibit spectra nearly identical to those in Fig.4.

Irrespective of the precise normalization of the data in Fig. 4, the interesting fact emerges that the spectral weights of the $4f_{5/2}$ features are more or less constant for all materials, or at least do not monotonically increase with $T_K$. While the 90 meV resolution precludes any firm conclusions about the widths and positions of the $4f_{5/2}$, it has little effect on the broad $4f_{7/2}$ peaks so that their peak amplitudes relative to $f^0$ need no deconvolution. One is forced to the conclusion that there is no monotonic scaling of the $4f_{5/2}$ spectral weights with $T_K$ relative to the $f^0$ peak. This lack of scaling cannot be blamed, as has been discussed above [Joyce and Arko, 1993a], on inclusion of $f^2$ intensity originating from finite $U_{ff}$ effects, since the $f^2$ DOS, if present, is broad and featureless and hence cannot change the relative amplitudes. Nor can one resort to a lowering of the $4f_{5/2}$ degeneracy to $N_f = 2$ by the introduction of two additional crystal field levels, since, as again shown by Joyce and Arko [1993a], even in the presence of these CF states the GS and NCA models predict a scaling of the $4f_{5/2}$ spectral weight with $T_K$. The nearly constant and much too broad (relative to GS) width of the $4f_{5/2}$ has been attributed by Patthey et al. [1993] to lifetime broadening. Even if this were the case, such broadening will not affect the spectral weight so that the inevitable conclusion of a lack of scaling with $T_K$ remains a reality.

To underscore the lack of scaling still further we point out the amazing similarity in Fig.4 between $CeBe_{13}$ (110) {$T_K \approx 400$K} and the spectrum for $CePt_{2+x}$ (111) {$T_K \approx 10$K}. In Fig. 20 the comparison is somewhat less striking than actuality because of the strong Pt-5d emission in the region of the $f^0$ peak. For a better comparison one must first subtract out the non-Ce portion of the $CePt_{2+x}$ (111) spectrum by subtracting the hν = 112 eV spectrum (anti-resonance) from the hν = 120 eV spectrum (resonance). For $CeBe_{13}$ the intensity at 112 eV is negligible. The two curves are overlaid in Fig. 5 (normalized at the $4f_{5/2}$ peak), with the secondary backgrounds also removed. With the exception of the shift in the $f^0$ peak the two curves overlay almost exactly. The region of the Ce-5d emission (between -0.5 eV and -2 eV) also overlays if the $f^0$ peaks are removed. This type of agreement is astounding when one considers that there is a factor of 40 difference in $T_K$'s. Moreover, the $CePt_{2+x}$(111) spectrum was taken at 120K where the KR should have been of negligible intensity. On the basis of the single crystal data, then, it would be difficult to reach a conclusion other than one of a total lack of scaling of the spectral weight with $T_K$.

The thin lined spectrum in Fig.5 is a GS calculation with parameters approximately matching $CeBe_{13}$ (i.e., W = 10 eV, $U_{ff}$ = 14 eV, $T_K$ = 400 K), smoothed to match the instrument resolution, and normalized in such a way that the total area of the calculated Lorentzian $f^0$ peak matches the total area of the measured Gaussian $f^0$ peak in $CeBe_{13}$. The mismatch is obvious. While better fits can be obtained with different parameters such as a smaller W, these are difficult to justify. The width of the conduction band used above (10 eV) may in fact be even conservative since experimentally it peaks at about -8 eV. Also, there are no Ce nearest neighbors in $CeBe_{13}$ so that the 4f electron can only hybridize with the Be conduction band.

**The f-level Occupancy, $n_f$**

It can already be seen in Fig. 5 that the spectral weight ratio $W_r=f^1/f^0$ in $CeBe_{13}$ is too large relative to a calculation; indeed, about a factor of 2.5 too large. The discrepancy must then be even larger for $CePt_{2.2}$. This indicates that the calculated f-occupancy $n_f$ of the bare f-level at ≈ -2.5 eV is much larger than measured. One can, of course vary the parameters (primarily increase the hybridization) until calculation matches experiment, but this will increase $T_K$ to values incommensurate with the thermodynamic quantities. Obtaining a good fit for one compound by varying parameters says very little about the necessary systematics. Basically, the lack of scaling seen in Fig.5 implies that the $n_f$ in Ce compounds does not change very dramatically with $T_K$. This is totally at odds with the SIM expectations.

The f-level occupancy is best studied in Yb compounds where it becomes the hole occupancy rather than electron occupancy, and where GS theory places the KR centroid on the occupied side of the Fermi energy. Here the integrated intensities of the $f^{13}$ (trivalent) and $f^{14}$ (divalent) portions of the photoemission spectrum are expected to be directly related to the f hole occupancy, $n_f$, by the relation

$$n_f = I(f^{13})/[I(f^{13})+13/14(f^{14})]$$

where $I(f^n)$ represents an integral spectral weight of the $f^n$ feature. One of the best compounds for this study is $YbAl_3$ where the Al states yield almost no photocurrent, thus making it relatively easy to subtract the background and to delineate the trivalent and divalent (bulk) intensities. An early study for this material [Oh, 1993] was an XPS study where direct integration yielded $n_f = 0.65$ which is far lower than expected for a material having $T_K \approx 400K$. Indeed this places it well into the mixed valent regime, outside the range of the validity of NCA. Tjeng et. al. [1993] later utilized the temperature dependence of the $4f_{7/2}$ level and obtained an $n_f$ value of 0.85 which is in exact agreement with NCA predictions [Bickers et al., 1987]. However, using the temperature dependence is difficult at best. The assumption is made that all of the temperature dependence of the $4f_{7/2}$ is due to Kondo effect. With no effort made to delineate the effects of a 300K Fermi function combined with thermal broadening, this assumption seems tenuous at best. Additionally the choice of backgrounds in that study is somewhat questionable. Since the non-$4f_{7/2}$ (bulk) background in the polycrystalline samples was assumed to comprise more than 80% of the signal, one can see how even in the case of correct assumptions regarding temperature dependence, small errors in the choice of background can yield huge errors in the temperature dependence.

The most reliable method for determining $n_f$ would seem to be direct integration of the divalent and trivalent features in a material where the background is relatively small. Toward

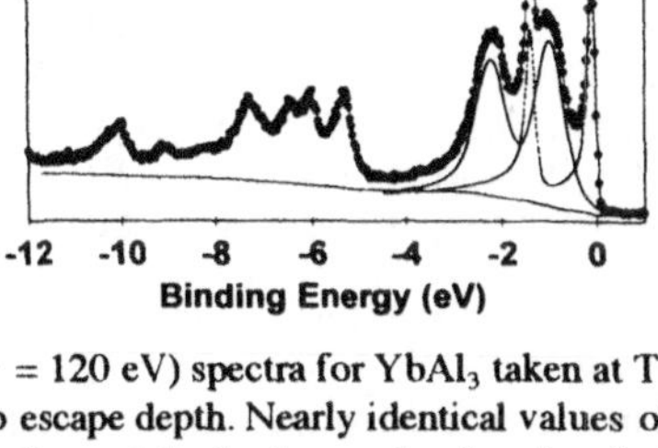


Fig. 6 XPS and UPS (hν = 120 eV) spectra for $YbAl_3$ taken at T = 20 K to determine any differences in signal due to escape depth. Nearly identical values of $n_f$ were obtained by direct integration after separating out the background and surface features as shown

this end Joyce et al. [1996] and Andrews et al. [1994] analyzed $YbAl_3$ spectra both from XPS as well as UPS and came up with values basically agreeing with the first investigation of Oh [1993]. This is shown in Fig.6 where the delineation into background and surface components is also illustrated. Similar analyses for $YbCu_2Si_2$ yielded [Andrews et al., 1994 and Joyce et al., 1996] nearly the identical values despite the fact that $YbCu_2Si_2$ has a $T_K$ of only 35K, an order of magnitude lower than $YbAl_3$. All of these values are totally out of the range of NCA validity and clearly belong to the mixed valence regime. While there remains some question as to whether the photoelectron cross sections for the $f^{13}$ and $f^{14}$ states are identical (a necessary condition in order to compare intensities), the fact that identical $n_f$ values are obtained at both XPS energies as well as near hv = 100 eV would suggest that the photoelectron cross sections of the $f^{13}$ and $f^{14}$ states are at least very similar if not identical. Thus direct integration of the spectra to obtain $n_f$ would seem preferable to the method of Tjeng et al. [1993].

PAM calculations indicate that the ratio of the $f^1$ to $f^0$ (or, the divalent to trivalent in Yb compounds) spectral weights is substantially larger when calculated within the PAM than when calculated within the SIM. It would seem that the trends predicted by the PAM are much more consistent with the experimental spectra obtained from single crystals.

There is no question about the fact that the $n_f$ values obtained from PES are dramatically in disagreement with those obtained from bulk properties as well as those obtained from X-ray absorption $L_{III}$ edge studies. While it is our contention that we must resort to a different model, such as the PAM, to resolve this discrepancy, a useful question to ask is whether photoemission measures the same material as is measured in $L_{III}$ studies. In particular, is it possible that owing to its surface sensitivity a PES study does not probe the bulk but rather only a surface or a sub-surface layer whose $n_f$ is dramatically different from the bulk. Such a scenario would render PES useless and would breathe new life into SIM.

But let us look at all possibilities. The existence of a surface layer is not in question. Moreover it is clear that in Ce compounds the surface tends to be trivalent while in Yb compounds it is divalent. But while it is easy to delineate a surface feature and separate it from bulk electronic structure, a thick subsurface has never yet been reported in any system known to us (thin subsurfaces may be preset in Yb compounds as evidenced by multiple surface features). If it were to exist in one system, the likelihood of its existence in other systems is high, making PES useless as a bulk probe. Fortunately, excellent agreement between calculated and PES-derived electronic structure is often reported. Indeed, most calculations suggest that only about 3 atomic layers below the surface are sufficient to obtain the bulk electronic structure.

But let us say for the sake of argument that a sub-surface layer may exist only in strongly correlated f-electron systems. Immediately we can rule it out in Ce compounds since studies at photon energies as high as 3 KeV show only bulk and surface states with no third layer. Such a sandwich would in any case be preposterous for Ce heavy fermion compounds since it would require a trivalent surfce, a strongly mixed valent sub-surface, and again a nearly trivalent bulk. We can also rule it out in purely trivalent as well as purely divalent Yb compounds since PES measurements yield the required purely trivalent ($YbBe_{13}$) or purely divalent ($YbSn_3$) spectra. It would be pathological that a sub-surface would exist only in mixed valent Yb compounds, but that is all that we are left with. Even here we see from the previous section that scraping of a surface only affect the divalent portion of the spectrum - i.e., the coherent bands, with little effect on the trivalen spectrum.

But let us go further and assume that nature is indeed that pathological. Since the $n_f$ values obtained at XPS energies are the same as those obtained at lower photon energies, this suggests that the desired sub-surface must be at least of the order 100 Å thick otherwise the larger escape depth at XPS energies would begin to probe the supposed true bulk and yield different $n_f$ values. On the other hand it must not be so thick as to constitute a significant percentage of the probing depth in $L_{III}$ edge studies where a sub-surface has never been observed (again pathological). But the real question to be asked is the following: If a single crystal compound is cleaved in-situ at about 20 K, what is the driving force which causes this ≈100Å thick layer to form instantly, and why is the LEED pattern representative of the bulk structure? Certainly there are no dangling bonds as on the surface. The answer remains elusive.

One is forced to the conclusion that there exists no sub-surface and that the measured $n_f$ values are real. The disagreement with $L_{III}$ studies remains unresolved. Perhaps we should we should look to problems associated with $L_{III}$ edge measurements rather than PES measurements, particularly in view of the problems uncovered recently with bandwidths.

**Temperature Dependence**

**General Considerations:** The determination of the temperature dependence of the near-$E_F$ feature in heavy fermion compounds is perhaps the most difficult experiment to accomplish correctly. Every material (even Cu, [Matzdorf, 1993]) displays a temperature dependence to some degree in its electronic structure, and it is difficult to determine whether the behavior observed is due to conventional effects of lattice contraction and phonon broadening, or whether it is due to the unconventional behavior predicted by the NCA.

The difficulty of the experiment has precluded much investigation in this arena outside the single crystal work of Joyce et. al. Indeed, in poly-crystalline materials the only PES studies are on $CeSi_2$ [Patthey et al., 1987], $YbAl_3$ [Tjeng et al., 1993] and $YbAgCu_4$ [Weibel et al., 1993], while some BIS data exist for $CeSi_2$ [Malterre et al., 1992a; Grioni et al, 1993] and $CePd_3$ [Malterre, et al., 1992b]. In each of the above studies it was determined that the temperature dependence of the near-$E_F$ feature follows the single impurity model exactly. However, they all have in common the fact that the surfaces are those of scraped poly-crystals. Moreover, there is the ever present problem of the choice of a background which can be varied by more than a factor of two thus leading to nearly any desired result. In none of the experiments above was the background determined from measurements of related La or Lu compounds to properly delineate first the non-f contribution, and second, the contribution due to the ever present surface components. In general, the background is grossly over estimated with the assumption that all the spectral intensity at the valley immediately below the near-$E_F$ peak (i.e., all spectral weight at ≈ -0.2 eV in Fig. 23) is due to background. By drawing a horizontal line between this valley and the Fermi energy and calling it background perhaps half of the $4f_{7/2}$ peak intensity is eliminated. Clearly this will exaggerate any temperature dependence.

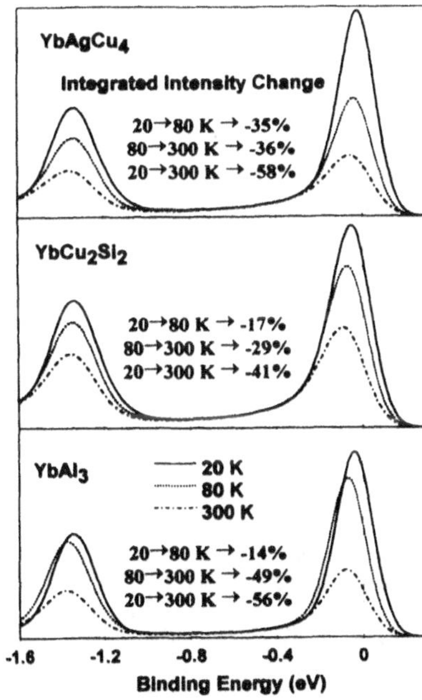


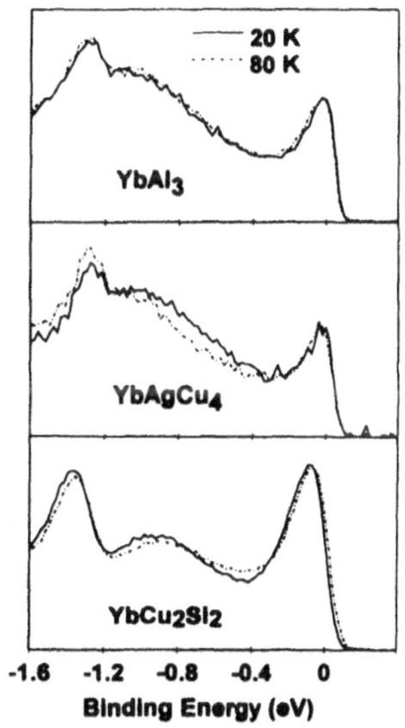


Fig. 7 NCA calculations at three different temperatures for three materials discussed here. Surface states are not modeled by NCA. The parameter used are: ($YbAgCu_4$) $T_K = 100$ K, $U_{ff} = \infty$, W =6.0 eV, $\Delta_{so} = 1.29$ eV, $e_f = 1.0$ eV, CF staes - none, $\Gamma = 0.05462$ eV. ($YbCu_2Si_2$) $T_K = 40$ eV, $U_{ff} = \infty$, W =6.0 eV, $\Delta_{so} = 1.29$ eV, $e_f = 1.0$ eV, $\Delta_{CF1} = 12$ meV, $\Delta_{CF2} = 30$ meV, $\Delta_{CF3} = 80$ meV, $\Gamma = 0.0606$ eV. ($YbAl_3$) $T_K = 400$ K, $U_{ff} = \infty$, W = 10.0 eV, $\Delta_{so} = 1.29$ eV, $e_f = 1.0$ eV, CF states - none, $\Gamma = 0.05856$ eV.

Fig. 8 Experimentally measured temperature dependence between 20 K and 80 K for the three materials calculated in Fig. 7.

**Yb Compounds - Comparison to NCA**: The NCA predictions (code provided by D. Cox) for the temperature behavior of these materials are shown in Fig. 7 [from Joyce et al., 1996]. Here the $4f_{7/2}$ and $4f_{5/2}$ features are shown without the omni-present surface states which cannot be modeled by NCA. The three temperatures for which calculations exist are those most easily accessible experimentally. The parameters are given in the figure caption, but are chosen so as to approximate the three materials and their respective $T_K$'s. The data has also been smoothed with a 190 meV Gaussian to match the resolution of the $YbCu_2Si_2$ data shown in the bottom frame of Fig. 8. In all cases a large temperature dependence is predicted. In particular, note the large decrease in intensity between 20 K and 80 K for $YbAgCu_4$, a material having a $T_K$ in the vicinity of 100 K.

By contrast, Fig. 8 shows that the measured temperature dependence between 20K and 80K in all three materials [Blyth et al., 1993] is almost nil, with the surface states usually showing larger effects than the bulk states. Here the $YbAgCu_4$ specimen was a coarse grained poly crystal cleaved in-situ. The results contrast sharply with the measurements of Malterre et al. [1992a] who reported a large temperature effect in $YbAgCu_4$ between 20K and 80K. Again, however, in view of the superiority of single crystal data (or at least cleaved vs scraped surfaces) more weight is given to results from cleaved surfaces, the scraped polycrystal data having too many uncertainties. Indeed, Arko et al. [unpublished] observed that scraping their $YbAgCu_4$ surface totally destroys the $4f_{7/2}$ and $4f_{5/2}$ bulk peaks which were present on the same specimen when the surface was prepared by cleaving. The intensities slowly recover with time and a temperature dependence can be observed below 80 K, but the surface composition prepared in this fashion surely is no longer representative of the starting material.

Above 80K a substantial temperature dependence is usually observed even in single crystal Yb heavy fermions, albeit not always. Spectral weight changes ranging from 0% to about 15% are common. Joyce et al. [1992c] suggested already in connection with $CeSi_2$ data of Patthey et al. [1987] that one should consider thermal broadening as a source of temperature dependence. In reply it was pointed out [Patthey et al., 1993] that while core level phonon broadening can be understood within the Franck-Condon model, the near-$E_F$

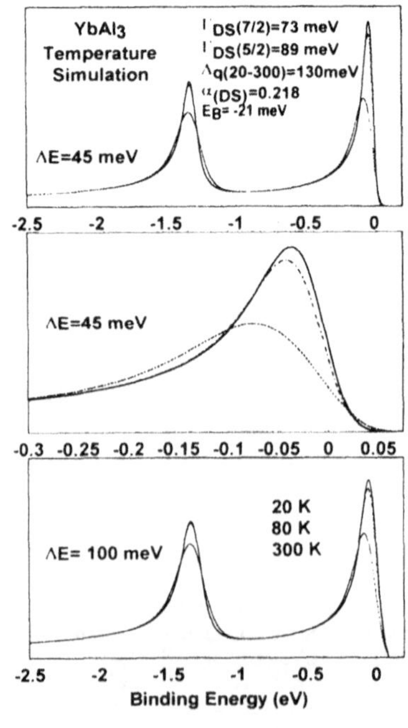


Fig. 9 Simulated temperature dependence of a fitted $YbAl_3$ spectrum A 12.5% spectral weight loss is simulated for the $4f_{7/2}$ peak and none for the $4f_{5/2}$ peak although the latter appears to have lost spectral weight due to broadening.

features in heavy fermions should not couple to the phonons owing to the well-screened condition of these states. Nevertheless, phonon broadening is observed in the valence bands of many materials, even in excellent metals like Cu [Matzdorf et al., 1993], where the screening is expected to be good. Thus, while not entirely understood, phonon broadening is an existing fact and must be considered. In $CeSi_2$ Joyce et al. [1992c] estimated the magnitude of the broadening (about 100 meV) from the broadening of the Si-2p core levels. While this is certainly not expected to be identical to the Ce states, it should serve as a first order approximation. In Yb compounds the phonon broadening was estimated from a measurement of pure Yb metal, and found to be of the order of 130 meV added in quadrature.

Assuming that a similar 130 meV broadening occurs in heavy fermion compounds where the $4f_{7/2}$ is within 20 to 30 meV of the Fermi energy, one can then simulate the effect of the Fermi function on the $4f_{7/2}$ intensity [Joyce et. al., 1996]. This is done in Fig. 9 where fits to the bulk divalent 4f peaks of Fig. 28 (i.e., the bulk $4f_{7/2}$ and $4f_{5/2}$) measured at 20K are broadened by 130 meV in quadrature and convoluted with a 300K Fermi function. The middle frame shows the effect near the Fermi energy using a 45 meV experimental resolution, while the bottom frame gives the same result with 100 meV experimental resolution. The calculated spectral weight loss of the Yb-$4f_{7/2}$ peak is about 15% which is actually larger than the observed result. Thus the broadening in compounds is perhaps somewhat less than 130 meV, indeed most likely closer to 100 meV as found [Joyce et. al., 1992c] in $CeSi_2$. It is important to note from the bottom frame of Fig. 9 that given some degree of data scatter one could never unambiguously distinguish between peak broadening and spectral weight loss, although a careful subtraction of the backgrounds often yields a zero weight loss for the Yb-$4f_{5/2}$ even in the presence of a 10% weight loss for the Yb-$4f_{7/2}$. In Fig. 27 no data analysis is needed to see this effect.

Thermal broadening and Fermi function effects work well for Ce compounds as well, but space does not permit a detailed examination.

**$YbInCu_4$: A Singular Test of the Single Impurity Model**

YbInCu4 provides the best opportunity to date for testing the predictions of the SIM. Not only is it an Yb compound where the Kondo features (in the Kondo limit) are on the occupied side of the Fermi energy, but also it undergoes an isostructural phase transition at $T_V = 42$ K with a 0.5% volume reduction below this temperature. It was initially suspected that this is the Yb analogue of the Ce Kondo volume collapse, but it quickly became apparent (Lawrence et al. ?) that subtle changes in the density of states were responsible for the phase transition. Nonetheless, within the SIM framework the Kondo temperature changes at the transition from a value of ≈ 25 K above $T_V$, to a value of ≈ 400 K below $T_V$. This more than an order of magnitude change in $T_K$ allows for easy testing of the NCA predictions within the same material. In particular one can measue: $n_f$, temperature dependence of spectral weight, linewidths and their changes predicted by NCA, peak position shifts, etc. In particular, $n_f$ is expected to change from a value of about 0.8 below $T_V$ to a value >0.9 above $T_V$, the intrinsic linewidth should narrow by an order of magnitude above $T_V$, the peak position of the spin-orbit peak (as well as the $K_R$ whose exact position is complicated by the Fermi function) should shift toward $E_F$ by 40 meV above $T_V$.

In Fig. 10 are shown a series of spectra at 45 meV resolution for temperatures ranging from 20 K to 200 K, thus covering adequately the $T_V$ region. Even a cursory visual inspection reveals that there cannot possibly be an order of magnitude change in the spectral weight of the bulk features above and below $T_V$. Joyce et al. (1988) have carefully measured the $n_f$ at 80 K (above $T_V$) and found a value of 0.5, far below the expected value of > 0.9 expected from NCA. Below $T_V$ this value decreases to about 0.4. Immediately, too, one can see that there is no shift at all in the SO peak position.

It is periodically suggested that that the surface sensitivity of PES measurements would render these results useless if a thick sub-surface layer prevents electron escape from the true bulk into the vacuum. In that instance we would be seeing perhaps a thick sub-surface whose $T_K$ is much larger than the bulk and where no transition occurs at $T_V$. We have already hinted that such a thick sub-surface layer would have to be pathological to Yb compounds. Further there is absolutely no evidence for its existence except for the disagreement with SIM. A thin (one atomic layer) sub-surface in Yb compounds is a possibility owing to the observation of several surface states, but in all cases the bulk is clearly visible. We suggest that it is not the PES measurement that is in error, but rather the SIM and possibly $L_{III}$ edge studies.

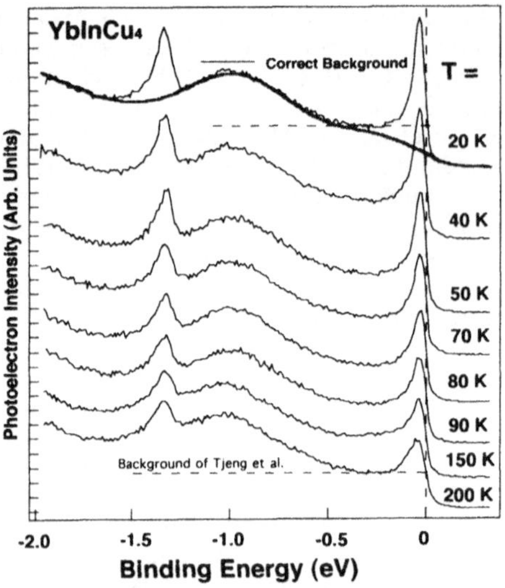


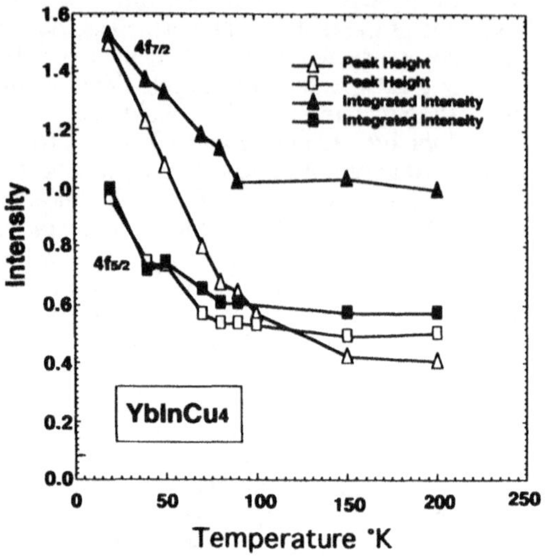


Fig. 10 Normal emission ARPES spectra for YbInCu4 at indicated temperatures. The SO intensity is T-independent above 50 K. Gray curve in 20 K spectrum accounts for surface features and non-f states, and is thus the best estimate of the background. Straight line background is shown for comparison.

Fig.11 Comparison of $4f_{5/2}$ and $4f_{7/2}$ intensities analyzed by two methods. Integrated intensities are T-independent above the bulk transition. Peak heights are measured from the straight line background.

It is easy to err when analyzing PES data. In order to measure the spectral weight of the near-$E_F$ features, a straight line background (used, e.g., by Tjeng et al., 1993) is often employed to eliminate the non-f portion of the spectral weight. This is shown as the dotted line in Fig. 10. But the real background can only be obtained by measuring a corresponding Lu compound under identical conditions. In that case the non-bulk, non-4f portion of the spectrum corresponds to the thick gray curve in the 20 K spectrum of Fig. 10. Thus the "Tjeng" background arbitrarily eliminates as much as 50% of the true bulk spectral weight of the $4f_{7/2}$ (shaded area) and leads to an erroneously enhanced temperature dependence. The error is compounded when no curve fitting employed to get the integrated spectral weight. Instead, only peak heights are measured from this background.

Even emplying curve fitting and a correct background, to get the true $4f_{7/2}$ spectral weight requires taking into account the effect of the Fermi function and theremal broadening. Every researcher, but we, has ignored this non-trivial effect. For this reason we suggest that it is actually easier to measure the weight of the $4f_{5/2}$ peak where thermal broadening effects are less significant, and there is no Fermi function to deal with. Indeed, simple peak height analysis is almost sufficient as shown in Fig.11. Here we present intensity data at various temperatures for both the $4f_{7/2}$ and the SO peaks, analyzed both from simple peak heights (using the Tjeng background) as well as by integrating fitted peaks. Clearly a peak height analysis using a Tjeng background overestimates the temperature dependence of the $4f_{7/2}$ by about 120%. Even the integrated intensity analysis for the $4f_{7/2}$ in Fig. 11 is not yet correct since we have not taken the Fermi function into account. It is best to consider the amplitude dependence of the $4f_{5/2}$ , or the SO peak, in Fig. 11 since this is not complicated by the Fermi function. Now we see that peak height analysis and integration yield very similar results. Indeed, we find that there is almost no temperature dependence above $T_V$, with all the temperature dependence below $T_V$ being due to the phase transition itself and not to SIM phenomena.

It is important to note that we are seeing the bulk transition in the $4f_{5/2}$ peak, so that the idea of a sub-surface with a single $T_K$ larger than the bulk, is of necessity eliminated. It was a preposterous idea in any case.

The high resolution of the data and narrow lineshapes of the $YbInCu_4$ data allow more comparisons to SIM. In Fig. 12 we overlay the 20 K and 200 K spectra after removing

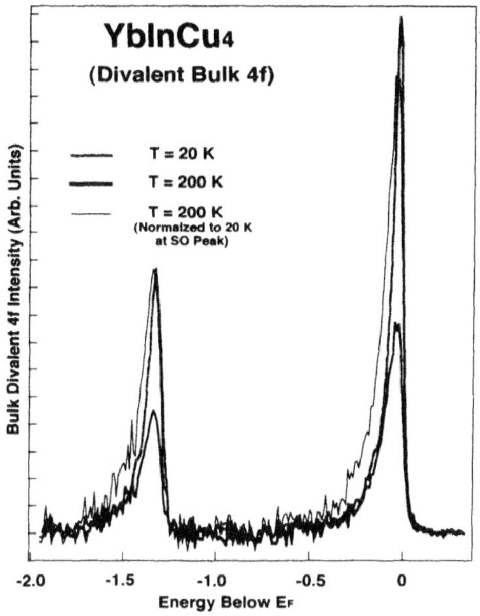


Fig. 12 Divalent bulk 4f peaks at 20 K and 200 K after subtracting off background. Thin-lined spectrum is 200 K spectrum normalized at peak of 20 K SO intensity. Note that 200 K spectrum is broader than 20 K spectrum, contrary to SIM expectations where thermal broadening is not included.

the surface and non-f contributions via the background shown in Fig. 10. We have already pointed out (Joyce et al., 1998) that the 200 K spectral weights are at least an order of magnitude too large relative to SIM predictions. But even more information is to be obtained. By normalizing the 200 K spectrum at the SO peak height of the 20 K spectrum we find that the 200 K data is actually substantially broader than the 20 K data. A convolution with a 90 meV Gaussian is needed to make the 20 K FWHM equal to the 200 K FWHM. This is exactly opposite to SIM predictions where a line narrowing of an order of magnitude is needed for the 200 K data. On the other hand, the 90 meV thermal broadening is almost exactly what we have been claiming all along for the $4f_{7/2}$ feature in Yb compounds, and the $4f_{5/2}$ feature in Ce compounds. Whether this broadening is due to phonons or to PAM effects is not delineated, but it is a fact of life.

To sum up, it is clear that in YbInCu4 we are seeing true bulk phenomena along with surface features. The measured $n_f$ for the low-$T_K$ region (above $T_V$) is $\approx$ 0.5 vs the expected 0.9, the spectral weight is at least an order of magnitude too large, there is no peak shift at the phase transition where a 40 meV shift is predicted, thermal broadening effects go opposite to SIM predictions but are precisely in line with our previous observations. Analysis of the $4f_{5/2}$ peak is preferred to the $4f_{7/2}$ owing to no interference from the Fermi function.

**Momentum Dependent Effects**

Perhaps the strongest argument in favor of a model which includes the periodic lattice is the observation of momentum dependent effects at temperatures far above the Kondo temperature. The PAM allows for such an electronic structure. By contrast, within GS and NCA the KR and its sidebands are a dispersionless feature [Bickers et al., 1987], indeed a thermodynamic quantity of a material, whose properties (i.e., the width, position in energy, spectral weight, etc.) are essentially determined by the value of $T_K$. True enough, at low temperatures it is agreed that there is band formation once coherence sets in (after all, the de Haas-van Alphen effect has been observed by numerous times ([Lonzarich [1988], Springford and Reinders [1988]), but the dispersion must be of the order $T_K$, and then only at $T << T_K$, or below the so-called coherence temperature. The manifestation of angle dependent effects, particularly if these exist at $T >> T_K$, then, would represent a serious disagreement with NCA.

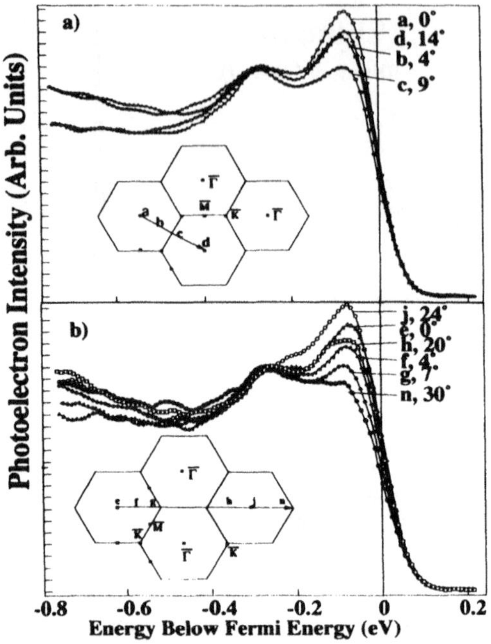


Fig. 13 $CePt_{2+x}$ ARPES spectra at the indicated point in the Brillouin zone. Note that the $4f_{5/2}$ amplitude shows amplitude variations exactly periodic with Brillouin zone.

**$CePt_{2+x}$ Amplitude Modulations:** Just such angle dependent effects have been reported by Andrews et. al first in $CePt_{2+x}$($0<x<1$) [Andrews et. al., 1995a], and later in $CeSb_2$ [Andrews et al., 1995b] and $CeBe_{13}$ [Andrews et al., 1996]. The data for $CePt_{2+x}$ are shown in Fig.13 where the amplitude of the near-$E_F$ feature is seen to undergo periodic amplitude modulation, the period of which coincides exactly with the inverse lattice. The insets indicate the positions in the lattice (surface projection of the Brillouin zone) where the various spectra were taken, while the angles are measured with respect to the surface normal (i.e., the Γ-point). Very large effects are observed, which bring into question the thermodynamic nature of the near-$E_F$ feature. The data were taken at resonance (photon energy = 120 eV) at a temperature of 120 K using a VSW HA50 angle resolved analyzer with a 2° acceptance cone and about 90 meV energy resolution. The data is highly reproducible.

The $CePt_{2+x}$ sample was prepared *in-situ* by evaporating Ce onto a Pt (111) single crystal. Analysis with LEED and XPS [Tang et. al., 1993] as well as comparison of Pt-4d intensities of the in-situ grown specimen to arc melted poly-crystals, determined the composition to be most likely $CePt_{2.2}$, with the value of x determined to within $0<x<1$. Bulk property measurements on arc melted specimens with compositions between $CePt_2$ and $CePt_3$ indicated very similar properties throughout the composition range so that the inexact determination of x is irrelevant to the physics. The Kondo temperatures were found to be less than 20K for all compositions, with the magnetic susceptibility showing Curie-Weiss behavior down to 20K. From bulk property measurements one would conclude that the materials are nearly trivalent so that it is all the more surprising that such an intense KR is observed at T = 120 K.

It is significant that the amplitude modulation effects are seen primarily in the KR. Indeed, in the figure the data have been normalized to the peak value of the spin-orbit peak, the Ce-$4f_{7/2}$. Because the amplitude modulation is seen primarily in the KR it rules out photoelectron diffraction as the source of the amplitude effect, since the orbital symmetry and the atomic point of origin of both features is identical, while the kinetic energies of the photoemitted electrons from the two features are likewise nearly identical. From photoelectron diffraction both peaks would be expected to undergo similar amplitude modulation. Moreover, photoelectron diffraction effects are not necessarily periodic with the inverse lattice, but are rather a geometric effect [Fadley, 1987] which yields maximum amplitude from forward scattering around masking atoms near the surface (the forward flashlight). The observation of dispersion in some systems in any case points to Bloch states.

In their first paper on the subject Andrews et al. [1995a] interpreted the amplitude effect as resulting from the dispersion of an extremely narrow band above $E_F$. They showed that the effect can easily be accounted for if a Lorentzian feature, derived from a very flat band about 20 meV below $E_F$, disperses to just above the $E_F$. The dispersion itself was not observable owing as much to the large natural linewidth (≈ 100 meV) as to instrument resolution. In subsequent materials studied there appears to be actual evidence for dispersion (see below) but surprisingly in a direction opposite to that assumed by Andrews et. al. [1995a].

The temperature at which the measurements were taken ($T_m$ = 120 K) is a most important factor in solving the heavy fermion puzzle. For $CePt_{2+x}$ we see that the measuring temperature is very high, $6T_K < T_m < 10T_K$. Based on NCA, the KR and any possible sidebands should not even exist at $10T_K$, much less be exhibiting Bloch state behavior which should have been observable only at $T_m << T_K$.

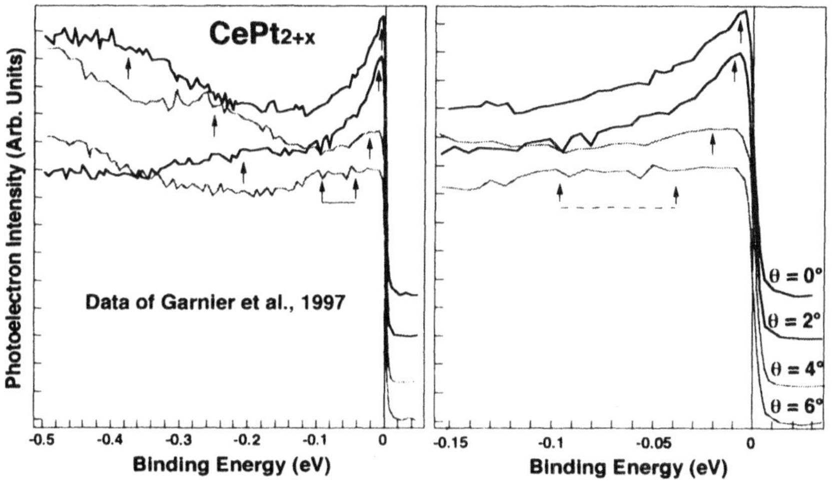


Fig. 14 Digitized and replotted data of Garnier et al. (1997) which strongly suggest (left panel) dispersion of both the $4f_{5/2}$ and $4f_{7/2}$ features. Arrows indicate apparent peak positions. The double arrow for θ = 6° derives from the uncertanity of the peak position. The right panel is a blowup of the near-$E_F$ region.

**Observation of Dispersion:** Superb confirmation of the above results was reported in a recent publication (Garnier, et al., 1997) in which the reported resolution was 5 meV and the data were taken at 40.8 eV photon energy. While the authors reported that no dispersion was evident, a replotting of the data in Fig. 14 suggests that one could reach different conclusions. Indeed, if the feature at -100 meV in the θ = 6° spectrum can be considered as 4f-derived, then there exists about 100 meV of dispersion. The double arrow on this feature at θ = 6° signifies the uncertanity of the exact peak position, but irrespective of the exact position it is hard to avoid the conclusion that the peak disperses. Indeed, not only the $4f_{5/2}$ but also the $4f_{7/2}$ is sees to strongly disperse in the left panel of Fig. 14. A blowup of the $4f_{5/2}$ dispersion is shown in the right panel. The observed dispersion is surprisingly close to the predictions of the PAM which suggest that the 4f band crosses $E_F$ away from a symmetry direction and loses f-character as it disperses below $E_F$. In Fig. 14 the most intense 4f peak is obtained at θ = 2° (away from (111)) and is nearly totally attenuated at θ = 6° where it is found below $E_F$. At normal emission the $4f_{5/2}$ intensity is consistent with a peak just above $E_F$. All these observations are entirely consaistent with the PAM predictions in which 4f-band dispersion is much larger than $T_K$, persist to very high temperatures, and crosses $E_F$ away from symmatry points in a cubic lattice.

**Dispersion in CeSb2:** Amplitude modulation of the Ce-$4f_{5/2}$ is seen to some degree in nearly every single crystal Ce heavy fermion investigated, but direct evidence for

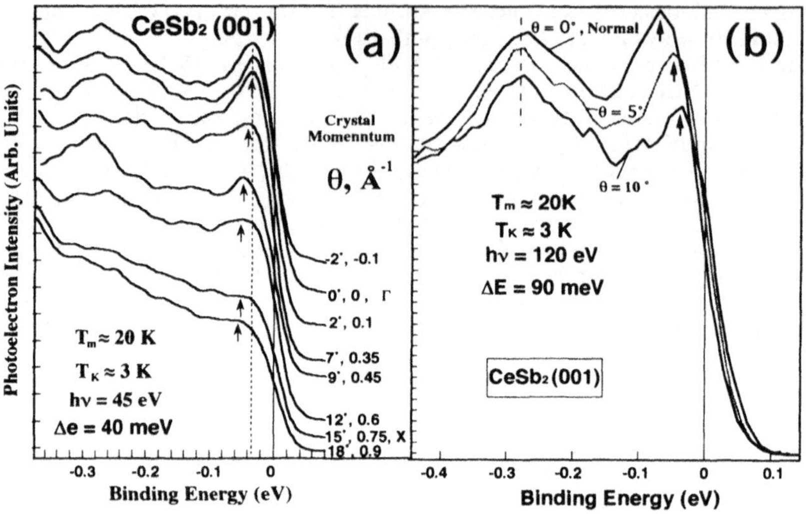


Fig. 15 $CeSb_2$ ARPES spectra along two directions in the Brillouin zone, but both starting at (001). In (a) the dispersion is below $E_F$ and the 4f intensity rapidly diminishes as the band takes on d-character. In (b) the dispersion is above $E_F$ with only minor diminnution of 4f intensity, consistent with the PAM.

dispersion requires better energy and momentum resolution. An especially crucial feature is a mirror-like cleaved surface without which all dispersion can be washed out. The intensity of the Ce-$4f_{5/2}$ as well as the strength of the amplitude modulations correlates more strongly with crystalline perfection than with $T_K$, with the most perfect single crystals yielding the most intense KR's. Indeed, an intense Ce-$4f_{5/2}$ peak is observed in $CeSb_2$, a layered material with a ferromagnetic transition at 10 K which implies that $T_K$ < 10 K. Amplitude modulation was first reported by Andrews et al. (1994). Subsequently (Arko et al., 1997), data at lower energies and T = 20K revealed about 25 meV of dispersion, with behavior amazingly similar to that found in Fig. 14. The left panel of Fig. 15 shows ARPES data at 45 eV photon energy, and we see that as the peak disperses away from $E_F$ it rapidly loses intensity as the f-character is diminished and replaced by d-character (according to the PAM). On the other hand, if the dispersion is toward $E_F$ (Fig. 41b), the $4f_{5/2}$ intensity is only slightly diminished as the 4f band disperses just above $E_F$ and flattens out, leaving the tail on the occupied side. This flattened region can be identified with the KR as in the SIM, though it persists far above $T_K$.

Similar dispersion has been observed also in $CeBe_{13}$ (Andrews et al., 1996) but is not shown here.

### Uranium Compounds

Measurements on highly correlated single crystal uranium compounds again seem to point to a band nature of 5f electrons at T >> $T_K$. Indications of this were already evident in early publications of Arko et al. [1983] in an angle resolved PES study of $UIr_3$ where evidence for slight dispersion of 5f states at 300 K was presented for the first time. In that paper dispersion was not considered unusual however, since, just as in $URh_3$, dHvA measurements [Arko et al., 1975, 1976] indicated a rather normal, transition-metal-like material with effective masses again no larger than 5 to 10. More intriguing were the indications of periodic ARPES structure [Arko et al., 1990] in the 5f features of the heavy fermion material $UPt_3$ measured at 300 K. Although in $UPt_3$ the near-$E_F$ 5f feature never showed measurable dispersion, it's width was resolution limited (≈ 200 meV) while it's intensity vanished near the M-point in the hexagonal lattice. Thus knowledge of the existence of sharp 5f features having intensities periodic with the inverse lattice in uranium compounds even at 300 K was available some time ago, but this information was mostly overlooked.

Recently single crystal uranium compounds have been investigated [Arko et al., 1998] both at the resonant photon energy as well as at much lower photon energies where momentum and energy resolution is substantially better. Evidence for dispersion is again

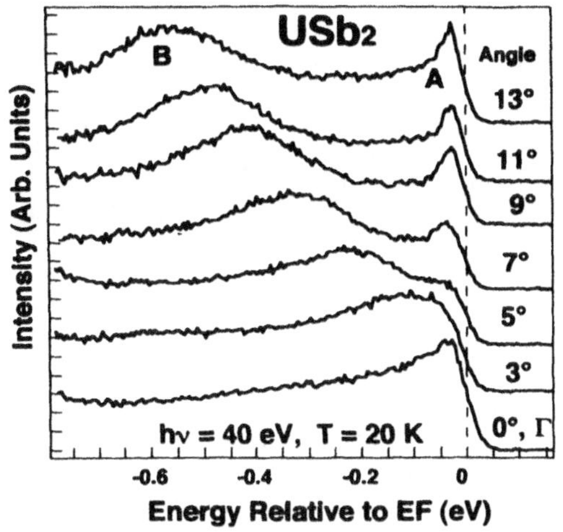


Fig. 16 ARPES spectra from a (100) surface of $USb_2$ along the Γ-X diraction. Peaks A and B are 5f-derived, based on hν-dependence.

found, primarily in materials having the most perfect cleaved surfaces. Here we will concentrate on $USb_2$, an antiferromagnet at 200 K, and $UPt_3$, a heavy Fermion with $T_K \approx 10$ K. Very recently 50 meV of dispersion was also found in UAsSe, a material claimed to display non-Fermi-liquid behavior, but this data is still too fresh to include here. All data shown were collected at $T \approx 20$ K, although data at higher temperatures exists.

In Fig. 16 we present ARPES data for $USb_2$ at lower photon energies. This material exhibits a layered tetragonal structure which is ideal for ARPES since the two-dimensional nature allows for band mapping without the complication from the third dimension. ARPES data at hν = 30 eV and T = 20K are shown as a function of angle [Arko et al., 1997] along

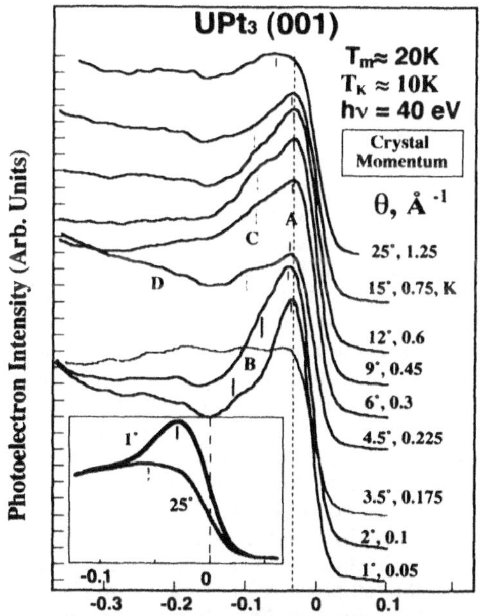


Fig. 17 ARPES spectra from $UPt_3$ (001) at T = 20 K, taken at hν = 40 eV ($T_K \approx 10$ K). Four dispersive peaks are evident. Note the vanishing of peak A between 2° and 3.5° as it disperses above $E_F$.

the Γ-X direction. Two dispersive peaks labeled A and B are evident. Both peaks are unambiguously 5f derived based on the photon energy dependence of the intensity, but peaks B appear to have substantial 6d-admixture, particularly as the binding energy increases. The dispersion of peak B is obvious and clearly 5f in nature, based on photon energy dependence. Indeed, this represents the first clear and unambiguous evidence of 5f dispersion.

The dispersion of peak A is smaller, of the order of 30 meV, though it is not obvious from Fig. 16. Indeed, peak A is nearly absent near Γ. Interestingly, band A disperses below $E_F$ as its intensity decreases, which is similar to the effect seen in Ce heavy fermions above. This sharp near-$E_F$ feature shows very little temperature dependence (except for Fermi function convolution), similar to the situation found in Ce and Yb compounds.

**Dispersion in Upt$_3$:** Since $USb_2$ is an anti-ferromagnet below 200K it leaves some room for argument regarding the band nature of heavy fermions despite the fact that one normally expects a magnetic state to exhibit greater localization. The next logical step is to look for evidence of dispersion and 5f band states in uranium compounds which, unlike antiferomagnetic $USb_2$, are unambiguously heavy fermions - e.g., $UPt_3$.

$UPt_3$ (001) was studied at lower photon energies, at temperatures of 20 K and 80 K, and the results are highly suggestive of dispersion of the near-$E_F$ peak as seen in Fig. 17. Here the data were taken at hν = 40 eV, resolution ΔE = 40 meV, and different angles relative to Γ, i.e., the surface normal in the Brillouin zone. . There is an obvious amplitude effect just as in Ce compounds. Note the complete vanishing of the near-$E_F$ peak between 2° and 3.5° analyzer angle. The drop in intensity corresponds reasonably well with band calculations which show a narrow band, located precisely at $E_F$ at the Γ-point, dispersing above $E_F$ at about 0.3Å$^{-1}$, or, just about 5° from Γ. The peak at 2° from Γ is dispersing toward the Fermi energy. The peaks beyond 5° are then probably due to different bands. The inset emphasizes the dispersion between the 1° and 25° data. It is not certain that improved resolution will be able to separate out the additional bands since the natural linewidths may be larger than the band separations, but it is nevertheless an experiment worth doing. Data taken at 80 K essentially reproduces the 20K data. This data clearly shows that most of the LDA-derived band structure exists already above $T_K$, except that the f-derived bands are flatter than predicted. This is totally consistent with the PAM predictions.

## CONCLUSIONS

Nearly every recent measurement performed on Ce or U based single crystal heavy fermion compounds seems to point to the existence of narrow f-electron bands for a wide range of temperatures. Yb compounds are less clear but are not totally inconsistent with this idea. The clearest evidence for bands is in uranium systems where actual dispersive bands have been observed in $USb_2$ and $UPt_3$, as well as suggestions of dispersion in $UBe_{13}$. Very recently 50 meV of 5f dispersion has also been found in UAsSe. In Ce compounds dispersion has been observed in $CePt_{2+x}$, $CeSb_2$, nad $CeBe_{13}$, at temperatures far above $T_K$. The feature commom to all these compounds is a mirror-like cleaved surface. The least obvious are Yb compounds where the 4f PES features for all intents and purposes are indistinguishable from core levels pinned at the Fermi energy, but with complications of surface states at lower photon energies. Some very recent data on $YbCdCu_4$, not included in this paper, does actually provide the first hint of dispersion even in Yb compounds so that it appears that all system may yet show some universal behavior. No mirror surfaces have been obtained in Yb compounds. In all cases the $n_f$ is far too low relative to SIM predictions.

By contrast, few systems unambiguously exhibit the behavior predicted by the single impurity model. While it is possible to fit a particular spectrum within NCA by adjusting parameters, the lack of systematics with $T_K$ are glaring. In particular, there is no clear systematic scaling of the near-$E_F$ spectral weight with $T_K$, the width of the $f^0$ peak is approximately constant with $T_K$, there is no temperature dependence which cannot be explained by conventional effects of thermal broadening and Fermi function effects (here, perhaps, the broadening may be attributed to the PAM), the spin-orbit feature does not shift toward $E_F$ as $T_K$ increases, and no clear-cut observation of a crystal field state has been reported despite sufficient resolution. Indeed, the temperature dependence in Ce compounds does not provide evidence for the existence of an intense 4f DOS feature just above $E_F$, otherwise one would observe a shift of the near-$E_F$ peak toward the Fermi energy (vs the observed shift in the opposite direction) as was first predicted by Gunnarsson and Schonhammer [1986]. In Yb compounds the spectral weight of the near-$E_F$ peaks (i.e., the $n_f$) is so totally inconsistent with SIM predictions as to be sufficient by itself to seriously question the model. The theory is only valid down to $n_f$ values of about 0.75, while typical $n_f$'s in Yb compounds are measured at 0.65 even for $T_K$'s of 35K. If the existence of the Kondo resonance with the stated properties is a necessary consequence of the SIM, as appears to be the case, then we must look beyond SIM to understand heavy electron behavior.

One the other hand, thermodynamic data are in better agreement with the predictions of SIM, with the recent bandwtdth extractions being the only serious disagreement. There is a consistent relationship between $T_K$ and, say, the peak in the magnetic susceptibility, the specific heat $\gamma$, and inelastic neutron scattering results. Actually, the most important parameter is in fact the hybridization from which $T_K$ and all other parameters follow within SIM. This has presented a dilemma prior to the emergence of new theories, many of which build on the SIM. Any successful theory must simultaneously be able to correctly predict the microscopic as well as the macroscopic properties. SIM and its extensions appear primarily successful with the latter, although, again as pointed out, the bandwidths obtained from a SIM analyis of susceptibility data in $YbXCu_4$ compounds are totally unphysical (0.1 eV vs measured values of $\approx$ 10 eV.). Nevertheless, at some level NCA contains some truth. Electron-electron correlations are indeed important, and formation of the singlet state with conduction electrons is not in dispute (except in the model of Liu (1993, 1997)). Somehow this must be reconciled with the existence of f-bands far above $T_K$ which are by clearly observed in ARPES measurements.

Various renormalized band approaches [Zwicknagel, 1992; Strange and Newns 1986] have succeeded in effectively reproducing the large electron masses in heavy fermions, but they are not in general able to accommodate the numerous magnetic states as well as superconductivity within the same model. A very interesting model that has been recently proposed is the two-electron band approach of Sheng and Cooper [1995], in which correlations are first included explicitly between the *f*-electrons and ligand conduction band states (*f-l* correlations) while *f-f* correlations are treated within an infinite-U approximation. While this is quite similar to the SIM approach, it differs in that a collective response of the ligand electrons which yields the Kondo resonance is not included. Instead, the singlet (or triplet) state is formed as an *f-l* pair of electrons which then form paired band states.

The nascent studies of the Periodic Anderson Lattice Model, do contain some of the ingredients crucial to the description of the photo-emission spectrum, such as substantially more intensity in the $f^1$ peaks vs the $f^0$ peak thus yielding much smaller $n_f$ values, the formation of dispersive quasi-particle bands, a much more mild temperature dependence than found in the SIM, broadening wich was initially attributed to phonons, and the weak hybridization at $E_F$ yielding flat bands at $E_F$ while strong hybridization away from $E_F$ allows f-spectral weight to disperse to energies far larger than $kT_K$ (as observed).
It is, of course, a long road from the first inception of any model to the final detailed proof of its applicability. In its early stages it appears to qualitatively fit the ARPES data, including the mild temperature dependence. Final proof - or discarding - of this theory, however, will occur only after successful - or unsuccessful - attempts to fit real spectra.

**Acknowledgements**
We would like to acknowledge useful conversations with J. Allen, J. Bonca, Barry Cooper, A. Chattopadhyay, D.L. Cox, P. van Dongen, J.K. Freericks, B. Goodman, M. Grioni, D.W. Hess, M. Hettler, H.R. Krishnamurthy, Jon Lawrence, D. McLaughlin, Th. Pruschke, R.N. Silver. Much of the experimental work was performed under the auspices of the U.S. Department of Energy, Office of Basic Energy Sciences, Division of Materials Science. Measurements were performed at the Los Alamos U3C beamline at the NSLS and at the Synchrotron Radiation Center of the Univ. of Wisconsin.

**References**

Allen, J.W., 1992, Synchrotron Radiation Research: Advances in Surface and Interface Science, Volume I: Techniques, edited by Z. bachrach, Plenum Press, New York.

Allen, J.W., S.-J. Oh, O. Gunnarson, K. Schonhammer, M.B. Maple, M.S. Torikachvili and I. Lindau, 1986, Advances in Physics **35**, 275.

Anderson, P.W., 1961, Phys. Rev. **124**, 41.

Andres, K., J.E. Graebner and H.R. Ott, 1975, Phys. Rev. Lett. **35**, 1779.

Andrews, A.B., J.J. Joyce, A.J. Arko, J.D. Thompson, J. Tang, J.M. Lawrence, and J.C. Hemminger, 1995a, Phys. Rev. B **5 1**, Rapid Comm., 3277.
Andrews, A.B., J.J. Joyce, A.J. Arko, J.D. Thompson, Z. Fisk, J. Tang, J.M. Lawrence, P. S. Riseborough, and P.C. Canfield, 1995b, Physica B **206 & 207**, 83.
Andrews, A.B., J.J. Joyce, A.J. Arko, Z. Fisk, and P.S. Riseborough, 1996, Phys. Rev. B **53**, 3317.
Andrews, A.B., R.I.R. Blyth, A.J. Arko, J.J. Joyce, Z. Fisk, J.D. Thompson, R.J. Bartlett, P.C. Canfield, C.G. Olson, P. Benning, and P.S. Riseborough, 1994, Physica B **199&200**, 15.
Arko, A,J., J.J. Joyce, A.B. Andrews, R.I.R. Blyth, C.G. Olson, P.C. Canfield, P.J. Benning, D.M. Poirier, J.H. Weaver, and P.S. Riseborough, 1994, in *Strongly Correlated Electronic Materials*, edited by K. Bedell, S.-Y. Huang, A. Balatsky, D.E. Meltzer, and E. Abrahams, (Addison Wesley), pg.101.
Arko, A.J., D.D. Koelling, and B. Reihl, 1983, Phys. Rev. **B 27**, 3955.
Arko, A.J., J.J. Joyce, A.B. Andrews, J.D. Thompson, J.L. Smith, D. Mandrus, M.F. Hundley, A.L. Cornelius, E. Moshopoulou, Z. Fisk, and P.C. Canfield, Alois Menovsky, 1996, Phys. Rev. B **56**, Rapid Com., R7041.
Arko, A.J., J.J. Joyce, A.B. Andrews, J.L. Smith, J.D. Thompson, Z. Fisk, E. Moshopoulou, A. Menovsky and P.C. Canfield, 1997, Physica B **230-232**, 16.
Arko, A.J., J.J. Joyce, L.E. Cox, L. Morales, J. Sarrao, J.L. Smith, Z. Fisk, A. Menovsky, A. Tahvildar-Zadeh, and M. Jarrell, 1998, J. Alloys and Comp.
Arko, A.J., J.J. Joyce, P.C. Canfield, J.D. Thompson, R.J. Bartlett, Z. Fisk, J. Lawrence, J. Tang, and P. Riseborough, 1993, in *Vacuum Ultraviolet Radiation Physics*, edited by F.J. Wuillemier, Y. Petroff, and I. Nenner, (World Scientific: Singapore), p. 291.
Arko, A.J., M.B. Brodsky, G.W. Crabtree, D. Karim, D.D. Koelling, L.R. Windmiller, and J.B. Ketterson, 1975, Phys. Rev. **B 12**, 4102.
Arko, A.J., M.B. Brodsky, G.W. Crabtree, D. Karim, L.R. Windmiller, and J.B. Ketterson, 1976, in *Plutonium 1975 and other Actinides*, H. Blank and R. Linder, eds. (North Holland, Amsterdam) p. 325.
Arko, A.J., P. Armstrong, and M. Wire, 1990, Physica **B163**, 167.
Bickers, N.E., 1987, Rev. Mod. Phys. **59**, 845.
Bickers, N.E., D.L. Cox and J.W. Wilkins, 1985, Phys. Rev. Lett. **54**, 230.
Bickers, N.E., D.L. Cox and J.W. Wilkins, 1987, Phys. Rev. B **36**, 2036.
Blyth, R.I.R., J.J. Joyce, A.J. Arko, J. Tang, and J.M. Lawrence, 1992, Phys. Rev. Lett.**69**, 3418.
Blyth, R.I.R., J.J. Joyce, A.J. Arko, P.C. Canfield, Z. Fisk, J.D. Thompson, R.J. Bartlett, J. Tang, and J.M. Lawrence, 1993, Phys. Rev. B **48,** 9497.
Fadley, C.F., 1987, Physica Scripta T17, 39.
Fisk, Z., J.L. Sarrao, J.L. Smith, and J.D. Thompson, 1995, *Proc. Natl. Acad. Sci. USA*, 92, 6663.
Galera, R.M., A.P. Murani, and J. Pierre, 1989, Physica B **156&157**, 801.
Garnier, M., D. Purdie, K. Breuer, M. Hengsberger, and Y. Baer, 1997, Phys. Rev. B **36**, R11399.
Grewe, N.and F. Steglich, 1991, *Handbook on the Physics and Chemistry of the Rare Earths*, K.A. Gschneidner, Jr. and L. Eyerling, editors, (North-Holland, Amsterdam), **Vol. 14,** pp. 343-474.
Grioni, M., D. Malterre, P. Weibel, B. Dardel, and Y. Baer, 1993, Physica **B 188**, 38.
Gunnarson, O., and K. Schonhammer, 1985, Phys. Rev. B **31,** 4815.
Gunnarson, O., and K. Schonhammer, 1986, in *Theory of Heavy Fermions and Valence Fluctuations*, T. Kasuya and T. Saso, editors (Springer-Verlag, Berlin, 1986) p. 110.
Hess, D.W., P.S. Riseborough, and J.L Smith, 1993, *Encyclopedia of Applied Physics,* Vol. 7, George L. Trigg., ed., (VCH Publishers, NY), pp.435-463.
Hirsch, J.E., and R.M. Fye, (1989), Phys. Rev. Lett. **56**, 2521.
Hufner, S., 1986, J. Phys. F **16**, L31.
Hufner, S., 1992, Z. Phys. B **86**, 241.
Hufner, S., L. Schlapbach, 1986, Z. Phys. B **64**, 417.
Jarrell, M., and J.E. Gubernatis, 1996, Physics Reports, Vol. 269, No.3, p133-195.
Joyce J.J., and A.J.Arko, 1993a, Phys. Rev. Lett. **70,** 1181.

Joyce, J.J., A. J. Arko, J. M. Lawrence, P. C. Canfield, Z. Fisk, R. J. Bartlett, and J. D. Thompson, 1992c, Phys. Rev. Lett. **68** 236.
Joyce, J.J., A. J. Arko, J. M. Lawrence, P. C. Canfield, Z. Fisk, R. J. Bartlett, J. D. Thompson, and J. L. Smith, 1992a, J. Less Comm. Met., **181,** 161.
Joyce, J.J., A.B. Andrews, A.J. Arko, R.I.R. Blyth, R.J. Bartlett, C.G. Olson, D.M. Poirier, P.C. Canfield, and P.J. Benning, 1996, Phys. Rev. B., to be published.
Joyce, J.J., A.J. Arko, A.B. Andrews, and R.I.R. Blyth, 1994a, Phys. Rev. Lett. **72**, 1774.
Joyce, J.J., A.J. Arko, A.B. Andrews, R.I.R. Blyth, R.J. Bartlett, J.D. Thompson, Z. Fisk, P.S. Riseborough, P.C. Canfield, C.G. Olson, and P.J. Benning, 1994b, Physica B **205**, 365.
Joyce, J.J., A.J. Arko, J.L. Sarrao, K.S. Graham, Z. Fisk, and P.S. Riseborough, 1998, submitted to Phys. Rev. Letters.
Joyce, J.J., A.J. Arko, J.M. Lawrence, P.C. Canfield, R.J. Bartlett, Z. Fisk, and J.D. Thompson, J. Tang, 1992b, Solid State Comm. **83**, 551.
Joyce, J.J., A.J. Arko, P.S. Riseborough, P.C. Canfield, J.M. Lawrence, R.I.R. Blyth, R.J. Bartlett, J.D. Thompson, and Z. Fisk, 1993b, Physica B **186-188,** 31.
Kakizaki, A., A. Harasawa, T. Ishii, T. Kashiwakura, A. Kamata, S. Kunii, 1995, J. Phys. Soc. Japan **64**, 302.
Kang, J.S., J.W. Allen, O. Gunnarsson, N.E. Christensen, and O.K. Anderson, Y. Lassailly, M.B. Maple and M.S. Torkachvilli, 1990, Phys. Rev. B **41**, 6610.
Lawrence, J.M., and D.M. Mills, 1991, Comments Cond. Mat. Phys., **15**, 163.
Lawrence, J.M., G.H. Kwei, P.C. Canfield, J.G. DeWitt, and A.C. Lawson, 1994, Phys. Rev. B49.
Lee, P.A., T.M. Rice, J.W. Serene, L.J. Sham, and J.W. Wilkins, 1986, Com. Cond. Mat. Phys. **12**, 99.
Liu, L.Z., J.W. Allen, O. Gunnarsson, N.E. Christensen, and O.K. Anderson, 1992, Phys. Rev. B **45**, 8934.
Liu, S.H., 1993, in *Handbook on the Physics and Chemistry of Rare Earths,* Vol. **17**, Ch. 111 (North Holland, Amsterdam, 1993) pp 87-148.
Liu, S.H., 1997, Physica B **240**, 49
Lonzarich, G.G., 1988, J. Magn. Magn. Mat. **76-77**, 1 .
Malterre, D., M. Grioni, P. Weibel, B. Dardel and Y. Baer, 1992a, Europhys. Lett.**20**, 445.
Malterre, D., M. Grioni, P. Weibel, B. Dardel and Y. Baer, 1992b, Phys. Rev. Lett. **68**, 2656.
Matzdorf, R., G. Meister, and A. Goldmann, 1993, Surf. Sci. **286**, 56.
Norman, M.R., 1985, Phys. Rev. B **31,** 6261.
Norman, M.R., D.D. Koelling and A.J. Freeman, 1985, Phys. Rev. B **31**, 6251.
Nozieres, P., 1985, Ann. de Phys. **10**, 19.
Oh, S.J., 1993, Physica **B 186-188**, 637.
Oliveira, L.N. and J. Wilkins (1981), Phys. Rev. Lett. **47**, 1553.
Onuki, Y., 1993, Physica B **186-188**, 92.
Ott, H.R., and Z. Fisk, 1987, in *Handbook on the Physics and Chemistry of the Actinides*, A.J. Freeman and G.H. Lander, editors, (North-Holland, Amsterdam), pp. 85-225.
Ott, H.R., H. Rudigier, Z. Fisk, and J.L. Smith, 1993, Phys. Rev. Lett. **50**, 1595.
Patthey, F., B. Delley, W.D. Schneider, and Y. Baer, 1985, Phys. Rev. Lett. **55**, 1518.
Patthey, F., J.-M. Imer, W.-D. Schneider, H. Beck and Y. Baer, 1990, Phys. Rev. B **42**, 8864.
Patthey, F., W.-D. Schneider, Y. Baer, and B. Delley, 1987a, Phys. Rev. Lett. **58**, 2810.
Patthey, F., W.D. Schneider, M. Grioni, D. Malterre, and Y. Baer, 1993, Phys. Rev. Lett. **70**, 1179.
Patthey, F., W.D. Schneider, Y. Baer, and B. Delley, 1986, Phys. Rev. B **34**, 2967.
Patthey, F., W.D. Schneider, Y. Baer, and B. Delley, 1987b, Phys. Rev. B **35**, 5903. Phys. Rev. Lett. **65**, 496-9 .
Pruschke, Th., and N. Grewe (1989), Z. Phys. B **74**, 439.
Sheng, Q.G. and B.R. Cooper, 1995, Phil. Mag. Lett. **72**, 123.
Silver, R.N., J.E. Gubernatis,D.S. Sivia, and Mark Jarrell, 1990

Springford, M., and P.H.P. Reinders, 1988, J. Magn. Magn. Mat. **76-77**, 11.
Steglich, F., J. Aarts, C.D. Bredl, W. Lieke, D. Meschede, W. Franz, and H. Schafer, 1979, Phys. Rev. Lett. **43**, 1892.
Steiner, M.M., R. C. Albers, and L. J. Sham, 1994, Phys. Rev. Lett. **72**, 2923.
Stewart, G. R., 1984, Rev. Mod. Phys. **56**, 755.
Stewart, G.R., Z. Fisk, J.O. Willis, and J.L. Smith, 1984, Phys. Rev. Lett. **52**, 679.
Strange, P., and D. M. Newns, J. Phys. F 16, 335.
Tahvildar-Zadeh, A.N., M. Jarrell and J.K. Freericks, 1997, Phys. Rev. B**55**, R3332.
Tahvildar-Zadeh, A.N., M. Jarrell and J.K. Freericks,1998, In preparation.
Takahashi, T., T. Morimoto, T. Yokoya, M. Kohgi, and T. Satoh, 1994, Phys. Rev. B **49**, 15688.
Tang, J., J.M. Lawrence, and J.C. Hemminger, 1993, Phys. Rev. B **48**, 15342.
Thompson, J.D. and J.M. Lawrence, 1994, *Handbook on the Physics and Chemistry of the Rare Earths*, K.A. Gschneidner, Jr. and L. Eyerling, editors, (North-Holland, Amsterdam), **Vol 19,** pp. 383-478.
Tjeng, L.H., S.-J. Oh, E.-J. Cho, H.-J. Lin, C.T. Chen, G.-W. Gweon, J.-H. Park, J.W. Allen, T. Suzuki, M.S. Makivic, and D.L. Cox, 1993, Phys. Rev. Lett. **71**, 1419.
Weibel, P., M. Grioni, D. Malterre, B. Dardel, Y. Baer, and M. Besnus, 1993, Z. Phys. B **91**, 337.
Weschke, E., C. Laubschat, T. Simmons, M. Domke, O. Strebel, and G. Kaindl, 1991, Phys. Rev. B 44, 8304.
Wilson, K.G., (1975), Rev. Mod. Phys. **47**, 733.
Zwicknagl, Gertrud, 1992, Advances in Physics **41**, 203.

# HEAVY ELECTRON PHENOMENA

H.R. Ott

Laboratorium für Festkörperphysik
ETH Hönggerberg
8093 Zürich
Switzerland

## INTRODUCTION

The low temperature features of simple metals and alloys are fairly well understood. The low energy excitations in the normal state are sufficiently well described by the quasiparticle concept of the Fermi liquid model[1] and the occurrence of superconductivity is very well mastered by the BCS model[2] and invoking the interaction of conduction electrons with the lattice of the solid as providing the necessary pairing of the charge carriers in the superconducting state. Still somewhat less transparent is the situation for describing the behaviour of d- and f-transition metals and alloys. The characteristic presence of incompletely filled electron shells in the atoms of the transition elements provides the possible formation of atomic magnetic moments. The interaction of these moments with the metallic environment has been investigated in the early work of Friedel[3], Blandin[4], Anderson[5] and Kondo[6]. All of this work dealt with only single magnetic moments embedded in a sea of conduction electrons and hence these models are at best sufficient for describing the properties of metals with dilute magnetic impurities. Very well established[7] and understood[8] is the Kondo effect where the conduction electrons tend to shield the impurity moment, leading to the characteristic increase of the electrical resistivity with decreasing temperature.

Much more difficult is the treatment of magnetic moments occupying regular sites of the crystal lattice of metallic materials, where the additional interaction among magnetic scattering centers poses substantial difficulties in the theoretical modelling of the situation. In the most simple cases, the interaction among the localized magnetic moments is provided by conduction electrons via the Ruderman-Kittel-Kasuya-Yosida (RKKY) interaction[9,10] which, at low enough temperatures, results in a cooperative transition to a magnetically ordered ground state. Complications arise if the above mentioned Kondo interaction tends to cancel the magnetic moments and thus competes with the RKKY interaction. In extremal situations, even a full compensation of the moments may occur which results in a complete transfer of the magnetic degrees of freedom to the conduction electron system. Since this doesn't occur abruptly, the entropy of the ensemble of conduction electrons is still anomalously large even

*Electron Correlations and Materials Properties*, edited by Gonis *et al.*
Kluwer Academic/Plenum Publishers, 1999.

at very low temperatures, providing a substantial enhancement of the effective mass of the conduction electrons. This happens because the specific heat of these electrons

$$c^{el}(T) = T(dS^{el}/dT) = \gamma T, \quad (1)$$

where $\gamma = (1/3)k_B{}^2\pi^2(k_F/h^2)m^*$ and $m^*$ is the effective electron mass, thus leading to the formation of heavy electrons in metals. Since large effective masses also imply low velocities, these electrons may also be regarded as slow. Another consequence is a small Fermi energy $E_F$ or a low Fermi temperature $T_F$. Nevertheless, assuming only a simple renormalization of the relevant parameters, even in this case the ground state of these metals would be regarded as that of a Fermi liquid as described by Landau[1)]. There is, however, increasing experimental evidence that such a simple view is, in many cases inadequate, hence suggesting that other models for describing the possible ground states of metals ought to be considered. As a further amusing detail we note that large mass enhancements may lead to the fundamentally intriguing situation of $T_F < \theta_D$, the Debye temperature

In this brief overview we mainly present and discuss various unusual features of materials containing elements with incompletely filled f-electron shells, where a rich variety of heavy electron phenomena due to extremely strong electron correlations has been found experimentally. As examples we may list i) the onset of magnetic order among strongly reduced magnetic moments, i.e., small moment ordering, ii) the coexistence of small moment ordering and itinerant electrons with substantially enhanced effective masses, iii) the onset of superconductivity involving the heavy electrons, iv) the coexistence of this superconductivity with small moment ordering and v) the observation of features in the physical properties of such materials which are not compatible with expectations of Landau's Fermi liquid model.

In addition we present some evidence for considerable mass enhancements of charge carriers in very narrow bands of electronic states in doped semiconductors and we also touch upon the question of electron mass enhancements in metals with rather low concentrations of itinerant charge carriers.

Yet another route to mass enhancement of electrons is discussed in the contribution of Peter Fulde[11)] to this volume.

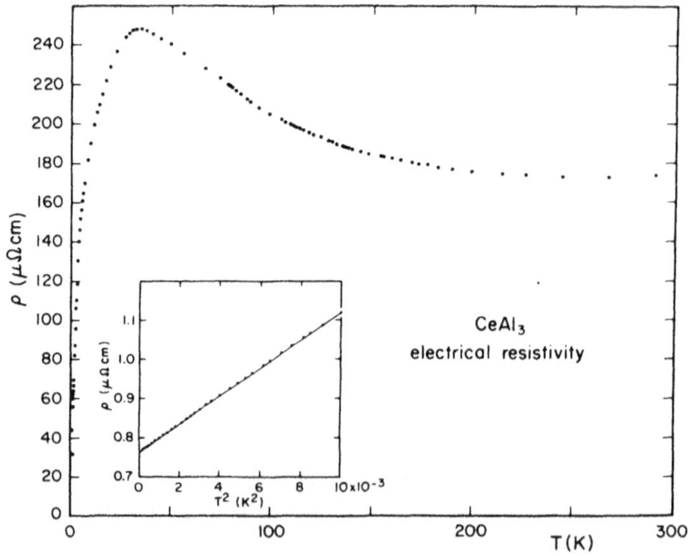


Fig. 1. Temperature dependence of the electrical resistivity of $CeAl_3$ between 1.5 and 300 K, and below 0.1 K. Data are from refs. 12 and 13.

## UNUSUAL LOW TEMPERATURE PROPERTIES OF $CeAl_3$

For a long time, $CeAl_3$ has been considered as the prototype example of a heavy electron compound whose itinerant electrons adopt a Fermi liquid type ground state with strongly renormalized parameters, such as a very high effective mass $m^*$. This view was mainly supported by the results of measurements of thermal and transport properties[12,13] down to millikelvin temperatures. The transition from an incoherent state at high temperatures to correlated electronic features at very low temperatures may well be seen from a diagram of the temperature dependence of the electrical resisitivity $\rho(T)$ of $CeAl_3$, shown in fig.1. The inset emphasizes the temperature range close to 0 K and demonstrates the $T^2$ variation of $\rho$, a typical prediction of the Fermi liquid model. The enormous prefactor of the $T^2$ term, $A = 35\mu\Omega cm$, is a manifestation of the large mass enhancement of the conduction electrons, otherwise established by the huge electronic specific heat $C^{el} = \gamma T$, with $\gamma = 1.6$ J/molK$^2$, roughly three orders of magnitude larger than the electronic specific heat of Al metal. None of this type of experiments gave any hint for a cooperative phase transition down to temperatures of 10 mK. The subsequent probing of magnetic properties of $CeAl_3$ using muon spin resonance ($\mu$SR) techniques, however, revealed the onset of frustrated and quasi-static magnetic correlations below about 2 K[14]. Below 0.7 K the appearance of a non-zero quasistatic internal magnetic field at the muon site revealed the onset of some, at least short range, magnetic order involving extremely reduced magnetic moments of the order of less than 0.1 $\mu_B$/Ce. The observed asymmetry of the $\mu$SR spectra indicated that this type of magnetic order developed in a spatially inhomogeneous way. At the lowest temperatures the sample was divided into two magnetically inequivalent portions of roughly equal size in a "sponge-like" fashion.

A more recent attempt of a microscopic characterization of the low temperature features of $CeAl_3$ was made via measurements of $^{27}$Al nuclear magnetic resonance (NMR) of $CeAl_3$ in low magnetic fields[15]. The properties of $CeAl_3$ are known to be very sensitive to chemical composition and homogeneity of the sample. Indeed, the onset of cooperative magnetic order has recently been claimed on the basis of results from measurements of thermal properties of single-crystalline material[16]. It is therefore of utmost importance to compare data from different sets of experiments only, if they have been obtained using the same sample. Therefore the NMR sample has been prepared from powder that has been obtained by grinding a piece of the material that previously had been characterized by transport and specific heat experiments and was used for the $\mu$SR study.

In fig.2, examples of low-field NMR spectra of partially oriented powder of $CeAl_3$, monitored at different temperatures, are displayed. The preferred orientation of the powder, achieved by applying a large magnetic field at low temperatures, proved to be very stable at all temperatures of our experiments. At temperatures exceeding 3 K, a rather sharp quadrupole split resonance pattern (see dotted line in fig.2) indicates the expected normal paramagnetic environment of the resonant $^{27}$Al nuclei. The narrow lines visible in the 3.5 K spectrum are due to the non-oriented part of the powder sample. Below 3 K, however, the resonance patterns broaden considerably and the growth of a broad unresolved line (broken lines in fig.2) of increasing intensity with decreasing temperature is observed. At the lowest temperatures reached in this series of experiments, the NMR response appears as a broad peak with little structure and a width (FWHM) of the order of a few hundred Gauss. The increasing integrated intensity of the mentioned broad structure with decreasing temperature is compensated by the decrease of the area of the rest of the signal.

From these experiments again no indication of a cooperative phase transition has been obtained. The NMR spectra below 3 K consist of two components whose relative weights change continuously with decreasing temperature. The broad structure is associated with regions of the sample containing correlated magnetic moments and the narrower lines are due to nuclei experiencing a paramagnetic environment. The width of the broad structure of the

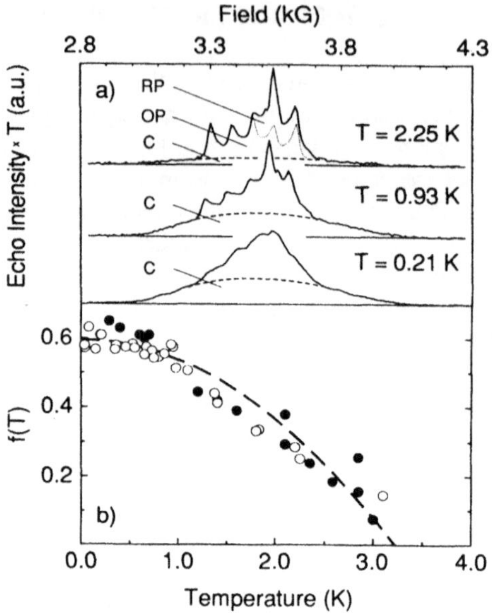


Fig. 2. (a) $^{27}$Al NMR spectra of partially oriented $CeAl_3$ powder. The solid line is the total signal, consisting of contributions from randomly oriented grains (RP) and oriented grains (OP) as indicated by the dotted line. The broken line estimates the broad signal from the correlated phase (C). (b) f(T) as obtained from the spectra (empty circles) and the analysis of the relaxation data using eq.(2) (full circles). The broken line indicates the $T^2$ dependence of f(T).

order of 0.5 kG reflects a distribution of internal fields of the same range which corresponds to localized Ce moments of less than 0.05 $\mu_B$. The width of the broad structure and consequently the magnitude of the internal field amplitude does not vary much with temperature. This observation is compatible with the almost temperature independent precession frequency which modulates the zero-field muon spin relaxation patterns at the lowest temperatures. Thus the temperature dependence of the NMR signals below 3 K to a large extent corroborate the earlier findings from the μSR experiments.

Additional information, confirming the conjecture of two coexisting, magnetically inequivalent phases in $CeAl_3$ at temperatures close to 0 K, has been obtained from measurements of NMR spin lattice relaxation rates. The experiments were performed in various external magnetic fields irradiating the whole spectrum with a special field sweep technique[17].The experimental renormalized magnetization recovery m(t,T) may quite accurately be fitted by

$$\frac{m(t,T)}{m(\infty,T)} = 1 - \left\{ (1-f(T))\exp\left(\frac{-t}{T_1(T)}\right) + f(t)\cdot\exp\left[-\left(\frac{t}{\tau(T)}\right)^{1/2}\right]\right\} , \qquad (2)$$

where f(T) is a temperature dependent weight factor that results from the fits and is shown in the lower half of fig. 2. Below 3 K the value of f(T) increases from zero with decreasing temperature almost exactly in the same way as does the integrated intensity of the broad part of the NMR signal, mentioned above. The choice of the stretched exponential in eq.(2), which may be due to non-random fluctuations in the magnetically correlated phase, has mainly been chosen to achieve the best possible fits. The real physical implications need further insight.

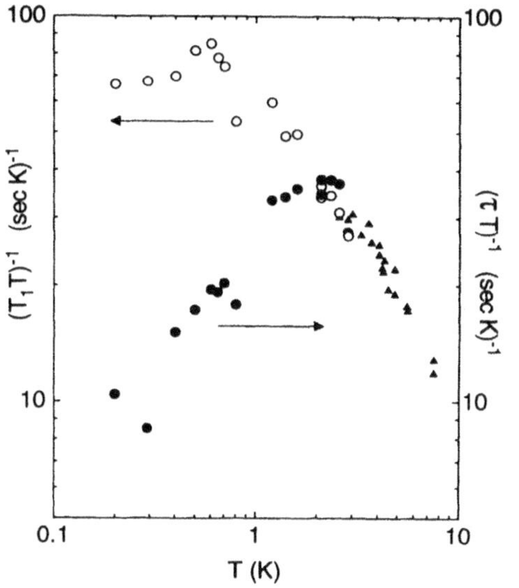


Fig. 3. Spin lattice relaxation rates in $CeAl_3$ as a function of temperature below 10 K. Data above 3 K (full triangles) are from fits to single exponential magnetization recoveries. Below 3 K, two contributions were considered. The full circles represent the parameter $\tau$ of the stretched exponential in eq. 2, the open circles denote $T_1$ in the same equation.

This fitting procedure provides the two temperature dependent parameters $T_1(T)$ and $\tau(T)$; both are shown in fig.3. The relaxation rate $T_1^{-1}$ which we associate with the relaxation in the paramagnetic environment joins smoothly the data at T > 3K where f(T) = 0, and agrees very well with previously reported results[18-20]. The Korringa-type behaviour below 0.6 K is what is expected from the features of the transport and thermal properties and the related Korringa constant is considerably enhanced from its value for, e.g., Al metal. The relaxation rate $\tau^{-1}$ is more difficult to interpret. The strong decrease of $(\tau T)^{-1}$ below 1 K may manifest a substantial coherence in the relaxing fluctuations. At any rate, the very different behaviours of two competing relaxation channels seem to confirm the coexistence of two magnetically and electronically different phases in $CeAl_3$ at very low temperatures. In this sense, the electronic subsystem of this compound may no longer be regarded as a simple physical realization of a Fermi liquid with strongly renormalized parameters.

## NON FERMI-LIQUID FEATURES

Non Fermi-liquid features are usually identified by anomalous temperature dependences of physical properties where the Fermi-liquid approximation provides clear predictions. A well documented case is $CeCu_{6-x}Au_x$ with x < 0.15. The binary compound $CeCu_6$ was one of the first examples where the low temperature properties suggested the presence of heavy-mass electrons[21], forming a strongly renormalized Fermi liquid at very low temperatures. No indication of a cooperative phase transition could be detected down to mK temperatures . Later it was shown that a magnetic phase transition may be triggered by replacing part of the Cu atoms by Au[22]. For concentrations x where still no phase transition was apparent, the temperature dependence of the enhanced electronic specific heat changed from $C^{el}/T$ being approximately constant to $C^{el}/T \sim \ln T$, a distinct deviation from Fermi-liquid type behaviour.

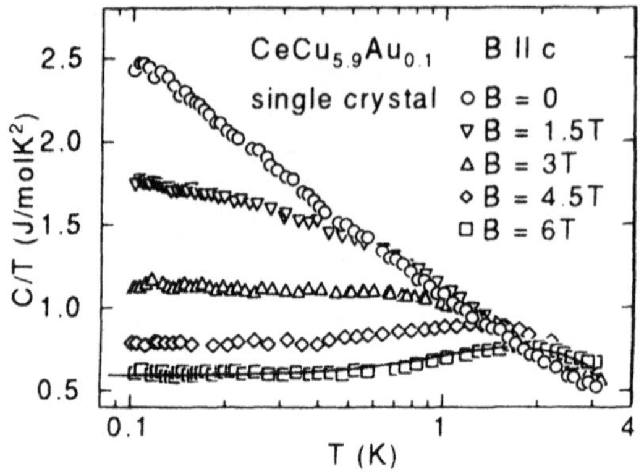


Fig. 4. C/T vs lnT of $CeCu_{5.9}Au_{0.1}$ in various external magnetic fields and below 3 K (from ref.23).

Concomitantly, the electrical resistivity revealed a linear-in-T variation instead of the expected $T^2$ dependence. Quite surprisingly, the Fermi liquid features may be restored, simply by applying an external magnetic field. An example of these features is shown in fig.4 for the specific heat of $CeCu_{5.9}Au_{0.1}$[23].

Theoretically, various routes may lead to non Fermi-liquid fix points. These include the most trivial case of a disorder induced distribution of Kondo temperatures[24], singular cases of multi-channel Kondo scattering scenarios[25] and the occurrence of a quantum critical point (QCP) in case of the onset of magnetic order at exactly T = 0 K[26]. It appears that the non Fermi-liquid features of $CeCu_{5.9}Au_{0.1}$ are a QCP phenomenon. Most of the other well documented cases of non Fermi-liquid type properties are related with U compounds, such as $U_{0.2}Y_{0.8}Pd_3$[27,28], $UCu_{3.5}Pd_{1.5}$[29,30], $U_{1-x}Th_xPd_2Al_3$[31], and possibly $UBe_{13}$[32]. Except for the last binary compound, the non Fermi-liquid behaviour is provoked by chemical tuning and unavoidable disorder.

## COMPETITION OF KONDO AND RKKY INTERACTIONS

As mentioned in the introduction, the main reason behind many of the unusual properties of f-electron metals is the competition between Kondo-type interactions involving itinerant electrons which tend to cancel magnetic moments and the RKKY-type coupling of magnetic moments, also mediated by the conduction electrons. This competition often results in the coexistence of long range magnetic order among substantially reduced magnetic moments and a subsystem of conduction electrons exhibiting a correspondingly enhanced density of electronic states at $E_F$. As an example we present and discuss the case of $CePd_2In$, an intermetallic compound that orders antiferromagnetically at $T_N$ = 1.23 K [33]. The result of specific-heat $C_p(T)$ experiments, shown in fig.5 not only demonstrate the cooperative phase transition by a distinct anomaly peaking at $T_N$, but they also reveal the anomalously large contribution to $C_p(T)$ varying linearly with T for T < 0.6 K. The corresponding prefactor $\gamma$ is 137 mJ/moleK$^2$, considerably larger than an estimate for the electronic specific heat parameter at temperatures T > 7 K, i.e., far above $T_N$, where $\gamma$ = 30 mJ/moleK$^2$. This and the data of the electrical resistivity $\rho(T)$ indicate that $T_K$ is of the same order of magnitude as $T_N$. Thus both the Kondo and the RKKY interaction are small but of similar strength, an intriguing situation per se.

Microscopic information on the low temperature state of $CePd_2In$ has subsequently been obtained via Indium nuclear magnetic resonance (NMR) studies[34]. Measurements both in zero (NQR) and non zero magnetic fields (NMR) have been made. In our context, one of the

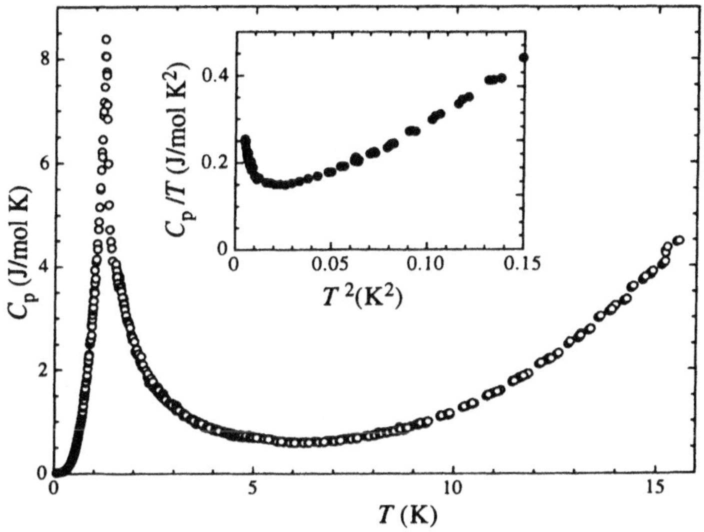


Fig. 5. Temperature dependence of the specific heat of $CePd_2In$ below 15 K, demonstrating the cooperative phase transition at 1.23 K. The inset emphasizes the region below 0.4 K, revealing a substantial C/T ratio and the onset of a contribution due to split nuclear levels.

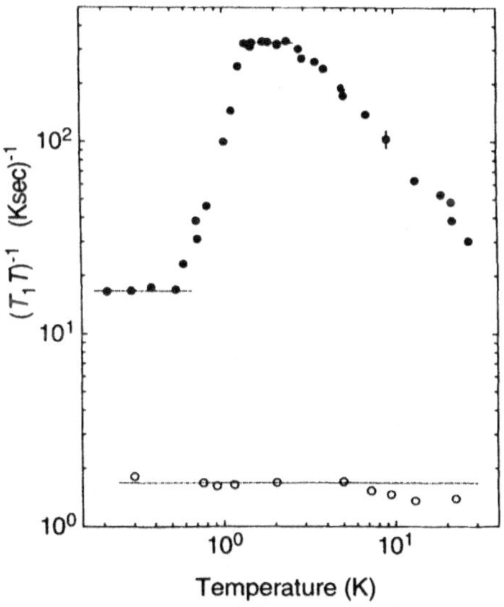


Fig. 6. In-NMR relaxation rates for $CePd_2In$ (full circles) and $LaPd_2In$ (open circles). The horizontal dotted lines indicate Korringa type behaviour.

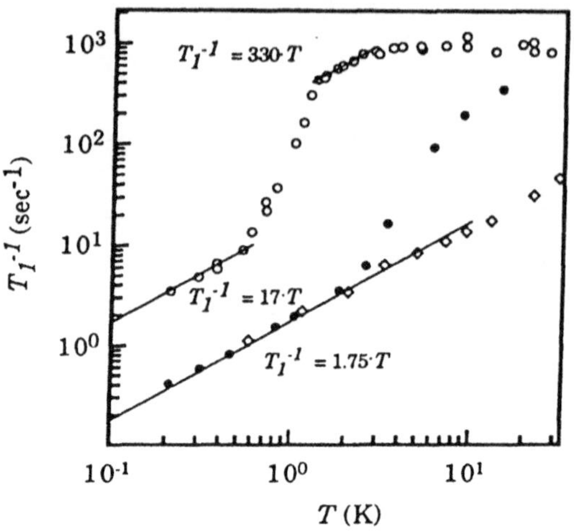


Fig. 7. Temperature dependence of the In-NMR spin lattice relaxation rate for $LaPd_2In$ (open diamonds), and $CePd_2In$ at zero magnetic field (open circles) and at 4T (closed circles). The solid lines indicate Korringa type behaviour.

main results is shown in fig.6, where the temperature dependence of the NQR spin-lattice relaxation rate is plotted in the form of $(T_1T)^{-1}$ versus T on logarithmic scales. For comparison the analogous data for $LaPd_2In$ is shown in the same figure. It may be seen that for the non magnetic compound, a simple Korringa-type behaviour is observed, as expected.

The temperature dependence of $(T_1T)^{-1}$ for $CePd_2In$ is less simple and a much higher relaxation rate is observed at all temperatures. The distinct drop of $(T_1T)^{-1}$ below 1.2 K is due to the onset of antiferromagnetic order. Below 0.5 K a Korringa-type behaviour, but with a much enhanced relaxation rate is again observed, confirming the presence of a high density of low energy excitations in the electronic subsystem, compatible with the above mentioned findings for the specific heat in the same temperature range.

The analysis of the NMR spectra and corresponding Knight shift data, combined with results of magnetisation measurements leads to the conclusion that the ordered moments are only of the order of 0.1 $\mu_B$/Ce, i.e., much reduced from the expectation of 1.3 $\mu_B$, the moment of the $|\pm 3/2>$ 4f electron ground state of the $Ce^{3+}$ ion. We thus have exactly the situation that reflects the competing influence of the two interactions described above.

That the strength of these two weak interactions may be altered rather easily via the application of moderate external magnetic fields H became apparent in NMR studies for H > 3.5 T. In fields of this magnitude, the antiferromagnetic ground state gives way to an induced ferromagnetic, i.e., paramagnetic state. From magnetization measurements it may be concluded that in this field range the full expected moment of 1.3 $\mu_B$ is restored. In fig. 7, it may be seen that in this field regime also the low temperature NMR relaxation rate is drastically reduced, reaching values which are very close to those of $LaPd_2In$, i.e., the influence of the enhancement mechanism, fully effective in zero field, is completely quenched. $CePd_2In$ seems as a very favourable candidate for achieving a reduction by external pressure of $T_N$ to T = 0 K, the situation that was briefly discussed in the previous section.

Large low temperature specific heats varying linearly with temperature do not necessarily imply a high density of electronic states. It has recently been shown[35)] that incommensurate magnetically ordered structures may lead to high densities of low energy excitations, mimicking heavy-electron behaviour. Experimentally, there is evidence for this situation from specific heat and NMR experiments on $CeAl_2$[36)].

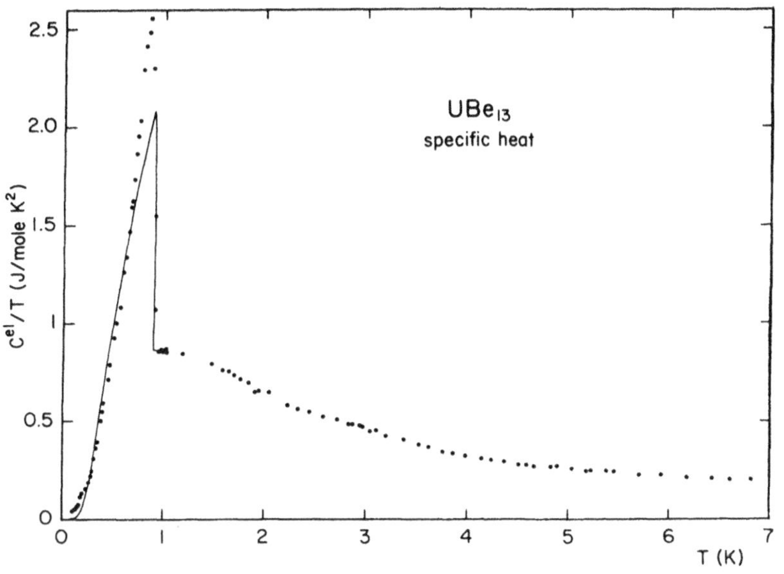


Fig. 8. The electronic specific heat of $UBe_{13}$ below 7 K. The anomaly reflects the superconducting transition, the solid line is calculated using the weak coupling BCS model prediction.

## EXOTIC SUPERCONDUCTIVITY

By now it is fairly well established that superconductivity involving heavy electrons in some rare-earth and actinide compounds is of a different kind than that of common metals like Al or Pb[37)]. Different varieties of order parameters have been suggested to characterize the superconducting states of heavy-electron superconductors[38)]. Consequently it also seems quite likely that the dominant mechanism provoking this superconductivity is rather of magnetic origin than being due to the usual electron-phonon interaction. Strong evidence for this has been obtained[39)] via the observation that some Ce intermetallics which, under ambient pressure conditions, order antiferromagnetically at low temperatures can be made superconducting by application of moderate hydrostatic pressure with a transition temperature $T_c < T_N$.

As one example of heavy-electron superconductivity we briefly discuss some features of $U_{1-x}Th_xBe_{13}$. The binary actinide intermetallic $UBe_{13}$ is superconducting below approximately 0.9 K[40)]. Based on measurements of the electrical resistivity it seems quite obvious that the superconducting state is preceded by a normal state that does not exhibit Fermi-liquid type features.

The fact that heavy mass electrons are involved is convincingly demonstrated by the large anomaly of the specific heat at and below $T_c$, shown in fig.8. Anomalous temperature dependences of the specific heat[41)], the $^{11}$B-NMR relaxation rate[42)], the ultrasound absorption[43)] and the penetration depth[44)] in the superconducting state have indicated nodes in the gap of electronic excitations, thereby implying an unusual symmetry of the order parameter. Quite anomalous is also the temperature dependence of the upper critical field $H_{c2}(T)$, leading to suggestions of the formation of a Fulde-Ferrell-Larkin-Ovchinnikov state of superconductivity in high magnetic fields[45)].

Rather intriguing is the $[T_c,x]$ phase diagram for $U_{1-x}Th_xBe_{13}$, shown in fig.9. The phases denoted by U and L are separated by a phase transition that manifests itself by a second sizeable anomaly of the specific heat. Ginzburg-Landau type model calculations[46)] have suggested that phase L might break the time reversal symmetry and the results of subsequent $\mu$SR experiments[47)] seem to confirm this conjecture.

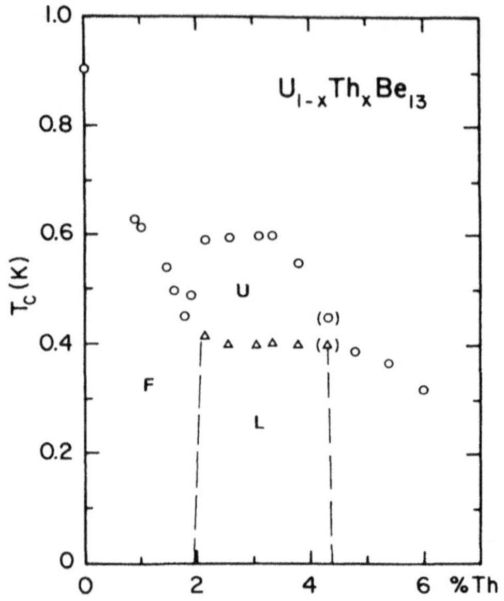


Fig. 9. [$T_c$,x] phase diagram for $U_{1-x}Th_xBe_{13}$, as obtained from specific heat data. The vertical broken lines indicate phase boundaries that are based on the non observations of a second phase transition (open triangles) in the alloys with $x = 0.019$ and 0.047.

The aspect of non Fermi-liquid behaviour of the normal state in these alloys also needs to be considered in the interpretation of the electronic specific heat of Th-doped $UBe_{13}$ below 1 K. The problem may be appreciated in fig.10. The entropy balance of $C_p(T)$ below $T_c$ cannot be achieved without assuming that $C^{el}/T$ of the underlying normal state increases dramatically towards T = 0 K[48].

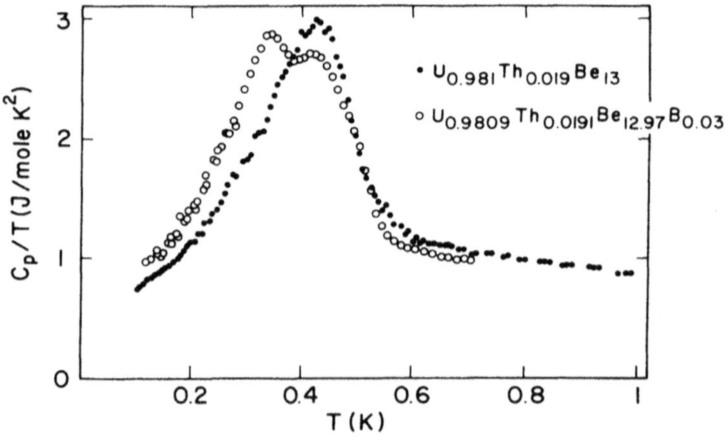


Fig.10. C/T vs T for Th and B doped $UBe_{13}$ below 1 K. The entropy balance can only be achieved if a considerable increase of C/T in the normal state for T approaching 0 K is assumed.

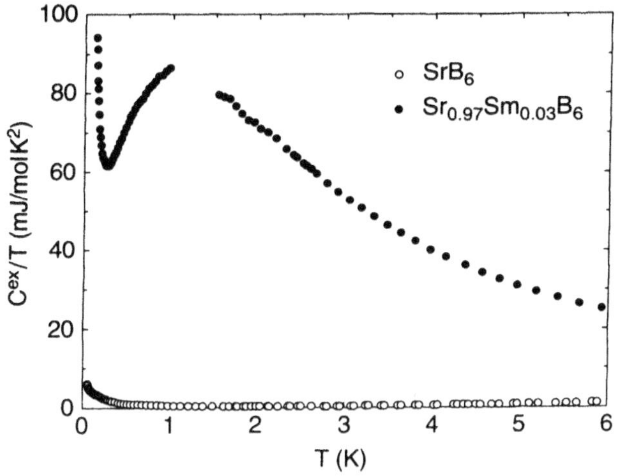


Fig.11. Comparison of the electronic specific heats of $SrB_6$ and $Sr_{0.97}Sm_{0.03}B_6$ below 6 K.

## MASS ENHANCEMENTS AT LOW ELECTRON DENSITIES

An obvious question that one may pose in the context of heavy-electron phenomena is that of how many electrons are needed to screen the moment of a magnetic impurity? This question may be addressed by investigating the case of introducing a magnetic moment into an essentally insulating environment and, at the same time, provide one extra electron. For this purpose a particularly well suited class of materials is provided by cubic hexaborides with divalent and trivalent cations.

$SrB_6$, in which the Sr ions adopt a divalent configuration, is known to be a poor conductor at low temperatures with a charge carrier concentration of the order of $10^{-3}$ electrons per unit cell[49]. Band structure calculations[50] indicate that the electronic features of divalent hexaborides are entirely determined by a small region in k-space around the X point of the Brillouin zone. Substituting a small amount of $Sr^{2+}$ by $Sm^{3+}$ will thus provide exactly the situation outlined above. In fig.11, we show one of the implications of this procedure in the form of a comparison of the low temperature specific heats of $SrB_6$ and $Sr_{0.97}Sm_{0.03}B_6$ below 6 K. The latter compound reveals metallic behaviour below room temperature and the considerable excess specific heat reflects the additional degrees of freedom. At very low temperatures this excess specific heat is well described by

$$C^{ex}T^2 = a + \gamma T^3 \tag{3}$$

where a and $\gamma$ are the prefactors of $C_p$ contributions proportional to $T^{-2}$ and T, respectively. If we interpret the second term on the right of eq.(3) as being due to the electronic density of states $D(E_F)$, the parameter $\gamma$ is compatible with 1.5 J/moleSmK$^2$, in line with metals exhibiting very strong electronic mass enhancements. Although the origin of the distinct $T^{-2}$ increase of $C_p$ at very low temperatures is not yet understood, it appears that electronic mass enhancements also occur if the number of moments to be shielded is the same as that of the available itinerant electrons[51].

Of course, the question of mass enhancements close to a metal-insulator transition is an old standing problem in the physics of transition metal oxides[52], a topic that is enjoying much renewed interest recently[53].

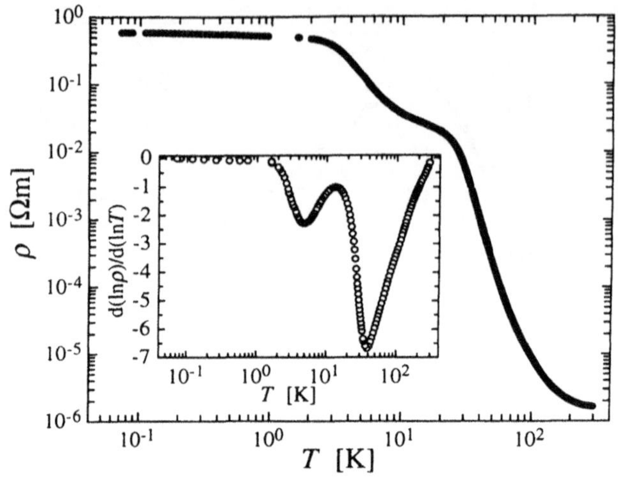


Fig.12. Temperature dependence of the electrical resistivity ρ(T) of FeSi below room temperature. The inset shows the logarithmic temperature derivative of ρ(T).

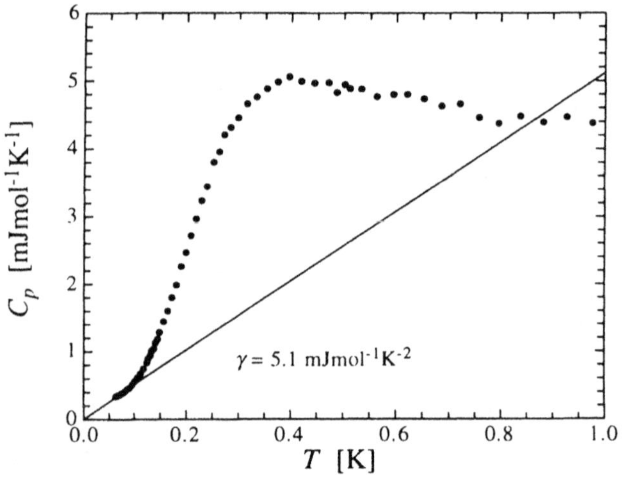


Fig.13. Low temperature specific heat of FeSi between 0.05 and 1 K. The solid line marks the slope of the data below 0.1 K.

## HEAVY ELECTRONS IN EXTREMELY NARROW IMPURITY BANDS

In recent studies of the cubic compound FeSi it has been found that this material may be regarded as a narrow gap semiconductor with unusual features of both the valence and the conduction band[54]. Although the electrical resistivity ρ(T) of FeSi grows by more than five orders of magnitude below room temperature, a low but non-zero electrical conductivity for T approaching 0 K, as shown in fig.12, is observed. At the lowest temperatures, i.e., below 0.2 K, $\rho = \rho_0 - AT^2$ with A = 0.27 $\Omega m/K^2$, a very large prefactor.

A discussion of all the features of the physical properties of FeSi is complicated[55-58] and still, certain interpretations are controversial. Here we only consider the region of very low temperatures. Since FeSi is not a so called line compound, defects in both the Fe and the Si sublattice are almost unavoidable and therefore, a very small concentration of charge carriers may occupy itinerant states, even in the absence of impurities. From various experiments it may be deduced that these defect states form a very narrow band of extended states close to the lower edge of the conduction band, with a width of less than 1 meV. Indeed, below 1 K, the specific heat has a very anomalous temperature dependence, as shown in fig.13. The overall shape of the anomaly is of Schottky type with a characteristic energy of 0.1 meV. The linear-in-T variation of $C_p$ below 0.1 K is compatible with a parameter $\gamma = 5.1$ mJ/moleK$^2$. Together with an estimate of the carrier concentration of $n = 2 \times 10^{22}$ m$^{-3}$, based on Hall voltage measurements, a mass enhancement for these electronic states of the order of 900 may be inferred. According to fig.13, this mass enhancement is clearly temperature dependent and obviously only effective at the very lowest temperatures.

## SUMMARY

In this very brief review of heavy-electron phenomena in metals and alloys we have only scratched the surface of a large body of available experimental and theoretical results. The current interest in these types of problems is distributed over the aspects that have been discussed above and some others not mentioned here. However, the issue concerning the validity of the Fermi-liquid model as the only description of the ground state of metals seems to be the most attractive item in the agenda. There is growing experimental and theoretical evidence that other models also have to be considered for describing the low temperature behaviour of new and exotic metals.

## ACKNOWLEDGEMENTS

I should like to thank many friends and colleagues who have contributed in various ways to the results that are presented above. It is mainly their lasting enthusiasm for the field of heavy-electron physics that made this review possible. In particular, my thanks go to E. Felder, J. Gavilano, J. Hunziker, P. Vonlanthen, B. Ambrosini, M. Chernikov, S. Paschen, L. Degiorgi, A. Schenck, Z. Fisk, J. Sarrao, J.L. Smith, M. Sigrist and T.M. Rice. Continuous financial support of the Schweizerische Nationalfonds zur Förderung der wissenschaftlichen Forschung is also gratefully acknowledged.

## REFERENCES

1. L.D. Landau, Zh. Eksp. Teor. Fiz. **30**, 1058 (1956) (Sov. Phys. JETP 3, 920 (1957)).
2. J. Bardeen, L.N. Cooper, and J.R. Schrieffer, Phys. Rev. **108**, 1175 (1957).
3. J. Friedel, Can. J. Phys. **34**, 1190 (1956).
4. A. Blandin and J. Friedel, J. Phys. Radium **19**, 573 (1958).
5. P.W. Anderson, Phys. Rev. **124**, 41 (1961).
6. J. Kondo, Progr. Theoret. Phys. (Kyoto) **32**, 37 (1964).
7. A.J. Heeger, Solid State Phys. **23**, 283 (1969).
8. N. Andrei, K. Furuya, and J.H. Lowenstein, Rev. Mod. Phys. **55**, 331 (1983).
9. T. Kasuya, Progr. Theoret. Phys. (Kyoto) **16**, 45 (1956).
10. K. Yosida, Phys. Rev. **106**, 893 (1957).
11. P. Fulde, this volume
12. K. Andres, J.E. Graebner, and H.R. Ott, Phys. Rev. Lett. **35**, 1979 (1975).
13. H.R. Ott, O. Marti, and F. Hulliger, Solid State. Comm. **49**, 1129 (1984).
14. S. Barth, H.R. Ott, F.N. Gygax, B. Hitti, E. Lippelt, A. Schenck, and C. Baines, Phys. Rev. B**39**, 11695 (1989).
15. J.L. Gavilano, J. Hunziker, and H.R. Ott, Phys. Rev. B**52**, R13106 (1995).

16. G. Lapertot, R. Calemczuk, C. Marcenat, J.Y. Henry, J.X. Boucherle, J. Flouquet, J. Hammann, R. Cibin, J. Cors, D. Jaccard, and J. Sierro, Physica B **186-188**, 454 (1993).
17. J. Hunziker, Doctoral Dissertation 11613, ETH Zürich, 1996, unpublished.
18. H. Nakamura, Y. Kitaoka, K. Asayama, and J. Flouquet, J. Phys. Soc. Jpn. **57**, 2644 (1988).
19. W.H. Wong and W.G. Clark, J. Magn. Magn. Mater. **108**, 175 (1992).
20. M.J. Lysak, and D.E. MacLaughlin, Phys. Rev. B **31**, 6963 (1985).
21. H.R. Ott, H. Rudigier, Z. Fisk, J.O. Willis, and G.R. Stewart, Solid State Commun. **53**, 235 (1985).
22. S. Mock, C. Paschke, A. Schröder, J. Sereni, M. Sieck, H. v. Löhneysen, Physica B **199-200**, 39 (1994).
23. H. v. Löhneysen, T. Pietrus, G. Portisch, H.G. Schlager, A. Schröder, M. Sieck, and T. Trappmann, Phys. Rev. Lett. **72**, 3262 (1994).
24. E. Miranda, V. Dobrosavljevic, and G. Kotliar, Physica B **230-232**, 569 (1997).
25. D.L. Cox, Phys. Rev. Lett. **59**, 1240 (1987).
26. A.M. Tsvelik and M. Reizer, Phys. Rev. B**48**, 9887 (1993).
27. M.B. Maple, M.C. de Andrade, J. Herrmann, Y. Dalichaouch, D.A. Gajewski, C.L. Seaman, R. Chau, R. Movshovich, M.C. Aronson, R. Osborn, J. Low Temp. Phys. **99,** 223 (1995).
28. B. Andraka and A.M. Tsvelik, Phys. Rev. Lett. **67**, 2886 (1991).
29. B. Andraka, Physica B **199-200**, 239 (1994).
30. L. Degiorgi and H.R. Ott, J. Phys.: Condens. Matter **8**, 9901 (1996).
31. Y. Dalichaouch and M.B. Maple, Physica B **199-200**, 176 (1994).
32. D.L. Cox and M. Makivic, Physica B **199-200**, 391 (1994).
33. A.D. Bianchi, E. Felder, A. Schilling, M.A. Chernikov, F. Hulliger, and H.R. Ott, Z. Phys. B **99**, 69 (1995).
34. P. Vonlanthen, J.L. Gavilano, B. Ambrosini, D. Heisenberg, F. Hulliger, and H.R. Ott, Z. Phys. B **102**, 347 (1997).
35. O. Hudak, J.L. Gavilano, and H.R. Ott, Z. Phys: B **99**, 587 (1996).
36. J.L. Gavilano, J. Hunziker, O. Hudak, T. Sleator, F. Hulliger, and H.R. Ott, Phys. Rev. B **47**, 3438 (1993).
37. H.R. Ott, Czechoslovak J. of Phys. **46**, Suppl. S6, 3131 (1996).
38. M. Sigrist and K. Ueda, Rev. Mod. Phys. **63**, 239 (1991).
39. N.D. Mathur, F.M. Grosche, S.R. Julian, I.R. Walker, D.M. Freye, R.K.W. Haselwimmer, and G.G. Lonzarich, Nature **394**, 39 (1998).
40. H.R. Ott, H. Rudigier, Z. Fisk, and J.L. Smith, Phys. Rev. Lett. **50**, 1595 (1983).
41. H.R. Ott, E. Felder, C. Bruder, and T.M. Rice, Europhys. Lett. **3**, 1123 (1987).
42. D.E. MacLaughlin, C. Tien, W.G. Clark, M.D. Lan, Z. Fisk, and J.L. Smith, Phys. Rev. Lett. **53**, 1833 (1984).
43. B. Golding, D.J. Bishop, B. Batlogg, W.H. Haemmerle, Z. Fisk, J.L. Smith, and H.R. Ott, Phys. Rev. Lett. **55**, 2479 (1985).
44. F. Gross, B.S. Chandrasekhar, D. Einzel, K. Andres, P.J. Hirschfeld, H.R. Ott, Z. Fisk, and J.L. Smith, Z. Phys. B **64**, 175 (1986).
45. J.P. Brison, A. Buzdin, L. Glémot, F. Thomas, and J. Flouquet, Physica B**230-232**, 406 (1997).
46. M. Sigrist and T.M. Rice, Phys. Rev. B**39**, 2200 (1989).
47. R.H. Heffner, J.L. Smith, J.O. Willis, P. Birrer, C, Baines, F.N. Gygax, B. Hitti, E. Lippelt, H.R. Ott, A. Schenck, E.A. Knetsch, J.A. Mydosh, and D.E. MacLaughlin, Phys. Rev. Lett. **65**, 2816 (1990).
48. H.R. Ott, E. Felder, Z. Fisk, R.H. Heffner, and J.L. Smith, Phys. Rev. B**44**, 7081 (1991).

49. H.R. Ott, M. Chernikov, E. Felder, L. Degiorgi, E.G. Moshopoulou, J.L. Sarrao, and Z. Fisk, Z. Phys. B**102**, 337 (1997).
50. S. Massidda, A. Continenza, T.M. Pascale, and R. Monnier, Z. Phys. B**102**, 83 (1997).
51. H.R. Ott, E. Felder, M. Chernikov, Z. Fisk, J.L. Sarrao, and D.P. Young, Phys. Rev. B**57**, 10644 (1998).
52. W.F. Brinkman and T.M. Rice, Phys. Rev. B**2**, 4302 (1970).
53. M. Imada, A. Fujimori, and Y. Tokura, Rev. Mod. Phys., in print.
54. S. Paschen, E. Felder, M.A. Chernikov, L. Degiorgi, H. Schwer, H.R. Ott, D.P. Young, J.L. Sarrao, and Z. Fisk, Phys. Rev. B**56**, 12916 (1997).
55. D. Mandrus, J.L. Sarrao, A. Migliori, J.D. Thopson, and Z. Fisk, Phys. Rev. B**51**, 4763 (1995).
56. Z. Schlesinger, Z. Fisk, H.T. Zhang, M.B. Maple, J.F. DiTusa, and G. Aeppli, Phys. Rev. Lett. **71**, 1748 (1993).
57. B.C. Sales, E.C. Jones, B.C. Chakoumakos, J.A. Fernandez-Baca, H.E. Harmon, J.W. Sharp, and E.H. Volckmann, Phys. Rev. B**50**, 8207 (1994).
58. L. Degiorgi, M.B. Hunt, H.R. Ott, M. Dressel, R.J. Feenstra, G. Gruner, Z. Fisk, and P.C. Canfield, Europhys. Lett. **28**, 341 (1994).

## Discussion

**Sawatzky:** How did you determine the electron density in FeSi?

**Ott:** From the Hall effect.

**Sawatzky:** How reliable is the Hall effect measurement for such a complicated Fermi surface? You also showed some optical conductivity measurements, what was the electron density as determined from the Drude weight?

**Ott:** At high temperatures the Drude weight yielded about 10**19 electrons per cubic centimeter which is of the same order of magnitude as that obtained from the Hall effect.

# LATTICE EFFECTS IN THE LIGHT ACTINIDES


A. C. Lawson, B. Cort, J. A. Roberts and B. I. Bennett,
T. O. Brun and R. B. Von Dreele

Los Alamos National Laboratory
Los Alamos, NM 87545 USA

J. W. Richardson, Jr.

Argonne National Laboratory
Argonne, IL 60439 USA


## INTRODUCTION

The light actinides show a variety of lattice effects that do not normally appear in other regions of the periodic table. Following the pioneering work of Johansson and his collaborators (Johansson et al., 1981) we know that these effects arise from interactions involving the 5f electrons. Progress in theory is bringing these effects within the range of detailed theoretical calculation: we have prepared this article to summarize the experimental situation.

The article will cover the crystal structures of the light actinides, their atomic volumes, their thermal expansion behavior, and their elastic behavior as reflected in recent thermal vibration measurements made by neutron diffraction. A discussion of the melting points will be given in terms of the thermal vibration measurements. Pressure effects will be only briefly indicated.

To indicate the range of exotic effects that are encountered in the light actinides, Fig. 1 shows the linear thermal expansion of Pu metal plotted vs. temperature. This figure shows a succession of six distinct crystallographic phases as the material is heated from room temperature to the melting point. There is a wide range of atomic volumes and thermal expansions, including a range of *negative* thermal expansion, for two of the phases, δ and δ'. The thermal expansion of Fe is shown for comparison: by this standard the thermal effects in Pu are quite large. The melting point of 641°C is anomalously low, either by rare earth or by transition metal norms.


*Electron Correlations and Materials Properties*, edited by Gonis *et al.*
Kluwer Academic/Plenum Publishers, 1999.

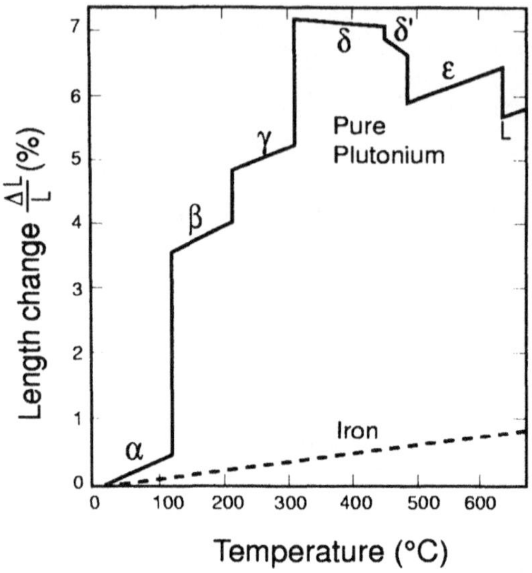


Fig. 1. Linear thermal expansion of the six phases of Pu metal. The thermal expansion of Fe is shown for comparison.

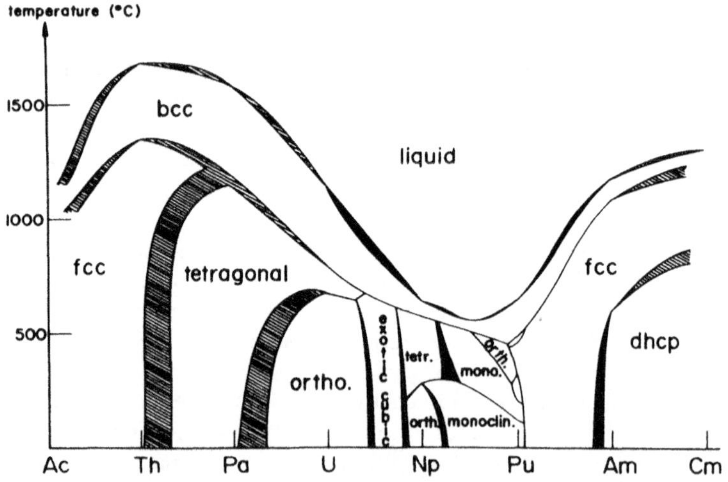


Fig 2. Schematic phase diagram of the light actinides. Reprinted from J. Less-Common Metals 90, "Magnetism and bonding: a nearly periodic table of transition elements," p. 83, Copyright (1983), with permission from Elsevier Science.

With so many distinct phases involved, it is necessary that some of the phases are crystallographically complex. This has been found to be the case, and the structures are in fact the most complex found for any metal. The situation is nicely presented in the schematic phase diagram of Kmetko and Smith, 1983, which is shown as Fig. 2. This diagram shows that as one proceeds from Ac through to Am, there is a progression of complex crystal structures that is correlated with a deep depression of the melting point. These features, together with other symptoms of dominant 5f electron influence on the materials properties, are the subject of this article.

## STRUCTURES

The light actinides are famous for their idiosyncratic structures. At normal pressures, Th has two polymorphs, U and Np each have three, while Pu has six. As one gets to Am in the periodic system there are again only two polymorphs. The structures of the various polymorphs are summarized in Table 1. The general trend with atomic number is from simple to complex to simple again. The general trend with temperature is from complex to simple. These trends are evident also in Fig. 2.

Table 1. Polymorphs of the light actinides

| Element | phase | crystal system | Stability Range (K) | space group | $N_{cell}$ | $N_{type}$ | ref. |
|---|---|---|---|---|---|---|---|
| Th | α | cubic | 0-1673 * | Fm3m | 4 | 1 | 1 |
| | β | cubic | 1673-2115 (M) | Im3m | 2 | 1 | 1 |
| Pa | α | tet. | ambient | I4/mmm | 2 | 1 | 1 |
| | β | cubic | high T | Fm3m | 4 | 1 | 1 |
| U | α' | CDW | 0-43 | | | | 2 |
| | α | ortho. | 43-935 | Cmcm | 4 | 1 | 3 |
| | β | tet. | 935-1045 | $P4_2/mnm$ | 30 | 5 | 4 |
| | γ | cubic | 1045-1405 (M) | Im3m | 2 | 1 | 1 |
| Np | α | ortho. | 0-553 | Pmcn | 8 | 2 | 5 |
| | β | tet. | 553-850 | P4/nmm | 4 | 2 | 6 |
| | γ | cubic | 850-910 (M) | Im3m | 2 | 1 | 1 |
| Pu | α | mono. | 0-395 | $P2_1/m$ | 16 | 8 | 7 |
| | β | mono. | 395-479 | I2/m | 34 | 7 | 8 |
| | γ | ortho. | 479-592 | Fddd | 8 | 1 | 9 |
| | δ | cubic | 592-724 | Fm3m | 4 | 1 | 1 |
| | δ' | tet. | 724-749 | I4/mmm | 2 | 1 | 10 |
| | ε | cubic | 749-913 (M) | Im3m | 2 | 1 | 1 |
| Am* | α | hex. | 339-931 | $P6_3/mmc$ | 4 | 2 | 1 |
| | β | cubic | 1066-1277 | Fm3m | 4 | 1 | 1 |
| | γ | cubic | 1323-1443 (M) | Im3m (?) | 1 | 1 | 11 |
| Cm | α | hex. | ambient | $P6_3/mmc$ | 4 | 2 | 1 |

References for Table 1
* more phases are possible for Am
1 Donohue (1982)
2 Lander et al. (1994)
3 Jacob and Warren (1937)
4 Lawson et al., (1988)
5 Zachariasen, (1952a)
6 Zachariasen, (1952b)
7 Zachariasen and Ellinger, (1963b)
8 Zachariasen and Ellinger, (1963a)
9 Zachariasen and Ellinger, (1955)
10 Ellinger, (1956)
11 Rose et al. (1979)

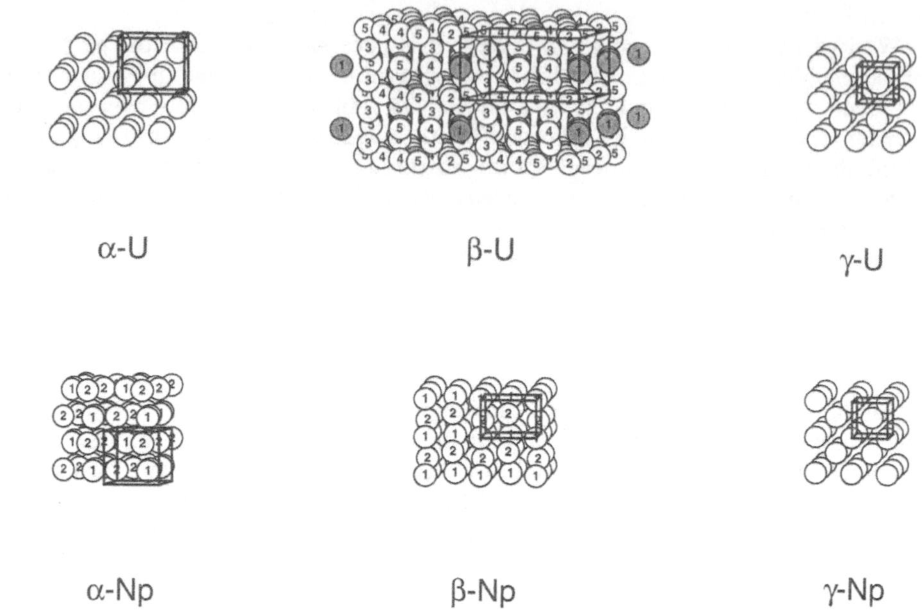


Fig 3. Crystal structures of the phases of U and Np. Blocks of 2x2x2 unit cells are shown. The atoms are numbered when there is more than one atom type. For β-U, the atom at the origin is shaded.

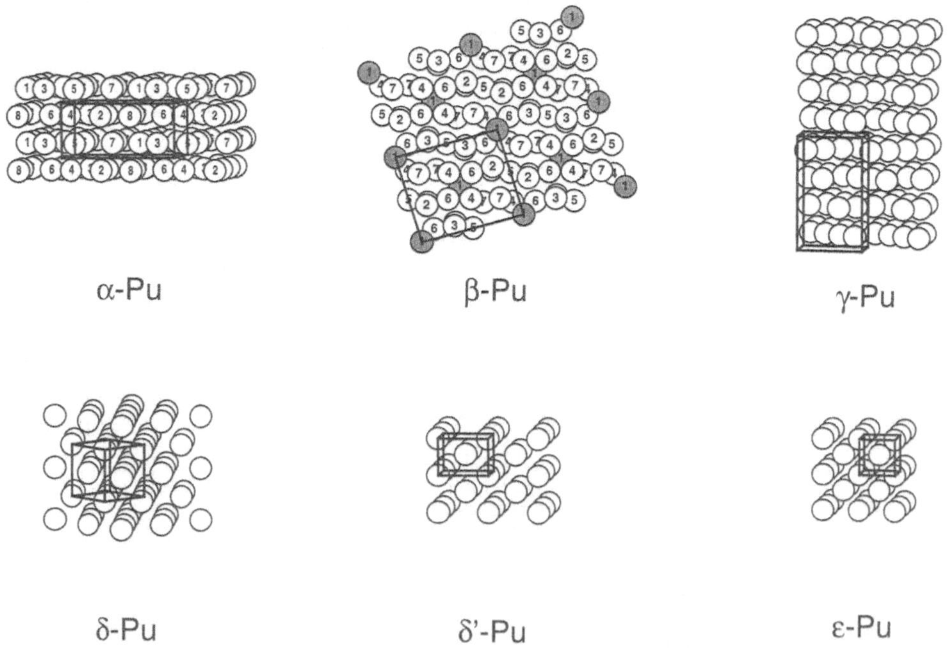


Fig 4. Crystal structures of the phases of Pu. Blocks of 2x2x2 unit cells are shown. The atoms are numbered when there is more than one atom type. For β-Pu, the atom at the origin is shaded

The main point of interest is the complexity of the low temperature structures of U, Np and Pu. The α-form of Pu metal is the most complex known for any metallic element: it is monoclinic, with 16 atoms per primitive cell, and there are 8 unique atom types. Zachariasen's solution of this structure from powder data is one of the outstanding achievements in all of crystallography. Figs. 3 and 4 illustrate this structural complexity. The figures, which are drawn to scale, show blocks of eight unit cells (2x2x2) in order to convey some idea of the packing.

Fig. 5 shows the atomic volumes of the light actinides plotted vs. atomic number. These volumes are obtained from crystallographic data, and there is a point shown for each polymorph. The points corresponding to the polymorph of lowest volume (always the low temperature structure) are connected to illustrate the trend. The wide range of values for Pu is taken by Johansson et al. (1981) as evidence for a Mott transition among the 5f electrons.

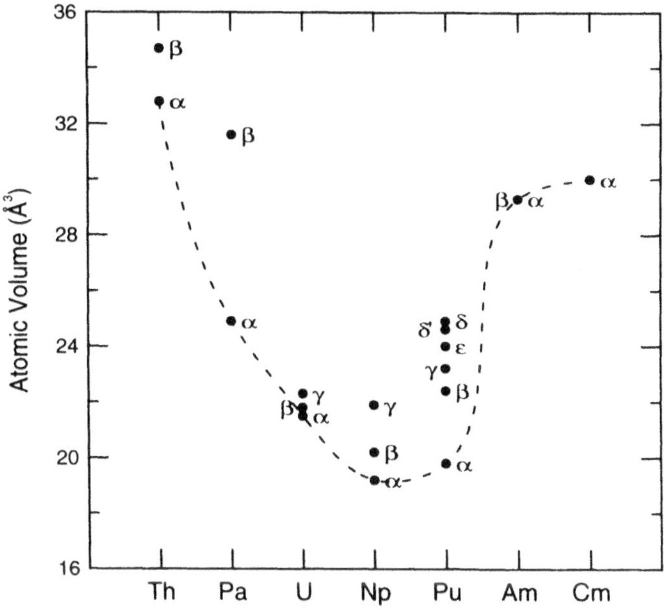


Fig. 5. Atomic volumes for the various light actinide phases.

## THERMAL EXPANSION

Thermal expansion can also be determined from crystallographic data. Fig. 6 shows the atomic volume plotted vs. temperature for the light actinides. To a good approximation, the volume vs. temperature behavior has a quadratic dependence, and the coefficient of thermal expansion can be obtained by numerical differentiation. A factor of 1/3 has been applied to obtain a coefficient of linear expansion, and the results are plotted in Fig. 7. Schonfeld and Tate (1996) give further information on a range of thermal expansion measurements.

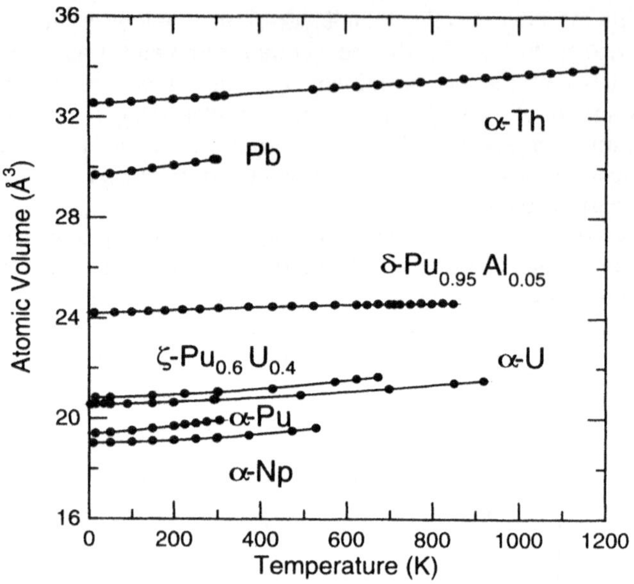


Fig 6. Atomic volume vs. temperature for some of the light actinide phases. Data are from Lawson et al. (1994a,b) and Lawson et al. (1996).

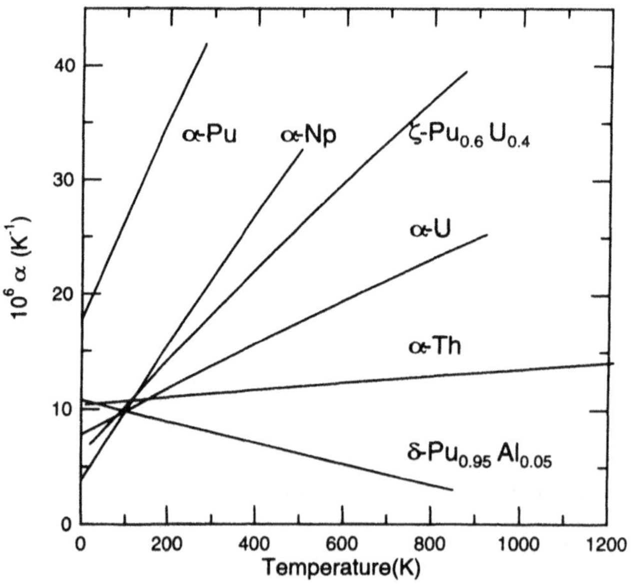


Fig. 7. Coefficient of linear thermal expansion for some of the light actinide phases.

The δ- and δ'-phases have negative thermal expansion. The temperature dependence of the sequence δ→ δ'→ ε is shown in detail in Fig. 8. The sequence is really quite unusual, as it connects two phases of cubic symmetry: the BCC ε-phase and the FCC δ-phase. The intermediate tetragonal δ'-phase can be construed as a distortion either of the ε-phase or of the δ-phase, and its space group, I4/mmm is a subgroup of both Im3m (ε) and Fm3m (δ). In Fig. 8 we have applied appropriate geometric factors to the lattice constants in order to clarify the metric relationships between the phases.

The approximate continuity of the normalized lattice constants of Fig. 8 suggests that the δ→ δ' transition could be second order, while the δ'→ ε transition is first order. But the situation is obscured by impurity effects: in actual experiments, the δ- and δ'-phases are separated by a two-phase region where δ and ε-phases are observed simultaneously. The behavior of this region – indeed whether δ' is observed at all – is extraordinarily sensitive to impurity content: 200 ppm of metallic impurities is known to play a significant role.

Ordinarily, the δ-phase of Pu is stable only over a narrow temperature range 592-724 K. The stability range can be considerably extended by the addition of certain solutes, including Al, Ga, Zn and Ce. The lattice constants of several Pu-Ga alloys plotted vs. temperature are shown in Fig. 9. This figure shows that the addition of only small amounts of solute gradually changes the high temperature thermal expansion from negative to positive. At the same time, the lattice constant is reduced. Similar x-ray data for a wide variety of solutes is available from Lee et al. (1960) and Elliott et al. (1960). Fig. 10 shows the lattice constants measured at 573 K for a variety of Pu alloys, and the effect of all solutes – except Ce – is to reduce the lattice constant. The effect of these solutes on the high-temperature thermal expansion (573-673 K) is to increase the thermal expansion coefficients from negative to positive, as shown in Fig. 11. Again, Ce is anomalous; otherwise the effect is roughly independent of the solute.

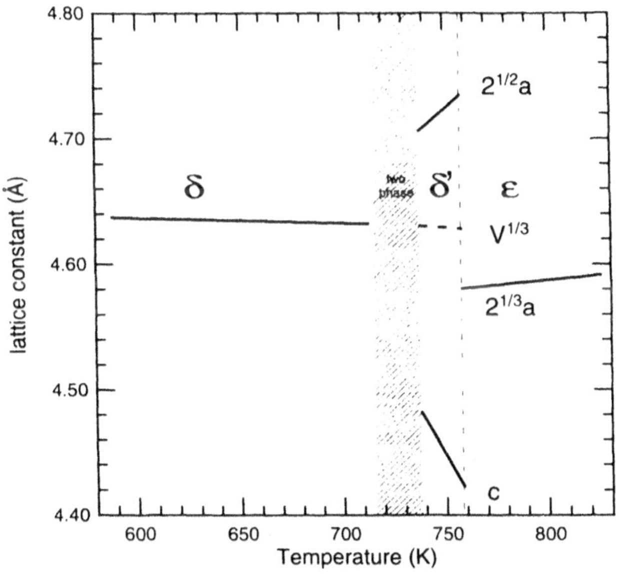


Fig. 8. Normalized lattice constants vs. temperature for δ-, δ'- and ε-phases of Pu metal. After Ellinger (1956).

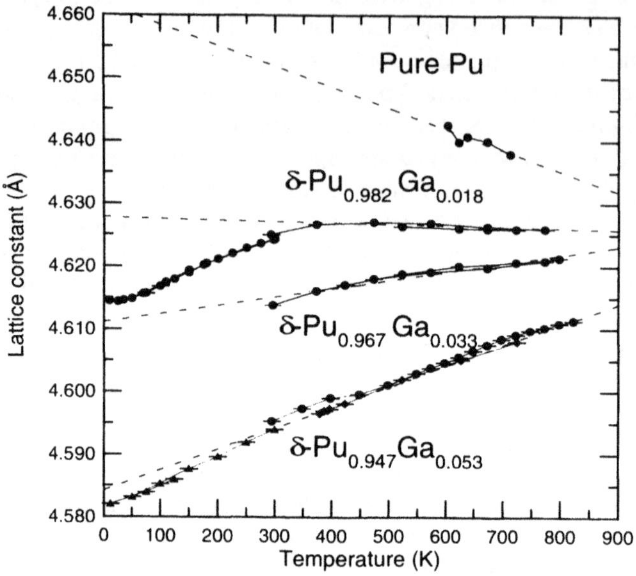


Fig. 9. Lattice constant vs. temperature for some Pu-Ga alloys.

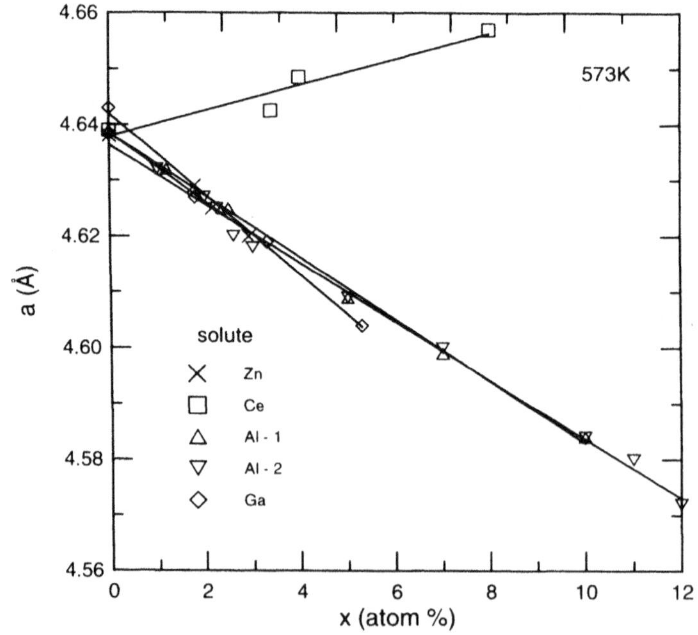


Fig. 10. Lattice constant at 573 K vs. concentration for some Pu-X alloys. "Al-2" data are from Lee et al. (1960), and the other data are from Elliott et al. (1960).

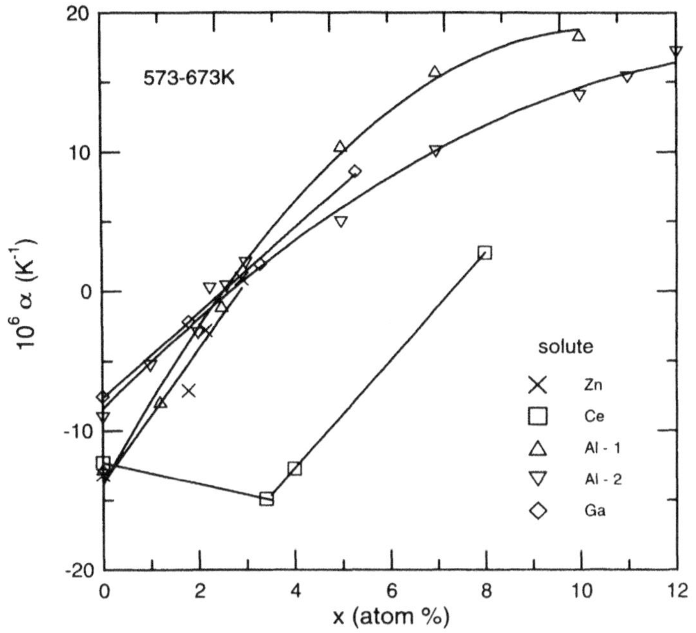


Fig. 11. Coefficient of linear thermal expansion over the temperature range 573-673 K. "Al-2" data are from Lee et al. (1960), and the other data are from Elliott et al. (1960).

## ELASTIC PROPERTIES

Neutron diffraction offers an interesting method for determining the elastic properties of materials via the Debye-Waller factors. The basic idea is that diffraction measures the mean-square vibrational amplitude of atomic vibration, $<u^2>$, and the atomic spring constants can be inferred from these quantities. The effect of increasing vibrational amplitude on the appearance of the diffraction pattern is quite strong: this is shown in Fig. 12. Precise numerical values for $<u^2>$ can be extracted from the diffraction patterns by Rietveld refinement (Young, 1993).

$<u^2>$ is a strong function of temperature. For an ideal Debye solid, $<u^2>$ would be a nearly linear function of temperature at high enough temperatures, and the slope of the $<u^2>$ vs. temperature curve would be the inverse of the atomic spring constant, $\kappa_{atom}$ according to

$$\kappa_{atom} < u^2 >= k_B T \tag{1}$$

where $k_B$ is the Boltzmann constant. $\kappa_{atom}$ is related to the Debye-Waller temperature by the formula

$$\kappa_{atom} = \frac{m k_B{}^2 \Theta_{DW}{}^2}{3\hbar^2} \tag{2}$$

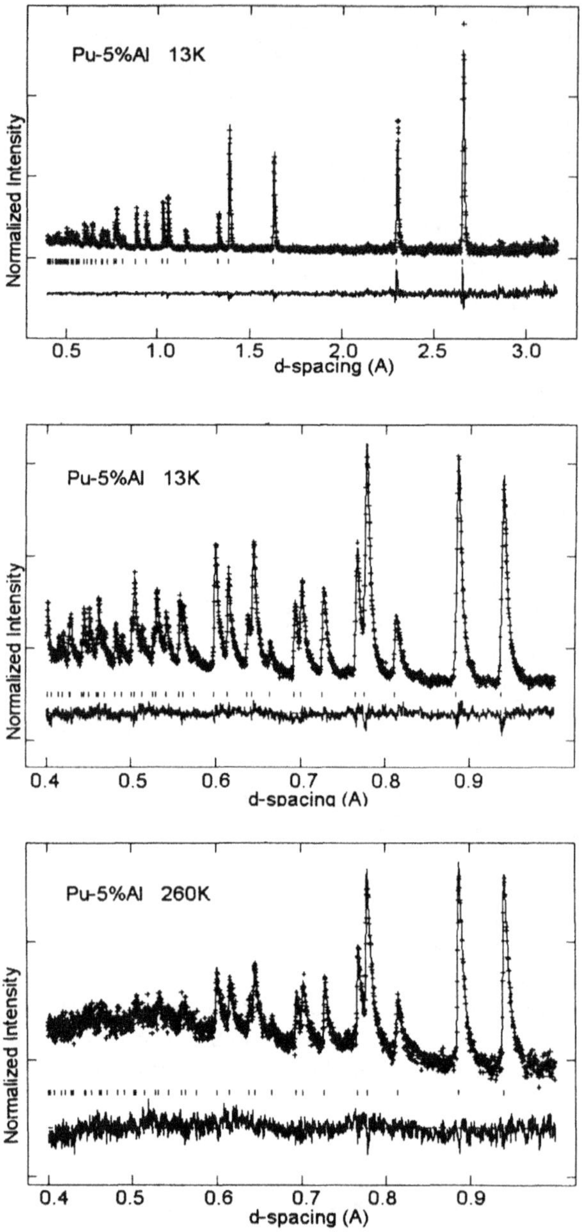


Fig. 12. Neutron diffraction patterns for $Pu_{0.95}Al_{0.05}$ at 13 and 260 K. The data have been fit by Rietveld analysis. +'s are the data, the line is the Rietveld fit, and the lower curve is the residual. The effect of thermal vibration is shown by the dramatic decrease in intensity at small d-spacings as the temperature increases from 13 to 260K in the second and third panels.

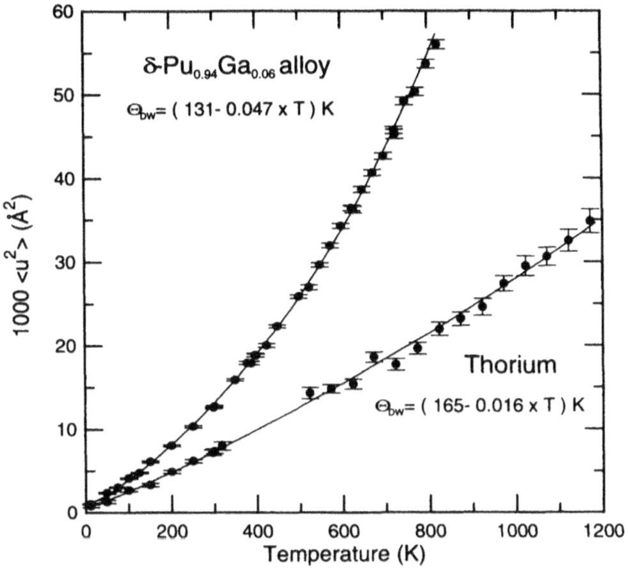


Fig. 13. $\langle u^2 \rangle$ vs. temperature for $Pu_{0.94}Ga_{0.06}$ and for Th. The data have been fit with $\Theta_{DW} = \Theta_{DW,0} + cT$.

This is just a definition of the Debye-Waller temperature in terms of atomic spring constants. For any temperature,

$$\langle u^2 \rangle = \frac{3\hbar^2 T}{mk_B\Theta_{DW}^2}\left\{\frac{\Theta_{DW}}{4T} + \frac{T}{\Theta_{DW}}\int_0^{\frac{\Theta_{DW}}{T}} \frac{udu}{e^u - 1}\right\} \tag{3}$$

Fig. 13 shows the measured values of $\langle u^2 \rangle$ plotted vs. temperature for thorium metal and for δ-stabilized $Pu_{0.94}Ga_{0.06}$, and the behavior of these two isostructural materials is quite different. The difference has two causes: first, Th and δ-phase Pu have different Debye-Waller temperatures, and second, the $\Theta_{DW}$'s have different temperature dependencies. Plotting the data in dimensionless form can separate these effects, and this is done in Fig. 14. In this figure the dimensionless displacement, $mk_B\Theta_{DW}\langle u^2 \rangle/\hbar^2$ is plotted versus the dimensionless temperature, $T/\Theta_{DW}$ for several δ-phase Pu alloys. The ideal Debye behavior (Eq. 2) is also shown in this graph. On this graph it is easier to see that the curvature of the $\langle u^2 \rangle$ vs. T plot is much greater for Pu than for Th, and that the deviation for Pu from the ideal behavior is quite significant. The data have been fitted by assuring the validity of Eq. 3, but with the additional model assumption that $\Theta_{DW}$ is linear in temperature:

$$\Theta_{DW} = \Theta_{DW,0} + cT. \tag{4}$$

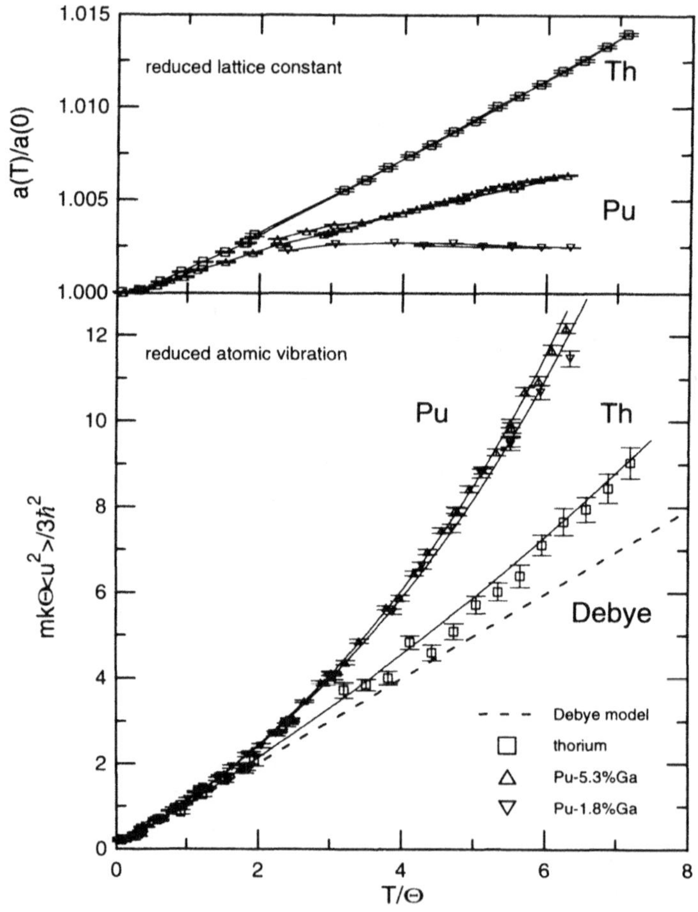


Fig. 14. Dimensionless plot of $<u^2>$ vs. temperature for some Pu-Ga alloys and for Th. The ideal Debye behavior is also shown. The upper curve shows the linear thermal expansion data, $\Delta a/a$ vs $T/\Theta_{DW,0}$.

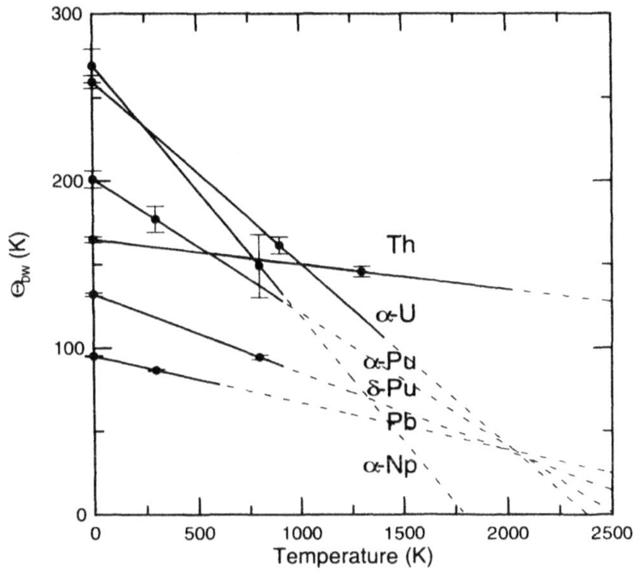


Fig. 15. $\Theta_{DW}$ vs. temperature for some of the light actinides and for Pb. The points with error bars show the limits of our measurements. The solid line changes to dashed at the limit of the solid phase.

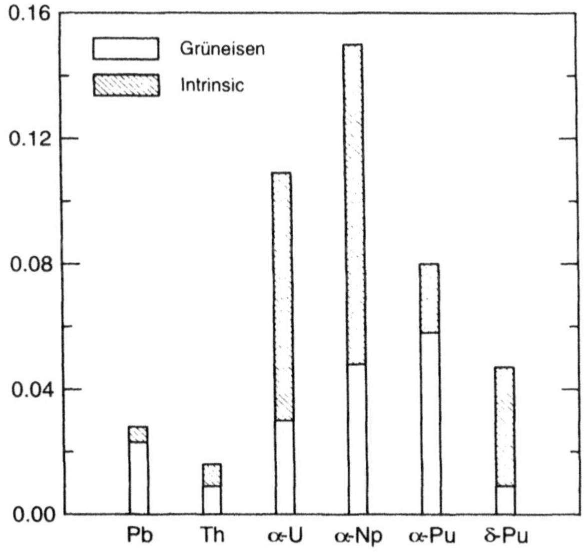


Fig. 16. Anharmonicity parameter, c, for the light actinides and for Pb. The anharmonicity has two parts, Grüneisen and intrinsic, as explained in the text.

In this equation, $\Theta_{DW}$ is the temperature dependent Debye-Waller temperature, $\Theta_{DW,0}$ is the fitted Debye-Waller temperature at T=0, and c is a fitted parameter we may call the anharmonicity parameter. For a given shape of the phonon spectrum, $\Theta_{DW}$ is essentially independent of temperature (Lovesey, 1984). This means that significant values of c imply that the shape of the spectrum is itself temperature dependent.

In the quasi-harmonic model of lattice dynamics, one assumes that $\Theta_{DW}$ changes according to the Grüneisen equation of state

$$\frac{\Theta}{\Theta_0} = \left[\frac{\Omega}{\Omega_0}\right]^{\gamma} . \tag{5}$$

In this equation, $\Omega$ is the atomic volume and $\gamma$ is a materials constant, the Grüneisen parameter. This would imply that the fitted constant c should be proportional to the thermal expansion. The data of Fig 14 are in direct contradiction to this conclusion, as the variation of thermal expansion with Ga concentration is quite large, while c hardly changes at all. In the case of Pu-Ga alloys, we conclude that $\Theta_{DW}$ is *explicitly* (or intrinsically) dependent on temperature (Grimvall, 1986).

We have completed neutron diffraction experiments on many of the light actinide phases. The temperature dependence of the $\Theta_{DW}$'s is shown in Fig. 15. The linear behavior is a consequence of the model assumption of Eq. 4. $\Theta_{DW,0}$ is the intercept and c the slope of the straight line plots of Fig. 15. Error bars of the fits are shown for the high and low temperature limits of the experimental data. The extrapolation of $\Theta_{DW}$ to the melting point is shown with a full line, and beyond the melting point with a dashed line. The size of these error bars depends strongly on the temperature range over which measurements can be performed, as $\Theta_{DW}$ and c are strongly correlated. The temperature range is of course limited by phase stability, and this is a serious limitation for the actinides. For example, we have not so far been able to say very much about the temperature dependence of $\Theta_{DW}$ in $\varepsilon$-phase Pu.

In the case of stabilized $\delta$-phase Pu, we can compare our results to the ultrasonic measurements of Taylor et al. (1967), and we find very good agreement. Their data also support the model assumption of Eq. 4; in particular, it seems that it is $\Theta_{DW}$, rather than $\kappa_{atom}$, that has a linear dependence on temperature. The linear variation of $\Theta_{DW}$ with temperature seems to be well established for $\delta$-phase Pu because of its large stability range; this is less well established experimentally for some of the other phases.

We have plotted the measured anharmonicities, c, for the light actinides in Fig. 16. We have estimated the relative contributions of Grüneisen and intrinsic anharmonicity by calculating the expected change in $\Theta_{DW}$ from thermal expansion data as assuming reasonable values for $\gamma$. If reliable experimental data are not available, we have assumed $\gamma$=2. The intrinsic contributions to the anharmonicity are very large for U and Np and somewhat smaller for Pu. For comparison, data for Pb are also shown on this graph.

## Diffuse Scattering

A curious feature of the neutron diffraction patterns of $\delta$-phase Pu is the diffuse background scattering that is barely perceptible in Fig 12. This scattering becomes quite pronounced at high temperature, as shown in Fig 17, which shows a diffraction pattern for *unalloyed* δ-phase Pu metal at 440°C. The observation of this scattering in unalloyed material, together with the determination that the scattering is proportional to the temperature suggests that the lattice vibrations alone cause the scattering. The inset to this feature shows the calculated thermal diffuse scattering (TDS) using a formula devised by Warren and modified by Borie (Warren, 1969). The scattering can be approximated by Rietveld refinement using a special background function of the form

$$B_{TDS}(k) = \sum_j C_j \frac{d^2}{dR^2}\left(\frac{\sin(kR_j)}{kR_j}\right)e^{-2W_j} \tag{6}$$

In this expression, k is the scattering vector, $R_j$ is an interatomic distance for the FCC structure, $C_j$ is a parameter fitted to the diffraction data, and $\exp(-2W_j)$ is a Debye-Waller factor. The function sin(kR)/(kR) is just the Debye scattering function that describes the diffraction patterns of liquids and amorphous materials. In practice, a successful fit can be made with a small number of interatomic distances (seven for Fig. 17), and the R's and W's are constrained to values obtained from the usual powder refinement. A comparison

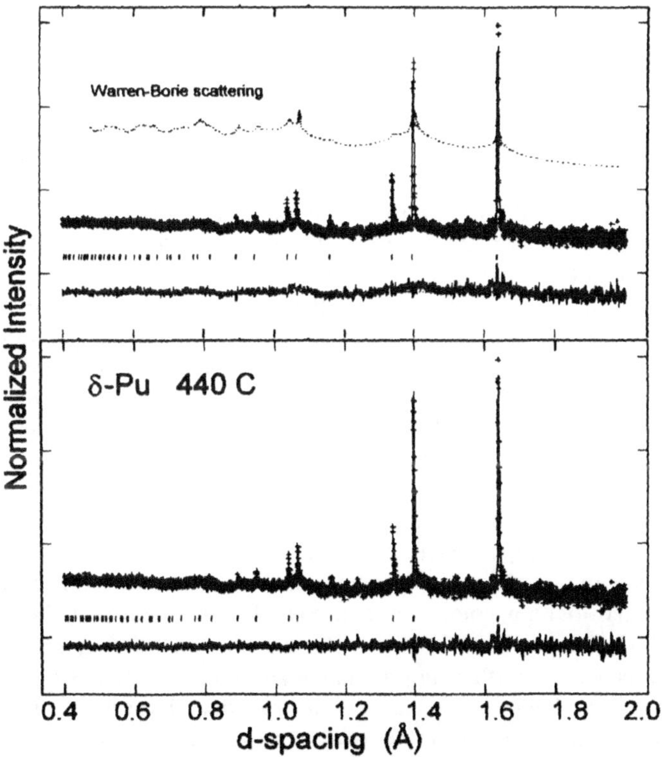


Fig. 17. Diffuse scattering observed in unalloyed δ-phase Pu at 440°C. In the upper curve, the diffuse scattering has not been fit. In the lower curve, the diffuse scattering has been fit as explained in the text. The inset shows the calculated Warren-Borie scattering for appropriate parameters.

of the fitted values of C shows that the observed scattering can be accounted for by the elastic continuum described by Debye's theory of lattice vibrations.

This type of scattering is not often observed with neutron powder diffraction, because the velocities of the scattered neutrons are much less than the sound speeds in most materials. Then the static approximation, on which the Warren-Borie formula depends, is invalid. Pu is an exception, as its low Debye temperature means that the static approximation is valid and the diffuse scattering can be observed and interpreted. The diffuse scattering is unusual but has no special significance.

### Elastic Anisotropy

Ledbetter and Moment (1976) measured single crystal elastic constants for δ-phase Pu. Their results show that δ-phase Pu is the most elastically anisotropic of all the face centered cubic metals. The anisotropy is characterized by the so-called Zener anisotropy defined by

$$Z = \frac{2c_{44}}{c_{11} - c_{12}}. \qquad (7)$$

For Z = 1, the surface representing the Young's modulus is spherical. The available data for the FCC metals are shown in Fig. 18. Among the FCC metals, δ-phase Pu is much more anisotropic than its nearest competitor, Pb.

## MELTING POINTS

As shown in Fig. 2, there is a dramatic depression of the melting point near Pu. This fact has long been regarded as one of the essential features of Pu physics and metallurgy and has received extensive discussion in the literature, for example, by Matthias et al. (1967) and by Kmetko and Hill (1976). It turns out that the melting behavior of the light actinides is explained rather naturally by the Lindemann rule if the temperature dependence of the $\Theta_{DW}$ shown in Fig. 14 is taken into account.

The Lindemann rule (Lindemann, 1910) was originally put forward to provide a means of estimating the Einstein temperature of a solid from its measured melting point. Lindemann's idea was that a solid should melt when the vibrational amplitude of the atoms is large enough that they bump into each other. Gilvarry (1958) reinterpreted the Lindemann rule to say that a solid melts when the vibrational amplitude is a fixed fraction of the interatomic distance, the Lindemann constant of about 10%. The Debye temperature (or the atomic spring constant) determines the vibrational amplitude, so we can use our measured values of $\Theta_{DW}$ to predict the melting point. However, it is necessary to take into account the temperature dependence of $\Theta_{DW}$, or the vibrations will be underestimated.

If the Lindemann rule is applied without correction for the temperature dependence of $\Theta_{DW}$, the estimated melting points are in serious disagreement with experiment, as shown in Fig. 19. If $\Theta_{DW}$ is evaluated at the melting point, then the agreement is much better. This is also shown in Fig 19, using a Lindemann constant of 8.3% for all the elements shown. With a look back at Fig. 2, one can say that depression of the melting points observed in the actinide series is not really anomalous, given the behavior of $\Theta_{DW}$.

The elastic properties and melting points (Emsley, 1998) of the actinides vs. position in the periodic system are shown in Fig. 20. The trend in $\Theta_{DW,0}$ is actually *opposite* to that

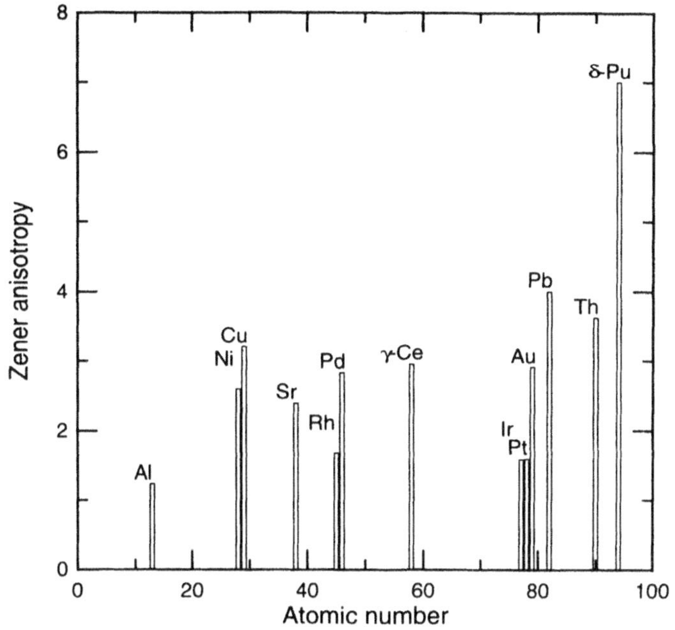


Fig. 18. Zener anisotropy ratio for various FCC metals. Data are from Hearmon (1979, 1984) and from Ledbetter and Moment (1976).

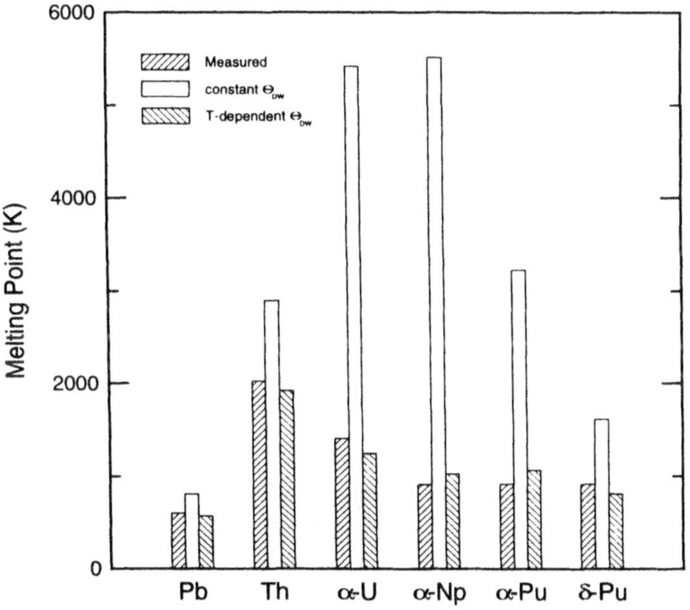


Fig. 19. Measured and calculated melting points for Pb and the light actinides. The calculations are made with the Lindemann rule, both including and neglecting the temperature dependence of $\Theta_{DW}$.

of the melting point. However, it is more than compensated by the trend in the anharmonicity, and this gives the observed depression of the melting points with the minimum at Pu.

## PRESSURE EFFECTS

We can only suggest the range of pressure effects in the light actinides by showing the pressure-volume-temperature diagram determined by Morgan (Fig. 21). Clearly this is a fertile area for exploration, but the experimental difficulties are considerable.

## SUMMARY AND DISCUSSION

The physical properties of the light actinides are dominated by 5f-electron interactions. These interactions are manifest in idiosyncratic structures, a singular dependence of volume on atomic number, and on anharmonic behavior that dominates the thermal expansion and elastic properties. The latter effects are large, and it is necessary to take the effect of temperature into account explicitly since the materials properties are highly temperature dependent.

Theoretical progress has been quite rapid during the past decade. There is now a good explanation of the structural trends, given in a series of papers by Wills and Eriksson (1992), Eriksson et al. (1993), and Söderlind et al. (1995, 1997, 1998). The general idea is

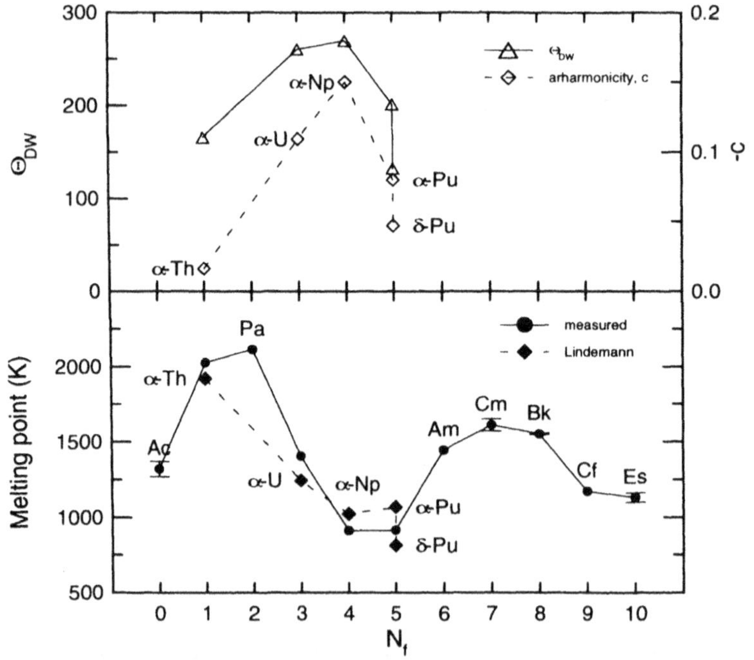


Fig. 20. Melting points for the actinides and rare earths and measured values for $\Theta_{DW}$ and anharmonicity, c.

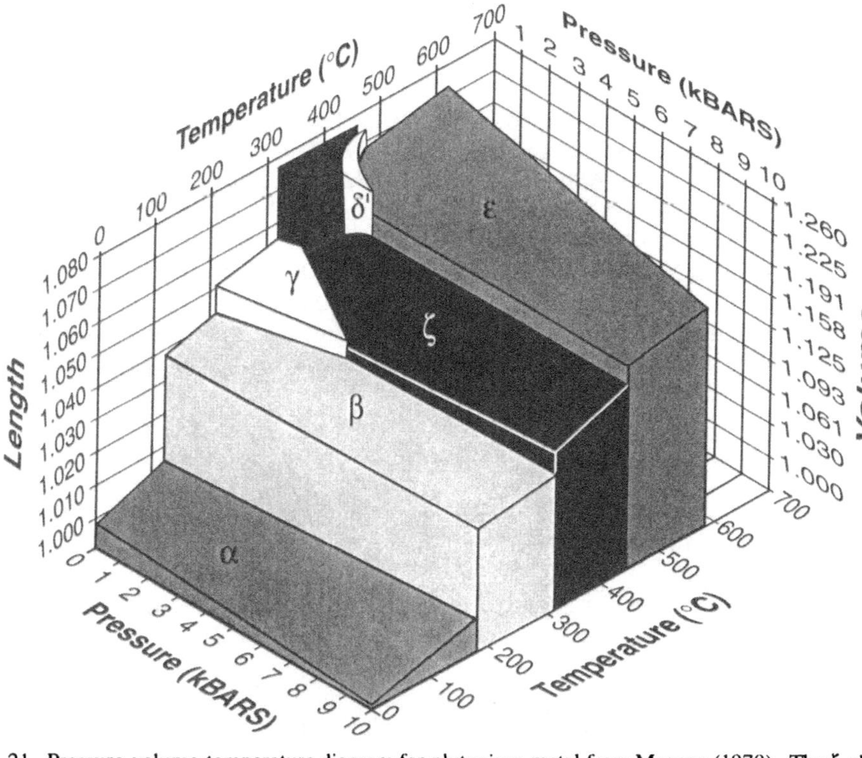


Fig. 21. Pressure-volume-temperature diagram for plutonium metal from Morgan (1970). The ζ-phase appears only at high pressure and its structure is not known. Reprinted with permission of the AIME.

that structural complexity is to be expected when narrow, delocalized f-bands cross the Fermi level as is believed to occur in the light actinides. These theories are not yet predictive of structural details, and they do not yet address the tendency for Pu to form "self-intermetallic compounds" in those cases where there is more than one crystallographically unique Pu atom (Lawson et al, 1996).

There have been some attempts to calculate the thermal expansion behavior of the light actinides by Söderlind et al. (1990). It is difficult to compare with experiment, because the experimental results are dominated by the temperature dependence of the expansion behavior (Fig. 7). In the same way, calculations of elastic constants (Söderlind et al., 1993) will also have to include the temperature dependence before a meaningful comparison with experiment (Fig. 15) can be made.

## ACKNOWLEDGMENTS

This research is sponsored by the United States Department of Energy. The work has benefited from the use of the Intense Pulsed Neutron Source at Argonne National Laboratory and of the Los Alamos Neutron Scattering Center at Los Alamos National Laboratory. IPNS is funded by the U.S. Department of Energy, BES-Materials Science, under Contract W-31-109-ENG-38, and LANSCE is funded by US Department of Energy under Contract W-7405-ENG-36. We are grateful to J. M. Wills for providing us with many reprints, to S. Hecker and J. L. Smith for providing us with the materials for Figs 1 and 2 and for permission to use them, to M. F. Stevens for a careful reading of the manuscript, and to J. L. Smith for some useful information on the behavior of americium.

## REFERENCES

Donohue, J. 1974, *The Structures of the Elements*, New York, John Wiley and Sons, reprint: Malibar FL, Robert E. Krieger Publishing Co.

Ellinger, F. H., 1956, Crystal structure of delta-prime plutonium and the thermal expansion characteristics of delta, delta-prime and epsilon plutonium, Trans. AIME, J. Metals 8:1256.

Emsley, J., 1998, *The Elements*, Oxford, Oxford University Press.

Elliott, R. O, Gschneidner, K. A., Jr., and Kempter, C. P, 1960, Thermal expansion of some delta plutonium solid solution alloys, in: *Plutonium 1960*, E. Grison, W. B. H. Lord and R. D. Fowler, eds., London, Cleaver-Hume Press.

Eriksson, Olle, Söderlind, Per, Wills, J. M., and Boring, A. M., 1993, First principles studies of the crystal structures of the f elecments, Physica B 190:5.

Gilvarry, J. J., 1956, The Lindemann and Grüneisen laws, Phys. Rev. 102:308.

Grimvall, G., 1986, *Thermophysical Properties of Materials*, Amsterdam, North-Holland.

Hearmon, R. F. S., 1979, The elastic constants of crystals and other anisotropic materials, in: *Landolt-Börnstein Numerical Data and Functional Relationships in Science and Technology*. Vol. III/11, K -H. Hellwege and A. M. Hellwege, eds., Berlin, Springer.

Hearmon, R. F. S., 1984, The elastic constants of crystals and other anisotropic materials, in: *Landolt-Börnstein Numerical Data and Functional Relationships in Science and Technology.* Vol. III/18, K -H. Hellwege and A. M. Hellwege, eds., Berlin, Springer.

Jacob, C. W. and Warren, B. E., 1937, J. Am. Chem. Soc. 59:2588.

Kmetko, E. A. and Hill, H. H., 1976, Anomalous melting of f electron metals (with attention to Pu), J. Phys. F. 6:1025.

Johansson, B., Skriver, H., L., and Anderson, O. K., 1981, "Electronic structure of the actinide metals," in: *Physics of Solids Under Pressure*, J. S. Shilling and R. N. Shelton, eds., North Holland, New York.

Lander, G. H. Fisher, E. S. and Bader, S. D., 1994, The solid-state properties of uraniuum, Adv. Physics 43:1.

Lawson, A. C., Olsen, C. E., Richardson, J. W., Jr., Mueller, M. H. and Lander, G. H., 1988, Crystal structure of beta-uranium, Acta Cryst. B 44:89.

Lawson, A. C., Goldstone, J. A., Cort, B., Sheldon, R. I and Foltyn, E., M., 1994a, Atomic thermal vibrations of the light actinide elements, J. Alloys and Compounds 213/214:426.

Lawson, A. C., Goldstone, J. A., Cort, B., Sheldon, R. I and Foltyn, E., M., 1994b, Debye-Waller factors of the light actinide elements, in: *Actinide Processing: Methods and Materials*, B. Mishra and W. A. Averill, eds. Warrendale PA, TMS.

Lawson, A. C., Goldstone, J. A., Cort, B., Martinez, R. J., Vigil, F. A., Zocco, T. G., Richardson, J. W. and Mueller, M. H., 1996, Structure of $\zeta$-phase plutonium-uranium, Acta Cryst B52:32.

Ledbetter, H. M. and Moment, R. L., 1976, Elastic properties of face-centered-cubic plutonium, Acta Met. 24:891.

Lee, J. A., Hall, R. O. A., King, E., and Meaden, G. T., 1960, Some properties of plutonium and plutonium-rich alloys, in: *Plutonium 1960*, E. Grison, W. B. H. Lord and R. D. Fowler, eds., London, Cleaver-Hume Press.

Lindemann, F. A., 1910, Über die Berechnung molecularer Eigenfrequenzen, Phys. Z. 11:609.

Lovesey, Stephan W., 1984, *Thoery of Neutron Scattering From Condensed Matter, Vol. 1*, Oxford, Clarendon Press.

Matthias, B. T., Zachariasen, W. H., Webb, G. W. and Engelhardt, J. J., 1967, Melting point anomalies, Phys. Rev. Lett. 18:781.

Morgan, James R., 1970, New temperature-pressure phase diagram of plutonium, in: *Plutonium 1970 and Other Actinides*, William N. Miner, ed., New York: Metallurgical Society of the American Institute of Mining, Metallurgical and Petroleum Engineers.

Rose, R. L., Kelley, R. E. and Lesuer, D. R., 1979, Dilatometry and differential thermal analysis of americium Metal, J. Nuc. Matl. 79:414.

Schonfeld, Fred W. and Tate, Raymond E., 1996, The thermal expansion behavior of unalloyed plutonium, Los Alamos National Laboratory, LA-13034.

Smith, J. L. and Kmetko, E. A., 1983, Magnetism and bonding: a nearly periodic table of transition elements, J. Less-Common Metals 90:83.

Söderlind, Per, Nordström, Lars, Yongming, Lou and Johanson, Börje, 1990, Relativistic effects on the thermal expansion of the actinide elements, Phys. Rev. B 42:4544.

Söderlind, Per, Eriksson, Olle, Wills, J. M. and Boring, A. M., 1993, Elastic constants of cubic f-electon elements: theory, Phys. Rev B. 48:9306.

Söderlind, Per, Eriksson, Olle, Johannson, Börje, Willis, J. M. and Boring, A.M. 1995, A unified picture of the crystal structures of metals, Nature 374:524.

Söderlind, Per, Wills, J. M., Johansson, B. and Eriksson, O., 1997, Structural properties of plutonium from first-principles theory, Phys. Rev. B. 55:1997.

Söderlind, Per, Wills, J. M. and Eriksson, O., Simple model for complex structures, 1998, Phys. Rev. B 57:1920.

Taylor, J. C., Loasby, R., G., Dean, J. D., and Linford, P. F., Some physical properties of plutonium at low temperatures, in: *Plutonium 1965*, A. E. Kay and M. B. Waldron, eds., London, Chapman and Hall.

Warren, B. E., 1969, *X-Ray Diffraction*, Reading, MA, Addison Wesley, reprint: Mineola, NY, Dover Publications.

Wills, J. M. and Eriksson, Olle, 1992, Crystal structure stabilities and electronic structure for the light actinides Th, Pa and U, Phys. Rev. B 45 13879.

Young, R. A., 1993, Introduction to the Rietveld method, in: The Rietveld Method, R. A. Young, ed., Oxford, Oxford University Press.

Zachariasen, W. H., 1952a, The crystal structure of neptunium metal, Acta Cryst. 5:680.

Zachariasen, W. H., 1952b, Crystal structure studies of neptunium metal at elevated temperatures, Acta Cryst. 5:664.

Zachariasen, W. H. and Ellinger, F. H., 1955, The crystal structure and thermal expansion of $\gamma$-plutonium, Acta Cryst. 8:431.

Zachariasen, W. H. and Ellinger, F. H., 1963a, The crystal structure of beta plutonium metal, Acta Cryst. 16:369.

Zachariasen, W. H. and Ellinger, F. H., 1963b, The crystal structure of alpha plutonium metal, Acta Cryst. 16:777.

# ANOMALOUS MAGNETIC AND RELATED ELECTRONIC PROPERTIES OF URANIUM INTERMETALLIC COMPOUNDS

Vladimír Sechovský,[1] Ladislav Havela,[1]
Karel Prokeš,[1] and Alexander V. Andreev[2]

[1]Department of Metal Physics
Charles University
Ke Karlovu 5, 121 16 Prague, The Czech Republic
[2]Institute of Physics
Academy of Sciences of the Czech Republic
Na Slovance 2, 180 40 Prague, The Czech Republic

## INTRODUCTION

The relatively large spatial extent of *5f*-electron wave-functions of actinide ions yields increased probability of the overlap with *5f* wave functions of actinide neighbors and wave functions of valence electrons of other ligands in alloys and intermetallic compounds. Consequently, the original actinide free-ion character of *5f* electron states is modified to smaller or larger extent depending on the geometry of the actinide ion neighborhood and the electronic states of ligands. The involved *5f*-ligand hybridization plays an underlying role in formation of a *5f* magnetic moments and correlations between them (Cooper et al., 1985; Hu and Cooper, 1993; Koelling et al., 1985). These attributes essentially determine the difference between magnetism and related electronic properties of light actinide (U÷Pu) based intermetallic materials and their lanthanide counterparts.

The *4f* electrons deeply embedded in the core of lanthanide ions yield by rule stable magnetic moments hardly affected by any direct interaction with ligand valence electrons and the only important modification of the free-ion *4f*-electron magnetic moment is observed due to crystal field lifted degeneracy of *4f*-electron energy levels. The only exceptions are usually observed in some compounds with Ce and some other anomalous lanthanides characterized as Kondo and valence-fluctuation systems. The really delocalized *4f* states appear only in few materials like $CeFe_2$ (Eriksson et al., 1988) due to strong *4f-3d* hybridization.

In contrast, hardly any uranium intermetallic with localized *5f* electron states can be found leaving $UPd_3$ as a prominent exceptional case (Baer et al., 1980; Sechovsky and Havela, 1988). Electronic structure of majority of uranium intermetallics is characterized by a *5f* band pinned at Fermi level with part of the *5f* states hybridized with the *s*, *p* and *d* states of ligands yielding substantial delocalization of *5f* electrons. Owing to very the strong spin-orbit interaction of energy comparable to the *5f* bandwidth, the uranium magnetic moment emerging in the itinerant *5f* electron systems bears, however, a substantial orbital component. The orbital magnetism together with participation of *5f* electrons in anisotropic covalent bonding are the principal sources of the huge magnetic anisotropy observed in uranium compounds even in as itinerant cases as $UNi_2$ having the uranium moment of $\mu_U = 0.08\ \mu_B$ only (Frings et al., 1986; Sechovsky and Havela, 1988). Such small moments are partly due to the strong delocalization of *5f* electrons and partly due to the mutual cancellation of spin and orbital component ($\mu_U = \mu_L - \mu_S$). In the limiting case of $\mu_L \approx \mu_S$, this may lead to a special "nonmagnetic case" (Brooks and Kelly, 1983; Lebech et al., 1989) analogous to certain extent to Eu case with J = 0. When the $\mu_L$ and $\mu_S$ become comparable anomalous magnetic form factors showing a maximum for $Q > 0$ are observed in the polarized neutron diffraction experiment. Such situation has been identified in experimental form factor in $UFe_2$ (Lebech et al., 1989) which exhibits almost zero uranium moment although the orbital $\mu_L = 0.23\ \mu_B$ and spin $\mu_S = 0.22\ \mu_B$ moments have considerable values (see Fig. 1).

A different sensitivity of spin or orbital moments to external variables like pressure, deduced in the original work already (Brooks and Kelly, 1983), led to the assumption that the ratio of orbital and spin moment should reflect the degree of the *5f*-delocalization. Indeed, the non-metallic compounds showing the least delocalised nature of the *5f* states like $UO_2$, USb, $NpAs_2$, PuSb have values of $\mu_L/\mu_S$ of corresponding to a free $5f^n$ ion. On the other hand, magnetically ordered materials with presumably the most itinerant *5f* states like $UFe_2$, $UNi_2$, $NpCo_2$, $PuFe_2$ have this ratio close to 1 (Lebech et al., 1991). The general validity of this scheme remains, however, questionable, as indicated by e.g. an almost free-ion like value of $\mu_L/\mu_S$ determined for a band metamagnet UCoAl (Wulff et al., 1990).

These characteristic features of *5f* electrons have important consequences for various electronic properties of U intermetallic compounds, particular in magnetism. In this paper we point out some typical phenomena, which, we believe, deserve attention of theorists.

## EVOLUTION OF MAGNETISM IN URANIUM INTERMETALLICS

It is widely accepted that the degree of *5f* electron delocalization responsible for magnetic moment washout is controlled mainly by the *5f*-ligand hybridization. The *5f-d*

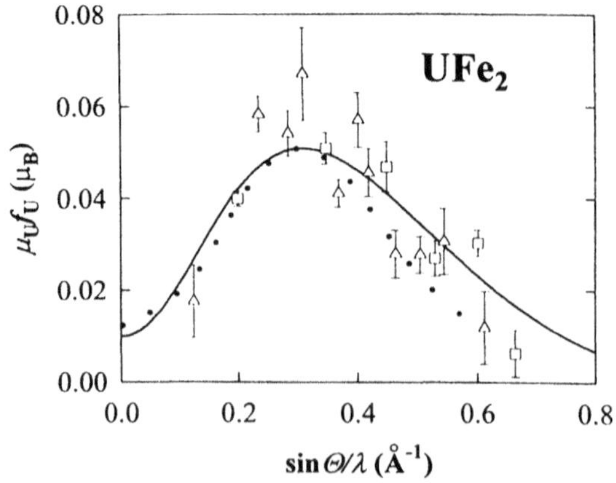


**Figure 1.** The U-magnetic form factor, $\mu_U f_U$, as function of $\sin\Theta/\lambda$ in $UFe_2$ - data from (Lebech et al., 1989). The squares are deduced from Bragg reflections from the U-sublattice only, the triangles represents rest of magnetic reflections. The solid curve represents the magnetic amplitude given by $\mu_L = 0.23\ \mu_B$ and $\mu_S = -0.22\ \mu_B$, $C_2 = 23$. The dotted curve is the magnetic amplitude calculated using the results of spin-polarised energy band calculations (Eriksson et al., 1990) re-scaled to match the experimental data.

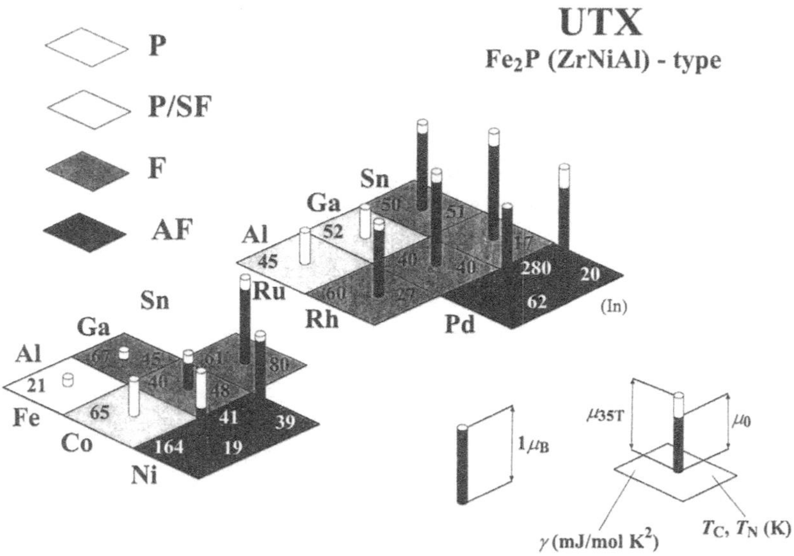


**Figure 2.** Schematic plot illustrating evolution of some electronic properties (type of magnetic ground state, magnetic ordering temperature, $\gamma$-value, spontaneous magnetic moment (black part of the columns - extrapolated from high magnetic field data to zero field), magnetic moment induced in field of 35 T (white top parts of the columns) of *UTX* compounds with the hexagonal ZrNiAl-type structure. Data from (Sechovsky and Havela, 1998) and references therein.

hybridization between uranium and transition metal neighbors is particularly effective in this process. Consequently, *5f* moment formation is strongly influenced by *d*-electron states of ligands, although essentially nonmagnetic. Besides delocalization of *5f* electrons, the hybridization plays also a substantial role in magnetic ordering by mediating the exchange interactions between uranium magnetic moments. The two competing effects of the hybridization then yield a large variety of magnetic ground states observed in uranium intermetallics. The evolution of magnetism with the varying *5f*-ligand hybridization may be best studied within families of isostructural compounds in which the variation of the geometry of the uranium atom neighborhood is not involved. Fig. 2 shows how magnetism develops across the series of U*TX* compounds (*T* represents a late transition metal from the 3d and 4d series, *X* is a p-electron metal Al, Ga, In and Sn) with the hexagonal crystalline structure of the ZrNiAl type. We can see that paramagnetic ground state is typically observed in compounds with Fe (UFeAl), Ru (URuAl, URuGa) and Co (UCoAl). Whereas the first compound is a Pauli paramagnet with very low susceptibility the latter three compounds show local moment behavior in high temperatures expressed by the Curie-Weiss $\chi$ vs. *T* dependence followed by spin-fluctuation anomalies in the low-temperature regime. The URh*X*, the rest of UCo*X* compounds and also URuSn exhibit ferromagnetism, whereas the materails with latest transition metals Ni and Pd order antiferromagnetically. The stability ranges of ferromagnetism and antiferromagnetism within this group of materials seems to depend on the occupation of the *3d* or *4d* band of the transition metal. This aspect will be a subject of further discussion. As for the magnetic moments which are predominantly residing on uranium sites (only weak induced magnetization can be associated with transition metal atoms as follows from neutron diffraction data of e.g. (Wulff et al., 1990)) a gradual increase of the moment is clearly seen when moving in the diagram towards compounds with latest transition metals and similar development is seen also in the series U*T*Al, U*T*Ga, U*T*Sn. Analogous pictures can be observed within various other groups of ternary isostructural compounds of various stoichiometry (e.g. U*TX*, U$T_2X_2$, $U_2T_2X$) involving transition metals (Sechovsky and Havela, 1998).

## MAGNETOCRYSTALLINE ANISOTROPY

The mechanism usually responsible for magnetocrystalline anisotropy in lanthanide intermetallics with well localized *4f* states is based on the single-ion crystalline electric field (CEF) interaction. The involvement of delocalised *5f* states in light actinide intermetallics in anisotropic covalent bonding implies an essentially different, two-ion (*5f-5f*) interaction, which plays a major role in the anisotropy. The strong spin-orbit interaction is then a necessary pre-requisite in both cases.

A relatively simple model, which leads to qualitatively realistic results, has been worked out by (Cooper et al., 1985), on the basis of Coqblin-Schrieffer approach to the mixing of ionic *f*-states and conduction-electron states (Coqblin and Schrieffer, 1969), in which the mixing term of the hamiltonian of Anderson type is treated as a perturbation, and the hybridization interaction is replaced by an effective *f*-electron-band-electron resonant exchange scattering. Considering an ion-ion interaction to be mediated by different covalent-bonding channels, each for a particular magnetic quantum number $m_l$, the strongest interaction is realized for the orbitals pointing along the ion-ion bonding axis, which represents the quantization axis of the system. The two *5f* ions maximize their interaction by compression of the *5f* charge towards the direction to the nearest *5f* ion. This has serious impacts on magnetic anisotropy, because it means a population of the *5f* states with orbital moments perpendicular to the bonding axis. This interaction prefers a strong ferromagnetic coupling of actinide atoms along the strong bonding direction, whereas there is no special general tendency to ferromagnetism or antiferromagnetism perpendicular to it, where the interaction is much weaker, and can be comparable to the RKKY-type isotropic exchange interaction ubiquitous in metallic systems. The anisotropy of the two-ion interaction in uranium intermetallics is also responsible for specific magnetic structures in antiferromagnets which will be further discussed.

A clear evidence of anisotropic hybridization has been provided by (Paixao et al., 1992) who performed experiments with polarized neutrons in order to measure the

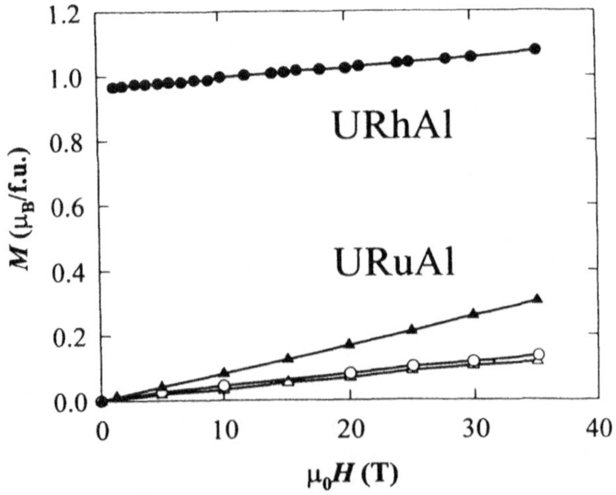


**Figure 3.** Magnetization $M$ vs. field $H$ measured at $T$ = 4.2 K on single crystals of URhAl (circles) and URuAl (triangles) in magnetic fields along (full symbols) and perpendicular (open symbols) to the $c$-axis. Data from (Veenhuizen et al., 1988).

magnetization density in URhAl. The magnetization density analysis shows that the total moment, as measured by magnetization, has four contributions, $\mu_U = 0.94\ \mu_B$, $\mu_{Rh(I)} = 0.28\ \mu_B$, $\mu_{Rh(II)} = 0.03$ and -0.11 $\mu_B$ as a conduction electron contribution. The Rh(I) and Rh(II) sites are located within the U-Rh and Rh-Al planes, respectively. The observation of a large moment at the Rh(I) site points to a much stronger *5f*(U)-*4d*(Rh) hybridization within the U-Rh basal plane than the *5f*(U)-*4d*(Rh) along the *c*-axis.

The fact, that the *5f* intermetallic compounds display much stronger magnetic anisotropy than their lanthanide counterparts has been known for a long time. The anisotropy field, estimated as the intercept field of the easy and hard magnetization curves extrapolated to high magnetic fields, frequently reaches values of the order of $10^2$ - $10^3$ T (Sechovsky and Havela, 1992; Sechovsky and Havela, 1988; Sechovsky et al., 1994). These values represent anisotropy energies of the same order (0.2 eV) as obtained for US by (Brooks et al., 1986) in the calculations done within LDA.

As discussed in the first section, the crystal environment of actinide atoms in different intermetallics is responsible for variable strengths of the *5f*-bonding and the *5f*-ligand hybridization. Resulting variations of the *5f*-electron delocalization and consequent washout of the *5f* magnetic moment lead to a wide spectrum of magnetic properties, ranging from Pauli paramagnetism to ferromagnetic or antiferromagnetic ordering of local *5f* moments. The strong magnetocrystalline anisotropy is, however, observed in all uranium compounds exhibiting any *5f* contribution to the magnetic moments, and is not thus restricted to magnetically ordered materials. Moreover, the strong anisotropy is clearly observed both in magnetically ordered and paramagnetic (except for Pauli paramagnetic) regime. This is well documented in Fig. 3 showing the magnetization curves measured on single crystals of URuAl and URhAl at 4.2 K in magnetic fields applied along and perpendicular to the *c*-axis. Whereas URuAl is paramagnetic down to lowest temperatures, URhAl orders ferromagnetically below $T_C = 27$ K. Nevertheless, both materials exhibit the uniaxial anisotropy.

Owing to different ground states, the magnetization curves along the *c*-axis (easy magnetization direction) differ drastically. The magnetization in a perpendicular direction, however, is nearly identical for both compounds irrespective to the ground state. Also at high temperatures, above $T_C$ of URhAl, the susceptibility measured in fields applied in the basal plane is nearly identical for both compounds whereas the *c*-axis signals are dramatically different (much stronger in URhAl).

The systematic occurrence of particular types of anisotropy over diverse groups of uranium intermetallics offers a more general conclusion on the vast majority of materials, in which the U-U coordination is clearly defined in the crystal structure, the easy-magnetization direction is perpendicular to the nearest U-U links. As follows from simple geometrical considerations, if the links are found in different directions in a plane, moments have to orient themselves perpendicular to the plane yielding the easy-axis anisotropy type. In an opposite case, if nearest uranium neighbor atoms form only linear chains (the intra-chain U-U separation is significantly smaller that the inter-chain one), an easy-plane anisotropy appears. Good number of well documented examples which manifest the relation between the layout of the uranium sublattice and the type of anisotropy can be found in the work of (Sechovsky and Havela, 1998).

## EXCHANGE INTERACTIONS MEDIATED BY ANISOTROPIC HYBRIDIZATION - EVOLUTION OF MAGNETIC ORDERING

An interesting excursion into physics of U*TX* compounds has been achieved by studies of the evolution of magnetism in some solid solutions of parent compounds included in Fig. 2. The relatively simple cases are represented in Fig. 4 by the $URu_{1-x}Rh_xAl$ and the $URu_{1-x}Rh_xGa$ pseudoternary compounds (Sechovsky et al., 1988; Sechovsky et al.,

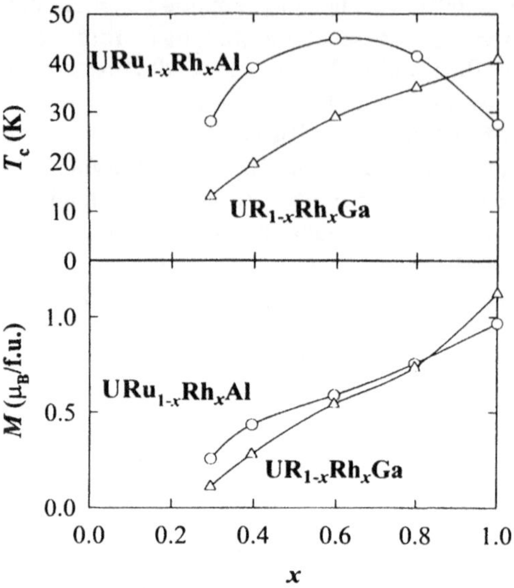


**Figure 4.** Concentration dependence of the ordering temperature $T_C$ and paramagnetic Curie temperature $\theta_p$ (a), and the spontaneous magnetic moment in $URu_{1-x}Rh_xAl$ and $URu_{1-x}Rh_xGa$ systems (b). Data from (Sechovsky et al., 1992).

1992; Veenhuizen et al., 1988) in which the development of the high-temperature susceptibility behavior characterized by a gradual reduction of negative values of the paramagnetic Curie temperature when substituting Rh for Ru. This probably reflects the suppression of spin-fluctuation effects. For $x \approx 0.3$, ferromagnetism is stabilized in both systems and the spontaneous moment increases monotonously with increasing Rh contribution. The same holds for $T_C$ in the latter system, whereas the ordering temperature of $URu_{1-x}Rh_xAl$ reaches a maximum for $x \approx 0.6$ and then decreases considerably with further increasing Rh concentration. The non-monotonous concentration dependence of $T_C$ has probably a close connection with the evolution of magnetism observed in solid solutions $URu_{1-x}Ni_xAl$ shown in Fig. 5 (Andreev et al., 1997) and $URu_{1-x}Pd_xGa$ (Tran et al., 1998), i.e. solid solutions of the paramagnet URuAl and the antiferromagnets UNiAl and

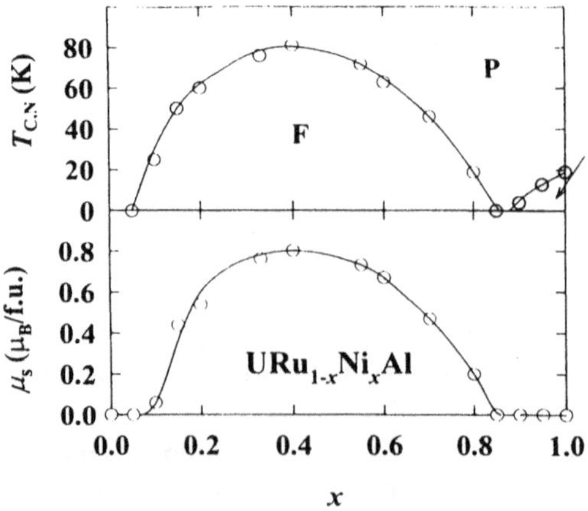


**Figure 5.** Concentration dependence of the ordering temperature $T_c$ and spontaneous magnetic moment $\mu_s$ in $URu_{1-x}Ni_xAl$ solid solutions. Data from (Andreev et al., 1997).

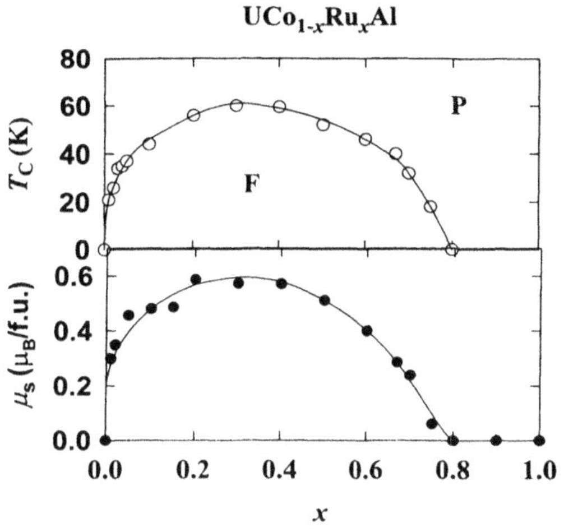


**Figure 6.** Concentration dependence of the ordering temperature $T_c$ and spontaneous magnetic moment $\mu_s$ in $UCo_{1-x}Ru_xAl$ solid solutions. Data from (Andreev et al., 1997).

UPdGa, respectively. In both these cases the substitution of Ru by Ni or Pd yields ferromagnetism for intermediate concentrations and antiferromagnetism on the Ru poor side of series.

A very surprising result has been obtained in (Andreev et al., 1997) by observing ferromagnetism in a wide concentration range of solid solutions of the type $UCo_{1-x}Ru_xAl$ although both the parent compounds, UCoAl and URuAl, have a non-magnetic ground state. As can be seen in Fig. 6, ferromagnetism is observed already for very small substitutions of Ru for Co. Both the spontaneous moment $\mu_s$ and the ordering temperature $T_C$ grow rapidly with increasing $x$, reach their maximum values for $x = 0.3$ and then decrease with increasing Ru content up to $x = 0.8$, where ferromagnetism in this systems disappears.

To account for these "striking" results, which seem to be in conflict with the standard picture of the general development of magnetism in *UTX* ternaries of the ZrNiAl-structure, we may consider the following simple model. As the prominent mechanisms affecting the size of U magnetic moments and their ordering, one should consider the overlap of neighboring U atom *5f* wave functions and the hybridization of the U *5f* states with the *s,p,d* valence states of ligands (*T* and *X* atoms). Both mechanisms induce two important strongly anisotropic effects:

- delocalization of the U *5f* states leading ultimately to the washout of magnetic moments
- exchange interactions correlating the U *5f* magnetic moments via direct *5f-5f* and indirect exchange involving non-*f* states.

The close packing of U and *T* atoms in the basal plane probably results both in a non-negligible *5f-5f* overlap and a strong *5f-d* hybridization involving the transition metal *d*-states, which compresses the *5f* charge density towards the basal plane. Since the *5f* states with strong spin-orbit interaction promote orbital magnetic moments, which are indeed observed experimentally (Wulff et al., 1990), a strong uniaxial anisotropy with magnetic moments locked along the *c*-axis is an inevitable element of magnetism leading to Ising like systems. Another consequence of the *5f-5f* overlap and the *5f-d* hybridization within the basal plane is a strong ferromagnetic coupling of the involved U magnetic moments. This situation leads to magnetic structures build up of ferromagnetic basal-plane layers.

Along the *c*-axis both, the *5f-d* (U-*T*) and *5f-3p* (U-Al) hybridization should be considered. The first type should dominate for earlier *d*-elements (Ru), but it ceases with additional filling of the *d*-shell leaving the *5f-3p* hybridization channel gradually more important. We may speculate that the former type leads preferably to a ferromagnetic type of U-U coupling along *c*, whereas the latter tends to an antiferromagnetic type of coupling.

Starting from URuAl, the *5f-d* hybridization becomes reduced when partly substituting Rh, Co, or Ni for Ru, which allows formation of the U *5f* magnetic moments (coupled ferromagnetically within the basal plane) for certain concentration of the substituting element, but the moments are relatively small and the interaction along *c* is ferromagnetic. When proceeding further with the substitution, the U-moments increase in magnitude, but the coupling along *c* weakens. Finally, after the decay of the *5f-d* exchange polarization the *5f-3p* channel overtakes the dominance leading to the relatively weak antiferromagnetic coupling along *c*. Thus the reduction of the exchange coupling along c leads to the final decay of the ordering temperature, despite of the fact that a strong coupling along the other two dimensions persists.

Not only the strength of the hybridization, but also the U-U exchange interaction as the second order effect, can influence the magnitude of U-moments. This can be deduced from the non-magnetic ground state of UCoAl. In the framework of our simple model this compound represents a delicate balance of ferromagnetic and antiferromagnetic components of the exchange interaction along *c*.

A proper theoretical background for the hybridization induced two-ion exchange interaction is being constantly developed (Hu and Cooper, 1993). *Ab initio* calculations using this approach are strongly desirable to explain the intriguing evolution of magnetic properties of U*TX* compounds.

## ELECTRICAL RESISTIVITY AND MAGNETIC STATE

Comparison of resistivity of uranium intermetallics with analogous rare-earth compounds, or with transition metals, shows that in actinide materials we often encounter values an order of magnitude higher at room temperature (resistivity values exceeding 200 μΩcm are not exceptional). Because similar high values are found in spin-fluctuators and magnetically ordered materials in the paramagnetic state, one may find its origin e.g. in the strong coupling of electrons to spin fluctuations as a consequence of the strong *5f*-ligand hybridization. Unlike standard metallic systems, in which the dominating resistance mechanism is the electron-phonon scattering, which leads to monotonously increasing $\rho(T)$, U-systems often show a clear fingerprint of a strong scattering on magnetic fluctuations, which are incoherent at high temperatures. In periodic systems the condensation from the high-temperature local-moment state to a Fermi liquid state manifests itself in a pronounced decrease of the resistivity below a characteristic temperature marking the onset of such coherent regime, $T_{coh}$.

One of the specific features of most uranium compounds is the strong non-additivity of partial contributions to the resistivity (the Matthiessen's rule does not hold) showing a fast suppression of the low-temperature decrease with practically any substitution destroying the coherence. At high temperatures, a more or less pronounced tendency to saturation of $\rho(T)$ appears due to a suppressed contribution of electron-phonon scattering, which, when unchanged, yields an increasing slope of $\rho(T)$ in conventional materials. In uranium intermetallics this behavior is different due to the strong electron spin-fluctuation scattering. The mentioned non-additivity means that the standard procedure of subtraction of the phonon contribution in the form of $\rho(T)$ of a non-magnetic analog can yield misleading results.

In the spirit of the analogy between electron gas and Fermi liquid systems, a simple $\rho \propto a\,T^2$ behavior due to electron-electron scattering is observed in the low-temperature limit in most of U-systems, too. The values of $a$ are many orders of magnitude higher than in simple metals, and an approximate scaling with $\gamma^2$ was observed experimentally (yielding the universal ratio $a/\gamma^2 \approx 10^{-5}$ μΩcm $mJ^{-2}K^2mol^2$ (Kadowaki, 1993). The validity of this universal behavior, which supposes that $a$ is determined by Fermionic excitations, can break down in magnetically ordered materials with different types of excitations, or generally in materials where the anisotropy of moment ordering and magnetic excitations leads to anisotropic $a$-parameters. The value of $a$ involves more complex cases not only the Fermi liquid interaction parameters, but also an electron scattering in magnetic excitations (e.g. spin waves). Finally, such a magnetic scattering can be affected by the existence of gaps in the electron energy or magnetic excitation spectra. The former one affects qualitatively the $\rho(T)$ dependencies mainly in the vicinity of $T_N$. The latter leads generally to exponential terms affecting $\rho(T)$ already from the low temperature limit. The contribution of scattering of electrons on magnon-like excitations can be derived in a simple model mentioned in the context of the contribution to specific heat (Andersen and Smith, 1979). It leads to the expression

$$\rho(T) = \rho_0 + aT^2 + bT\,(1 + 2\,T/\Delta)\,\exp(-\Delta/T), \qquad (1.1.)$$

the last term accounting for electron scattering on magnetic excitations with a gap $\Delta$. Such term was identified e.g. in UNiGe, where it yields $\Delta \approx 40$ K (Prokes et al., 1994) similar to the value obtained from the specific heat data.

In several cases a power law with an exponent lower than 2 has been observed. Exponents of 3/2 or 5/3 are expected in the spin fluctuation theories to replace the value of 2 for the situation very close to the onset of magnetic order (Mathon, 1968; Moriya, 1991; Ueda, 1977). Such behavior was found for UNiAl, in which inter-site antiferromagnetic spin fluctuations along the $c$-axis apparently survive into the low temperature limit (Brück et al., 1994). The lower $\rho(T)$ curve for current along the $c$-axis is rapidly increasing with temperature, and can be described up to $T = 3.5$ K by a power law $\rho \sim T^{5/3}$. This behavior is attributed to spin fluctuations close to the onset of magnetic ordering and is observed in weak itinerant magnets, and contrasts with the simple $aT^2$ type of behavior for current along the basal plane.

A strong anisotropy of the electrical resistivity in actinide intermetallics is not at all exceptional, and is mostly related to the strong anisotropy of the exchange coupling of $5f$-moments. It is easily observed in single crystal samples of tetragonal, hexagonal or orthorhombic materials. There can be, in particular, a very striking difference in type of the temperature dependence of the resistivity for different directions of the electrical current. In certain materials, antiferromagnetic correlations can be suppressed by an external magnetic field, and we observe a metamagnetic transition at a critical field which is generally accompanied by a giant magnetoresistance effect. Thus measurements of the $\rho(T)$ dependencies with current along various crystal directions and in different magnetic fields are usually very helpful for understanding the magnetic contribution to the resistivity. What is much less clear, is how the anisotropy in the underlying electronic structure (mainly the anisotropy of hybridization, which must strongly affect conduction-electron states) is reflected in various types of $\rho(T)$ behavior.

The residual resistivity $\rho_0$ reaches typically smaller values in ferromagnets. The much larger $\rho_0$ values observed in antiferromagnets can be explained by several models, which are based either on the notion of gapping due to additional periodicity in the antiferromagnetic state, or, more generally, on spin dependent scattering and $k$-space restrictions (Plumer, 1994).

A strong anisotropy of the resistivity has been observed for numerous single crystals of non-cubic structures (for example in $UT_2X_2$ and $UTX$ compounds). The essential features can be illustrated on the case of UNiGa. As seen in Fig. 7, for the electrical current in the basal plane ($i \perp c$) containing ferromagnetically coupled U magnetic moments, the

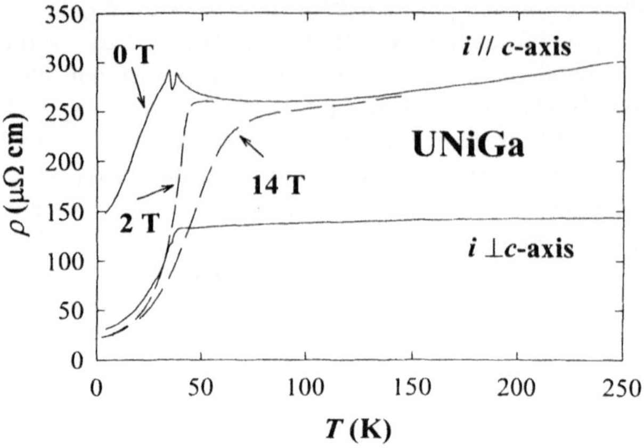


**Figure. 7** Electrical resistivity $\rho$ vs. temperature $T$ for single-crystalline UNiGa with current $i$ along the $c$-axis in various magnetic fields applied in the same direction. The zero field state is antiferromagnetic with several magnetic phase transitions between T = 35 and 40 K. In fields higher than 1 T, a ferromagnetic state is induced. The comparison with the resistivity behavior for current within the basal plane ($i \perp c$, in zero field) illustrates the effect of ferromagnetic coupling of the uranium moments within the basal plane. Data from (Havela et al., 1996).

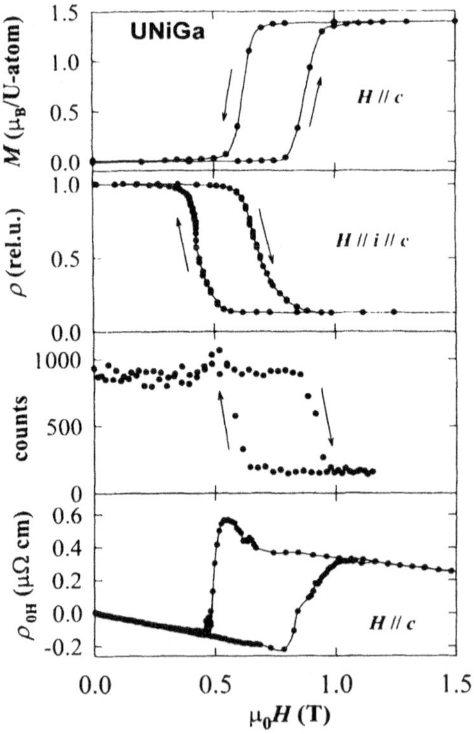


**Figure. 8** Field dependence of the magnetization $M$ (a), reduced electrical resistivity $R$ for current $i$ along and perpendicular to the $c$-axis (b) - data from (Jirman et al., 1992), the peak intensity of the neutron diffraction reflection 1,1,4/3 (c) - data from (Prokes et al., 1996), and Hall resistivity $R_{0H}$(d) measured on UNiGa single crystal at $T$ = 4.2 K in magnetic field $H$ applied along the $c$-axis - data from (Kobayashi et al., 1996).

temperature dependence of the resistivity mimics a ferromagnet with a dramatic drop below the ordering temperature and a gradual saturation in the low temperature limit.

The flat part of the $\rho(T)$ curve above $T_N$ demonstrates that the contribution of the electron-phonon scattering, which is normally linearly increasing in this temperature range, plays only a minor role compared to the spin-disorder scattering (which is of the order of 100 μΩcm). Qualitatively a different resistance behavior is observed for the current along the $c$-axis ($i \parallel c$), which senses the antiferromagnetic coupling of uranium moments. Below 80 K, the resistivity increases with decreasing temperature, becoming more pronounced approaching $T_N$. In the range 35÷40 K, several anomalies emerge due to magnetic phase transitions, and finally the resistivity decreases at low temperatures. However, the residual resistivity $\rho_0$ is much higher (140 μΩcm) for $i \parallel c$ than for $i \perp c$. The high resistivity is removed in the ferromagnetic state. The negative derivative $d\rho/dT$, which develops gradually below 80 K for $i \parallel c$, is unambiguously due to incipient antiferromagnetic correlations (Havela et al., 1994) and can be suppressed in 14 T.

In most of the materials mentioned, a simple $aT^2$ law characterizes the low temperature behavior both in the antiferromagnetic and metamagnetic states. UNiGa can be taken as an example, in which the coefficient $a$ decreases by a factor of 5 (from 0.24 μΩcm $K^{-2}$ to 0.051 μΩcm $K^{-2}$) at the metamagnetic transition, but the basic form of $\rho(T)$ does not change with temperature up to 20 K. Since the field dependence of the specific-heat $\gamma$-value is weak, the coefficient $a$ then does not scale with $\gamma^2$.

## GIANT MAGNETORESISTANCE AT METAMAGNETIC TRANSITIONS

A ferromagnetic state in UNiGa is achieved via a spin-flip metamagnetic transition at a magnetic field strength of 1 T (at 4.2 K) applied along the $c$-axis (Sechovsky et al., 1991). The magnitude of the U magnetic moment is not affected by this transition, but a large drop of the resistivity (for i // c) is apparent amounting to 87 % (see Fig. 8), and the $\rho(T)$ curve measured in the metamagnetic state mimics the behavior of a ferromagnet (Fig. 7).

When the magnetic field is applied along the basal plane in UNiGa and other antiferromagnets of the ZrNiAl-structure family, do not lead to any moment reorientation and have practically no influence on the resistivity. This demonstrates that standard magnetoresistance effects connected with the evolution of electron trajectories in a magnetic field have only little influence compared to "giant" effects due to spin reorientation. The magnitude of the resistivity changes due to magnetic structure variations in actinides is often larger than so called Giant Magnetoresistance (GMR) effects in magnetic multilayers. Therefore the notion of the GMR effect is commonly used in actinides, too.

The case of UNiGa provides clues to the mechanism for the GMR effects in this class of compounds. The change of the $\gamma$-coefficient, reflecting the total density of electronic states at the Fermi level, at the transition from the AF to F state (from 43 mJ/mol $K^2$ to 48 mJ/mol $K^2$) goes in the right direction (Aoki et al., 1996), but is too small to give an unambiguous proof of substantial Fermi surface reconstruction. However, a complementary information pointing to the reconstruction can be obtained from additional experiments of Hall effect and thermoelectric power (Kobayashi et al., 1996). The Hall resistivity in magnetic materials is described as a sum of two terms:

$$\rho_H(H) = R_H H = R_0 H + 4\pi R_s M, \qquad (1.2)$$

where $R_0$ and $R_s$ are the normal and the extraordinary Hall coefficient, respectively, and $M$ is the volume magnetization.

The normal Hall effect contains information on the Fermi surface, such as carrier concentration and/or the *k*-dependence of the conduction-electron-scattering. The extraordinary Hall effect provides information on the left-right asymmetry of the scattering (Hurd, 1972). In UNiGa the magnetization at $T = 4.2$ K is practically field-independent both below and above the metamagnetic transition (see Fig. 8). The same holds for normal resistivity (see Fig. 7), which means that the anomalous Hall effect contribution can be taken as field independent in both antiferromagnetic and ferromagnetic phase. As can be seen from Fig. 8, the field dependence of the Hall resistivity at $T = 4.2$ K is indeed linear in both phases, and therefore the change of the normal Hall coefficient $R_0$ at the transition can be evaluated from the slope of $\rho_H(H)$. We have thus obtained $R_0 = (-2.9\pm0.1)*10^{-9}$ m$^3$/C) for the antiferromagnetic state and $(-1.7\pm 0.1)*10^{-9}$ m$^3$/C for the ferromagnetic state. The difference in $R_0$ in the two phases reflects a considerable change of the Fermi surface resulting from the spin-reorientation, suggesting that an electronic structure variation is responsible for the GMR effect.

Another complementary information is provided by the field dependence of thermoelectric power. The change of the sign of $S$ at the metamagnetic transition in UNiGa probably reflects the sign change of the energy derivative of the density of state (DOS) at the Fermi level, which is again due to the Fermi surface reconstruction resulting from the moments reorientation (Kobayashi et al., 1996).

The phenomena described here illustrate the fact that in actinide intermetallics the type of magnetic structure, especially the type of magnetic coupling along a particular direction, plays a dominant role both in size and temperature dependence of the electrical resistivity. The impact is much stronger than in analogous lanthanide-based materials (Fournier and Gratz, 1993). The reason can be found in the presence of a narrow *f*-band at the Fermi surface, and in the stronger hybridization of the *5f*-states with other electronic states, which leads to an extreme sensitivity of electronic structure to the orientation of the *5f* magnetic moments. The question remains whether it is predominantly a gapping effect which leads to the high resistivity in antiferromagnetic state due to the enlargement of magnetic unit cell with respect to the crystallographic one. The large bulk of experimental data on various compounds show that there is a reduction of resistivity accompanying ferromagnetic

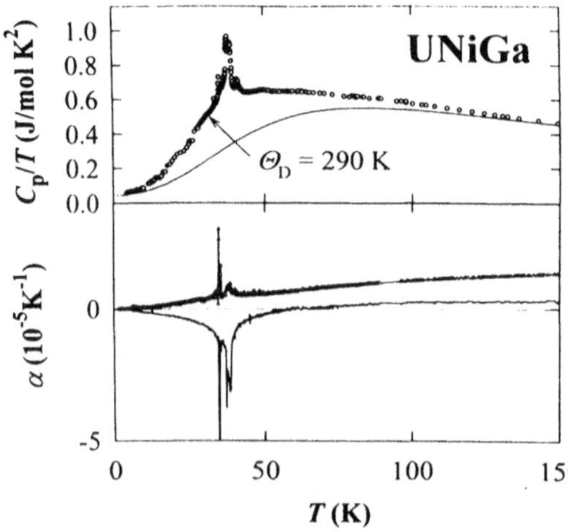


**Figure 9.** Temperature dependence of: a) the specific heat - $C_p/T$ vs. $T$ plot. The Debye function with $\theta_D$ = 290 K is shown by the solid line, and b) the linear thermal expansion coefficient of UNiGa measured on single crystals along the *a*- and *c*-axis. Data from (Prokes, 1997).

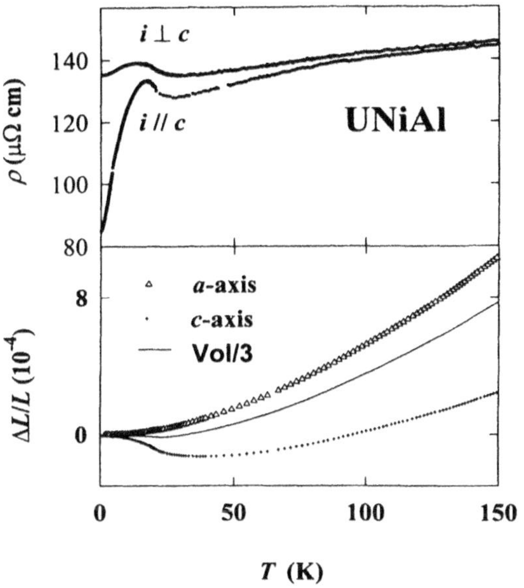


**Figure 10.** Temperature dependence of: a) the electrical resistivity $\rho$ with current $i$ along and perpendicular to the $c$-axis; data taken from (Brück et al., 1994), and b) the linear ($\Delta L/L$ - along the $a$- and $c$-axis) and volume ($\Delta V/V = 2\Delta L_a/L_a + \Delta L_c/L_c$ - full line) thermal expansion of UNiAl measured on single crystals. Data from (Prokes, 1997; Prokes et al., 1996).

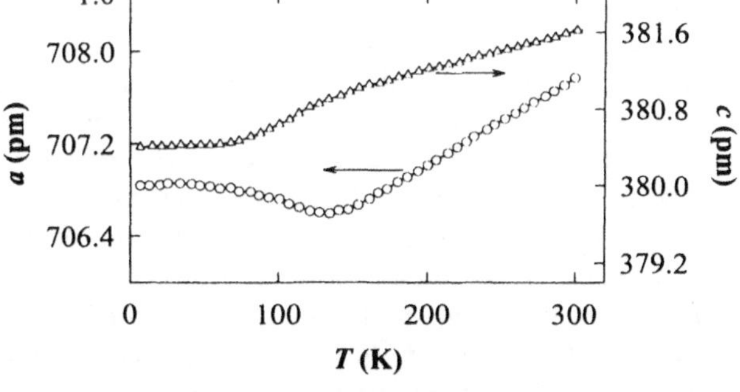


**Figure 11.** Temperature dependence of a) the electrical resistivity $\rho$, and b) lattice parameters $a$ and $c$ measured on polycrystalline URuGa. Data from (Havela et al., 1983).

coupling in the majority of cases, consistent with the action of a general superzone boundary effect.

The fact that the GMR effect can be understood on the basis of variations of the electronic structure only (i.e., without any spin-dependent scattering) is corroborated by electronic structure calculations for UNiGa performed using the fully relativistic, spin-polarized LDA approximations in both ferromagnetic and simple antiferromagnetic (+ –) states (Antonov et al., 1996). It is especially interesting that the calculations reproduce the magnitude of the GMR effect without any spin dependent scattering, because notion of spin diffusion length (used widely in the context of magnetic multilayers) cannot be simply applied to actinides with a strong spin orbit coupling, where *S* cannot be taken as a good quantum number.

Thus, the entire GMR effect in UNiGa can be explained by variations of topology of the Fermi surface. It would be interesting to test additional theoretical predictions of GMR effects for actinide materials with identical magnetic and crystallographic unit cells in antiferromagnetic state. One example is the calculation of magnetoresistance effects in $U_2Pd_2In$ and $U_2Pd_2Sn$ (Richter et al., 1996). The latter material is expected to exhibit a large positive magnetoresistance at the metamagnetic transition. This would mean that the magnetic superzone effect is not a necessary ingredient for the occurrence of a GMR effect.

## ANISOTROPY OF MAGNETOELASTIC PROPERTIES

Similar to anisotropy of transport properties we often find in antiferromagnetic compounds strikingly anisotropic thermal expansion, and magnetostriction, especially near magnetic-phase transitions. The anomalies are frequently of opposite sign for directions along which the antiferromagnetic and ferromagnetic coupling propagates, respectively and compensate to such extent, that the effect on volume is negligible if any. Anisotropy is also a quite common feature as concerns compressibility. Relevant data in this field are, unfortunately, still very scarce, especially due to lack of large enough single crystals. Systematic effort is highly desirable. Experimental results available in literature by end of 1997 can be found in (Sechovsky and Havela, 1998). In this paper we like to demonstrate the above mentioned effects on the thermal expansion behavior of 3 compounds, antiferromagnets UNiGa (Fig. 9) and UNiAl (Fig. 10) and a spin fluctuator URuGa (Fig. 11).

Specific-heat data in the low- and the high-temperature limits (see Fig. 9) can be approximated by a Debye function with $\theta_D = 290$ K when considering $\gamma = 44$ mJ/mol K$^2$. Closer inspection of fig. 5.12 also shows that the $Cp/T$ vs. $T$ dependence gradually deviates from a Debye function with decreasing temperature below 180 K which may be correlated with the anomalous behavior of the thermal expansion. In Fig. 9 one can see the anisotropy of the linear thermal expansion coefficient, which is strongly negative along the *c*-axis and shows pronounced anomalies not only in the vicinity of magnetic phase transitions and in antiferromagnetic region but also far above $T_N$. The positive thermal expansion anomalies within the basal plane compensate for the negative *c*-axis effects yielding the volume thermal expansion looking almost normal, as there were no magnetic phenomena in UNiGa. The electrical resistivity with current along the *c*-axis shows the onset of an upturn below 150 K (see Fig. 7). These anomalies, which can be removed by applying sufficiently large magnetic field along *c*, are attributed to the short range antiferromagnetic correlations propagating along the *c*-axis. Future efforts in theory correlating these effects with electronic structure and exchange interactions are highly desirable.

The temperature dependencies of the relative length changes, $(\Delta L/L)_a$ and $(\Delta L/L)_b$ with respect to the length of the sample along a particular direction at $T = 1.5$ K are displayed in Fig. 10, together with the relative change in volume in UNiAl. It is evident that the thermal expansion also in this case is highly anisotropic, similar to most of other bulk properties (Prokes et al., 1996). Along the *a*-axis, the lattice monotonously expands with increasing temperature ($\alpha_a$ is always positive) whereas the *c*-axis shrinks considerably with increasing temperature up to $T \approx 35$ K. Around 35 K, $\alpha_c(T)$ changes sign and the lattice expands at higher temperatures. The sharp anomaly observed near 19.3 K in both $\alpha_a(T)$ and $\alpha_c(T)$ clearly marks the magnetic phase transition at $T_N$. Note that the relatively small value of $\alpha_V$, especially at low temperatures, is a consequence of mutual cancellation of significantly large coefficients $\alpha_a$ and $\alpha_c$, one positive and the other negative. This is the main reason why the electronic term of the effective Grüneissen parameter in the case of UNiAl is only $\Gamma_e = 6.8$, which is about one order of magnitude smaller than that of typical heavy-fermion compounds, although it is considerably enhanced with respect to normal metals ($\Gamma_e \approx 2$).

The thermal expansion data may be better understood in conjunction with results of elastic and inelastic neutron scattering experiments on single crystals (Prokes, 1997), which reveal the presence of strong antiferromagnetic fluctuations in UNiAl. The temperature development of fluctuations can be followed from low temperatures up to at least 1.5 $T_N$. However, no particular dispersion can be assigned to magnetic fluctuations. Instead of that, a wide spread in energy transfer of scattering intensity in the inelastic spectrum is present in both spectra, below and above $T_N$.

The magnetic fluctuations have entirely different behavior along and perpendicular to the *c*-axis. The onset of magnetic fluctuations is visible around 35 - 40 K, where a deviation from the high-temperature behavior of the thermal-expansion coefficient $\alpha_c$ can be observed, accompanied by a contraction along the hexagonal axis with lowering temperature. Magnetic fluctuations along the *c*-axis are clearly of antiferromagnetic type. The behavior along the *a*-axis is rather regular, except for the critical region around $T_N$.

As the last selected case, demonstrating need of a further concerted action in experiment and theory is the spin-fluctuator URuGa (see Fig. 11). The susceptibility follows a modified Curie-Weiss behavior above 100 K. With decreasing temperature below 100 K, the susceptibility is leveling off and forms a plateau below 50 K, followed by an upturn with further lowering temperature. As can be seen from the temperature dependence of lattice parameters, the thermal expansion is anisotropic with $\alpha_c$ considerably larger than $\alpha_a$. Below approximately 150 K, a positive (negative) anomaly in the *c*- (*a*-) axis evolves with lowering temperature, which may be tentatively attributed to magnetic correlations. In this context we recall the above discussed thermal expansion anomalies accompanying antiferromagnetic correlations propagating along *c*.

## REFERENCES


Andersen, N.H., and Smith, H., 1979, Electron-magnon interaction and the electrical resistivity ofTb, *Phys.Rev. B, 19:* 384

Andreev, A.V., Havela, L., Sechovsky, V., Bartashevich, M.I., Dremov, R.V., and Kozlovskaya, I.K., 1997, Ferromagnetism in the $UCo_{1-x}Ru_xAl$ quasiternary intermetallics, *Phil.Mag. B, 75:*827.

Andreev, A.V., Sechovsky, V., Havela, L., Bartashevich, M.I., Goto, T., Dremov, R.V., and Kozlovskaya, I.K., 1997, Magnetic properties of $UCo_{1-x}Ru_xAl$ and $UNi_{1-x}Ru_xAl$ solid solutions, *Physica B, 237-238:*224.

Antonov, V.N., Perlov, A.Y., Oppeneer, P.M., Yaresko, A.N., and Halilov, S.V., 1996, Mechanism of the giant magnetoresistance in UNiGa from first-principles calculations, *Phys Rev. Lett., 77:* 5253.

Aoki, Y., Kobayashi, Y., Sato, H., Sugawara, H., Sechovsky, V., Havela, L., Prokes, K., Mihalik, M., and Menovsky, A., 1996, Superzone Gap Formation Evidenced by Specific-Heat in UNiGa, *J.Phys. Soc.Japan, 65:*3312.

Baer, Y., Ott, H.R., and Andres, K., 1980, 5f electron localization in metallic UPd3, *Solid State Commun., 36:*387.

Brooks, M.S.S., Johansson, B., Eriksson, O., and Skriver, H.L., 1986, Relativistic Effects in Heavy Elements, *Physica B, 144:* 1.

Brooks, M.S.S., and Kelly, P.J., 1983, Large Orbital-Moment Conribution to 5f Band Magnetism, *Phys. Rev. Lett.*, *51:* 1709.
Brück, E., Nakotte, H., de Boer, F.R., de Châtel, P.F., van der Meulen, H.P., Franse, J.J.M., Menovsky, A.A., Kim Ngan, N.H., Havela, L., Sechovsky, V., Perenboom, J.A.A.J., Tuan, N.C., and Sebek, J., 1994, Electronic properties of UNiAl in high magnetic fields, *Phys.Rev. B*, *49:* 8852.
Cooper, B.R., Siemann, R., Yang, D., Thayamballi, P., and Banerjea, A., 1985, Hybridization-induced anisotropy in cerium and actinide systems, In A.J. Freeman and G.H. Lander (Eds.), *Handbook on the Physics and Chemistry of the Actinides* (pp. 435-500), Amsterdam: Noth-Holland.
Coqblin, B., and Schrieffer, J.R., 1969, *Phys Rev*, *185:*847
Eriksson, O., Brooks, M.S.S., and Johansson, B., 1990, *Phys.Rev. B*, *41:*9087
Eriksson, O., Nordstrom, L., Brooks, M.S.S., and Johansson, B., 1988, *Phys.Rev.Lett.*, *60:*2523
Fournier, J.M., and Gratz, E., 1993, K.A. Gschneidner, Jr., L. Eyring, G.R. Choppin, and G.H. Lander (Eds.), *Handbook on the Physics and Chemistry of Rare Earths* (pp. 409-537), Amsterdam: North-Holland.
Frings, P.H., Franse, J.J.M., Menovsky, A., Zemirli, S., and Barbara, B., 1986, Magnetic properties of single-crystalline $UNi_2$, *J.Magn.Magn.Mater.*, *54-57:*541.
Havela, L., Hrebik, J., Zeleny, M., and Andreev, A.V., 1983, Magnetic properties of UGaX (Ni, Co, Ru), *Acta Phys Polonica A*, *68:* 493.
Havela, L., Sechovsky, V., Nakotte, H., Brück, E., and de Boer, F.R., 1994, Anisotropy of magnetic and transport properties in actinide intermetallics, *IEEE Trans.Magn.*, *30:*1130.
Havela, L., Sechovsky, V., Prokes, K., Nakotte, H., Fujii, H., and Lacerda, A., 1996, Giant Magnetoresistance Effects in 5f-Materials, *Physica B*, *223-224:* 245.
Hu, G.J., and Cooper, B.R., 1993, Theory of Anisotropic Hybridization-Broadened Magnetic Response in Cerium and Actinide Systems, *Phys.Rev. B*, *48:* 12743.
Hurd, C.M., 1972, *The Hall Effect in Metals and Alloys*, New York: Plenum Press.
Jirman, L., Sechovsky, V., Havela, L., Ye, W., Takabatake, T., Fujii, H., Suzuki, T., Brück, E., and de Boer, F.R., 1992, Magnetic and transport properties of UNiGa, *J.Magn.Magn.Mater.*, *104-107:*19.
Kadowaki, K., 1993, Electrical-Resistivity of Some Uranium-Compounds, *Physica B*, *186-188:* 727.
Kobayashi, Y., Aoki, Y., Sugawara, H., Sato, H., Sechovsky, V., Havela, L., Prokes, K., Mihalik, M., and Menovsky, A.A., 1996, Hall effect and thermoelectric power in UNiGa, *Phys Rev.B*, *54:*15330.
Koelling, D.D., Dunlap, B.D., and Crabtree, G.W., 1985, f-electron hybridization and heavy-fermion compounds, *Phys.Rev. B*, *31:* 4966.
Lebech, B., Wulff, M., and Lander, G.H., 1991, Spin and Orbital Moments in Actinide Compounds, *J.Appl.Phys.*, *69:* 5891.
Lebech, B., Wulff, M., Lander, G.H., Rebizant, J., Spirlet, J.C., and Delapalme, A., 1989, Neutron diffraction studies of the crystalline and magnetic properties of $UFe_2$, *J Phys : Condens Matter*, *1:*10229.
Mathon, J., 1968, Magnetic and electrical properties of ferromagnetic alloys near the critical concentration, *Proc.Roy.Soc. A*, *306:*355.
Moriya, T., 1991, Theory of itinerant electron magnetism, *J.Magn.Magn.Mater.*, *100:*261.
Paixao, J.A., Lander, G.H., Brown, P.J., Nakotte, H., de Boer, F.R., and Brück, E., 1992, Magnetization Density in URhAl - Evidence for Hybridization Effects, *J Phys : Condens Matter*, *4:* 829.
Plumer, M.L., 1994, Mean-Field Phase-Diagrams of $UT_2X_2$ Compounds, *Phys.Rev. B*, *50*, 13003.
Prokes, K., 1997, Magnetism in uranium intermetallic compounds, University of Amsterdam;
Prokes, K., Brück, E., de Boer, F.R., Mihalik, M., Menovsky, A., Burlet, P., Mignot, J.M., Havela, L., and Sechovsky, V., 1996, Giant Magnetoresistance and Magnetic Phase-Diagram of UNiGa, *J.Appl.Phys.*, *79:* 6396.
Prokes, K., de Visser, A., Menovsky, A.A., Brück, E., de Boer, F.R., Sechovsky, V., and Gortenmulder, T.J., 1996, Thermal-Expansion of Single-Crystalline UNiAl, *J.Appl.Phys.*, *79:* 6358.
Prokes, K., Nakotte, H., Brück, E., de Boer, F.R., Havela, L., Sechovsky, V., Svoboda, P., and Maletta, H., 1994, Anisotropic magnetic and transport properties of UNiGe, *IEEE Trans.Magn.*, *30:* 1214.
Richter, M., Zahn, P., Divis, M., and Mertig, I., 1996, Giant magnetorestance in uranium intermetallics: Ab initio calculations for $U_2Pd_2In$ and $U_2Pd_2Sn$, *Phys.Rev. B*, *54*, 11985.
Sechovsky, V., and Havela, L., 1988, Intermetallic compounds of actinides, In E.P. Wohlfarth and K.H.J. Buschow Eds.), *Handbook of Magnetic Materials* (pp. 309-491). Amsterdam: Elsevier Science B.V.
Sechovsky, V., and Havela, L., 1992, Anisotropic Hybridization and Magnetism in Actinide Intermetallics, *Physica Scripta*, *Vol T45:*99.
Sechovsky, V., and Havela, L., 1998, Magnetism of ternary intermetallic compounds of uranium, In K.H.J. Buschow (Ed.), *Handbook of Magnetic Materials* (pp. 1-289). Amsterdam: Elsevier Science B.V.
Sechovsky, V., Havela, L., de Boer, F.R., Veenhuizen, P.A., and Brück, E., 1988, Magnetic properties and hybridization effects in UTX compounds, *J.de Phys*, *49:* 497.
Sechovsky, V., Havela, L., de Boer, F.R., Veenhuizen, P.A., Sugiyama, K., Kuroda, T., Sugiura, E., Ono, M., Date, M., and Yamagishi, A., 1992, Hybridization and magnetism in U(Ru,Rh)X, X = Al, Ga, *Physica B*, *177:*164.

Sechovsky, V., Havela, L., Jirman, L., Ye, W., Takabatake, T., Fujii, H., Brück, E., de Boer, F.R., and Nakotte, H., 1991, Giant Magnetoresistance Effects in UNiGa, *J.Appl.Phys.*, *70:* 5794.
Sechovsky, V., Havela, L., Nakotte, H., de Boer, F.R., and Brück, E., 1994, Anisotropy of magnetism in actinide and lanthanide intermetallics, *J Alloys and Compounds*, *207&208:*221.
Tran, V.H., André, G., Bourée, F., Troc, R., and Noël, H., 1998, Ac-susceptibility ansd neutron diffraction studies of the $URu_{1-x}Pd_xGa$ system, *J Alloys and Compounds*, *271-273:*503.
Ueda, K., 1977, Electrical Resistivity of Antferromagnetic Metals, *J Phys Soc Japan*, *43:* 1497.
Veenhuizen, P.A., de Boer, F.R., Menovsky, A.A., Sechovsky, V., and Havela, L., 1988, Magnetic Properties of URuAl and URhAl Single Crystals, *J Phys (Paris)*, *49-C8:* 485.
Veenhuizen, P.A., Klaasse, J.C.P., de Boer, F.R., Sechovsky, V., and Havela, L., 1988, The role of hybridization in U(Ru,Rh)Al, *J.Appl.Phys.*, *63:*3064.
Wulff, M., Fournier, J.M., Delapalme, A., Gillon, B., Sechovsky, V., Havela, L., and Andreev, A.V., 1990, Study of Orbital and Spin Magnetism in UCoAl by Polarized Neutron Diffraction, *Physica B*, *163:*331.

# THE ROLE OF SELECTED f IONS IN THE SUPPRESSION OF HIGH-$T_C$ SUPERCONDUCTIVITY

L. Soderholm[1] and U. Staub[2]

[1]Chemistry Division, Argonne National Laboratory, Argonne IL 60439, USA
[2]Swiss Light Source, CH-5232 Villigen PSI, Switzerland

## INTRODUCTION

The initial observations of superconductivity at temperature above 77K in copper-oxide based materials was surprising from a variety of different perspectives. Among the unexpected findings were reports of superconductivity for the series $RBa_2Cu_3O_7$ where R is a rare earth(Y,Nd,Tm)[1,2], which may carry a large, local magnetic moment. Superconductivity was subsequently demonstrated for all 4f analogs in this series except Ce, Pr, and Tb. In addition to the $RBa_2Cu_3O_7$ series, there are several other CuO based series of superconductors that are formed by substituting R ions. The most studied of these are listed in Table I, together with the f ions that form isostructural compounds and their superconducting critical temperatures ($T_c$). The presence of an R ion with a large magnetic moment does not significantly influence the superconductivity. In contrast, even the presence of small concentrations of magnetic impurity ions in a conventional superconductor inhibits superconductivity by interfering with the formation of Cooper pairs.

Most R ions substitute into an isostructural series with no observable effect on the superconducting properties of the material. As can be seen from Table I, there are notable exceptions to this observation. In particular, the rare-earth ions Ce, Pr, and Tb, together with the 5f-actinide ions Am and Cm, sometimes behave anomalously. These ions either do not form an isostructural phase, or if the phase forms it may not be superconducting. The fact that an f-ion can suppress superconductivity is clearly demonstrated in the isostructural series $Y_{1-x}Pr_xBa_2Cu_3O_7$ $(0<x<1)$[3], in which the critical temperature decreases with increasing x, such that for concentrations of Pr greater than $x \approx 0.5$, the samples no longer superconduct.

There have been at least four mechanisms proposed to account for the suppression of superconductivity by selected f ions: (1) Hole filling. The suppression of $T_c$ in the $RBa_2Cu_3O_7$ series was initially attributed to the presence of a tetravalent ion at the R site that would effectively transfer charge density according to:

$$R^{3+}Ba_2Cu_2{}^{2+}Cu^{3+}O_7 \rightarrow R^{4+}Ba_2Cu_3{}^{2+}O_7,$$

thereby filling the holes on the CuO planes that are responsible for superconductivity. Whereas this mechanism does not play a significant role in the $T_c$ suppression observed in $PrBa_2Cu_3O_7$, it has been subsequently used to understand other Ce and Am compounds. (2) Hole localization, (3) Hybridization of 4f with oxygen 2p orbitals. The latter two mechanism are sometime combined[4,5], and (4) Magnetic-pair breaking[6]. Whereas each of these theories can explain some aspects of the available data, a comprehensive understanding of $T_c$ suppression by selected f ions has yet to emerge.

The ability of select f-ions to sometimes suppress superconductivity is the subject of this article. In an effort to clearly delineate the effects of R substitution on $T_c$, we have undertaken a systematic study of f-ion substitutions into three different high-$T_c$ series, $RBa_2Cu_3O_7$, $R_2CuO_4$ and $Pb_2Sr_2RCu_3O_8$. Herein we present a summary of our results. We focus the discussion on results from the R ions that have shown to behave anomalously, that is Ce, Pr, Tb, Am and Cm. A more general review of the overall aspects of our work is presented elsewhere [7]. We begin by reviewing relevant chemical and electronic properties of these f-ions, then we present results from the relevant analogs of the three systems that we studied, and follow these results this with a discussion of the systematics that have emerged from this study.

## CHEMICAL AND ELECTRONIC PROPERTIES OF f-IONS

The f-ion has three characteristics of importance relative to it's influence on superconductivity in high-$T_c$ oxides: (1) the chemical behaviour of the ions, which includes ion size and valence stabilities, (2) electronic behaviour, including both single ion and bonding properties and (3) the single-ion and collective magnetic properties. These will be discussed in turn.

The rare-earth ions, excluding Y, La and Lu, have partially filled 4f-states. The f-orbitals are not effectively screened from the nucleus, therefore their ionic radii decrease across the lanthanide series, as shown in Table II. This decreasing size with increasing atomic number, the lanthanide contraction, effectively buries the valence f-orbitals in the s-, p-, and d-electron density. As a result, the f-electrons are localized, and the f-orbitals are not readily available for bonding or interactions with surrounding near-neighbour ions. Their electronic properties are predominantly influenced by electrostatic interactions and they can be treated within single-ion models. $Ce^{3+}$ has the most exposed f-orbitals, which are thought to sometimes hybridize with ligand orbitals resulting in novel electronic properties, including heavy-fermion behaviour and the Kondo effect[8,9]. The ions heavier than Nd are generally considered to behave as trivalent ionic spacers, and the f-states are often treated theoretically as core states. For the most part, the chemistry of these elements is determined by the size of the ion.

**Table I.** Superconducting CuO series based on R substitutions. Single phase isostructural compounds form except where listed as a parenthetical exception, in which case isostructural member does not form.

| Series | R | Exceptions | Approx. $T_c$ (K) |
|---|---|---|---|
| | | | |
| $RBa_2Cu_3O_7$ | Y, La-Yb, Cm | (Ce), Pr, (Tb) | 92 |
| $RBa_2Cu_4O_8$ | Nd-Er | (Tb) | 51-84 |
| $R_{1.85}M_{.15}CuO_4$ | Pr-Eu, Cm; $M=Ce^{4+}, Th^{4+}$ | Cm | 24 |
| $Pb_2Sr_2R_{.5}Ca_{.5}Cu_3O_8$ | La-Tm, Am, Cm | Ce, Am, Cm | <84 |
| $Bi_2Sr_2R_{.5}Ca_{.5}Cu_2O_8$ | Y, Nd-Tm | | <85 |
| $(R_{1.5}Ce_{.5})Sr_2Cu_2NbO_{10}$ | Pr-Eu | Pr | 51-84 |

**Table II.** The standard electrode potentials, $E^0$, for the reduction of the tetravalent to the trivalent valence state[10]. The higher the number, the more stable the trivalent relative to the tetravalent state. Also listed are the trivalent ionic radii for a variety of lanthanide[11] and actinide[12] ions.

| 4f Element | Radius 3+ (A) | Reduction Potential (eV) | 5f Element | Radius 3+ (A) | Reduction Potential (eV) |
|---|---|---|---|---|---|
| | | | | | |
| La | 1.160 | | | | |
| Ce | 1.143 | 1.7 | | | |
| Pr | 1.126 | 3.1 | | | |
| Nd | 1.109 | 5.0 | U | | -0.52 |
| | | | Np | | 0.15 |
| Sm | 1.079 | | Pu | | 1.0 |
| Eu | 1.066 | | Am | 1.11 | 2.4 |
| Gd | 1.053 | | Cm | 1.10 | 3.2 |
| Tb | 1.040 | 3.2 | Bk | 1.08 | 1.7 |
| Dy | 1.027 | 5.2 | Cf | 1.08 | 3.2 |
| (Y) | 1.019 | | | | |
| Ho | 1.015 | | | | |
| Er | 1.004 | | | | |

Whereas the predominant oxidation state of the lanthanide series is trivalent, selected f-ions are redox active under favourable conditions. Ce, Pr and Tb have energetically accessible tetravalent states in oxidic systems. Ce is the most easily oxidized with a comparatively stable 4+, tetravalent state. A variety of complex oxides containing $Ce^{4+}$ are known[13]. Pr and Tb also form complex oxide compounds in their tetravalent states, the latter of which is isoelectronic with $Gd^{3+}$, and has a half-filled f shell. The relative stabilities of the tetravalent states of lanthanides are compared in Table II. These data are obtained electrochemically in non-complexing solutions. We argue that although the exact reduction potentials may not be valid in our complex oxides, the trends established by these data are followed. In other words, within an isostructural series, Ce is the most likely to be tetravalent, with Pr and Tb exhibiting similar behaviour.

The actinide series is the 5f counterpart of the lanthanides series, with partially filled 5f states. Whereas an actinide contraction exists across this series for the same reasons as the lanthanide contraction, the effect is not as important. The 5f orbitals are not as buried in the core-electron density. This is demonstrated by the comparison of $Pr^{3+}$ ($4f^2$) with $Cm^{3+}$ ($5f^7$) shown in Figure 1. It is clear that although the ionic radius of $Cm^{3+}$ is slightly smaller than that of $Pr^{3+}$, the 5f valence orbitals on Cm extend radially further from the nucleus than do their 4f counterparts. The result is that $Cm^{3+}$ and $Am^{3+}$ are expected to have valence f orbitals that are more influenced than the 4f ions by their surrounding ligands. The 5f orbitals are more able to hybridize effectively with ligand orbitals than the valence 4f orbitals in the lanthanide series.

The chemistry of the actinide elements is much more complex than that of the lanthanides[10]. The lighter actinides have multiple oxidation states accessible in both solution and in the solid state. Whereas stable trivalent and tetravalent states are reported for all the ions from Th-Bk in solid oxides, Am and Cm are the only two ions that are expected to form isostructural members of a R-CuO series. Elements lighter than Am are expected to reduce Cu during synthesis, and therefore not form the desired compounds. Higher Z elements are expected to show effects of radiation damage that vitiate property characterization.

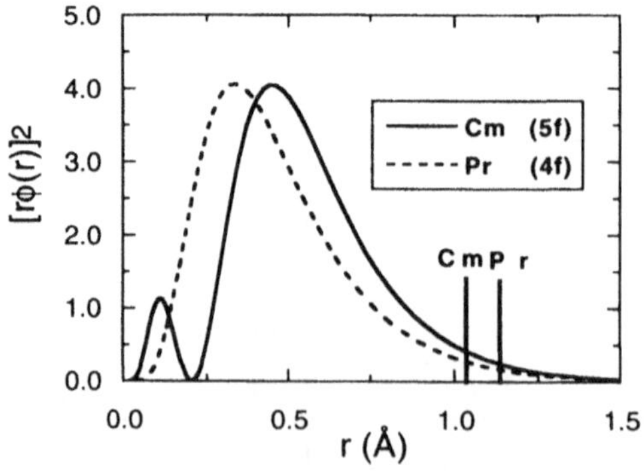


**Figure 1** The radial distribution of the f-electron density of $Pr^{3+}$ ($4f^2$) is compared with that of $Cm^{3+}$ ($4f^7$). The vertical lines at 1.10 and 1.226 Å represent the ionic radii of Cm and Pr respectively (see Table II)

## RESULTS

### $RBa_2Cu_3O_7$

This series has been extensively studied. The substitution of R results in isostructural compounds for R=Y, La-Yb (except Ce,Pr, Tb) and Cm. The Ce and Tb analogs do not form single-phase materials, and studies of substitutions into bulk $YBa_2Cu_3O_7$ show very limited stability ranges for both ions. The range of phase stabilities of Ce and Tb in $R_{1-x}Y_xBa_2Cu_3O_7$ increase when samples are made as thin films, into which it is possible to substitute to concentrations of x < 0.3 for Ce and x < 0.7 for Tb[14]. Resistivity data from these thin films show that $T_c$ is suppressed with increasing x for both Ce and Pr, whereas $T_c$ remains constant with x for the Tb samples through the substitutional range. These data are summarized in Figure 2.

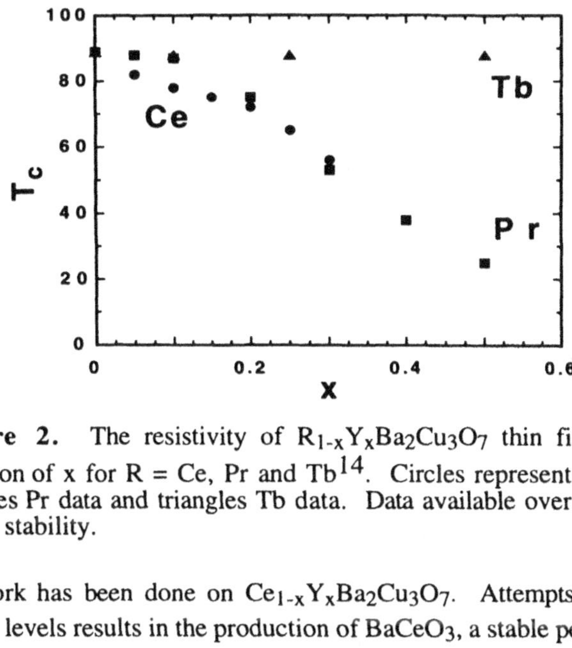


**Figure 2.** The resistivity of $R_{1-x}Y_xBa_2Cu_3O_7$ thin films as a function of x for R = Ce, Pr and Tb[14]. Circles represent Ce data, squares Pr data and triangles Tb data. Data available over range of phase stability.

Very little work has been done on $Ce_{1-x}Y_xBa_2Cu_3O_7$. Attempts to prepare samples even at low doping levels results in the production of $BaCeO_3$, a stable perovskite phase that forms easily under the synthetic conditions required to form the superconductor. The work on thin film samples[14] provides evidence of $T_c$ suppression, but there is little characterization on the oxidation state of Ce in this sample. Bulk samples of $Ce_{.2}Y_{.8}Ba_2Cu_3O_7$ have been studied using x-ray absorption near-edge structure (XANES)[15] spectroscopy, and have been shown to contain primarily $Ce^{4+}$. Further work to characterize Ce in this series would be useful.

A variety of experiments have been centered on the Tb oxidation state and hybridization in $Tb_{1-x}Y_xBa_2Cu_3O_7$[16,17]. An analysis of neutron diffraction data on a x=0.1 sample reveals that Tb is incorporated into the sample, substituting at the Y site. There are no anomalies observed in the lattice constants or in the planar-oxygen thermal parameters determined from the neutron data. This result would be expected if Tb is trivalent because their $Y^{3+}$ and $Tb^{3+}$ have very similar ionic radii, as shown in Table II. However the low concentration prohibits the determination of a Tb oxidation state from Tb-O distances alone. Paramagnetic susceptibility data also suggest that Tb is trivalent. An effective moment of $\mu_{eff}=(9.8\pm0.3)\mu_B$ was determined by fitting experimental data. This measured moment compares with free-ion effective moments of $9.72\mu_B$ and $7.94\ \mu_B$ for the $Tb^{3+}$ and $Tb^{4+}$ respectively. Further support for trivalent Tb in this material is obtained from a calculation of the single-ion paramagnetic susceptibility for comparison with experiment. Assuming a sublattice of non-interacting Tb ions with a valence of 3+ and no significant hybridization, the effect of the crystal-field generated by the ions surrounding the Tb was modeled using crystal-field parameters interpolated from other $RBa_2Cu_3O_7$ compounds[17]. Including J-mixing and intermediate coupling effects, the single-ion susceptibilities were calculated as a function of temperature, and agree very well with the measured values. For comparison, $Tb^{4+}$ has a spherically symmetric $^8S_{7/2}$ ground term that is not affected to first order by the crystal field, and therefore the free ion susceptibility is the best value to compare with experiment. It is clear that the magnetic data favor model in which Tb is trivalent in $Tb_{1-x}Y_xBa_2Cu_3O_7$. This result is further confirmed by the x-ray absorption near-edge (XANES) data shown in Figure 3. A comparison of the Tb $L_3$-edge data from $Tb_{1-x}Y_xBa_2Cu_3O_7$ with appropriate trivalent and tetravalent standards demonstrates that Tb is essentially trivalent when substituting for Y in this structure, confirming the analysis of the diffraction and the magnetic susceptibility results.

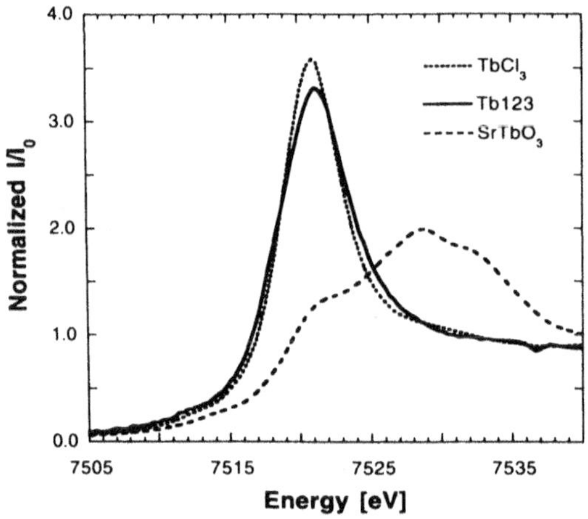


**Figure 3.** Tb $L_3$-edge x-ray absorption data from $Tb_{1-x}Y_xBa_2Cu_3O_7$ (solid line) are compared with a $Tb^{3+}$ standard, $TbCl_3$, and a $Tb^{4+}$ standard, $SrTbO_3$. These data were obtained through electron-yield detection at ambient temperature [17].

The incorporation of increasing amounts of Tb into $Y_{1-x}Tb_xCu_3O_7$ does not influence the superconducting properties of the material. The suppressed $T_c$ previously reported for $TbBa_2Cu_3O_7$ arose because the sample is not single phase. The stability of $BaTbO_3$ interferes with the formation of the superconducting material. However, when Tb is incorporated into the $RBa_2Cu_3O_7$ structure, it is trivalent, and it does not suppress superconductivity.

The mechanism involved in the suppression of superconductivity by Pr in $PrBa_2Cu_3O_7$ is complex and is a central focus of the issue of $T_c$ suppression by f-ions in high-$T_c$ oxides

[3,18,19]. Initially it was argued that $T_c$ suppression occurred through hole filling, and that Pr was mixed valent or tetravalent. This argument is supported by the chemistry of Pr, which is known to have stable trivalent and tetravalent states in oxidic systems. Initially magnetic susceptibility data were also used in support of mixed-valent Pr. Pr in $PrBa_2Cu_3O_7$ has a measured paramagnetic susceptibility that is intermediate between the free ion values of the trivalent and tetravalent ions.

Inelastic neutron scattering (INS) has been used extensively to characterize the electronic properties of the Pr f states in $PrBa_2Cu_3O_7$[7,20-25]. Whereas there are minor discrepancies about some of the peak assignments, the overall conclusions are clear. The data clearly show that Pr is trivalent in this compound. The linewidths of the magnetic peaks are unusually broadened in the spectra, indicating either static or dynamic processes are influencing the Pr electronic states. This effect is clearly demonstrated by a comparison of the INS data from $PrBa_2Cu_3O_7$ with those obtained form $Pr_2CuO_4$, shown in Figure 4. Despite the line broadening, it is clear that the INS data are consistent with a trivalent Pr ion that has a $^3H_4$ electronic ground level. This ground level is split by the orthorhombic symmetry into 9 singlets that are well represented by a crystal-field model similar to that found for other R in this compound[7,20]. The crystal-field splitting determined by fitting these data is somewhat unusual, in that there are two excited-state singlets within 4 meV of

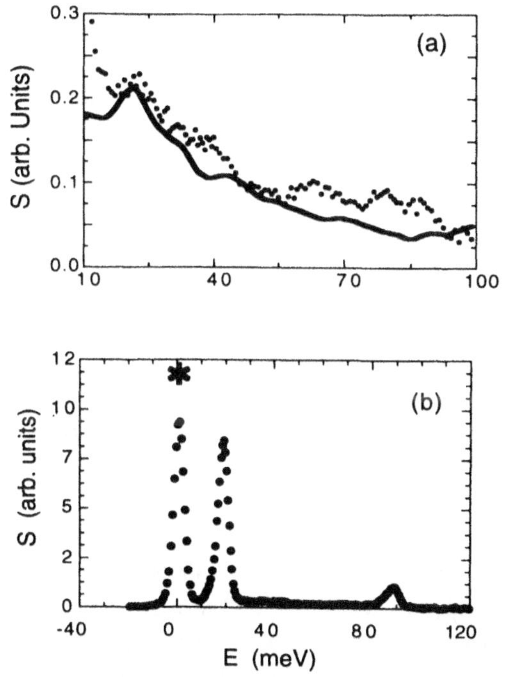


**Figure 4.** A comparison of the INS energy loss spectra obtained from (a) $PrBa_2Cu_3O_7$ and (b) $Pr_2CuO_4$. Both data sets were taken on the same spectrometer with the same experimental parameters. Solid line in (a) are data from $YBa_2Cu_3O_7$ to show the contribution from phonons over the same energy range. The * at zero energy transfer in (b) indicates elastic scattering.

the ground state, with the next excited state located about 45 meV higher in energy. This splitting is depicted in Figure 5. The splitting of the ground level into a "pseudo-triplet" ground state, which is isolated from the higher states, results from the nearly cubic R-site symmetry. The magnitude and temperature dependence of the magnetic susceptibility can be entirely accounted for by this electronic structure of $Pr^{3+}$ [20]. The calculated effective moment of 2.8$\mu_B$ agrees well with that measured experimentally, and is significantly reduced from the free ion value of 3.58$\mu_B$. There is no need to invoke mixed-valent behaviour to understand the susceptibility data. The reduced effective moment, together with the saturation moment of 0.79$\mu_B$, measured in the ordered state[26,27], are understood in terms of a renormalization of the pseudo-triplet state[22].

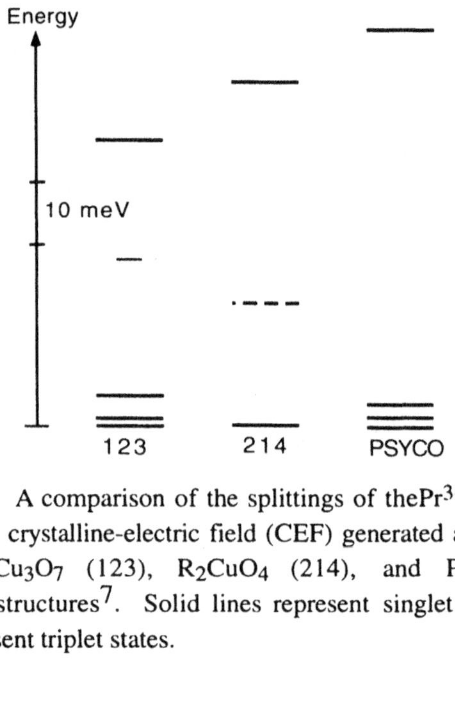


**Figure 5.** A comparison of the splittings of the$Pr^{3+}$ $^3H_4$ ground term by the crystalline-electric field (CEF) generated at the R site in the $RBa_2Cu_3O_7$ (123), $R_2CuO_4$ (214), and $Pb_2Sr_2RCu_3O_8$ (PSYCO) structures[7]. Solid lines represent singlet states, dotted lines represent triplet states.

The majority of data suggest that Pr is trivalent in $PrBa_2Cu_3O_7$[7,19], thereby arguing against a charge-transfer, hole-filling mechanism for the suppression of $T_c$. The strongest supporting evidence in favour of a charge transfer model to explain suppressed a $T_c$ in $PrBa_2Cu_3O_7$ comes from studies involving Ca substitution. A study of the superconducting behaviour of the solid-solution series $Y_{1-x-y}Ca_yPr_xBa_2Cu_3O_7$ shows that the substitution of $Ca^{2+}$ for Pr restores superconductivity[28]. The argument presented is that $Pr^{4+}$ removes holes from the CuO planes, which are subsequently restored by the addition of $Ca^{2+}$. Even more convincing is the report of superconductivity in the thin film $Pr_{.5}Ca_{.5}Ba_2Cu_3O_7$ at 43 K[29]. As we will point out our discussion of results from $Pb_2Sr_2Pr_{.5}Ca_{.5}Cu_3O_8$, there is an alternate explanation to hole filling that is consistent with the presence of $Pr^{3+}$ [30].

The only other isostructural member of the $RBa_2Cu_3O_7$ series that can be made single phase but is not superconducting is the R=Cm analog[31]. X-ray diffraction data on $CmBa_2Cu_3O_7$ can be indexed on the Pmmm orthorhombic structure seen for the Y analog.

The lattice constants are the same, within experimental error, as those obtained from $PrBa_2Cu_3O_7$. The inverse susceptibility is plotted as a function of temperature in Figure 6. There is no evidence of superconductivity over the temperature range studied, that is down to 10 K. The effective moment, determined from the slope of the inverse susceptibility, is (8.9 ± 0.4) $\mu_B$. This high moment implies the full $Cm^{3+}$ valence because $Cm^{4+}$ has no magnetic moment, assuming Russell Sanders coupling (L=3; S=3; J=L-S=0)[32]. An expanded view of the low temperature behaviour is shown as an inset in Figure 6. The inflection in this data has a very similar appearance to that observed for the Pr analog[26] where neutron diffraction was used to show that the Pr moments order. Thus, the shape of the curve, together with the linearity of the M vs. H curve at 12 K in fields up to 5 T, suggests antiferromagnetic ordering of the Cm moments at 22 K.

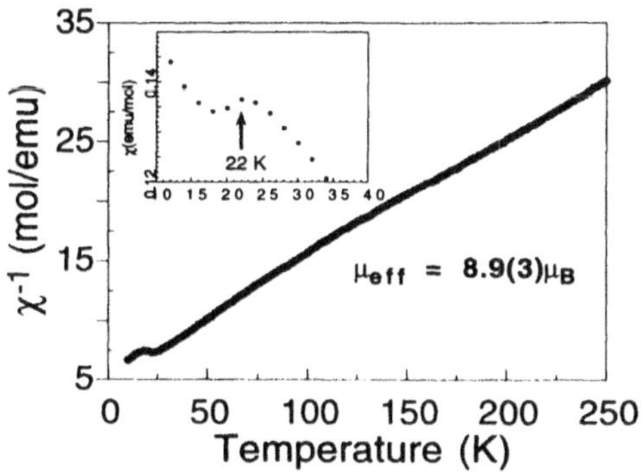


**Figure 6.** The inverse magnetic susceptibitiy of $CmBa_2Cu_3O_7$ plotted versus temperature (a Curie plot). There is no evidence of superconductivity down to 4 K. An expanded view of the susceptibitiy in the low temperature range 10-50 K is shown in the inset. We interpret the cusp at 22 K as an indication of long-range magnetic ordering of the Cm moments.

Although $CmBa_2Cu_3O_7$ forms a single-phase member of the $RBa_2Cu_3O_7$ series, it does not superconduct. It is clear that Cm is trivalent in this compound, ruling out a charge transfer model. In addition, the Cm magnetic moments order at an unusually high temperature compared to the other isostructural $RBa_2Cu_3O_7$ compounds, except the R=Pr analog.

## $R_2CuO_4$

The $R_2CuO_4$ series of compounds is isostructural for R=Pr-Gd and Cm. They crystallize with the $K_2NiF_4$-related T'-structure[33,34], which provides a significantly different crystal-field environment at the R site than found in the $RBa_2Cu_3O_7$ or the $Pb_2Sr_2RCu_3O_8$ series[7]. In addition, the magnetic transitions between states within the ground multiplet are markedly different for Pr in $Pr_2CuO_4$ compared with Pr in $PrBa_2Cu_3O_7$, as observed by inelastic neutron scattering, As shown in Figures 4, the transitions for the former compound are narrow, resolution limited, whereas those seen for the latter compound are very broadened. The sample qualities, as judged by x-ray and neutron diffraction, are comparable. The line broadening seen for the $PrBa_2Cu_3O_7$ is attributed to a dynamic, lifetime effects, perhaps resulting from interactions with the Cu spins[7]. In addition to differences in linewidth, the magnetic transitions in these two materials occur at much different energies, as seen by the comparison of the splitting of the $Pr^{3+}$ ground term in the R-site environments provided by the different structures. $Pr^{3+}$ in the $R_2CuO_4$ compound has an isolated-singlet ground state, in contrast to the $RBa_2Cu_3O_7$ and $Pb_2Sr_2RCu_3O_8$ materials, in which Pr has an isolated pseudo-triplet ground state. This difference in ground state electronic properties manifests itself in the magnetic behaviour of Pr in these three structure types. $Pr_2CuO_4$ has only a very small moment at low temperature, $0.08\mu_B$ observed by neutron diffraction [35]. This moment is induced by the ordering on the Cu sublattice. There is no evidence of long range ordering of the Pr sublattice, in contrast to the other two series, in which the Pr analog orders at an anomalously high temperature with respect to other R in the same series. The Nd moments in $R_2CuO_4$ order at about $T_N$=1.3 K[35], $T_N$=5.95 K for the R=Sm[36], $T_N$=6.5 K for the R=Gd[37] and $T_N$=25 K for the R=Cm[38,39] analogs.

Superconductivity is achieved by replacing $R^{3+}$ by $Ce^{4+}$ (or $Th^{4+}$) to form $R_{1.85}Ce_{.15}CuO_4$[34], such that the superconducting carriers are electrons. The R ions Pr-Sm all form single-phase superconductors with $T_c$'s of about 24 K. The solubility limit of Ce in the Gd analog is less than that required to induce superconductivity, which accounts for the absence of a transition in this material. The Pr analog has superconducting properties indistinguishable from other members of the series. In contrast, the Cm analog $Cm_{1.83}Th_{.17}CuO_4$ does not superconduct, although it is isostructural with the other T' compounds[38,39]. Instead, the antiferromagnetic ordering of the Cm moments is depressed from 25 K in the parent compound to 13 K for the Th-doped material, as shown in Figure 7. The effective moment obtained for Cm in both the parent and the doped compound is $(7.89\pm0.5)\mu_B$, which compares well with the free ion moment of $Cm^{3+}$.

In summary, $R_{1.83}Ce_{.17}CuO_4$ superconducts when R=Pr, but not when R=Cm. Both Pr and Cm are trivalent in this series. Pr has no local magnetic moment, whereas Cm as the full $7.89\mu_B$ associated with the trivalent s-state ion. These Cm moments order at anomalously high temperatures.

## $Pb_2Sr_2RCu_3O_8$

The Pb-based series of compounds, $Pb_2Sr_2RCu_3O_8$, form for R=Y,La-Lu, Am, and Cm. This series is of considerable interest because the Ce, Pr and Tb analog all form, and therefore provide an ideal opportunity to study the effects of selected f-ions on superconducting properties. These compounds are essentially isostructural [40], with only slight distortions from the basic orthorhombic structure[41,42]. The R environment in $Pb_2Sr_2RCu_3O_8$ is very similar to that found in the $RBa_2Cu_3O_7$ series[43]. These materials are made superconducting by replacing 50% of the $R^{3+}$ by $Ca^{2+}$. The excess holes are transferred to the CuO planes thought to be responsible for superconductivity. Superconductivity in these materials are reported as high as 84 K in single crystal samples,

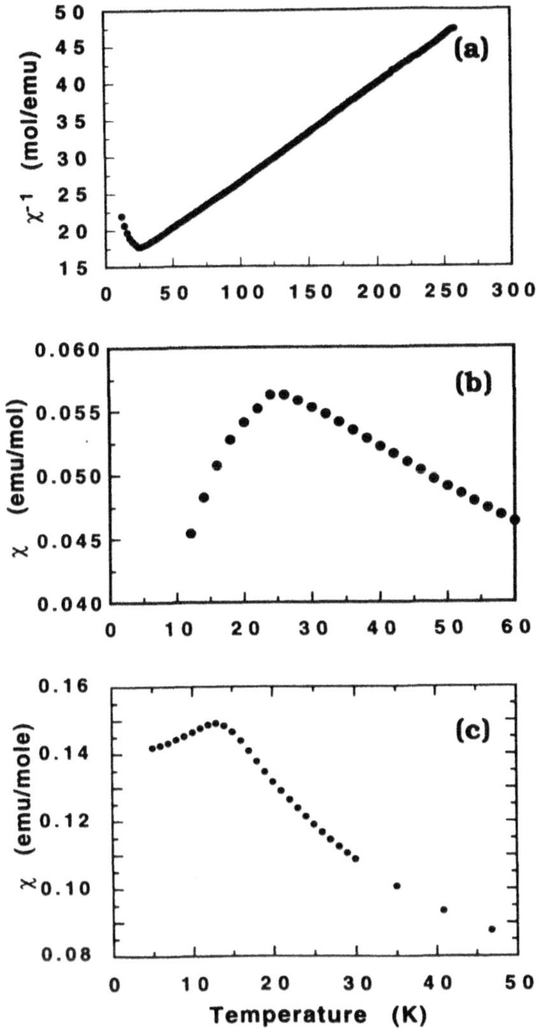


**Figure 7.** (a) A Curie plot of the magnetic susceptibilty data obtained from $Cm_2CuO_4$. The effective moment obtained from these data is consistent with that expected for the $Cm^{3+}$ free ion. (b) An expanded view of the low temperature susceptiblity. The cusp at 25 K is indicative of antiferromagnetic ordering of the Cm moments. This ordering has been confirmed by neutron diffraction. (c) The low temperature magnetic susceptibility of $Cm_{1.87}Th_{.13}CuO_4$. There is no evidence of superconductivity in this sample. The cusp in the susceptibility has moved down to 13 K from 25 K in the parent compound with the introduction of 8.5 % diamagnetic $Th^{4+}$.

with somewhat lower temperatures reported for powder samples synthesized by solid-state techniques. All the samples, including the R=Pr and Tb analogs can be made superconducting except for the R=Ce and Am analogs, which do not superconduct down to the lowest temperatures measured (4 K)[41,44]. The magnetic responses, as a function of temperature, for the R=Ce, Pr, Tb, and Am analogs are shown in Figure 8. The Tb analog has a $T_c$(onset)=71 K, which is typical of other R analogs including R=Gd, Ho, and Er samples that have been made by the same synthetic route. Whereas the R=Pr analog is also superconducting, its $T_c$ is depressed by 10 K relative to the other superconductors in this series.

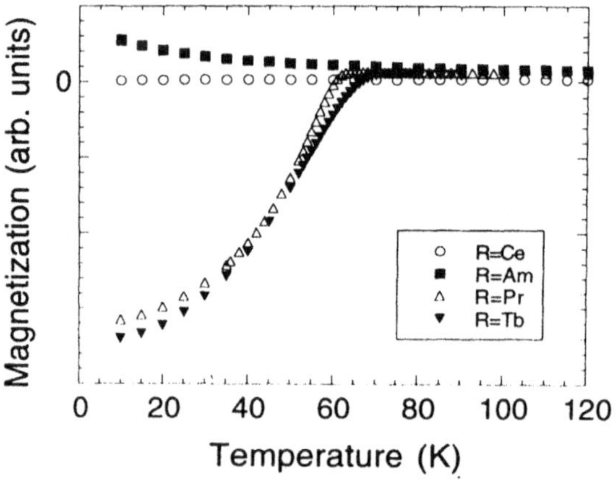


**Figure 8.** The low-field magnetization versus temperature data obtained from polycrystalline compounds of $Pb_2Sr_2R_{.5}Ca_{.5}Cu3O_8$ for R= Ce, Am, Pr, and Tb. These data show superconducting transitions for the R=Pr and Tb samples. There is no evidence of superconductivity in the R=Ce and Am samples.

The Am and Ce compounds do not superconduct down to 5 K, the lowest temperature measured. The structural and electronic behaviours of both of these compounds have been studied in detail [41,44]. A comparative list of the lattice constants and R-O bond distances is provided in Table III. Both the Ce and Am analogs have R-O distances that are shorter than expected by comparison of other ions with similar trivalent ionic radii, suggesting that these

**Table III.** Lattice parameters (a, b, and c), cell volumes, and R-O near neighbour distances in $Pb_2Sr_2RCu_3O_8$. Structural data (except for R=Am) were obtained from refinement of powder neutron diffraction data based on the space group Cmmm[41]. The lattice parameters for the R=Am data were obtained from analysis of powder x-ray data, and the R-O bond distance was obtained from fitting EXAFS spectra[44].

| R | a(Å) | b(Å) | c(Å) | Volume (Å$^3$) | R-O(Å) |
|---|---|---|---|---|---|
| Y | 5.3943(1) | 5.4337(1) | 15.7251(2) | 460.93(1) | 2.3989(1) |
| Ce | 5.4312(1) | 5.4741(2) | 15.8238(4) | 470.45(2) | 2.4857(2) |
| Pr | 5.4449(1) | 5.4836(1) | 15.7982(3) | 471.70(1) | 2.4909(1) |
| Tb | 5.4033(1) | 5.4433(1) | 15.7325(3) | 462.72(1) | 2.4213(1) |
| Am | 5.411(3) | 5.459(4) | 15.822(8) | 467.4(5) | 2.22(7) |

ions may be tetravalent in this host. This suggestion is confirmed by an analysis of the XANES data shown in Figure 9. These edge spectra clearly show that Ce and Am are tetravalent. Magnetic susceptibility data obtained from $Pb_2Sr_2AmCu_3O_8$ support this conclusion. Neither of these samples show any indication of magnetic ordering down to the lowest temperature measured.

The absence of superconductivity in $Pb_2Sr_2Ce_{.5}Ca_{.5}Cu_3O_8$ and $Pb_2Sr_2Am_{.5}Ca_{.5}Cu_3O_8$ can then be understood in terms of charge transfer from Ce or Am to the CuO lattice. The reduced charge on the CuO planes, confirmed by Cu K edge XANES[41] (not shown), can account for the absence of superconductivity. $Ca^{2+}$ substitution, which increases charge on the CuO sublattice, cannot be accomplished to high enough concentrations to sufficiently increase the charge on the CuO planes. It is argued that the

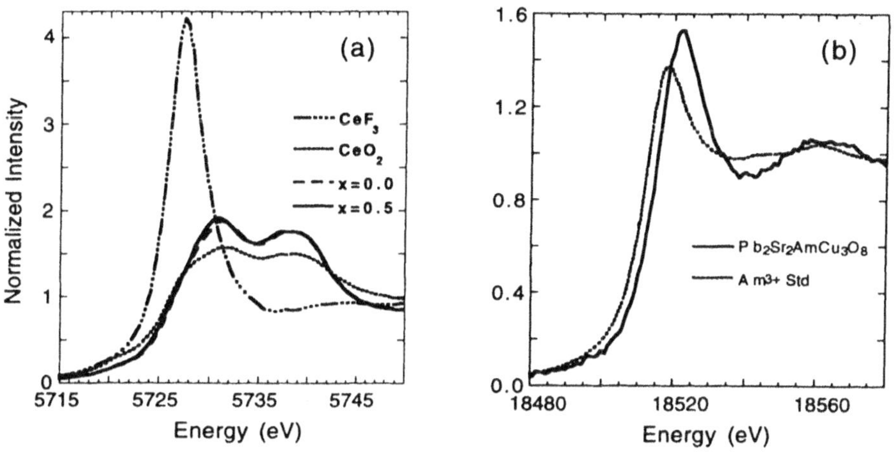


**Figure 9.** (a) A comparison of the Ce $L_3$-edge XANES from a trivalent standard ($CeF_3$) and a tetravalent standard ($CeO_2$) with data from$Pb_2Sr_2Ce_{1-x}Ca_xCu3O_8$ for x=0 and 0.5. The similarity of the edge features of the $CeO_2$ and the two Pb-based samples reveals that Ce is tetravalent in the latter compounds.

concentration of holes on the CuO planes is not sufficient to permit superconductivity. $T_c$ in these two compounds is suppressed by a simple charge-transfer mechanism.

As seen from Figure 8, $Pb_2Sr_2Tb_{.5}Ca_{.5}Cu_3O_8$ is superconducting below 71 K. $L_3$-edge XANES measurements, shown in Figure 10, demonstrate that Tb is trivalent. Inelastic neutron scattering data support this conclusion. Crystal-field modeling of the data, using the crystal field scheme derived from work on the R=Ho and Er analogs[43,45], has been used to determine the splitting of the $Tb^{3+}$ $^7F_6$ ground multiplet[46]. The wavefunctions, energies and symmetries of the crystal-field states are used to calculate the magnetic properties of the Tb analog. The strong anisotropy in the magnetic response from a single crystal of $Pb_2Sr_2TbCu_3O_8$[46] is well reproduced by the calculations. However, the observed $\chi$(parallel) deviates considerably for the calculation at temperatures below about 120 K, as does the powder data, shown in Figure 11. For comparison, the figure also includes data obtained from a 5%Tb in Y sample, $Pb_2Sr_2Tb_{.05}Y_{.95}Cu_3O_8$. It can be seen that the susceptibility data obtained from the 5% Tb compound are very different from those obtained from the pure R=Tb sample, and that the data from the magnetically dilute sample agree very well with the calculated susceptibility. We interpret this behaviour to the effects of Tb-Tb spin interactions that persist to very high temperatures. These interactions are strongly suppressed in the diluted system. Support for this observation comes from the unusual temperature dependence of the INS elastic line [46]. Both the magnetic susceptibility and the INS data show deviations from expected single-ion behaviour at temperatures less than 120 K. Neutron diffraction experiments[47] on the parent compound reveal a quasi two dimensional (2D) phase with finite antiferromagnetic correlations along the c-direction together with a three dimensional phase with ferromagnetic correlations along the c-direction. The co-existence of the two phases is likely to be related to structural imperfections such as stacking faults, strains, oxygen disorder or cation vacancies. The $Pb_2Sr_2Tb_{.5}Ca_{.5}Cu_3O_8$ sample, which has a superconducting critical temperature of 71 K, also exhibits a quasi 2D magnetic ordering of the Tb sublattice with a finite ferromagnetic correlation along the c-direction. These magnetic correlations are more than an order of magnitude higher than seen in any related system, as well as much higher than the two-dimensional long-range ordering temperature of $T_N$=5.3 K[48] for the parent compound. This unusual behaviour is understood in terms of the highly two dimensional nature of the lattice combined with the strongly Ising-like magnetic properties.

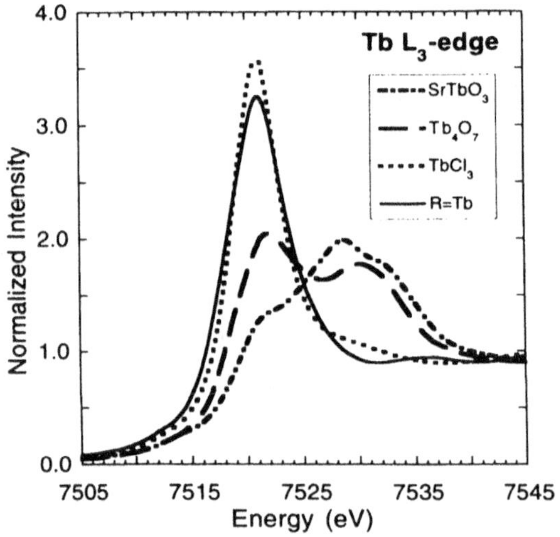


**Figure 10.** Tb $L_3$-edge XANES from $Pb_2Sr_2TbCu_3O_8$ are compared with a trivalent standard ($TbCl_3$) as well as mixed trivalent and tetravalent ($Tb_4O_7$) and tetravalent ($SrTbO_3$) standards. The comparison demonstrates that Tb is trivalent in the Pb-based sample.

The Tb moments in $Pb_2Sr_2TbCu_3O_8$ order at a high temperature relative to the other members of the series, except the Pr analog. For example, the Gd moments order at $T_N$=2.3 K[49]. The relatively high Neel temperature found for the Tb moments in $Pb_2Sr_2TbCu_3O_8$ is understood in terms of the effects of a small molecular field on the quasi-doublet ground state[47]. A large longitudinal transition strength strongly enhances the susceptibility in the c-direction and suppresses the perpendicular susceptibility, relative to an isotopic ion such as Gd, in a similar environment. There appears to be no need in the Tb case for any additional exchange interactions or hybridization to explain its unusually high $T_N$ in these double-layer cuprates.

The Pr analogs of the $Pb_2Sr_2R_{1-x}Ca_xCu_3O_8$ series provide unique information on the role of Pr in suppressing high $T_c$ superconductivity. $Pb_2Sr_2PrCu_3O_8$ is isostructural with the other members of this series. Pr in this compound sits in a site structurally very similar to that of Pr in $PrBa_2Cu_3O_7$. The parent compound exhibits semiconducting behaviour and has a magnetically ordered Pr sublattice below $T_N \approx 5.8K$[50,51], which is a higher $T_N$ than that observed for $Pb_2Sr_2GdCu_3O_8$, but lower than the $T_N$=17K observed for $PrBa_2Cu_3O_7$[26]. The magnetic ordering temperatures of the Gd analogs of these two compounds are comparable. The magnetic structure at low temperature, determined from neutron diffraction experiments[52]. is similar to that observed for the Tb analog The saturation moment of $1.4\pm0.4\mu_B$ for Pr is slightly larger than the $0.79\pm.04\mu_B$ found in $PrBa_2Cu_3O_7$.

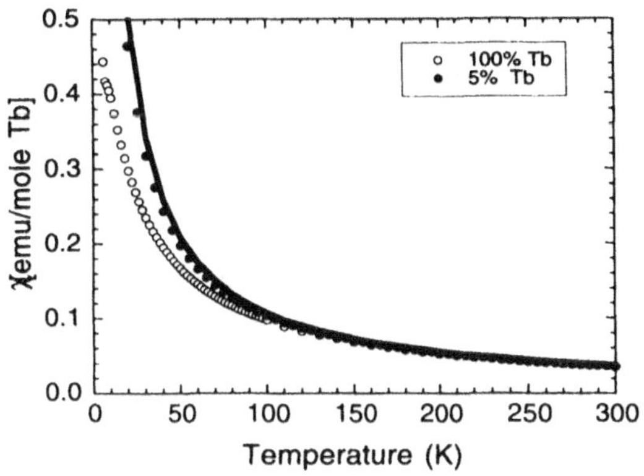


**Figure 11.** A comparison of the susceptibilities measured from powdered samples of $Pb_2Sr_2TbCu_3O_8$ (open circles) and the magnetically dilute sampe $Pb_2Sr_2Tb_{.05}Y_{.95}Cu_3O_8$ (full circles). These data are compared with single-ion CEF calculations (solid line).

The INS data from $Pb_2Sr_2PrCu_3O_8$ reveal the expected similarities for the electronic properties of Pr in this and the $PrBa_2Cu_3O_7$ structure[52]. The magnetic-scattering lines are significantly broadened over that expected from the experimental resolution, as is demonstrated in Figure 4 for $PrBa_2Cu_3O_7$. Overall, the splitting of the $Pr^{3+}$ $^3H_4$ ground level is similar in both materials, including the splitting of the lowest three singlets. The splittings of the $Pr^{3+}$ ground levels for the various compounds of interest herein are depicted schematically in Figure 5. There overall splitting is about 4 meV in both materials, and they are well isolated from higher energy states, resulting in a quasi-triplet ground state, that renormalizes at low temperatures to result in a local, ordered moment on Pr [22,52].

In order to induce superconductivity, it is necessary to replace approximately 20-80% of the $Pr^{3+}$ in $Pb_2Sr_2PrCu_3O_8$ with $Ca^{2+}$. The holes introduced by this substitution reside in the CuO planes, and are the superconducting carriers. $Pb_2Sr_2Pr_{.5}Ca_{.5}Cu_3O_8$ is a superconductor similar to the other R members of this series, except that $T_c$ is about 60K, reduced by 10K from the other members of the series. The long-range ordering of the Pr moments is suppressed by this substitution.

In addition to the effect that the $Ca^{2+}$ substitution has on the electronic properties of the CuO sublattice, it also has an effect on the single-ion electronic properties of the R ions. This has been explicitly demonstrated for Ca substitution in $Pb_2Sr_2ErCu_3O_8$[45], where the Er states are shown to broaden and slightly shift in energy with Ca substitution. This change in the Er states is understood to result from a statistical distribution of Ca second near neighbours, which slightly modifies the CEF potential. A comparison of low energy INS data obtained from undoped and doped $Pb_2Sr_2PrCu_3O_8$, shown in Figure 12, reveals a similar effect in this case. The effect of Ca substitution on the Pr INS data is to broaden the magnetic peaks, and to shift the magnetic scattering of the quasi triplet to higher energies. This increased splitting of the quasi triplet with Ca substitution could be interpreted as effectively reducing the number of magnetic Pr ions, or as an increase of the average CEF splitting and therefore a reduction in the average magnetic moment on the Pr ions. This result is extremely important because it demonstrates that the role of Ca substitution is more complex than initially understood. Not only does Ca substitution alter the number of carriers in the CuO planes, it also alters the single ion magnetic properties of the R ions, which we argue can have a significant influence on the bulk properties exhibited by the material.

Preliminary data are available on $Pb_2Sr_2CmCu_3O_8$. X-ray powder diffraction data show this compound to form as a single phase material, with lattice constants very similar to those obtained from the Pr analog under similar conditions[53]. Susceptibility data as a function of temperature show a cusp at 15 K, which is attributed to antiferromagnetic ordering of the Cm sublattice. Neutron diffraction measurements are planned to confirm this finding.

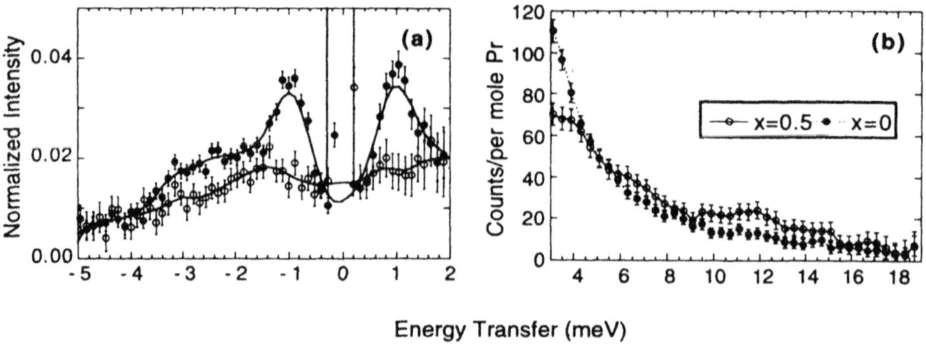


**Figure 12.** A comparison of detail-balance corrected INS energy spectra obtained from $Pb_2Sr_2Pr_{1-x}Ca_xCu_3O_8$ for x=0 and 0.5. Data are normalized per mole Pr ions at 20 K. (a) Low energy transfer region shows sharply lower intensity for the x=0.5 sample. (b) Magnetic scattering of the x=0.5 sample is seen to extend out to higher energies. These data indicate that the ground state splitting of Pr in this materials is influenced by the incorporation of Ca. The energy splitting of the pseudo-triplet ground state, about 4 meV for the x=0 compound, is expanded to about 14 meV in the Ca-doped sample

## DISCUSSION AND CONCLUSIONS

The results of this systematic study on the Ce, Pr, Tb, Am and Cm analogs of the three superconducting series $RBa_2Cu_3O_7$, $R_{2-x}M_xCuO_4$, and $Pb_2Sr_2R_{1-x}Ca_xCu_3O_8$ are summarized in Table IV. These results can be divided into 4 main categories: (1) compounds that do not form, i.e. $CeBa_2Cu_3O_7$ and $TbBa_2Cu_3O_7$; (2) compounds in which R is tetravalent, i.e. $Pb_2Sr_2Ce_{1-x}Ca_xCu_3O_8$ and $Pb_2Sr_2Am_{1-x}Ca_xCu_3O_8$. $T_c$ is suppressed in these compounds; (3) compounds which form the correct phase, R is essentially trivalent, and $T_c$ is suppressed, i.e. $PrBa_2Cu_3O_7$, $CmBa_2Cu_3O_7$, $Cm_2CuO_4$ and $Pb_2Sr_2CmCu_3O_8$; and (4) superconducting compounds, i.e. $Pr_{1.82}Ce_{.18}CuO_4$, $Pb_2Sr_2Pr_{.5}Ca_{.5}Cu_3O_8$, and $Pb_2Sr_2Tb_{.5}Ca_{.5}Cu_3O_8$. It is informative to compare and contrast these categories.

**Table IV**. A summary of the work presented herein. For full references, see text.

| | **$Y_{1-x}R_xBa_2Cu_3O_7$**<br>**$T_c \approx 92$ K** | **$R_{1.85}M_{0.15}CuO_4$**<br>**$T_c \approx 20$ K** | **$Pb_2Sr_2R_{1-x}Ca_xCu_3O_8$**<br>**$T_c \approx 80$ K** |
|---|---|---|---|
| **Ce** | 4+ (x < 0.1)<br>suppresses $T_c$ | Does not form | 4+<br>suppresses $T_c$ |
| **Pr** | 3+ (x ≤ 1)<br>suppresses $T_c$<br>$T_N = 17$ K | 3+<br>normal $T_c$ | 3+<br>lower ($T_c$=60K)<br>$T_N = 6$ K |
| **Tb** | 3+ (x = 0.3)<br>normal $T_c$ | Does not form<br>isostructural phase | 3+<br>normal $T_c$<br>$T_N = 5.3$ K |
| **Am** | $BaAmO_3 + Cu_2O$ | $AmO_2 + Cu_2O$ | 4+<br>suppresses $T_c$ |
| **Cm** | 3+ (x ≤ 1)<br>suppresses $T_c$<br>$T_N = 21$K | 3+<br>suppresses $T_c$<br>$T_N = 25$ K | 3+<br><br>($T_N = 15$ K) |

There are several compounds in which $T_c$ appears to be suppressed by the f-ion. For those compounds in which R is tetravalent, $Pb_2Sr_2Ce_{1-x}Ca_xCu_3O_8$ and $Pb_2Sr_2Am_{1-x}Ca_xCu_3O_8$, the mechanism for $T_c$ suppression is expected to be one of simple charge transfer and resultant CuO hole filling. This proposed mechanism is supported by experimental results for $Pb_2Sr_2Ce_{1-x}Ca_xCu_3O_8$[41].

For the case in which R is trivalent but $T_c$ is suppressed, the situation is much more complex, but several trends emerge. This category is restricted to samples that contain either Pr or Cm. These ions have similar ionic radii, similar 4+/3+ redox couples (see Table II), and they both have f-orbitals whose radial extent permits limited hybridization with neighbouring ligand states. $Cm^{3+}$ ($5f^7$) is an s-state ion and therefore it has a large magnetic moment that is unaffected to first order by crystal-field effects[32]. Cm suppresses superconductivity in all the compounds in which it is incorporated. In addition to suppressing superconductivity, the local moments on Cm also exhibit long-range ordering at temperatures an order of magnitude higher than superconducting members of the same series, as shown in Table IV.

$Pr^{3+}$ also suppresses superconductivity. It has a $4f^2$ configuration with single-ion magnetic properties that depend significantly on the details of the crystal field magnitude and symmetry. For orthorhombic or lower site symmetry the $^3H_4$ ground level is split into 9 singlets. The details of this splitting play a central role in determining the single-ion and collective properties of samples containing Pr. The relative energies of the singlet states in $PrBa_2Cu_3O_x$ and $Pb_2Sr_2PrCu_3O_8$ are such that there is an effective ground-state triplet that is isolated from other, higher energy singlets. The result is that $Pr^{3+}$ has an effective moment that is reduced from its free-ion value, and an ordered saturation moment at lower temperatures. Like Cm, the Pr analogs of these two compounds have Neel temperatures that are nearly an order of magnitude higher than those for the superconducting members of these series. In contrast to this behaviour, $Pr_2CuO_4$ has an isolated singlet ground state, no local moment at low temperature, and it is superconducting when appropriately doped. These results suggest that Pr suppresses superconductivity only when it has a local moment.

The influence of a Pr local moment on superconductivity is explored with the work on $Pb_2Sr_2Pr_{1-x}Ca_xCu_3O_8$. The data demonstrate that when x=0, the Pr ground multiplet is similar to that found in $PrBa_2Cu_3O_x$. The Pr-Pr magnetic interactions are sufficiently strong that magnetic ordering is induced at 6K in the undoped x=0 compound; an ordering temperature comparable to that found for $PrBa_2Cu_3O_6$ with $T_N \approx 10K$[54]. When Ca replaces some Pr in the Pb-based compound, the average splitting of the $Pr^{3+}$ pseudo-triplet is increased, as evidenced by INS data shown in Figure 12. We argue that $Pb_2Sr_2Pr_{0.5}Ca_{0.5}Cu_3O_8$ superconducts, but $T_c$ remains slightly suppressed because the pseudo-triplet splitting is not large enough to remove all moment from the ground state.

This analysis of the electronic and magnetic properties of Pr in the $RBa_2Cu_3O_7$, $R_{2-x}Ce_xCuO_4$, and $Pb_2Sr_2R_{1-x}Ca_xCu_3O_8$, taken together with the data on the Cm analogs of these materials provide a platform on which to build a fundamental understanding of the selective suppression of high-$T_c$ superconductivity. The relevance of the magnetic ground state, specifically the size of the quasi-triplet ground-state splitting, to $T_c$ suppression may affect the f-state hybridization with the CuO band states. It is possible that an increased quasi-triplet splitting may inhibit an effective hybridization between the 4f- and the CuO states. A non-magnetic singlet may not be able to hybridize effectively with the CuO states. Such effects are well known in Kondo systems, in which the CEF energy splitting can compete with the Kondo hybridization[55]. Of the f-ions that form superconducting or related phases, Pr and Cm have the most extended occupied f-orbitals, relative to the outer p- and d-orbitals, which allows them to most effectively hybridize with the CuO states. However, this alone is not sufficient to suppress superconductivity, as evidenced by the superconductivity observed for $Pr_{1.85}Ce_{.15}CuO_4$. In addition, there must be unpaired spins in the hybridizing states. We argue that it is the magnetic moments in hybridizing states that interact with the CuO states to suppress $T_c$. Whereas to date there is no direct information on the hybridization strength in these systems, this argument is strongly supported by the high magnetic ordering temperatures observed for all Pr and Cm compounds that do not superconduct.

The theory of hole localization proposed by Fehrenbacher and Rice[4] can explain selected x-ray absorption and optical reflectivity data. However, the proposed admixture of 50% $Pr^{+3}$ ($4f^2$) and $Pr^{+4}$ ($4f^1$) is not consistent with the overall INS results. The magnetic data from Pr is well understood in terms of $Pr^{3+}$ ions only. The Cm data are even more definitive in this regard ($\mu_{eff}(Cm^{3+}) = 7.94\mu_B$; $\mu_{eff}(Cm^{4+}) = 0\mu_B$). Certainly it has not been demonstrated how a mixing of $Pr^{3+}$ and $Pr^{4+}$ can explain the high ordering temperatures of the Pr and Cm moments. In contrast, the theory developed by Guo and Temmerman[6] is consistent with the results discussed herein.

Of the 4 mechanisms often proposed to explain the suppression of superconducting in $PrBa_2Cu_3O_7$, it is clear that simple charge transfer is not sufficient Although a full understanding of the fundamental mechanism for the suppression of $T_c$ in these systems is yet to be attained, the current results clearly point toward the central role played by magnetic interactions. Further progress on this problem may necessitate an understanding of interactions responsible for high temperature superconductivity.

## ACKNOWLEDGMENTS

The authors wish to thank C.W. Williams, S. Skanthakumar, G.L. Goodman, M.R. Antonio and C.-K. Loong for their contributions to aspects of this project. This work is supported by the U.S. DOE - Basic Energy Sciences, Chemical Sciences, under contract W31-109-ENG-38.

## REFERENCES

1L. F. Schneemeyer, J. V. Waszczak, S. M. Zahurak, R. B. van Dover, and T. Siegrist, Mater. Res. Bull. **22**, 1467 (1987).

2S.E. Brown, J.D. Thompson, J.O. Willis, R.M. Aikin, E. Zirngiebl, J.L. Smith, Z. Fisk, and R.B. Schwarz, Phys. Rev. B **36**, 2298 (1987).

3L. Soderholm, K. Zhang, D. G. Hinks, M. A. Beno, J. D. Jorgenson, C. U. Segre, and I. K. Schuller, Nature **328**, 604 (1987).

4R. Fehrenbacher and T. M. Rice, Phys. Rev. Lett. **70**, 3471 (1993).

5A. I. Liechtenstein and I. I. Mazin, Phys. Rev. Lett. **74**, 1000 (1995).

6G. Y. Guo and W. M. Temmermann, Phys. Rev. B **41**, 6372 (1990).

7U. Staub and L. Soderholm, _Handbook on the Physics and Chemistry of Rare Earths_ (in the press).

8M.B. Maple, L.E. DeLong, and B.C. Sales, "Kondo Effect: alloys and compounds," in _Handbook on the Physics and Chemistry of the Rare Earths_, edited by Karl A. Gschneidner and LeRoy Eyring (Elsevier, Amsterdam, 1978), Vol. 1, pp. 797.

9Peter Fulde, _Electron Correlations in Molecules and Solids_, 3rd ed. (Springer, Berlin, 1995).

10L R Morss, "Comparative Thermochemical and Oxidation-Reduction Properties of Lanthanides and Actinides," in _Handbood on the Physics and Chemistry of Rare Earths_, edited by K A Gschneidner, Jr., L Eyring, G R Choppin, and G H Lander (Elsevier Science, Amsterdam, 1994), Vol. 18, pp. 239.

11R.D. Shannon, Acta Crystallogr. A **32**, 751 (1976).
12Darleane Hoffman, private communication.
13Barry T. Kilbourn, _A Lanthanide Lanthology_ (Molycorp, Inc., White Plains, 1993).
14C. R. Fincher, Jr., and Graciela B. Blanchet, Phys. Rev. Lett. **67**, 2902 (1991).
15U. Staub, L. Soderholm, S. Skanthakumar, and M.R. Antonio, J. Phys. IV France 7 **C2**, 1077 (1997).
16K. I. Gnanasekar, A. S. Tamhane, R. Pinto, R. Nagarajan, M. Sharon, L. G. Gupta, and R. Vijayaraghavan, Physica C **219**, 183 (1994).
17U. Staub, Mark R. Antonio, L. Soderholm, M. Guillaume, W. Henggeler, and A. Furrer, Phys. Rev. B **50**, 7085 (1994).
18Y. Dalichaouch, M. S. Torikachvili, E. A. Early, B. W. Lee, C. L. Seaman, K. N. Yang, H. Zhou, and M. B. Maple, Solid State Comm. **65**, 1001 (1988).
19H.B. Radousky, J. Mater. Res. **7**, 1917 (1992).
20L. Soderholm, C.-K. Loong, G. L. Goodman, and B. D. Dabrowski, Phys. Rev. B **43**, 7923 (1991).
21S. Skanthakumar, W-H. Li, J. W. Lynn, A. Kebede, J. E. Crow, and T. Mihalsin, Physica B **163**, 239 (1990).
22G.L. Goodman, C.-K. Loong, and L. Soderholm, J. Phys.: Condens. Matter **3**, 49 (1991).
23H.-D. Jostarndt, U. Walter, J. Harnischmacher, J. Kalenborn, A. Severing, and E. Holland-Moritz, Phys. Rev. B **46**, 14872 (1992).
24V. Nekvasil, E. Holland-Moritz, H.-D. Jostarndt, U. Walter, and G. Hilscher, J. Magn. Magn. Mater. **117**, 11 (1992).
25G. Hilscher, E. Holland-Moritz, T. Holubar, H.-D. Jostarndt, V. Nekvasil, G. Schaudy, U. Walter, and G. Fillion, Phys. Rev. B **49**, 535 (1994).
26W.-H. Li, J. W. Lynn, S. Skanthakumar, and T. W. Clinton, Phys. Rev. B **40**, 5300 (1989).
27S. Skanthakumar, J.W. Lynn, N. Rosov, G. Cao, and J.E. Crow, Phys. Rev. B **55**, R3406 (1997).
28J.J. Neumeier, T. Bjornholm, M.B. Maple, and Ivan K. Schuller, Phys.Rev. Lett. **63**, 2516 (1989).
29David P. Norton, D.H. Lowndes, B.C. Sales, J.D. Budai, B.C. Chanourmakos, and H.R. Kerchner, Phys.Rev.Lett. **66**, 1537 (1991).
30U. Staub, L. Soderholm, S. Skanthakumar, R. Osborn, and F. Fauth, Europhys. Lett. **39**, 663 (1997).
31L. Soderholm, G.L. Goodman, U. Welp, C.W. Williams, and J. Bolender, Physica C **161**, 252 (1989).
32G.K. Liu, J.V. Beitz, and Jin Huang, J. Chem. Phys. **99**, 3304 (1993).
33R. Saez Puche, M. Norton, T.R. White, and W.S. Glaunsinger, J. Solid State Chem. **50**, 281 (1983).
34H. Takagi, S.Uchida, and Y. Tokura, Phys.Rev.Lett **62**, 1197 (1989).
35M. Matsuda, K. Yamada, K. Kakurai, H. Kadowaki, T.R. Thurston, Y. Endoh, Y. Hidaka, R.J. Birgeneau, M.A. Kastner, P.M. Gehring, A.H. Moudden, and G. Shirane, Phys.Rev.B **42**, 10098 (1990).
36I.W. Sumarlin, S. Skanthakumar, J.W. Lynn, J.L. Peng, Z.Y. Li, W. Jiang, and R.L. Greene, Phys.Rev.Lett. **68**, 2228 (1992).

37J.D. Thompson, S.-W. Cheong, S.E. Brown, Z. Fisk, S.B. Oseroff, M. Tovar, D.C. Vier, and S. Schultz, Phys.Rev. B **39**, 6660 (1989).
38L. Soderholm, C.W. Williams, and U. Welp, Physica C **179**, 440 (1991).
39L. Soderholm, C.W. Williams, and S. Skanthakumar, (to be published).
40S. Skanthakumar and L. Soderholm, unpublished .
41S. Skanthakumar and L. Soderholm, Phys. Rev. B **53**, 920 (1996).
42U. Staub, L. Soderholm, S. Skanthakumar, P. Pattison, and K. Conder, Phys.Rev.B **57**, 5535 (1998).
43L. Soderholm, C.-K. Loong, J. S. Xue, J.P. Hammonds, J.E. Greedan, and M. Maric, J. Appl. Phys. **73**, 6314 (1993).
44L. Soderholm, Clayton Williams, S. Skanthakumar, Mark R. Antonio, and S. Conradson, Z. Phys. B **101**, 539 (1996).
45L. Soderholm, C.-K. Loong, U. Staub, S. Skanthakumar, J. S. Xue, J. P. Hammonds, J. E. Greedan, and M. Maric, Physica C **246**, 11 (1995).
46U. Staub, L. Soderholm, S. Skanthakumar, and Mark. R. Antonio, Phys. Rev. B **52**, 9736 (1995).
47U. Staub, L. Soderholm, S. Skanthakumar, S. Rosenkranz, C. Ritter, and W. Kagunya, Z. Phys. B. **104**, 37 (1997).
48C. R. Shih, T. H. Meen, Y. C. Chen, and H. D. Yang, Phs. Rev. B **50**, 9619 (1994).
49C.C. Lai, J.H. Shieh, B.S. Chiou, J.C. Ho, and H.C. Ku, Phys. Rev. B **49**, 1499 (1994).
50W. T. Hsieh, W-H. Li, K. C. Lee, J. W. Lynn, J. H. Shieh, and H. C. Ku, J. Appl. Phys **76**, 7124 (1994).
51J. H. Shieh, H. C. Ku, and J. C. Ho, Phys. Rev. B **50**, 3288 (1994).
52U. Staub, L. Soderholm, S. Skanthakumar, R. Osborn, F. Fauth, and C. Ritter, (to be published).
53L. Soderholm, C.W. Williams, and S. Skanthakumar, (unpublished).
54A. Kebede, J.P. Rodrigqquez, I. Perez, T. Mihalsisin, G. Myer, J.E. Crow, P.P. Wise, and P.Schlottmann, J. Appl. Phys. **69**, 5376 (1991).
55S. Watanabe and Y. Kuramoto, Z. Phys. B **104**, 535 (1997).

## Discussion

**Sawatzky**: Since it seems that Pr influences the superconducting transition in the high Tc's in very different ways in the different structures, it would be interesting to try to relate this to the local oxygen coordination and the role in this played by the CuO plane Oxygen ions.

**Solderholm:** Well, that may be so. We have not looked at this in detail, but all the crystallographic information is available to do this.

# AN INVESTIGATION OF THE MAGNETIC FLUCTUATIONS ABOVE AND BELOW $T_c$ IN THE HEAVY FERMION SUPERCONDUCTOR $UPd_2Al_3$

N. Bernhoeft[#]

Institut Laue Langevin, BP 156X, F-38042 Grenoble, France

**Abstract**
The heavy fermion superconductor $UPd_2Al_3$ exhibits the unusual combination of an antiferromagnetic phase transition, at $T_N = 14.3$ K, followed by a superconducting phase transition below 2 K without destruction of the ordered magnetic moment. In the superconducting phase the low energy magnetic inelastic response is qualitatively of a different form from that above $T_c$. We review this low energy response.

## Introduction

The interest in strongly correlated electron metals has been heightened by the discovery of the co-existence of magnetism and superconductivity in selected members of this class. In particular, the advent of both heavy-fermion and high-$T_c$ superconducting materials has given rise to the active debate that internal (dynamic) magnetic fields may play a crucial role in the formation of novel superconducting phases. Inelastic neutron scattering techniques permit one to investigate spatial and temporal magnetic correlations in a wide variety of materials, the principal limitation being the available signal to noise ratio. The heavy fermion superconductor, $UPd_2Al_3$ [1], characterised by an initial phase transition, at $T_N = 14.3$ K, into an antiferromagnetic state and a second transition, at $T_c$= 2K, into a state of coexisting superconductivity and antiferromagnetism, appears to be a favourable case in which to study the interplay of superconductivity and magnetism through neutron scattering techniques on three accounts. First, it has a significant magnetic moment at low temperatures, ~ 0.85 $\mu_B$ per uranium ion; second, $T_c$ sets the energy scale in a region accessible to high resolution neutron spectrometers, and third, $UPd_2Al_3$ is congruently melting which allows sizable single crystals to be made, for example, by the Czochralski method.

The compound crystallises in the hexagonal $PrNi_2Al_3$ structure (space group *P6/mmm*) with lattice constants $a$ = 5.350Å and $c$ = 4.185Å at room temperature, Fig. 1. Characterisation by specific heat measurements reveals not only a strong enhancement of the linear term at low temperatures, characteristic of the heavy fermion state, but also two

---

[#] Present address: CEA Grenoble, F-38054 Grenoble, France

distinct anomalies associated with the antiferromagnetic and superconducting phase transitions respectively[1]. The large specific heat jump at $T_c$, $\Delta C = 1.2\gamma T_C$ ($\gamma = 140\ mJ / mole\ K^2$)[1,2] suggests that the superconducting ground state evolves out of interactions between quasiparticles located in strongly renormalised states in a low energy shell around the Fermi surface.

For the appearance of a collective paired state, as in the generic BCS theory, a primary ingredient is that the quasiparticles present a sufficiently high density of states such that the energy gain of condensing into the superconducting state is not offset by a cost in kinetic energy. We analyse results of inelastic neutron scattering experiments carried out at the Institut Laue-Langevin, Grenoble, France, (ILL) that suggest the primary fluctuations implicit in the low frequency thermodynamics of this material to be of antiferromagnetic character. The major experimental results on the antiferromagnetic fluctuation spectrum thus far established by high resolution neutron scattering techniques, and arrived at independently at the ILL[3], Grenoble and the Advanced Science Research Centre, JAERI, Tokai, Japan[4], may be summarised as follows: first, in the normal antiferromagnetic phase, the existence of a spin wave gap at the magnetic zone centre, $\mathbf{Q}_o$ = (0 0 .5), together with a distinct low energy mode which lies in this spin gap. Second, in the superconducting antiferromagnetic phase, that the spin wave pole changes little on passing below $T_c$ whilst the low energy mode changes in a qualitative manner.

There have been many investigations to find signs of a connection between antiferromagnetism and superconductivity in this compound since its discovery. Experiments on the pressure dependence of the heat capacity in the superconducting phase have been interpreted as implying the co-existence of two different subsystems of more or less localised quasiparticles of $5f$ symmetry[5]. Roughly speaking, one sub-system is the more itinerant in character and gives rise to superconducting properties whilst the other behaves, in some sense, as a more localised component. There is assumed to exist an (unspecified) interaction between the two parts. The idea, that the properties of the $5f$ quasiparticles be considered in terms of two interacting sub-systems, has also been raised based on the analysis of, for example, muon spin rotation experiments. These experiments have been used to assess the reduction of the $5f$ susceptibility in the superconducting state. An approximately isotropic reduction is found which is to be distinguished from the *an*isotropic total susceptibility[6]. The isotropic component has been associated with the itinerant / superconducting behaviour and the residual anisotropic part with a more localised behaviour. Similar conclusions have been drawn from $Al^{27}$ Knight shift data[7]. These empirical ideas have found support in calculations of the $UPd_2Al_3$ Fermi surface by the local spin density[8] and augmented plane wave approximations[9] used in the interpretation of de Haas-van Alphen spectra. This dual magnetic character of an electronic state has its precedents; for example, the anomalous properties of iron in the vicinity of the Curie temperature have been interpreted in analogous terms of a 'long lived local moment' in the presence of an itinerant contribution, the two parts representing different facets of the same electronic states[10,11]. In heavy electron systems the idea has been advanced by, for example, Kuromoto and Miyake[12], and most recently by Pépin and Lavagna[13].

Microscopic details of the magnetic phase, revealed through scattering experiments, show that $UPd_2Al_3$, in addition to its elevated $T_c$, has an unusually large low temperature ordered moment when compared with the other uranium based heavy fermion superconductors such as $UPt_3$ ($0.02\mu_B$)[14], $URu_2Si_2$ ($0.03\mu_B$)[15] or $UNi_2Al_3$ ($0.1\mu_B$)[16]. The ordered moments in $UPd_2Al_3$, which are associated with the uranium ions, are coupled ferromagnetically and lie in the basal hexagonal plane, Fig. 1. These ferromagnetic planes are then stacked antiferromagnetically along the $c$-axis with a wave vector $\mathbf{Q}_o$ = (0 0 0.5)[17-19]. Changes in antiferromagnetic Bragg peak intensities on passing through $T_c$ were reported[20], however, these results were questioned by independent measurements using both neutron diffraction and resonant magnetic $x$-ray scattering[4,18,21]. There are no reported studies of the width of the antiferromagnetic Bragg peaks on passing below $T_c$ and it is taken that a long range antiferromagnetic state coexists with the superconducting phase. This conclusion is corroborated by μSR data[22] where the $\mu^+$ depolarisation rate and lineshape are unchanged below $T_c$ indicating that the magnetisation density distribution is not affected by the onset of superconductivity. That is to say, within the experimental resolution, neither the magnitude of the local magnetisation density at the muon site is changed nor is there a volume segregation of magnetism and superconductivity, since this would produce two, inequivalent spin relaxation sites. Concerning dynamic features of the magnetic susceptibility, pioneering inelastic neutron scattering experiments under relatively coarse energy resolution revealed the existence of defined, although heavily damped, spin waves with no apparent changes in the spectral form around the antiferromagnetic zone centre on cooling through $T_c$[23]. More recent work, referred to above, has raised the possibility of a gap in the spectrum of antiferromagnetic excitations at low temperatures in the superconducting state[3,4].

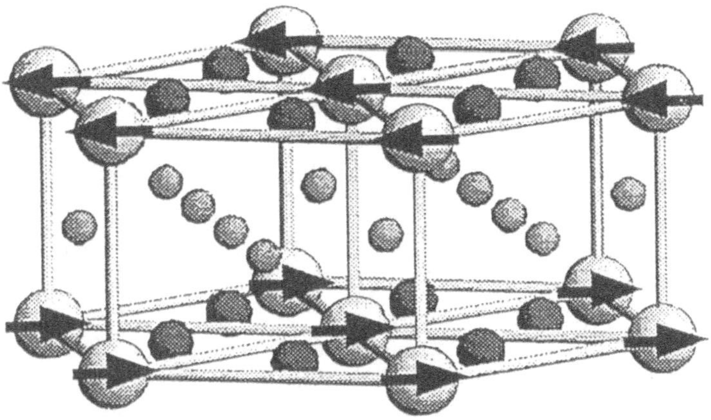

**Figure 1.** The crystal structure of $UPd_2Al_3$. The large circles represent the positions of uranium ions with the bold arrows marking the relative directions of the magnetic moments. The smaller circles in the same planes represent the positions of the palladium ions, and, the smallest circles in the intercalating plane represent the aluminium ions.

## Experimental

The sample was pulled from a carefully homogenised melt of high purity elements by the Czochralski method with a nominal composition of $UPd_{2.02}Al_{3.03}$[24,25], and had a superconducting transition temperature of $1.90 \pm 0.07K$. It was mounted with the c*-a* axes in the scattering plane of the IN14 cold source spectrometer at the ILL, Grenoble, operated variously with graphite monochromator, graphite analyser and a beryllium filter or using a graphite monochromator with polarising bender mirror and Heusler analyser in the polarisation studies. The spectrometer was optimised in all cases for low energy scattering around $\mathbf{Q}_0$ with a typical resolution half width half maximum (*hwhm*) of 45 μeV in energy and 0.025 $Å^{-1}$ in wave vector. Uncertainties in calibration, made with a standard vanadium sample, lead to errors in estimation of the absolute values of the parameters used to characterise the spectra of the order of 20%.

An overview of the spectral response (taken with an unpolarised neutron beam) at $\mathbf{Q}_{0}$ as a function of neutron energy transfer, $\hbar\omega$, is given in Fig. 2. The four temperatures in the left hand frame represent, on a logarithmic vertical scale, the interval around $T_N$ and the normal antiferromagnetic state. For clarity of presentation, data at different temperatures are drawn with displaced axes indicated by the dashed lines. Starting at 14 K one has, within the resolution of the spectrometer, a quasielastic response of Lorentzian form centered around

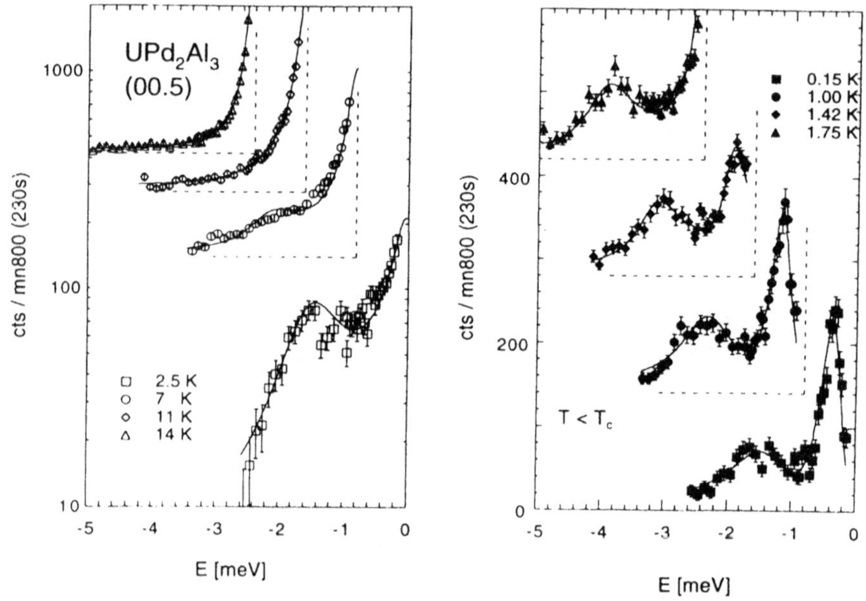


**Figure 2.** The temperature dependence of scattering at (0 0 .5) as a function of neutron energy transfer. In the left hand frame, the scattering in the normal antiferromagnetic phase is dominated by a strong low energy mode which, within our resolution is quasielastic. The solid lines are calculated within the framework outlined in the text. Note the logarithmic vertical scale. The data have been displaced by 0.8 meV horizontally and 140 cts vertically for successive temperatures above 2.5 K. In the right hand frame, the scattering in the superconducting antiferromagnetic phase. Again the spectra have been displaced in a similar manner to the left hand frame. The strong, low energy, feature below $T_c$ is clearly seen to have a qualitatively different form from the response in the normal antiferromagnetic phase. The solid lines are model calculations outlined in the text including the experimental resolution in the energy momentum plane of the experiment. As described in the text the model includes a Lorentzian centred at finite frequency for both the spin wave and the pair breaking response together with a quasielastic mode which only becomes significant near $T_c$.

those reciprocal lattice vectors which, in the antiferromagnetic phase, become resolution limited Bragg peaks. The solid lines are fits to a model to be described later. On cooling, the response broadens out in energy as the spin wave mode develops, however a strong residual quasielastic intensity below 0.5 meV energy transfer remains in the normal state. The separation of the spin wave and quasielastic modes becomes more distinct at lower temperatures, for example at 2.5 K. This arises on two accounts, the spin wave gap increases with decreasing temperature and, at the same time, the quasielastic response becomes tighter in energy transfer on account of the falling Bose factor. In the right hand frame, also with displaced axes but on a linear vertical scale, we show scattering in the neighbourhood of $T_c$ and the low temperature superconducting antiferromagnetic phase. The scattering at 1.75 K, just below $T_c$, is very similar to that at 2.5 K. However, on further cooling into the superconducting phase, a distinct, qualitative change in the low energy response occurs with a peaking of the response at finite energy transfer.

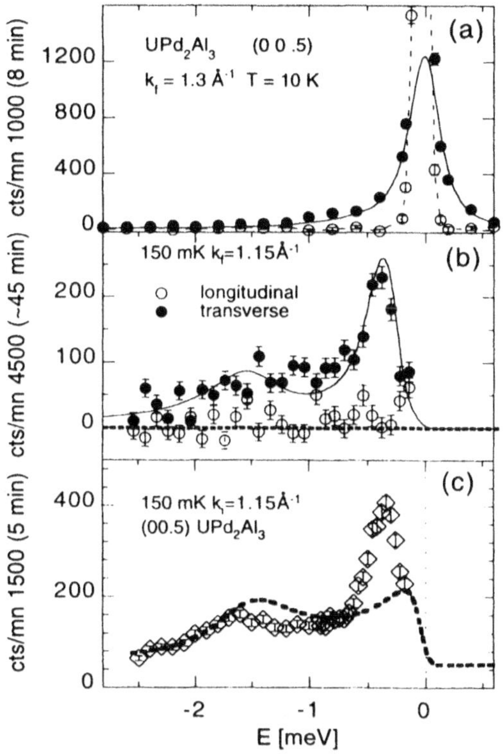


**Figure 3.** Experimental data from $UPd_2Al_3$ at the antiferromagnetic wave vector $Q_0$. The value of the fixed outgoing wave vector is given in each frame as is the temperature. (a) and (b) are taken with polarised neutrons and the transverse (longitudinal) response is shown as solid (open) circles. A background of 6 cts/mn 1000 has been subtracted and the data corrected for the measured flipping ratio. In (b) the solid line is a scaled plot of the unpolarised data at 150 mK (frame c). The dashed line is the zero level. The lower panel (open diamonds) is taken with an unpolarised incident beam. The dashed line is the calculated response in the normal antiferromagnetic state at this temperature.

To make an analysis of the data, it is important to establish that the low energy and spin wave parts of the spectral form may reasonably be taken to represent two aspects of the 5*f* response at all temperatures. This we have done by (a) noting that they have a common polarisation, see top two frames Fig. 3[26,27] together with (b) the observation that the scattering of the two modes scales in an identical manner around different magnetic reciprocal lattice vectors, Fig. 4. This latter scaling would not be anticipated if the low energy feature were associated with a lattice mode or an electronic state of different symmetry.

For the polarisation studies, reported in Fig. 3, the sample was field cooled at 2.5 Tesla into the antiferromagnetic phase after which the field was lowered to 0.8 Tesla. Under these conditions the antiferromagnetic peak at $\mathbf{Q}_o$ had an intensity 15 times lower than that observed in the similar configuration without polarisation analysis. The flipping ratio at the antiferromagnetic peak was 2.5 in agreement with the magnetic phase diagram[18] in which the sub-lattice moments lie at 30 degrees to the horizontal under the given conditions of temperature and applied field. The geometric constraints allow one to identify the fluctuations as being transverse to the sub-lattice magnetisation and lying in the crystallographic (*a-b*) plane. Data are shown at 10 K, Fig. 3*a*, with a background of 6 cts/mn 1000 subtracted and corrected for the measured flipping ratio. In the superconducting state at 150 mK the results indicate that the new, low energy, feature maintains the same polarisation. This is shown in frame 3*b* where the strong low energy peak can be seen to join smoothly with the spin wave at higher energies. From the solid line, which is a scaled copy of data collected without polarisation analysis (no applied field) in the same spectrometer configuration, *e.g.* frame 3*c*,

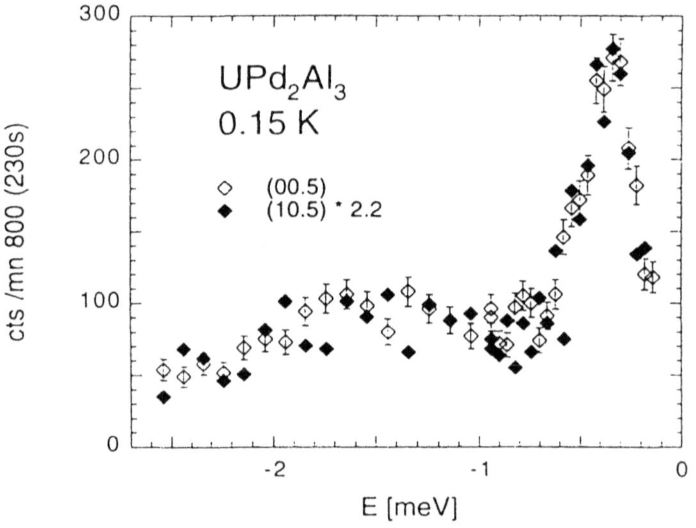


**Figure 4.** The scattered intensity at 150 mK versus neutron energy transfer at (0 0 .5) and (1 0 .5). The scattering at (1 0 .5) having been multiplied by a constant factor matches that at (0 0 .5) over both the spin wave and low energy feature. This, together with the results of polarisation analysis (Fig. 3) indicate that it is reasonable to associate the low energy scattering with the 5*f* electronic response.

one sees that the field necessary for polarisation analysis does not significantly distort the spectrum. The polarisation of the new low energy mode is important. The continuous decay of scattering intensity which occurs on cooling in the normal antiferromagnetic phase is replaced by a qualitatively new feature at the lowest temperatures. This scattering, having the same polarisation and form factor as the spin wave and the (vanishing) quasielastic mode, can reasonably be associated with the $5f$ electronic response. Having established the nature of the scattering, high resolution studies to monitor the temperature evolution from 150 mK to 20 K were carried out with the full intensity of the unpolarised beam.

**Discussion T > T$_c$**

The analysis follows a phenomenological approach. We consider the $f$ response to be represented by two subsystems with starting susceptibilities, $\chi_1$ and $\chi_2$, in which all internal interactions have been included apart from their mutual coupling which is effected through a mean field parameter $\lambda$[28].

$$M_1 = \chi_1[H + \lambda M_2] \tag{1a}$$

$$M_2 = \chi_2[H + \lambda M_1] \tag{1b}$$

$$M = M_1 + M_2 \tag{2}$$

and hence

$$\chi = \frac{\chi_1 + \chi_2 + 2\lambda\chi_1\chi_2}{1 - \lambda^2\chi_1\chi_2} \tag{3}$$

For the dissipative component of the total susceptibility, as $\omega \to 0$, this yields a denominator

$$1 - \lambda^2 \operatorname{Re}[\chi_1]\operatorname{Re}[\chi_2]. \tag{4}$$

Since, in the limit of low frequencies, the real part of $\chi$ tends to a finite value[29] the mutual coupling leads to a low frequency enhancement of the total susceptibility and a concomitant renormalisation of the effective low energy linewidth. At high frequencies the bare susceptibilities vanish[29] and the denominator goes to unity.

The low frequency part, $\chi_1$, in the normal antiferromagnetic state is represented as a quasielastic excitation.

$$\chi_1^{-1}(\bar{\mathbf{q}},\omega) = \chi_1^{-1}(\bar{\mathbf{q}})\left[1 - \frac{i\omega}{\Gamma_1(\bar{\mathbf{q}})}\right] \tag{5}$$

with

$$\bar{\mathbf{q}} = \mathbf{Q}_o + \mathbf{q}, \tag{6}$$

$$\chi_1^{-1}(\bar{\mathbf{q}}) = \chi_1^{-1}(\mathbf{Q}_o) + c_1 q^2 \tag{7}$$

and

$$\Gamma_1(\bar{\mathbf{q}}) = u_1\chi_1^{-1}(\bar{\mathbf{q}}). \tag{8}$$

In this formulation $\chi(\mathbf{Q}_o)$ is the static magnetic susceptibility at the antiferromagnetic wave vector, $c$ is a torsional stiffness coefficient, $q$ is $|\mathbf{q}|$ as measured from the antiferromagnetic wave vector and $u$ is a measure of the relaxation rate of the fluctuations. This is coupled together with $\chi_2$, a damped spin wave at pole $\omega(q)$.

$$\chi_2^{-1}(\bar{\mathbf{q}},\omega) = \chi_2^{-1}(\bar{\mathbf{q}})\left[1 - \frac{\omega(\omega + i\Gamma_2(\bar{\mathbf{q}}))}{\omega^2(q) - i\Gamma_2(\bar{\mathbf{q}})(\omega + i\Gamma_2(\bar{\mathbf{q}}))}\right] \tag{10}$$

$$\hbar\omega(q) = \hbar\omega_o + Dq^2 \tag{11}$$

where $\chi_2^{-1}(\bar{q})$ and $\Gamma_2(\bar{q})$ take the same functional form as above, equations (7) and (8). The spin wave pole gap at the antiferromagnetic zone centre is given by $\hbar\omega_o$ and $D$ is the stiffness.

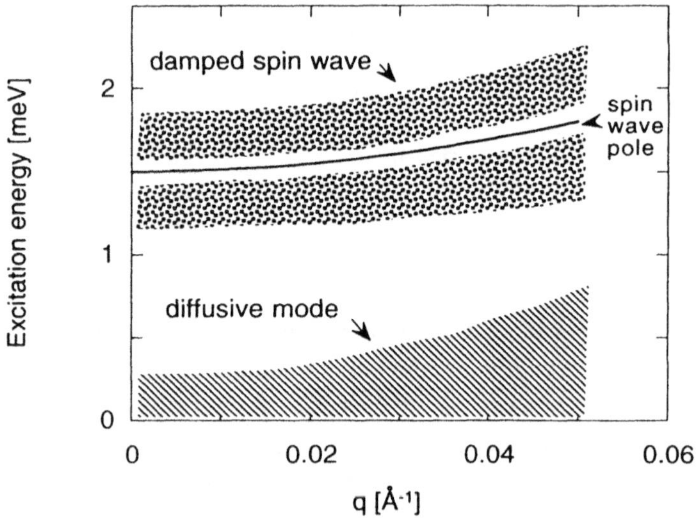


**Figure 5.** The form of the two parts of the 5*f* response, in the range $T_c < T < T_N / 2$, as determined from the measured spectra. The spin wave dispersion is centered around 1.5 meV and the *hwhm* is typically 0.5 meV as indicated by the pebbled region on either side of the spin wave pole. The quasielastic mode, with its pole at zero frequency, has a characteristic *hwhm* of 0.5 meV indicated by the shaded area. This is the spectral form which was used to estimate the low temperature scattering in Fig. 2*c*, given as dashed line.

In the neighbourhood of $T_N$ there remains only one response, the quasielastic scattering. Taking into account the experimental resolution we have derived parametric fits to the data, represented by the solid lines in the left hand frame of Fig. 2. In this regime one may use antiferromagnetic spin fluctuation theory[30,31] to estimate the Néel temperature as 16 K in reasonable agreement with the experimental value. On cooling below $T_N$, in the range $T_c < T < T_N / 2$, the microscopic parameters characterising the dynamical susceptibility are found to change little, and the evolution of scattering intensity is accounted for principally by the falling temperature. The typical form of the excitation spectrum over this temperature interval, before mean field coupling, is depicted in Fig. 5. From the low energy part of this excitation spectrum, renormalised by the mean field coupling $\lambda$, one may estimate a contribution to the linear coefficient of the low temperature heat capacity[30-33]. Extrapolated to 0 K, one obtains approximately 0.1 $J / mole\ K^2$. Although it is difficult to have a precise figure, since it depends on a cut off parameter of integration[30-33], it is close to, and certainly within order of magnitude of, the experimental value. This strongly supports the idea that the dominant renormalisation, yielding the massive quasiparticles which give rise to the superconducting ground state, may be considered as arising from the low frequency exchange field.

**Discussion T < $T_c$**

The observation that the linear coefficient of heat capacity falls to 9 $mJ/mole\,K^2$ [1,2] at base temperature implies that a major part of the low energy spin excitation spectrum must be gapped in the superconducting state. That the spectrum is strongly modified below $T_c$ is evident from Fig. 2. We may underline the change by plotting the anticipated scattering that would have been seen at the lowest temperatures if the spectral form of the normal state, Fig. 5, had been unmodified on passing below $T_c$. This is given as the dashed line in the bottom frame of Fig 3. The opening of a gap, to satisfy the loss of a major part of the low temperature heat capacity, alone is clearly seen not to be sufficient to explain the observed scattering. There is also a strong increase in intensity at finite energy transfer. In the approximation that the magnetic response is a measure of a two site correlation function, given the superconducting gap occurs in the (single) quasiparticle spectrum, it becomes model dependent to relate the observed resonance to the superconducting gap.

Whilst we are unaware of any detailed forms for the dynamical susceptibility in the antiferromagnetic superconducting state, on the basis of a general pairing mechanism we have been able to approach this problem in the following manner. First, we recognise that at the lowest temperatures, neutron energy loss excitations above the superconducting ground state will consist principally of pair annihilation events, since there is a negligible density of excited quasiparticles (low heat capacity) the density of quasiparticle - quasiparticle scattering will be small. As the temperature is raised, beyond about $T_c$ / 2, the quasiparticle - quasiparticle scattering will increase since $k_B T$ becomes comparable with the (falling) gap excitation energy. The spectra at the lowest temperatures will thus be characterised by a paucity of scattering around zero energy transfer whilst the scattering above $T_c$ / 2 can have an additional component. This second component may, at low energies, qualitatively resemble that above $T_c$. Second, for the calculation of the full dynamical susceptibility one is to couple, as in the normal antiferromagnetic state, the low energy and spin wave modes with the mean field parameter, $\lambda$.

We proceed in the calculation as follows, first we calculate the bare spin susceptibility above the superconducting ground state and then make an analytical approximation to the spectral form which is inserted into the second stage of the calculation. To set the framework of the calculation we assume a spin singlet superconducting ground state[34]. The coherence factor for creation of a pair of quasiparticles is, in the case where the perturbation changes sign under time reversal of the electronic states, expected to be of the form[35],

$$\left[1-\frac{\xi(\mathbf{k}+\mathbf{q})\xi(\mathbf{k})+\cos[\Phi(\mathbf{q})]|\Delta(\mathbf{k}+\mathbf{q})||\Delta(\mathbf{k})|}{E(\mathbf{k}+\mathbf{q})E(\mathbf{k})}\right] \tag{12}$$

where $\xi(\mathbf{k})=\varepsilon(\mathbf{k})-\varepsilon_F$ is the normal state single particle energy relative to the Fermi energy, $\Delta(\mathbf{k})$ the gap and $E(\mathbf{k})=\sqrt{\xi(\mathbf{k})^2+|\Delta(\mathbf{k})|^2}$ the quasiparticle excitation energy above the superconducting state. The factor $\Phi(\mathbf{q})$ is the phase difference between $\Delta(\mathbf{k})$ and $\Delta(\mathbf{k+q})$. At minimum excitation energy $\xi(\mathbf{k})$ = 0 and thus, for a finite response, $\Delta(\mathbf{k+q})$ must be the negative of $\Delta(\mathbf{k})$ at least over a sizable portion of the zone. The observation of scattering around the antiferromagnetic reciprocal lattice vectors is compatible with an order parameter displaying sign inversion on translation by $\mathbf{Q}_{//}$ over a significant part of the zone. This is in general accord with the given interpretations of the temperature dependence of the NQR data[36], the μSR reduction in spin susceptibility[6], the angular dependence of the upper critical field[34] and tunneling data[37], the interpretation of which invoke a spatially asymmetric (*d*-wave) spin singlet superconducting state. The unrenormalised spin susceptibility above the superconducting ground state is then calculated in the following approximation[38,39],

$$\begin{aligned}\chi(\mathbf{q},\omega) = &\sum_{\mathbf{k}}\frac{1}{2}\left[1+\frac{\xi(\mathbf{k}+\mathbf{q})\xi(\mathbf{k})+\cos[\Phi(\mathbf{q})]|\Delta(\mathbf{k}+\mathbf{q})||\Delta(\mathbf{k})|}{E(\mathbf{k}+\mathbf{q})E(\mathbf{k})}\right]\frac{f(\mathbf{k}+\mathbf{q})-f(\mathbf{k})}{\omega-[E(\mathbf{k}+\mathbf{q})-E(\mathbf{k})]+i\Gamma}\\ &+\sum_{\mathbf{k}}\frac{1}{4}\left[1-\frac{\xi(\mathbf{k}+\mathbf{q})\xi(\mathbf{k})+\cos[\Phi(\mathbf{q})]|\Delta(\mathbf{k}+\mathbf{q})||\Delta(\mathbf{k})|}{E(\mathbf{k}+\mathbf{q})E(\mathbf{k})}\right]\frac{1-f(\mathbf{k}+\mathbf{q})-f(\mathbf{k})}{\omega-[E(\mathbf{k}+\mathbf{q})+E(\mathbf{k})]+i\Gamma}\\ &-\sum_{\mathbf{k}}\frac{1}{4}\left[1-\frac{\xi(\mathbf{k}+\mathbf{q})\xi(\mathbf{k})+\cos[\Phi(\mathbf{q})]|\Delta(\mathbf{k}+\mathbf{q})||\Delta(\mathbf{k})|}{E(\mathbf{k}+\mathbf{q})E(\mathbf{k})}\right]\frac{1-f(\mathbf{k}+\mathbf{q})-f(\mathbf{k})}{\omega+[E(\mathbf{k}+\mathbf{q})+E(\mathbf{k})]+i\Gamma}\end{aligned} \quad (13)$$

The first term arises from scattering between quasiparticle levels above the superconducting ground state, the second term corresponds with the creation of pairs of quasiparticles (neutron energy loss scattering) and is the leading contribution at low temperatures, whilst the final term corresponds with Cooper pair condensation (neutron energy gain scattering). Each magnetic sub-lattice is taken to have finite population of spin up and spin down quasiparticles on surfaces which are degenerate in energy (the lattice has inversion symmetry)[40], this suggests that the approximate form, equation (13), whilst clearly lacking in details, may serve to orient the calculations. The quasiparticle dispersion has been approximated by a Fermi surface with cylindrical symmetry about the c*-axis[8,9].

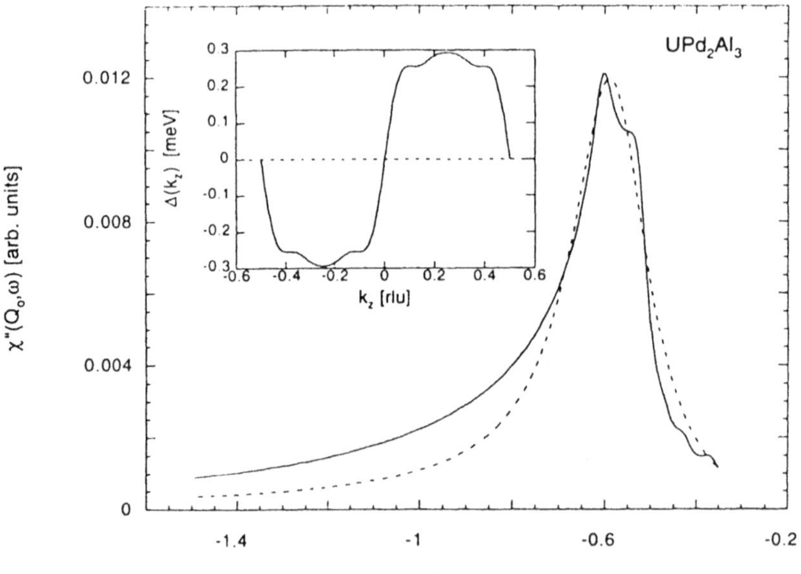


**Figure 6.** The low temperature form of the spin susceptibility divided by frequency at $\mathbf{Q}_o$ as a function of neutron energy loss calculated from equation (13) in the text. The form for the superconducting energy gap as a function of $k_z$ is given in the insert.

On heating towards $T_c$, the observation of an increasing quasielastic term in the response around $\mathbf{Q}_o$, Fig. 2 right hand panel, indicates the mounting contribution from the quasiparticle term in the susceptibility (first term in equation (13) above). From the resonant energy denominator we have, for a quasielastic process, $E(\mathbf{k}+\mathbf{Q}_o)=E(\mathbf{k})$, or since $\xi(\mathbf{k}+\mathbf{Q}_o)=\xi(\mathbf{k})$ on account of inversion symmetry, $|\Delta(\mathbf{k}+\mathbf{Q}_o)|=|\Delta(\mathbf{k})|$. Taken in conjunction with the sign inversion of the order parameter on translation by $\mathbf{Q}_o$ this suggests an *antiferromagnetic* form of $\Delta$ be taken, $\Delta(\mathbf{k}) = -\Delta(\mathbf{k}+\mathbf{Q}_o)$. That is, $\Delta(\mathbf{k})=\sum a_n \sin((2n+1)ck_z)e^{i\phi}$, where $\varphi$ is the angle in the $k_x$-$k_y$ plane, a typical form for $\Delta(k_z)$ is given as in insert of Fig. 6. For such a gap parameter one may, in gross outline, represent the imaginary part of the spin susceptibility divided by frequency by a Lorentzian centred at a finite energy as shown in Fig. 6. In this case, the peak energy of the response corresponds, approximately, with twice the maximum single particle gap, with the width being determined by the given forms of $\varepsilon(\mathbf{k})$ and $\Delta(\mathbf{k})$. Neither the peak response nor the measured 'gap' are simply related to the superconducting quasiparticle excitation spectrum in general, however, as an *a posterei* motivation for the given form $\Delta(\mathbf{k})$, we will find that the parameter values, extracted from the following data analysis, for the superconducting energy gap and its temperature dependence are in reasonable accord with other estimates[34,36-37].

For the second stage of calculation, the coupling of the low and high frequency modes, we use the Lorentzian approximation to the low frequency dynamics together with a damped spin wave similar to that determined in the normal state. In this manner we have parameterised the spectra around $\mathbf{Q}_o$ as a function of temperature, solid lines in Fig. 2 right hand panel.

We find the following results: (i) a resonant energy for the low frequency susceptibility ~ 0.6 meV ($3.7k_BT_c$) at low temperatures with an effective band width ~ 0.15 meV and an evolution of the pole position of the Lorentzian with temperature of approximately BCS form as shown in Fig. 7; (ii) a coupling constant $\lambda$ close to that found in

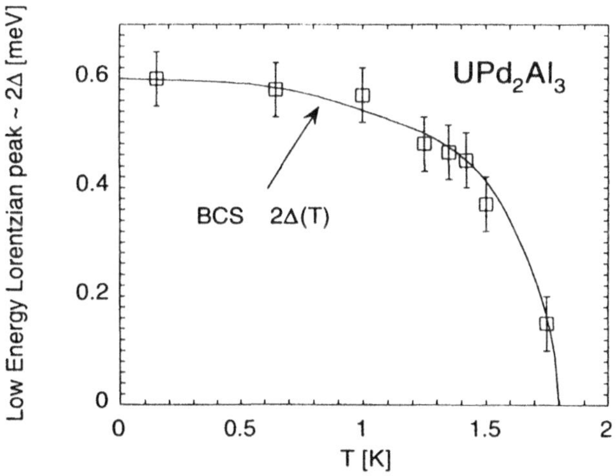


**Figure 7.** The thermal evolution of the low energy Lorentzian pole in the superconducting state which is approximately given by twice the maximum energy gap. The solid line corresponds with the BCS form normalised to the value at low temperatures, $2\Delta_o = 3.86k_BT_o$, with $T_o$ = 1.8 K lying close to the estimated bulk $T_c$ = 1.90 ± 0.07 K.

the normal antiferromagnetic state; (iii) a quasielastic component, which sets in above 1 K, with similar parametric form to that found above $T_c$; (iv) some changes in the spin wave pole and width at the lowest temperatures, which however demand a more refined theory in order to assess their significance. Extension of the analysis to encompass the wave vector dependence of the scattering intensity in the superconducting antiferromagnetic phase is in progress and will be reported elsewhere.

Interestingly, a model of superconducting order with antiferromagnetic periodicity has been advanced previously on the basis of general symmetry arguments in heavy electron systems[41-43]. Within such a scenario the width of the response around a reciprocal lattice vector gives an inverse measure of the spatial coherence of the superconducting order parameter. This is estimated to be around 100Å in agreement with values extracted from bulk measurements.

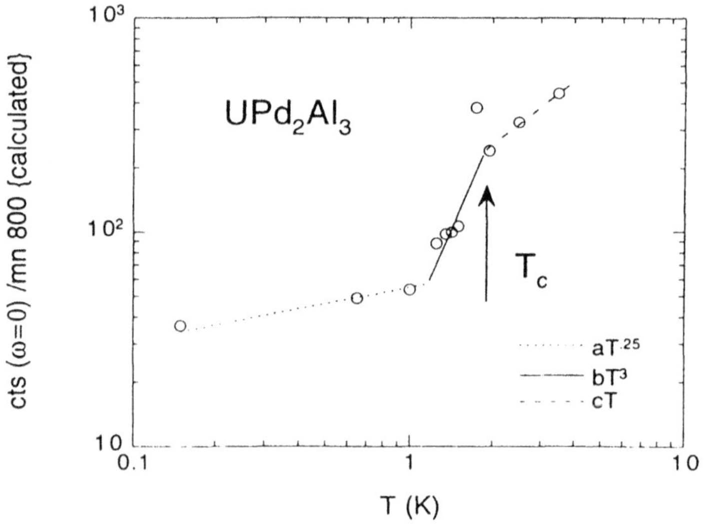


**Figure 8.** The nuclear relaxation process is a measure of the very low frequency local magnetic fluctuations. In the approximation that these are dominated by $\chi''(Q_0)$, the thermal evolution of the computed response at $Q_0$ in the elastic channel should mimic the measured relaxation rate.

As a control of the low frequency part of the calculated scattering cross-section, we note that, since the magnetic response is strongly focused around the antiferromagnetic wave vector, we may compare the computed response around the elastic position at $\mathbf{Q}_0$ with the thermal evolution of the nuclear resonance relaxation frequency, $1/T_1 \propto k_B T \sum_q \chi''(q,\omega_{NMR})/\omega_{NMR} \approx k_B T \chi''(Q_0,\omega_{NMR})/\omega_{NMR}$. As shown in Fig. 8, and in common with the resonance data[36], we obtain three regions of approximate power law behaviour. At the lowest temperatures, where the susceptibility is dominated by thermal excitation of quasiparticle pairs, one has the slowest increase with temperature; between ~ 1 K and $T_c$ the dominant contribution comes from the quasiparticle scattering above the superconducting ground state and the increase is approximately $T^3$; finally above $T_c$, in the presence of quasielastic normal state excitations, the relaxation rate is linear in temperature. We recall that the low energy susceptibility is not only dependent on the quasiparticle excitations associated with the superconducting ground state, and hence $\Delta(\mathbf{k})$, within our model, it is also strongly influenced by the presence of, and coupling to, the spin wave mode. In comparison with the $^{105}$Pd NQR data[44], the rate of fall of calculated intensity below 1 K is much slower. In this temperature regime, the calculation is dominated by pair annihilation processes(strong inelastic peak in neutron data), suggesting that this part of the response may be local to the uranium ions whilst the quasiparticle-quasiparticle term, which gives rise to the $T^3$ relaxation above the superconducting ground state and the linear in T relaxation in the normal antiferromagnetic state, is of a more delocalised nature.

## Conclusion

As the result of wave vector and temperature dependent studies, polarisation analysis and rudimentary calculations, it appears reasonable to suggest the following features concerning the superconducting antiferromagnetic state in $UPd_2Al_3$. First, the strongly correlated electronic state from which the superconductivity evolves at low temperatures may be considered to arise from the presence of intense antiferromagnetic fluctuations of low energy in the normal state. Second, whilst the role of the superconducting energy gap anisotropy and finite coupling, $\lambda$, between the low and intermediate frequency modes preclude a simple interpretation of the gap which appears in the magnetic fluctuation spectrum, the wave vector and polarisation specificity of the neutron magnetic cross-section may allow one to extract valuable information on the symmetry of the ground state from the enhanced magnetic scattering observed at low temperatures. We hope the work reported here will stimulate further theoretical and experimental investigations of superconducting heavy fermion materials. In particular, it is necessary to improve this work and construct a realistic model of the antiferromagnetic superconducting state.


## Acknowledgments

I would like to thank my colleagues for help and inspiration. Also I thank the ILL, Grenoble, and ITU, Karlsruhe, for their generous support which made this work possible.


## References


1) C Geibel, C Schank, S Thies, H Kitazawa, CD Bredl, A Bohm, M Rau, A Grauel, R Caspary, R Helfrich, U Ahlheim, G Weber and F Steglich, Heavy fermion superconductivity at $T_c$ = 2K in the antiferromagnet $UPd_2Al_3$, *Z. Phys.* **B84:** 1-2 (1991)

2) F Steglich, P Gegenwart, C Geibel, R Helfrich, P Hellmann, M Lang, A Link, R Modler G Sparn, N Buttgen and A Loidl, New observations concerning magnetism and superconductivity in heavy fermion metals, *Physica* **B223:** 1 (1996)

3) N Sato, N Aso, GH Lander, B Roessli, T Komatsubara and Y Endoh, Spin fluctuations in the heavy fermion superconductor $UPd_2Al_3$ studied by neutron inelastic scattering, *J. Phys. Soc. Japan.* **66:** 1884 (1997)

4) N Metoki , Y Haga, Y Koike, N Aso and Y Onuki, Coupling between magnetic and superconducting order parameters and evidence for the spin excitation gap in the superconducting state of a heavy fermion superconductor $UPd_2Al_3$, *J . Phys. Soc. Japan* **66:** 2560 (1997); N Metoki , Y Haga, Y Koike, N Aso and Y Onuki, Neutron scattering study of the correlation of magnetism and superconductivity in the heavy fermion superconductor $UPd_2Al_3$, *J Magn. & Magn. Mat.* **177-181:** 449 (1998); N Metoki , Y Haga, Y Koike, and Y Onuki, Superconducting energy gap observed in the magnetic excitation spectra of a heavy fermion superconductor, *Phys. Rev. Lett.* **80**: 5417 (1998)

5) R Caspary, P Hellmann, M Keller, G Sparn, C Wassilew, R Kohler, C Geibel, C Schrank and F Steglich, Unusual ground state properties of $UPd_2Al_3$: Implications for the coexistence of heavy fermion superconductivity and local moment antiferromagnetism, *Phys Rev Lett* **71:** 2146 (1993)

6) R Feyerherm, A Amato, FN Gygax, A Schenck, C Geibel, F Steglich, N Sato and T Komatsubara, Coexistence of local moment magnetism and heavy fermion superconductivity in $UPd_2Al_3$, *Phys Rev Lett* **73:** 1849 (1994)

7) M Kyogaku, Y Kitaoka, K Asayama, C Geibel, C Schank and F Steglich, NMR and NQR studies of magnetism and superconductivity in the antiferromagnetic heavy fermion superconductors $UM_2Al_3$ (M = Ni and Pd), *J. Phys. Soc. Japan* **62:** 4016 (1993)

8) K Knopfle, A Mavromaras, LM Sandratskii and J Kubler, The Fermi surface of $UPd_2Al_3$, *J. Phys. Condens. Matter*, **8:** 901 (1996)
9) Y Inada, A Ishiguro, J Kimura, N Sato, A Sawada, T Komatsubara and H Yamagami, Fermi surface of the heavy fermion superconductor $UPd_2Al_3$, *Physica* **B206-207:** 33 (1995)
10) D Edwards, Some current problems in itinerant electron magnetism, *J. Magn. Magn. Mat.* **15-18:** 262 (1980); D Edwards, The paramagnetic state of itinerant electron systems with local magnetic moments: Static properties, *J. Phys.* **F12:** 1789 (1982)
11) H Hasegawa, Single-site functional integral approach to itinerant electron ferromagnetism, *J. Phys. Soc. Japan* **46:** 1504 (1979)
12) Y Kuromoto and K Miyake, Quantum phenomenology for heavy fermion systems. I. Formation of the model, *J. Phys. Soc. Japan* **59:** 2831 (1990)
13) C Pépin and M Lavagna, Dynamic spin correlation function near the antiferromagnetic quantum phase transition of heavy fermions, *preprint submitted Phys. Rev.* **B** (1998)
14) G Aeppli, D Bishop, C Broholm, E Bucher, K Siemensmeyer, M Steiner and N Stusser, Magnetic order in the different superconducting states of $UPt_3$, *Phys. Rev. Lett.* **63:** 676 (1989)
15) JP Brison, N Keller, P Lejay, A Buzdin, A Huxley, JL Tholence, B Fak, AN Stepanov, J Flouquet, N Bernhoeft, L Schmidt, RA Fisher, N Phillips and C Vettier, Magnetism and superconductivity in heavy fermion materials, *J. Low Temp. Phys.* **95:** 145 (1994)
16) JG Lussier, A Schroder, BD Gaulin, JD Garrett, WJL Buyers, L Rebelsky and SM Shaprio, Incommensurate magnetic order in $UNi_2Al_3$, *Physica* **B199-200:** 137 (1994)
17) A Krimmel, P Fisher, B Roessli, H Maletta, C Geibel, C Schank, A Grauel, A Loidl and F Steglich, Neutron diffraction study of the heavy fermion superconductors $UM_2Al_3$ (M=Pd, Ni), *Z. Phys.* **B86:** 161 (1992)
18) H Kita, A Donni, Y Endoh, K Kakurai, N Sato and T Komatsubara, Single crystal diffraction study of the magnetic phase diagram of the heavy fermion superconductor $UPd_2Al_3$, *J. Phys. Soc. Japan* **63:** 726 (1994)
19) L Paolasini, JA Paixao, GH Lander, P Burlet, N Sato and T Komatsubara, Field dependence of magnetic structure of $UPd_2Al_3$ in the normal state, *Phys. Rev.* **B49:** 7072 (1994)
20) A Krimmel, A Loidl, P Fisher, B Roessli, A Donni, H Kita, N Sato, Y Endoh, T Komatsubara, C Geibel and F Steglich, Single crystal neutron diffraction studies of the heavy fermion superconductor $UPd_2Al_3$, *Solid State Comm.* **87:** 829 (1993)
21) BD Gaulin, ED Isaacs, JG Lussier, JN Reimers, D Gibbs, Pzschack, A Schroder, L Taillefer and JD Garrett, Resonant magnetic x-ray scattering studies of heavy fermion superconductors, *Phys. Rev. Lett.* **73** 890-893 (1994)
22) A Amato, R Feyerherm, FN Gygax, A Schenk, M Weber, R Caspary, P Hellmann, C Schank, C Geibel, F Steglich, DE MacLaughlin, Knetsch and RH Heffner, Magnetic and superconducting properties of the heavy fermion superconductor $UPd_2Al_3$, *Europhys. Lett.* **19:** 127 (1992)
23) T Petersen, TE Mason, G Aeppli, AP Ramirez, E Bucher and RN Kleiman, Magnetic fluctuations and the superconducting transition in the heavy fermion material $UPd_2Al_3$, *Physica* **B199-200:** 151 (1994)
24) N Sato, T Sakon, N Takeda, T Komatsubara, C Geibel and F Steglich, Anisotropy in a heavy fermion superconductor: $UPd_2Al_3$, *J. Phys. Soc. Japan.* **61:** 32 (1992)
25) N Sato, Y Inada, T Sakon, K Imamura, A Ishiguro, J Kimura, A Sawada and T Komatsubara, Magnetic, electrical and thermal properties of heavy fermion superconductor $UPd_2Al_3$, *IEEE Trans. Magn.* **30:** 1145 (1994)
26) N Bernhoeft, B Roessli, N Sato, N Aso, A Hiess, GH Lander, Y Endoh and T Komatsubara, $UPd_2Al_3$: An analysis of the inelastic neutron scattering spectra, in *Itinerant Electron Magnetism: Fluctuation Effects,* D Wagner, W Brauneck and A Solontsov, eds., Kluwer (1998)
27) N Bernhoeft, N Sato, B Roessli, N Aso, A Hiess, GH Lander, Y Endoh and T Komatsubara, Enhancement of magnetic fluctuations on passing below $T_c$ in the heavy fermion superconductor $UPd_2Al_3$, *Phys. Rev. Lett.* **81:** 4244 (1998)
28) N Bernhoeft and GG Lonzarich, Scattering of slow neutrons from long wavelength magnetic fluctuations in $UPt_3$, *J. Phys.* **C7:** 7325 (1995)
29) W Marshall and SW Lovesey, *Theory of Thermal Neutron Scattering,* Oxford University Press, (1971)

30) K Nakayama and T Moriya, Quantitative aspects of the theory of weak itinerant antiferromagnetism, *J. Phys. Soc. Japan* **56:** 2918 (1987)
31) T Moriya and T Takimoto, Anomalous Properties around magnetic instability in heavy fermion systems, *J. Phys. Soc. Japan* **64:** 960 (1995)
32) GG Lonzarich, The magnetic equation of state and heat capacity in weak itinerant ferromagnets, *J. Magn. Magn. Mat.* **54:** 612 (1986)
33) GG Lonzarich, The magnetic electron, in *Electron* M Springford, ed., Cambridge University Press (1997)
34) J Hessert, M Huth, M Jourdan, H Adrian, CT Rieck and K Scharnberg, Temperature and angular dependence of the upper critical field of $UPd_2Al_3$ thin films, *Physica* **B230-232:** 373 (1997)
35) JR Schrieffer, *Theory of superconductivity*, Benjamin Cummings, London (1964)
36) K Matsuda, Y Kohori and T Kohara, Observation of $^{105}Pd$ NMR and NQR signals in the heavy fermion superconductor $UPd_2Al_3$, *Phys. Rev.* **B55:** 15223 (1997)
37) M Jourdan, M Huth, J Hessert and H Adrian, Superconducting tunneling spectroscopy on epitaxial $UPd_2Al_3$ thin films, *Physica* **B230-232:** 335 (1997)
38) JP Lu, Neutron scattering as a probe for unconventional superconducting states, *Phys. Rev. Lett.* **68:** 125 (1992)
39) M Lavagna and G Stemmann, Spin excitations of two-dimensional-lattice electrons: Discussion of neutron scattering and NMR experiments in high-$T_c$ superconductors, *Phys. Rev.* **B49:** 4235 (1994)
40) C Herring, *Magnetism Volume IV*, GT Rado and H Suhl eds., Academic Press, New York and London, (1966)
41) PW Anderson, Further consequences of symmetry in heavy fermion superconductors, *Phys. Rev.* **B32:** 499 (1985)
42) MR Norman, Solutions of the magnetic Eliashberg equations for heavy fermion superconductors, *Phys. Rev.* **B37:** 4987 (1988)
43) AI Liechtenstein, I Mazin and OK Andersen, s-wave superconductivity from an antiferromagnetic spin fluctuation model for bilayer materials, *Phys. Rev. Lett.* **74:** 2303 (1995)
44) K Matsuda, Y Kohori and T Kohara, Observation of 105 Pd NMR and NQR signals in the heavy fermion superconductor, *Phys. Rev.* **B55**: 15223 (1997)

## Discussion

**Sawatzky:** Was the quasi-elastic structure also present on the gain side of the spectrum?

**Bernhoeft:** Yes, it was all-consistent with a quasi elastic interpretation including the different temperature dependence of the structure on the loss and the gain side.

**Sawatzky:** What is the symmetry of the order parameter and what do you mean by saying that along the c axis it is like that of the magnetic order parameter?

**Bernhoeft:** By this we mean that there is a node along the c axis and although we can't say what it is in general, we think it may result from dxz, dyz-like symmetries.

**Sawatzky:** But then there would be lines of nodes on the Fermi surface and the inelastic neutron scattering, which averages over k in scattering a particle from k to k+q would not exhibit a clear gap.

**Bernhoeft:** Yes, that is correct. We do not know the k dependence of the scattering cross section, so we can't really say if the data is consistent with such a d wave scenario.

## NON-FERMI-LIQUID PROPERTIES AND EXOTIC SUPERCONDUCTIVITY IN $CeCu_2Si_2$ AND (UTh)$Be_{13}$


M. Lang[1], P. Gegenwart[1], R. Helfrich[1], M. Köppen[2], F. Kromer[1], C. Langhammer[1], C. Geibel[1], F. Steglich[1,2], J.S. Kim[3], G.R. Stewart[3,4]

[1]Max-Planck-Institut für Chemische Physik fester Stoffe, D-01187 Dresden Germany
[2]Institut für Festkörperphysik, SFB 252, TU Darmstadt, D-64289 Darmstadt Germany
[3]Department of Physics, University of Florida, Gainesville, Florida 32611, USA
[4]Institut für Physik, Universität Augsburg, D-86159 Augsburg, Germany


## INTRODUCTION

Tetragonal $CeCu_2Si_2$ [1] and cubic $UBe_{13}$ [2] belong to the class of strongly correlated f-electron materials where a periodic lattice of partially filled f shells is embedded in a metallic environment. Below a characteristic "Kondo temperature" $T_K$, typically of the order of 10K, quasiparticles composed of both local f degrees of freedom and itinerant conduction-electron degrees of freedom form. As has been inferred from the giant Sommerfeld coefficient $\gamma$ of about 1 J/$K^2$mole, the huge effective quasiparticle masses $m^*$ (100 - 1000 $m_{el}$) are governed by the f degrees of freedom. The discontinuity of the specific heat at $T_c$ which was found to scale with the large $\gamma$ [1,2] proved that superconductivity is, indeed, formed by those heavy fermions (HF).

In the years that followed the discovery of HF superconductivity, the number of systems showing similar properties has grown. Compared to their later discovered counterparts $UPt_3$ [3], $URu_2Si_2$ [4], $UNi_2Al_3$ [5] and $UPd_2Al_3$ [6], it appears that $CeCu_2Si_2$ and $UBe_{13}$ reveal a more complex phenomenology and are less understood. For $CeCu_2Si_2$, an as yet unidentified, presumably magnetically ordered "phase A" has been found which is almost degenerate with HF superconductivity. Further on, the low-temperature normal state is characterized by strong violations of the behavior expected for a Landau Fermi liquid. This holds true also for the normal-conducting state of $UBe_{13}$. For the latter, particular interest arose, however, because of its highly anomalous superconducting-state properties, notably the occurrence of a double phase transition for weakly Th doped $U_{1-x}Th_xBe_{13}$ [7].

In this paper we summarize our recent works [8-10] on these two canonical HF superconductors. Two main aspects will be addressed: In Sects. 2 and 3 we will explore their


*Electron Correlations and Materials Properties*, edited by Gonis *et al.*
Kluwer Academic/Plenum Publishers, 1999.

non-Fermi-liquid properties. We will investigate whether this behavior is related to the nearness of a quantum critical point, defined by a magnetic ordering temperature $T_m \to 0$. If so, we will study how heavy fermions behave in its vicinity. In Sect. 4 we will address our recent discovery of a "new line of anomalies" in the superconducting state of $UBe_{13}$ and its thoriated variant $U_{1-x}Th_xBe_{13}$. Our results call for a revision of the interpretation of the low-T phases in the complex T-x phase diagram that had been established for this system.

## 2. $CeCu_2Si_2$

### 2.1 Different ground-state behaviors

By means of a thorough thermodynamic investigation of $CeCu_2Si_2$ single crystals it was found that distinctly different ground-state properties exist which depend sensitively on the composition of the starting melt and/or the subsequent heat treatment [11,12]. This can be most clearly visualized by comparing the coefficient of thermal expansion, $\alpha(T) = l^{-1}dl/dT$, for such differently prepared single crystals (Fig. 1). The transition from the paramagnetic into the superconducting state manifests itself in a positive jump in $\alpha$ measured along the a-axis [11]. This is shown in Fig. 1a for an annealed single crystal with $T_c = 0.63$K. In an overcritical field of B = 3 T superconductivity is suppressed and no indication for any further anomaly can be resolved in $\alpha(T)$ down to 0.05 K. For a number of unannealed crystals it was found, however, that instead of bulk superconductivity they showed a transition into phase A with an onset temperature $T_A \simeq 0.7$ K. This is accompanied by a large negative discontinuity in $\alpha(T)$, cf. Fig.1b. Measurements in B = 3 T demonstrate that compared to superconductivity, phase A is much more robust against magnetic fields. For other crystals, an intermediate behavior was found [12] (Fig. 1c): Upon cooling an incipient A-phase transition is replaced by the superconducting transition at $T_c \simeq 0.67$ K. In an overcritical field of B = 1.5 T, superconductivity is suppressed and phase A fully recovers. These astonishing differences in the ground-state properties of $CeCu_2Si_2$ are most likely related to variations in site occupation within the single crystals. However, no significant difference in the lattice constants could be resolved by X-ray diffractometry. Using polycrystalline $Ce_{1+x}Cu_{2+y}Si_{2+z}$ samples with compositions deliberately chosen to be slightly off stoichiometric it was possible to map those different ground-state behaviors onto separate sectors of the narrow homogeneity range of the primary $CeCu_2Si_2$ phase within the ternary Ce-Cu-Si phase diagram [13]. While samples prepared with slight Cu excess show superconductivity without phase A, a transition into phase A is found in those samples with Cu/Ce deficiency. Both the "S" and "A" sectors are separated from each other by the "AS" sector containing the true stoichiometric point. The thermal expansion of a polycrystalline sample out of the "AS" sector is displayed in Fig.1d.

### 2.2 On the nature of phase A

The nature of phase A is still unknown. Neutron diffractometry has so far failed to resolve magnetic Bragg reflections. Different assignments spanning the whole range from spin-glass [14] to dynamical [15] and unconventional spin-density-wave [16] order have been made. Partial Ge substitution for Si was found to stabilize phase A and to support strong evidence for an antiferromagnetic transition at $T_A$ [17]. For all $CeCu_2Si_2$ samples studied so far, an additional "phase B" [12] phenomenologically related to phase A, was found to form at fields B > 6T. Below we focuss on single crystals of the "AS" sector where phase A can be studied when

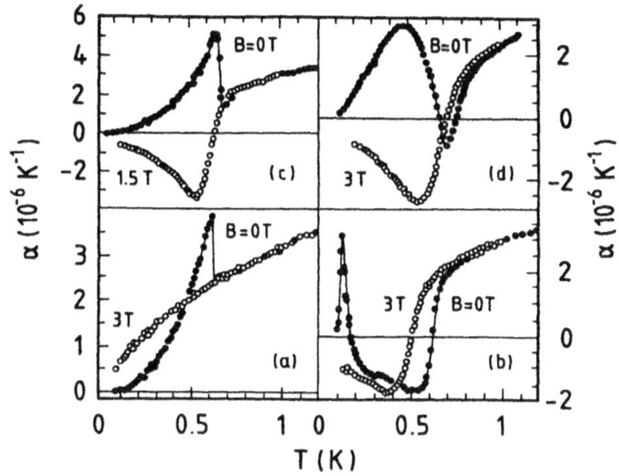


Fig. 1. Coefficient of thermal-expansion measured parallel to the a axis as a function of temperature at both B = 0 and B > $B_{c2}(0)$, with B∥a, for three $CeCu_2Si_2$ single crystals: a) superconducting crystal identical to that studied in [11]; b) non-bulk-superconducting crystal ("as grown") [11]. c) same crystal as studied in [12] showing a superconducting transition in B = 0 but a transition into "phase A" at B > $B_{c2}$; In d) one third of the volume-expansion coefficient, $\alpha(T) = 1/3 \cdot \beta(T)$, at B = 0 T and 3 T is shown for a polycrystalline sample from the same type as the single crystal in c).

superconductivity has been suppressed by a sufficiently large magnetic field. Fig. 2 shows the normalized resistivity measured along the respective a- and c-axes in an overcritical field B = 5T. The additional scattering contribution visible at low temperatures marks the transition from the paramagnetic phase into phase A. As will be discussed in Sect 2.3 in more detail, phase A (and B) develops out of a state where the resistivity varies as $\rho(T) = \rho_0 + aT^2$ ($\rho_0$: residual resistivity) with a giant coefficient a ≈ 10 $\mu\Omega cmK^{-2}$. The resistivity contribution $\delta\rho = \rho(T) - (\rho_0 + aT^2)$ associated with the transition at $T_A$ is shown in the insert. While $\delta\rho$ increases below $T_A$ for current j∥a, no change is found in $\delta\rho$ for j∥c at this temperature. This is consistent with a

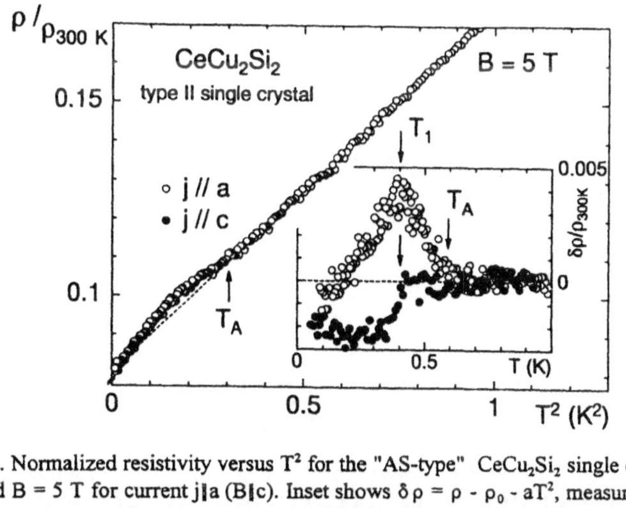


Fig. 2. Normalized resistivity versus $T^2$ for the "AS-type" $CeCu_2Si_2$ single crystal at a field B = 5 T for current j∥a (B∥c). Inset shows $\delta\rho = \rho - \rho_0 - aT^2$, measured along the respective a- and c- axes at the same field (applied perpendicular to the current).

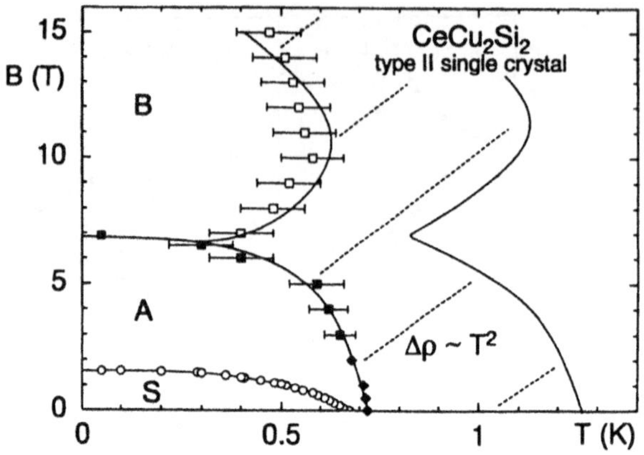


Fig.3. B-T diagram, based upon resistivity results, for the "AS-type" $CeCu_2Si_2$ single crystal. Validity range for the $T^2$-law is indicated.

transition into a SDW state with a nesting vector lying within the basal plane, see, e.g. [18]: The increase in $\delta\rho$ for j$\|$a reflects the reduction of the effective carrier number, while the isotropic decrease at somewhat lower temperatures may indicate the freezing out of incoherent scattering. The resistively determined B-T phase diagram is shown in Fig. 3 which also includes the high-field phase B as well as the validity range of the $\rho \propto T^2$ behavior.

In order to study the phase boundary between the A and B phase in great detail, we have performed low-temperature magnetostriction measurements. Fig. 4 shows relative length changes, $\Delta l/l$, as a function of field at T = 0.25 K. From the discontinuous behavior of $\Delta l/l$ at 6.8T - the A-B phase boundary - the first-order character of this transition can be inferred. Similar observations [19] have been made on the "AS" crystal studied in [12]. These observations are consistent with a field-induced transition between different SDW states as has been established for the quasi-onedimensional organic salts [20]. In addition, Fig. 4 provides

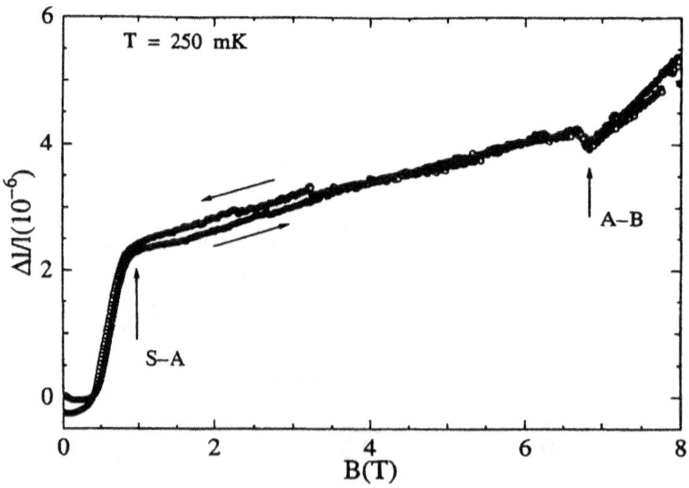


Fig. 4. Relative length changes versus magnetic field at T = 0.25 K for the "AS-type" $CeCu_2Si_2$ single crystal. Arrows indicate the respective phase transitions between the superconducting (S) and A phase (S-A) as well as A and B phase (A-B).

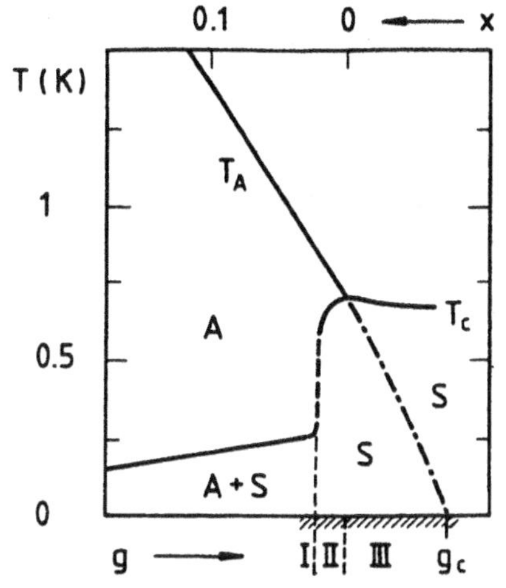


Fig. 5. Schematic phase diagram for $CeCu_2Si_2$ at zero field, indicating existence ranges for phase A (A), superconductivity (S) and coexistence range (A+S). For samples labeled "type I" ("A" type), "II" ("AS" type) and "III" ("S" type): $T_A > T_c$, $T_A \geq T_c$ and $T_A < T_c$, respectively (see text). On the abscissa an effective coupling constant g is used. The phase boundaries $T_A(g)$ [17,23] and $T_c(g)$ [22,23] are determined from B = 0 measurements (solid lines) or extrapolated from data taken at $B > B_{c2}$ [22,23] (dash-dotted line).

evidence that as a function of field the transition from the superconducting state at small fields to the A phase at intermediate fields is of first order, too. This reflects the field-induced

Taken together, our results indicate that $T_A$ marks the transition into a SDW state with a nesting vector lying within the basal plane.

## 2.3 Break-up of heavy fermions on the brink of "phase A"

As mentioned before, phase A can be stabilized by partial Ge-substitution for Si [17]. Polycrystalline $CeCu_2(Si_{1-x}Ge_x)_2$ samples with $0.02 < x < 0.15$ and undoped samples of type A exhibit an A-phase transition between $T_A \approx 0.8$ K and 1.75 K, followed by a bulk HF-superconducting transition between $T_c \approx 0.3$ K and 0.15 K, respectively, cf. the phase diagram in Fig. 5. Phase A and (thermodynamically weak) superconductivity coexist on a microscopic scale [21]. In "AS" type $CeCu_2Si_2$ samples of near stoichiometric composition and with $T_A \geq T_c$, (thermodynamically strong) HF superconductivity expells phase A [12]. Bulk measurements of "S"-type $CeCu_2Si_2$ polycrystals reveal A-phase signatures only at magnetic fields sufficient to suppress superconductivity. From the B-T phase diagrams collected for such polycrystals yielding fictitious transition temperatures $T_A(B{=}0) \approx 0.5$ K and 0.35 K [22] we infer that via suitable composition the A phase can be tuned to $T_A \rightarrow 0$. In the phase diagram in Fig. 5, an effective coupling constant g measuring the strength of the 4f hybridization with the conduction electrons is used on the abscissa. g is expected to be a complicated function of the composition in homogeneous $CeCu_2Si_2$ samples [24] and to be proportional to the Ge-content, x, in $CeCu_2(Si_{1-x}Ge_x)_2$. Fig. 5 suggests the existence of a critical coupling constant $g_c$ at which $T_A \rightarrow 0$. From the absence of any A-phase signatures for our single crystals of type "S" a coupling constant slightly in excess of $g_c$ can be expected. In the following, we will discuss, whether $g_c$ defines a quantum critical point (QCP) and, if so, how heavy fermions behave in its vicinity.

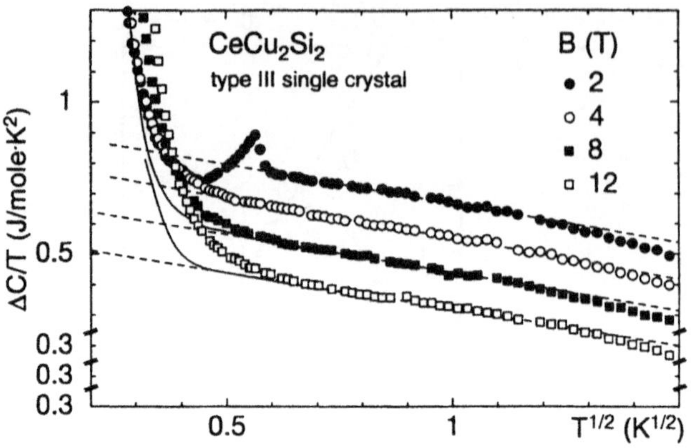


Fig.6. Ce-increment to the specific heat as $\gamma = \Delta C/T$ vs $T^{1/2}$ at varying fields for the "AS-type" $CeCu_2Si_2$ single crystal. Dashed lines indicate $(-T^{1/2})$ dependence of $\gamma(T) - \gamma_0$. Solid lines display $\Delta C(T)/T$ data after subtraction of nuclear hyperfine contributions due to the applied fields. For B = 2 T, the superconducting transition anomaly at $T_c \approx 0.3$ K is seen.

It has been predicted that close to an antiferromagnetic (AF) QCP, low-lying and spatially extended spinfluctuations with wave vector $\mathbf{q} \approx \mathbf{Q}$, the afm ordering wavevector, give rise to strongly T-dependent quasiparticle masses and quasiparticle-quasiparticle cross sections [25-27]. For threedimensional (3D) systems these should manifest themselves in coefficients $\gamma = C/T$ and $a = (\rho - \rho_0)/T^2$ in the specific heat and resistivity which are not constant as in a Landau Fermi liquid, but obey the following asymptotic T dependences [26,27]:

$$\gamma(T) = \gamma_0 - \alpha T^{1/2} \quad (1)$$

and [25-27]

$$a(T) = \beta T^{-1/2} \quad (2a)$$

corresponding to

$$\Delta\rho = \rho(T) - \rho_0 = \beta T^{3/2} \quad (2b).$$

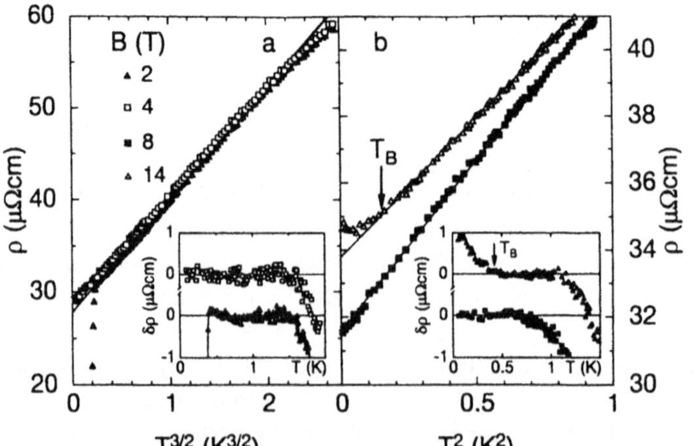


Fig.7. Resistivity for the same "AS-type" crystal as in Fig. 6 as $\rho$ vs $T^{3/2}$ for B = 2 T and 4 T (a) as well as $\rho$ vs $T^2$ for B = 8 T and 14 T (b). Insets show $\delta\rho$ vs T with $\delta\rho = \rho - \rho_0 - \beta T^{3/2}$ (a) and $\delta\rho = \rho - \rho_0 - \beta T^2$ (b), respectively. Arrows in (b) indicate B-phase transition for B = 14 T.

Since the singular scattering expressed by Eq. 2a is associated with the AF wavevector **Q**, i.e. occurs only along certain "hot lines" on the Fermi surface, all other quasiparticle-quasiparticle scattering events ought to give rise to the ordinary Fermi liquid term $\Delta\rho \propto T^2$ (a=const) which, consequently, must short-circuit the anomalous $\beta T^{3/2}$ term at sufficiently low temperatures [28]. In Figs. 6 and 7 we show results of $\gamma(T)$ and $\Delta\rho(T)$ for an "S"-type single crystal which lacks any A-phase signature. Let us first focus on the data taken at sufficiently low fields ($B < 6T$) and intermediate temperatures, cf. Figs. 6 and 7a. For $T > 0.2$ K and $B < 6$ T, both $\gamma(T)$ and $\Delta\rho(T)$ obey Eqs. 1 and 2b for $T < 1.7$ K suggesting the existence of an AF-QCP. The same conclusion has been drawn on the basis of specific-heat and resistivity results on an "A"-type $CeCu_2Si_2$ polycrystal which has been tuned through $T_A \rightarrow 0$ via hydrostatic pressure [29].
It is straightforward to relate this QCP to the disappearance of phase A at the critical coupling constant $g_c$ (Fig. 5). However, upon further approaching the QCP by cooling the single crystal to below 0.2 K, $\rho(T)$ and C(T) behave very disparately: While the resistivity (at sufficiently low fields) keeps varying as $\Delta\rho \propto T^{3/2}$ down to 20 mK, the lowest temperature of our experiment, the n-state specific-heat coefficient $\gamma(T)$ does <u>not</u> follow the corresponding $T^{1/2}$ dependence anymore (Fig. 6). Rather, it shows a steep upturn at low temperatures. Since this upturn cannot be ascribed to the Zeeman splitting of nuclear $^{63}Cu$, $^{65}Cu$ or $^{29}Si$ spin states through the external field (cf. solid lines in Fig. 6), an anomalous enhancement of the hyperfine coupling, i.e. an (average) finite "internal magnetic field" has to be invoked to account for this anomalous T dependence. The origin of this internal field is, however, not clear.
Upon increasing the magnetic field to $B > 6$ T one recognizes another surprising low-T property of our $CeCu_2Si_2$ single crystal. While the gross T dependence of $\gamma(T)$ remains unaffected (Fig.6), the $\Delta\rho(T)$ dependence becomes qualitatively changed into $\rho(T) = \rho_0 + aT^2$ (Fig. 7b). In addition, the B = 14 T data display the broadened transition into phase B which is not visible in $\gamma(T)$ measured at, e.g. B = 12 T (Fig. 6).
The $\rho(T) \propto T^2$ dependence suggests that phase B (as well as phase A, cf., e.g. Fig. 2 and [8]) forms out of a heavy Landau-Fermi-liquid phase - a notion which is, however, in conflict with the pronounced T dependences of $\gamma(T)$ precursive to both the B- and A-phase transition. The striking different T dependences of the resistivity for the $CeCu_2Si_2$ single crystal below and above $B \approx 6$ T are shown as $a(T) = \Delta\rho(T)/T^2$ vs T in Fig. 8a, along with the resistively determined B-T phase diagram in Fig 8b. We note that the field dependence of the limiting temperature $T_{lim}$ below which the $\Delta\rho = aT^2$ dependence is obeyed tracks that of the phase transition temperatures $T_B(B)$ and $T_A(B)$ for "A" and "AS"-type $CeCu_2Si_2$ [8]. From these observations one might be inclined to ascribe the $\Delta\rho = aT^2$ dependence preceding the A/B transitions to some critical fluctuations. However, assuming the A/B phases to be of itinerant nature, $\Delta\rho \propto T$ is predicted [30] in the critical regime $T > T_{A/B}$.

To summarize, n-state resistivity and specific-heat measurements performed in sufficiently low fields and at intermediate temperatures suggest the existence of a QCP of antiferromagnetic nature where $T_A \rightarrow 0$. This is concluded from the agreement of experimental data with theoretical predictions for a nearly antiferromagnetic Fermi liquid (NAFFL) in a one-band system of itinerant fermions [25-27]. However, upon approaching the QCP sufficiently closely, there are two striking observations that strongly violate this NAFFL scenario: 1) the absence of a crossover in $\Delta\rho(T)$ to a Landau-Fermi-liquid-type $T^2$ behavior (at least above 20mK). This suggests that singular scattering occurs on the <u>whole</u> Fermi surface rather than along some "hot lines" only. 2) $\Delta\rho(T)$ and $\gamma(T)$ behave very disparately. This indicates a breakdown of the concept of heavy quasiparticles. Rather it appears that, near the A-phase transition, their itinerant component (probed by $\Delta\rho(T)$) and the local 4f component (probed by C(T)) are decoupled from each other.

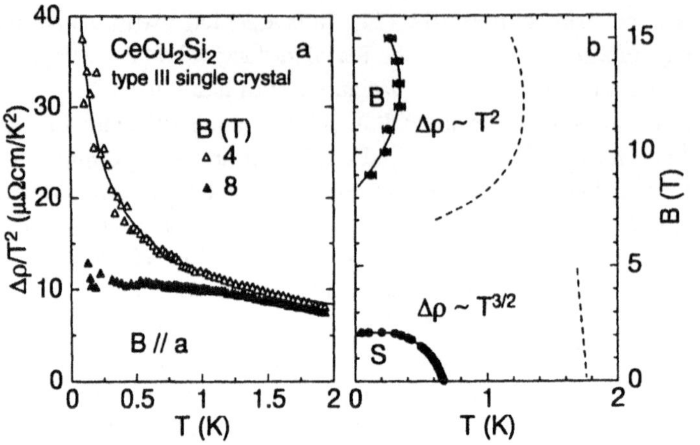


Fig.8. a: $\Delta\rho/T^2$ vs T ($\Delta\rho = \rho - \rho_0$) at B = 4 T and 8 T for the same "AS-type" $CeCu_2Si_2$ single crystal as in Figs. 6 and 7. Solid line marks $T^{-1/2}$ dependence of a(T) = $\Delta\rho(T)/T^2$.
b: B-T phase diagram for the same crystal derived from ρ(T) measurements. Existence ranges for superconductivity (S) and phase B are indicated along with limiting temperatures, $T_{lim}$, for $T^{3/2}$ and $T^2$ dependences of $\Delta\rho(T)$ (dashed lines).

### 3. $UBe_{13}$

Two variants of $UBe_{13}$ with markedly different sc- and n-state properties have been recently identified [31,32]: While "H-type $UBe_{13}$ exhibits $T_c$ values between 0.85 K and 0.9 K, "L-type" $UBe_{13}$ is characterized by $T_c \approx 0.75$ K. Most of the polycrystalline samples reported are of type H, while all L-type samples are single crystals.
In the following we discuss low-T properties of high-quality $UBe_{13}$ single crystals of the "H-type" variant.

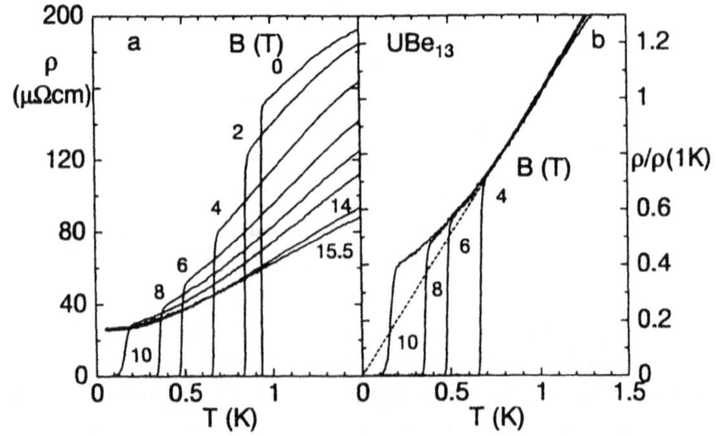


Fig. 9. a: ρ vs T for single crystalline $UBe_{13}$ at differing magnetic fields. b: Same data as in a, normalized to the respective ρ value at T = 1K. Dashed straight line is an extrapolation to T = 0 of the data for T ≥ 0.8K.

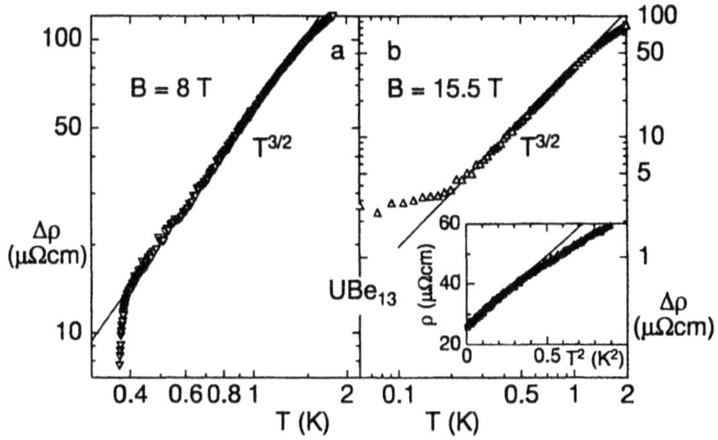


Fig. 10a,b. $\Delta\rho = \rho - \rho_0$ vs T (on logarithmic scales) for the same $UBe_{13}$ single crystal as in Fig. 9 at B = 8T (a) and B = 15.5T (b), respectively. Inset shows low-T data of (b) as ρ vs $T^2$.

### 3.1 Non-Fermi-liquid normal state

Amongst HF compounds the superconductor $UBe_{13}$ [2] is one of the most prominent examples of a Non-Fermi-liquid-type normal state.
Besides a characteristic (Kondo) scale $T^*$ ranging from 8 K [33] to 30 K [34] which accounts for the large effective carrier masses another low-energy scale exists in this compound: At $T \simeq 2K$ both thermodynamic properties, i.e. specific heat and thermal expansion as well as electrical resistvity [2] reveal more or less pronounced maxima (shoulders). From the maximum value of the resistivity, an inelastic mean free path as short as a few lattice spacings can be inferred. This anomaly has been commonly attributed to some, as yet unidentified, itinerant afm spin fluctuations [34]. On the other hand, as will be discussed below, the results of thermal-expansion measurements rather indicate these "2K fluctuations" to be of local (Kondo) type. As demonstrated in Fig. 9, already moderate fields are apt to suppress these fluctuation contributions efficiently: In a wide field range, $4\ T \leq B \leq 10\ T$, we are able to scale the various ρ(T) curves within $T_c(B) < T \leq 1.1K$ to a universal curve, by normalizing ρ(T) by its respective value at T = 1K (Fig. 9b).
Above $T \approx 0.8$ K, a linear ρ(T) dependence is found, that can be extrapolated to $\rho \simeq 0$ for $T \to 0$. At lower temperatures the data follow a $\rho = \rho_0 + \beta T^{3/2}$ dependence. This is demonstrated in Fig. 10a for the B = 8 T data. Apparently, this NFL behavior is in full accord with the predictions for the 3D NAFFL [25-27]. A $T^{3/2}$ dependence is found even up to the highest fields applied, i.e. 15.5 T. However, the low-T data for $T \leq 0.3$ K taken at high fields of B = 14 T and 15.5 T are better described by $\Delta\rho = \rho - \rho_0 = aT^2$, see inset Fig. 10b. The gigantic coefficient a is found to decrease with increasing field from 52 $\mu\Omega cmK^{-2}$ at B = 14 T to 45 $\mu\Omega cmK^{-2}$. We cannot decide whether this asymptotic $T^2$ term also exists at low fields, where it is masked by the sc transition, or whether it is induced by the high magnetic field. We note that already at low fields a crossover from a $T^{3/2}$ to $T^2$ at sufficiently low temperatures is required for a NAFFL [28]. Such a crossover, however, cannot be expected on the basis of isothermal n-state magnetoresistance results yielding $\Delta\rho(B) < 0$ below T = 2 K in the whole field range, B < 14 T. A negative sign of $\Delta\rho(B)$ is typical for NFL, while $\Delta\rho(B) > 0$ is a hallmark of the Landau FL. In fact, for $B \geq 14$ T $\Delta\rho(B)$ becomes positive at the same low temperatures where $\Delta\rho(T)$ follows the $T^2$ law.

To summarize, the NFL properties found for $UBe_{13}$ are consistent with the nearness of a QCP, though there is no clear-cut evidence for the presence of afm order at ambient pressure. A more detailed analysis of the NFL n-state properties is, however, hampered by the large value of its upper critical field. On the other hand, from the appearance of a positive peak in the thermoelectric power in pressurized (p > 23 Kbar) $UBe_{13}$ [35], a high-pressure AF ground state has been inferred which might be related to the present observations.

### 3.2 Revision of the phase diagram in $U_{1-x}Th_xBe_{13}$

First indications for the unconventional nature of the superconducting order parameter in $UBe_{13}$ stem from the observation of a $T^3$ dependence of the low-T specific heat [36]. Even more intriguing was the subsequent discovery of a non-monotonic $T_c(x)$ dependence for weakly Th doped $U_{1-x}Th_xBe_{13}$ [7], notably the occurrence of two subsequent phase transitions above a critical Th concentration $x_{cr} \approx 0.019 < x < 0.045$. So far, the origin of the lower of the subsequent transitions at $T_{c2} < T_{c1}$ is unclear. While early ultrasonic-attenuation measurements reveal indications for an AF transition [37], a superconducting nature of the transition has been claimed from measurements of the lower critical field [38]. In addition, the existence of very small magnetic moments of $\mu_s \approx 10^{-3}\mu_B/U$ for $T < T_{c2}$ has been deduced from muon-spin-relaxation ($\mu$SR) studies [39].

Attempting to explain the non-monotonic $T_c(x)$ dependence theoretical models started from the assumption of different anisotropic sc states for $x < x_{cr}$ and $x > x_{cr}$ as arising from the crossing of two different representations of the cubic symmetry group [40,41]. Studying various combinations of two respective representations Sigrist and Rice achieved a qualitative description of the above experimental results [41]. In their model, $T_{c2}$ marks the transition from a state belonging to a single representation for $T_{c2} < T < T_{c1}$ into a state formed by a combination of two representations. Since the latter is nonunitary it possesses magnetic

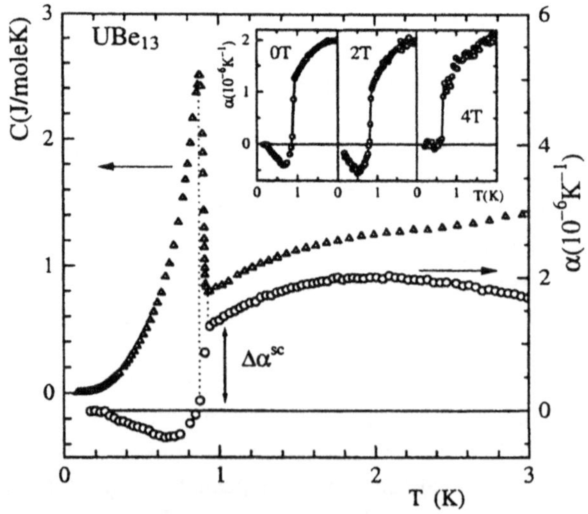


Fig. 11. Low-temperature specific heat (left scale) and thermal expansion (right scale) of single crystalline $UBe_{13}$ plotted on the same temperature scale. Width of superconducting transition is indicated by vertical dotted lines. Inset shows thermal expansion at varying fields.

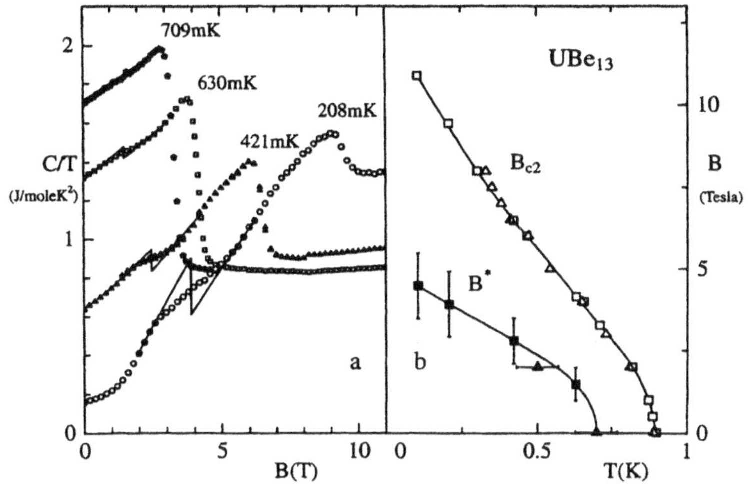


Fig. 12. a: Low-temperature specific heat as C/T vs B at varying temperatures of the same $UBe_{13}$ single crystal as in Fig. 11. b: Corresponding B-T phase diagram including the upper critical field, $B_{c2}(T)$, as determined by specific heat (open squares) and thermal expansion, (open triangles) as well as positions of anomalies observed as a function of temperature in $\alpha(T,B=const)$ (solid triangles) and magnetic field in $C(B,T=const)/T$ (solid squares) giving rise to a line of anomalies $B^*(T)$.

properties which then might account for the magnetic signatures found in the μSR experimens. In Fig. 11 we compare results of specific-heat and thermal-expansion measurements on the same high-quality $UBe_{13}$ single crystal by plotting the data sets on the same temperature scale. Compared to C(T), the $\alpha(T)$ data in the superconducting state display a rather non-monotonic T dependence suggesting the existence of a further anomaly below $T_c = 0.9$ K. In fact, by projecting the width of the sc transition from C(T) on the $\alpha(T)$ data (vertical lines in Fig.11) we find that the superconducting transition manifests itself in the steep decrease in $\alpha(T)$ at $T_c = 0.9K$, thus leaving the broadened minimum as an independent anomaly. This assignment is corroborated by (i) a thermodynamic analysis of the sc transition and (ii) an investigation of the field dependence of $\alpha(T)$: (i) Employing the construction as indicated in Fig. 11 to extract the $\alpha(T)$ discontinuity at $T_c$, $\Delta\alpha^{sc}$, we are able to calculate the initial hydrostatic-pressure dependence of $T_c$ by means of the Ehrenfest relation, i.e. $(\partial T_c/\partial p)_{p\to 0} = T_c \cdot V_{mol} \cdot 3 \cdot \Delta\alpha^{sc}/\Delta C^{sc}$, where $V_{mol} = 81.3$ cm³/mole is the molar volume. The so-derived pressure coefficient of $(\partial T_c/\partial p)_{p\to 0} = -(13\pm4)$ mK/kbar is in perfect agreement with the $T_c$ shift observed under hydrostatic-pressure conditions of -(13±4) mK/kbar [42]. (ii) The $\alpha(T)$ minimum is almost completely suppressed by a field of 4 T which has, however, only little effect on the sc transition (inset Fig. 11). From these $\alpha(T)$ measurements at zero field, or at constant, finite fields as well as isothermal field scans of the specific heat (Fig. 12a) a "new line of anomalies", $B^*(T)$, has been established in the B-T phase diagram of $UBe_{13}$ (Fig. 12b).

In Fig. 13 we follow the evolution of the $\alpha(T)$ minimum as a function of increasing Th doping. Upon increasing x to 0.01, 0.017 and 0.0185 (Fig. 13a), being still subcritical, the minimum becomes grows and shifts to lower temperatures. Using the corresponding specific-heat results, we can separate from the $\alpha(T)$ data the sc jump anomaly, leaving behind the $\alpha(T)$ minimum. For lack of any other well founded criterion, the (steeper) high-T flank of this minimum is replaced by an "equal-areas construction" by an idealized sharp jump. This defines the positions of the anomaly $T_L(x)$ in Fig. 12. A comparison of the data for subcritical Th concentrations with those for $x > x_{cr}$ (Fig. 13b) strongly suggests that the lower phase transition anomaly at $T_{c2}$ for $x > x_{cr}$ evolves out of the $\alpha(T)$ minimum in the subcritical concentration range: Being still similar to its outward appearance (i.e. sign and asymmetry) to that for $x < x_{cr}$, the minimum has become progressively deeper and sharper upon increasing x to 0.022 and 0.03, now revealing the character of a true phase transition. The respective transition temperatures as derived from an "equal-areas construction" in an $\alpha/T$ vs T plot are in good agreement with literature results, cf. Fig. 14.

The T-x phase diagram is supplemented by the "new line of anomalies" $T_L(x)$. As the central result of this study we propose that the anomaly at $T_L(x)$ (for $x < x_{cr}$) marks the precursor of the second phase transition at $T_{c2}(x)$ (for $x > x_{cr}$).

To interpret the broadened anomalies for $x < x_{cr}$ we have two possibilities: They manifest either an (inhomogeneously) broadened phase transition or some short-range ordering effects.

An anisotropic sc transition at the $T_L(x)$ - $T_{c2}(x)$ line is discarded for two reasons: (i) the sharpening of the anomaly upon increasing the impurity (Th) concentration, and (ii) the isotropic lattice response associated with the anomaly at $T_L(x)$ [10] both of which are in contrast to the expectation for a transition into a state with a gap anisotropy [40]. In particular, a combined-representation state below $T_{c2}$ as proposed in [41] must be ruled out, since this requires the crossing of two phase boundaries at $x_{cr}$.

Rather the broad shape as well as the negative sign of the $\alpha(T)$ anomaly at $T_L$ would be consistent with short range AF correlations. The sharpening of these features for Th concentrations in excess of $x_{cr}$ then indicates the onset of true long-range order.

Finally, we like to address the line denoted $T_{max}$ in the T-x phase diagram in Fig. 14. $T_{max}(0)$ denotes the position of the "2K anomaly" in the coefficient of thermal expansion. To follow its evolution as a function of Th doping we plot in Fig. 15 $\alpha(T)$ data for various x over a somewhat extended temperature range. Compared to resistivity and specific heat, this anomaly

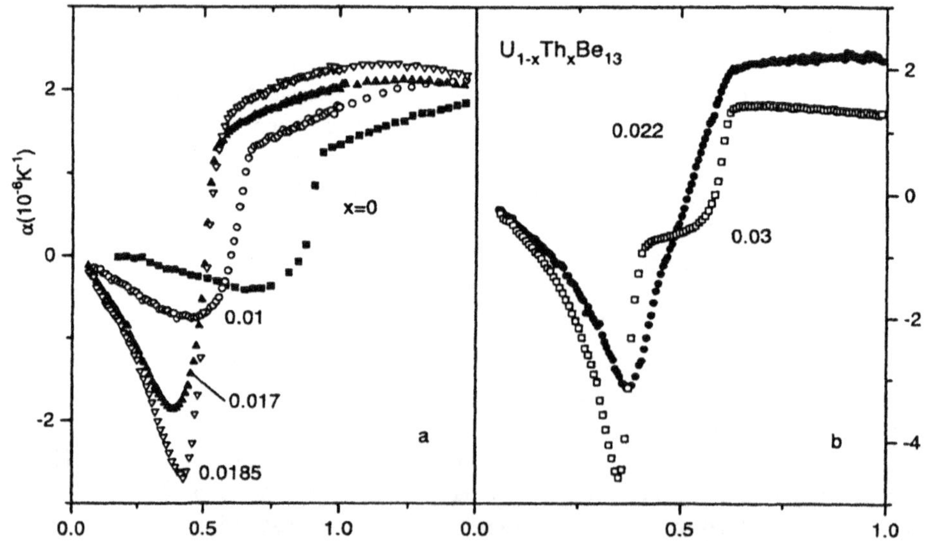


Fig. 13. Coefficient of thermal expansion of single crystalline $UBe_{13}$ and polycrystalline $U_{1-x}Th_xBe_{13}$ for x = 0.01, 0.017 and 0.0185 ($< x_{cr}$) (a) as well as x = 0.022 and 0.03 ($x > x_{cr}$) (b).

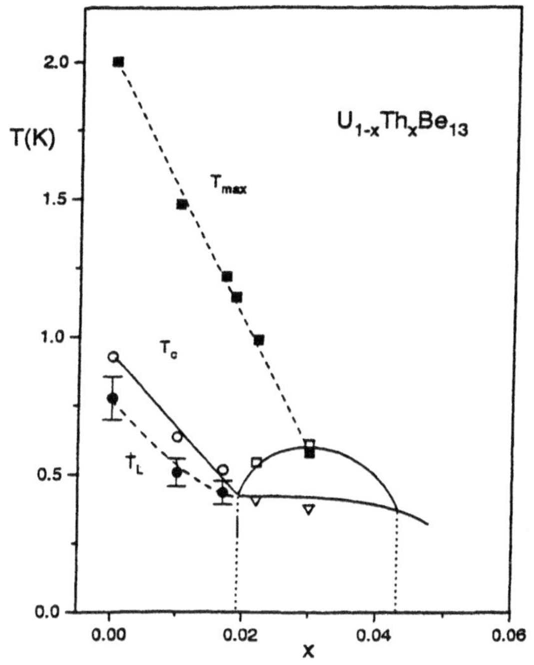


Fig. 14. T-x phase diagram of $U_{1-x}Th_xBe_{13}$ including results from literature [39] (solid lines) and the present study. Open symbols indicate second-order phase transitions, while closed symbols mark anomalies at $T_L$ and $T_{max}$ as described in the text. Vertical bars indicate uncertainties in determining $T_L$.

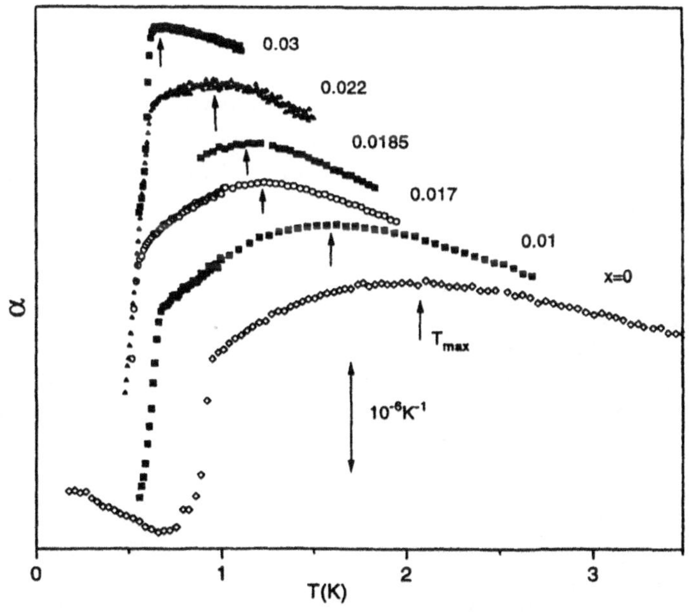


Fig. 15. Low-temperature thermal expansion of single crystalline $UBe_{13}$ and polycrystalline $U_{1-x}Th_xBe_{13}$. A vertical shift of each data set have been employed for clarity. Arrows mark the positions of the $\alpha(T)$ maximum.

is most strongly pronounced in $\alpha(T)$ which renders a background correction unnecessary. Most interestingly, and in contrast to the results of resistivity measurements [34], we find that $T_{max}(x)$ shows a linear reduction with x, hitting $T_{c1}$ at x = 0.03, i.e. right at its maximum value.
As already mentioned in Sect. 3.1 both the positive sign of $\alpha$ at the "2K anomaly" as well as the large corresponding Grüneisenparameter $\Gamma_{eff} \approx 50$ indicate the gradual freezing out of local Kondo-type spin fluctuations. This suggests an effective two-band scenario for $U_{1-x}Th_xBe_{13}$: HF superconductivity is carried by "less localized 5f states" with a characteristic temperature $T_{K1}$ = 8 - 30 K [33,34]. These states are scattered by Kondo fluctuations of the "more localized 5f states" with $T_{K2} \approx T_{max}(x)$. As long as $T_{max} > T_c$, these fluctuations are practically frozen out, i.e. harmless for superconductivity. Once $T_{max}$ becomes smaller than $T_{c1}$, however, an additional pair-breaking channel might open. This could be the case at x = 0.03 above which the pair-breakingcharacter of these fluctuations causes $T_{c1}(x)$ to decrease, i.e. destabilizes the sc state.

To summarize, thermodynamic evidence is provided for a new line of anomalies in the T-x diagram of sc $U_{1-x}Th_xBe_{13}$. Although its nature cannot be established unambiguously by the present thermodynamic studies, our results rule out the models applied to this system so far and further constrain possible explanations for the various low-T phases of this canonical HF system.

## 4. Conclusions

The prototypical HF materials $CeCu_2Si_2$ and $UBe_{13}$ display significant NFL phenomena in their normal-conducting states. In either case our experimental findings indicate the existence of a "nearby" quantum critical point of antiferromagnetic type.
For $CeCu_2Si_2$, our investigations of different variants prove that the NFL normal state is neither necessary nor sufficient for the occurrence of HF superconductivity. Upon approaching the quantum critical point by further cooling the different response from resistivity on the one hand and specific heat on the other highlights a break-up of the itinerant and local (4f) parts of the heavy quasiparticles.
For $UBe_{13}$, our resistivity data indicate that the incoherent normal state is caused only to a smaller part by the so-called "2K-fluctuations". In addition, thermodynamic measurements provide evidence for a new line in T-x diagram of $U_{1-x}Th_xBe_{13}$ that marks the precursor of the lower transition at $T_{c2}$ for x > 0.019.

**Acknowledgements**
We are grateful to W. Assmus for supplying the $CeCu_2Si_2$ single crystals. We gratefully acknowledge stimulating discussions with M. Sigrist and P. Thalmeier. Work in Darmstadt was supported by the BMBF grant NO. 13N6608/1. Work in Gainesville was supported by the DOE, Contract No. De-FG05-86ER45268.

## References

[1] F. Steglich, J. Aarts, C.D. Bredl, W. Lieke, D. Meschede, W. Franz and H. Schäfer, Phys. Rev. Lett. **43**, 1892 (1997).
[2] H.R. Ott, H. Rudigier, Z. Fisk and J.L. Smith, Phys. Rev. Lett. **50**, 1595 (1983).

[3] G.R. Stewart, Z. Fisk, J.O. Willis and J.L. Smith, Phys. Rev. Lett. **52**, 679 (1984).
[4] W. Schlabitz, J. Baumann, B. Pollit, U. Rauchschwalbe, H.M. Mayer, U. Ahlheim and C.D. Bredl, Abstr. ICVF5, Köln 1984 (unpublished); Z. Phys. B **62**, 171 (1986).
[5] C. Geibel, S.Thies. D. Kaczorowski, A. Mehner, A. Grauel, B. Seidel, U. Ahlheim, R. Helfrich, K. Petersen, C.D. Bredl and F. Steglich, Z. Phys. B **83**, 305 (1991).
[6] C. Geibel, C. Schank, S. Thies, H. Kitazawa, C.D. Bredl, A. Böhm, M. Rau, A. Grauel, R. Caspary, R. Helfrich, U. Ahlheim, G. Weber and F. Steglich, Z. Phys. B **84**, 1 (1985).
[7] H.R. Ott, H. Rudigier, Z. Fisk, J.L. Smith, Phys. Rev. B **31**, 1651 (1985).
[8] F. Steglich, P. Gegenwart, R. Helfrich, C. Langhammer, P. Hellmann, L. Donnevert, C. Geibel, M. Lang, G.Sparn, W. Assmus, G.R. Stewart, A. Ochiai, Z. Phys. B **103,** 235 (1997).
[9] P. Gegenwart, L. Donnevert, C. Langhammer, R. Horn, A. Link, C. Geibel, R. Helfrich, M. Lang, G. Sparn and F. Steglich (submitted).
[10] F. Kromer, R. Helfrich, M. Lang, F. Steglich, C. Langhammer, A. Bach, T. Michels, J.S. Kim and G.R. Stewart (submitted).
[11] M. Lang, R. Modler, U. Ahlheim, R. Helfrich, P.H.P. Reinders, F. Steglich, W. Assmus, W. Sun, G. Bruls, D. Weber and B. Lüthi, Physica Scripta T **39**, 135 (1991).
[12] G. Bruls, B. Wolf, D. Finsterbusch, P. Thalmeier, I. Kouroudis, W. Sun, W. Assmus, B. Lüthi, M. Lang, K. Gloos, F. Steglich and R. Modler, Phys. Rev. Lett. **72**, 1754 (1994).
[13] F. Steglich, P. Gegenwart, C. Geibel, R. Helfrich, P. Hellmann, M. Lang, A. Link, R. Modler, G. Sparn, N. Büttgen and A. Loidl, Physica B **223&224**, 1 (1995).
[14] Y. Uemura, W.J. Kossler, X.H. Yu, H.E. Schone, J.R. Kempton, C.E. Stronach, S. Barth, F.N. Gygax, B. Hitti, A. Schenck, C. baines, W.F. Lankford, Y. Onuki, T. Komatsubara, Phys. Rev. B **39**, 4726 (1989).
[15] H. Nakamura, Y. Kitaoka, H. Yamada, K. Asayama, J. Magn. Magn. Mat. **76&77**, 517 (1988).
[16] P. Thalmeier, Z. Phys. B **95**, 39 (1994).
[17] O. Trovarelli, M. Weiden, R. Müller-Reisener, M. Gómez-Berisso, P. Gegenwart, M. Deppe, C. Geibel, J.G. Sereni, F. Steglich, Phys. Rev. B **56**, 678 (1997).
[18] Y. Miyako, T. Takeuchi, T. Taniguchi, Y. Yamamoto, S. Kawarazaki, M. Acet, G. Dumpich, E.F. Wassermann, Z. Phys. B **101**, 339 (1996).
[19] R. Modler, Dissertation TH Darmstadt (1995) (unpublished).
[20] Kwak et al. Phys. Rev. Lett. **46**, 1296 (1981), D. Jérome, Science **252**, 1509 (1991).
[21] C. Geibel, Phys. Bl. **53**, 689 (1997).
[22] P. Gegenwart, M. Lang, A. Link, G. Sparn, C. Geibel, F. Steglich, W. Assmus, submitted to SCES '98, Paris, France.
[23] P. Gegenwart, Dissertation, TU Darmstadt (unpublished).
[24] R. Müller-Reisener, Diploma Thesis, TH Darmstadt (1995), unpublished.
[25] A.J. Millis, Phys. Rev. B **48**, 7183 (1993).
[26] G.G. Lonzarich, College on Quantum Phases, ICTP Trieste 1994, unpublished.
[27] T. Moriya and T. Takimoto, J. Phys. Soc. Jpn. **64**, 960 (1995).
[28] R. Hlubina and T.M. Rice, Phys. Rev. B **51**, 9253 (1995).
[29] F. Steglich, B. Buschinger, P. Gegenwart, M. Lohmann, R. Helfrich, C. Langhammer, P. Hellmann, L. Donnevert, S. Thomas, A. Link, C. Geibel, M. Lang, G. Sparn and W. Assmus, J. Phys.: Condens. Matter **8**, 9909 (1996).
[30] K. Ueda, J. Phys. Soc. Jpn. **43**, 1497 (1977).
[31] R. Helfrich, Dissertation, TU Darmstadt (1996), unpublished.

[32] C. Langhammer, R. Helfrich, A. Bach, F. Kromer, M. Lang, T. Michels, M. Deppe, F. Steglich, G.R. Stewart, J. Magn. Magn. Mat. **177-181**, 443 (1998).
[33] R. Felten, F. Steglich, G. Weber, H. Rietschel, F. Gompf, B. Renker, J. Beuers, Europhys. Lett. **2**, 323 (1986).
[34] E.A. Knetsch, Dissertation, University of Leiden (1993), unpublished.
[35] D. Jaccard, J. Sierro, J.P. Brison, and J. Flouquet, J. Physique **49**, C8-741 (1988).
[36] H.R. Ott, H. Rudigier, T.M. Rice, K. Ueda, Z. Fisk, J.L. Smith, Phys. Rev. Lett. **52**, 1915 (1984).
[37] B. Batlogg, D. Bishop, B. Golding, C.M. Varma, Z. Fisk, J.L. Smith, H.R. Ott, Phys. Rev. Lett. **55**, 1319 (1985).
[38] U. Rauchschwalbe, F. Steglich, G.R. Stewart, A.L.Giorgi, P. Fulde, K. Maki, Europhys. Lett. **3**, 751 (1987).
[39] R.H. Heffner, J.L. Smith, J.O. Willis, P. Birrer, C. Baines, F.N. Gygax, B. Hitti, E. Lippelt, H.R. Ott, A. Schenck, E.A. Knetsch, J.A. Mydosch, D.E. MacLaughlin, Phys. Rev. Lett. **65**, 2816 (1990).
[40] R. Joynt, T.M. Rice and K. Ueda, Phys. Rev. Lett. **56**, 1412 (1986).
[41] M. Sigrist and T.M. Rice, Phys. Rev. B **39**, 2200 (1989).
[42] S.E. Lambert, Y. Dalichaouch, M.B. Maple, J.L. Smith and Z. Fisk, Rev. Lett. **57**, 1619 (1986).

## DISCUSSION

**Sawatzky:** If I may make a comment. In RKKY theory the exchange interaction is usually described by delta function in real space and for rare earths the exchange is mostly due to 4f - 5d exchange interactions and the 5d character in the conduction electron sea is determined by the d-conduction electron hybridization. This mechanism uses the conduction electron states of local d symmetry. In the Kondo-like coupling derived from an Anderson like impurity Hamiltonian the exchange is derived starting from a hybridization of the f orbitals with the conduction electron states using therefore states of local f symmetry and then using a Schrieffer-Wolff transformation to get to the Kondo Hamiltonian. Note that these mechanisms use different symmetries of conduction electron states and are therefore not directly related. In addition one has the superexchanged terms I have mentioned before which destroy the simple relation between RKKY and Kondo-like terms.

# ONSET OF MAGNETISM AND NON-FERMI LIQUID BEHAVIOUR IN UTX COMPOUNDS

L. Havela and V. Sechovský

Department of Metal Physics, Charles University, Ke Karlovu 5, 121 16 Prague, The Czech Republic

## INTRODUCTION

Magnetism based of the *5f*-states in light actinides can be very vaguely characterized as a phenomenon spanning the range between the *3d*-band magnetism of transition metals and the magnetism of localized *4f* states in regular lanthanides. The tunability of the degree of delocalization of the *5f*-electronic states, which depends on the overlap of the *5f* wave functions centered on neighbouring sites, and on the strength of the *5f*-ligand hybridization, makes indeed possible to reach, depending on crystal structure, stoichiometry, and type of constituents, a large variety of types of the ground state. The narrow *5f*-band regime, into which most the materials fall, means much higher complexity of the problem due to the prominent role of the *5f*-electron correlations. Additional non-standard feature is the crucial role of relativistic phenomena. The last ingredient leads, for example, to a strong orbital polarization even in strongly itinerant systems, and to a strong coupling of the spin system to the crystal lattice, both being the consequence of strong spin-orbit interaction.

In this overview we plan to describe a recent development in several particular issues of the magnetism of uranium intermetallics, which are studied on the background of equiatomic ternary compounds of the UTX type, with a transition metal T and a non-transition metal X. First issue is the maximum possible localization of the *5f* states. In another words, whether the *5f*-ligand hybridization can be weak enough to allow the localization of *5f* states, withdrawing them from the metal bonding. The necessity to minimize the hybridization with the transition metal *d*-states means that we should survey materials with the very late transition metals (noble metals), which form the structure of the $CaIn_2$ type. In particular, we concentrate on UPdSn, with possible dilution of the U sublattice.

On the other hand, a lot of interest is devoted to the region of vanishing magnetic order, with interesting non-Fermi liquid features often found around the verge. The critical assessment leads however to a hesitation to attribute the observed anomalous power laws to quantum critical point in case of substitutional disorder. Thus we target also the materials with stronger *5f-d* hybridization, crystallizing in the ZrNiAl structure type.

*Electron Correlations and Materials Properties*, edited by Gonis *et al.*
Kluwer Academic/Plenum Publishers, 1999.

In this survey we concentrate in more detail on the concetntartion range around UCoAl, where a clear example of the *5f*-band metamagnetism appears. We demonstrate that certain non-Fermi liquid features occur here even in absence of substitutional disorder.

Finally we will describe results of a dilution study for such material, which corroborates the strong band character of such type of magnetism.

## DILUTION OF U SUBLATTICE IN UPdSn

From the UTX compounds with the $CaIn_2$ structure type or its derivative (so far known UPdSn, UAuSn, UCuSn, UCuGa, and UPbSb) the most attention was paid to UPdSn, which is the real ordered ternary compound (GaGeLi structure type), which was synthesized in the single crystal form. This compound is an antiferromagnet with $T_N = 40$ K which undergoes another magnetic phase transition at 27 K. On the basis of the magnitude of effective moment (3.3 $\mu_B$/U), ordered moment ($\approx 2$ $\mu_B$/U) and a relatively large magnetic entropy estimated as 1.5*Rln2 (de Boer et al., 1992) it can be definitely classified as a local *5f*-moment system. The low value of the electronic specific heat $\gamma = 5$ mJ/mol K$^2$ (similar value as recorded for $UPd_3$) can be attributed to the *5f* states removed from the Fermi surface. Some uncertainty about the *5f* localization remains even after electronic structure calculations (Trygg et al. 1994) and photoelectron spectroscopy experiments (Havela et al., 1992a, 1992b), which have revealed a fingerprint of the *5f* states at the Fermi level. Most recently, the electronic structure calculations performed for the realistic magnetic structure by Sandratskii and Kübler (1997) have demonstrated that assuming the itineracy of the *5f* states a low value of $\gamma = 7.5$ mJ/mol K$^2$ can be obtained as a result of a deep minimum of the density of electronic states at the Fermi energy. To learn more about the nature of the *5f* states in UPdSn, we have undertaken a dilution study, with the U-sublattice diluted by Lu.

As cast samples prepared by arc melting of stoichiometric amounts of pure elements show a single-phase character up to $x = 0.25$. The sample with $x = 0.30$ shows already some spurious phases. The concentration dependence of lattice parameters within the homogeneity range shows a weak monotonous contraction.

Fig. 1 displays results of magnetic susceptibility measurements performed on randomly oriented powder samples. All $\chi(T)$ data can be in the paramagnetic range fitted to a modified Curie-Weiss law: $\chi(T) = C/(T - \Theta_p) + \chi_0$. The value of the effective moment per 1 U atom $\mu_{eff}$ decreases weakly with increasing Lu concentration from 3.1 to 2.7 $\mu_B$/U atom, whereas the paramagnetic Curie temperature $\Theta_p$ becomes progressively more negative (developing from -5 K for UPdSn to -40 K for 30% Lu substituted sample). One has to keep in mind that the data are affected by magnetic anisotropy (*c*-axis is the hard magnetization direction in UPdSn), which leads to $\mu_{eff}$ values somewhat smaller than obtained for single crystal with field along the easy axis. The values of the temperature independent term $\chi_0$ remains of the order of $3*10^{-9}$ m$^3$/mol for all the concentrations studied. Inspection of the critical region around the magnetic phase transition shows a substantial change of the behaviour induced already by a 2% Lu substitution. The lower-temperature magnetic phase transition, which represents an onset of a new order parameter in the pure UPdSn [5,6], disappears completely in Lu substituted sample and the low temperature susceptibility becomes substantially higher compared to the parent compound. The Néel temperature $T_N$ decreases systematically with increasing Lu content, but for higher concentrations this tendency saturates, as seen in Fig. 2. Although we cannot be sure without a neutron diffraction experiment that the low temperature magnetic phase from UPdSn is suppressed with Lu substitutions, the shape of the maximum on the $\chi(T)$ curve, which is more or less preserved throughout, makes the impression that the high-temperature

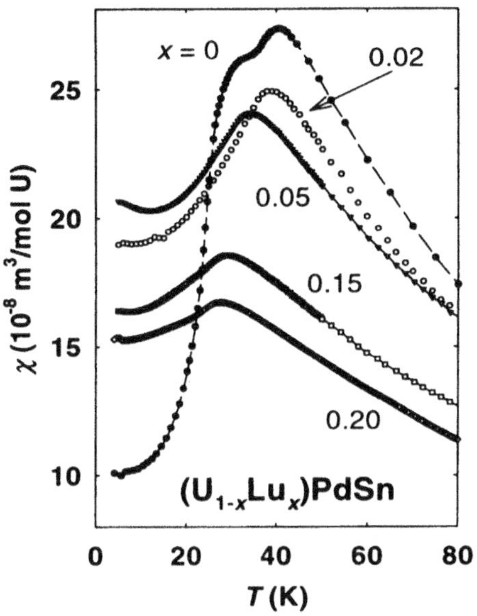


**Figure 1.** Detail of the temperature dependence of magnetic susceptibility $\chi(T)$ measured in the field of 2 T for various Lu concentrations. (After Havela et al. 1998).

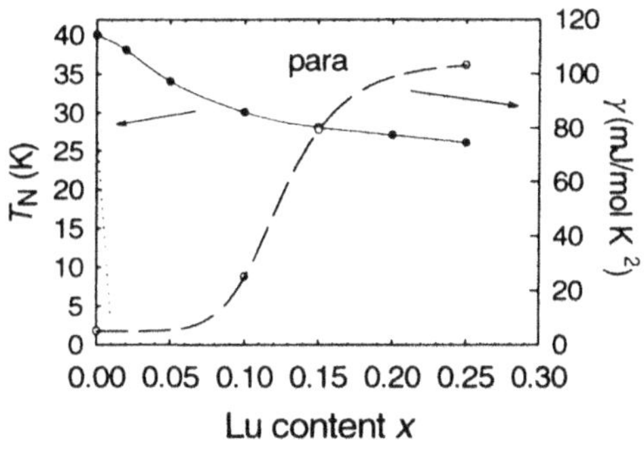


**Figure 2.** Tentative magnetic phase diagram of the $(U_{1-x}Lu_x)PdSn$ system. The dotted line represents schematically the fast supression of the low temperature magnetic structure of UPdSn. The temperatures of the order-disorder magnetic phase transitions were determined from maxima in $\chi(T)$. The empty symbols show the concentration development of the low temperature $\gamma$-coefficient. The dashed line is guide to the eye. (After Havela et al. 1998; Dremov 1998).

antiferromagnetic phase of UPdSn is spread down to lowest temperatures in all Lu substituted samples.

Fig.2 shows also the increase of $\gamma$ with increasing Lu content. It is obvious that already few percent of Lu increase $\gamma$ above the limit of about 10 mJ/mol $K^2$, below which the *5f*-localization can be speculated. All that happens without any dramatic modification of ordering temperature, effective moment or paramagnetic Curie temperature. Thus we incline to understanding of the low $\gamma$ in UPdSn as due to "accidental" position in a minimum of the Fermi level in a deep minimum of the density of states, and not due to the *5f*-localization.

## METAMAGNETISN IN UCoAl

Similar to the materials with the $CaIn_2$ structure type, also the compounds having the ZrNiAl structure type and contain late transition metals display a magnetic (mainly AF) ground state. The fact that magnetism is lost for the earlier *d*-metals (like in UFeAl) is a tendency more or less general in all intermetallics of U with *d*-metals, and is due to a stronger *5f-d* hybridization. (For an overview see the chapter of Sechovsky et al. (1998) in the same volume). If we, however, study the loss of magnetism in detail in the course of substitution of, e.g., Ni in UNiAl by Co or Fe, we see that the loss of magnetism is non-monotonous. The antiferromagnetism of UNiAl lost very fast, but close to UCoAl a new island of stability of ferromagnetism appears. One should keep in mind that also here we deal with the *5f*-magnetism: only weaker (exchange-induced) transition-metal moments were observed by neutron diffraction in materials like URhAl (Paixao et al., 1992). And finally, also this ferromagnetism is lost if we progress further to the left, e.g. from URhAl to URuAl.

The interesting behaviour of band metamagnetism is traditionally asociated with several transition-metal based materials as, for example, $YCo_2$ (see Goto et al., 1994). The ground state is non-magnetic, with more or less pronounced spin fluctuation features. A metastable ferromagnetism can be induced by high magnetic fields, which can be understood as a simple consequence of fulfillment of the Stoner criterion under the influence of the field. A clear distinction from "standard" metamagnetism on the basis of antiferromagnets makes absence of any magnetic phase transition in zero magnetic field. The fact that the entropy at a particular temperature is lower in the metamagnetic state leads to an increase of the critical metamagnetic field $H_C$ with increasing temperature.

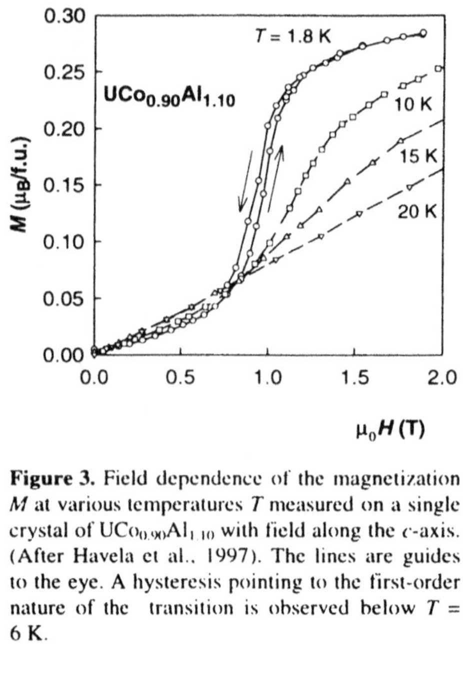


**Figure 3.** Field dependence of the magnetization $M$ at various temperatures $T$ measured on a single crystal of $UCo_{0.90}Al_{1.10}$ with field along the $c$-axis. (After Havela et al., 1997). The lines are guides to the eye. A hysteresis pointing to the first-order nature of the transition is observed below $T$ = 6 K.

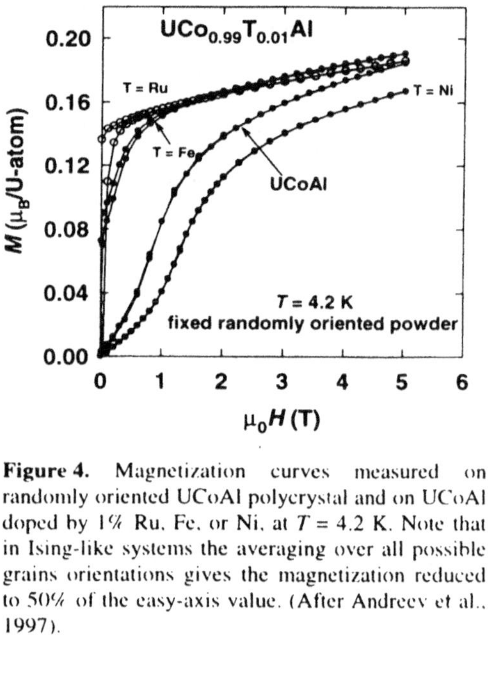


**Figure 4.** Magnetization curves measured on randomly oriented UCoAl polycrystal and on UCoAl doped by 1% Ru, Fe, or Ni, at $T$ = 4.2 K. Note that in Ising-like systems the averaging over all possible grains orientations gives the magnetization reduced to 50% of the easy-axis value. (After Andreev et al., 1997).

The metamagnetism in UCoAl has one specific feature, namely the uniaxial character, which conforms to the general Ising-like character of all U-compounds with the ZrNiAl structure type. Its critical field is rather low (about 1 T) (see Fig.3) and can be varied by deviations from the ideal 1:1:1 stoichiometry (Havela et al., 1997). The excess of Co, as well as doping by Fe or Ru, leads to a decrease of the critical field and stabilization of ferromagnetism (see Figs. 4 and 5). On the other hand, substitution of e.g. Ni for Co or Co-defficiency increases the critical field. Thus the metamagnetic region appears on the verge of the island of stability of ferromagnetism. Furthermore, the metamagnetism is strongly affected by external pressure (see Fig.5), which increases the critical field and reduces the high-field magnetization, corroborating thus the band character of the observed phenomenon. Extrapolation of the pressure variation of the magnetization increase at $H_C$, which is linear, to zero leads to estimated suppression of metamagnetism in $p$ = 22—24 kbar (Andreev et al., 1997).

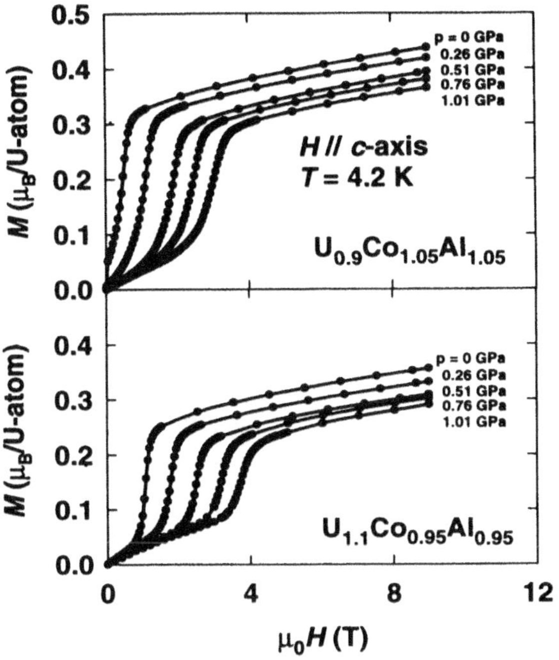


**Figure 5.** Field dependencies of magnetization $M$ at $T = 4.2$ K, measured for two off-stoichiometric single crystals with field along the $c$-axis. Applied hydrostatic pressure is shown at each curve. (After Havela et al., 1997).

Fortunately, pure stoichiometric UCoAl falls into the metamagnetic region, which allows to study transport properties in a material of reasonable perfection, i.e. without substitutional disorder. In particular, we were curious whether deviations from the Fermi liquid picture, expected due to spin fluctuations around the onset of magnetic ordering (Mathon, 1968, Ueda, 1977, Moriya and Takimoto 1995) can be found in a material, in which the proximity to magnetic ordering is manisfet by band metamagnetism with small critical field.

Fig.6 shows that the temperature dependence of electrical resistivity $\rho(T)$ does not follow indeed the standard Fermi liquid dependence of the $aT^2$ type expected for a standard electron quasiparticle scattering. Instead, the exponent is lower, and we can fit the experimental data up to about 20 K to the $bT^{3/2}$ type of dependence. This power law should appear in case of antiferromagnetic spin fluctuations, but the exponent $n = 5/3$ derived spin fluctuations of the ferromagnetic type does not show much worse fit quality. In magnetic fields $H > H_c$ the $aT^2$ power law is restored, as seen in Fig. 7, which means that the exponent $n < 2$ is indeed due to magnetic fluctuations in the low-field state.

The effect of spin fluctuations on temperature dependence of various quatities in the vicinity of a quatum critical point for different dimensionalities and different types of coupling was recently reviewed by Mishra and Sreeram (1998). It was shown that in case of 3D ferromagnets one can indeed expect the resistivity exponent $n$, normaly equal 2 in the Fermi liquid case, to be somewhat lower. It is further reduced in the case of antiferromagnetic coupling ($n \leq 3/2$) or lower dimensionality ($n \leq 4/3$) of the ferromagnetic coupling.

The necessity to distinguish small differences in the resistivity exponent motivated a new resistivity study on a crystal with an order of magnitude lower residual resistance $\rho_0$. Preliminary data show in this clean case that the dependence of the $bT^{5/3}$ is obeyed

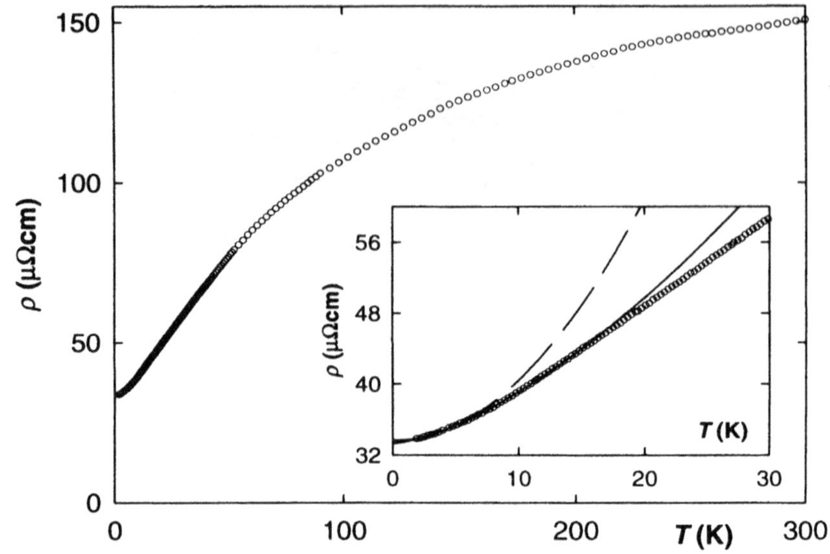


**Figure 6.** Temperature dependence of electrical resistivity $\rho$ (current $i$ along the $c$-axis) for $UC_{0.95}Al_{1.05}$ single crystal in zero magnetic field. The inset shows the low temperature detail as well as the fits to $\rho \sim bT^{3/2}$ and $\rho \sim aT^2$, the latter yielding somewhat lower $\rho_0$ (After Kolomiets et al., 1998).

significantly better than the $T^{3/2}$ type of dependence both or current along the $c$-axis and in the basal plane, which conforms to the situation in the close vicinity of long-range *ferromagnetism*. Remaining conclusions of the previous experiment remain valid in this case, too (Kolomiets, 1998). It is interesting to note that the power law with $n$ = 5/3 was indicated also for another compound of this type, namely mid-weight heavy-fermion antiferromagnet UNiAl with strongly reduced U-moments and $T_N$ = 19 K (Brück et al., 1994), but only for the $c$-axis direction of the current. For current along the basal plane the standard $aT^2$ type of dependence appears.

## DILUTION OF U SUBLATTICE IN UCoAl

Up to 40% non-magnetic Y can be substituted for U while preserving the ZrNiAl structure type. The substitution expands slightly the lattice. The lattice volume increases by 1.8% for the ultimate Y concentration, which comes from the 2.2% linear expansion in the basal plane and 2.6% shrinking along the $c$-axis. The expansion of the critical U-U spacing in the basal plane, which is just of the critical Hill limit $d_{U-U} \approx 350$ pm, is one of possible reasons why the metamagnetic behaviour is suppressed and long-range ferromagnetism is established for small Y concentrations (Andreev et al. 1998). At the lowest Y concentration studied (2%), the metamagnetism and ferromagnetism coexist. This development can be understood in terms of the observred strong pressure sensitivity of the meagnetic properties of UCoAl. Based on the compressibility data and the pressure derivative of the critical metamagnetic field, we estrimate that 0.5% volume expansion stabilizes ferromagnetism by reducing the critical metamagnetic field to zero. In the course of the Y substitution, it is already 0.3% volume expansion for 6% Y suppressing the metamagnetsim. This difference can be qualitatively explained by the dominating influence of the basal-plane U-U spacing, which is much more affected by the Y substitution than the volume itself.

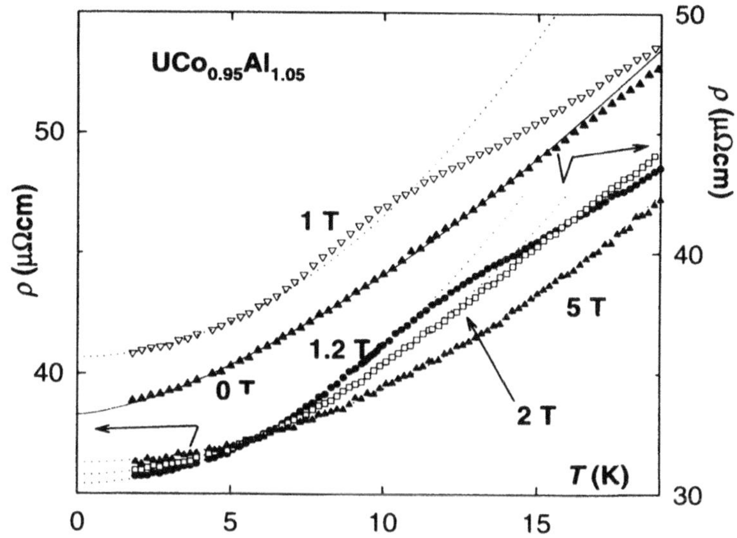


**Figure 7.** Temperature dependencies of resistivity $\rho$ for $UCo_{0.95}Al_{1.05}$ in various fields. The dotted lines represent the fits to the $T^2$ law, the full lines fits to the $T^{3/2}$ law. Note two different $\rho$-scales (left and right). (After Kolomiets et al., 1998).

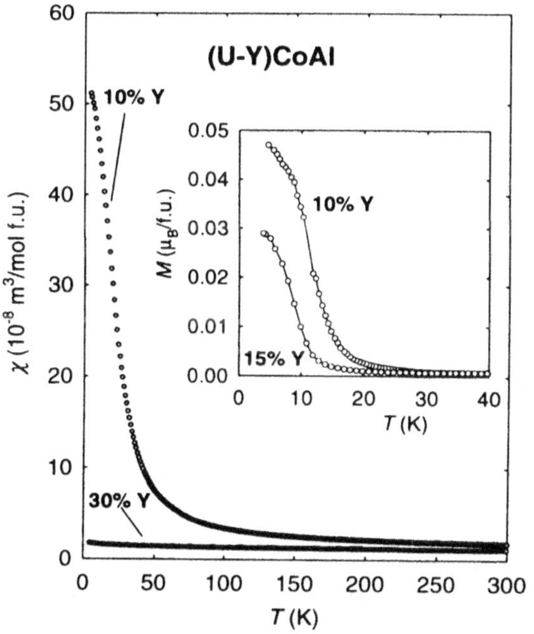


**Figure 8.** Temperature dependence of magnetic susceptibility for polycrystalline materials $U_{1-x}Y_xCoAl$, measured in $\mu_0 H = 2$ T. The inset shows the measurements done in $\mu_0 H = 0.01$ T, which reflect a residual magnetization. Again, the magnetization along the easy axis should be twice larger.

The long-range ferromagnetism does not survive, however, more Y substitution than 10%. It is suppressed in a way not yet entirely understood, and weak Pauli paramagnetism is observed for 30% Y. (See Fig.8). This is qualitatively different behaviour comparing to what was found at more local-moment analogs like UCoSn, studied in the course of Th substituting for U. Here ferromagnetism disappears around 60 % Th, but for 80% Th, U still preserves features of the Curie-Weiss behaviour and the paramagnetic Curie temperature tends to 0 K (Sechovsky et al 1988). Nevertheless, it seems that in the U-dilution limit U-moments vanish systematically even in compounds of the latter type, which brings us the question about importance of intersite exchange coupling for the formation of U-moments.

## Acknowledgments

This work was supported by the Grant Agency of the Czech Republic (Grant No. 202/96/0207).

## References


Andreev, A.V., L. Havela, V. Sechovsky, M.I. Bartashevich, T. Goto, and K. Kamishima, 1997, 5f band metamagnetism in UCoAl: Effects of alloying and pressure, *J.Magn.Magn.Mater.* 169: 229

Andreev, A.V., I.K. Kozlovskaya, N.V. Mushnikov, T. Goto, V. Sechovsky, L. Havela, Y. Homma, and Y. Shiokawa, 1998, Magnetic properties of $U_{1-x}Y_xCoAl$, to be published

Brück, E., H. Nakotte, F.R. de Boer, P.F. de Châtel, H.P. van der Meulen, J.J.M. Franse, A.A. Menovsky, N.H. Kim-Ngan, L. Havela, V. Sechovsky, J.A.A.J. Perenboom, N.C. Tuan, J. Sebek, 1994, Electronic Properties of UNiAl in High Magnetic Fields, *Phys.Rev. B* 49: 8852

de Boer, F.R., E. Brück, H. Nakotte, A.V. Andreev, V. Sechovsky, L. Havela, P. Nozar, C.J.M. Denissen, K.H.J. Buschow, B. Vaziri, M. Meissner, H. Maletta, and P. Rogl, 1992, Magnetic, electrical, and specific-heat properties of UPdSn and UAuSn, *Physica B* 176: 275.

Dremov, R.V., 1998, unpublished results

Goto, T., H. Aruga Katori, T. Sakakibara, H. Mitamura, K. Fukamichi, and K. Murata, 1994, Itinerant electron metamagnetism and related phenomena in Co-based intermetallic compounds, *J.Appl.Phys.* 76: 6682

Havela, L., T. Almeida, J.R. Naegele, V. Sechovsky, and E. Brück, 1992a, High resolution photoemission on UPdSn and UNiAl, *J.Alloys Comp.* 181: 205.

Havela, L., V. Sechovsky, J.R. Naegele, T. Almeida, E. Brück, H. Nakotte, and F.R. de Boer, 1992b, 5f electron localization in UTX compounds, *J.Magn.Magn.Mater.* 104: 23.

Havela, L., A.V. Andreev, V. Sechovsky, I.K. Kozlovskaya, K. Prokes, P. Javorsky, M.I. Bartsahevich, T. Goto, and K. Kamishima, 1997, 5f-band metamagnetism in UCoAl, *Physica B* 230-232: 98

Havela, L., R.V. Dremov, A.V. Andreev, V. Sechovsky, J. Sebek, and K. Prokes, 1998, Electronic properties of UPdSn diluted by Lu, *J.Alloys Comp.*, in press

Kolomiets, A.V., 1998, unpublished data

Kolomiets, A.V., L. Havela, V. Sechovsky, L.E.DeLong, D.B. Watkins, and A.V. Andreev, 1998, Evidence for an extended critical region near the metamagnetic transition of UCoAl, *J.Appl.Phys.* 83: 6435

Mathon, J., 1968, Magnetic and electrical properties of ferromagnetic alloys near the critical concentration, *Proc.Roy.Soc. A* 306: 355

Mishra, S.G., and P.A. Sreeram, 1998, Fluctuation-induced non-Fermi-liquid behavior near a quantum phase transition in itinerant electron systems, *Phys.Rev.B* 57: 2188

Moriya, T. and T. Takimoto, 1995, Anomalous properties around magnetic instability in heavy electron systems, *J.Phys.Soc.Japan* 64: 960

Nakotte, H., 1994, Magnetism in cerium and uranium intermetallics, PhD Thesis, University of Amsterdam

Paixao, J.A., G.H. Lander, A. Delapalme, H. Nakotte, F.R. de Boer, and E. Brück, 1992, Magnetization density in URhAl - evidence for hybridization effects, *J.Phys.:Condens.Matter.* 4: 829

Robinson, R.A., A.C. Lawson, J.W. Lynn, and K.H.J. Buschow, 1992, Temperature dependence of magnetic order in UPdSn, *Phys.Rev.B* 45: 2939

Sandratskii, L.M. and J. Kübler, 1997b, Electronic and magnetic properties of UPdSn: the itinerant 5f electrons approach, *J.Phys.: Condens.Matter* 9: 1

Sechovský, V., L. Havela, G. Hilscher, N. Pillmayr, A.V. Andreev, P.A. Veenhuizen, F.R. de Boer, 1988, 5f magnetism in (U,Th)CoSn, *J.Appl.Phys.* 63: 3070

Sechovsky, V., L. Havela, K. Prokes, and A.V. Andreev, Anomalous magnetic and related electronic properties of U-intermetallics, in the same volume

Trygg, J., B. Johansson, and O. Eriksson, 1994, Photoelectron spectra and magnetic properties in UPdSn: Theory, *Phys.Rev.B* 49: 7165

Ueda, K., 1977, Electrical resistivity in antiferromagnetic Metals. *J.Phys.Soc.Japan* 43: 1497

# NON-FERMI LIQUID BEHAVIOR IN $U_{3-x}Ni_3Sn_{4-y}$ SINGLE CRYSTALS

L. Shlyk,[1,2] J. C. Waerenborgh,[3] P. Estrela,[4] L. E. De Long,[1]
A.de Visser,[4] and M. Almeida[3]

[1] B. Verkin Institute for Low Temperature Physics and Engineering
47 Lenin Ave., 310164 Kharkov, Ukraine
[2] Department of Physics and Astronomy University of Kentucky
Lexington, KY, USA 40506-0055
[3] Departamento Quimica, Instituto Tecnologico e Nuclear
P-2686 Sacavem Codex, Portugal
[4] Van der Waals-Zeeman Institute, University of Amsterdam
Valckenierstraat 65, NL-1018 XE Amsterdam, The Netherlands

## INTRODUCTION

The behavior of an increasing number of metallic, paramagnetic 4f- and 5f-electron systems are known to be inconsistent with simple Fermi liquid theory. Although magnetic order is the typical groundstate for rare earth and actinide intermetallics, the ordering temperature of some materials is sensitive to f-state hybridization and can be driven to zero by varying composition or pressure. The suppression of magnetic order toward zero temperature may result in an extended "Non-Fermi-Liquid" (NFL) regime in which critical fluctuations induce non-analytic temperature dependences of physical properties at temperatures $T<T_0$, where $T_0$ is a characteristic scaling temperature[1].

The $U_3T_3X_4$ compounds (T = transition metal, X = metalloid) with the cubic $Y_3Au_3Sb_4$ structure (a filled $Th_3P_4$-type) are good candidates for NFL behavior, since they exhibit a range of unusual properties related to variations of 5f-ligand hybridization[2,3]. The magnetic character of the uranium 5f-electrons in these compounds may be suppressed by increasing 5f-conduction electron hybridization via filling of the vacancies on the cation sites. For example, the stannides, $U_3T_3Sn_4$ (where T=Ni, Pt, Au)[3], do not exhibit long-range magnetic order down to 1.5 K, but are metallic and display heavy fermion properties, except for $U_3Cu_3Sn_4$ which is antiferromagnetic ($T_N$ = 12 K). In particular, an $U_3Ni_3Sn_4$ *polycrystal* was classified[3] as a moderate heavy fermion compound with an electronic heat capacity

*Electron Correlations and Materials Properties*, edited by Gonis *et al.*
Kluwer Academic/Plenum Publishers, 1999.

coefficient $\gamma$ = 92 mJ/moleU-$K^2$, accompanied by saturating susceptibility and quadratic temperature dependence of the electrical resistivity below 10 K, characteristic of a Fermi liquid ground state.

In contrast, recent observations[4] reveal that the magnetic susceptibility $\chi$ of *single-crystal* $U_3Ni_3Sn_4$ has no tendency towards saturation as T→0 K. This is not surprising, since the physical properties of many uranium compounds are sensitive to metallurgical factors such as impurity concentration, phase distributions and structural defects. Therefore, we have grown single-crystal $U_3Ni_3Sn_4$ samples having slightly different starting compositions for extensive characterization via X-ray diffraction, transmission electron microscopy, magnetic and heat capacity measurements.

## SAMPLE PREPARATION AND X-RAY CHARACTERIZATION

Samples were prepared by induction-melting two different charges with 3:3:4 or 3:2.8:4 atomic ratios of pure U (depleted), Ni and Sn, respectively. Single crystals were obtained by slow cooling the charges in a levitated cold crucible. The resulting ingots were about 2-2.5 cm in diameter, and contained many single crystals with a maximum size of 3x3x3 $mm^3$.

Single-crystal X-ray diffraction (XRD) data were collected at room temperature on an Enraf Nonius CAD-4 diffractometer using graphite-monochromated $MoK_\alpha$ radiation ($\lambda$ = 0.71069 Å) and an $\omega$-2$\theta$ scan mode. Measured intensities were corrected for Lorentz-polarization effects[5] and absorption by an empirical method based on $\psi$-scans[6] . Data, consistent with space-group $I\bar{4}3d$, were refined assuming a $Y_3Au_3Sb_4$-type structure by a full-matrix, least-squares method[7] based on the squares of the structure factors. In the later stages of refinement, site occupation factors (SOF) were allowed to vary. Best agreement factors R = 0.0548 and $R_W$ = 0.1297, or R = 0.0565 and $R_W$ = 0.1114, were obtained in correspondence to estimated compositions $U_{2.9}Ni_{3.0}Sn_{2.9}$ and $U_3Ni_3Sn_4$ (hereafter referred to as Sample 1 and Sample 2, respectively). These stoichiometries are consistent with ideal composition, within experimental error (standard deviations of calculated SOF are ≤ 4 %). Unit-cell parameters (9.3577(4) or 9.3524(5) Å for Sample 1 or 2, respectively) were obtained by least-squares refinement of 25 reflections with 11° < 2θ < 45°, and demonstrate that estimated deviations from perfect stoichiometry correlate with an increase in lattice parameter, so that ideal $U_3Ni_3Sn_4$ probably has the minimum unit-cell volume.

Specimens for transmission electron microscopy (TEM) were prepared from a larger single crystal of nominal composition $U_{2.9}Ni_{3.0}Sn_{3.9}$. Flakes of thickness ≈ 120 μm were produced by crushing and placed on a Cu grid, and several electron diffraction patterns were taken over selected areas of a few $\mu m^2$. Careful examination of several samples revealed no foreign phases or splitting of the diffraction spots due to twinning. As point defects usually have dimensions smaller than the resolution of the electron microscope, we were not able to observe single vacancies. There was no evidence of superlattice reflections due to vacancy ordering.

## MAGNETIC MEASUREMENTS

DC magnetization and susceptibility measurements were performed in the temperature interval 1.7-300 K in applied fields up to 5 T using a Quantum Design MPMS5S SQUID

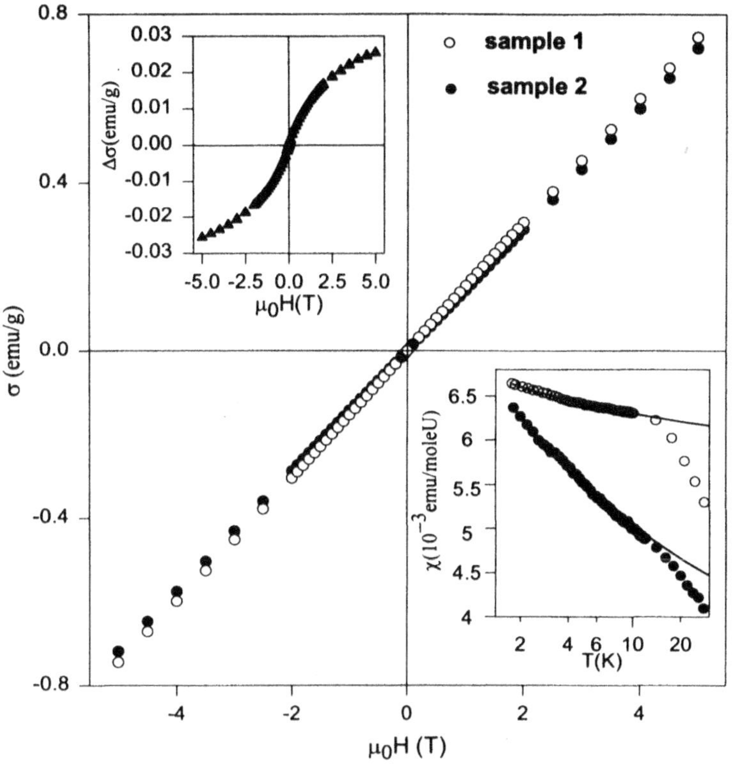


Fig.1. Magnetization σ(H) versus magnetic field H at T = 1.8 K for $U_{2.9}Ni_{3.0}Sn_{3.9}$ (Sample 1) and $U_3Ni_3Sn_4$ (Sample 2). Upper Inset: Difference in magnetizations Δσ of Samples 1 and 2 versus H. Lower Inset: Low temperature susceptibilities; the lines are fits to $\chi(T) \propto T^{-0.3}$.

Magnetometer. The temperature dependences of the magnetic susceptibilities of Samples 1 and 2 follow a $T^{-0.3}$ dependence in the range 1.7 - 5 K (Fig.1, lower inset). Above 25 K, $\chi^{-1}$ vs. T data approximately obey a modified Curie-Weiss law with Curie-Weiss temperature $\theta_p \approx -50\ (-60)$ K, effective moment $\mu_{eff} \approx 2.0\ (1.8)\ \mu_B/U$, and constant term $\chi_0 \approx 1.1\ (0.95) \times 10^{-3}$ emu/mole U for Sample 1 (2), in agreement with polycrystal results [3]. The reduction of $\mu_{eff}$ from 3.62 or 3.58 $\mu_B$ expected for $U^{3+}$ or $U^{4+}$ free ions, respectively, may indicate the combined effects of the crystalline electric field and significant hybridization between 5f and itinerant electron states.

## HEAT CAPACITY MEASUREMENTS

The heat capacity of Sample 1 was measured in a $^3He$ cryostat using the relaxation-time method. The results are similar to polycrystal data[3] above 1.5 K (see Fig. 2a), but lower temperature single-crystal data reveal a rapid increase in C/T between 0.3 and 0.7 K. The heat capacity of Sample 1 can be fitted (see Fig.2a) with the following expression:

$$C = (\gamma_0 - \alpha\sqrt{T})T + \beta T^3 + D/T^2, \qquad (1)$$

where $C_E = (\gamma_0 - \alpha\sqrt{T})T$ is the electronic contribution, $C_L = \beta T^3$ is the lattice contribution, and $C_N = D/T^2$ represents the high-temperature form of a nuclear Schottky term[8]. The best-fit coefficients are $\gamma_0 = 0.124$ J/moleU-K$^2$, $\alpha = 0.0151$ J/moleU-K$^{2.5}$, $\beta = 2.071 \times 10^{-3}$ J/moleU-K$^4$, and $D = 4.622 \times 10^{-4}$ J-K/moleU.

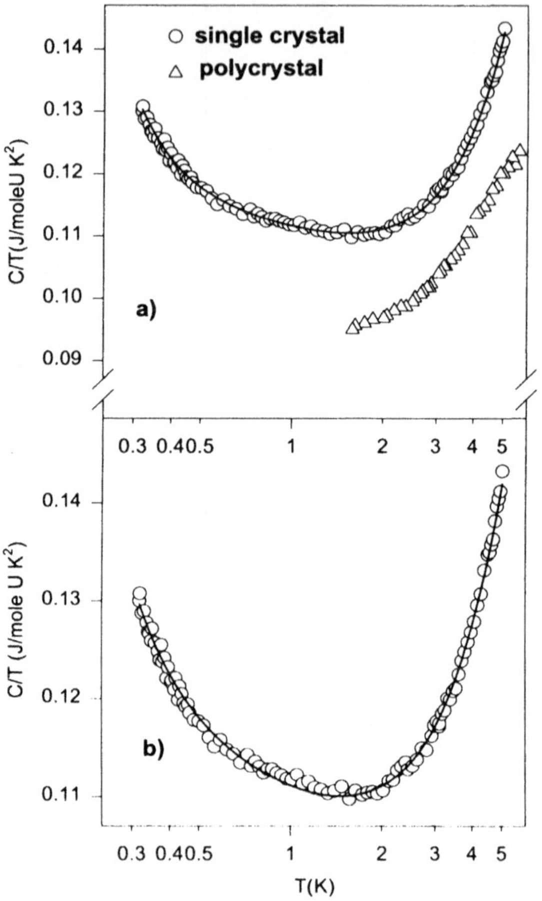


Fig. 2. a) Heat capacity C divided by temperature T for $U_{2.9}Ni_{3.0}Sn_{3.9}$ (Sample 1) and polycrystalline sample data[3], vs. log(T). The solid line is a fit of the data to Eq. 1. b) Heat capacity C divided by temperature T for $U_{2.9}Ni_{3.0}Sn_{3.9}$ vs. Log (T). The solid line is a fit of the data to Eq. 3.

The nuclear heat capacity anomaly in paramagnetic, cubic $U_3Ni_3Sn_4$ can be attributed to atoms located on the $\bar{4}$- and 3-sites having axial symmetry; hence, isotopes having a non-zero quadrupole moment (i.e., having nuclear spin I >1/2) may interact with the electric field gradient of neighboring atoms. In addition, unpaired d- or f-electrons of Ni and U, respectively, can generate magnetic hyperfine fields if the atomic spin relaxation is slower than the Larmour frequency of the nuclear spin[8-11]. Considering the natural abundances of relevant isotopes, only 1.2% of the Ni ($^{61}$Ni) and much less than 0.7% ($^{235}$U) of the depleted U may contribute to the nuclear Schottky term, assuming it reflects both quadrupolar and hyperfine interactions. Alternatively, $^{119}$Sn comprises 9% of natural Sn, and could yield a dominant hyperfine contribution to the nuclear heat capacity. As a check on the credibility of our numerical fits, we estimate an *upper limit* (neglecting all nuclear contributions of the

$^{61}Ni$ and $^{235}U$ isotopes) of the Sn-site hyperfine magnetic field $|H_{eff}| = 240$ kG, which is several times larger than that determined from Mössbauer experiments on non-magnetic heavy fermion materials[10,11].

Renormalization group theory[12] predicts $\gamma \propto \gamma_0 - \alpha\sqrt{T}$ near a zero-temperature antiferromagnetic instability. The fitted value of $\alpha = (15/64)k_B N_A N[2/\pi T_0]^{3/2}\zeta(5/2)$ allows us to estimate the characteristic temperature $T_0 \approx (10\text{ K}) \times N^{2/3}$. Assuming the N-dimensional bosonic order parameter N = 1, the best-fit value of $T_0$ corresponds to the onset temperature of the nonanalytic behavior of $\chi(T)$ shown in the lower inset of Fig.1. Alternatively, self-consistent renormalized spin fluctuation (SCR) theory[13] also predicts a square-root form of the electronic heat capacity at low temperatures that evolves into a logarithmic behavior, $C/T \sim \ln(T_0/T)$, and a good fit of the data was obtained using

$$C = \gamma_0 T\ln(T_0/T) + \beta T^3 + D/T^2\,, \qquad (2)$$

with $\gamma_0 = 7.74 \times 10^{-3}$ J/moleU-K$^2$, $T_0 = 1.34 \times 10^6$ K, $\beta = 1.8 \times 10^{-3}$ J/moleU-K$^4$, and $D = 3.79 \times 10^{-4}$ J-K/moleU. Multi-channel Kondo[14] or Kondo disorder[15] models also predict a logarithmic dependence for $\gamma(T)$. The Kondo disorder model demands $\chi(T) \propto -\ln T$; whereas, the multichannel Kondo model can yield $\chi(T) \propto -T^{0.5}$ or $-\ln T$ at low temperatures. Fits to a $T^{-0.3}$ dependence yield slightly better results than the $-T^{0.5}$ or $-\ln T$ forms; however, the best-fit value of $T_0 \approx 10^6$ K is extremely high, and probably does not correspond to a physically significant spin fluctuation energy. On the other hand, the $T \rightarrow 0$ behavior of $\gamma(T)$ is also predicted to be of square-root form as a consequence of a zero-temperature quantum transition from a paramagnetic metal to a spin-glass[16,17], but the predicted[16] low temperature susceptibility $\chi \propto -T^{3/4}$, in conflict with our data.

The assumed slight increase in defect concentration of Sample 1 evidently leads to a small nonlinear field dependence of the difference in magnetization $\Delta\sigma \equiv \sigma_1 - \sigma_2$ of the two samples (see upper inset to Fig. 1), which begins to saturate above $H_{sat} = 6$ T. This is consistent with a characteristic temperature $T_0 \approx \mu_B H_{sat}/k_B > 4$ K, and corroborates the $\chi(T)$ and C(T) data analysis.

Very recent experimental[18] and theoretical[19] work proposes that NFL behavior is caused by competition between RKKY and Kondo interactions in the presence of atomic disorder, leading to a Griffiths phase (large magnetic clusters) close to a quantum critical point. We find that the NFL behavior of C(T) and $\chi(T)$ in $U_{2.9}Ni_{3.0}Sn_{3.9}$ can be described by divergent power law predicted by this model, i.e. $C(T)/T \propto \chi(T) \propto T^{-1+\lambda}$ with $\lambda = 0.7$. We have obtained a good fit of the specific heat of Sample 1 using :

$$C = (\gamma_0 + \alpha T^{-0.3})T + \beta T^3 + D/T^2\,, \qquad (3)$$

where $C_E = (\gamma_0 + \alpha T^{-0.3})T$ is the electronic contribution, $C_L = \beta T^3$ is the lattice contribution, and $C_N = D/T^2$ represents the high-temperature form of a nuclear Schottky term[8] (Fig.2b). The best-fit coefficients are $\gamma_0 = 0.0932$ J/moleU-K$^2$, $\alpha = 0.0159$ J/moleU-K$^{2.5}$, $\beta = 1.6421 \times 10^{-3}$ J/moleU-K$^4$, and $D = 2.5968 \times 10^{-4}$ J-K/moleU. These values show that the heat capacity contributions at T = 5 K (or T = 0.3 K) are 72% (92%) fermionic, 28 % (less than 1%) lattice, and less than 1% (7.6%) nuclear. Note that the calculated contributions are slightly different from those obtained above from renormalization group theory; for example, the *upper limit* (neglecting all nuclear contributions of the $^{61}Ni$ and $^{235}U$ isotopes) of the Sn-site hyperfine magnetic field $|H_{eff}| = 146$ kG, 1.6 times less than the previous estimate. We note that the renormalization group analysis yields slightly better agreement with the experimental heat capacity data in temperature interval 0.6 - 1.7 K (compare Figs. 2a and

2b); nevertheless, the Griffiths phase model provides a self-consistent description of both the thermal and magnetic data.

## CONCLUSION

We have undertaken a thorough analysis of the heat capacity and susceptibility of the $U_{2.9}Ni_{3.0}Sn_{3.9}$, and conclude that several other NFL models (e.g., multichannel Kondo[14], Kondo disorder[15]) commonly considered in the contemporary literature do not describe the entire data set known for this material. Two models (Millis and Griffiths phase), both of which depend upon the existence of a low temperature QCP between a paramagnetic and magnetic order regime, appear to be consistent with experiments so far. Millis's treatment does not include the effects of disorder, which might be present in our sample and in any other real system, while Griffiths phase model includes disorder as a crucial ingredient. Recent predictions[20] for the susceptibility behavior near antiferromagnetic QCP for the 3-dimensional case yield $\chi \propto -T^{1.5}$, in conflict with our data, and may indicate that the low temperature behavior of the $\chi$ (T) is strongly affected by a small amount of disorder . Whether either of these two models is valid for the NFL properties of $U_3Ni_3Sn_4$ is a question for further experimental investigation.

## Acknowledgments

We thank Prof. Elizabeth Dickey for assistance in the transmission electron microscopy studies. The single-crystal samples were grown by L. S. during her stay in Portugal under the support of a NATO fellowship. Research at the University of Kentucky was supported by NSF Grant #INT-9515504. P.E. acknowledges the European Commission for a grant within the TMR programme. We thank Prof. M. B. Maple and Dr. B. A. Jones for sending us preprints of unpublished work.

## REFERENCES

1. M. B. Maple, R. P. Dickey, J. Herrmann, M. C. de Andrade, E. J. Freeman, D. A. Gajewski, R. Chau, *J. Phys.: Condens. Matter* 8: 9773 (1996).
2. T. Endstra, G. J. Nieuwenhuys, J. A. Mydosh and K. H. J. Buschow, *J. Magn. Magn. Mat.* 89: L273 (1990).
3. T. Takabatake, S. Miyata, H. Fujii, Y. Aoki, T. Suzuki, T. Fujita, J. Sakurai, T. Hiraoko, *J. Phys. Soc. Japan* 59: 4412 (1993).
4. L. Shlyk, R. Troc, J. C. Waerenborgh, J. A. Paixao, M. M. R. Costa and M. Almeida, in *Proceedings 26ièmes Journées des Actinides*, Szklarska Poreba, Poland, P. 214 (1996).
5. C. K. Fair, *MOLEN*, Enraf-Nonius, Delft, The Netherlands, (1990).
6. A. C. T. North, *Acta Cryst.* A24:351 (1968).
7. G. M. Sheldrick , University of Göttingen, Germany, SHELXL-93: Program for Crystal Structure Refinement (1993).
8. O. V. Lounasmaa, in *Hyperfine Interactions*, Academic Press, New York, (1967)
9. F. Steglich B. Buschinger, P. Gegenwart, M. Lohmann, R. Helfrich, C. Langhammer, P. Hellmann, L. Donnevert, S. Thomas, A. Link, C. Geibel, M. Lang, G. Sparn and W. Assmus, J. Phys.: *Condens. Matter* 8: 9909 (1996).

10. B. Chevalier, L. Fournes and J. Etourneau, in *Proceedings 26ièmes Journées des Actinides,* Szklarska Poreba, Poland, p. 66 (1996).
11. M. A. Polikarpov, V. M. Cherepanov, M. A. Chuev, S. S. Yakimov, *Physica B* 199/200:45 (1994).
12. A. J. Millis, *Phys. Rev. B* 48: 7183 (1993); *Phys. Rev. B* 51: 8996 (1995).
13. T. Moriya and T. Takimoto, *J. Phys. Soc. Japan* 64: 960 (1995).
14. D. L. Cox and M. Jarrell, *J. Phys.: Condens. Matter* 8: 9825 (1996).
15. E. Miranda, V. Dobrosavljevic, G. Kotliar, *J. Phys.: Condens. Matter* 8: 9871 (1996).
16. A. M. Sengupta and A. Georges *Phys. Rev. B* 52: 10295 (1995)
17. S. Sachdev, N. Read and R. Oppermann, *Phys. Rev. B* 52: 10286 (1995).
18. M. C. de Andrade, R. Chau, R. P. Dickey, N. R. Dilley, E. J. Freeman, D. A. Gajewski, M. B. Maple, R. Movshovich, A. H. Castro Neto, G. E. Castilla, B. A. Jones, *cond-mat/9802081* (unpublished).
19. A. H. Castro Neto, G. Castilla and B. A. Jones, *cond-mat/9710123* (unpublished).
20. A. J. Millis, Privite Communication.

# Part 2

# Phenomenological Studies of Correlation Effects

# INTRODUCTORY OVERVIEW AND HEAVY-FERMION PHENOMENOLOGY

P. Fulde

Max Planck Institute for the
Physics of Complex Systems,
Nöthnitzer Straße 38,
01187 Dresden (Germany)

## INTRODUCTION

Approximately 20 years ago it was found [1] that there are metallic systems which behave at low temperatures like ordinary metals except that the quasiparticles appear to have a very large effective mass, i.e., one which is hundreds of times larger than the free electron mass. Therefore these systems are often called heavy-fermion or heavy-electron systems. In the meantime research on those systems has developed into an own branch of metal physics and it has become clear over the years that their basic properties are strong electron correlations. Therefore it is no surprise that heavy-fermion systems contain in almost all cases rare-earth ions, in particular $Ce$ or $Yb$ or actinide ions such as $U$ or $Np$. Typical representatives are $CeAl_3$, $CeCu_2Si_2$, $CeCu_6$, $YbAl_3$, $YbCu_2Si_2$, $UBe_{13}$, $UPt_3$ and $NpBe_{13}$. But also $Nd_{2-x}Ce_xCuO_4$ $(0.1 \leq x \leq 0.2)$ or $Yb_4As_3$ belong into that category, the latter being a semimetal at low temperatures.

From the observed low-temperature properties of the heavy-fermion systems we must conclude that there exists obviously to good approximation a one to one mapping between the low-energy quasiparticle excitations of the heavy-fermion systems and those of a weakly interacting electron system, provided the effective mass of the latter is properly renormalized.

Therefore we speak of a heavy-fermion system when a metal meets the following conditions:

(a) The low temperature specific heat $C = \gamma T$ has an enhanced coefficient $\gamma$ of order 1 $Jmol^{-1}K^{-2}$, rather than 1 $mJmol^{-1}K^{-2}$ as, e.g., is found for sodium metal;

(b) the Pauli paramagnetic susceptibility $\chi_s$ is similarly enhanced as $\gamma$;

*Electron Correlations and Materials Properties*, edited by Gonis *et al.*
Kluwer Academic/Plenum Publishers, 1999.

(c) the ratio $R = \pi^2 k_B^2 \chi_s / (3\mu_{eff}^2 \gamma)$ (Sommerfeld-Wilson ratio) is of order unity. Here $\mu_{eff}$ is the effective magnetic moment of the quasiparticles. The quantities $\gamma$ and $\chi_s$ are both proportional to the quasiparticle density of states (per spin direction) at the Fermi level $N^*(0)$. The latter is proportional to $m^*$, i.e., the effective mass of the low-energy fermionic excitations. When the Sommerfeld Wilson ratio is calculated, the density of states $N^*(0)$ drops out. For free electrons $R = 1$. In the presence of quasiparticle interactions $R = (1 + F_0^a)^{-1}$. The Landau parameter $F_0^a$ relates to the interactions and enters $\chi_s$.

The high density of states of excitations is limited to a small energy interval around the Fermi energy. This is seen from the specific heat which levels off at a characteristic temperature $T^*$. Also the spin susceptibility changes from Pauli- to Curie- like behaviour around that temperature. Therefore the high effective mass $m^*$ disappears when $k_B T$ becomes larger than this energy interval.

Obviously the low-energy scale is characterized by $T^*$. As the temperature of the system increases to values above $T^*$, the rare earth or actinide ions behave more and more like ions with well-localized $f$ electrons.

One key problem is to understand the physical origin of the low-energy scale. Until few years ago, it was commonly believed that the Kondo effect is the sole source of heavy-fermion behaviour. The physics associated with the Kondo effect is extensively described in a monograph [2] and a number of reviews [3-8]. However, more recently it has been found that heavy quasiparticles may result from rather different physical effects. In all cases a lattice of $4f$ (or $5f$) ions is involved. It is coupled by hybridization with the conduction electrons. The latter are either weakly correlated as in $CeAl_3$ or strongly correlated as in the high-$T_c$ cuprates. Strongly correlated conduction electrons may influence substantially the physical properties of the system. This situation is encountered, e.g., in $Nd_{2-x}Ce_xCuO_4$ and it is shown below that here the Zeeman effect is responsible for the formation of heavy quasiparticles [9-11]. In the semimetal $Yb_4As_3$ a third physical effect is responsible for the high density of states of low energy excitations, [12-17]. Here the $4f$ holes of the $Yb$ ions from a lattice superimposed on the atomic lattice and it is the chain like structure of that superimposed lattice which results in heavy-fermion behaviour of the low-energy excitations. Thus, instead of having one single physical origin, heavy fermions may have a variety of effects responsible for their existence.

In all cases, the low-lying excitations characterizing heavy-fermion systems involve predominantly spin degrees of freedom. Direct evidence is given by the amount of entropy associated with the excess specific heat. The latter is associated with an entropy of order $S \simeq k_B \ell n \nu_f$ per $f$ site, where $\nu_f$ denotes the degeneracy of the crystal-field ground state of the atomic $f$ shell. It is pretty safe to state that in all likelihood yet unknown mechanisms will add to the presently known ones (see, e.g., Ref. [18]). In the following, a discussion is given of the three routes to heavy-fermion behaviour just outlined.

## KONDO LATTICE SYSTEMS

The essence of the single-site Kondo effect is the formation of a singlet ground-state due to a weak hybridization of the incomplete $4f$ shell of a rare-earth ion with the conduction electrons. A specific form of the singlet wavefunction is obtained by starting from the Anderson impurity Hamiltonian

$$H = \sum_{km} \epsilon(k) c^{+}_{km} c_{km} + \epsilon_f \sum_{m} n^f_m + \frac{U}{2} \sum_{m \neq m'} n^f_m n^f_{m'} + \tag{1}$$
$$+ \sum_{km} V(k)(f^{+}_m c_{km} + c^{+}_{km} f_m) + \tilde{H}_0.$$

Here $f^{+}_m$ denotes the creation operator of an $f$ electron in state $m$ of the lowest $J$ multiplet and $n^f_m = f^{+}_m f_m$. The $f$-orbital energy is $\epsilon_f$ and $U$ is the $f - f$ Coulomb repulsion. The $c^{+}_{km}$ create conduction electrons with momentum $| \mathbf{k} |= k$ and the three quantum numbers $\ell = 3$, $J$ and $m$. The hybridization between the $f$-and conduction electrons is given by the matrix element $V(k)$. Finally, $\tilde{H}_0$ contains all those degrees of freedom of the conduction electrons which do not couple to the impurity. The following ansatz for the singlet ground-state wave function is due to Varma and Yafet [19]

$$| \psi_0 >= A(1 + \frac{1}{\sqrt{\nu_f}} \sum_{km} \alpha(k) f^{+}_m c_{km}) \mid \phi_0 > \tag{2}$$

where $| \phi_0 >$ represents the Fermi sea of the conduction electrons. The ansatz (2) is closely related to the one suggested by Yoshida [20] for the ground state of the Kondo Hamiltonian. The variational parameters $A$ and $A\alpha(k)$ are obtained by minimizing the energy. The energy $E_0$ of $| \psi_0 >$ is always lower than the one of the multiplet $| \psi_m > \ = f^{+}_m \mid \phi_0 >$. The difference $\epsilon$ is found to be

$$\epsilon = - D exp[- \mid \epsilon_f \mid /(\nu_f N(0) V^2)] \tag{3}$$

and denotes the energy gain due to the formation of the singlet. Here $D$ is half of the bandwidth of the conduction electrons and $N(0)$ is their density of states per spin direction at the Fermi energy. It is customary to associate with this energy gain a temperature $T_K$, i.e., the Kondo temperature. The singlet-triplet excitation energy is often of the order of a few meV only, and provides a low-energy scale. When a lattice of $f$ ions is considered like, e.g., $CeAl_3$ the Anderson lattice Hamiltonian is replacing Eq.(1). The energy scale $k_B T_K$ is here replaced by $k_B T^*$ which takes into account modifications in the presence of the lattice, i.e., due to interactions between different $f$ sites. The energy gain due to singlet formation competes with the one due to the RKKY interaction between $f$ sites [21]. In the limit of small hybridization $V$ the RKKY interaction energy always wins out because it is proportional to $V^4$ while the singlet-formation energy depends exponentially on $V$ (see Eq. (3)) and therefore is smaller. This seems to be the case in systems like $CeAl_2, CePb_3$ and $NpBe_{13}$ which become antiferromagnets at low temperatures.

In addition to $T^*$ there exists another characteristic temperature $T_{coh} < T^*$ below which the local singlet-triplet excitations lock together and form coherent quasiparticles with large effective masses. The details of this coupling are not yet understood, but de Haas - van Alphen measurements demonstrate convincingly that the $f$ electrons behave like delocalized electrons. At the Fermi surface they show strong anisotropies in the effective mass. It is somewhat surprising that one can calculate the Fermi surface of a heavy-electron system and determine the anisotropic masses with one adjustable parameter only. This is achieved by renormalized band-structure calculations [22-27, 6]. Thereby the effective potential seen by a quasiparticle is described by energy-dependent phase shifts $\eta^A_\ell(\epsilon)$ of the different atoms $A$. Thereby the index $\ell$ refers to the different angular momentum channels.

In the following we consider $CeRu_2Si_2$ as an example [6,26,27]. The essential point is to use for the phase shifts the ones computed by application of the local-density approximation (LDA) to density functional theory, with the exception of the $\ell = 3$ phase shift of the $Ce$ ion. Note that this approximation neglects the coupling of conduction electrons to different configurations of the $4f$ or $5f$ shell with *fixed f* electron number. The mass enhancement of the conduction electrons of $Pr$ metal falls into that category. It results from the virtual transitions between different crystal-field eigenstates of the $4f^2$ system caused by the coupling between conduction and $4f$ electrons [28]. Thus, only the $\eta^{Ce}_{\ell=3}(\epsilon)$ phase shift remains undetermined. According to Hund's rules the ground-state multiplet of the $4f^2$ configuration of $Ce$ is $j = 5/2$ with the $j = 7/2$ multiplet being much higher in energy. Therefore, we may set $\eta^{Ce}_{j=7/2}(\epsilon_F) = 0$. Of the $j = 5/2$ multiplet, only the Kramers-degenerate crystal-field ground state is considered, because it is the only state occupied at low temperatures. Let $\tau$ denote the degeneracy index of that ground state. Only the phase shift $\eta^{Ce}_{\tau}(\epsilon_F)$ among the different $\ell = 3$ channels is different from zero. It must contain the strong correlations of the $4f$ electrons and is unknown. In the spirit of Landau's Fermi liquid theory we expand this unknown function around the Fermi energy and write

$$\eta^{Ce}_{\tau}(\epsilon) = \eta^{Ce}_{\tau}(\epsilon_F) + a(\epsilon - \epsilon_F) + O((\epsilon - \epsilon_F)^2). \tag{4}$$

Of the two parameters one, i.e., $\eta^{Ce}_{\tau}(\epsilon_F)$ is fixed by requiring that the number of $4f$ electrons $n_f = 1$ . According to Friedel's sum rule this implies $\eta^{Cu}_{\tau}(\epsilon_F) = \frac{\pi}{2}$. The remaining parameter $a$ fixes the slope of the phase shift at $\epsilon_F$. The latter determines the density of states and hence the effective mass. We set $a = (k_B T^*)^{-1}$ and determine $T^*$ by requiring that the linear specific heat coefficient $\gamma$ calculated from the resulting quasiparticle dispersions agrees with the experimental one. The different computational steps are summarized in Fig. 1.

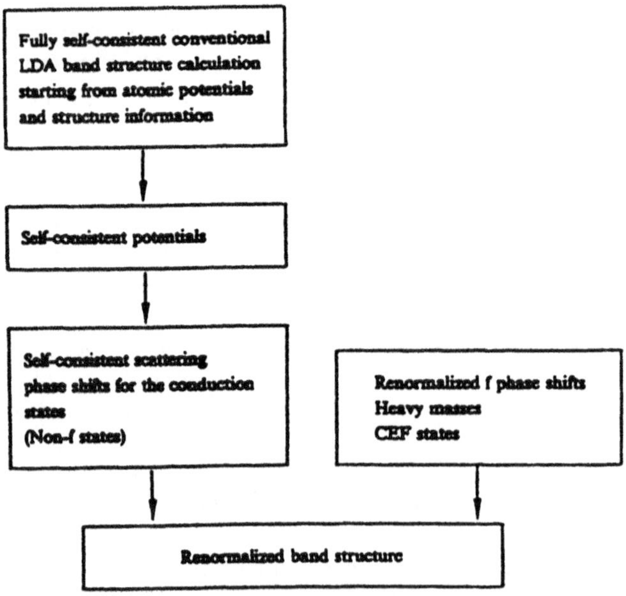


**Fig. 1.** *Different computational steps of renormalized band structure calculation. (From Ref. [6])*

Calculations of this form have explained and partially predicted [6,26,27] the large mass anisotropies in $CeRu_2Si_2$ [29]. For more details on renormalized band theory we refer to comprehensive reviews which are available [6-30].

When the temperature increases beyond $T_{coh}$ the excitations lose their coherence properties and the problem reduces to that of independent impurities. In that regime the specific heat contains large contributions from the incoherent part of the $f$ electron excitations. The noncrossing approximation (NCA) is a valuable tool for treating the coupled $4f$ and conduction electrons in that temperature regime [31-33]. It leads to a system of coupled equations of the form

$$\begin{aligned} \Sigma_0(z) &= \frac{\Gamma}{\pi}\sum_m \int_{-\infty}^{+\infty} d\zeta \rho_m(\zeta) K_+(z-\zeta) \\ \Sigma_m(z) &= \frac{\Gamma}{\pi} \int_{-\infty}^{+\infty} d\zeta \rho_0(\zeta) K_-(z-\zeta). \end{aligned} \tag{5}$$

Here $\Gamma = \pi N(0) V^2$ and $K_\pm(z)$ are defined by

$$K_\pm(z) = \frac{1}{N(0)} \int_{-\infty}^{+\infty} d\epsilon \frac{N(\pm\epsilon) f(\epsilon)}{z+\epsilon} \tag{6}$$

where $f(\epsilon)$ is the Fermi energy and $N(\epsilon)$ is the energy-dependent conduction-electron density of states. The function $\Sigma_\alpha(z)$ and $\rho_\alpha(z)$ ($\alpha = 0; m$) are related to each other through

$$\begin{aligned} \rho_\alpha(z) &= -\frac{1}{\pi} Im \left\{ R_\alpha(z) \right\} \\ R_\alpha(z) &= \frac{1}{z - \epsilon_\alpha - \Sigma_\alpha(z)} \end{aligned} \tag{7}$$

with $\epsilon_{\alpha=0} = 0$, $\epsilon_{\alpha=m} = \epsilon_{fm}$. The NCA equations have to be solved numerically [34]. However, one can find simple, approximate solutions which have the virtue that crystal-field splittings can be explicitly included, a goal which has not been achieved yet by numerical methods [35]. Once the $\rho_\alpha(\epsilon)$ are known, one can determine, e.g., the temperature dependence of the $f$-electron occupancies $n_{fm} =< f_m^+ f_m >$ through

$$n_{fm}(T) = \frac{1}{Z_f} \int_{-\infty}^{+\infty} d\epsilon \rho_m(\epsilon) e^{-\beta(\epsilon-\mu)}, \tag{8}$$

where $\mu$ is the chemical potential and

$$Z_f = \int_C \frac{dz}{2\pi i} e^{-\beta z} (R_0(z) + \sum_m R_m(z)) \tag{9}$$

is the partition function of the $f$ electrons. Knowing the $n_{fm}(T)$ enables us to compute quantities like the temperature dependence of the quadrupole moment of the $f$ sites [36].

When $T >> T^*$, the $f$ electrons can be treated as being localized. Their moment is weakly interacting with that of the conduction electrons and perturbation theory can be applied to study the resulting effects. The different temperature regimes are shown in Fig. 2.

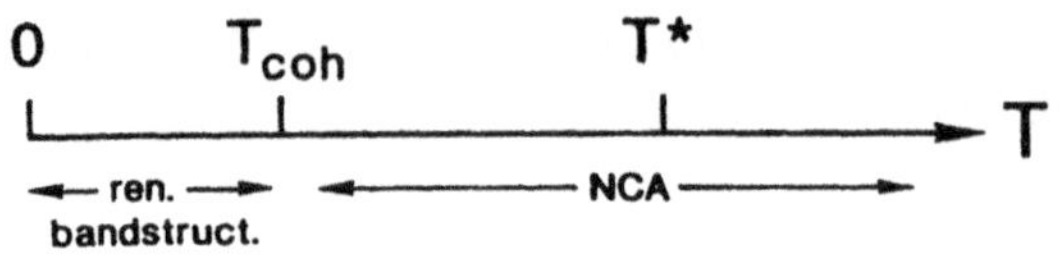


**Fig. 2.** *Different temperature regimes and corresponding theoretical methods for describing the low-energy excitations of a Kondo-lattice*

A beautiful demonstration of the above scenario is the experimentally observed difference in the Fermi surfaces of $CeRu_2Si_2$ and $CeRu_2Ge_2$ [37]. When $Si$ is replaced by $Ge$ the distance between $Ce$ and its nearest neighbors is increased. This implies a decrease in the hybridization of the $4f$ electrons with the valence electrons of the neighboring ions. While in $CeRu_2Si_2$ the characteristic temperature is $T^* \simeq 15K$, it is practically zero in $CeRu_2Ge_2$. De Haas-van Alphen experiments are performed at a temperature $T \simeq 1K$ implying $T << T^*$ for $CeRu_2Si_2$ while $T >> T^*$ for $CeRu_2Ge_2$. Therefore, the $4f$ electron of $Ce$ contributes to the volume of the Fermi sea in the former case, but not in the latter. Indeed, experiments show that the two Fermi surfaces have similar features but differ in volume by one electron [37] (see Fig. 3).

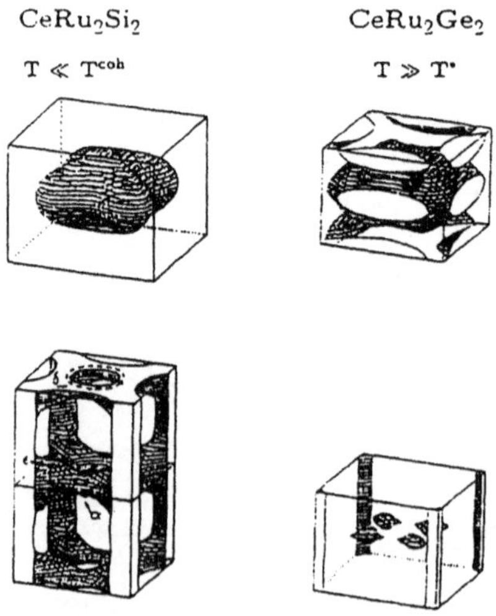


**Fig. 3.** *Fermi surface of $CeRu_2Si_2$ and $CeRu_2Ge_2$ as derived from de Haas - van Alphen measurements [37]. The $4f$ electron of $Ce^{3+}$ is well localized in $CeRu_2Ge_2$ but participates in the Fermi surface in $CeRu_2Si_2$. Therefore the volume enclosed by the Fermi surface differs by one electron in the two cases. In $CeRu_2Ge_2$ the electronic part is diminished and the hole part is enlarged as compared with $CeRu_2Si_2$.*

## HEAVY FERMIONS IN $Nd_{2-x}Ce_xCuO_4$

Low-temperature measurements of the specific heat and magnetic susceptibility demonstrate the existence of heavy-quasiparticle excitations in the electron-doped system $Nd_{2-x}Ce_xCuO_4$ [9]. For $x = 0.2$ and temperatures $T \leq 1K$ the linear specific heat coefficient is $\gamma = 4J/(mol \cdot K^2)$. The magnetic susceptibility $\chi_s$ is approximately $T$-independent in that temperature regime and the Sommerfeld-Wilson ratio is $R \simeq 1.8$ (see Fig. 4).

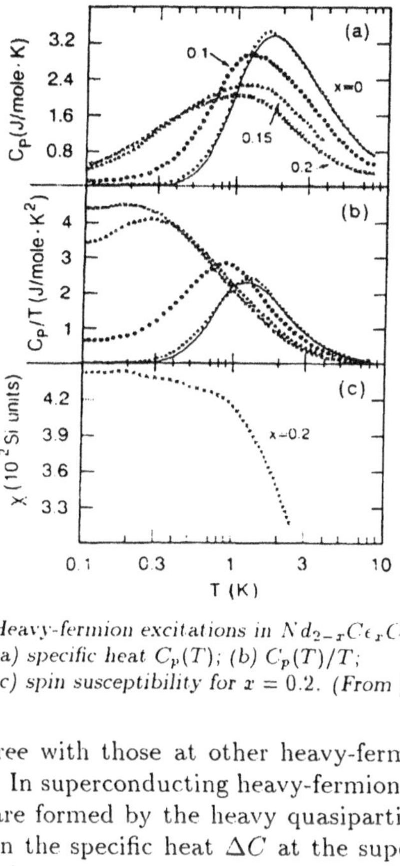


**Fig. 4.** *Heavy-fermion excitations in $Nd_{2-x}Ce_xCuO_4$: (a) specific heat $C_p(T)$; (b) $C_p(T)/T$; (c) spin susceptibility for $x = 0.2$. (From [9])*

While these features agree with those at other heavy-fermion systems, there are also pronounced differences. In superconducting heavy-fermion systems like $CeCu_2Si_2$ or $UPt_3$ the Cooper pairs are formed by the heavy quasiparticles. This is evidenced by the fact that the jump in the specific heat $\Delta C$ at the superconducting transition temperature $T_c$ is directly related to the large $\gamma$ coefficient, i.e., $\Delta C(T_c)/(\gamma T_c) \approx 2.4$. The low-energy excitations are therefore strongly reduced below $T_c$. In superconducting $Nd_{1.85}Ce_{0.15}CuO_4$ the formation of Cooper pairs has no noticeable effect on the heavy-fermion excitations. They remain uneffected by superconductivity.

A crucial difference between $Nd_{2-x}Ce_xCuO_4$ and, e.g., $CeCu_2Si_2$ are the strong electron correlations between the conduction electrons which are present in the former, but not in the latter material. In the two-dimensional $Cu-O$ planes of $Nd_{2-x}Ce_xCuO_4$ with $x \geq 0.1$ we have to account for antiferromagnetic fluctuations which are very slow at low temperatures. There is considerable experimental evidence for that feature. Consider undoped $Nd_2CuO_4$, which is an antiferromagnet with a Néel temperature of $T_N \simeq 270K$. Since the exchange interactions between a $Nd$ ion and its nearest-neighbor $Cu$ ions cancel because of the antiferromagnetic alignment of the $Cu$ spins, one is left with the next-nearest neighbor $Cu - Nd$ spin interaction. The latter is of the form $\alpha s_{Cu} S_{Nd}$ and is larger than the $Nd - Nd$ interaction. The Schottky peak in the specific heat seen in Fig. 4 results from the spin flips of the $Nd$ ions in the staggered effective field $\alpha < s_{Cu} >$ set up by the $Cu$ spins (Zeeman effect). It is also present in doped systems like $Nd_{1.8}Ce_{0.2}CuO_4$ where antiferromagnetic long-range order is destroyed by doping. This can only be understood if the changes in the preferred direction of the $Cu$ spins occur sufficiently slowly, i.e., slower than $10^{-10}$ sec in the present case, so that the $Nd$ spins can follow adiabatically these motions. Only then is a similar energy required to flip a $Nd$ spin as in $Nd_2CuO_4$. This physical picture has been confirmed by $\mu SR$ experiments [38] which demonstrate the presence of slow magnetic fluctuations. Spin-glass behaviour can be excluded. Also a recent inelastic neutron scattering supports this point of view [39]. Another paper on magnetically ordered samples misses somewhat the key issue [40].

Due to an effective valency of $Ce$ of approximately +3.5 the $Cu-O$ planes are doped with electrons, i.e., a corresponding number of $Cu$ sites are in a $3d^{10}$ configuration. Since these sites have no spin they do not interact with the $Nd$ ions. The extra electrons move freely in the $Cu - O$ planes and therefore, the interaction of a $Nd$ ion with the next-nearest $Cu$ site is repeatedly turned off and on. It is this feature which results in heavy-quasiparticle excitations.

Two model descriptions have been advanced in order to explain the low-energy excitations of $Nd_{2-x}Ce_xCuO_4$. One is based on a Hamiltonian in which the $Nd - Cu$ interaction is treated by a hybridization between the $Nd$ $4f$ and $Cu$ $3d$ orbitals. Usually it is much easier to extract heavy quasiparticles out of such a Hamiltonian than from one with an effective spin-spin interaction obtained after the $4f$ charge fluctuations have been eliminated by a Schrieffer-Wolff transformation. The slow antiferromagnetic fluctuations of the $Cu$ spins are replaced by a static staggered field acting on them. This symmetry-breaking field also accounts for the strong correlations in the $Cu - O$ planes because charge fluctuations between $Cu$ sites are strongly reduced this way (unrestricted Hartree-Fock). Thus the Hamiltonian $H$ reads

$$\begin{aligned} H &= -t \sum_{<ij>\sigma} (a^+_{i\sigma} a_{j\sigma} + h.c.) + h \sum_{i\sigma} \sigma \, e^{i\mathbf{QR}_i} a^+_{i\sigma} a_{i\sigma} \\ &+ V \sum_{i\sigma} (a^+_{i\sigma} f_{i\sigma} + h.c.) + \tilde{\epsilon}_f \sum_{i\sigma} f^+_{i\sigma} f_{i\sigma} . \end{aligned} \tag{10}$$

Here $\mathbf{Q} = (\pi, \pi)$ is a reciprocal lattice vector, $\mathbf{R}_i$ denotes the positions of the $Cu$ ions and $h$ is the staggered field. The operators $a^+_{i\sigma}, f^+_{i\sigma}$ create an electron in the $Cu$ $3d_{x^2-y^2}$ and $Nd$ $4f$ orbital, respectively. For simplicity, only one $Nd$ site per $Cu$ site is considered and one $4f$ orbital with energy $\tilde{\epsilon}_f$ is assumed instead of seven. The energies $\tilde{\epsilon}_f$ and $V$ are strongly renormalized quantities because of the $4f$ electron correlations.

The Hamiltonian (10) is easily diagonalized. Four bands are obtained, two of which are $d$-like ($Cu$) and two which are $f$-like ($Nd$). The dispersions of the four bands are given by

$$E_\nu(\mathbf{k}) = \frac{\tilde{\epsilon}_f \pm \epsilon_{\mathbf{k}}}{2} \pm \frac{1}{2}\sqrt{(\epsilon_{\mathbf{k}} \mp \tilde{\epsilon}_f)^2 + 4V^2} \; , \; \nu = 1, ..., 4 \tag{11}$$

where $\epsilon_{\mathbf{k}} = (\epsilon_0^2(\mathbf{k}) + h^2/4)^{1/2}$ and $\epsilon_0(\mathbf{k}) = -2t(cosk_x + cosk_y)$. A schematic plot is shown in Fig. 5.

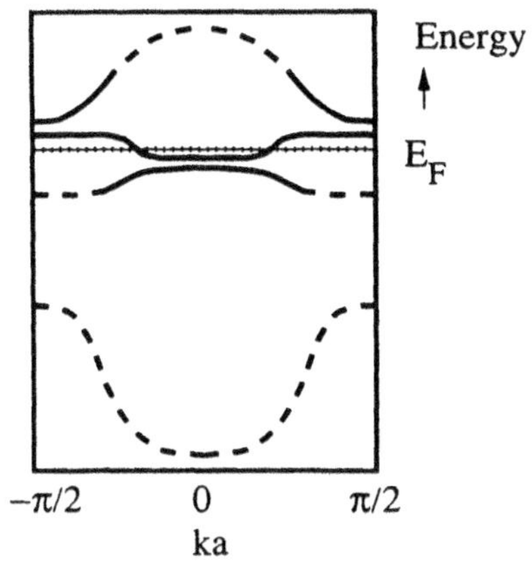


**Fig. 5.** *Schematic drawing of the hybridized Cu 3d and Nd 4f bands. The effective hybridization V and the f-orbital energy $\tilde{\epsilon}_f$ are strongly affected by conduction electron correlations. Solid and striped lines indicate f- and d-like excitations, respectively. (From [10])*

At half-filling only the lower $f$ band is filled and the Schottky-peak contributions to $C(T)$ are due to transitions from the filled lower to the empty upper $f$ band. When the planes are doped with electrons the upper $f$ band becomes partially filled resulting in low-energy excitations with large effective mass. The latter follows from the quasiparticle dispersion

$$E_{qp}(\mathbf{k}) \simeq \tilde{\epsilon}_f + \frac{V^2}{(\tilde{\epsilon}_f + \epsilon(\mathbf{k}))}. \tag{12}$$

It is noticed that here it is the Zeemann splitting of the $f$ states which is responsible for the occurrence of heavy-electron behaviour. The effect of superconductivity on the heavy quasiparticles can be studied by adding an attractive interaction part $H_{attr}$ for the charge carriers in the $Cu - O$ planes to the Hamiltonian [41]. For $V = 0$ the conventional BCS spectrum is recovered for the electrons in the upper $Cu$ band. When $V \neq 0$ the lower $Cu$ band hybridizes with one of the dispersionless $f$ bands. The lower $d$ band remains uneffected by superconductivity because pairing occurs in the upper $d$ band. Therefore superconductivity has no effect here. The upper $d$ band hybridizes with the second $f$ band. When $H$ is diagonalized one finds a BCS gap in the $Cu$ band while the $f$ band remains virtually unchanged as compared with a vanishing superconducting order parameter.

The second model description of the $Nd$ spins coupled to the $Cu$ spin is based on stochastic forces acting on the latter [11]. They mimic the interaction of the $Cu$ spin with its environment, i.e., with the other $Cu$ spins. In that case we start from the Hamiltonian

$$H_{int} = \alpha\, \mathbf{s}_{Cu} \cdot \mathbf{S}_f \quad , \quad \alpha > 0 \tag{13}$$

for the $Nd - Cu$ interaction. For simplicity, both spins are assumed to be of magnitude $S$. We treat the vector $\mathbf{\Omega} = \mathbf{s}_{Cu}/S$ like a classical variable, subject to a stochastic force. We assume a Gauss-Markov process in which case the distribution function obeys a Fokker-Planck equation. Then the correlation function is of the form

$$< \mathbf{\Omega}(0)\mathbf{\Omega}(t) > \;=\; e^{-2D_r t} \tag{14}$$

where $D_r$ can be obtained from the nonlinear $\sigma$ model [42]. Because there is no long-range order $< \mathbf{\Omega}(t) >= 0$. The motion of the $Nd$ spin is governed by the equation

$$\frac{d}{dt}\,\mathbf{n}(t) \;=\; \omega_0(\mathbf{\Omega}(t)\mathrm{x}\mathbf{n}(t)) \tag{15}$$

where $\mathbf{n}(t) = \mathbf{S}_f$ and $\omega_0 = \alpha S$. The spectral function

$$I(\omega) \;=\; \frac{1}{2\pi}\int_{-\infty}^{+\infty} dt\; e^{i\omega t} < \mathbf{n}(0)\mathbf{n}(t) > \tag{16}$$

is evaluated by making use of the corresponding stochastic Liouville equation. We find that $I(\omega)$ is of the form

$$I(\omega) \;=\; \frac{1}{3\pi}\,\frac{4D_r}{\omega^2 + (4D_r)^2} \;+\; (\text{side peaks at } \omega_0). \tag{17}$$

While $D_r(T)$ vanishes as $T \to 0$ in the presence of long-range order, it remains finite when the latter is destroyed by doping. A linear specific-heat contribution of the $4f$ spin is obtained from

$$C(T)_{imp} \;=\; \frac{d}{dT} < H_{int} >= \frac{S(S+1)}{T^2}\int_0^{\infty} d\omega\omega^2 \frac{I(\omega)}{cosh^2(\omega/2T)} \tag{18}$$

when $D_r(T=0) \neq 0$. The side peaks in $I(\omega)$ give raise to a Schottky-type contribution. The calculated specific heat reproduces the experiments reasonable well. One shortcoming of the theory in its present form is the low-temperature spin susceptibility which follows from

$$\chi_{imp}(T) \;=\; \frac{4}{3}(g\mu_B)^2 S(S+1)\int_0^{\infty} d\omega\,\frac{I(\omega)}{\omega}\,\tanh\frac{\omega}{2T}. \tag{19}$$

We find $\chi_{imp}(T) \sim ln(Dr/T)$ at low $T$. Possibly this is due to the neglect of $Nd - Nd$ interactions.

However, when evaluated for $T = 0.4\ K$ one finds for $Nd_{1.8}Ce_{0.2}CuO_4$ a Sommerfeld-Wilson ratio of $R \simeq 1.4$.

## ELECTRONIC CHARGE ORDER AND HEAVY FERMIONS

It was first suggested by Wigner that electrons will form a lattice when their mutual Coulomb repulsion exceeds their kinetic energy [43]. He considered a homogeneous electron gas. In this case one needs rather dilute systems in order that the repulsive energy is larger than the kinetic one, a consequence of Pauli's principle. This suggest considering systems in which the conduction electrons are strongly localized at atomic sites so that the overlap of atomic wavefunctions of neighboring sites is small. Then the energy gain due to delocalization is small and therefore electron crystallization or charge ordering is expected to take place already at high densities. The most localized electrons are $4f$ electrons. Therefore the question arises whether one can find intermetallic rare-earth systems in which $4f$ electrons form a superlattice on the underlying atomic lattice. We find that $Yb_4As_3$ is such an example and we want to discuss this system in more detail.

The intermetallic compound $Yb_4As_3$ has an anti-$Th_3P_4$ structure. The $Yb$ ions are situated on chains with directions along the diagonals of a cube. Thus we are dealing with a system of four sets of interpenetrating chains (see Fig. 6).

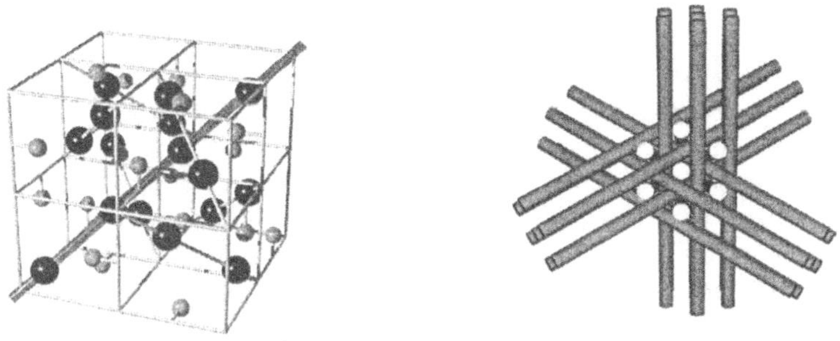

(a) (b)

**Fig. 6.** *(a) Structure of $Yb_4As_3$: large spheres represent $Yb$ and small spheres $As$ ions. (b) four families of interpenetrating chains on which the $Yb$ ions are located.*

We want to draw attention to the fact that the distance between neighboring $Yb$ ions on a chain exceeds the one between neighboring ions on different chains.

Because $As$ has a valency of $3-$, three of the four $Yb$ ions have a filled $4f$ shell, i.e., a valency $2+$, while one ion is in a $4f^{13}$ configuration (valency $3+$). Since all $Yb$ sites are equivalent, the hole in the $4f$ shell is shared between four $Yb$ ions and the system is metallic. However, at a temperature $T_s \simeq 300$ $K$ the $4f$ holes start to order by avoiding nearest neighbor positions. At $T = 0$ they occupy one of the four families of chains only, while the remaining three families remain empty. The phase transition at $T_s$ is of weak first order and the system goes over into a trigonal distorted structure [12-17]. This is so because the $Yb^{3+}$ ions collecting, e.g., along [1,1,1] have a smaller radius than the $Yb^{2+}$ ions. Therefore the chains in that direction contract while the chains in the other three directions elongate a bit so that the volume of unit cell remains unchanged.

Measurements of the Hall constant reveal a dramatic increase below $T_s$, implying a sharp drop in the charge-carrier density with decreasing temperature. At low $T$ one is left with one carrier per $10^3$ $Yb$ ions. The resistivity increases below $T_s$ with decreasing temperature until it reaches a maximum of approximately 10 $m\Omega$ $cm$. At low $T$ it is of the form $\rho(T) = \rho_0 + AT^2$ and therefore shows Fermi-liquid behaviour. The linear specific-heat coefficient $\gamma$ is found to be of order $\gamma \simeq 200\ mJ/(mol \cdot K^2)$. The spin susceptibility is Pauli like and equally enhanced as $\gamma$, giving raise to a Sommerfeld-Wilson ratio of order unity. No indication of magnetic order is found down to $T = 0.045K$, but below $2K$ the susceptibility increases again which indicates the presence of another low-energy scale [15]. The above findings strongly suggest heavy-fermion behaviour which is further confirmed by the observation that the ratio $A/\gamma^{\nu}$ ($\nu \simeq 2$) compares well with that of other heavy-fermion systems [14]. One should appreciate that despite the low-carrier concentration the $\gamma$ value exceeds that of, e.g., $Na$ by a factor of more than $10^2$. This demonstrates that the high density of low-energy excitations must clearly involve spin-degrees of freedom of the $Yb^{3+}$ ions. The Kondo effect can be ruled out as a source of heavy quasiparticles. The low-energy scale which corresponds to the observed $\gamma$ value is $T^* \simeq 40K$. Inelastic neutron scattering shows a well resolved crystal-field excitation of $Yb^{3+}$ at a comparable energy which would be impossible if local singlets would form with a binding energy of similar size.

Although the anti-$Th_3P_4$ structure is not a one-dimensional structure, the crystallization or charge order of the $4f$ holes changes it into one, because the different chains, e.g., along [1,1,1] are well separated from each other. Therefore one may describe the structural phase transition at $T_s$ in terms of a collective band Jahn-Teller (CBJT) effect of strongly correlated electrons or holes [17]. The Coulomb repulsion which is responsibly for the charge ordering takes here the form of a strong deformation-potential coupling, a feature which is quite common in mixed-valence systems.

The $CBJT$ transition splits the fourfold degenerate quasi-$1D$ density of states into a nondegenerate one corresponding to the short chains and a threefold one due to the long chains. The nondegenerate one is lower in energy and would be half filled if charge ordering were perfect and the holes were uncorrelated fermions. Instead, the holes are strongly correlated. Two holes on a site would imply a $4f^{12}$ configuration for $Yb$ and that has a much too high energy. Therefore, we are dealing with an almost filled lower (hole) Hubbard band rather than with an almost half-filled band. The ideal system with a full lower Hubbard band should therefore be an insulator.

The effective Hamiltonian for the description of the $CBJT$ is of the form [44]

$$\begin{aligned} H &= -\sum_{\mu}\sum_{<ij>\sigma} t_\mu f^+_{i\mu\sigma} f_{j\mu\sigma} + U\sum_{\mu}\sum_{i} n_{i\mu\uparrow}n_{i\mu\downarrow} \\ &+ \epsilon_\Gamma \sum_{\mu}\sum_{i\sigma} \Delta_\mu n_{i\mu\sigma} + N_L c_\Gamma \epsilon_\Gamma^2. \end{aligned} \tag{20}$$

The subscript $\mu = 1, \ldots 4$ refers to the four different chain directions and $< ij >$ denotes neighbouring sites on a chain. The first term describes the kinetic energy of holes moving in a chain. Hopping between chains is neglected here. In the distorted phase the $t_\mu$ may differ for the elongated and the contracted chains and an ansatz of the form

$$\begin{aligned} t_\mu &= t_+\delta_{\mu 1} + t_-(1-\delta_{\mu 1}) \\ t_+ &= t\ e^{\lambda\epsilon_\Gamma}, \qquad\qquad t_- = t\ e^{-\lambda\epsilon_\Gamma/3} \end{aligned} \tag{21}$$

seems appropriate with $\epsilon_\Gamma$ defined below. From LDA calculations [45] one can extract a band width of $4t = 0.2eV$. The second term takes the on-site Coulomb repulsions of holes into account. The third term is due to a volume-conserving coupling of the $f$ bands to the trigonal strain $\epsilon_\Gamma$ with $\Gamma = \Gamma_5$. This coupling is described by a deformation potential of the form

$$\Delta_\mu = \frac{\Delta}{3}(4\delta_{\mu 1} - 1). \tag{22}$$

The basis for this is the Coulomb repulsion of holes on neighbouring sites. It is treated here as an effective attraction between holes on next-nearest neighbour sites, i.e., nearest sites on a chain. Here the tacit assumption has been made that the ground state is one with the holes positioned in on type of chain only. The last term in Eq. (20) is the elastic energy associated with the lattice deformation. Here $N_L$ is the number of sites in a chain and $c_\Gamma$ is the bulk elastic constant for which we choose $c_\Gamma/\Omega = 4 \cdot 10^{11}$ erg/$cm^3$ with $\Omega$ denoting the volume of a unit cell. Note that the lattice constant is $a_0 = 8.789\mathring{A}$.

The Hamiltonian (20) is suitable for describing a band Jahn-Teller transition. It lifts the four-fold degeneracy of the $f$ bands by the spontaneous appearance of a trigonal strain which lowers the symmetry. In order to demonstrate this we consider first the case of $U = 0$. This neglects the effect of the strong electron correlations on the CBJT effect which prevent occupancies of sites by two holes. Near the transition, where each of the chains is still nearly quarter-filled this is not a too bad approximation. However, it is not acceptable for low temperatures where in the distorted phase all holes are ensembled in one family of chains so that there is nearly one hole per site. The condition for a CBJT transition is $\Delta^2/(t c_\Gamma) > 3$ [17]. Choosing $\Delta = 5eV$ leads to a transition temperature of $T_c \simeq 250K$ which is close to the experimental value of $T_c^{exp} \simeq 300K$. The Grüneisen parameter $\Omega_G = \Delta/(4t)$ is 25 in this case which seems reasonable when compared with other mixed valence systems. The temperature dependence of the spontaneous strain as well as the associated occupation numbers $n_\mu$ of the chains are shown in Fig. 7.

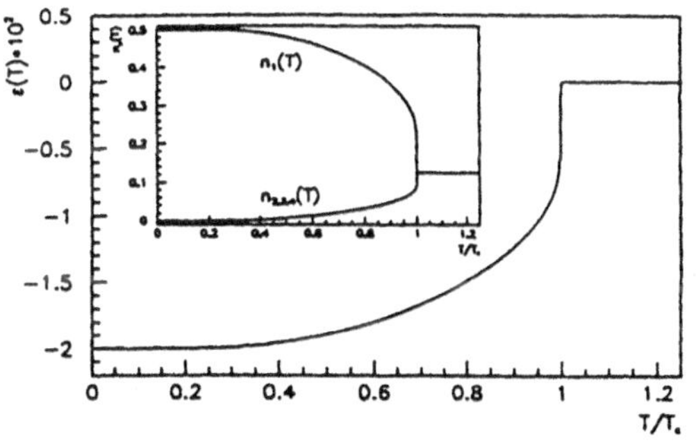


**Fig. 7.** *Temperature dependence of the trigonal strain $\epsilon_\Gamma(T)$ below $T_c$. Insert: $T$ dependence of the occupation numbers of the four one-dimensional $f$-bands. (From Ref. [17])*

As a consequence of the trigonal strain the centers of the fourfold degenerate one-dimensional $f$ bands split into three (upper) and one (lower). The energy difference of their centers is $(4/3) \mid \epsilon_\Gamma \Delta \mid$ with the equilibrium strain given by $\epsilon_\Gamma = -\Delta/(2c_\Gamma) \approx 0.02$. It is noted that band refilling by a CBJT transition and hole crystallization are equivalent pictures, in particular when large $U$ values exclude double occupancies of sites by holes. This brings us to the effects of finite values of $U$.

Since the work of Lieb and Wu [46] it is well known that the one-dimensional Hubbard model can be solved exactly. The same solution can be adapted to the present Hamiltonian because the chains are not directly coupled. Only an indirect coupling exists via the strain $\epsilon_\Gamma$ which enters $t_\mu$ and the chemical potential. The Lieb-Wu type of equations based on the Bethe ansatz must therefore be solved self-consistently [44]. While for $U = \infty$ analytic solutions can be obtained one must resort to numerical solutions when $U$ is arbitrary. For $U = \infty$ and $U = 0$ one finds for the ground-state energy per site [44]

$$\begin{aligned}\frac{E_s}{N_L} = -\frac{2}{\pi}[sin(n_1\frac{\pi}{q})e^{\lambda\epsilon_\Gamma} &+ 3sin(n_2\frac{\pi}{q})e^{-\lambda\epsilon_\Gamma/3}] \\ &- \epsilon_\Gamma(n_1 - n_2) + c_\Gamma\epsilon_\Gamma^2\end{aligned} \tag{23}$$

where $q = 1$ for $U = \infty$ and $q = 2$ for $U = 0$. Furthermore, $n_1, n_2$ are the hole concentrations in the elongated chains and in the other three chains, respectively. Here and in the following the energy scale is set by taking $t = 1$. In order to find the equilibrium strain $\epsilon_\Gamma$ we must minimize $E_s$ with respect to $\epsilon_\Gamma$ under the condition that the chemical potential is the same in all chains. At $T = 0$ this implies that $\delta E_1/\delta n_1 = \delta E_2/\delta n_2$, and therefore we must know $E_1$ and $E_2$ separately. One finds

$$\frac{E_i}{N_L} = -\frac{2q}{\pi}t_i sin(n_1\frac{\pi}{q}) - \epsilon_\Gamma\Delta_i n_i + \frac{c_\Gamma}{4}\epsilon_\Gamma^2 \tag{24}$$

which leads to the constraint

$$\epsilon_\Gamma = 1.5[cos(n_2\frac{\pi}{q})e^{-\lambda\epsilon_\Gamma/3} - cos(n_1\frac{\pi}{q})e^{\lambda\epsilon_\Gamma}]. \tag{25}$$

After minimization of $E_s/N_L$ with respect to $\epsilon_\Gamma$ (or $n_1$) the hole concentrations in the different chains are obtained. Without the CBJT transition, i.e., for $\epsilon_\Gamma = 0$ it is $n_1 = n_2$. But when $c_\Gamma$ is sufficiently small so that $\Delta^2/(c_\Gamma t) \geq 3$ we have $n_1 > n_2$.

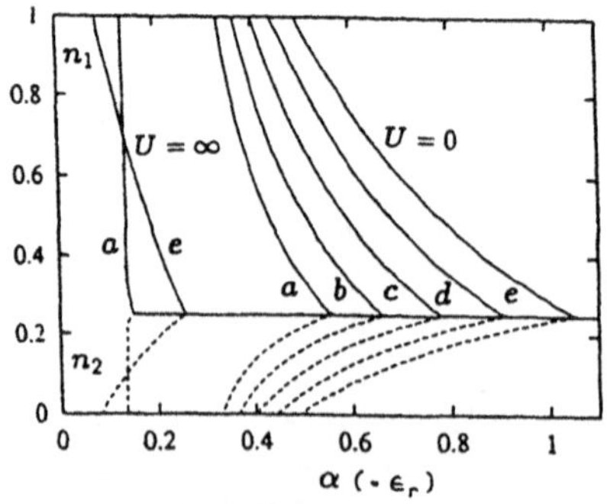


**Fig. 8.** *Hole concentration $n_1$ (solid lines) and $n_2$ (dashed lines) as function of $c_\Gamma (= \alpha\Delta^2)$ for different values of $\lambda$ and $U = 0, \infty$. Curves $a - e$ correspond to $\lambda/\Delta = 0, 0.05, 0.10, 0.15$ and 0.2 respectively. (From Ref. [44])*

Thereby two different cases must be distinguished. For the largest part of the parameter space we find $n_1 = 1$ and $n_2 = 0$, i.e., all holes collect in the long chains while the short chains do not contain holes. This is the same situation as shown in Fig. 7. However, depending on the size of $\lambda, U$ and $c_\Gamma$ there is also a parameter regime for which $n_1 \neq n_2 \neq 0, \epsilon_\Gamma \neq 0$.. This is demonstrated in Fig. 8.

When $n_1 \neq n_2 \neq 0$ holds the system is a conductor rather an insulator and the situation corresponds to self-doping, but here without interband hopping as we want to stress. The above estimate of $\alpha \simeq 0.25$ suggests that $Yb_4As_3$ could well be in the regime of $n_1 \neq 0$, but the origin of the low carrier concentration is very likely to be another one. This is seen by performing an LDA+U bandstructure calculation [47]. Thereby a gap is found to exist between $As$ $4p$ and $Yb$ $4f$ bands on one side and the $Yb$ $4d$ bands on the other. Therefore the Fermi energy is pinned to the $4f$ hole level which itself is rather rigidly pinned to the top of the $As$ $p$ band. The few charge carriers are therefore pnictide $p$ holes, in agreement with cyclotron resonance experiments which have found small effective masses for the particles involved.

The physical picture of $Yb^{3+}$ ions accumulating in one family of chains was convincingly confirmed by beautiful inelastic neutron scattering experiments of Kohgi and coworkers [16]. The dispersion of the magnetic excitations found in these experiments agrees very well with the known work of des Cloizeaux and Pearson [48] on antiferromagnetic Heisenberg spin chains. Only near $q \simeq 0$ are some deviations found which may have their origin in crystal-field effects.

When the low-temperature specific heat is calculated from the observed excitation spectrum one finds $C = \gamma T$ with a $\gamma$ coefficient of the observed size. This proves that the large value of $\gamma$ is due to spin fluctuations in chains.

From the above findings we can conclude that in the low-temperature phase of $Yb_4As_3$ the conventional Landau description of Fermi liquids is inapplicable. Basic assumption of that description is the existence of a one to one correspondence between the excitations of a given metallic (or semimetallic) system and those of a nearly free electron system *provided* certain quantities like the effective mass are properly renormalized. This assumption cannot be made here. Since the $\gamma$ coefficient is due to the spin degrees of freedom of the Heisenberg chains there is no relation between the effective masses as they would show up in de Haas - van Alphen experiments and the size of $\gamma$. This is quite different from, e.g., $CeRu_2Si_2$ where the measured Fermi surface contains parts with very high effective masses from which the observed large $\gamma$ value can be derived [6, 27, 29]. The formation of a lattice of $4f$ holes with some remaining itinerant electrons invalidates the conventional Fermi liquid description. We can still save this concept if we introduce in addition to the *charged* quasiparticles a second system of *neutral* fermionic quasiparticles. They describe in a phenomenological way the spin excitations of the Heisenberg chains. Usually spin excitations like magnons are bosons, but in one dimension they can be transformed into fermions. The neutral Fermi liquid dominates the specific heat and spin susceptibility at low temperatures. Of course, it does not contribute to the electric current but it takes part in the heat current.

## ACKNOWLEDGEMENT

I would like to thank Dr. B. Schmidt, Dr. P. Thalmeier and Dr. G. Zwicknagl for a long standing cooperation on strongly correlated electrons.

# References


[1] K. Andres, J.E. Graebner and H.R. Ott, *Phys. Rev. Lett.* 35:1779(1975)

[2] A.C. Hewson, The Kondo Problem to Heavy Fermions, Cambridge University Press, Cambridge (1993)

[3] P.A. Lee, T.M. Rice, J.M. Serene, L.J. Sham and J. W. Wilkins, Comments in *Cond. Matt. Phys.* 12:99(1986)

[4] P. Fulde, J. Keller and G. Zwicknagl, *Solid State Phys.* eds. H. Ehrenreich and D. Turnbull, Academic Press, San Diego 41:p.1(1988)

[5] P. Schlottmann, *Phys. Rep.* 181:1(1984)

[6] G. Zwicknagl, *Adv. Phys.* 41:203(1993)

[7] P. Wachter, *Handbook Phys. and Chem. Rare Earths*, eds. K.A. Gschneidner, Jr., L. Eyring, G.H. Lander and G.R. Chappin, Elsevier, Amsterdam 19:177(1994)

[8] N. Grewe and F. Steglich, *Handbook Phys. and Chem. Rare Earths*, eds. K.A. Gschneidner, Jr. and L. Eyring, Elsevier, Amsterdam 14:343(1991)

[9] T. Brugger, T. Schreiner, G. Roth, P. Adelmann and G. Czjzek, *Phys. Rev. Lett.* 71:2481(1993)

[10] P. Fulde, V. Zevin and G. Zwicknagl, *Z. Phys. B* 92:133(1993)

[11] J. Igarashi, K. Murayama and P. Fulde, *Phys. Rev. B* 52:15966(1995)

[12] A. Ochiai, T. Suzuki and T. Kasuya, *J. Phys. Soc. Jpn.* 59:4129(1990)

[13] T. Suzuki. *Physica B* 186-188:347(1993)

[14] A. Ochia, D.X. Li, Y. Haga, O. Nakamura and T. Suzuki, *Physica B* 186-188:437[1993)

[15] P. Bonville, A. Ochiai, T. Suzuki and E. Vincent, *J. Phys.* I 4:595(1994)

[16] M. Kohgi, K. Iwasa, J.-M. Mignot, A. Ochiai and T. Suzuki, *Phys. Rev. B* 56:R11 388(1997)

[17] P. Fulde, P. Thalmeier and B. Schmidt, *Europhys. Lett.* 35:109(1995)

[18] Q.G. Sheng and B.R. Cooper, *Phil. Mag. Lett.* 72:123(1995)

[19] C.M. Varma, Y. Yafet, *Phys. Rev. B* 13:2950(1976)

[20] K. Yoshida, *Phys. Rev.* 147:223(1966); see also K. Yoshida and A. Yoshimori, *Magnetism*, ed. by G.T. Rado and H. Suhl, Academic Press, New York 5:253(1973)

[21] S. Doniach, *Physica B* 91:231(1971)

[22] H. Razafimandimby, P. Fulde, J. Keller, *Z. Phys. B* 54:111(1984)

[23] N. d'Ambrumenil and Fulde P, *J. Magn. Magn. Mater.* 47+48:1(1985)

[24] J. Sticht, N. d'Ambrumenil, J. Kübler, *Z. Phys. B* 65:149(1986)

[25] P. Strange and D.M. Newns, *J. Phys. F* 16:335(1986)

[26] G. Zwicknagl, *J. Magn. Magn. Mater.* 76+77:16(1988)

[27] G. Zwicknagl, E. Runge and N.E. Christensen, *Physica B* 163:97(1990)

[28] J. Jensen and P. Fulde, *Phys. Rev. B* 27:4085(1983), see also R.: White and P. Fulde, *Phys. Rev. Lett.* 47:1540(1981)

[29] H. Aoki, S. Uji, A. Albessand and Y. Onuki, *Phys. Rev. Lett.* 71:2120(1993)

[30] M.N. Norman and D. Koelling D, *Handbook on the Physics and Chemistry of Rare Earths*, ed. by K.A. Gschneidner, L. Eyring. G.H. Lander, G.R. Choppin, Elsevier, Amsterdam 17:1(1993)

[31] H. Keiter and J.C. Kimball, *Int. J. Magn.* 1:233(1971)

[32] H. Kojima, Y. Kuramoto and M. Tachiki, *Z. Phys. B* 54:293(1984)

[33] N.E. Bickers, *Rev. Mod. Phys.* 59:845(1987)

[34] N.E. Bickers, D.L. Cox and J.W. Wilkins, *Phys. Rev. Lett.* 54:230(1985)

[35] V. Zevin, G. Zwicknagl and P. Fulde, *Phys. Rev. Lett.* 60:2331(1988)

[36] K. Thomala, G. Weschenfelder, G. Czjzek and E. Holland-Moritz, *J. Magn. Magn. Mat.* 89:143(1990)

[37] C.A. King and G.G. Lonzarich, *Physica B* 171:161(1991)

[38] M. Hillberg, M. de Melo, H.H. Klauß W. Wagner, F.J. Litterst, P. Adelmann and G. Czjzek, *Hyperfine Interact.* 104:221(1997)

[39] M. Loewenhaupt, P. Fabi, S. Horn, P.v. Aken and A. Severing, *J. Magn. Magn. Mat.* 140-144:1293(1995)

[40] W. Henggeler, B. Roessli, A. Furrer, P. Vorderwisch and T. Chatterji, *Phys. Rev. Lett.* 80:1300(1998)

[41] P. Fulde and V. Zevin, *Europhys.* 24:791(1993), see also S. Tornow, V. Zevin and G. Zwicknagl, *Ann. Physik* 5:501(1996)

[42] S. Chakravarty, B.I. Halperin and D.R. Nelson, *Phys. Rev. B* 39:2344(1989), S. Chakravarty and R. Orbach, *Phys. Rev. Lett.* 64:224(1990)

[43] E.P. Wigner, *Phys. Rev.* 46:1002(1934)

[44] Y.M. Li, N. d'Ambrumenil and P. Fulde, *Phys. Rev. Lett.* 78:3386(1997)

[45] K. Takegahara and Y. Kaneta, *J. Phys. Soc. Jpn.* 60:4009(1991)

[46] E.H. Lieb and F.Y. Wu, *Phys. Rev. Lett.* 20:1445(1968)

[47] V.N. Antonov, A.N. Yaresko, A.Ya. Perlov, P. Thalmeier, P. Fulde, P.M. Oppeneer and H. Eschrig (submitt. for publ.)

[48] J. des Cloizeaux and J.J. Pearson, *Phys. Rev.* 128:2131(1962)

# MAGNETIC AND THERMODYNAMIC PROPERTIES OF THE 3-D ANDERSON LATTICE HAMILTONIAN

C. Huscroft,[1] R.T. Scalettar,[1] A.K. McMahan,[2] and E.L. Pollock[2]

[1]Physics Department
University of California
Davis, CA 95616

[2] Lawrence Livermore National Laboratory
P. O. Box 808
Livermore, CA 94551

## INTRODUCTION

Tight binding models capture many of the qualitative features of interaction induced effects in solids. For example, the simplest such model, the single band Hubbard Hamiltonian, [1] describes the "Mott" insulating phase which occurs in correlated systems, despite the fact that the one electron band is nominally only half-filled, as well as the tendency towards magnetic order. Both phenomena occur in the transition metal oxides. The Periodic Anderson Model (PAM) is a step towards incorporating more complex orbital structure. [2] It contains a pair of orbitals on each site– a delocalized conduction band and a set of highly correlated, localized states. The PAM successfully describes conditions for transitions between antiferromagnetic order of the local moments and phases in which these moments are quenched into singlets paired with conduction electrons. [3] These phenomena are central to heavy fermion systems. [4] The pressure-induced volume collapse in Ce has also been attributed to Kondo-like quenching of the local f moments in this metal, as has been discussed in the context of the impurity Anderson Model. [5]

We will describe Quantum Monte Carlo (QMC) calculations of the magnetic and thermodynamic properties of the PAM in three dimensions. Previous QMC studies have been reported in one and two dimensions. [6,7] A focus of our attention will be on the density of states and the specific heat. The organization of this paper is as follows. We first introduce the PAM and outline some of its properties. Next, a brief presentation of the Quantum Monte Carlo, Maximum Entropy, and Hartree–Fock methods is given. We then show the equilibrium magnetic properties of the PAM, including the spin correlations between conduction and localized orbitals, and antiferromagnetic correlations in the localized band, before turning to the thermodynamics and the density of states. A concluding section describes connections of this work to the problem of the rare earth volume collapse transitions.

*Electron Correlations and Materials Properties*, edited by Gonis *et al.*
Kluwer Academic/Plenum Publishers, 1999.

## THE ANDERSON LATTICE HAMILTONIAN

The Periodic Anderson Model is,

$$
\begin{aligned}
H = & -t_{dd} \sum_{\langle ij \rangle \sigma} (d^\dagger_{i\sigma} d_{j\sigma} + d^\dagger_{j\sigma} d_{i\sigma}) - t_{fd} \sum_{\langle ij \rangle \sigma} (d^\dagger_{i\sigma} f_{j\sigma} + f^\dagger_{j\sigma} d_{i\sigma}) \\
& + U_f \sum_i (n_{if\uparrow} - \frac{1}{2})(n_{if\downarrow} - \frac{1}{2}) + \epsilon_f \sum_{i\sigma} n_{if\sigma} - \mu \sum_{i\sigma} (n_{if\sigma} + n_{id\sigma}) \,. \qquad (1)
\end{aligned}
$$

Here $d^\dagger_{i\sigma}$ and $f^\dagger_{i\sigma}$ represent the creation operators for $d$ and $f$ electrons of spin $\sigma$ on site $i$. $n_{id}$ and $n_{if}$ are the total number of up and down spin electrons on the $d$ and $f$ orbitals at lattice site $i$. $t_{dd}$ and $t_{fd}$ are the intersite valence and $f$–valence hybridizations, $U_f$ is an on–site repulsion for the $f$ electrons, $\epsilon_f$ is the position of the $f$ level, and $\mu$ is the chemical potential. The interaction is written in particle–hole symmetric form which has the convenience that at $\mu = \epsilon_f = 0$, both orbitals are precisely half–filled, $n^f_i = n^d_i = 1$, for any choice of $t_{dd}$, $t_{fd}$, $U_f$ or temperature $T$. The choice $\epsilon_f = 0$ is often referred to as the "symmetric" limit of the PAM. We will present results mostly for this case.

Writing the one–body terms in momentum space, the PAM becomes,

$$
\begin{aligned}
H = & \sum_{k\sigma} \epsilon_k d^\dagger_{k\sigma} d_{k\sigma} + \sum_{k\sigma} V_k (d^\dagger_{k\sigma} f_{k\sigma} + f^\dagger_{k\sigma} d_{k\sigma}) \\
& + U_f \sum_i (n_{if\uparrow} - \frac{1}{2})(n_{if\downarrow} - \frac{1}{2}) + \epsilon_f \sum_{k\sigma} n_{kf\sigma} - \mu \sum_{k\sigma} (n_{kf\sigma} + n_{kd\sigma}) \,. \qquad (2)
\end{aligned}
$$

In this paper, we will study the properties of the PAM on a simple cubic structure for which,

$$
\begin{aligned}
\epsilon_k &= -2t_{dd} \left[\cos k_x a + \cos k_y a + \cos k_z a\right], \\
V_k &= -2t_{fd} \left[\cos k_x a + \cos k_y a + \cos k_z a\right], \qquad (3)
\end{aligned}
$$

where $a$ is the lattice constant. Often the PAM is defined with an $f$–valence hybridization on the same site, leading to a momentum independent $V_k$. With that choice, even in the non–interacting limit $U_f = 0$ the PAM is an insulator at half–filling owing to the opening of a hybridization gap at the Fermi surface. We believe our choice of near-neighbor intersite hybridization in Eq. 1 is physically well motivated, as strictly onsite one–body terms in the Hamiltonian do not couple different angular momenta. However, it is likely that, while the positions of critical coupling and temperatures are somewhat different, the underlying physics is similar between the two choices.

In numerical solutions of Eq. 2, we have assumed parameters which are roughly characteristic of the rare earths. Specifically, we have taken $t_{dd} = 1$ eV, $U_f = 6$ eV, and considered a range of $t_{fd}$ values. The expression for $\epsilon_k$ in Eq. 3 yields a $d$ band of width $12t_{dd}$, and for convenience we shall similarly define an $f$ band width, $W_f = 12t_{fd}$. Parameter calculations show the ratio $U_f/W_f$ has roughly quadratic dependence on atomic volume, due primarily to the variation of $t_{fd}$ with volume. [8]

As we shall describe in detail below, the central issue in numerical and analytic work on the PAM concerns magnetic and thermodynamic properties. As the on–site repulsion $U_f$ increases, double occupancy of the $f$ orbitals is suppressed, and local moments form. This phenomenon is most marked in the symmetric limit where there is on average one electron per $f$ orbital, so that as double occupancy is reduced, each $f$ orbital individually has a single electron and hence a spin. In addition, the Fermi wavevector of the conduction band is $k_F = (1, 1, 1)\pi/a$ which is optimal for staggered order. The central question is then whether the $f$ moments align. In the absence of any direct $f$–$f$ hybridization, ordering can occur via an indirect RKKY interaction mediated by the conduction electrons. However, there is also a competing tendency. As $t_{fd}$ is increased, $d$ and $f$ electrons can form local singlets, suppressing long range antiferromagnetic order. The local moments are not destroyed through the magnetic transition, instead they remain rather well–formed, but simply lose their coherence.

We will present new QMC results for magnetic order in the three dimensional case. However, we will also focus on the thermodynamics of the transition. We expect that as singlets form, the entropy of the system will be significantly reduced, a phenomenon which might show up as a prominent feature in the energy, specific heat, and free energy. As we will see, observing this is complicated in the symmetric case since one must disentangle the effects of singlet formation from those of magnetic ordering.

## CALCULATIONAL TECHNIQUES

Over the last fifteen years, Quantum Monte Carlo techniques have made great progress in simulations of the equilibrium charge, spin, and pairing correlations of two–dimensional correlated electron systems. [9] The usefulness of the approach is that it treats interactions *exactly*. The main limitation is that only finite size systems, typically a few hundred electrons, can be treated. Equivalently, the orbital complexity of models that can be studied is highly constrained. Systems of a few hundred sites or orbitals are, however, an order of magnitude larger than can be studied by exact diagonalization, and large enough so that finite size scaling can often be performed to analyze the thermodynamic limit. A second limitation is that, unless symmetries prevent it, the "sign problem" restricts the accessible parameter range and precludes studies of low temperatures and strong coupling. This, however, does not occur in the symmetric limit of the PAM. In this section we will discuss a particular approach, the determinant QMC algorithm. [10]

We will also describe the "Maximum Entropy" method which has recently been developed to take QMC data and extract *dynamical* information like the density of states, [11] our approach for obtaining the specific heat, and analytic Hartree Fock calculations with which we will compare our results.

### The Determinant Quantum Monte Carlo Method

The central quantity in classical or quantum statistical mechanics is the partition function,

$$Z = \mathrm{Tr} e^{-\beta H}. \tag{4}$$

Just as a trace over any potential energy which is a quadratic form in the position operators can be done in a classical statistical mechanics problem, the trace over a quantum mechanical Hilbert space can be performed analytically for Hamiltonians which are quadratic in the fermion operators. However, the interaction term $U_f$ is quartic, and prevents us from doing the trace. The determinant QMC method is a technique which develops an exact reformulation of the original problem so that only quadratic terms appear.

We can isolate the interaction term by discretizing the inverse temperature, $\beta = L\Delta\tau$. The exponential of the full Hamiltonian $H = K+V$, now has a small parameter in its argument, and we can approximate the full exponential by the product of individual exponentials of the kinetic and potential energies.

$$Z = \mathrm{Tr}[e^{-\Delta\tau H}]^L \approx \mathrm{Tr}[e^{-\Delta\tau K}e^{-\Delta\tau V}]^L. \quad (5)$$

This "Trotter" approximation [12] becomes exact as $\Delta\tau \to 0$. In practice, simulations are conducted at several finite $\Delta\tau$ and the limit then extracted.

The interaction term can then be decoupled by introducing a discrete "Hubbard–Stratonovich" field, [13]

$$e^{-\Delta\tau U(n_{i\uparrow}-1/2)(n_{i\downarrow}-1/2)} = \frac{1}{2}e^{-\Delta\tau U/4} \sum_{S_{i\tau}=\pm 1} e^{\lambda S_{i\tau}(n_{i\uparrow}-n_{i\downarrow})}, \quad (6)$$

with coupling constant $\lambda$ given by $\cosh(\lambda)=e^{\Delta\tau U/2}$. The Hubbard–Stratonovich field is indexed both by a spatial variable $i$ labeling which of the interaction terms in the Hamiltonian (one for each site) is being decoupled, and an "imaginary time" variable $\tau$ which specifies which of the $L$ exponentials of the potential energy is being considered.

The right hand side of Eq. 6 is now quadratic in the fermion operators, so the trace can be performed analytically. The physical content of the procedure is that the set of interacting electrons has been replaced by independent electrons moving in an appropriate fluctuating classical field. The result for the partition function is the product of two determinants, one for each spin species.

$$Z = \sum_{\{S_{i\tau}\}} \det M_\uparrow \det M_\downarrow. \quad (7)$$

These determinants depend on the particular values of the Hubbard–Stratonovich field. The matrices $M_\sigma$ have dimension the number of spatial sites, $N$, in the original lattice.

One is really interested in evaluating ratios of traces which represent specific expectation values. Traces which involve additional operators reduce to a similar product of determinants, but with additional matrix elements of the inverse of $M$. Thus, for example, the equal time fermion Green's function is given by,

$$G_{\uparrow ij} = \langle c_{i\uparrow}c^\dagger_{j\uparrow}\rangle = Z^{-1}\,\mathrm{Tr}[c_{i\uparrow}c^\dagger_{j\uparrow}e^{-\beta H}] = Z^{-1}\sum_{\{S_{i\tau}\}} M^{-1}_{\uparrow ij}\det M_\uparrow(S)M_\downarrow(S). \quad (8)$$

Here $c$ is a generic Fermi operator. This takes precisely the form of a classical Monte Carlo problem: If the classical variables $S_{i\tau}$ are generated with a "Boltzmann weight" given by the product of the two determinants, for example with the Metropolis algorithm, the Green's function is given by simply accumulating a simple average of the values of $M^{-1}_{\uparrow ij}$ in the sequence so generated. From $G$ come all quantities of physical interest. Setting $i=j$ allows us to evaluate the occupation $n_{\uparrow i} = 1 - G_{\uparrow ii}$. Making $i,j$ near neighbors determines the kinetic energy. More complicated expectation values of, for example, two particle Green's functions like spin and density correlations can be reduced, via Wick's theorem, to appropriate products of the single particle $G$.

The Boltzmann weight, the product of the two determinants, is extremely complicated. Specifically, the computation a change in its value, required to do the Monte Carlo, scales as the square of the number of sites in the lattice, $N^2$. Thus to update all the $NL$ variables requires a computation time which scales as $N^3L$. It is this fact that limits the size of the simulations that can be performed.

While there are many refinements to this method, [14,15] the above description contains the essence of the methodology.

## The Maximum Entropy Method

Above, we described how the determinant QMC method allows the measurement of equal time Green's functions $G$. It is straightforward, but time consuming, to measure imaginary time displaced Green's functions:

$$G_{\uparrow ij}(\tau) = \langle c_{i\uparrow}(\tau)c_{j\uparrow}^{\dagger}(0)\rangle = \langle e^{\tau H}c_{i\uparrow}e^{-\tau H}c_{j\uparrow}^{\dagger}\rangle. \tag{9}$$

Such quantities are needed to measure susceptibilities, where the generalization of the classical fluctuation-dissipation result, $\chi_A = \beta[\langle A^2\rangle - \langle A\rangle^2]$, to quantum problems is $\chi_A = \int_0^\beta d\tau[\langle A(\tau)A(0)\rangle - \langle A\rangle^2]$.

Unfortunately, one is also very much interested in the Green's functions in *real* time and frequency, but operators like $e^{-iHt}$ are difficult to deal with numerically because of their rapidly fluctuating phases. Thus a crucial task is to relate the imaginary time Green's functions which can be measured in the simulations to real frequency response functions. In principle such relationships are known. For example, the single particle Green's function $G(\mathrm{p},\tau) = \langle c(\mathrm{p},\tau)c^\dagger(\mathrm{p},0)\rangle$, is given by an appropriate convolution of the spectral weight $A(p,\omega)$,

$$G(\mathrm{p},\tau) = \int_{-\infty}^{\infty} d\omega \frac{e^{-\tau\omega}}{1+e^{-\beta\omega}} A(\mathrm{p},\omega). \tag{10}$$

We want to invert this expression and get the spectral weight from the known Green's function. This inversion, or analytic continuation, is difficult to perform numerically because the kernel is small at large absolute values of the integrand $\omega$. Additionally, the Green's function values obtained from QMC are known only to within statistical error bars. These are small, a tenth of a percent of less typically, but nevertheless because of the ill conditioned nature of the kernel, significantly different $A(\mathrm{p},\omega)$ will be consistent with the data.

The method of Maximum Entropy [11] circumvents the problem of the ill defined integrand and noisy Green's function data by utilizing an analytic model of the high energy behavior in combination with a fit to the Green's function data obtained from QMC. The approach has had considerable success for the Hubbard and periodic Anderson Hamiltonians in two dimensions. Here we will apply it to our three dimensional simulations.

## Fitting E(T) to Obtain the Specific Heat

The specific heat $C = dE/dT$ can be evaluated from determinant QMC by calculating the energy at two closely spaced temperatures and doing the derivative numerically, or else by measuring $\beta[\langle H^2\rangle - \langle H\rangle^2]$. In practice, both approaches run into serious difficulties. The former has large fluctuations as typically occurs when one takes a numerical derivative of data with statistical errors. The latter approach involves high order Green's functions which are rather noisy, and also some non-trivial considerations concerning the Trotter approximation. [16]

A more promising technique is to take the QMC data for the energy, fit to a reasonable analytic form, and then differentiate the resulting expression. This approach has been applied to the single band Hubbard model in two dimensions to obtain the peaks in $C$ associated with charge and spin fluctuations. [17] Here we will use a similar idea, with, however, a rather different analytic form for the fitting. Specifically, we chose,

$$E(T) = E(0) + \sum_{n=1}^{N} c_n e^{-\beta n\Delta}, \tag{11}$$

where $\Delta$ and $c_n$ are free parameters. The number of terms $N$ was chosen to be about half of the number of QMC data points fit by this expression. We have verified that other fitting forms give similar results. Various further checks of the procedure come out of the physics, and are described below.

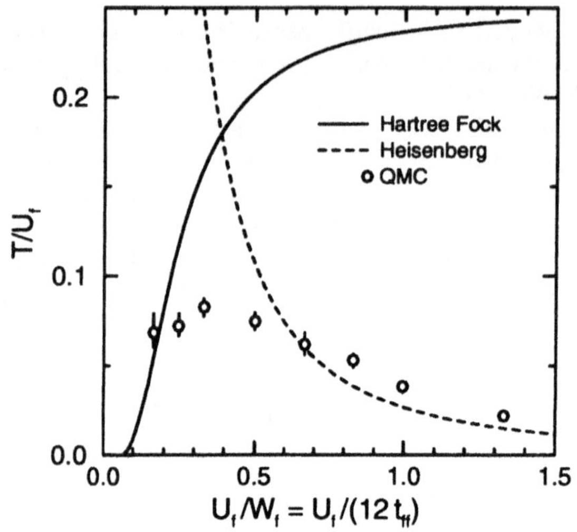


FIG. 1. The Néel temperature of the single band Hubbard model as calculated within Hartree–Fock (solid line), Heisenberg (dashed line), and Quantum Monte Carlo simulations (open circles).

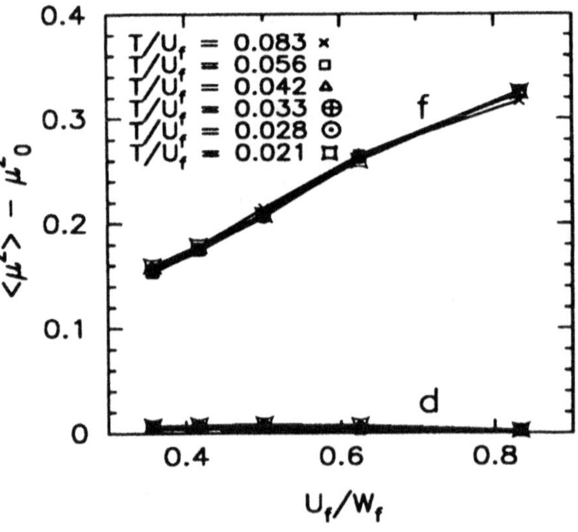


FIG. 2. Enhancement of the local moments on the $f$ and $d$ orbitals over their non-interacting values. Local moment formation occurs at a relatively high temperature, so that the results shown here are independent of $T$.

### Hartree–Fock Approach

The Hartree–Fock (HF) approach is the simplest analytic technique to incorporate interaction effects. HF takes an interaction term like $U_f n_{if\uparrow} n_{if\downarrow}$ in Eq. 1 and expresses it in the decoupled form $U_f[n_{if\uparrow}\langle n_{if\downarrow}\rangle + \langle n_{if\uparrow}\rangle n_{if\downarrow} - \langle n_{if\uparrow}\rangle\langle n_{if\downarrow}\rangle]$. Typically, one makes an *ansatz* for the form of the expectation values, $\langle n_{i\sigma}\rangle = \rho + \sigma m_u + (-1)^i \sigma m_s$. Here $\rho$ is the electron density per spin, $m_u$ represents a uniform (ferromagnetic) difference in up and down spin occupation, and $m_s$ represents a staggered (antiferromagnetic) difference in occupations. The resulting Hamiltonian is quadratic in the fermion operators and can be diagonalized by going to momentum space. The expectation values $\rho, m_u$, and $m_s$ are computed self-consistently by filling the resulting single particle levels. One can either compute a global minimum of the free energy by allowing $m_u$ and $m_s$ to take arbitrary values, or restrict them to special values, for example $m_u = m_s$ to get the paramagnetic solution.

HF has a number of well-known drawbacks, some of which are quite disastrous. This is illustrated in Fig. 1 where we show the Néel temperature $T_N$ as a function of the ratio $U_f/W_f$ for a single band Hubbard model corresponding to a simple cubic lattice. [18] The Hamiltonian is like that in Eq. 1 except with a single $f$ orbital, which has a non-zero $f$–$f$ hybridization $t_{ff}$, and an $f$-band width, $W_f = 12\,t_{ff}$. The on-site repulsion is fixed at $U_f = 6$. $T_N$ is grossly over-estimated at large $U_f/W_f$. This is a consequence of the fact that the sole magnetic energy scale in HF is set by $U_f$, the energy for the formation of moments. However, the exchange energy $J = t_{ff}^2/U_f$ is the appropriate scale of the Néel temperature at strong coupling. HF does not know about $J$ and hence at large $U_f/W_f$ has a $T_N$ which goes into the temperature axis at $U_{ff}/4$ instead of going to zero as $1/U_{ff}$. As can be seen in Fig. 1, QMC accurately describes this strong coupling (Heisenberg) limit, the weakly coupled Hartree-Fock limit, and crosses over naturally between these two regimes. [18,19]

There are, of course, many analytic treatments which improve over HF. In subsequent work, we plan on comparing QMC with those approaches. In this manuscript, however, we will focus on the results of QMC, and compare only with HF calculations.

Despite the cautionary message of Fig. 1, when properly interpreted, HF is a useful guide to the effects of interactions. Furthermore, it is worth emphasizing that while HF theory gets $T_N$ qualitatively wrong at strong coupling, one can get at the energy scale $J$ by examining the energy difference between the ferromagnetic and antiferromagnetic HF solutions. This difference does come down as $1/U_{ff}$ at strong coupling. We will exploit this fact later when we suggest further directions for QMC, in particular when suggesting what might occur in the non-symmmetric PAM, where we have yet to carry through detailed QMC simulations.

## MAGNETIC CORRELATIONS OF THE SYMMETRIC ANDERSON LATTICE HAMILTONIAN

We first describe the nature of the local moments in the PAM. In Fig. 2 we show the enhancement of the square of the local moment $\mu^2 = \langle (n_{\uparrow i} - n_{\downarrow i})^2\rangle$ over its value $\mu_0^2$ in the absence of interactions. The $x$-axis is the ratio of the on-site interaction $U_f$ to the $f$ bandwidth $W_f = 12t_{fd}$. Moment formation occurs at a temperature set by the "charge fluctuation" energy scale $T/U_f \sim 1/4$ so that in the range shown here, the moment enhancement is temperature independent. Thus, features we will subsequently identify for these temperatures in the specific heat are not caused by entropy associated with moment formation.

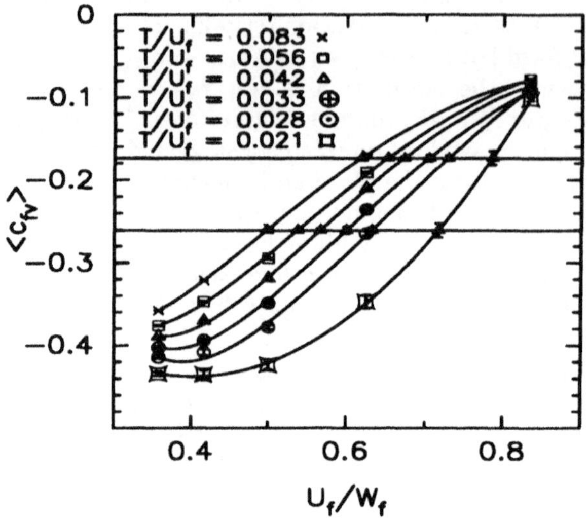


FIG. 3. The spin correlations between the $d$ and $f$ electron moments is shown as a function of $U_f/W_f$ for different temperature values. The horizontal lines give an approximate range for the cross–over to a singlet phase.

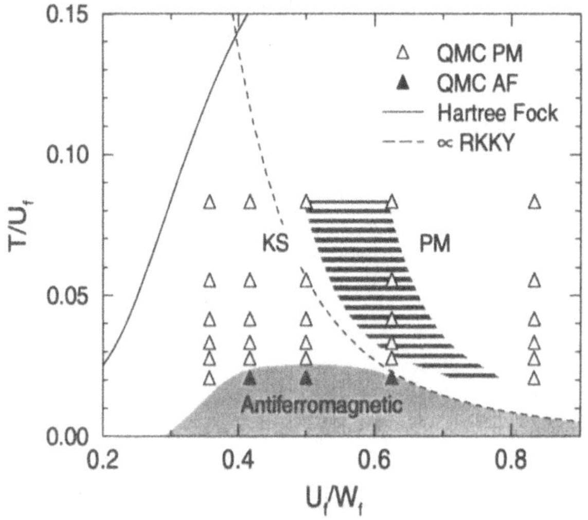


FIG. 4. The magnetic phase diagram of the 3D periodic Anderson model. Open triangles are the paramagnetic phase, while solid triangles are antiferromagnetic (see comment in text). The hatched region shows the region of cross–over to a singlet phase. Also shown are the Hartree–Fock and RKKY results.

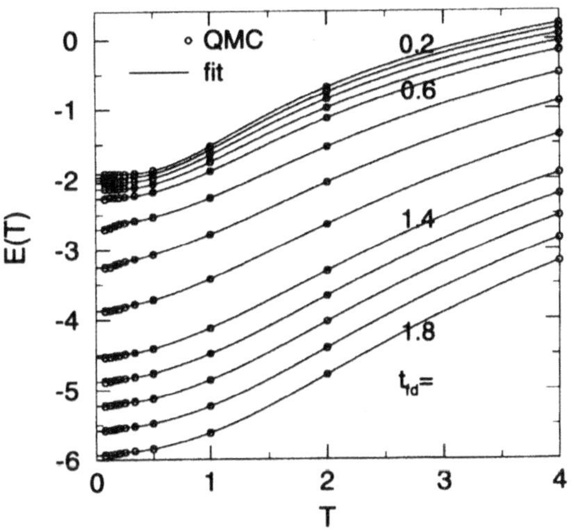


FIG. 5. Energy as a function of temperature. The open circles are the results of QMC simulations. The interpolating lines are the fit described in the text.

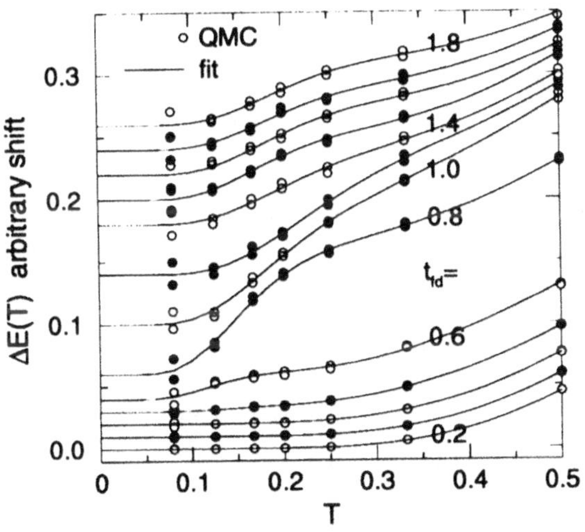


FIG. 6. The data of the preceding figure are reproduced with a higher resolution for the low temperature points to emphasize how the fit tracks the QMC energy data. The doubling of each QMC data point denotes uncertainty in the measured energy. Points on adjacent curves are alternately filled or open.

As described above, one of the features of the physics of local moments interacting with a delocalized band is the extent to which they form singlets. In the language of the PAM this is measured by the correlation function,

$$c_{fd} = \langle \vec{S}_i^f \cdot \vec{S}_j^d \rangle / [\langle (S_i^f)^2 \rangle \langle (S_i^d)^2 \rangle] \tag{12}$$

Here $i$ and $j$ are near neighbors, and $\vec{S}$ is the vector whose components are $\vec{S}_i = (c_{i\uparrow}^\dagger c_{i\downarrow}^\dagger)\vec{\sigma}\binom{c_{i\uparrow}}{c_{i\downarrow}}$. That is, $c_{fd}$ is just the spin–spin correlation function normalized to the size of the magnetic moment on the two sites. In Fig. 3 we show this quantity for various interaction strengths and temperatures. As the system is cooled at fixed $W_f$, we see that singlets form. The most rapid variation in $c_{fd}$ is at $U_f/W_f \approx 0.6$.

Measurements of the spin correlations between the $f$ electrons on different sites

$$c_{ff}(l) = \langle \vec{S}_i^f \cdot \vec{S}_{i+l}^f \rangle \tag{13}$$

allow us to determine whether long range antiferromagnetic order is present. In Fig. 4 we show the phase diagram which results from a consideration of the behavior of $c_{fd}$ and $c_{ff}$. Note, however, that a rigorous treatment of magnetic long range order, such as has been done in two dimensions, [7,20,15] requires simulations on a range of system sizes and an appropriate finite size scaling analysis, Here we have looked only at $4^3$ lattices, so our antiferromagnetic phase boundary is only approximate.

## THERMODYNAMIC PROPERTIES OF THE SYMMETRIC ANDERSON LATTICE HAMILTONIAN

The results of fitting the energy to the form Eq. 11 for various values of the $f$–$d$ hybridization are shown in Fig. 5, and, with higher resolution at low temperatures, in Fig. 6. It is useful to compare these results with those of a Hartree–Fock calculation. On the rather coarse energy scale of Fig. 7 it is seen that the antiferromagnetic HF solution appears fairly good. Indeed, it is guaranteed to be exact both at $t_{fd} = 0$ and at $U_f = 0$. We show the difference between the antiferromagnetic HF and QMC energies in Fig. 8. With the resulting better resolution, we see the two break apart at large $t_{fd}$ (small $U_f/W_f$) as already suggested by Fig. 7.

The specific heat obtained by differentiating the analytic fits in Fig. 5 is shown in Fig. 9. The points of divergence of the antiferromagnetic HF and QMC energies correlate fairly well with the positions of the sharp peaks in the specific heat of Fig. 9. This is consistent with the notion that HF accurately reflects the charge fluctuation energy scale $U_f$, but is less good at picking up smaller energy scales associated with more subtle magnetic ordering, whether antiferromagnetism or singlet formation.

The entropy of the system can be computed by evaluating the integrated area,

$$\int_0^\infty dT \frac{C(T)}{T} = \sum_{i=1}^{N} \frac{c_n}{n\Delta} = 4\ln 2 - S_0 \,. \tag{14}$$

Here the term $S_0$ arises from the fact that on a finite periodic cluster characterized by the dispersion, Eq. 3, the entropy does not go to zero at $T = 0$. For our $4^3$ PAM, $S_0 = 0.4332$, as may be directly calculated using HF, since this is a one-body effect. As shown in Fig. 10, we find the sum rule on the total entropy, Eq. 14 is obeyed to within less than 2 percent for $t_{fd} \geq 0.8$, which is a good check both on our QMC simulation and on our fitting procedure for the energy. For smaller $t_{fd}$ the integral instead approaches $3\ln 2 - S_0$, which implies the presence of some missing entropy. That is, for small $t_{fd}$, the system orders magnetically at temperatures below those

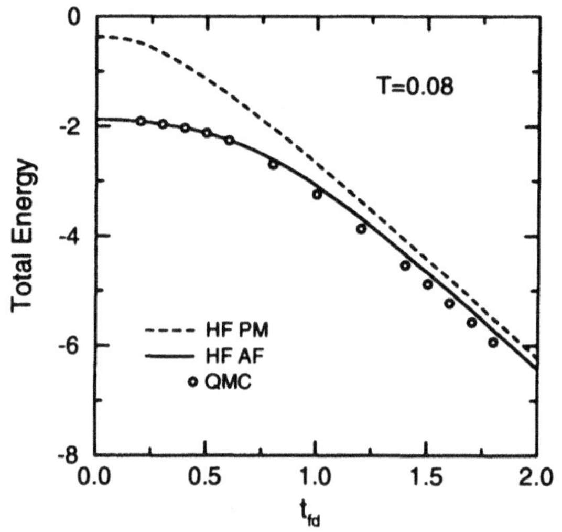


FIG. 7. Hartree-Fock and QMC energies are compared as a function of $f$-valence hybridization. The dashed curve is the paramagnetic solution, and the solid curve the antiferromagnetic solution. HF is exact at zero hybridization, and shows apparently small deviations at large $t_{fd}$.

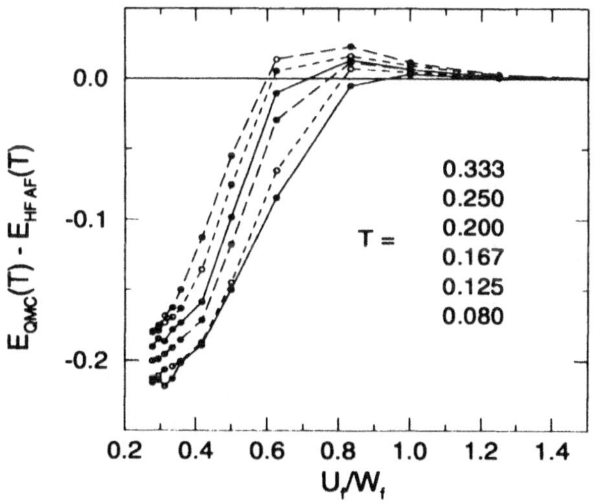


FIG. 8. Difference between antiferromagnetic Hartree-Fock and Quantum Monte Carlo energies. When the ratio $U_f/W_f$ is small, the QMC energies break away rather suddenly and fall below the HF energies.

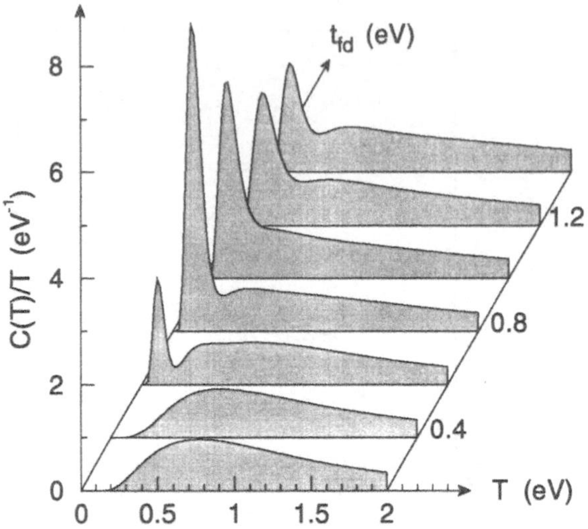


FIG. 9. Specific heat of the PAM for different choices of interband hybridization. The wide structure at $T \sim 1$ is associated with moment formation, while the sharper low $T$ structure is associated with magnetic ordering (either singlet formation or antiferromagnetism). As $t_{fd}$ is decreased, the sharp feature moves to lower temperature and, eventually, is no longer captured by the temperature range accessed in our simulations.

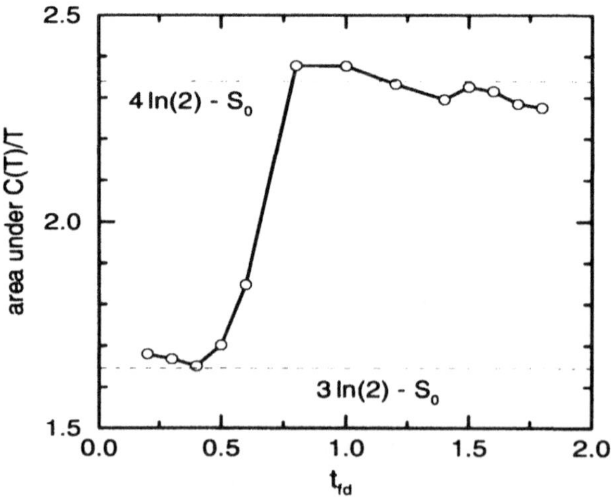


FIG. 10. The integrated areas under the curves of the preceding figure are shown. This measure captures the full entropy $4\ln 2$, adjusted for the effects of the finite lattice, for $t_{fd}$ above 0.75. Below this value there is missing entropy, indicating that there is indeed a peak in $C/T$ at lower temperatures. $S_0 = 0.4332$ for a $4^3$-site periodic lattice.

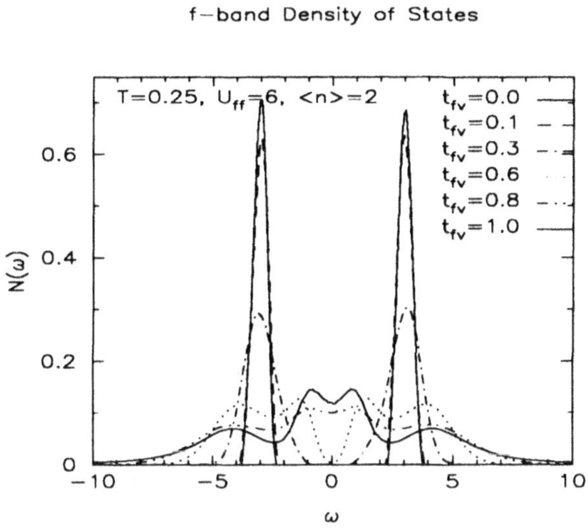


FIG. 11. The $f$ electron density of states for different choices of $t_{fd}$ at $T = t_{dd}/4$. At small $t_{fd}$, $N_f(\omega)$ consists of the two narrow bands at $\pm U_f/2$ expected for a localized orbital. At larger $t_{fd}$, $N_f(\omega)$ broadens and develops a Kondo resonance near $\omega = 0$.

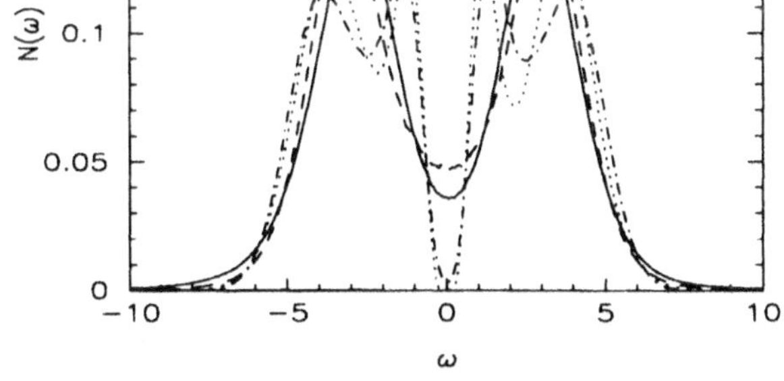


FIG. 12. The $f$ electron density of states for $t_{fd} = 0.6$ at different temperatures. As $T$ is lowered, the broad Mott peaks develop sharper Kondo resonances.

accessed in the simulations, $T \geq 0.1t_{dd}$, which would account for the missing $\ln 2$ in the entropy.

## DYNAMIC PROPERTIES OF THE SYMMETRIC ANDERSON LATTICE HAMILTONIAN– THE DENSITY OF STATES

The evolution of the $f$ density of states with changing $f$–valence hybridization is given in Fig. 11. For $t_{fd} = 0$ we have completely isolated localized sites, and therefore expect to have two delta function peaks at $\pm U_f/2 = \pm 3$. That is, we have two infinitely narrow Mott bands, separated by a Mott gap $U_f$. Our choice of the form of the interaction places these symmetrically about $\omega = 0$. As $t_{fd}$ is made non–zero, the $f$ electrons increasingly hybridize with the conduction band and their density of states widens. Eventually a structure appears which has both broad remnants of the Mott bands, but also Kondo resonance peaks near $\omega = 0$.

It is also interesting to examine the temperature dependence of the $f$ density of states at fixed $t_{fd}$. We show this in Fig. 12. At high temperatures there is only a pair of Mott peaks associated with charge fluctuations on the $f$ orbital. As $T$ is lowered, sharp Kondo resonances split off of these broad structures. The suppression of the $f$ density of states at $\omega = 0$ which separates the sharp peaks can arise either as a result of the development of an antiferromagnetic gap, or as a Kondo gap associated with the energy required to break a singlet pair.

The Kondo resonance in $N_f(\omega)$ is a further signature of singlet formation in the PAM. It develops at a temperature which coincides with that for which the static spin–spin correlation function $c_{fd}$ shows antiferromagnetic correlations are developing between the $f$ and $d$ electron spins. Similarly, the resonance develops at a temperature in rough coincidence with the peak in the specific heat shown in Fig. 9. However, because in the same temperature range antiferromagnetic correlations among the $f$ moments are also developing, we cannot as yet unambiguously attribute the thermodynamic signature to singlet formation.

We have seen that the Hartree–Fock technique overestimates $T_N$ and also fails to capture the thermodynamics of spin ordering at low temperatures $T \sim J$. It is interesting to ask what HF theory predicts for the density of states. The answer is that HF theory can describe the strong coupling case where two Mott bands exist at $\omega = \pm U_f/2$ as well as the weak coupling metallic limit where a single broad feature in $N_f(\omega)$ is centered about $\omega = 0$. However, it cannot correctly capture the intermediate case where a Kondo resonance develops.

## CONCLUSIONS

The periodic Anderson Hamiltonian is the fundamental model of interacting local bands and delocalized conduction electrons. It contains the essential competition between the tendency towards singlet formation of the local electron and conduction electron spin, and antiferromagnetic order in the local band, and has been used as a qualitative model of heavy fermion and other systems. Here we have presented new results for its magnetic and thermodynamic properties in three dimensions.

In this concluding section of the manuscript we turn to a more specialized application of the PAM, namely to the problem of volume collapse transition in rare earth systems.

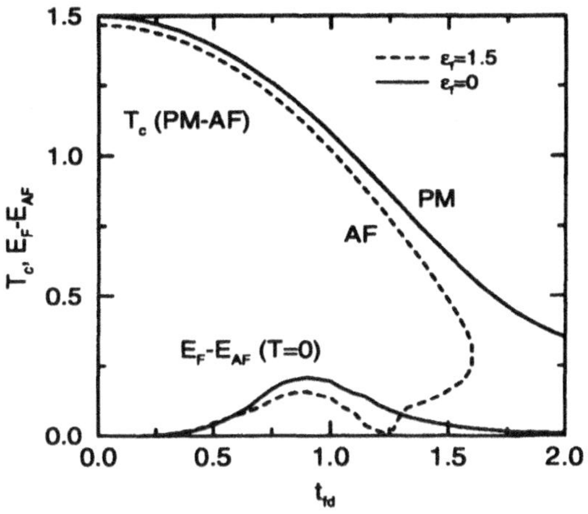


FIG. 13. The Hartree-Fock phase diagram in the non-symmetric limit is compared to the symmetric case. The region of antiferromagnetism is reduced for non-zero $\epsilon_f$. The real HF Néel temperatures is shown (curves labelled $T_c$), as well as the ferromagnetic–antiferromagnetic energy differences (curves labelled $E_{\mathrm{F}} - E_{\mathrm{AF}}$). For the symmetric case ($\epsilon_f = 0$), this energy difference is much closer to the actual Néel temperature seen in Fig. 4. The HF calculations were carried out with Brillouin zone sampling corresponding to $12^3$ sites, for $t_{dd} = 1$ and $U_f = 6$.

Several Lanthanides exhibit phase transitions under pressure characterized by abnormally large volume changes (14% for Cerium and 9% for Praseodymium), as has recently been reviewed. [8] The physical mechanisms responsible for these transitions have been debated since discovery of the Cerium phenomenon over 50 years ago. Currently, the two main viable conjectures [21] are a Mott transition of the 4f electrons accompanied by magnetic ordering, [22] and a "Kondo volume collapse" due to rapid change in the $4f$–valence electron coupling. [5] While some estimates of the free energy changes in the two scenarios have been made which indicate that the size may be in the right ball–park to drive the transition, it is useful to have exact treatments of specific correlated electron Hamiltonians to test the ideas more quantitatively.

We have demonstrated sharp thermodynamic signatures in both the low-temperature specific heat and in the $t_{fd}$ (or atomic volume) dependence of the total energy for the PAM, and correlated these features with structure in the $f$ electron density of states. These features are not reproduced in simpler Hartree–Fock calculations. The former is absent also in more sophisticated approaches [23] we have attempted, where, however, there is some suggestion of similar $t_{fd}$ dependence in the total energy. Whether these effects are sufficiently large, and sufficiently abrupt, to drive the volume collapse transitions is still unclear. In the symmetric limit that we have studied to date, the situation is also complicated by the presence of antiferromagnetic ordering in the model.

Hartree–Fock calculations, despite their weaknesses, can give us some qualitative insight into the probable effect of going to the non–symmetric limit on the antiferromagnetic ordering, either by shifting $f$ level $\epsilon_f$ away from zero or else by adjusting the chemical potential $\mu$ to go off half–filling. In Fig. 13 we show the results of HF calculations in the case of a shifted $f$ level. We see that the tendency to antiferromagnetic order is reduced. We plan, therefore, to carry out simulations in the non–symmetric case where we can be more sure that the thermodynamic signatures are not associated with long range magnetic order.

It remains to be established that this transition in the PAM occurs in a reasonable place on the experimental phase diagram. This requires making a determination of the volume dependence of the PAM parameters through, for example, a connection to LDA calculations. [8] This connection involves either developing a mapping of the full, many $f$ and conduction band structure to the simpler PAM, or else QMC simulations of more complicated multi-band models.

## ACKNOWLEDGEMENTS

This work was performed under the auspices of the U.S. Department of Energy by Lawrence Livermore National Laboratory under Contract No. W-7405-Eng-48, and the Accelerated Strategic Computing Initiative. We are also grateful to the hospitality of the Lawrence Livermore National Laboratory Materials Research Institute, where part of this work was performed.

## REFERENCES

[1] J. Hubbard, Proc. R. Soc. London **A276**, 238 (1963); *ibid* **A277**, 237 (1964); *ibid* **A281**, 401 (1964). For a collection of recent papers, see, "The Hubbard Model," A. Montorsi (ed), World Scientific, 1992.

[2] P.W. Anderson, Phys. Rev. **124**, 41 (1961).

[3] P. Fulde, J. Keller, and G. Zwicknagel, Solid State Physics **41**, 1 (1988); D.W. Hess, P.S. Riseborough, and J.L. Smith, Ency. Appl. Phys., **7**, 435 (1993).

[4] G.R. Stewart, Rev. Mod. Phys. **56**, 755 (1984); P.A. Lee, T.M. Rice, J.W. Serene, L.J. Sham, and J.W. Wilkins, Comm. on Cond. Mat. Phys. **12**, 99 (1986).

[5] J.W. Allen and R.M. Martin, Phys. Rev. Lett. **49** 1106 (1982); J.W. Allen and L.Z. Liu, Phys. Rev. **B46**, 5047 (1992); M. Lavagna, C. Lacroix, and M. Cyrot, Phys. Lett. **90A**, 210 (1982); J. Phys. F **13**, 1008 (1983).

[6] R.M. Fye, Phys. Rev. **B41**, 2490 (1990).

[7] M. Vekic, J.W. Cannon, D.J. Scalapino, R.T. Scalettar, and R.L. Sugar, Phys. Rev. Lett. **74**, 2367 (1995).

[8] A. McMahan, C. Huscroft, R.T. Scalettar, and E.L. Pollock, Journal of Computer-Aided Materials Design, to appear.

[9] For reviews see: D.J. Scalapino, in *Proceedings of the International School of Physics* (July 1992), R. A. Broglia and J.R. Schrieffer, eds. (North-Holland, New York, 1994); and E. Dagotto, Rev. Mod. Phys. **66**, 763 (1994).

[10] R. Blankenbecler, D.J. Scalapino, and R.L. Sugar, Phys. Rev. **D24**, 2278 (1981).

[11] M. Jarrell and J.E. Gubernatis, Phys. Rep. **269**, 135 (1996).

[12] H. F. Trotter, Proc. Am. Math. Soc. **10**, 545 (1959); M. Suzuki, Comm. Math. Phys. **51**, 183 (1976).

[13] R.L. Stratonovich, Dokl. Akad. Nauk. SSSR **115**, 1097 (1957); J.E. Hirsch, Phys. Rev. **B31**, 4403 (1985).

[14] G. Sugiyama and S.E. Koonin, Ann. Phys. **168**, 1 (1986); S. Sorella, S. Baroni, R. Car, and M. Parrinello, Europhys. Lett. **8**, 663 (1989); S. Sorella, E. Tosatti, S. Baroni, R. Car, and M. Parrinello, Int. J. Mod. Phys. **B1**, 993 (1989).

[15] S.R. White, D.J. Scalapino, R.L. Sugar, E.Y. Loh, Jr., J.E. Gubernatis, and R.T. Scalettar, Phys. Rev. **B40**, 506 (1989).

[16] R.M. Fye and R.T. Scalettar, Phys. Rev. **B36**, 3833 (1987).

[17] D. Duffy and A. Moreo, Phys. Rev. **B55**, 12918 (1997).

[18] R.T. Scalettar, D.J. Scalapino, R.L. Sugar, and D. Toussaint, Phys. Rev. **B39**, 4711 (1989)

[19] The Néel temperature for the single band Hubbard Hamiltonian has also been computed within "dynamical mean field theory." The results are in much better agreement with QMC. In particular, the strong–coupling Heisenberg limit is correctly picked out.

[20] J.E. Hirsch and S. Tang, Phys. Rev. Lett. **62**,591 (1989).

[21] Interestingly, there have been some recent suggestions, that in fact the two mechanisms may be related, with the sharp quasiparticle peak in the single band Hubbard model playing a similar role to that of the Kondo resonance in a multi–band model.

[22] B. Johansson, Philos. Mag. **30**, 30 (1974).

[23] M.M. Steiner, R.C. Albers, D.J. Scalapino, and L.J. Sham, Phys. Rev. **B43**, 1637 (1991).

## Discussion

**Sawatzky:** Was this calculation done for one conduction electron per site and 1 f electron?

**Scalettar:** Yes.

**Sawatzky:** Yes, the couplings are indeed large and there is a gap at $E_F$ indicative of a Kondo insulator.

**Scalettar:** I will now give what I retained from the first discussion evening. I think this is fairly accurate but I may have missed some points and also I may have colored the discussion a bit according to personal preference. But perhaps it is something you could use or perhaps not. Decide for yourself.

# NARROW-BAND EFFECTS IN RARE-EARTHS AND ACTINIDES : INTERACTION BETWEEN THE KONDO EFFECT AND MAGNETISM

**B. Coqblin[1], B.H. Bernhard[2-3], J. R. Iglesias[4], C. Lacroix[3], and K. Le Hur[1]**

[1] L.P.S., Bât. 510, Université Paris-Sud, 91405 Orsay, France
[2] Departamento de Fisica, U.E.S.C., Joinville, S.C., Brazil
[3] Laboratoire L. Néel, CNRS, 38042 Grenoble Cedex 09, France
[4] Instituto de Fisica, U.F.R.G.S., 91501-Porto-Alegre, R. S., Brazil

**Abstract .** Many Cerium, Ytterbium and Uranium systems exhibit a heavy fermion behaviour due to the Kondo effect and, in the case of a lattice, there is a strong interaction between the Kondo effect and the Ruderman-Kittel-Kasuya-Yosida (RKKY) indirect interaction. In this paper, we discuss the competition between the Kondo effect and magnetism in both the Kondo-lattice and Anderson-lattice models and we present a "revisited" version of the Doniach diagram with a drastically reduced Kondo temperature for the lattice. Then, we discuss very briefly the multichannel Kondo effect and the case of Uranium compounds where a coexistence between the Kondo effect and magnetism seems to occur.

## 1. INTRODUCTION TO STRONGLY CORRELATED ELECTRON SYSTEMS

The physics of narrow band effects or strongly correlated electron systems is an extensively studied subject, which has been presented in many review papers[1-7] and discussed in many international conferences as for example in the series of International Conferences on Strongly Correlated Electron Systems (SCES), including the previous ones in Amsterdam[8], Goa[9], Zurich[10] and obviously the last one in Paris (SCES98)[11].

*Electron Correlations and Materials Properties*, edited by Gonis *et al.*
Kluwer Academic/Plenum Publishers, 1999.

On our side, we have made several reviews and the present one can be considered as a continuation of our two most recent ones[12,13]; in particular, two tables with roughly 50 Cerium Kondo compounds are given in ref. 12. For a detailed knowledge of these compounds, the reader is sent back to these review papers[6,12,13].

Many rare-earth metals, alloys and compounds show an anomalous behaviour which has been firstly characterized by a departure from the normal $4f^n$ (with integer n values) configuration at normal or high pressures[14,15]. Then, anomalous rare-earth systems have been separated in two groups : the "intermediate valence" systems, where the rare-earth has clearly a valence different from 3 and the "Kondo" systems where the rare-earth has a valence close to 3 and presents, however, a Kondo effect due to the proximity of the 4f and Fermi levels ; the valence is defined here as the average number of conduction electrons and, in Cerium compounds, it is equal to 4 - $n_f$, where $n_f$ is the average number of 4f electrons[15]. However, the separation between these two regimes is not always very sharp and moreover some experimental features, such as for example the well known anomalous phase diagram of Cerium metal, have been theoretically accounted for by both intermediate valence or Kondo models[15-18].

The Cerium Kondo compounds have been extensively studied from a theoretical and an experimental point of view. At "high temperatures" ($T >> T_k$), i. e. for temperatures larger than the Kondo temperature $T_k$, the perturbation theory holds[19] and the properties of Cerium Kondo compounds are well accounted for by the Coqblin-Schrieffer Hamiltonian[20] including crystal field effects[21]. For example, the decrease of the magnetic resistivity in LogT, either only at sufficiently high temperatures after passing through a maximum corresponding approximately to the crystalline field overall splitting $\Delta$ or at both low and high temperatures, on the two sides of $\Delta$ with LogT slopes characteristic of the degeneracies of the different levels of the crystalline field multiplet, is now considered as a clear "signature" of the Kondo effect in magnetic Cerium compounds, according to the predictions of the old theoretical model given in ref. 21. The other transport properties, such as the thermoelectric power[22] or the thermal conductivity[23], can also be explained by the Kondo effect and allow to determine the crystalline field splitting. The list of Cerium Kondo compounds is always increasing up to roughly one hundred and the LogT decrease of the magnetic resistivity has been observed in many new Cerium compounds as for instance recently in $Ce_5Ge_3$[24], $Ce_2Pd_3Ge_5$[25] or $CeNi_{1-x}Cu_x$[26] alloys among other examples.

At “low temperatures” ($T << T_k$), a so-called heavy fermion beahviour has been observed in many Cerium Kondo compounds, such as $CeAl_3$, $CeCu_6$ or $CeCu_2Si_2$[6,12] and we will briefly summarize their main properties. First of all, the problem of a single Kondo impurity has been exactly solved at T = 0 by Wilson[27] using the renormalization group technique on the exchange Hamiltonian $H = -\Gamma\, \vec{s}_c \cdot \vec{S}_f$ with a $S_f = \frac{1}{2}$ spin and $\Gamma < 0$. Wilson was the first to show numerically that the exact T = 0 solution is a non magnetic singlet. An analytical exact solution of the single Kondo impurity problem has then been obtained by the Bethe-Ansatz method for the classical exchange Hamiltonian and the Coqblin-Schrieffer one[28-30].The exact solutions have given a clear description of the Kondo effect, but since it is difficult to use them for computing the different physical properties, sophisticated "mean field"-type methods have been introduced, such as the "slave boson" method[31,32] or the method using correlators[33,34]. A dynamical mean-field theory which takes the limit of infinite dimension (or very large lattice coordination) has been recently introduced because this limit can yield an exact mapping of lattice models onto quantum impurity models[35]. We can also mention many works to describe dynamical effects or spectroscopic properties and both the impurity Kondo and Anderson problems including dynamical properties have been solved by different theoretical techniques and in particular by the renormalization group technique[36-38].

The low temperature properties of a Kondo impurity are those of a "Fermi liquid" [39] and can be summarized as follows :

- The electronic specific heat constant $\gamma$ and the magnetic susceptibility $\chi$ are much larger than those corresponding to a normal magnetic behaviour, but the Wilson ratio $R=\chi/\gamma$ remains constant between 1 and 2 in suitable units. The values of $\gamma$ are typically of order 100 mJ /mole $K^2$ in the case of compounds which order magnetically and much larger, of order 1000 mJ/mole $K^2$, in the case of compounds which remain non magnetic at low temperatures as explained in ref. 12. The name of "Heavy fermions" originates from these high values of $\gamma$ and $\chi$.

- A direct evidence of the heavy fermion behaviour has been obtained by de Haas van Alphen effect experiments in many Cerium compounds, such as $CeCu_6$, $CeB_6$, $CeRu_2Si_2$ and some Uranium compounds [7]. An effective mass of order $80m_e$ (where $m_e$ is the electron mass) has been typically observed in Cerium compounds.

- The transport properties of Cerium Kondo compounds are described by the Fermi liquid formalism at low temperatures. In particular, the electrical resistivity follows a $T^2$ law and the coefficient A of this term is particularly large in heavy fermion systems, as given in ref. 12.

The regular Kondo effect corresponds in fact either to a S = 1/2 spin screened by only one conduction electron channel (n = 1) or to the Coqblin-Schrieffer Hamiltonian, where the 4f electron is screened by the l = 3, s = 1/2, j = 5/2 component of the conduction electrons ; in the two preceding cases, there is an equal number of 4f electrons and conduction electrons. However, in several cases such as the $5f^2$ configuration of Uranium ions or the multichannel p- or d-like conduction electrons, the Kondo effect is underscreened ($n < 2S$, where n is the number of conduction electron channels and S the spin value) or overscreened ($n > 2S$) and the behaviour can be different from the Fermi liquid one[40].

Thus, we have presented here a brief summary of the single-impurity properties of Cerium Kondo systems. These properties are essentially those of Cerium Kondo compounds which do not order magnetically at low temperatures. On the other hand, there exists, in a lattice of Cerium atoms, a strong competition between the Kondo effect and the RKKY (Ruderman-Kittel-Kasuya-Yosida) interaction, which has been previously described within the Doniach diagram[41]. In the next section, we present a "revisited" version of the Doniach diagram[42], which can yield a rather flat variation of the Kondo temperature, in better agreement with some new experimental results in Cerium compounds. Then, in section 3, we present a new model[43] starting from the Anderson lattice Hamiltonian which is able to recover the previous results of the revisited Doniach diagram. Finally in section 4, we describe the case of Uranium compounds where there seems to exist rather a coexistence between the Kondo effect and magnetism[44] than a competition which occurs in Cerium compounds.

## 2. THE "REVISITED" DONIACH DIAGRAM

The competition between the Kondo effect on each atom which tends to suppress the magnetic moment with decreasing temperature and the RKKY interaction which, on the contrary, tends to give a magnetic ordering between different rare-earth atoms is well described by the well-known "Doniach diagram" [41], as shown in figure 1. Let us call $T_{KO}$ the Kondo temperature for a single impurity and $T_{NO}$ the Neel (or Curie in a few cases) temperature, as if there were no Kondo effect. For the exchange Hamiltonian $H = - J_K \mathbf{s_c} \mathbf{S}$, $T_{KO}$ is proportional to $\exp(-1/|J_K\rho|)$ and $T_{NO}$ to $|J_K\rho|^2$, where $\rho$ is here the density of states for the conduction band at the Fermi energy. Thus, for small $|J_K\rho|$ values, $T_{NO}$ is larger than $T_{KO}$ and the system tends to order magnetically, with often a reduction of the magnetic moment due to the Kondo effect[5]; on the contrary, for large $|J_K\rho|$, $T_{KO}$ is larger than $T_{NO}$ and the system tends to become non magnetic and does not order magnetically. The real ordering temperature $T_N$, therefore, increases initially with increasing $|J_K\rho|$, then passes through a maximum and tends to zero at a critical value $|J_K\rho|_c$ corresponding to point I of Fig. 1.

Such a behaviour of $T_N$ has been experimentally observed with increasing pressure in several Cerium Kondo compounds, namely $CeAl_2$[45,46], CeAg[47], $CePd_2Al_3$[48], CePtGa[49], $CeIn_3$[50], $CePd_2Si_2$[51], $CeAu_2Si_2$[52], $CeCu_2$[52] and $CeRh_2Si_2$[53]. A maximum has been observed in some of them, such as in CeAg[47] or $CePd_2Al_3$[48], but in all these

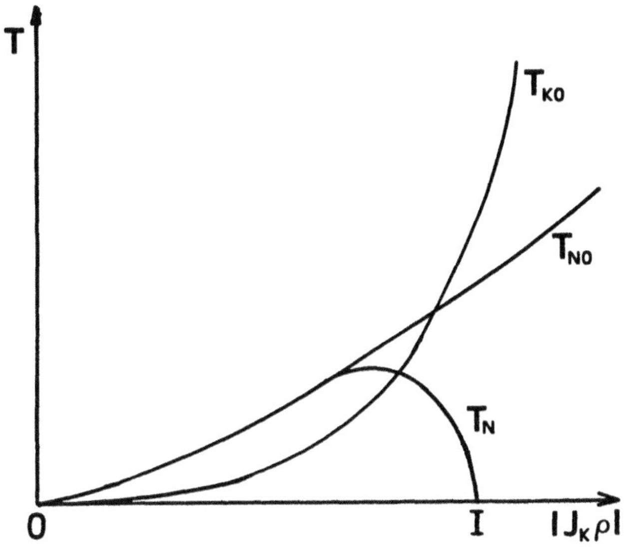


**Figure 1** . Plot of the Neel and Kondo temperatures according to the Doniach diagram, as explained in text.

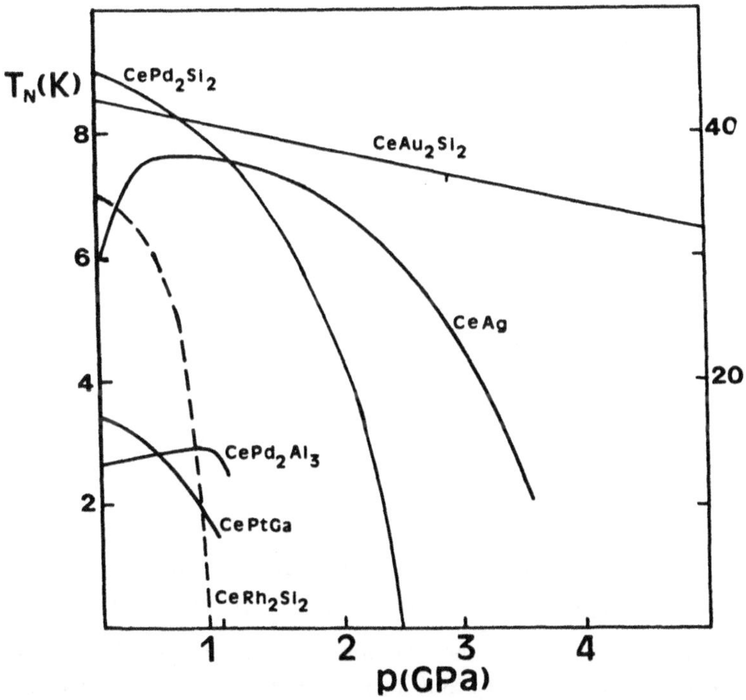


**Figure 2** . Experimental curves of the pressure dependence of the Neel temperature for some Cerium compounds. The full lines correspond to the left scale for the Neel temperature, while the dotted line of the $CeRh_2Si_2$ compound corresponds to the right scale.

compounds, the Neel temperature decreases and tends to zero at high pressures as shown in Fig. 2. A similar variation of $T_N$ has been also observed in $Ce(Ru_xRh_{1-x})_2Si_2$[54,56], $CeSi_x$[57] or $CeNi_xPt_{1-x}$[58] with a variation of the relative concentration x. We will discuss in detail the first case later on, but the two other cases, namely $CeSi_x$[57] or $CeNi_xPt_{1-x}$[58], are not really relevant for our present study since the experimentally observed transition to non magnetism corresponds probably here to a change of valence[59]. This type of transition to an intermediate valence has been also observed in alloys $Ce(Cu_xPd_{1-x})_2Si_2$[60] or $CeM_xPd_{1-x}$ (with M=Ni or Rh) [61,62].

In conclusion, the variation of the Neel temperature which is predicted by the Doniach diagram is well followed in many Cerium Kondo compounds. The argument has been already used to separate Cerium Kondo compounds between those which magnetically order and those which do not order magnetically[12].

However, the preceding description appears to be too much simplified for the real Kondo temperature $T_k$ which is experimentally observed in Cerium Kondo compounds. In fact, compounds with large $|J_K\rho|$ values have clearly much larger $\gamma$ values than those which have $|J_K\rho|$ values smaller than the critical value $|J_K\rho|_c$, as previously observed[6]. From the proportionality relation between $\gamma$ and $1/T_k$, we would deduce that the corresponding $T_k$ values for compounds with large $|J_K\rho|$ values must be much smaller than the $T_k$ values for compounds with small $|J_K\rho|$ values. For example, if one tries to estimate the value of $|J_K\rho|$ for both $CeAl_3$ and $CeAl_2$ compounds, which are on the two sides of the magnetic-non magnetic transition, one finds from the magnetic resistity curves a ratio of 1.43 for the $|J_K\rho|$ values[14], while the estimation of $T_k$ from the specific heat results would give a $T_k$ for $CeAl_3$ ($\gamma$ = 1600mJ/mole $K^2$) roughly smaller by a factor of 10 than that for $CeAl_2$ ($\gamma$ = 135mJ/mole $K^2$)[12)].

Recent measurements[53] of the resistivity and specific heat under high pressure up to 17 kbar at low temperatures have been performed in $CeRh_2Si_2$ compound. At normal pressure, it orders antiferromagnetically with a relatively large value of the Neel temperature $T_N$ = 35 K and a relatively small value of the electronic specific heat constant $\gamma$ = 22.8 mJ/mole $K^2$. The Neel temperature decreases rapidly with pressure and disappears at 9 kbar. On the other hand, the electronic specific heat constant increases rapidly and goes through a broad maximum around 14 kbar with a value of $\gamma$ = 80 mJ/mole $K^2$, which is much larger than the normal pressure value. Thus, if we apply the relationship $\gamma = 1/T_k$ which is in principle valid only in the non magnetic domain, we obtain a clear decrease of $T_k$, by a factor of roughly 3, in the pressure range where the compound changes from antiferromagnetic to non magnetic and then a slow increase of $T_k$ above 14 kbar; it is clear that $T_k$ is still decreasing in the non magnetic region above 9 kbar. It is interesting to note also that the low temperature resistivity appears to vary as T at high pressure and that the Fermi liquid picture is still valid around the magnetic instability 9 kbar pressure. These experiments on $CeRh_2Si_2$ compound under pressure show clearly that the variation of the Kondo temperature $T_K$ for the lattice is absolutely not exponentially increasing up to 17 kbar.

We will go back now to the case of $Ce(Ru_{1-x}Rh_x)_2Si_2$[54-56] alloys. It is well established that one goes from a non magnetic case for x = 0 to a magnetically ordered case with increasing x and the transition itself occurs at x = 0.08. Recent studies[63] of these alloys have shown that the experimentally deduced Kondo temperature $T_K$ remains roughly constant of order 20 K when the relative concentration x is varying from x = 0 to x = 0.2, in the concentration range where the Neel temperature disappears; let us finally notice that Cerium does not undergo any valence change in this concentration range, so that our present study will be applicable.

Moreover, short-range magnetic correlations have been observed in Cerium compounds which do not present any long-range magnetic order. Extensive neutron diffraction experiments have been performed in $CeCu_6$[64], $CeInCu_2$[65], $CeRu_2Si_2$[56,64,66] or $Ce_{1-x}La_xRu_2Si_2$[56,66-68] single crystals at low temperatures; the last system is particularly interesting, since it is non magnetic and shows short-range magnetic correlations for x=0, while antiferromagnetic order appears for approximately x=0.08 with increasing Lanthanum concentration. The neutron scattering spectra can be considered as sums of a q-independent, single site, quasi-elastic contribution associated with the single-site Kondo effect and of an inter-site q-dependent contribution associated with short-range magnetic correlations between different Cerium sites. Thus, short-range magnetic correlations between neighbouring Cerium atoms develop in $CeCu_6$, $CeInCu_2$, $Ce_{1-x}La_xRu_2Si_2$ and other similar compounds at low temperatures ; the temperature $T_{cor}$ at which antiferromagnetic short-range correlations appear is clearly larger than the Kondo temperature $T_K$ ($T_{cor} \simeq$ 60-70 K and $T_K \simeq$ 14-23 K in $CeRu_2Si_2$[64,66], $T_{cor} \simeq$ 10 K and $T_K \simeq$ 5 K in $CeCu_6$). Finally, these correlations are also destroyed in $CeCu_6$ compound[64] and $Ce_{1-x}La_xRu_2Si_2$ alloys[66] by the application of a strong magnetic field.

Thus, we will present now theoretical models describing the Kondo and Anderson lattices which can yield a lattice Kondo temperature much smaller than the single impurity one.

Firstly, we have derived a theoretical model which has included short-range magnetic correlations in the Kondo lattice. Detailed calculations can be found in reference 42 and we just present here a summary of the main results. The corresponding Hamiltonian is given by :

$$H = E_0 \sum_{i\sigma} n^f_{i\sigma} + \sum_{k\sigma} \varepsilon_k c^*_{k\sigma} c_{k\sigma} - J_K \sum_i s_i . S_i - J_H \sum_{i\delta} S_i . S_{i+\delta} \quad (1)$$

In (1), we take a zero width f band, a conduction band with a width 2W and a constant density of states and a classical s-f exchange term. But, here we add an Heisenberg interaction between neighbouring f magnetic moments; $J_H$ is the Heisenberg interaction coupling constant and $J_K(<0)$ is the Kondo coupling constant. We take firstly here $J_H$ and $J_K$ independent from each other, although in fact it is not the case, since $J_H$ is proportional to $\rho|J_K|^2$, where $\rho$ is the density of states of the conduction band ; finally, we will assume a relationship between $J_H$ and $J_K$ in order to derive a "revisited" version of the Doniach diagram.

In our model, we use a simple mean-field approximation, by introducing the two following parameters for describing the non-magnetic phase :

$$\lambda_{i\sigma} = \langle c^*_{i\sigma} f_{i\sigma} \rangle = \langle f^*_{i\sigma} c_{i\sigma} \rangle$$

$$\Gamma_{i,i+\delta} = \langle f^*_{i\sigma} f_{i+\delta,\sigma} \rangle = \langle f^*_{i+\delta,\sigma} f_{i,\sigma} \rangle \quad (2)$$

The first mean value describes the formation of the singlet state between the f and conduction electrons[33], while the second mean value gives a description of the nearest neighbour f-f correlation[34]. Indeed, this method is an approximate one, which is very useful in absence of exact calculations.

We consider here only spin independent and translation invariant solutions for which we have $\lambda_i = \lambda$ and $\Gamma_{i,i+\delta} = \Gamma$ and we describe the conduction band in the tight binding approximation and write:

$$\varepsilon_k = \frac{W}{z} \sum_R \cos k.R \quad (3)$$

where z is the number of neighbours of a given atom.

The Hamiltonian (1) can be easily diagonalized if we use the mean field approximation given by (2) and the resulting eigenvalues are given by :

$$E^{\pm}(k) = \frac{1}{2} \left[ \varepsilon_k(1+B) + E_0 \pm \sqrt{[\varepsilon_k(1-B) - E_0]^2 + 4J_K^2\lambda^2} \right] \quad (4)$$

with :

$$B = \frac{z\ J_H \Gamma}{W} \tag{5}$$

$E_0$ and the Fermi energy $E_F$ have to be determined self-consistently in order to obtain :

$$\left\langle \sum_\sigma f^*_{i\sigma} f_{i\sigma} \right\rangle = 1$$
$$\left\langle \sum_\sigma c^*_{i\sigma} c_{i\sigma} \right\rangle = n \tag{6}$$

The calculation of ref. 42 has been performed for a half-filled conduction band (n = 1) with $E_0 = E_F = 0$ in the non magnetic case.

The T = 0 results depend on the relative values of $\lambda$ and $\Gamma$ ; in the limit of small $J_K/W$ values, we have obtained for $\lambda$ :

$$\lambda \approx \left( \frac{W\ (1-B)}{J_K} \right) \exp \left( \frac{W\ (1-B)}{J_K} \right) \tag{7}$$

$\lambda$ has the same dependence as in earlier calculations without correlations[23], but the bandwidth is renormalized by the correlations : $W_{eff} = W\ (1 - B)$. Thus, $W_{eff}$ can be larger or smaller than W and consequently $\lambda$ can be smaller or larger depending on whether $J_H$ is negative or positive.

Thus, if exchange Heisenberg interaction is antiferromagnetic ($J_H < 0$), $\lambda$ is reduced and the non-magnetic phase is stabilized by the correlations. On the other hand, if $J_H$ is positive, the correlations increase the Kondo effect according to (7), but according to the calculation of the energy, the non-magnetic phase is found to be less stable. This result is in good agreement with the experimental fact that the heavy fermion behaviour is much more often observed in nearly antiferromagnetic systems than in nearly ferromagnetic systems.

Then, we have performed calculations at finite temperatures in the only case of antiferromagnetic correlations ($J_H < 0$), for the non magnetic solution and the half-filled conduction band $E_0 = E_F = 0$. The different results are given in ref. 42. The mean field parameters $\lambda$ and $\Gamma$ are obtained by minimizing the total energy.

We have derived the $\lambda$ and $\Gamma$ curves as a function of temperature for different $J_H$ and $J_K$ values. The results depend essentially on the ratio $J_K/J_H$ rather than on the

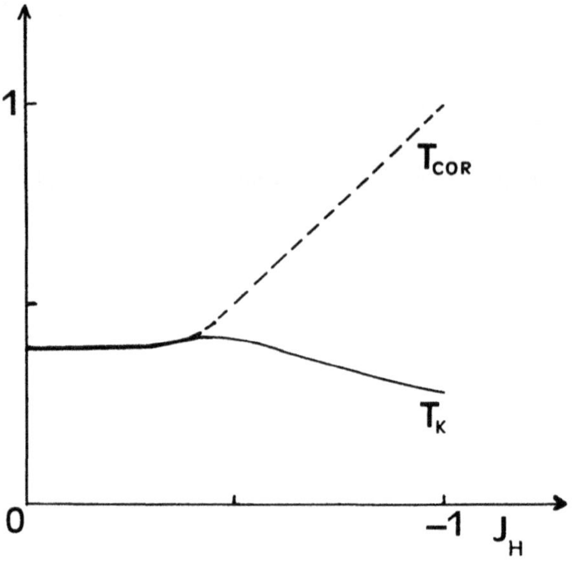


**Figure 3** . Plot of the Kondo temperature $T_k$ (full line) and the correlation temperature $T_{cor}$ (dashed line) versus $J_H$, for $J_K$ = -1.

individual values of $J_K$ and $J_H$. The $\lambda$ versus T curves start from a finite value at T = 0 and finally decrease down to zero at a temperature corresponding in the present mean field approximation to the Kondo temperature $T_k$. The $\Gamma$ versus T curves decrease also down to zero at a temperature $T_{cor}$ corresponding to the disappearance of short-range antiferromagnetic correlations.

Figure 3 gives the plot of these two temperatures $T_k$ and $T_{cor}$ versus $J_H$ for a given value $J_K$ = - 1. The Kondo temperature remains constant for small $J_H$ values and equal to the $J_H$=0 Kondo temperature $T_{ko}$ ; after a very weak maximum, $T_k$ decreases slowly, indicating that the short-range antiferromagnetic correlations tend to decrease $T_k$. The temperature $T_{cor}$ remains equal to $T_k$ up to the $J_H$ value of the maximum of $T_k$ and then $T_{cor}$ increases linearly with $J_H$ according to the very simple law $T_{cor}$ = $J_H$, predicted in the region where $\lambda$ = 0 with a number of neighbours z = 6. Thus, it is remarkable to notice that there are essentially two regions : $T_{cor}=T_k$ for small $|J_H|$ values compared to $|J_K|$ and $T_{cor}>T_k$ when the effect of antiferromagnetic correlations overcomes the Kondo effect. A similar effect had been obtained in the Monte Carlo study of the two-impurity Kondo Hamiltonian[69].

But in fact, $J_K$ and $J_H$ are not independent from each other and it is necessary to take into account the relationship between these two parameters in order to get the real dependence of the Kondo temperature $T_k$ versus $|J_K|$.

Thus, we take in the following :

$$J_H = - \alpha \ |J_K|^2 \qquad (8)$$

which can well represent the quadratic dependence of the $J_H$ parameter on $J_K$ in the RKKY interaction.

Figures 4 and 5 give the plots of the Kondo temperature $T_k$ and the correlation temperature $T_{cor}$ as a function of $J_K$ for a small value $\alpha = 0.25$ and a large value $\alpha = 2$ of the $\alpha$ parameter. We have plotted for comparison the single impurity (obtained for $J_H = 0$) Kondo temperature $T_{ko}$.

For the small value $\alpha = 0.25$, the three temperatures $T_k$, $T_{ko}$ and $T_{cor}$ are equal to each other, since the f-f correlation has almost no real importance. On the other hand, for $\alpha = 2$, we see that the real Kondo temperature $T_k$ is firstly increasing with $|J_K|$ and almost equal to $T_{ko}$, goes through a maximum and decreases for large $|J_K|$ values. On the other hand, the correlation temperature $T_{cor}$ is much larger than $T_k$ and increases rapidly with $|J_K|$.

The curve showing $T_k$ in figure 5 provides, therefore, a "revisited" version of the Doniach diagram for the non magnetic domain when the Neel temperature has disappeared : antiferromagnetic short-range correlations decrease drastically the Kondo temperature with respect to that defined for the single impurity and yield slowly decreasing or roughly constant values of $T_k$ versus $|J_K|$. This theoretical result accounts for the experimental decreases of $T_k$ observed in $CeRh_2Si_2$ under pressure[53] since pressure makes $|J_K|$ increase. A similar behaviour has been observed in $Ce(Ru_{1-x}Rh_x)_2Si_2$ alloys[54-56], as previously explained. Resistivity measurements under pressure in $CeAu_2Si_2$ compound seem to yield a similar behaviour[52].

Moreover, figure 3 gives a clear explanation of the occurrence of short-range correlations in $CeRu_2Si_2$[64,66] or $CeCu_6$[64] at a temperature clearly larger than $T_k$ ; these compounds must correspond here to quite large $|J_H|$ values, while other compounds corresponding to smaller $|J_H|$ values would have to be characterized by equal $T_k$ and $T_{cor}$ values.

It would be very interesting to perform new studies in almost antiferromagnetic systems, in order to observe this rather constant value of $T_k$ in a given range of $|J_K|$ in the non magnetic domain.

On the contrary, $T_k$ is predicted to have a quite different behaviour in almost ferromagnetic systems and to increase rapidly with $|J_K|$. Such a behaviour is observed in alloys such as $CeNi_{1-x}Cu_x$[26], but however further experiments are necessary to better account for a difference of behaviour in almost ferro- or antiferromagnetic systems.

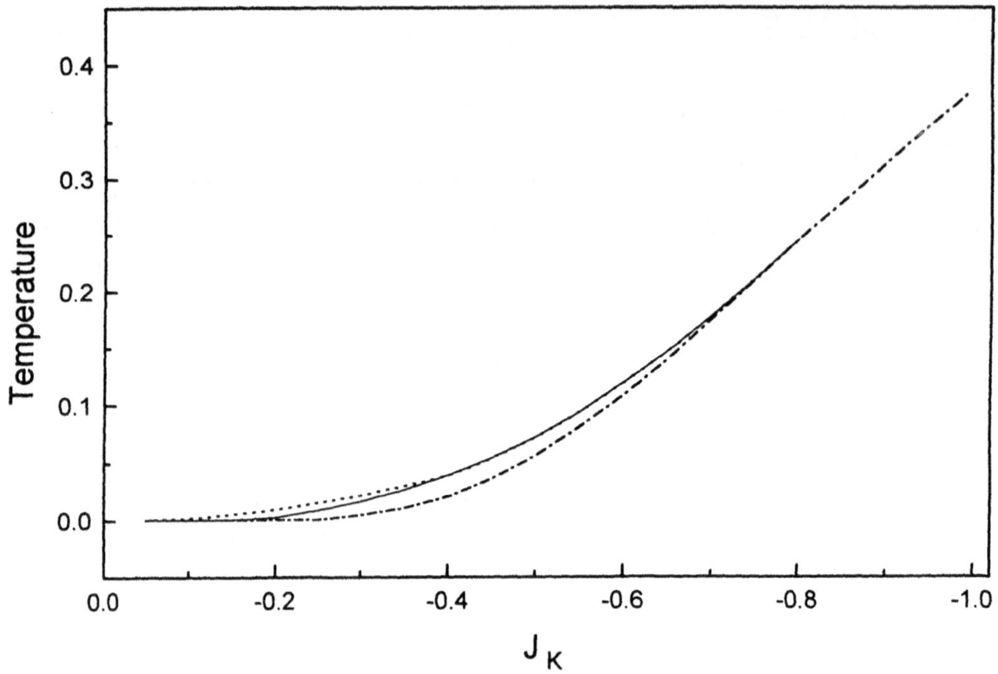


**Figure 4** . Plot of the single-impurity Kondo temperature $T_{KO}$ (dashed-dotted line), the real Kondo temperature $T_K$ (full line) and the correlation temperature $T_{cor}$ (dashed line) as a function of $J_K$ using the formula (8) with $\alpha = 0.25$.

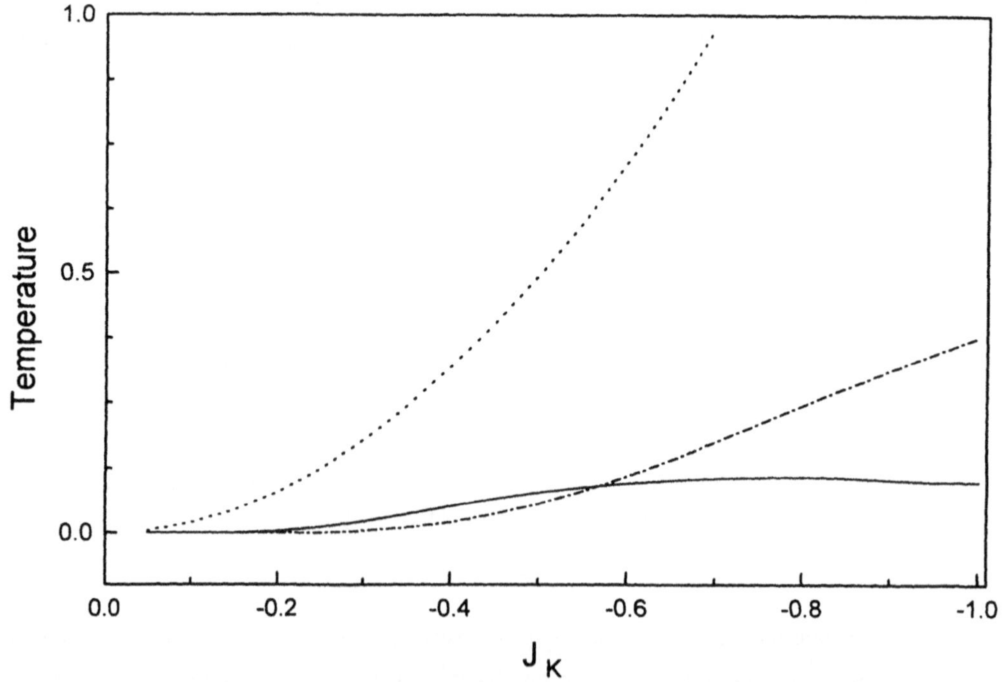


**Figure 5** . Plot of the one-impurity Kondo temperature $T_{KO}$ (dashed-dotted line), the real Kondo temperature $T_K$ (full line) and the correlation temperature $T_{cor}$ (dashed line) as a function of $J_K$ using the formula (8) with $\alpha = 2$.

Finally, it is important to remark that our theoretical model is obviously based on the "Fermi liquid" description of the Kondo effect and we have applied it to Cerium compounds, such as $CeRh_2Si_2$, which show a clearly well established Fermi liquid experimental behaviour.

Ytterbium-based compounds offer an alternative way to test the Doniach diagram because pressure makes the exchange integral $|J_K|$ decrease and favors the magnetic state. At present, a large decrease of the Kondo temperature has been observed in $YbCu_4Ag$[70] and in $YbCu_2Si_2$[52] under pressure, as well as in $YbCu_{5-x}Al_x$ alloys with increasing Aluminium concentration[71].

A similar theoretical result yielding a much smaller Kondo temperature for the Kondo lattice ($T_k$) than for the single impurity ($T_{ko}$) has been established independently by Nozières[72] using the so-called "exhaustion" problem[73]. In the case of a single impurity, there are always enough conduction electrons to screen the localized electron spins. On the opposite, in the case of a Kondo lattice, only conduction electrons within an energy width $T_{ko}$ around the Fermi energy are able to screen the localized spins : this is the exhaustion problem introduced by Nozières[73] and thus only a fraction $p \doteq T_{ko}/D$ of the conduction electrons screens the localized spins, where $T_{ko}$ and D are respectively here the single impurity Kondo temperature and the conduction electron band width. Then, it results that each conduction electron has to screen a large number 1/p of localized spins and finally, using an argument of visit time on each site by the conduction electron in the spirit of real space renormalization, Nozières[72] has deduced that the Fermi liquid behaviour can occur only below the "lattice" Kondo temperature $T_k$ given by :

$$T_k \approx p.T_{ko} = \frac{(T_{ko})^2}{D} \qquad (9)$$

which is clearly much smaller than $T_{ko}$.

Thus, the argument of Nozières does not involve the introduction of Heisenberg-type interaction between spins on different sites. However, we can think that the real wavefunction of the problem involves singlets formed with conduction and localized electrons on one side and with localized electrons on different sites on the other side. The Kondo lattice model has been also studied in the one-dimension case[74] and similar results have been obtained.

We will study in the next section the Anderson lattice model and finally we will discuss some recent results obtained with the two Kondo and Anderson lattice models.

## 3. THE ANDERSON LATTICE MODEL

The Anderson lattice model is able to account for different cases, such as the intermediate valence one (when the number of 4f electrons $n_f$ is clearly different from an integer) and obviously the Kondo limit (with $n_f \simeq 1$) where the Kondo and the Anderson Hamiltonians have in principle the same behaviour. The single impurity Anderson Hamiltonian has been extensively studied since the pionnering work of Anderson in 1961.

The Anderson lattice model has been recently studied by several authors in order to better describe the situation of rare-earth compounds and concentrated alloys. Here, we will firstly describe our recent calculation[43] and then we will give a brief review of some other works.

In our recent work[43], the periodic Anderson Hamiltonian is written as follows :

$$H = H_0 + H' \tag{10}$$

where :

$$H_0 = \sum_{i\sigma}\left\{E_f n^f_{i\sigma} + \frac{1}{2}U n^f_{i\sigma} n^f_{i\sigma} - V\left(f^*_{i\sigma}d_{i\sigma} + \text{h.c.}\right) - \mu\left( n^f_{i\sigma} + n^d_{i\sigma} \right)\right\} \tag{11}$$

$$H' = \sum_{\langle ij\rangle}\left\{- t \sum_{\sigma}\left(d^f_{i\sigma}d_\sigma + \text{h.c.}\right) - 2J_H S^f_i . S^f_j\right\} \tag{12}$$

Here $c^*_{i\sigma}$ $\left(c_{i\sigma}\right)$ is the operator that creates (annihilates) one electron of spin $\sigma$ in the c = d or f orbital at the i-th site, $n^c = c^*_{i\sigma}\ c_{i\sigma}$ is the number operator in that orbital and $S^f_i$ is the f-electron spin at site i. The parameters involved are the hopping t of the d-electrons between nearest neighbor sites, the local hybridization V between f and d electrons, and the exchange $J_H$ between f-electron spins at nearest-neighbor sites (indicated as $< ij >$ in Eq. 12).

Then, we consider here, as in our previous work[42] on the Kondo lattice Hamiltonian, both the Anderson lattice Hamiltonian and an Heisenberg-type interaction between f-electron spins. Then, the adopted separation of the total Hamiltonian is similar to the one applied in ref. 75 to the pure Periodic Anderson model. The first term, $H_0$, is a

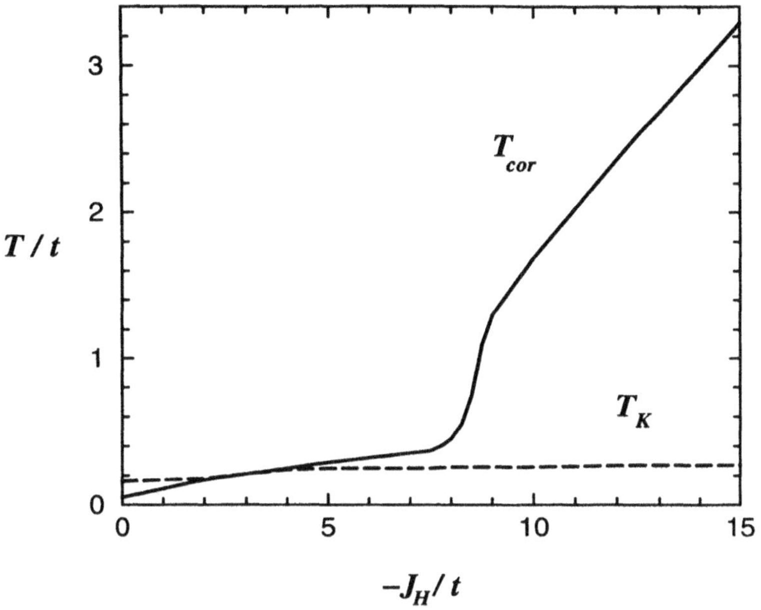


**Figure 6** . Crossover temperatures $T_k$ and $T_{cor}$ as a function of the parameter $J_H$, for E = -3t and V = t.

pure atomic Hamiltonian and is treated exactly, while the perturbation term, H', is approximated and is treated by means of a Dyson equation which involved only chain-like diagrams. Thus, we transform the second term of (12) into an one-particle term by applying a mean-field approximation which involves only the average $\left\langle f_{i\sigma}^* \; f_{j\sigma} \right\rangle$ in the studied non magnetic case.

The averages $\left\langle d_{i\sigma}^* \; f_{i\sigma} \right\rangle$ and $\left\langle f_{i\sigma}^* \; f_{j\sigma} \right\rangle$ are calculated as a function of temperature in the simple case of both half-filled (particle-hole symmetry case) conduction band and non magnetic localized electrons, which implies $n^f = 1$, $n^d = 1$ and $\mu = 0$.

The averages $\left\langle d_{i\sigma}^* \; f_{i\sigma} \right\rangle$ and $\left\langle f_{i\sigma}^* \; f_{j\sigma} \right\rangle$ correspond respectively to the Kondo effect and to the short-range antiferromagnetic correlations, as previoulsy established in the previous work on the Kondo lattice[42]. However, the curves giving these averages go smoothly to zero with increasing temperature; thus, the Kondo temperature $T_k$ and the short-range antiferromagnetic correlation temperature $T_{cor}$ are defined by the temperature at which the derivative of the corresponding curves reaches a minimum. In figure 6, the temperatures $T_k$ and $T_{cor}$ are plotted as a function of the exchange parameter $J_H$ and the results are qualitatively similar to those shown in figure 3 for the Kondo lattice. Moreover, we have to recall that the conduction band width is given here by W = 12t; the temperature values that we obtain here for $T_k$ and $T_{cor}$, are, therefore, one order of magnitude smaller and consequently closer to the experimental values. We can also remark that the correlation functions decrease smoothly with temperature, which yields a better description than the abrupt decrease found in ref. 42 for the Kondo lattice.

In conclusion, Figure 6 gives both a rather flat variation of $T_k$ with $J_H$ and a value of $T_{cor}$ much larger than $T_k$ for large $J_H$ values, as previously established in ref. 42.

The Kondo temperatures have been also computed for both the single impurity Anderson model (SIAM) and the periodic Anderson model (PAM), by studying the infinite dimension case with a quantum Monte-Carlo numerical simulation[76]. The Kondo temperatures are computed as a function of the conduction band filling $n_d$ for a fixed half-filled f band $n_f = 1$. In the symmetric half-filled case, $n_d = n_f = 1$, the PAM Kondo temperature is slightly enhanced compared to the SIAM Kondo temperature, as it has been found earlier[77]. On the opposite, far from the symmetric limit, i. e. for $n_d$ values well smaller than 1, the PAM Kondo temperature is strongly suppressed compared to the SIAM one. Thus, the lattice Kondo temperature is found by these numerical calculations to be much smaller than the single impurity one only for $n_d$ clearly smaller than 1, but not for $n_d = 1$, in contrast to our calculations.

Thus, the Kondo temperatures for both the Kondo and Anderson lattices are clearly different from those of the single impurity case. For example, recent photoemission experiments[78] in strongly correlated f-electron systems with Cerium, Ytterbium or Uranium show clear departures from the single impurity model[79] and ask for a need of a new approach including lattice effects.

There are different approaches including lattice effects which are presently under development and there are some disagreements between the different theoretical results. However, the Kondo temperature for a lattice is clearly different from the single impurity one and our calculations given in ref. 42 and ref. 43 show a Kondo lattice temperature much smaller than the single impurity one, in good agreement with some recent experiments in Cerium compounds.

## 4. THE MULTI-CHANNEL KONDO EFFECT AND THE CASE OF THE URANIUM COMPOUNDS

The preceding models imply a "Fermi liquid" behaviour appropriate to describe the physics of the Kondo effect. We have shown that the obtained lattice Kondo temperature $T_k$ is roughly constant as a function of $|J_K|$ in the non-magnetic domain above the point I corresponding to the magnetic-non magnetic instability (or to the disappearance of the Neel temperature $T_N$).

On the other hand, many departures from a simple Fermi liquid behaviour, characterized in particular by a constant $\gamma$ value and a $T^2$ law for the resistivity $\rho_m$, have been observed near the magnetic-non magnetic instability point I. This behaviour is called "Non Fermi Liquid" (NFL) and excellent reviews, such as those of Coleman[80], Von Löhneysen[81] or Steglich[82] are available on that subject. There are several different NFL behaviours, in contrast to the obviously unique Fermi liquid (FL) behaviour and also several theoretical explanations for such experimentally observed NFL behaviours.

An experimental NFL behaviour has been observed in many Cerium compounds at normal or under high pressure, such as $CeCu_{5.9}Au_{0.1}$ or $CeCu_{5.8}Au_{0.2}$ studied in detail by von Löhneysen et al.[81,83,84], $CePtSi_{0.9}Ge_{0.1}$[85], $Ce_{1-x}La_xRu_2Si_2$[86], the disordered alloys $Ce_{1-y}Cu_{2+x}Si_2$[81] or similar alloys with Ge substituting Si or another transition metal substituting Cu[81,87] or finally $Ce_7Ni_3$[88]. Such a behaviour can be theoretically accounted for by the self-consistent renormalized (SCR) theory of spin fluctuations[89] which predicts a LnT law for $\gamma$ and a $T^{3/2}$ law for $\rho_m$ when one is very close to the magnetic instability and a FL behaviour when one goes farther from the magnetic instability point I.

A NFL behaviour has been also observed in dilute alloys with Uranium, such as $Y_{1-x}U_xPd_3$[90] at normal or high pressure[91], $Th_{1-x}U_xRu_2Si_2$[92] or $UCu_xAl_{5-x}$[93]. There is still an effect of the disorder[94,95], but in addition, Uranium compounds show a different behaviour from that of Cerium compounds and are rather more described by the so-called "multichannel Kondo model"[40,96-98].

The regular Kondo effect, studied up to now, corresponds to an exact compensation of the impurity spin S by the conduction electrons ; if we call n the number of channels for conduction electrons, the regular or "screened" Kondo effect corresponds to $n = 2S$. If $n < 2S$, the impurity spin is only partially compensated, which gives the "underscreened" Kondo effect. If $n > 2S$, there are too many conduction electrons to compensate the impurity spin ; the spin S is completely compensated and in addition there remain delocalized conduction electrons, which gives the "overscreened" Kondo effect[40,98] The multi-channel Kondo model for a single impurity has been completely solved analytically by the Bethe ansatz method[99,100] and numerical solutions[101,102] have been obtained to describe the different physical properties; a review on these solutions can be found in ref. 98.

In the underscreened Kondo effect where $2S > n$, a part n/2 of the impurity spin is compensated, while the part $S' = S - n/2$ is not compensated. It results that the magnetic susceptibility is diverging in 1/T as for a Curie law, which is obviously different from the constant magnetic susceptibility obtained at low temperatures for the regular screened

Kondo effect (2S = n). A ferromagnetic coupling between the resulting spin S' and the conduction electron spin is finally obtained in the underscreened Kondo model.

In the overscreened Kondo effect where 2S < n, it remains (n - 2S) delocalized conduction electrons and the spin S' given by S' = n/2 - S is antiparallel to S. The magnetic susceptibility $\chi$ and the electronic specific heat constant $\gamma$ have been numerically computed and are given in ref. 101. As an example for S = 1/2, constant values of $\chi$ and $\gamma$ are obtained at low temperatures for the screened Kondo effect corresponding to n = 1. On the contrary, for S = 1/2, n = 2, a logarithmic dependence of $\chi$ and $\gamma$ is obtained, while for S = 1/2 and n > 2, the temperature dependence of $\chi$ and $\gamma$ is given by a power law $T^n$. As a conclusion, a FL behaviour is obtained for n = 2S and a NFL behaviour for n > 2S ; we can finally note that there are several possible NFL behaviours, in contrast to the unique FL one. A particular model within the overcompensated case is the quadrupolar Kondo effect introduced by Cox[103], corresponding to the two-channel case ; a logarithmic dependence of $\gamma$, a linear one of the magnetic resistivity and an unusual value for the entropy (corresponding to half the value of the entropy for the regular Kondo effect) have been obtained and these theoretical results have been used to account for the non-Fermi liquid behaviour observed in $Y_{1-x}U_xPd_3$[90].

The underscreened Kondo effect has been studied again recently in order to better understand the apparent coexistence observed in some Uranium compounds between the heavy fermion character due to the Kondo effect and a weak magnetic order resulting from the RKKY interaction. The Uranium compound $UPt_3$ has an outstanding behaviour, since it undergoes a transition to an antiferromagnetic-like ordering with a tiny ordered magnetic moment below the Neel temperature $T_N \simeq$ 5K[104] and becomes superconducting below $T_c \simeq$ 0.5 K[105]; a heavy-fermion behaviour characterized by a large electronic specific heat constant $\gamma \simeq$ 0.4 J/mole $K^2$ [106] and a $T^2$ term of the resistivity[107] is observed in $UPt_3$ at low temperatures. A similar behaviour has been observed in other Uranium compounds, namely $UBe_{13}$[108], $URu_2Si_2$[109], $UNi_2Al_3$[110,111], $UPd_2Al_3$[110,111]. In these Uranium compounds, when temperature decreases, there is first a transition to an antiferromagnetically ordered state and then a transition to a superconducting state ; as an example, for $UNi_2Al_3$ and $UPd_2Al_3$, there are a magnetic transition at $T_N$= 4.6 and 14 K and a superconducting one at $T_c$ = 1 and 2K respectively.

The underscreened Kondo effect has been studied within a model[44] of two impurities with a spin S = 1 interacting with the conduction band and via an interimpurity coupling $-K\vec{S}_1.\vec{S}_2$. For a small antiferromagnetic (K<0) coupling, a completely asymmetric situation is obtained with one s = 1/2 component strongly screened by the Kondo effect and the other one almost free to yield magnetism, which shows finally a possible coexistence between a RKKY interaction and a local Kondo effect. This theoretical result could account for the apparently observed coexistence in Uranium compounds such as $UPt_3$. However, several questions have to be solved : firstly, it is not obvious that the underscreened model (with $n < 2S$) applies to Uranium compounds more likely than the overscreened one (with $n > 2S$) which has been previously applied[90]. Second, the experimental situation of Uranium compounds showing a possible coexistence between a weak magnetic order and the heavy fermion character has to be established experimentally more clearly. Finally, the difference of behaviour between Uranium compounds showing rather a coexistence and Cerium compounds presenting a competition has to be understood in more detail.

## 5. CONCLUSIONS

Thus, we have discussed, in the preceding sections, the problem of the competition or coexistence between the Kondo effect and magnetism and in particular the behaviour of heavy fermions close to the magnetic instability.

Another remarkable and unsolved problem occuring in heavy fermions is the experimental evidence of low temperature superconductivity in several Cerium and Uranium compounds. After the discovery of superconductivity at $T_c$=0.6K in $CeCu_2Si_2$[112], superconductivity has been recently found at very high pressures in several compounds, namely $CeCu_2Ge_2$[113] at $T_c$=0.6 K above 77 kbar, $CePd_2Si_2$[114] at $T_c$=0.5 K above 23 kbar, $CeRh_2Si_2$[53] at $T_c$=0.4 K above 9 kbar and possibly $CeCu_2$ around 60 kbar[115]. The first compound $CeCu_2Si_2$ is non magnetic at normal pressures, while the other ones present a magnetic order which disappears at high pressures and superconductivity appears approximately when the Neel temperature disappears. Thus, this peculiar heavy fermion superconductivity seems to appear around the magnetic - non magnetic transition.

On the other hand, superconductivity exists in several Uranium compounds, namely $UPt_3$[105], $UBe_{13}$[108], $URu_2Si_2$[109], $UNi_2Al_3$[110,111], $UPd_2Al_3$[110,111], with both a transition to an antiferromagnetically ordered state and then a transition to a superconducting state . Thus, the heavy fermion superconductivity appears to be different in Cerium and Uranium compounds, since it occurs in a non magnetic state in the first case and in a magnetically ordered case in the second one. A possible explanation of this difference could arise from the fact that there is a classical Kondo effect for Cerium systems and a multichannel one for Uranium compounds. In fact, both the question of the nature of the heavy fermion superconductivity and that of the symmetry of the order parameter are not really solved at present.

At last, another well studied subject concerns the so-called "Kondo insulators", which have, at low temperatures, a semiconducting behaviour characterized by a small gap value. The compounds $SmB_6$[116], SmS[117], TmSe[118], $YbB_{12}$[119] and $Ce_3Bi_4Pt_3$[120] are already well known and their corresponding gap, which is typically ranging from 30 to 80 K, arises from the hybridization between 4f and conduction electrons. A new class of "Kondo semiconductors"[121-123] such as CeNiSn, CePdSn or CeRhSb is presently studied and both their properties and the origin of the gap are not completely well understood at present.

In conclusion, the behaviour near the magnetic instability, the difference of behaviour between Cerium and Uranium compounds, the heavy fermion superconductivity and the nature of the Kondo semiconductors are certainly at present very interesting and not completely understood problems of the physics of heavy fermion compounds.

## REFERENCES


(1) A. C. Hewson. "The Kondo Problem to Heavy Fermions", Cambridge University Press (1992).

(2) B. Coqblin, in "Magnetism of Metals and Alloys", ed by M. Cyrot, North Holland Publishing Company, Chap 3, pp. 295-377 (1982).

(3) P. Fulde, J. Keller and G. Zwicknagl, in "Solid State Physics", ed. by H. Ehrenreich and D. Turnbull, Academic Press, New-York, vol **41**:1 (1988).

(4) N. Grewe and F. Steglich in "Handbook on the Physics and Chemistry of Rare-earths", ed. by K. A. Gschneidner, Jr. and L. Eyring, Elsevier Science B.V., vol **14**:343 (1991).

(5) J. G. Sereni, in "Handbook on the Physics and Chemistry of Rare-earths", ed. by K. A. Gschneidner, Jr. and L. Eyring, Elsevier Science B.V., vol. **15**:1 (1991).

(6) B. Coqblin, J. Arispe, A. K. Bhattacharjee and S. M. M. Evans, in "Frontiers in Solid State Physics", ed. by L. C. Gupta and M. S. Multani, World Scientific, p. 75 (1993); S.M.M. Evans, A.K. Bhattacharjee and B. Coqblin, Physica B **171**:293 (1991); C. Lacroix, J. Magn. Magn. Mater. **100:**90 (1991).

(7) Y. Onuki and A. Hasegawa, in "Handbook on the Physics and Chemistry of Rare-earths", ed. by K. A. Gschneidner, Jr. and L. Eyring, Elsevier Science B.V., vol. **20**:1 (1995).

(8) Proceedings of the SCES'94 Conference, Amsterdam, Physica B, volumes **206-207**, North Holland Elsevier (1995).

(9) Proceedings of the SCES'95 Conference, Goa, Physica B, volumes **223-224**, North Holland Elsevier (1996).

(10) Proceedings of the SCES'96 Conference, Zurich, Physica B, volumes **230-232**, North Holland Elsevier (1997).

(11) Proceedings of the SCES98 Conference, Paris (15-18 July 1998), Physica B, volumes **259-261**, North Holland Elsevier (1999).

(12) B. Coqblin, J. Arispe, J. R. Iglesias, C. Lacroix and Karyn Le Hur, J. Phys. Soc. Japan, suppl. B **65**:64 (1996).

(13) B. Coqblin, Karyn Le Hur, J. R. Iglesias, C. Lacroix and J. Arispe, Molecular Physics Reports **17**:11 (1997).

(14) B. Coqblin. "The Electronic Structure of Rare-Earth Metals and Alloys : the Magnetic Heavy Rare-Earths", Academic Press, New York, (1977).

(15) B. Coqblin and A. Blandin, Adv. in Phys. **17**:281 (1968).

(16) R. Ramirez and L. M. Falicov, Phys. Rev. B **3**:2425 (1971).

(17) M. Lavagna, C. Lacroix and M. Cyrot, J. Phys. F **13**:1007 (1983); J.W. Allen and R. M. Martin, Phys. Rev. Lett. **49**:1106 (1982).

(18) A. Svane, Phys. Rev. B **53**:4275 (1996) and references therein.

(19) J. Kondo, Theoret. Phys. (Kyoto) **32**:37 (1964).

(20) B. Coqblin and J. R. Schrieffer, Phy. Rev. **185**:847 (1969).

(21) B. Cornut and B. Coqblin, Phys. Rev. B **5**:4541 (1972).

(22) A. K. Bhattacharjee and B. Coqblin, Phys. Rev. B 13:3441 (1976) ; A. Guessous, Thesis, Grenoble (1987).

(23) A. K. Bhattacharjee and B. Coqblin, Phys. Rev. B **38**:338 (1988).

(24) M. Kurisu, T. Mitsumata and I. Oguro, Physica B **259-261**:96(1999).

(25) B. Becker, S. Ramakrishnan, D. Groten, S. Sullow, C.C. Mattheus, G. J. Nieuwenhuys and J.A. Mydosh, Physica B **230-232**:253 (1997)

(26) J. C. Gomez Sal., J. Garcia Soldevilla, J. A. Blanco, J. I. Espeso, J. Rodriguez Fernandez, F. Luis, F. Bartolome and J. Bartolome, Phys. Rev. B **56**:11741 (1997) ; J. I. Espeso, J. Garcia Soldevilla, J. C. Gomez Sal, M. Reiffers and J. A. Blanco, J. Appl. Phys. **79**:6352 (1996)

(27) K. G. Wilson, Rev. Mod. Phys. **47**:773 (1975).

(28) N. Andrei, Phys. Rev. Lett. **45**:379 (1980) ; N. Andrei, K. Furuya and J. H. Lowenstein, Rev. Mod. Phys. **55**:331 (1983).

(29) A. M. Tsvelick and P. B. Wiegmann, J. Phys. C **15**:1707 (1982).

(30) P. Schlottmann, Z. Phys. B **49**:109 (1982).

(31) J. W. Rasul, N. Read and A. C. Hewson, J. Phys. C **16**:L1079 (1983); N. Read and D. N. Newns, J. Phys. C **16**:3273 (1983).

(32) P. Coleman, Phys. Rev. B **29**:3035 (1984).

(33) C. Lacroix and M. Cyrot, Phys. Rev. B. **20**:1969 (1979).

(34) P. Coleman and N. Andrei, J. Phys. : Condens. Matter **1**:4057 (1989).

(35) A. Georges, G. Kotliar, W. Krauth and M. J. Rozenberg, Rev. Mod. Phys. **68**:13 (1996).

(36) J. W. Allen, S. J. Oh, O. Gunnarsson, K. Schönhammer, M. B. Maple, M. S. Torikachvili and I. Lindau, Adv. in Phys. **35**:275 (1986).

(37) O. Gunnarsson and K. Schönhammer, in "Handbook on the Physics and Chemistry of Rare-earths", ed. by K. A. Gschneidner, Jr., L. Eyring and S. Hüfner, Elsevier Science B.V., vol.**10**:1 (1987).
(38) N. E. Bickers, Rev. Mod. Phys. **59**:845 (1987).
(39) P. Nozières, J. Low Temp. Phys. **17**:31 (1974).
(40) P. Nozières and A. Blandin, J. Physique **41**:193 (1980).
(41) S. Doniach, Physica B **91**:231 (1977).
(42) J. R. Iglesias, C. Lacroix and B. Coqblin, Phys. Rev. B **56**:11820 (1997)
(43) B.H. Bernhard, C. Lacroix, J. R. Iglesias and B. Coqblin, Physica B **259-261**:227(1999).
(44) Karyn Le Hur and B. Coqblin, Physica B, **230-232**:454 (1997) and Phys. Rev. B **56**:668 (1997).
(45) B. Barbara, H. Bartholin, D. Florence, M. F. Rossignol and E. Walker, Physica B **86-88**:177 (1977).
(46) M. C. Croft, R. P. Guertin, L. C. Kupferberg and R. D. Parks, Phys. Rev. B **20**:2073 (1979).
(47) A. Eiling and J. S. Schilling, Phys. Rev. Lett. **46**:364 (1981).
(48) S. A. M. Mentink , N. M. Bos, G. J. Nieuwenhuys, A. A. Menovsky and J. A. Mydosh, Physica B **186-188**:497 (1993) ; F. Nolting, A. Eichler, S. A. M. Mentink and J. A. Mydosh, Physica B **199-200**:614 (1994).
(49)Y. Uwatoko, T. Ishii, G. Oomi and S. Malik, Physica B **206-207**:199 (1995).
(50) J. Flouquet, P. Haen, P. Lejay, P. Morin, D. Jaccard, J. Schweizer, C. Vettier, R. A. Fisher and N. E. Phillips, J. Magn. Magn. Mater. **90-91**:377 (1990).
(51) P. Link and D. Jaccard, Physica **B 223-224**:303 (1996).
(52) P. Link and D. Jaccard, Physica B **230-232**:31 (1997) ; D. Jaccard, P. Link, E. Vargoz and K. Alami-Yadri, Physica B **230-232**:297 (1997).
(53) R. Movshovich, T. Graf, D. Mandrus, M. F. Hundley, J. D. Thompson, R. A. Fischer, N. E. Phillips and J. L. Smith, Physica **B 223-224**:126 (1996) ; T. Graf, J. D. Thompson, M. F. Hundley, R. Movshovich, Z. Fisk, D. Mandrus, R. A. Fisher and N. E. Phillips, Phys. Rev. Lett. **78**:3769 (1997).
(54) P. Haen, J. Flouquet, F. Lapierre, P. Lejay and G. Remenyi, J. Low Temp. Phys. **67**:391 (1987).
(55) M. J. Besnus, P. Lehmann and A. Meyer, J. Magn. Magn. Mater. **63-64**:323 (1987); C. Sekine, T. Sakakibara, H. Amitsuka, Y. Miyako and T. Goto, J. Phys. Soc. Japan **61**:4536 (1992).
(56) J. Flouquet, S. Kambe, L. P. Regnault, P. Haen, J. P. Brison, F. Lapierre and P. Lejay, Physica B **215**:77 (1995).
(57) W. H. Lee, R. N. Shelton, S. K. Dhar and K. A. Gschneidner Jr., Phys. Rev. B **35**:8523 (1987).
(58) D. Gignoux and J. C. Gomez-Sal, Phys. Rev. B **30**:3967 (1984).

(59) J. G. Sereni, Physica B **206-207**:246 (1995).
(60) M. Weiden, O. Trovarelli, M. Gomez-Berisso, R. Müller-Reisener, C. Geibel and J. G. Sereni, Physica **B223-224**:299 (1996).
(61) J. G. Sereni, E. Beaurepaire and J. P. Kappler, Phys. Rev. B **48**:3747 (1993).
(62) J. G. Sereni and J. P. Kappler, J. Phys. : Cond. Matter. **6**:4771 (1994).
(63) Y. Miyako, T. Takeuchi, T. Taniguchi, S. Kawarazaki, K. Marumoto, R. Hamada, Y. Yamamoto, M. Ocio, P. Pari and J. Hammann, J. Phys. Soc. Japan, suppl. B, **65**:12 (1996).
(64) J. Rossat-Mignod, L. P. Regnault, J. L. Jacoud, C. Vettier, P. Lejay, J. Flouquet, E. Walker, D. Jaccard and A. Amato, J. Magn. Magn. Mater. **76-77**:376 (1988).
(65) J. Pierre, P. Haen, C. Vettier and S. Pujol, Physica B **163**:463 (1990).
(66) L. P. Regnault, W. A. C. Erkelens, J. Rossat-Mignod, P. Lejay and J. Flouquet, Phys. Rev. B **38**:4481 (1988).
(67) J. M. Mignot, J. L. Jacoud, L. P. Regnault, J. Rossat-Mignod, P. Haen, P. Lejay, Ph. Boutrouille, B. Hennion and D. Petitgrand, Physica B **163**:611 (1990).
(68) L. P. Regnault, J. L. Jacoud, J. M. Mignot, J. Rossat-Mignod, C. Vettier, P. Lejay and J. Flouquet, Physica B **163**:606 (1990).
(69) R. M. Fye and J. E. Hirsch, Phys. Rev. B **40**:4780 (1989).
(70) E. Bauer, R. Hauser, E. Gratz, M. Maikis, Le Tuan, A. Indinger, R. Hatzl, G. Oomi and T. Kagayama, Physica B **199-200**:527 (1994)
(71) E. Bauer, R. Hauser, L. Keller, P. Fisher, O. Trovarelli, J. G. Sereni, J. J. Rieger and G. R. Stewart, Phys. Rev. B **56**:711 (1997).
(72) P. Nozières, Eur. Phys. J. B**6**:447 (1999).
(73) P. Nozières, Ann. de Phys. (France)**10**:19 (1985).
(74) Karyn Le Hur, Phys. Rev. B **58**:10261 (1998) and Thesis, Orsay (1998).
(75) B. H. Bernhard and J. R. Iglesias, Phys. Rev. B **47**:12408 (1993).
(76) A. N. Tahvildar-Zadeh, M. Jarrell and J. K. Freericks, Phys. Rev. **B55**:R3332 (1997).
(77) T. M. Rice and K. Ueda, Phys. Rev. B **34**:6420 (1986); M. Jarrell, Phys. Rev. B **51**:7429 (1995).
(78) A. J. Arko, J. J. Joyce, J. Sarrao, Z. Fisk, J. L. Smith, J. D. Thompson, M. Hundley, A. Menovsky, A. N. Tahvildar-Zadeh and M. A. Jarrell, to be published in these Proceedings.
(79) O. Gunnarsson and K. Schönhammer, Phys. Rev. B **28**:4315 (1983).
(80 P. Coleman, Physica B **259-261**:353 (1999).
(81) H. v. Löhneyssen, F. Huster, S. Mock, A. Neubert, T. Pietrus, M. Sieck, O. Stockert and M. Waffenschmidt, Physica B **230-232**:550 (1997).
(82) F. Steglich, P. Gegenwart, C. Geibel, R. Helfrich, P. Hellmann, M. Lang, A. Link, R. Molder, G. Sparn, N. Büttgen and A. Noidl, Physica **B223-224**:1 (1996).
(83) H. v. Löhneyssen, T. Pietrus, G. Portisch, H. G. Schlager, A. Schröder, M. Sieck and T. Trappmann, Phys. Rev. Lett. **72**:3262 (1994).

(84) M. Sieck, F. Huster and H. v. Löhneysen, Physica B **230-232**:583 (1997).
(85) F. Steglich, C. Geibel, K. Gloos, G. Olesch, C. Schank, C. Wassilew, A. Loidl, A. Krimmel and G. R. Stewart, J. Low Temp. Phys. **95**:3 (1994).
(86) S. Kambe, S. Raymond, H. Suderow, J. Mc Donough, B. Fak, L.P. Regnault, R. Calemczuk and J. Flouquet, Physica **B223-224**:135 (1996).
(87) J. G. Sereni, C. Geibel, M. G. Berisso, P. Hellmann, O. Trovarelli and F. Steglich, Physica B **230-232**:580 (1997).
(88) K. Umeo, H. Kadomatsu and T. Takabatake, J. Phys. : Condens. Matter **8**:9743 (1996) and Phys. Rev. B **55**:R692 (1997).
(89) T. Moriya and T. Takimoto, J. Phys. Soc. Japan **64**:960 (1995).
(90) C. L. Seaman, M. B. Maple, B. W. Lee, S. Ghamaty, M. S. Torikachvili, J. S. Kang, L. Z. Liu, J. W. Allen and D. L. Cox, Phys. Rev. Lett. **67**:2882 (1991).
(91) Y. Sakurai, T. Kagayama, G. Oomi, Y. Aoki, H. Sato, Y. Onuki, H. Takahashi and N. Mori, Physica B **230-232**:596 (1997).
(92) H. Amitsuka, T. Sakakibara, A. de Visser, F. E. Kayzel and J. J. M. Franse, Physica B **230-232**:596 (1997).
(93) H. Nakotte, K. H. J. Buschow, E. Brück, J. C. P. Klaasse, K. Prokes, F. R. de Boer, A. V. Andreev and A. Lacerda, Physica B **230-232**:616 (1997).
(94) E. Miranda, V. Dobrosavljevic and G. Kotliar, Phys. Rev. Lett. **78**:290 (1997) and Physica B **230-232**:569 (1997).
(95) E. Miranda and V. Dobrosavljevic, Physica B **259-261**:359 (1999).
(96) D. M. Cragg and P. Lloyd, J. Phys. C **11**:L597 (1978).
(97) D. M. Cragg and P. Lloyd, J. Phys. C **12**:L-215 and 3301 (1979).
(98) P. Schlottman and P. D. Sacramento, Adv. in Phys. **42**:641 (1993).
(99) N. Andrei and C. Destri, Phys. Rev. Lett. **52**:201 (1984).
(100) A. M. Tsvelick and P. B. Wiegmann, Z. Phys. **54**:201 (1984) and J. Statist. Phys. **38**:125 (1985).
(101) P. D. Sacramento and P. Schlottman, Phys. Rev. B **43**:13294 (1991).
(102) P. D. Sacramento and P. Schlottman, J. Phys. : Condens. Matter **3**:9687 (1991).
(103) D. L. Cox, Phys. Rev. Lett. **59**:1240 (1987).
(104) G. Aeppli, E. Bucher, C. Broholm, J. K. Kjems, J. Baumann and J. Hufnagl, Phys. Rev. Lett. **60**:615 (1988).
(105) G. R. Stewart, Z. Fisk, J. O. Willis and J. L. Smith, Phys. Rev. Lett. **52**:679 (1984); L. Taillefer, K. Behnia, K. Hasselbach, J. Flouquet, S. M. Hayden and C. Vettier, J. Magn. Magn. Mater. **90-91**:623 (1990).
(106) G. E. Brodale, R. A. Fisher, N. E. Phillips, G. R. Stewart and A. L. Giorgi, Phys. Rev. Lett. **57**:234 (1986).
(107) J. O. Willis, J. D. Thompson, Z. Fisk, A. de Visser, J. J. M. Franse and A. Menovsky, Phys. Rev. B **31**:1654 (1985).
(108) H. R. Ott, H. Rudigier, Z. Fisk and J. L. Smith, Phys. Rev. Lett. **50**:1595 (1983); L. Taillefer, J. Flouquet and G. G. Lonzarich, Physica B **169**:257 (1991).

(109) T. T. Palstra, A. A. Menovsky, J. van den Berg, A. J. Dirkmaat, P. H. Kes, G. J. Nieuwenhuys and M. A. Mydosh, Phys. Rev. Lett. **55**:2727 (1985).

(110) C. Geibel, S. Thies, D. Kaczorowski, A. Mehner, A. Grauel, B. Seidel, R. Hefrich, K. Petersen, C. D. Bredl and F. Steglich, Z. Phys. B **83**:305 (1991) ; C. Geibel, C. Schank, S. Thies, H. Kitazawa, C. D. Bredl, A. Bohm, M. Rau, R. Caspary, R. Hefrich, U. Ahlheim, G. Weber and F. Steglich, Z. Phys. B **84**:1 (1991).

(111) M. Kyogaku, Y. Kitaoka, K. Asayama, C. Geibel, C. Schank and F. Steglich, J. Phys. Soc. Japan **62**:4016 (1993).

(112) F. Steglich, J. Aarts, C. D. Bredl, W. Lieke, D. Meschede, W. Franz and H. Schafer, Phys. Rev. Lett **43**:1892 (1979).

(113) D. Jaccard, K. Behnia and J. Sierro, Phys Lett A **163**:475 (1992).

(114) S. J. S. Lister, F. M. Grosche, F. V. Carter, R. K. W. Haselwimmer, S. S. Saxena, N. D. Mathur, S. R. Julian and G. G. Lonzarich, Z. Phys. B**103**:263 (1997)

(115) E. Vargoz, P. Link and D. Jaccard, Physica B **230-232**:182 (1997).

(116) A. Jayaraman, V. Narayanamurti, E. Bucher and R. G. Maines, Phys. Rev. Lett. **25**:1430 (1970).

(117) M. B. Maple and D. Wohlleben, Phys. Rev. Lett **27**:511 (1971).

(118) P. Haen, F. Lapierre, J. M. Mignot, R. Tournier and F. Holtzberg, Phys. Rev. Lett **43**:304 (1979).

(119) M. Kasaya, F. Iga, M. Takigawa and T. Kasuya, J. Magn. Magn. Mater. **47-48**:429 (1985).

(120) M. F. Hundley, P. C. Canfield, J. D. Thompson and Z. Fisk, Phys. Rev. B **42**:6842 (1990).

(121) T. Ekino, T. Takabatake, H. Tanaka and H. Fujii, Phys. Rev. Lett. **75**:4262 (1995).

(122) Y. Uwatoko, T. Ishii, G. Oomi, H. Takahai, N. Mori, J. D. Thompson, J. L. Shero, D. Madru and Z. Fisk, J. Phys. Soc. Japan **65**:27 (1996).

(123) K. Nakamura, Y. Kitaoka, K. Asayama, T. Takabatake, G. Nakamoto, H. Tanaka and H. Fujii, Phys. Rev. **53**:1 (1996).

## Discussion

**Sawatzky:** In the sample models of the competition between long-range magnetic order via RKKY and the Kondo effect, I would have thought that the details of the Fermi surface of the conduction electron system or, if you like, the host in which the localized spin carrying orbitals are placed would be extremely important. Has this been considered?

**Coqblin**: Up to now, most theories assume a simple free electron-like Fermi surface.

**Sawatzky:** Rare earth intermetallics form heavy Fermion systems but also the hard ferromagnetic systems. In the later the 4f-5d atomic exchange interactions play a very important role and are of order 1 eV. Even though the 5d bands may be located well above $E_F$ they, none the less, play a very important role in the exchange mechanisms between rare earths and the transition metals. This suggests that not only the itinerant electron states close to $E_F$ play are important, but also states located well above and below $E_F$, such as the d bands of the transition metals. These bands located up to 2 or 3 eV from $E_F$ will contribute to interactions similar to the superexchange interactions in the magnetic insulators, like the transition metal oxides. In superexchange states located as far as 10eV from $E_F$ contribute strongly. This would indicate that the Kondo-like coupling and the interatomic exchange interaction are not driven by the same mechanisms as assumed when considering only RKKY like mechanisms.

# CONSEQUENCES OF HAVING TWO KINDS OF *f*-ELECTRONS FOR SCE SYSTEMS AS TREATED BY A SYNTHESIS OF MANY-BODY THEORY AND ELECTRONIC STRUCTURE

Bernard R. Cooper and Yeong-Lieh Lin

Dept. of Physics
West Virginia University
Morgantown, WV 26506-6315, U.S.A.

## ABSTRACT

For strongly correlated electron (SCE) systems, one has to treat at least two-electron dynamics and to include both *p/d* band-*f* hybridization and *p/d* band-*f* coulomb exchange. However, the two great simplifications possible when considering *magnetic ordering* for *uranium* SCE systems have enabled us to make absolute material-specific predictions of alloying or high-pressure effects using Local Density Approximation (LDA) input into many-electron dynamics with only one *non-ab-initio* number (the baseline percentage division, for unalloyed material at ambient pressure, between occupied correlated-electron states having localized (magnetic) and those having delocalized (nonmagnetic) 5*f* electrons). Experimentally, the alloying effects can be dramatic, e.g. in $U_xLa_{1-x}S$ the magnetic ordering abruptly disappears at about 55% uranium. The theory is quite successful in its detailed absolute predictions, and this has important implications for the overall understanding of electronic behavior in SCE systems including heavy fermion systems. The key conclusion is that the hybridization, *as kinematically restricted by exchange symmetry*, leads to a chemical-environment-dependent sharp threshold for a drastic restructuring of the ground state in SCE systems with dramatic observable consequences. This restructuring is driven by the elimination of the localized-magnetic transition-shell electrons (*f* electrons for light actinide and cerium-based SCE materials, *d* electrons for transition-metal-oxide-based SCE materials).

## 1. Introduction

The need to include *f*-"ligand" (*p/d*) as well as *f-f* correlations, makes the development of *ab initio* (materially-predictive) theory for electronic and magnetic behavior difficult for

*Electron Correlations and Materials Properties*, edited by Gonis *et al.*
Kluwer Academic/Plenum Publishers, 1999.

Strongly Correlated Electron (SCE) systems. This is because of the computational complexity of going beyond refinements of LDA (one-electron dynamics) theory. One has to treat at least two-electron dynamics and to include both ligand-*f* hybridization and ligand-*f* coulomb exchange.

Starting from the first principles hamiltonian embodying the dynamics of the $\sim 10^{25}$ *f*-transition-shell and *p/d* valence electrons for the lattice, where $V_0$ is the periodic potential from the fixed background of the nuclei and filled-shell electrons,

$$H = \sum_i \nabla_i^2 + V_0(r_i) + \frac{1}{2}\sum_{i \neq j}\frac{1}{r_{ij}} \tag{1}$$

one can derive a model hamiltonian describing *f* electrons interacting with non-*f* band electrons, where all the quantities in it can either be matched to the output of a LMTO (linear combination of muffin-tin orbitals) LDA (local density approximation) band calculation or be calculated separately using the information given by a LMTO band calculation.

$$H = H_0 + H_1, \tag{2a}$$

$$H_0 = \sum_k \varepsilon_k b_k^\dagger b_k + \sum_{Rm} E_f c_m^\dagger(R) c_m(R) + \frac{U}{2}\sum_{R, m \neq m'} n_m(R) n_{m'}(R), \tag{2b}$$

$$H_1 = \sum_{km\sigma R}[V_{km\sigma} e^{-ik\cdot R} b_{k\sigma}^\dagger c_{m\sigma}(R) + H.c.] - \sum_{kk'}\sum_{mm',R} J_{mm'}(\mathrm{k},\mathrm{k}') e^{-i(\mathrm{k}-\mathrm{k}')\cdot R} \times b_k^\dagger b_{k'} c_m^\dagger(R) c_{m'}(R). \tag{2c}$$

The nature of the terms and sources of the parameters in this five-term Hamiltonian are summarized as follows. The first term is the non-*f* band energy and comes directly from the LDA calculation; the second term is the *f*-state energy, which is not a LDA energy but can be calculated indirectly from the LDA; the third term is the intra-atomic *f-f* coulomb interaction (correlation energy) which can be calculated indirectly from the LDA; the fourth term is the LDA hybridization; and the fifth term is the band-*f* exchange coulomb interaction.

Two great simplifications are possible when considering *magnetic ordering* for *uranium* SCE systems. (1) Magnetic ordering involves the electronic behavior *globally* (i.e. involves the spectral density in an integrated sense). (2) In a number of uranium-based compounds and alloys thereof the chemical environment is such that the behavior of the 5*f* electrons is neither closely atomiclike nor significantly influenced by direct overlap with 5*f* electrons from other lattice sites. In this situation, the $5f$ role in the electronic dynamics, and in consequent magnetic ordering, is determined[1] by hybridization with band electrons of non-$f$ atomic origin. Therefore, *as long as the hybridization is treated as acting between properly exchange-symmetrized two-electron wave functions*, the effects of exchange can be incorporated adequately by its inclusion in the one-electron exchange-correlation potential. This is not so simply the situation in cerium-based systems[2] where the $f$-electron behavior is more atomiclike. This is a very important simplification because coulomb exchange is a true two-body interaction (involves two creation and two annihilation operators), while hybridization is a one-body interaction (involves one creation and one annihilation operator) [see Eq (2.10) of Ref. 2]. This has enabled us to make absolute material-specific predictions of alloying or high-pressure effects using LDA input into many-electron dynamics with only one non-*ab-initio* number (the baseline percentage separation between localized and delocalized $f$ electrons---*why, as a consequence of the necessary exchange-symmetrization, there are two kinds of f electrons is discussed below*). This one number is chosen for the pure (unalloyed) material at ambient pressure. Experimentally, the alloying effects can be dramatic, e.g. in $U_xLa_{1-x}S$, the magnetic ordering abruptly disappears[3] at about 55% uranium. Thus magnetic ordering behavior provides a powerful tool for diagnosing the nature of the correlated (two-electron) electronic structure; and this leads to recognition of, and a measure of, the quantitative division in occupied composition between the two kinds of two-electron states. The theory is highly successful in its detailed absolute predictions, and this has important implications for the overall understanding of electronic behavior in SCE systems including heavy fermion systems.

## 2. Recognition and Quantification of Key Physics Concepts: Development of the Model

Through a combination of first principles model calculations[4] and *ab-initio-based* materially predictive calculations,[1,2,5-7] we have come to understand that there are two kinds of many-electron states involving $f$-electrons. To see this we consider that prior to including the coulomb interaction between the $4f$ (cerium) or $5f$ (uranium) electrons and the band ("ligand") electrons of $p/d$ character (both coulomb-induced hybridization and the coulomb exchange interaction), the $f$ electrons are quite sharp in energy and well localized in space around a single (cerium or uranium) nucleus, i.e. there is very little direct overlap of $f$ electrons centered around different nuclei. To consider the effects of hybridization we locally (with regard to a uranium nucleus) form properly (exchange) symmetrized two-electron states, i.e., we form $f$-ligand two-electron singlet and triplet states. The locally strong hybridization for uranium then acts only through the $f$-$l$ singlet.[4] To see this, we consider the following points. Because the ligand states are itinerant, their orbital angular momentum is quenched,[8] and thus they can be treated as $s$ states, i.e., the incoming ligand electrons do not establish residence in the local spherical environment for sufficiently long to have well defined orbital angular momentum. (The resonant scattering formulation of Coqblin and Schrieffer[9] corresponds to this physics.) Therefore the two-electron purely ligand state involved in hybridization, because it is itinerant and hence effectively involves one-electron $s$ states, must be a singlet.[4] Since hybridization does not mix spin, this state can only mix with the $f$-$l$ singlet; and the hybridization gives configuration interaction, i.e. locally gives valence fluctuations and correspondingly enhanced Pauli susceptibility, only by acting through the singlet two-electron states. Thus, the $f$ spectral density in magnetic (triplet $f$-$l$) states remains highly localized, hence sharp in energy, while the $f$ spectral density in nonmagnetic (singlet $f$-$l$) states is of mixed local and itinerant character. (The generalization of the treatment in Ref. 4, which involve only one $f$ electron, i.e. the cerium situation, to that involving more than one $f$ electron, i.e. the actinide situation, is given in

Ref. 6. Also, it should be noted that, given the permutation symmetry of all electrons, the magnetic *f-l* states cannot remain totally isolated from the nonmagnetic states, and thus cannot be perfectly sharp in energy; nevertheless we expect this strong distinction in character of the spectral distribution to be essentially preserved, at least so long as the *f* electron role in the overall dynamics is neither significantly atomistic nor significantly direct overlap.) It should be emphasized that the hybridization is strongly reduced by the strong onsite *f-f* coulomb repulsion energy (the *f-f* correlation energy, *U*, for uranium is between 3 and 4 eV). It should also be emphasized that higher order hybridization effects give intersite singlet-triplet mixing and thereby give intersite *f-f* coupling.

This intersite *f-f* coupling[10] gives highly anisotropic[11] magnetic ordering. Highly anisotropic ferromagnetic ordering, as discussed in Refs. 11, is a characteristic feature of the Coqblin-Schrieffer interaction both in its original form[9] or in its generalized form.[12,13] In its pure manifestation (without the presence of any significant antiferromagnetic kinetic superchange interaction of the type discussed in Ref. 13), this gives very strongly anisotropic ferromagnetism. This is a consequence of the situation that hybridization acts to give orbitally-driven magnetism, while exchange acts to give spin-driven magnetism. (The two types of magnetic ordering, orbitally-driven and spin-driven, are coupled together by the strong spin-orbit coupling for *f* electrons.) Both the generalized Coqblin-Schrieffer (C-S) and the kinetic superexchange interaction arise from a combination of hybridization and exchange effects. As one gets further into the intermediate *f* delocalization (hybridizing) regime, the generalized C-S interaction becomes relatively stronger. When the generalized Coqblin-Schrieffer interaction is overwhelmingly dominant, one expects extremely anisotropic ferromagneticsm. Thus the extraordinarily anisotropic strong ferromagnetic ordering behavior of the uranium monochalcogenides offers a benchmark series of compounds for testing the theory and predictions presented here. (For a discussion of the magnetic ordering of these compounds presented in the overall context of actinide and cerium-based magnetically ordered systems, see References 14 and 15. When the generalized Coqblin-Schrieffer interaction is dominant, but there is a relatively weak antiferromagnetic kinetic superexchange, the dominant C-S interaction manifests itself in the form of strongly ferromagnetic planes weakly antiferromagnetically coupled[10] as is characteristic of high-temperature copper-oxide-type superrconductors[16] and collosal magnetoresistance perovskites as well as cerium and many light actinide correlated-electron magnetically-ordered systems.

Thus, the key physics is the recognition and quantification of the concept that there are two kinds of *5f-p/d* (band) two-electron (correlated) states. The composition of one of these includes (one-electron) localized uranium 5*f* states and is magnetic in character. (The 5*f* spectral weight resides stably around a single site.) The composition of the other kind of *5f-p/d* two-electron states includes a mixture of (one-electron) uranium 5*f* states of localized and delocalized nature and is nonmagnetic in character. (The 5*f* spectral weight does not remain stably around a single site. Viewed from a single site, such a state fluctuates rapidly between valences.) Both pressure and dilution alloying (substituting a lanthanum for a uranium) increase the 5*f*-with-*p/d* (band) hybridization driving 5*f* electron occupation from the well-localized (stable *f* configuration) to the itinerant (rapidly fluctuating *f* configuration) two-electron states. This gives a *decrease in the low-temperature ordered moment*, and interestingly we expect a *correspondingly enhanced Pauli paramagnetism*. Thus both the pressure and dilution alloying effects act in the same qualitative way to increase the average number of "holes" in the magnetic lattice; however, increasing dilution alloying has a more strongly highly nonlinear effect on increasing the number of holes. Thus for dilution alloying, at a certain point the well-localized component falls abruptly below a critical value necessary for magnetic ordering, and a catastrophic collapse in magnetic ordering occurs as[3] in US.

A schematic drawing of typical behavior for the *f* spectral distribution in energy as provided by an LDA (one-electron dynamics) calculation is shown in Fig. 1. The true many-electron states (well approximated for this purpose by two-electron states, as discussed above) are expected to give spectral densities consisting of (1) a very sharp peak corresponding to the localized magnetic states, and (2) a distribution that is similar in appearance to that provided by the LDA calculation, with a broadened peak and diffuse tails, corresponding to the nonmagnetic locally fluctuating plus itinerant states. The broadened peak corresponds to the period of circulation around a given nucleus, and the diffuse tail corresponds to the period of itinerancy between lattice sites. Thus when projected onto the LDA one-electron states, we would expect the magnetic component to be wholly contained within the LDA broadened peak and the nonmagnetic component to be contained in the remainder of that peak plus the diffuse tail. This leads to the *critical ansatz* of our LDA *ab-initio-based* materially specific prediction of dilution alloying and pressure effects on magnetic ordering. This ansatz is that the *LDA calculation of the shift in f-electron spectral weight from the broadened peak to the diffuse tail with dilution alloying or pressure accurately approximates the true shift of spectral density from magnetic to nonmagnetic states.*

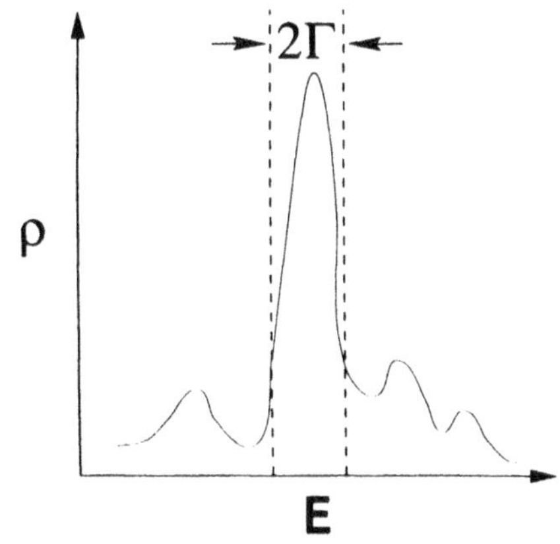


Fig. 1 Schematic drawing of typical behavior for the *f* spectral distribution in energy as provided by an LDA (one-electron dynamics) calculation.

With this ansatz, our *ab initio* LDA calculations allow us to predict the pressure and uranium-dilution dependence of transfer of *f* spectral weight from the localized magnetic *f* states to the hybridizing nonmagnetic *f* states. When used as input into our phenomenological theory of magnetic ordering, including correlation effects,[5] this allows us to predict concentration and pressure dependence of ordered moments and Curie temperatures with remarkable absolute accuracy. The approximation that in practice probably limits this accuracy is the relatively simple method used to evaluate the intersite coupling giving magnetic ordering. This is chosen for the ideal hybridization-mediated coupling limit. Thus this approximation will fail both at the limit where direct overlap effects are significant for the *f*-electrons, i.e. where the *f*-electrons become too itinerant, and at the limit where the *f*-electrons become too localized, so that[2] the role of the *f*-ligand coulomb exchange in the intersite magnetic coupling becomes more complicated.

## 3. Predictive Calculation of Low-Temperature Ordered Moment and of Curie Temperature

As stated above, our *ab-initio-based* model gives absolute material-specific predictions using input from the LDA paramagnetic uranium *f*-electron-projected density of states plus the *ab-initio* calculated value of the correlation energy U (3.4 eV for uranium in US). Figure 2 shows the density of states behavior for a superlattice LDA calculation for $U_xLa_{1-x}S$ (4 uranium-or-lanthanum, 4 sulfur sites per unit cell) using the full-potential linear combination of muffin-tin-orbitals technique.[12,17,18] The shaded area shows the uranium *f*-projected density of states. To calculate the LDA shift in *f*-electron spectral weight from the broadened peak to the diffuse tails with dilution alloying or pressure, we have to specify an objective (non-adjustable) quantitative criterion to set the energy separating the localized and delocalized *f* spectral weight, i.e. defining the "peak" by the selection of energies such as $E_l$ and $E_2$ shown in Fig. 1. We adopt a boundary defining the energy where, upon decreasing energy from the *j*=5/2 peak in the density of states, there is a distinct difference between the total and the *f*-projected density of states. (In effect, this is the energy where significant hybridization begins to occur between the *f* electrons and the non-*f* band electrons. Operationally, we define this as the energy where the density of states first stops dropping monotonically from the peak, i.e. the first local minimum in the density of states.) This gives $E_l$; and $E_2$ is chosen as the energy on the high-energy side of the peak where the density of states has the same value as at $E_l$. Once this is done, there is only one *non-ab-initio* number that must be provided[5] for a given material, e.g. US. This number sets the starting (baseline) division between uranium sites where an $f^3$ configuration exists for long enough for those sites to be able to magnetically order and those sites which fluctuate between an $f^3$ configuration and an $f^2$ configuration sufficiently rapidly to act as magnetic holes in much the same way as a La site. We choose that single number (i.e., the baseline fraction of uranium sites with stable $f^3$ configuration) for the pure (unalloyed) material (e.g. US) at ambient pressure.

The catastrophic ordered magnetic moment collapse that occurs at 55% per cent uranium for $U_xLa_{1-x}S$ is then a consequence of three effects providing an extremely nonlinear magnetic response to the dilution: most importantly, the increased delocalization acts wholly to reduce the stable $f^3$ component, secondarily that $f^3$ component becomes more localized, hence giving less coupling for a given $f^3$-$f^3$ separation, and in addition the $f^3$ sites are spread further apart with increasing lanthanum dilution.

While probably secondary in triggering the ordered moment collapse, the extreme localization, as it disappears, of the remaining $f^3$ component may be important for the observation of heavy fermion effects in the reconstructed ground state after moment collapse discussed in our concluding discussion in Parts 4. This concentrating of spectral density may be significant for specific heat and susceptibility enhancement in that reconstructed ground state.

The way in which *f* electrons are counted is a key step in implementing the LDA part of the physics. Given the physical context, we do this counting by integrating the 5*f* charge within the uranium muffin-tin spheres (where in the full-potential calculation, space is divided between muffin-tin spheres centered about nuclei and interstitial). There is a difference between the way the uranium muffin-tin radius is chosen in calculating pressure effects and is calculating dilution alloying effects. For the pressure effects, the muffin-tin radius is chosen as part of the self-consistent procedure that determines the lattice constant through total-energy minimization, and thus corresponds to the lattice constant used at a specified pressure. (The muffin-tin radius is chosen to give continuity of the self-consistent potential at the radius, i.e. between muffin-tin sphere and interstitial. In the pressure calculation, obviously this sphere radius decreases with increasing pressure.) This is not what is done for the dilution alloying. There the lattice constant is chosen from the experimental behavior for US or for USe and UTe using Vegard's law (linear dependence

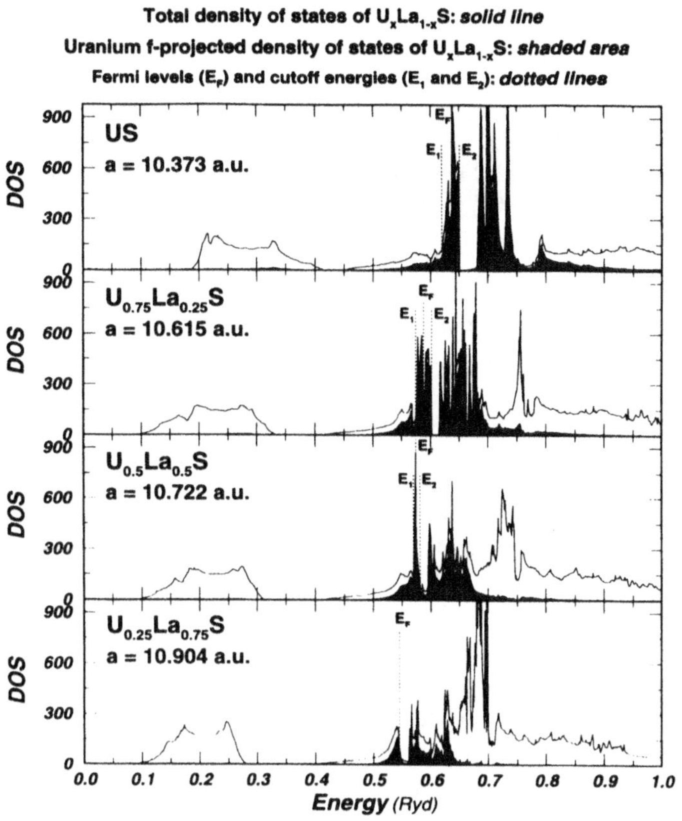


Fig. 2 Density of states for a superlattice LDA Calculation for $U_xLa_{1-x}S$ (4 uranium-or-lanthanum, 4 sulfur sites per unit cell) from a full-potential LMTO calculation. The shaded area shows the uranium *f*-projected density of states.

on concentration) for the lattice constant, i.e. substituting lanthanum gives an expansion of the lattice. (Experimentally for US there is evidence for a departure from Vegard's law behavior near the ordered moment collapse [see Bourdarot et al, Ref. 3].) This means that one is faced with using some physical criterion to specify the way the muffin-tin radius is chosen. If one uses the same, now inconsistent, choice as in the pressure calculations (continuity of the self-consistent potential), this will artificially increase the uranium muffin-tin sphere, thereby altering the LDA selection of *f*-electrons. This changes the nature of the LDA spectral distribution. A better choice, consistent with Vegard's law, is to use fixed uranium and lanthanum muffin-tin sphere radii corresponding to the end points of the concentration variation. This is what we do. (For comparison, we have also used the first, inconsistent choice and compared the behavior found. In fact, there is very little change in the predicted magnetic ordering behavior.) A self-consistent choice of lattice constant and muffin-tin radius could be done since the supercell calculation is, in effect, the calculation for an ordered intermetallic compound, e.g. $U_3LaTe_4$. This would give a departure from Vegard's law. It may be that the experimental departure from Vegard's law close to the moment collapse is associated with some degree of ordering of the lattice; and it could therefore be interesting to also compare results using this way of choosing the muffin-tin radius.

Using the Hubbard model, following the procedure described in Reference 5, the coupling (J) between localized moments on an *fcc* lattice is obtained from the width of the localized peak in the uranium *f*-electron-projected density of states and the calculated correlation energy (U). If $t$ is the inter-atomic interaction, then for an *fcc* lattice, the localized peak width Γ is related to $t$ through $\Gamma = 1.04t$ and $J = t^2/U$. The magnetic ordering is then calculated exactly as in Reference 5 *for an Ising lattice with randomly-located holes*. (An Ising lattice is used to mimic the very large magnetic anisotropy of these ferromagnets. A detailed discussion of the way in which this extremely anisotropic ferromagnetic interaction results from the correlated electronic behavior is given in Ref. 12 (this treatment was incorporated into the calculations in Refs. 1 and 2).) A very complete derivation and discussion of this physics is provided in the appendix of Ref. 13. In the final summary paragraph on page 12,518 of Ref. 13, it is pointed out that the particle-hole (generalized Coqblin-Schrieffer) interaction is ferromagnetic at short ranges.

The key point is to identify the number of holes, i.e. the number of sites not occupied by stable $f^3$ uranium. As stated above, we do this[5] by first establishing a baseline fraction of stable $f^3$ sites for the pure unalloyed material at atmospheric pressure. This is the one point at which we either insert a physically-motivated guess, or use one piece of experimental data, or use a value chosen to achieve some balance between fitting several pieces of data. The obvious choices for this purpose are the low-temperature ordered moment, the Curie temperature, or the LDA calculated *f*-electron count (which will vary slightly depending on the choice of technical method used in determining it). For US, in fact, all three of these indicate very closely the same baseline value for the fraction of stable $f^3$ sites. The experimental low-temperature ordered moment per uranium of US from the recent neutron measurements of Bourdarot et al[3] is $1.60\mu_B$, and from the magnetization measurements of Schoenes et al[3] is $1.55\mu_B$ per uranium. This is slightly less than half the ordered moment expected for $U^{3+}$ ($f^3$ uranium). (For atomic $U^{3+}$, this is $3.27\mu_B$; in the solid state, because of the breaking of spherical symmetry, one expects a slightly different value.) This would give a baseline fraction of stable $f^3$ of 0.49. There are several ways to technically define the *f*-count from a band calculation which give slightly different values; but these all give a value of $n_f$ close to 2.5 for US, indicating a baseline $f^3$ fraction of 0.5. In addition, for US as discussed below with that choice of 0.5, the calculated $T_C$ is 183K compared to the experimental value[3] of 180K. So we take the baseline value of holes as 50% for pure US at atmospheric pressure. (We note that this is a different baseline than was used in Reference 5, and there are also some changes in the technical details of the LDA calculation and of extracting the desired input information from them. Thus the predicted variation of Curie temperature with pressure shown below differs somewhat from that of Reference 5.)

For USe and UTe, the choice of the starting baseline fraction of stable $f^3$ uranium is not so clearcut. The reason for this is that our way of determining the coupling (J) between localized moments tends to underestimate that coupling with increasing degree of localization. Matching the low-temperature experimental ordered moment[15] from neutron measurements ($2.0\mu_B$ for USe and $2.25\mu_B$ for UTe) would give 0.61 and 0.69, respectively, for the baseline $f^3$ fractions, which are close to those expected from the *f*-count from band calculations. On the other hand, matching the Curie temperatures[15] (160K and 104K, respectively) requiries 0.72 and 0.91, respectively. Guided by these values, we have chosen 0.68 and 0.79 as the baseline fractions of $f^3$ for USe and UTe, respectively. It is this greater degree of localization, of having more stable $f^3$, on going from US to USe to UTe that gives rise to the dramatically different variation of $T_C$ with pressure.

Then the key input information from the LDA calculations is how this baseline number of holes increases with pressure and dilution alloying as the hybridization with the *p/d* electrons increases. We do this by calculating the decrease in integrated *f*-spectral weight per uranium contained in the localized peak of the LDA density of states. Thus for US, for pressure effects the increase in holes comes wholly from the size of that spectral transfer, which decreases the 0.5 electron per uranium that on average defines the baseline fraction of well ordered $f^3$ uranium.

For dilution alloying, there are two ways in which the number of holes increases. One is by the shift of spectral weight per uranium from the localized peak to the itinerant tail, i.e. the same effect as caused by pressure. In addition, substitution of a lanthanum by itself gives a hole. Thus dilution alloying has a much enhanced effect on magnetic ordering. The actual LDA calculations for dilution alloying have been carried out using supercells containing four uranium/lanthanum sites and four anion sites. Thus, at present, we can predict alloying effects only at intervals of 25% dilution.

## 4. Results and Discussion

The predicted variation of Curie Temperature with pressure and dilution alloying is compared to experiment[19] for US in Figure 3 and Fig. 4, respectively; while the nonmonotonic variation with pressure predicted in close agreement with experiment for USe and UTe is shown in Figs. 5 and 6. We emphasize that the predictions are absolute. All input information is *ab initio* except for the baseline specification of the fraction of stable $f^3$ uranium for unalloyed US at atmospheric pressure. In Fig. 7, we show the comparison between the predicted and experimental [Bourdarot et al, Ref. 3] low temperature ordered moment for $U_xLa_{1-x}S$. The remarkable agreement of prediction with experiment shown in Figures 3, 4 and 7, including the collapse of magnetic ordering somewhat above 50% uranium, provides strong confirmation of the correctness of the physical picture used.

Curie temperature and ordered moment calculations of the sort shown in Figs. 4 and 7 for the diluted US system have also been performed for the diluted USe and UTe systems, and our predictions for the Curie temperature and ordered moment in diluted UTe are shown in Figures 8 and 9, respectively. While the present calculations at intervals of 25% dilution are consistent with the experimental[20] observation that the moment collapse in diluted USe and UTe is similar to that in US in occurring at uranium concentrations not too far above 50%, calculations at 62.5% uranium (8 uranium sites in the supercell) are necessary to confirm that the detailed trend in behavior predicted between materials for the ordered moment collapse is consistent with experiment.[20] It will be interesting to have neutron diffraction results for the UTe system of the sort already available and shown for the US system in Figs. 4 and 7. The very substantial increase in $T_c$ predicted (see Fig. 8) at 75% uranium in the diluted UTe is a consequence of the more localized behavior for the pure material (baseline $f^3$ fraction of 0.79 for pure UTe at ambient pressure) as compared to US (baseline $f^3$ fraction of 0.50). If not substantiated by experiment, this rise could be indicative of the detailed quantitative limitations in the simplified treatment of the intersite coupling. In fact, the predicted order moment behavior (see Fig. 9) for UTe is almost flat between 100% and 75% uranium, with only a very slight rise at 75% uranium. Since the Curie temperature varies quadratically with the ordered moment and linearly with the magnetic coupling, it is a much more sensitive quantitiy to calculate than the ordered moment. Thus the likeliest explanation for the substantial increase in the predicted Curie temperture and the very slight rise in the predicted ordered moment at 75% uranium is that the boundary of the LDA peak has been chosen at somewhat too low an energy. Since this determines both and counting of states, and hence the change in ordered moment, and the LDA *f*-peak width, and hence the intersite magnetic coupling, any error in this choice would show up much more strongly in the Curie temperature than in the ordered moment. At a more fundamental level, the predicted behavior for UTe, if not supported by experiment, could indicate the limitations in the simplified treatment of exchange as the $f$ behavior becomes more localized.

For the baseline stable $f^3$ fractions used (0.50, 0.68 and 0.79 for US, USe and UTe, respectively), the predicted Curie temperature and ordered moment are 183K and 1.64$\mu_B$ (experimentally 180K and 1.6$\mu_B$) for US, 149K and 2.19$\mu_B$ (experimentally 160K and 2.0$\mu_B$) for USe, and 91K and 2.6$\mu_B$ (experimentally 104K and 2.25$\mu_B$) for UTe.

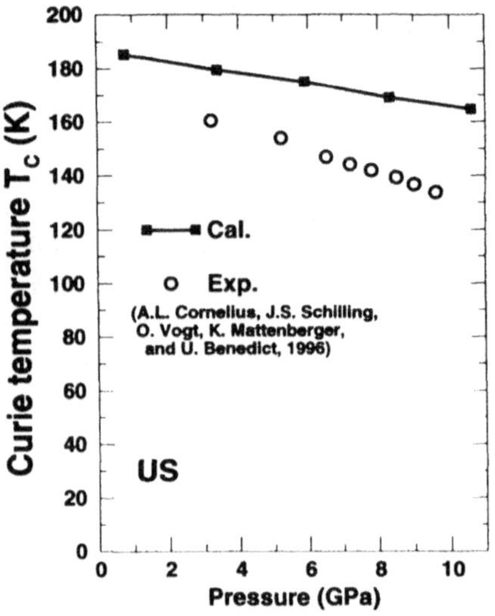


Fig. 3 Predicted variation of Curie temperature with Pressure for US compared to experiment.[18]

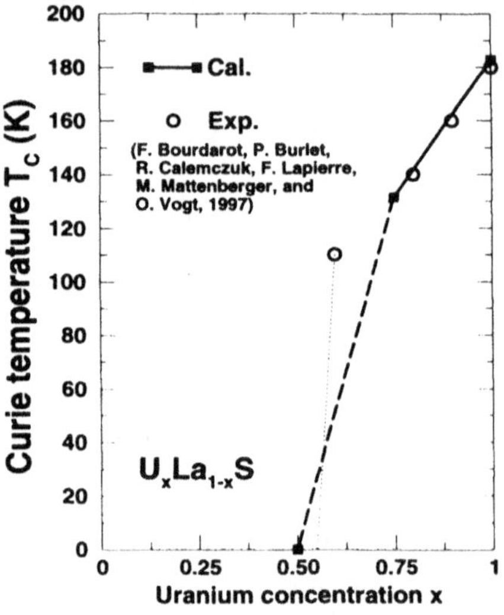


Fig. 4 Predicted variation of Curie temperature with dilution alloying For US compared to experiment.[3]

In conclusion, we want to emphasize that the collapse of magnetic ordering in US is a catastrophic consequence of the two-component, stable $f^3$ and fluctuating $f^3/f^2$ mixed valence, character of the uranium, and of the situation that the fluctuating component is increased at the expense of the localized component by increasing the $f$ delocalization with pressure or dilution alloying. The key conclusion is that the hybridization, *as kinematically restricted by exchange symmetry*, leads to a chemical-environment-dependent sharp threshold for a drastic restructuring of the ground state in SCE systems with dramatic observable consequences. This conclusion applies not only to $f$-electron (cerium and light actinide based) SCE systems but also to $d$-electron (e.g. $d$ transition-metal-oxide based) SCE systems, e.g., to the behavior of heavy fermion systems, high-temperature cuprate superconductors, and collosal magnetoresistance perovskites.

The question arises as to the relationship between the picture presented here and that provided by the competing nonmagnetic and magnetic correlated ultra-narrow band picture of References 4 and 6. In Reference 4, the relationship of a two-liquid ultra-narrow correlated band picture to the characteristic specific heat, susceptibility and Wilson-Sommerfeld ratio behavior for heavy fermion systems is discussed. The existence of these two liquids, or the corresponding ultra-narrow correlated bands, follow from treating two-body electronic correlations, i.e. corresponds again as in the discussion within the present paper to singlet (nonmagnetic) and

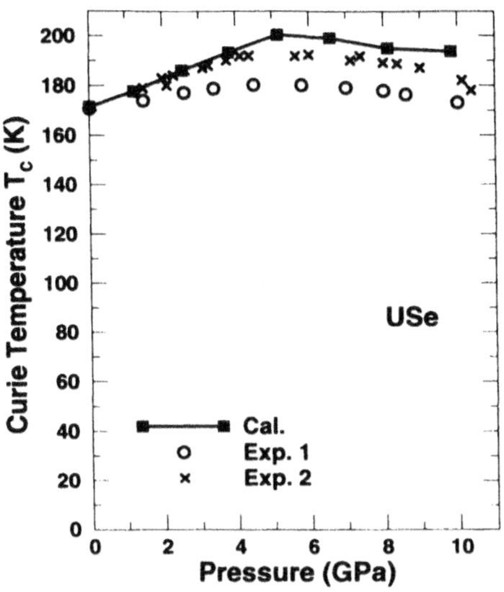


Exp. 1: Cornelius et al., J. Magn. Magn. Mater. **161** (1996) 169.
Exp. 2: Link et al., Physica B **190** (1993) 68.

Fig. 5 Predicted variation of Curie temperature With pressure for USe compared to experiment.[18]

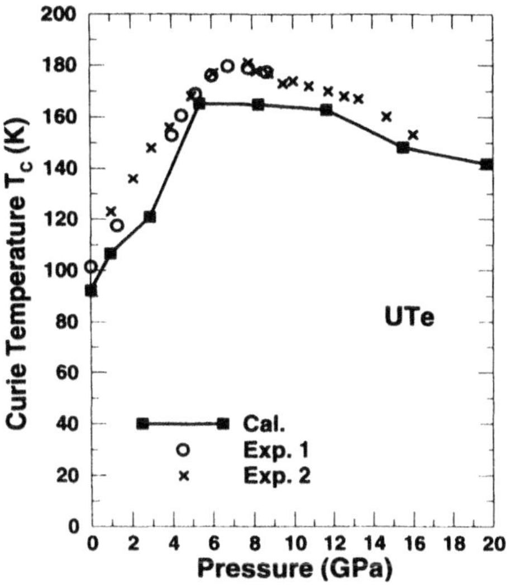


Exp. 1: Cornelius et al., J. Magn. Magn. Mater. **161** (1996) 169.
Exp. 2: Link et al., J. Phys: Condens. Matter **4** (1992) 5585.

Fig. 6 Predicted variation of Curie temperature with pressure for UTe compared to experiment.[18]

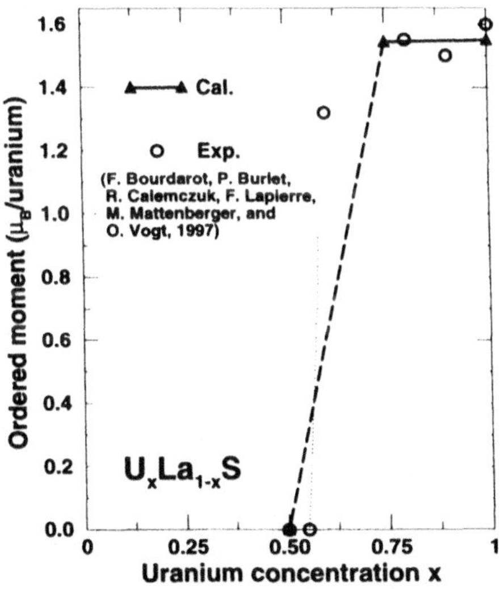


Fig. 7 Predicted variation of low-temperature ordered moment with dilution alloying for US compared to experiment.[3]

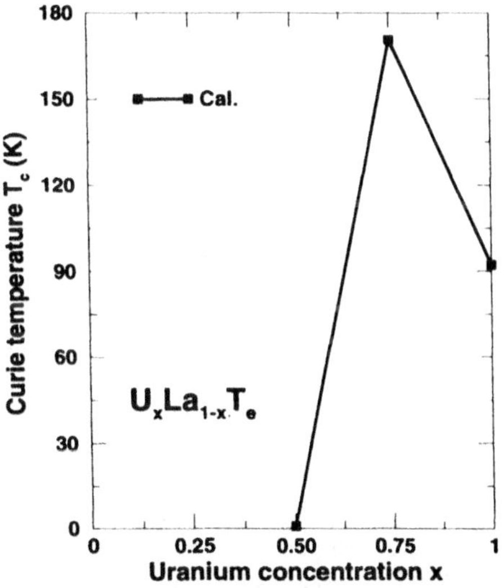


Fig. 8 Predicted variation of Curie temperature with dilution alloying for UTe.

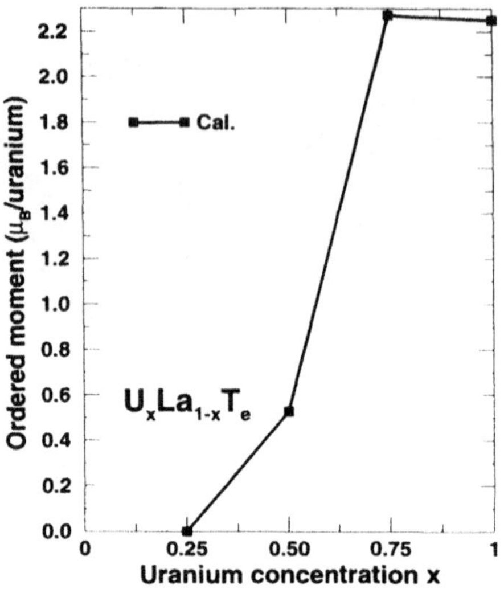


Fig. 9 Predicted variation of low-temperature ordered moment with dilution alloying for UTe.

triplet (magnetic) two-electron correlations. In Reference 6, a recipe is presented in outline form for obtaining these ultra-narrow bands through ab initio calculations. Furthermore, in Reference 6 it is noted that the resonance widths and hybridization dressings in Tables II and III of Ref. 12 and Tables I and II of Ref. 21 correspond to widths of the 1-10 mRy range desired for understanding observed heavy fermion behavior. Our overall picture then is that this two-liquid ultra-narrow band picture describes the behavior *after the moment collapse and restructuring of the ground state* that is the topic of the present paper.

What then is the unifying theme of this overall physics? *Why does the drastic restructuring of the ground state occur*? To answer this question, we must recognize that prior to the moment collapse, even without any alloying, the SCE system behaves like a *random alloy*. The stable $f^3$ sites break up the ability of the rapidly fluctuating $f^2/f^3$ sites to establish a coherent lattice state. Once the increase of hybridization destroys all of the stable $f^3$ sites, the correlated ultra-narrow band lattice state abruptly sets in. A significantly enhanced electronic specific heat and Pauli susceptibility will occur, which in a specific case may be sufficient to be labeled heavy fermion behavior. As hybridization increases further these bands will broaden, perhaps quite nonlinearly.

Once we recognize this *"random alloy" versus "correlated lattice banding (bonding) state" competition*, the winner of which is decided by the magnitude and nature of the *f*-band hybridization mechanism as treated via our generalization of the original Coqblin-Schrieffer picture, we can ask whether there is evidence for the wide applicability of this picture to SCE phenomenology. Three important instances come to mind. One is found in the behavior of the 1·2·3 copper-oxide-based high-temperature superconductors when praseodymium or cerium is substituted for yttrium (see the paper of L. Soderholm in this volume). The second is found in the behavior of the colossal magnetoresistance (CMR) perovskite manganese oxides. The third instance is found in the highly unusual thermo-structural behavior of elemental plutonium and the stabilization of its nominally face-centered-cubic δ phase at low temperatures, in preference to the monoclinic α phase, by the addition of small amounts of a trivalent III B additive such as gallium.

For the 1·2·3 superconductors, as discussed in Reference 16, the disappearance of superconductivity and the appearance of Pr magnetic ordering points to competing hybridization mechanisms determining the choice between magnetic and superconducting behavior. Thus the orbitally driven praseodymium-praseodymium magnetic coupling, resulting from the hybridization of Pr *f* electrons with oxygen-*p*-derived band electrons, can explain the relatively high magnetic ordering temperature of the 100% Pr system and the low ordered moment. Depriving the copper *d* electrons from hybridizing with these same oxygen *p* electrons then kills the superconductivity. Interestingly, this picture lends some support to one class of the proposed superconductivity mechanisms. This is the class of mechanism, such as the interlayer tunneling (ILT) model[22,23] of Anderson, depending on interlayer interactions. This is because the ability to control the process via the yttrium site implies the centrality of interlayer coupling.

A central feature of ideas and theories for understanding CMR behavior in doped manganese oxides is the ferromagnetic double exchange interaction between $Mn^{3+}$ and $Mn^{4+}$ ions. In fact, the motivation for the original ideas of Zener and the development of these ideas into the theory of double exchange[24] by de Gennes was provided by the behavior of systems such as $(La_{1-x}Ca_x)(Mn_{1-x}^{3+}Mn^4_x)O_3$. At both ends of the composition range, the manganite series $(La_{1-x}Ca_x)(Mn_{1-x}^{3+}Mn^4_x)O_3$ is an antiferromagnetic insulator. However, at intermediate compositions, both a dramatic increase in conductivity and the appearance of ferromagnetism occur. This occurrence is associated with the presence of mixed valence behavior, i.e., $Mn^{3+}$ and $Mn^{4+}$ occurring at crystallographically equivalent sites to compensate for the presence of both trivalent lanthanum and divalent calcium. This double exchange mechanism,[24] associated with mixed valence situations in the doped manganese oxides and other such systems, is a kinetic ferromagnetic interaction mediated through charge transfer, e.g. between $Mn^{3+}$ and $Mn^{4+}$ via a $Mn^{3+}$ $-O^{2-}$ $-$ $Mn^{4+}$ coupling. Double exchange provides a ferromagnetic interaction between two different valances within a static random distribution. The generalized Coqblin-Schrieffer interaction may add a dynamic enhancement of this coupling.

In the case of the δ Pu stabilization, presumably the few percent of randomly located trivalent additive nucleates the breakup of the coherent *f* bonding that stabilizes the monoclinic α structure. This breakup of coherence would result from the Pu *f*-bonding possibilities being tied up in a geometrically random pattern of hybridization with non-*f* electrons from the additive species, e.g., Ga. Given the highly nonlinear nature of the effects of alloying in driving the restructuring of the ground state, this nucleation is likely to give rise to an avalanche effect, with the formation of randomly located stable $f^5$ Pu sites further breaking up the *f* bonding coherency between the fluctuating $f^5/f^4$ Pu sites. (For many years there has been a fruitless attempt to find ordered magnetism in the stabilized δ Pu alloys. Presumably, by randomly breaking up the hybridization-mediated coupling, the bonding to the trivalent impurities destroys the coherent magnetic coupling between the stable $f^5$ sites.) This avalanche effect may relate to the behavior found in the experimental extended x-ray absorption fine structure (EXAFS) and x-ray diffraction work of Faure et al[25] which provides both Ga-Pu and Pu-Pu interatomic distances as Ga content varies over a range between 1.89 and 10.43 at %. Each substitutional gallium atom creates a Ga-Pu cluster where the 12 nearest neighbor Ga-Pu bonds are shortened. The observed bond shortening has a minimum near or at the 7.7% Ga corresponding to one Ga for every 12 Pu's. The x-ray diffraction data and the Pu-Pu EXAFS data provide an identical negative deviation from Vegard's law (defined by the straight line between the end points defined by the extrapolation to zero Ga and $Pu_3Ga$) over the composition range to 10.43 at % Ga. Faure et al state that increasing correlation between Ga-Pu clusters offers a qualitative explanation for the decrease of the Ga-Pu bond shortening up to 7.7 at % Ga, but state that they cannot understand why there is then a substantial increase in bond shortening on going to 10.4 at % Ga. It is possible that the avalanche effect provides an explanation for the increase in bond shortening beyond 7.7 at % by providing a reinforcement by the effect of the Ga nucleation centers in breaking up the Pu-Pu bonding.

A very important consequence of our *"random alloy" picture* is that it *explains the stabilization of the δ phase with increasing temperature for pure elemental plutonium*, i.e. this transition to δ phase occurs because pure δ plutonium is a solid solution (an inhomogenous mixture) of stable $f^5$Pu and fluctuating $f^4/f^5$ Pu sites. In fact, since we are dealing with quantum mechanical probability densities, this distribution is spatially time varying. The configurational entropy associated with the randomness then provides a free energy decrease that by the temperature of 319 C stabilizes the δ phase.

**Acknowledgments**

This research was supported by National Science Foundation Grant No. DMR-91-20333. We have benefited from discussion of their data with F. Bourdarot, P. Burlet, and O. Vogt, and we appreciate their permission to quote their results. Invaluable support was provided by the association of one of us (BRC) with Los Alamos National Laboratory through the Seaborg Institute and the Center for Materials Science and especially by A.C. Lawson, J.L. Smith, and M.F. Stevens of LANL.

**References**

1. Q.G. Sheng, B.R. Cooper, and S.P. Lim, Phys. Rev. B 50, 9215 (1994).
2. Q.G. Sheng and B.R. Cooper, Phys. Rev. B 50, 965 (1994).
3. J. Schoenes, O. Vogt, J. Löhle, F. Hulliger, and K. Mattenberger, Phys. Rev. B 53, 14,987 (1996); F. Bourdarot, P. Burlet, R. Calemczuk, F. Lapierre, K. Mattenberger, and O. Vogt, J. de Actinides, Dijon, France, April, 1997.
4. Q.G. Sheng and B.R. Cooper, Phil. Mag. Lett. 72, 123 (1995).
5. Q.G. Sheng and B.R. Cooper, J. Magn. Magn. Mater. 164, 335 (1996).
6. B.R. Cooper, Y.-L. Lin, and Q.G. Sheng, J. Appl. Phys. 81, 3856 (1997).
7. B.R. Cooper and Y.-L. Lin, J. Appl. Phys. 83, 6432 (1998).
8. H. Brooks, Ph.D. thesis, Harvard University, 1940.
9. B. Coqblin and J.R. Schrieffer, Phys. Rev. 185, 847 (1969).
10. See B.R. Cooper, R. Siemann, D. Yang, P. Thayamballi, and A. Banerjea, Chapter 6 in *Handbook on the Physics and Chemistry of the Actinides*, edited by A.J. Freeman and G.H. Lander, Elsevier Science Publishers, Amsterdam, 1985.
11. B.R. Cooper, G.-J. Hu, N. Kioussis, and J.M. Wills, J. Magn. Magn. Mater. 63-64, 121 (1987); B.R. Cooper, J. Less-Common Metals 133, 31 (1987); B.R. Cooper, J.M. Wills, N. Kioussis, and Q.G. Sheng, J. de Phys. 49, C8-463 (1988).
12. J.M. Wills and B.R. Cooper, Phys. Rev. B 36, 3809 (1987).
13. C. Sanchez-Castro, B.R. Cooper, and K.S. Bedell, Phys. Rev. B 51, 12506 (1995).
14. J.-M. Fournier and R. Troc, Chapter 2 "Bulk properties of the actinides" in *Handbook on the Physics and Chemistry of the Actinides*, Volume 2 (eds A.J. Freeman and G.H. Lander, Elsevier Science Publishers, Amsterdam, 1985).
15. O. Vogt and K. Mattenburger, Chapter 114 "Magnetic measurements on rare earth and actinide monopnictides and monochalcogenides" in volume 17 of the *Handbook on the Physics and Chemistry of Rare Earths* (eds K.A. Gschneidner, Jr., L. Eyring, G.H. Lander, and G.R. Chopin, Elsevier Science Publishers, Amsterdam, 1993).
16. B.R. Cooper, "Magnetism versus high temperature superconductivity—the role of competing hybridization" in *Proceedings of High-Tc Superconductors: Magnetic Interactions*, pp 7-16 (eds. L.H. Bennett, Y. Flom, G.C. Vezzoli, World Publishing, Singapore, 1989).
17. D.L. Price and B.R. Cooper, Phys. Rev. B 39, 4945 (1989).

18. D.L. Price, J.M. Wills, and B.R. Cooper, Phys. Rev. B 48, 15,301 (1993).
19. P. Link, U. Benedict, J. Wittig, and H. Wühl, J. Phys.: Condens. Matter 4, 5585 (1992). P. Link, U. Benedict, J. Wittig, and H. Wühl, Physica B 190, 68 (1993). A.L. Cornelius, J.S. Schilling, O. Vogt, K. Mattenberger, and U. Benedict, J. Magn. Magn. Mater. 161, 169 (1996).
20. O. Vogt, private communication (1998).
21. J.M. Wills and B.R. Cooper, Phys. Rev. B 42, 4682 (1990).
22. P.W. Anderson, *The Theory of Superconductivity in the High $T_c$ Cuprate Superconductors* (Princeton U. Press, Princeton, 1997); Science 279, 1196 (1998).
23. A.J. Leggett, Science 279, 1157 (1998).
24. C. Zener, Phys. Rev. 82, 403 (1951); P.G. DeGennes Phys. Rev. 18, 141 (1960).
25. Ph. Faure, B. Deslanders, D. Bazen, C. Tailland, R. Doukhan, J.M. Fournier, and A. Falanga, J. of Alloys and Compounds 244, 131 (1996).

# EFFECT OF DISORDER IN THE PERIODIC ANDERSON MODEL

Feng Chen and Nicholas Kioussis

Department of Physics and Astronomy
California State University
Northridge, CA 91324-8268

## ABSTRACT

We have studied the one-dimensional periodic, symmetric Anderson model at half filling in the presence of disorder using finite-temperature quantum Monte Carlo techniques. We have examined for the first time the disorder both in hybridization between the local *f*-orbitals and the conduction electrons and in the local *f*-site energy, using a uniform distribution. We find that the disorder in hybridization has a dramatic effect on the low-temperature magnetic properties giving rise to a non-Fermi-liquid behavior, and that the magnetic susceptibility can be scaled by a power law with an exponent that is comparable with recent experiments. On the other hand, disorder in the local *f*-orbital energy level does not show a non-Fermi liquid behavior.

The last several years have witnessed dramatic growth in both the experimental and theoretical activity in a new class of heavy fermion materials that exhibit non-Fermi-liquid (NFL) behavior in their physical properties at low temperatures. In several cases of heavy-fermion compounds, such as $Y_{1-x}U_xPd_3$[1] and $Ce_{1-x}Th_xRhb$[2], NFL behavior occurs only when the *f*-electron materials, consisting primarily of Ce or U intermetallics, have been alloyed with a nonmagnetic element. Several models have been proposed to explain these experimental results[3-6]. All these calculations strictly apply to systems with a *dilute* collection of Kondo centers. The inclusion of lattice effects in the presence of disorder presents an additional challenge that has only recently started to be addressed[7-9].

We present here the first investigation of the effects of disorder on the magnetic properties of the one-dimensional periodic Anderson model (PAM) at half filling using quantum Monte Carlo techniques[10]. The quantum Monte Carlo calculations treat both the RKKY and the Kondo interactions on an equal footing, and allow one to treat both the weak and strong disorder regimes.

The Hamiltonian for the one-dimensional PAM with disorder is

*Electron Correlations and Materials Properties*, edited by Gonis *et al.*
Kluwer Academic/Plenum Publishers, 1999. 

$$H = -t\sum_{i,\sigma}(c^{+}_{i,\sigma}c_{i+1,\sigma} + H.c.) + E^{c}\sum_{i,\sigma} n^{c}_{i\sigma} + \sum_{i,\sigma} V_i(f^{+}_{i,\sigma}c_{i,\sigma} + H.c.)$$
$$+U_f\sum_{i}(n^{f}_{i\uparrow} - \tfrac{1}{2})(n^{f}_{i\downarrow} - \tfrac{1}{2}) + \sum_{i,\sigma} E^{f}_{i} n^{f}_{i\sigma}. \qquad (1)$$

The important thing to notice in Eq.(1) is that, unlike the usual PAM, the local *f*-site parameters $V_i$ or $E^f_i$ are taken here to be random numbers distributed according to uniform distributions of width 2Δ centered at $V_0$ or $E^f_0$ respectively. In the following we will set t=1 and consider the half filling ($E_c$=0) and symmetric ($E^f_0$=-U/2) case.

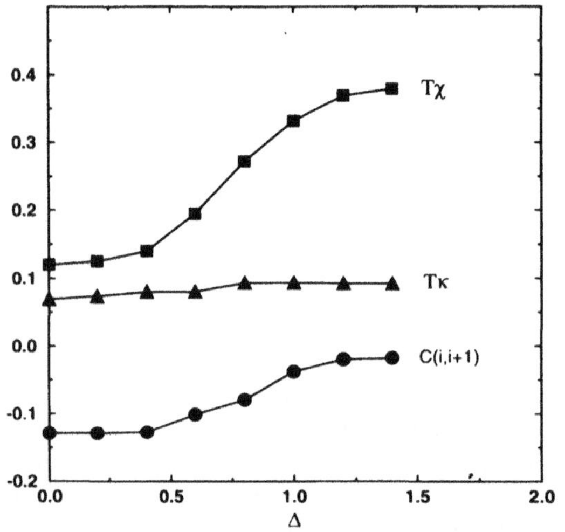


**Fig.1** The *f* magnetic susceptibility *Tχ* (solid squares), the charge susceptibility *Tκ* (sold triangles) and the nearest-neighbor *f-f* spin correlation function *C(i,i+1)* (solid triangles) as a function of the hybridization-disorder width *Δ* at temperature *T/t=1/8* with *U/t=1* and *V/t=1*.

We have employed a finite-temperature quantum Monte Carlo technique with an exact updating procedure. The calculated physical observables are averaged over a sufficient number of disorder realizations. We first study the case of disorder in hybridization. In Fig.1 we plot the impurity magnetic susceptibility Tχ, the charge compressibility Tκ, and the nearest-neighbor spin correlation function *C(i,i+1)* as a function of the disorder width Δ at low temperatures, β=8. We find that Tχ, which measures the "effective" magnetic moment, increases with increasing disorder width. The low value of the effective moment for Δ=0 at low temperatures indicates that the magnetic correlation act to screen the *f*-site moment leading to a singlet ground state. Turning on the disorder in hybridization results in a destruction of the singlet ground-state and a reduction of the spin gap between the singlet ground state and the lowest lying excited triplet state. Fig.1 shows also that *C(i,i+1)* increases monotonically with Δ. These short-range *f-f* antiferromagnetic correlations have been found to be as important as the *f-c* correlations in the formation of the singlet state in the PAM[11]. Consistent with the above results, we find that increase of disorder in hybridization tends to suppress these short-range antiferromagnetic *f-f* correlations, resulting in non-compensated *f* moments. On the other hand, Tκ is almost unchanged by disorder in hybridization. This suggests that the charge gap might be due to correlation effects rather than hybridization effects.

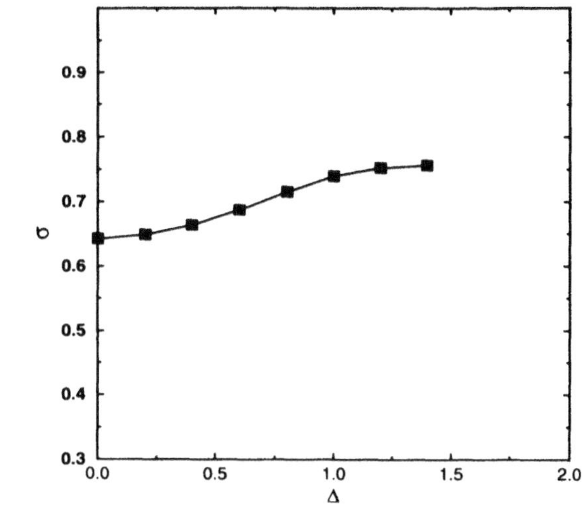


**Fig.2** The square of the *f*-orbital local moment σ as a function of the hybridization-disorder width Δ at temperature $T/t=1/8$ with $U/t=1$ and $V/t=1$.

In Fig.2 we plot the impurity magnetic moment σ as a function of disorder width Δ at low temperature, β=8. We find that σ increases as Δ increases. This result shows that the number of the *unquenched* local spins at low temperatures increases with disorder.

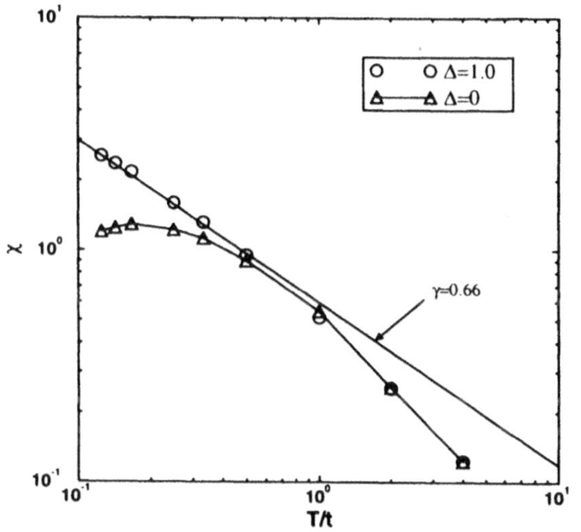


**Fig.3** The uniform *f* magnetic susceptibility as a function of temperature for the hybridization-disorder width $\Delta=0$ (triangles) and $\Delta=1$ (circles) for $U/t=1$ and $V/t=1$. For $T/t\leq 1$, χ can be scaled by the power law, χ~T-γ, with γ=0.66, in comparable with experiment[1].

We present in Fig.3 the *f* magnetic susceptibility χ as a function of temperature for Δ=0 (PAM) and Δ=1. In contrast to the non-disordered case, χ exhibits a non-Fermi-liquid behavior, diverging at low temperatures. We find that χ can be scaled at low temperatures by a power law, $\chi\sim T^{\gamma}$, with the exponent γ=0.66, which is comparable with recent experimental results[1]. This result might be interpreted that the hybridization-disorder causes some of the local $f$ sites to possess low Kondo temperatures and consequently to behave as nearly free magnetic moments, leading to a NFL behavior.

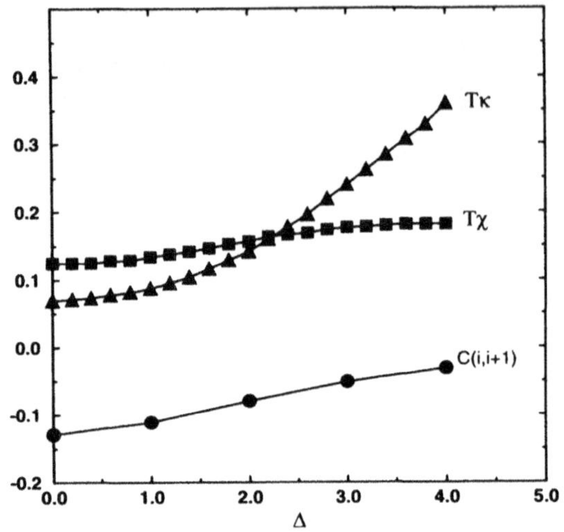


**Fig.4** The $f$ magnetic susceptibility $T\chi$ (solid squares), the charge susceptibility $T\kappa$ (sold triangles) and the nearest-neighbor $f$-$f$ spin correlation function $C(i,i+1)$ (solid triangles) as a function of the energy-level-disorder width $\Delta$ at temperature $T/t=1/8$ with $U/t=1$ and $V/t=1$.

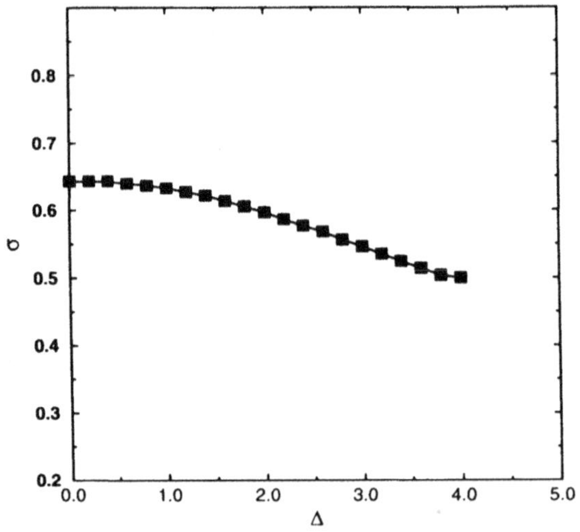


**Fig.5** The square of the $f$-orbital local moment σ as a function of the energy-level-disorder width Δ at temperature $T/t=1/8$ with $U/t=1$ and $V/t=1$.

We have also studied the effect of disorder in the $f$-electron energy level, $E_f$, on the magnetic properties. In Fig.4, we plot the impurity magnetic susceptibility $T\chi$, the charge compressibility $T\kappa$, and the nearest-neighbor spin correlation function $C(i,i+1)$ as a function of the disorder width Δ at low temperatures, β=8. In Fig.5, we plot the impurity magnetic moment σ as a function of disorder width Δ. We find that in the weak-disorder regime, $\Delta/t \leq 1$, all quantities are almost unchanged by the disorder in $E_f$, in contrast to the case of disorder in hybridization. This is due to the fact that the majority of the *f* sites are occupied, since the average *f*-electron energy level is below the Fermi energy ($E^f_0$=-U/2).

**Acknowledgments** This work was supported through the National Science Foundation under Grant No.DMR-9531005 and the Office of Research and Sponsored Projects at California State University Northridge.

**References**

1. M.C. de Andrade *et al.*, cond-mat/9802081, (1998).
2. B.Andraka and G.R. Stewart, *Phys. Rev.* **B** 47:3208(1993); B.Andraka, *Phys. Rev.* **B** 49:348(1994).
3. T.S. Kim and D.L. Cox, *Phys. Rev. Lett.* 75:1622(1995);
4. R. Bulla, A.C. Hewson, *Physica* **B** 230-232:627(1997).
5. R.N. Bhatt and D.S. Fisher, *Phys. Rev. Lett.* 68:3072 (1992)
6. V. Dobrosavljevic, T.R. Kirkpatrick, and G. Kotliar, *Phys. Rev. Lett.* 69:1113 (1992).
7. E. Miranda, V. Dobrosavljevic and G. Kotliar, *Phys. Rev. Lett.* 78:290(1997).
8. A. Chattopadhyay and M. Jarrell, *Phys. Rev.* **B** 56:2920 (1997)
9. A.H. Castro Neto, G. Castilla, and B.A. Jones, *cond- mat/9710123.*
10. J.E. Hirsch, and R.M. Fye, *Phys. Rev. Lett.* 56:2521 (1986); J.E. Gubernatis, J.E. Hirsch, and D.J. Scalapino, *Phys. Rev.* **B** 35:8478 (1987).
11. R. Blankenbecler, J.R. Fulco, W. Gill, and D.J. Scalapino, *Phys. Rev. Lett.* 58:411(1987).

# Discussion

**Sawatzky:** I don't understand how you define a Fermi liquid since a 1D system as in your model calculation can never be a Fermi liquid in the sense of having a discontinuity in the n(k) (the one electron occupation number) at the Fermi surface, which in this case, would consist of two points. If we define a Fermi liquid as being a system in which the Imaginary part of the self energy goes to zero at the Fermi surface and that there is a discontinuity in n(k) (required to define the Fermi surface in the first place), then your 1D system can never be a Fermi liquid.

**Kioussis** We have defined a Fermi liquid as one which has the same temperature dependent transport properties like the electrical conductivity (~T squared), the magnetic susceptibility (T ingredient), and the specific heat (linear in T) as a free electron gas. This is the definition used by most in the field of heavy Fermions.

# DYNAMICAL ELECTRON CORRELATIONS IN METALS: TB-LMTO AND MULTIBAND HUBBARD HAMILTONIAN

Václav Drchal, Václav Janiš, and Josef Kudrnovský

Institute of Physics, Academy of Sciences of the Czech Republic
Na Slovance 2, CZ-180 40 Praha 8, Czech Republic

## INTRODUCTION

A general scheme for determination of the electronic structure of solids with correlated electrons is still missing. The density functional formalism within the local density approximation (LDA), in which the electron-correlation part is guessed on the basis of the homogeneous electron gas model, has proved to be a highly reliable method for the evaluation of the ground state properties[1]. On the other hand, there are many examples of the failure of the LDA, e.g., the study of the excitation spectra of solids or the evaluation of the gap in insulators and semiconductors. Another example of the failure of the LDA are solids whose electronic structure is better described in terms of atomic-like electronic states rather than in terms of the homogeneous electron gas model on which the LDA is based.

The electronic properties of strongly correlated materials which cannot be adequately described within the LDA are usually studied in the framework of simplified models like, e.g., the single-band Hubbard model (for a recent review see Ref. 2). The correlation effects in solids are usually classified in terms of the ratio of the on-site Coulomb energy $U$ to the bandwidth $w$. One distiguishes three regimes, namely, (i) the weak interaction case ($U/w < 1$, transition metals), (ii) the intermediate interaction case ($U/w \approx 1$, metal-insulator transition regime and Kondo systems), and (iii) the strong interaction case ($U/w > 1$, rare-earth systems and wide gap solids). It should be mentioned that a general approach bridging all the above mentioned cases is still missing even at the model level.

Recently some progress has been made in the reformulation of the LDA (the time-dependent LDA[3]) which is useful in the study of excitation energies. The so-called GW approximation (GWA)[4] was successfully applied to the gap problem in a number of semiconductors and insulators. An alternative approach to the gap problem which is based on a straightforward generalization of the LDA is the so-called LDA+U method[5]. The GWA was also applied to metallic systems. Although computationally highly demanding, it yielded band narrowing in the photoemission spectra of nickel[6],

*Electron Correlations and Materials Properties*, edited by Gonis *et al.*
Kluwer Academic/Plenum Publishers, 1999.

but it failed to produce the well-known satellite structure below the main peak[7]. The presence of the satellite in Ni was succesfully explained in the framework of the $T$-matrix formalism developed recently on *ab initio* level (see Ref. 8 and references therein for previous semiempirical approaches to this problem). An alternative approach to the satellite and the band narrowing in nickel is based on a three-body scattering approximation that employs the Faddeev equations[9,10].

In a recent paper[11] the so-called LDA++ method was introduced which extends the LDA+U method by accounting for dynamical electron correlations. The starting point of the LDA++ method is a multiband Hubbard Hamiltonian whose one-electron part is identified with the LDA Hamiltonian with a subtracted double counting correction for the average Coulomb interactions in the LDA. The LDA++ approximates solutions in different correlation limits by different many-body approximations, namely it employs the so-called Hubbard I solution[12] in the strong-interaction limit, the iterated perturbation theory (IPT) within the dynamical mean-field theory (DMFT)[2] for the intermediate-interaction case, and the so-called fluctuation exchange approximation (FLEX)[13] in the weak-interaction limit.

In the present paper we follow the basic approach of the LDA++ but with some conceptual as well as computational differences. In particular, the main effort is directed towards a formulation of the problem in a way suitable for future generalizations to random alloys and their surfaces and interfaces. This means that the formalism is strictly based on Green functions which allow for configurational averaging in the case of alloys. The starting point is a multiband Hubbard Hamiltonian (MBHH) whose parameters are naturally determined from the corresponding tight-binding linear muffin-tin orbital (TB-LMTO) Hamiltonian (for a review of the TB-LMTO method and its applications to random surfaces and interfaces see a recent book, Ref. 14). The basic approximation adopted here is the assumption on the local (site-diagonal, or wave-vector independent) selfenergy which is reasonably well justified for transition metals and their alloys[15]. The local approximation for the selfenergy is also required from the formal point of view for the extension of the theory to the case of random alloys within the coherent potential approximation (CPA)[16]. We employ the single-channel approximations (FLEX) of the canonical perturbation theory to solve the many-body part of the problem.

## THEORY

The electronic structure determined within the LDA is described by a TB-LMTO Hamiltonian with the overlap matrix equal to unity

$$H^{\mathrm{LMTO}} = \sum_{\mathbf{R}\lambda,\mathbf{R}'\lambda'} |\mathbf{R}\lambda\rangle \, H^{\mathrm{LMTO}}_{\mathbf{R}\lambda,\mathbf{R}'\lambda'} \, \langle \mathbf{R}'\lambda'| \,, \tag{1}$$

where

$$H^{\mathrm{LMTO}}_{\mathbf{R}\lambda,\mathbf{R}'\lambda'} = C_{\mathbf{R}\lambda\lambda'} \, \delta_{\mathbf{R}\mathbf{R}'} + \Delta^{1/2}_{\mathbf{R}\lambda} \, S^{\gamma}_{\mathbf{R}\lambda,\mathbf{R}'\lambda'} \, \Delta^{1/2}_{\mathbf{R}'\lambda'} \,. \tag{2}$$

Here, $\mathbf{R}$ is the site index, $\lambda = (L\sigma) = (\ell m \sigma)$ is the spinorbital index, $L = (\ell m)$ is the orbital index, $\sigma$ is the $z$-projection of the spin, $C$, $\Delta$, and $\gamma$ are site-diagonal matrices of potential parameters and $S^{\gamma}$ is the matrix of structure constants in the orthogonal LMTO representation

$$S^{\gamma}_{\mathbf{R}\lambda,\mathbf{R}'\lambda'} = [S^0(1-\gamma S^0)^{-1}]_{\mathbf{R}\lambda,\mathbf{R}'\lambda'} = \left[ S^{\beta}\left(1-(\gamma-\beta)S^{\beta}\right)^{-1} \right]_{\mathbf{R}\lambda,\mathbf{R}'\lambda'} . \tag{3}$$

Here $S^0$ is a matrix of canonical structure constants, $S^{\beta}$ is a matrix of screened structure constants, and $\beta$ is a site-diagonal matrix of screening constants.

The parameters of a multiband Hubbard Hamiltonian in second quantization with creation ($a^+_{\mathbf{R}\lambda}$) and destruction $a_{\mathbf{R}\lambda}$ operators

$$H^{\rm Hubb} = \sum_{\mathbf{R}\lambda,\mathbf{R}'\lambda'} t_{\mathbf{R}\lambda,\mathbf{R}'\lambda'}\, a^+_{\mathbf{R}\lambda}\, a_{\mathbf{R}'\lambda'} + \sum_{\mathbf{R},\lambda,\lambda'} U_{\mathbf{R}\lambda\lambda'}\, n_{\mathbf{R}\lambda}\, n_{\mathbf{R}\lambda'} \qquad (4)$$

are found from the LDA calculations. The density operator $n_{\mathbf{R}\lambda} = a^+_{\mathbf{R}\lambda}\, a_{\mathbf{R}\lambda}$. The hopping integrals

$$t_{\mathbf{R}\lambda,\mathbf{R}'\lambda'} = \Delta^{1/2}_{\mathbf{R}\lambda}\, S^{\gamma}_{\mathbf{R}\lambda,\mathbf{R}'\lambda'}\, \Delta^{1/2}_{\mathbf{R}'\lambda'}\,, \quad \mathbf{R} \neq \mathbf{R}' \qquad (5)$$

are identified with the site off-diagonal elements of (1), while the atomic levels $t_{\mathbf{R}\lambda,\mathbf{R}\lambda'} = \epsilon_{\mathbf{R}\lambda\lambda'} = \epsilon_{\mathbf{R}\lambda}\delta_{\lambda,\lambda'}$ are determined from the condition that the average occupancy $\bar{n}_{\mathbf{R}\lambda}$ of the state $|\mathbf{R}\lambda\rangle$ as calculated within the LDA and within the Hartree-Fock approximation for multiband Hubbard Hamiltonian (4) are identical. It means

$$\epsilon_{\mathbf{R}\lambda} = C_{\mathbf{R}\lambda} - \Sigma^{\rm HFA}_{\mathbf{R}\lambda} = C_{\mathbf{R}\lambda} - \sum_{\lambda'} U_{\mathbf{R}\lambda\lambda'}\, \bar{n}_{\mathbf{R}\lambda'}\,. \qquad (6)$$

The Hubbard interaction parameter $U_{\mathbf{R}\lambda\lambda'}$ is determined within the LDA usually as a second derivative of $E_{tot}$ with respect to occupancies of the interacting states $\lambda, \lambda'$. Due to the Pauli principle, $U_{\mathbf{R}\lambda\lambda'} = U_{\mathbf{R}\lambda\lambda'}(1 - \delta_{\lambda\lambda'})$. Here we consider translationally invariant solids with one atom per unit cell. The hopping integrals then depend only on the difference $\mathbf{R} - \mathbf{R}'$. The on-site levels $\epsilon_{\mathbf{R}\lambda}$ as well as the pair interactions $U_{\mathbf{R}\lambda\lambda'}$ are independent of $\mathbf{R}$.

We study four different approximations of the canonical many-body perturbation theory, namely, (i) the second order perturbation theory (SOPT), and three approximations of infinite order that correspond to electron-electron (hole) scatterings in a single channel, namely, (ii) the T-matrix approximation (TMA) which describes electron-electron scatterings, (iii) the random-phase approximation (RPA) corresponding to electron-hole scatterings, and (iv) the GW-approximation (GWA) in which the bare interaction $U$ is replaced by a (complex and energy-dependent) interaction $W(E)$ (double wavy line) renormalized by repeated excitations of electron-hole pairs (polarization bubbles). The graphs for one-particle selfenergy are shown in Fig. 1. We assume

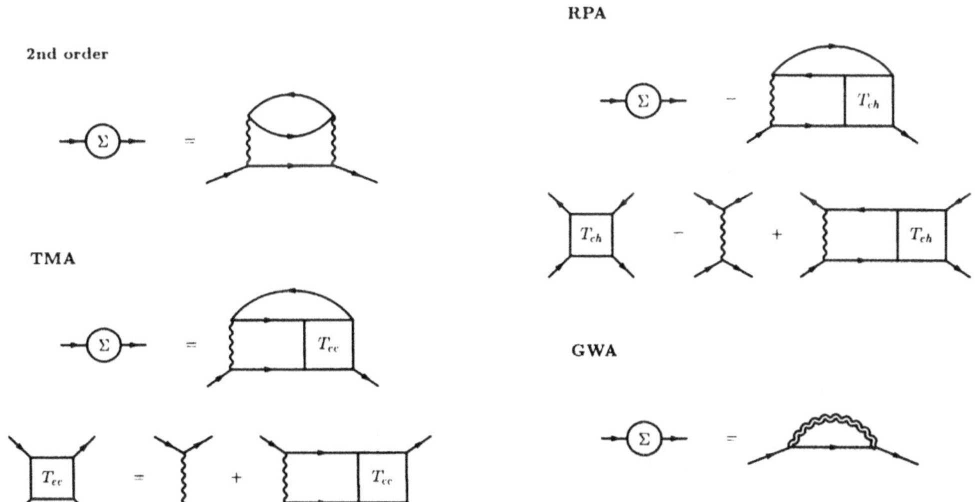


**Figure 1.** Graphs for one-particle selfenergy in second order perturbation theory (SOPT), T-matrix approximation (TMA), random-phase approximation (RPA), and GW-approximation (GWA).

a local approximation, i.e., the selfenergy is independent of the **k**-vector, which means that it is diagonal in the site representation, $\Sigma_{\mathbf{R}\lambda,\mathbf{R}'\lambda'} = \Sigma_{\mathbf{R},\lambda\lambda'}\delta_{\mathbf{R},\mathbf{R}'}$. We use the technique of causal Green functions at $T = 0$ in our calculations. It follows from the adopted construction of the on-site levels in (6) that the Hartree-Fock part of the one-electron selfenergy is already contained in the atomic levels $\epsilon_{\mathbf{R}\lambda}$. Consequently, the expansions of the selfenergy start from the terms quadratic in the Hubbard $U$. The unperturbed one-particle Green function corresponding to the Hartree-Fock approximation (HFA) applied to $H^{\mathrm{Hubb}}$ is identical with the Green function of $H^{\mathrm{LMTO}}$.

We assume a cubic symmetry of the lattice and confine ourselves to $s$, $p$, and $d$ states. In addition, we neglect the pair correlations in $s$ and $p$ states and for simplicity we assume $U_{\lambda\lambda'} = U(1-\delta_{\lambda\lambda'})\delta_{\ell,2}$, i.e., we neglect exchange interactions and consider only the pair interactions between the $d$-states. The Green functions and selfenergy are then diagonal in the spinorbital indices $\lambda$. The selfenergy can acquire only two values that correspond to $e_g$ and $t_{2g}$ representations of the cubic point group.

As an example we give here the derivation of basic equations for the TMA which is an extension of the derivation given in Ref. 17 to the multiband case. The basic equations of the TMA in the language of causal quantities (superscript $c$) read

$$i\Psi^{(c)}_{\lambda\lambda'}(E) = -\int_{-\infty}^{\infty}\frac{d\omega}{2\pi}\,G^{(c)}_{\lambda}(E-\omega)\,G^{(c)}_{\lambda'}(\omega)\,, \tag{7}$$

$$T^{(c)}_{\lambda\lambda'}(E) = \frac{U}{1-U\Psi^{(c)}_{\lambda\lambda'}(E)} - U\,, \tag{8}$$

$$-i\Sigma^{(c)}_{\lambda}(E) = \sum_{\lambda'}(-1)\int_{-\infty}^{\infty}\frac{d\omega}{2\pi}\,T^{(c)}_{\lambda\lambda'}(E+\omega)\,G^{(c)}_{\lambda'}(\omega)\,, \tag{9}$$

where $\Psi^{(c)}_{\lambda\lambda'}(E)$ is the two-particle propagator, $T^{(c)}_{\lambda\lambda'}(E)$ is the electron-electron T-matrix, and $\Sigma^{(c)}_{\lambda}(E)$ is the one-electron selfenergy. The factor $(-1)$ in (9) is due to a closed fermion loop. In numerical calculations, it is advantageous to replace the causal quantities by the retarded ones (without superscript) as follows

$$\Psi^{(c)}_{\lambda\lambda'}(E) = \mathrm{Re}\Psi_{\lambda\lambda'}(E) + i\,\mathrm{sgn}(E-2\mu)\,\mathrm{Im}\Psi_{\lambda\lambda'}(E)\,, \tag{10}$$

$$T^{(c)}_{\lambda\lambda'}(E) = \mathrm{Re}T_{\lambda\lambda'}(E) + i\,\mathrm{sgn}(E-2\mu)\,\mathrm{Im}T_{\lambda\lambda'}(E)\,, \tag{11}$$

$$\Sigma^{(c)}_{\lambda}(E) = \mathrm{Re}\Sigma_{\lambda}(E) + i\,\mathrm{sgn}(E-\mu)\,\mathrm{Im}\Sigma_{\lambda}(E)\,, \tag{12}$$

$$G^{(c)}_{\lambda}(E) = \mathrm{Re}G_{\lambda}(E) + i\,\mathrm{sgn}(E-\mu)\,\mathrm{Im}G_{\lambda}(E)\,. \tag{13}$$

By inserting (10)-(13) into (7)-(9) we find equations for the retarded quantities which are holomorphic in the upper halfplane of the complex energy $z$ and vanish for $z \to 0$. It is therefore sufficient to calculate only their imaginary parts because the real parts can be found using a dispersion relation

$$X(z) = \int_{-\infty}^{\infty}\frac{d\omega}{\pi}\,\frac{\mathrm{Im}X(\omega)}{\omega - z}\,, \tag{14}$$

which allows to determine the real and imaginary parts of $X(z)$ for complex $z$ as well as for real values of the energy. The equations for the retarded quantities read

$$\mathrm{Im}\Psi_{\lambda\lambda'}(E) = -\mathrm{sgn}(E-2\mu)\int_{E-\mu}^{\mu}\frac{d\omega}{\pi}\,\mathrm{Im}G_{\lambda}(E-\omega)\,\mathrm{Im}G_{\lambda'}(\omega)\,, \tag{15}$$

$$T_{\lambda\lambda'}(E) = \frac{U}{1-U\Psi_{\lambda\lambda'}(E)} - U\,, \tag{16}$$

$$\mathrm{Im}\Sigma_\lambda(E) = -\sum_{\lambda'} \int_{2\mu-E}^{\mu} \frac{\mathrm{d}\omega}{\pi} \, \mathrm{Im}T_{\lambda\lambda'}(E+\omega)\, \mathrm{Im}G_{\lambda'}(\omega)\,. \tag{17}$$

In the derivation of (15)-(17) we have used analytic properties of the retarded quantities from which follow useful relations between convolutions of real and imaginary parts, as, for example

$$\int_{-\infty}^{\infty} \frac{\mathrm{d}\omega}{2\pi} \, \mathrm{Re}G_\lambda(E-\omega)\, \mathrm{Re}G_{\lambda'}(\omega) = -\int_{-\infty}^{\infty} \frac{\mathrm{d}\omega}{2\pi} \, \mathrm{Im}G_\lambda(E-\omega)\, \mathrm{Im}G_{\lambda'}(\omega)\,. \tag{18}$$

The equations for the other approximations can be derived in a similar way. An expression for the renormalized pair interaction $W(E)$ requires matrix inversion even in the simplified case $U_{\lambda\lambda'} = U(1-\delta_{\lambda\lambda'})$ assumed here, since

$$\sum_{\lambda''} \left[\delta_{\lambda,\lambda''} - U_{\lambda\lambda''}\Phi_{\lambda''}(E)\right] W_{\lambda''\lambda'}(E) = U_{\lambda\lambda'}\,, \tag{19}$$

where $\Phi_\lambda(E)$ is the electron-hole bubble and

$$\mathrm{Im}\Phi_\lambda(E) = -\int_{\mu-E}^{\mu} \frac{\mathrm{d}\omega}{\pi} \, \mathrm{Im}G_\lambda(E+\omega)\, \mathrm{Im}G_\lambda(\omega)\,. \tag{20}$$

The retarded one-electron Green function of the interacting system is given by the resolvent

$$G(z) = [z - H^{\mathrm{eff}}(z)]^{-1} \tag{21}$$

of an effective one-electron Hamiltonian with the matrix elements

$$H^{\mathrm{eff}}_{\mathbf{R}\lambda,\mathbf{R}'\lambda'}(z) = H^{\mathrm{LMTO}}_{\mathbf{R}\lambda,\mathbf{R}'\lambda'} + \Sigma_\lambda(z)\, \delta_{\mathbf{R}\mathbf{R}'}\, \delta_{\lambda\lambda'}\,. \tag{22}$$

The density of states (DOS) is defined as

$$\rho(E) = -\frac{1}{\pi} \, \mathrm{Im} \sum_\lambda G_{\mathbf{R}\lambda,\mathbf{R}\lambda}(E+i0)\,. \tag{23}$$

The set of equations has to be completed by an equation for the chemical potential $\mu$

$$\int_{-\infty}^{\mu} \mathrm{d}E\, \rho(E) = n\,, \tag{24}$$

where $n$ is the number of electrons per one site.

The spectral density

$$A(\mathbf{k},E) = -\frac{1}{\pi} \mathrm{Im} \sum_\lambda G_{\lambda\lambda}(\mathbf{k}, E+i0) \tag{25}$$

is expressed in terms of the lattice Fourier transform of $G_{\mathbf{R}\mathbf{R}'}(z)$.

## NUMERICAL IMPLEMENTATION

The formalism developed in the previous section was applied to the case of the paramagnetic fcc-Ni and the paramagnetic bcc-Fe which belong to weakly interacting systems. There is no exact way of determining the Hubbard interaction parameter $U$ for the interaction among the $d$-electrons. One can find a broad variety of values in the literature. In the present calculations we have adopted $U$ = 0.18 Ry for the paramagnetic fcc-Ni which emphasizes its atomic-like character, while the value $U$ =

0.1 Ry used for the paramagnetic bcc-Fe is close to the value derived from experiment (see Ref. 15). The ratio $U/w$ is larger for Ni as compared to Fe because the Ni bandwidth is smaller and the corresponding $U$ is also larger. Therefore many-body effects will be more pronounced in Ni metal as compared to Fe. The smaller value of $U/w$ in iron is, however, partly offset by a larger number of holes so that one can still expect non-negligible influence of many-body effects on one-particle states. The parameters of the one-particle part of the MBHH were obtained from the TB-LMTO assuming experimental lattice constants and the Ceperley-Alder form of the exchange-correlation potential. We note that the TB-LMTO method employs the so-called atomic sphere approximation (ASA) which gives an accurate description of ground state properties of transition metals, their alloys and surfaces. For more details we refer to a recent book, Ref. 14.

The elements of the site-diagonal Green function corresponding to the one-particle part of the MBHH, $G_{\mathbf{RR}}(E)$, needed for the many-body calculations, are obtained from the site-diagonal elements of the auxiliary Green function, $g_{\mathbf{RR}}(z), z = E + i0$, as

$$\begin{aligned} G_{\mathbf{R}\lambda,\mathbf{R}\lambda'}(z) &= [(z - H^{\mathrm{LMTO}})^{-1}]_{\mathbf{R}\lambda,\mathbf{R}\lambda'} \\ &= \lambda_{\mathbf{R}\lambda}(z)\, \delta_{\lambda\lambda'} + \mu_{\mathbf{R}\lambda}(z)\, g_{\mathbf{R}\lambda,\mathbf{R}\lambda'}(z)\, \mu_{\mathbf{R}\lambda'}(z)\,, \end{aligned} \tag{26}$$

where

$$g_{\mathbf{R}\lambda,\mathbf{R}\lambda'}(z) = \frac{1}{N} \sum_{\mathbf{k}} [(\mathcal{P}(z) - S(\mathbf{k}))^{-1}]_{\lambda\lambda'}\,. \tag{27}$$

In (27), the sum runs over the Brillouin zone, $N$ denotes the number of lattice sites, and the site-diagonal quantities $\lambda_{\mathbf{R}\lambda}(z)$ and $\mu_{\mathbf{R}\lambda}(z)$ are functions of the potential parameters $C_{\mathbf{R}}$, $\Delta_{\mathbf{R}}$, and $\gamma_{\mathbf{R}}$ (see Ref. 14 for more details). Finally, $\mathcal{P}_{\mathbf{R}\lambda}(z)$ is the potential function which is expressed in terms of potential parameters as

$$\mathcal{P}_{\mathbf{R}\lambda}(z) = \frac{z - C_{\mathbf{R}\lambda}}{\Delta_{\mathbf{R}\lambda} + (\gamma_{\mathbf{R}\lambda} - \beta_{\mathbf{R}\lambda})(z - C_{\mathbf{R}\lambda})}\,. \tag{28}$$

The calculations are performed first along a line in the complex energy plane ($z = E + i\epsilon$, $|\epsilon| = 0.01$ Ry), and the results are analytically deconvoluted back to the real axis[18]. In this way, the poles present in Eq. (27) are avoided. Once the imaginary part of $G_{\mathbf{R}\lambda,\mathbf{R}\lambda}(E + i0)$ is known, the selfenergy is determined from Eqs. (15)-(17). This procedure is non-selfconsistent with respect to the Green function and the selfenergy. However, we determine the Fermi energy from (6) in a selfconsistent manner in order that the number of particles is conserved. In the last step we evaluate the quantities of interest, in particular, the density of states $\rho(E)$, Eq. (23), and the spectral density $A(\mathbf{k}, E)$, Eq. (25). To this end we need to perform calculations indicated above but now for the Hamiltonian $H^{\mathrm{eff}}$, Eq. (22), rather then for $H^{\mathrm{LMTO}}$. In other words, the potential function (28) is substituted by a function of the same form in which $C_{\mathbf{R}\lambda}$ is replaced by

$$\tilde{C}_{\mathbf{R}\lambda}(z) = C_{\mathbf{R}L} + \Sigma_{\mathbf{R}\lambda}(z)\,. \tag{29}$$

The calculations are again performed along the line $z = E + i\epsilon$ in the complex energy plane and then deconvoluted to the real axis. The selfenergy in the complex energy plane is obtained from its imaginary part on the real axis determined above using dispersion relation (14).

We have used 2480 and 1360 **k**-points in the irreducible Brillouin zone to perform the integrations in Eq. (27) for the fcc and bcc cases, respectively. We note that such a high number of **k**-points is only needed to reduce the oscillations in the DOS tails which are due to the discretization of **k**-space, while for the determination of the selfenergy it is sufficient to use much smaller number of **k**-points. We have verified that the

Fermi energy and the selfenergy obtained using 280 and 240 **k**-points give essentially the same result. The step in the energy was 0.005 Ry. A typical run for 750 energy points and 2500 **k**-points requires about 5 minutes on a medium workstation and the selfconsistency with respect to the Fermi energy is achieved typically after 5 iterations in the weak-interaction case.

## RESULTS AND DISCUSSION

In Fig. 2 we compare the DOS of the paramagnetic fcc-Ni evaluated in the non-selfconsistent SOPT and in the non-selfconsistent TMA. In each case the results are compared with the corresponding LDA-DOSs. The following points are to be mentioned: (i) we observe a narrowing of the one-particle (LDA) bands, which is in agreement with experimental data[7]; (ii) we note sharp structures close to the Fermi energy which are only weakly influenced by electron interactions. The DOS at the Fermi energy is thus rather similar to that obtained in the LDA. Note that a large value of the DOS at the Fermi level is a precursor of a magnetic state at $T = 0$ K (the Stoner criterion) which does not seem to be influenced by electron interactions;

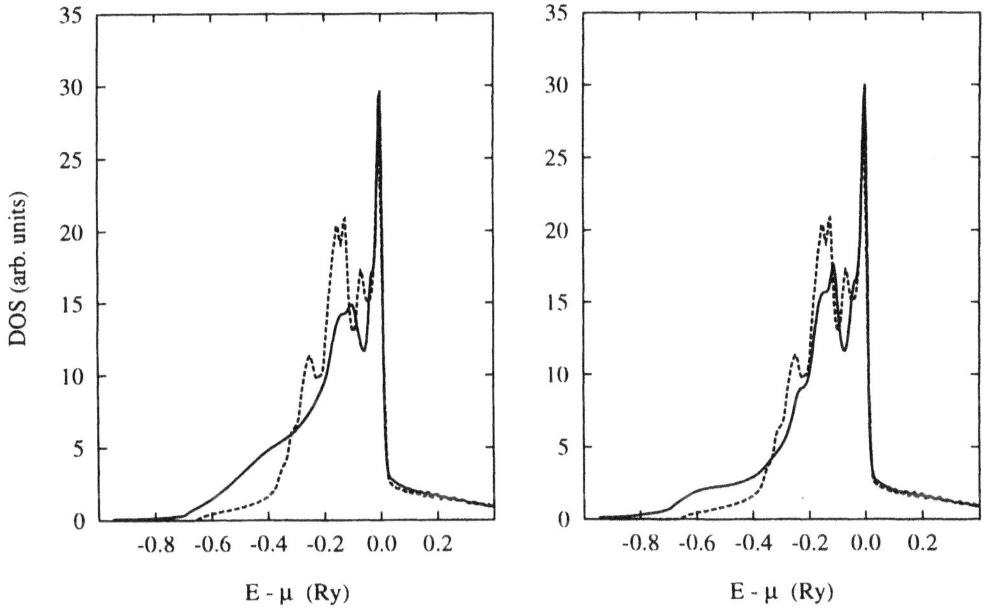


**Figure 2.** Densites of states for a paramagnetic fcc-Ni within the LDA (dashed lines) and non-selfconsistent 2nd-order perturbation theory (left frame, full line) and non-selfconsistent T-matrix approximation (right frame, full line) for the Hubbard parameter $U = 0.18$ Ry.

(iii) the $3d - 3d$ electron interactions in a partly filled band also give rise to a broad dispersionless satellite which appears below the Fermi energy. This satellite is positioned at about 6 eV below the Fermi energy in the SOPT in agreement with experiment[7]. It should be noted, however, that its position depends on the value of the interaction parameter $U$ used in the calculations. The results for the SOPT are in a reasonable agreement with similar calculations of Steiner et al.[15]; (iv) the effect of electron interactions for the same value of the Hubbard parameter $U$ is stronger in the TMA. In particular, the satellite structure is more pronounced and shifted deeper below the Fermi energy. This is due to the denominator of the T-matrix (see Eq. 16) which is missing in the SOPT expression.

We have performed similar calculations also for the RPA and the GWA. The results for the RPA are very similar to the results of the SOPT. The reason is that the $k$-order diagram of the RPA is of order of $O(n_h^k)$, where $n_h$ is the number of holes per site which is small for Ni so that the first term ($k = 2$) in the perturbation series, which is identical to the SOPT diagram, dominates. On the other hand, the effects of electron interactions (band narrowing and the shift of the satellite below the Fermi energy) are strongest within the GWA.

A deeper insight into the nature of the electron states influenced by electron-electron interactions could be obtained from spectral densities, or **k**-resolved densities of states. We note that the DOS and the spectral densities are, with exception of the transition matrix elements, proportional to the angle-integrated and angle-resolved photoemission spectra, respectively. Despite the fact that the effect of the transition matrix elements may be important, such quantities are still very helpful in understanding the influence of electron interactions on the electron states. In Fig. 3 we present the spectral densities of Ni metal calculated within the SOPT for $\mathbf{k}$=(0,0,0) ($\Gamma$-point) and for $\mathbf{k} = \frac{2\pi}{a}(1,0,0)$ (X-point) in the BZ of the fcc-lattice together with the LDA spectral

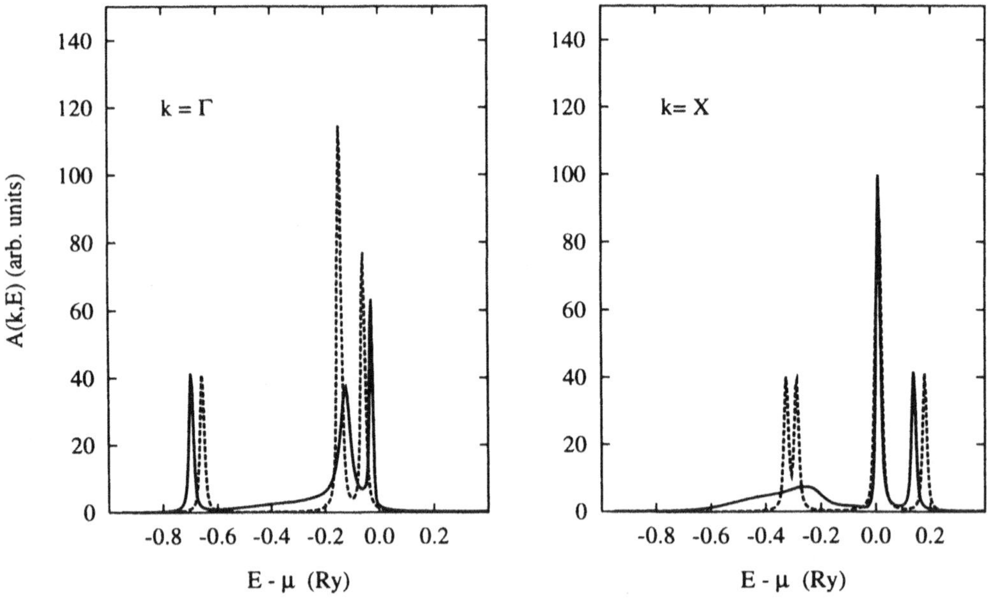


**Figure 3.** Spectral densities for a paramagnetic fcc-Ni within the LDA (dashed lines) and non-selfconsistent 2nd-order perturbation theory (full lines) at $\mathbf{k} = \Gamma$ (left frame) and at $\mathbf{k}$=X (right frame) for Hubbard parameter $U = 0.18$ Ry.

densities. In calculating the spectral densities we added a small imaginary part (0.005 Ry) to the energy to avoid too sharp peaks, in particular for the LDA curves with no damping due to the selfenergy. Three LDA peaks for $\mathbf{k} = \Gamma$ correspond to three bands, one of the $s$-character at $E \approx -0.75$ Ry and two other corresponding to the $d$-states around $E \approx -0.2$ Ry. At the X-point we have in the LDA two peaks around $E \approx -0.3$ Ry corresponding to the $d$-states of which the lower one has a strong admixture of the $s$-states, the other two $d$-bands at the Fermi energy (not resolved here because of their closeness and the finite imaginary part of the energy used in the calculations), and one $sp$-band above the Fermi energy (see Fig. 4). The influence of electron interactions results in a shift and broadening of the electron states, but the effect is anisotropic with respect to $\mathbf{k}$-vector and it depends also strongly on the energy region.

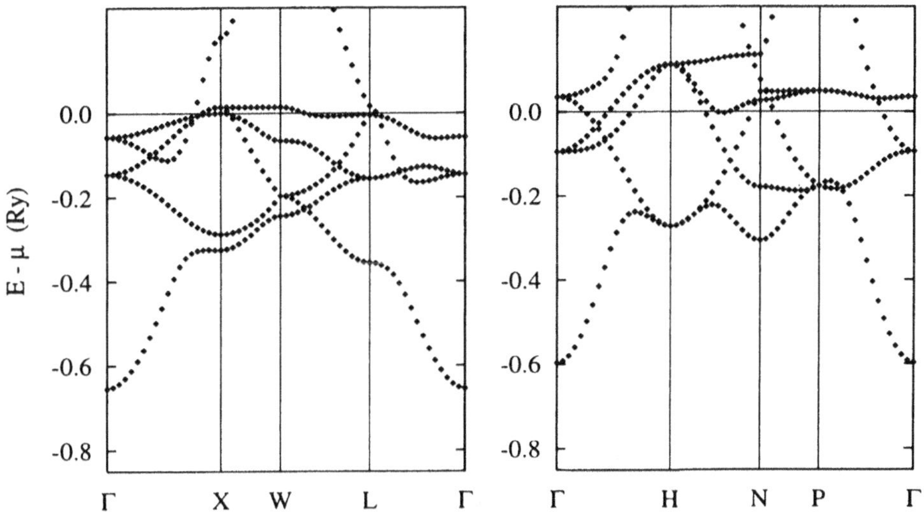


**Figure 4.** The LDA-bandstructures of paramagnetic fcc-Ni (left frame) and of paramagnetic bcc-Fe (right frame) along the lines of high symmetry in reciprocal space.

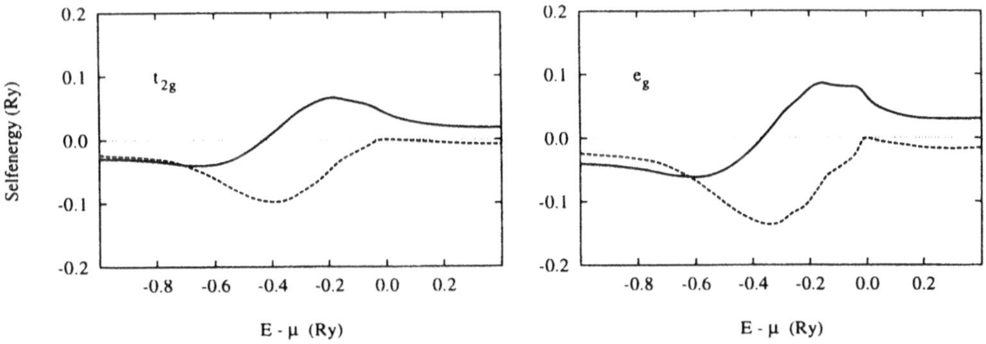


**Figure 5.** Real (full lines) and imaginary (dashed lines) parts of the selfenergy for a paramagnetic fcc-Ni for the non-selfconsistent 2nd-order perturbation theory: $t_{2g}$-symmetry (left frame) and $e_g$-symmetry (right frame) for Hubbard parameter $U = 0.18$ Ry.

To understand such a behavior in more detail, we plot in Fig. 5 the real and imaginary parts of the selfenergy $\Sigma_\lambda(E)$. Note that our approximations obey the Luttinger theorem, i.e., $\mathrm{Im}\Sigma_\lambda(E) \propto (E-\mu)^2$ in the neighborhood of the Fermi energy. Due to the cubic symmetry there are two different selfenergies, namely, $\Sigma_{t_{2g}}(E)$ and $\Sigma_{e_g}(E)$, corresponding to the three- and two-dimensional representations for the $d$-states, respectively. Another point to be mentioned is that $|\text{Im}\,\Sigma_{t_{2g}}(E)| < |\text{Im}\,\Sigma_{e_g}(E)|$, because the $e_g$ band is narrower than the $t_{2g}$ band, so that the ratio $U/w$ is larger for $e_g$-states and hence the influence of electron interactions is also stronger for $e_g$-states. The states above the Fermi energy are only weakly damped by electron interactions and are shifted almost rigidly, i.e., $\Sigma_\alpha(E) \approx \text{Re}\,\Sigma_\alpha(E) \approx$ const., ($\alpha = t_{2g}, e_g$). The same concerns the low-lying $s$-states at the bottom of the band ($\mathbf{k} = \Gamma$) which are influenced by electron-electron interactions among the $d$-electrons only indirectly, via a hybridization with the $d$-states. In both cases, the peaks have a Lorentzian shape indicating well-behaved quasiparticles. The same holds for the $d$-states close to the Fermi energy both for $\mathbf{k} = \Gamma$ and $\mathbf{k}$ =X. On the other hand, the $d$-states well below the Fermi energy, particularly in the energy region where $|\text{Im}\,\Sigma_\alpha(E)|$ have their maxima ($E \approx -0.4$ Ry), are strongly influenced by electron interactions.

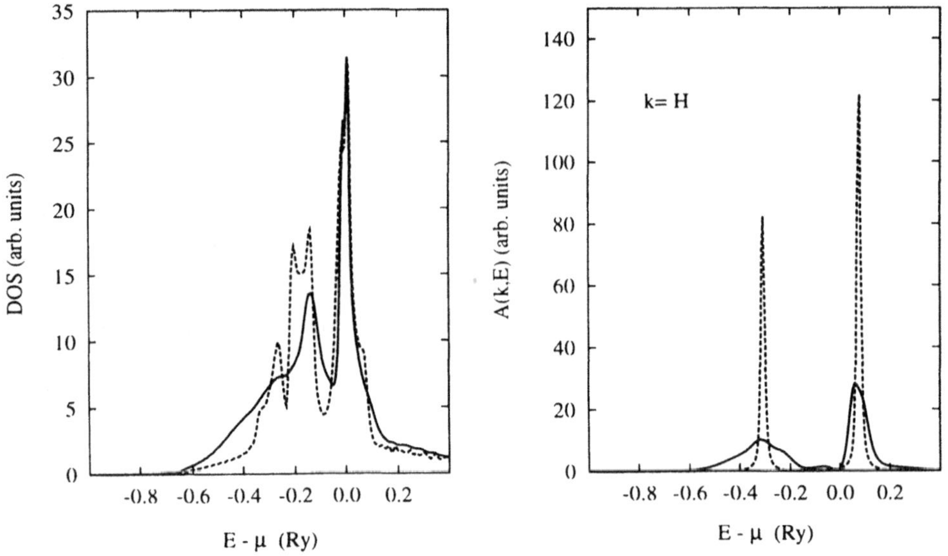


**Figure 6.** Densites of states (left frame) and spectral densities at $\mathbf{k}$=H (right frame) for a paramagnetic bcc-Fe within the LDA (dashed lines) and the non-selfconsistent 2nd-order perturbation theory (full lines) for Hubbard parameter $U = 0.1$ Ry.

This is illustrated by spectral densities for $\mathbf{k}$=X at energies around $E \approx -0.3$ Ry. The extra spectral weight in the energy region $E \in (-0.6, -0.2)$ Ry which is missing within the LDA, gives rise to a satellite structure observed in the DOS. A strongly non-Lorentzian behavior of spectral densities in this energy region indicates a breakdown of the quasiparticle behavior. Note a particularly strong influence on the lower $d$-band at $\mathbf{k} = X$ and a weaker influence on states at $\mathbf{k} = \Gamma$, where the $d$-states are closer to the Fermi energy. We note that along the line Γ-X, the band-like features move with $\mathbf{k}$-vector, while the extra spectral weight connected with the satellite structure, remains localized in the BZ. The change of the sign of $\text{Re}\Sigma_\alpha(E)$ means that the states below and above $E \approx -0.4$ Ry are shifted in opposite directions. The band narrowing can be explained by a negative slope of $\text{Re}\,\Sigma_\alpha(E)$. The results for $\mathbf{k}$ =X agree reasonably well with the results of the SOPT calculations[15] despite the fact that in the cited paper the authors studied a ferromagnetic Ni. Its exchange splitting is rather small. We have found that behavior of spectral densities in other approximations (TMA, RPA, and GWA) is qualitatively similar and thus need not be discussed separately here.

We now briefly discuss the results for a bcc Fe to see differences due to a different structure, smaller $U/w$ ratio, and a lower electron concentration. The results for the DOS and the spectral density at $\mathbf{k}$=(1,0,0) corresponding to the H-point (see Fig. 4) are plotted in Fig. 6 for the SOPT. The results are in a reasonable agreement with those of paper[15], although one has to keep in mind that our value of $U$ is slightly larger and calculations are performed for a paramagnetic rather than for a ferromagnetic state. In the LDA, a bcc Fe exhibits well-known bonding and antibonding peaks with a high value of the DOS at the Fermi energy. This feature indicating a strong tendency of a paramagnetic Fe towards a magnetic instability (Stoner criterion) is preserved in the SOPT. On the other hand, the peaks below the Fermi energy are smoothed by many-body effects (finite imaginary part of the selfenergy) and due to a larger number of holes (partly filled $d$-band) also states above the Fermi energy are now influenced. This is better seen in the spectral density evaluated at $\mathbf{k}$=H. In particular, the LDA peak above the Fermi energy (see Fig. 6) is now much more broadened than similar peaks in the case of Ni. The other peak below the Fermi energy is influenced by electron interactions more strongly showing a weak satellite feature around $E < -0.4$ Ry.

## CONCLUSIONS AND OUTLOOK

In summary, we have developed a scheme that allows to determine electronic properties of correlated solids which could be described by a multiband Hubbard Hamiltonian and which belong to the weak-interaction case, $U/w < 1$. Our approach describes the one-electron part of the many-body Hamiltonian in terms of a TB-LMTO Hamiltonian. The TB-LMTO is the LDA-based technique which describes reasonably well the ground-state properties of solids. The present formalism is particularly suitable for future extensions to random alloys and their surfaces. This represents a great advantage of the present formulation in comparison to similar constructions found in the literature. The many-body part is described by a non-selfconsistent FLEX-type method, but the Fermi energy is determined selfconsistently. There are strong arguments[19] that such an approach is in fact more correct than simple selfconsistent treatments using dressed rather than bare propagators, because within an internally consistent theory one has to renormalize simultaneously propagators and vertices. We are working on such a generalization of single-channel approximations to include a properly selected subclass of parquet diagrams[20]. In the future, we plan to implement it into the present formalism. The basic approximation adopted in the present paper is a local character of the selfenergy. This limitation which is reasonably justified for transition metals is also motivated by future applications to alloys and their surfaces as it is consistent with the treatment of the substitutional disorder using the CPA. On the other hand, the assumptions concerning the paramagnetic state, and, more importantly, the simplified form of the electron-electron interaction reduced here to a single value of the Hubbard parameter $U$ is not essential and can easily be lifted. We employ retarded quantities evaluated along a line in the complex energy plane coupled with an analytic deconvolution to the real axis and combined with the use of dispersion relations for the determination of the analytic function from its imaginary part. This is a powerful alternative to the conventional approach based on the use of the causal Green functions in which the energy integrals are replaced by finite sums over Matsubara energies. The knowlegde of the electron Green functions then allows one not only to evaluate one-electron properties, but also a future extension to transport properties of alloys and multilayers.


## Acknowledgements

Financial support for this work was provided by the Grant Agency of the Czech Republic (Project 202/98/1290), and the Grant Agency of the Academy of Sciences of the Czech Republic (Project A 1010829).


## REFERENCES


1. P. Hohenberg and W. Kohn, Phys. Rev. **136**, B864 (1964).
2. A. Georges, G. Kotliar, W. Krauth, and M. Rozenberg, Rev. Mod. Phys. **68**, 13 (1996).
3. M. Petersilka, U.J. Gossmann, and E.K.U. Gross, Phys. Rev. Lett. **76**, 1212 (1996).
4. L. Hedin, Phys. Rev. **139**, A796 (1965).
5. V. Anisimov, J. Zaanen, and O.K. Andersen, Phys. Rev. B **44**, 943 (1991).
6. F. Aryasetiawan and O. Gunnarsson, Phys. Rev. Lett. **74**, 3221 (1995).
7. S. Hüfner, G.K Wertheim, N.V. Smith, and M.M Traum, Solid State Commun. **11**, 323 (1972).
8. M. Springer, F. Aryasetiawan, and K. Karlsson, Phys. Rev. Lett. **76**, 1212 (1996).
9. J. Igarashi, P. Unger, K. Hirai, and P. Fulde, Phys. Rev. B **49**, 16181 (1994).
10. F. Manghi, V. Bellini, and C. Arcangeli, Phys. Rev. B **56**, 7149 (1997).
11. A.I. Liechtenstein and M.I. Katsnelson, Phys. Rev. B **57**, 6884 (1998).
12. J. Hubbard, Proc. Roy. Soc. London, Ser. A **276**, 238 (1963).
13. N.E. Bickers and D.J. Scalapino, Ann. Phys. (N.Y.) **193**, 206 (1989).
14. I. Turek, V. Drchal, J. Kudrnovský, M. Šob, and P. Weinberger, *Electronic Structure of Disordered Alloys, Surfaces and Interfaces* (Kluwer, Boston-London-Dordrecht, 1997)
15. M.M. Steiner, R.C. Albers, and L.J. Sham, Phys. Rev. B **45**, 13272 (1992).
16. B. Velický, H. Ehrenreich, and S. Kirkpatrick, Phys. Rev. **175**, 747 (1968).
17. V. Drchal, J. Phys. Chem. Solids **40**, 393 (1979).
18. K.C. Hass, B. Velický, and H. Ehrenreich, Phys. Rev. B **29**, 3697 (1984).
19. N.E. Bickers, D.L. Cox, and J.W. Wilkins, Phys. Rev. B **36**, 2036 (1986).
20. V. Janiš, J. Phys. Condens. Matter **10**, 2915 (1998).

# Part 3

# *Ab initio* Studies of Correlation Effects

# EXCHANGE AND CORRELATION IN ATOMS, MOLECULES, AND SOLIDS: THE DENSITY FUNCTIONAL PICTURE

John P. Perdew

Department of Physics and Quantum Theory Group
Tulane University
New Orleans, LA 70118
USA

## OUTLINE



## I. INTRODUCTION

The exchange-correlation energy is "nature's glue." It is largely responsible for the binding of atoms to form molecules and solids. Density functional theory (DFT)[1, 2, 3, 4, 5] provides a simple and powerful description of exchange and correlation, and makes successful predictions for moderately-correlated electronic systems.

Density functional theory predicts ground-state total energies and energy differences, such as cohesive energies of solids and atomization energies of molecules, surface energies, energy barriers, and structural energy differences. The theory also predicts ground-state total and spin densities, nuclear positions and vibrational frequencies, and static linear response functions.

*Electron Correlations and Materials Properties*, edited by Gonis *et al.*
Kluwer Academic/Plenum Publishers, 1999.

Questions to be answered in this article are: (a) What is DFT? (2) What are exchange and correlation? (3) What are the common DFT approximations for the exchange-correlation energy as a functional of the density, such as the local spin density (LSD) and generalized gradient approximation (GGA)? Why do they work for moderate correlation, and how can they fail for strong correlation? (4) What are the simple limits and physical effects of exchange and correlation? (5) What new approximations might lie beyond GGA?

## II. Kohn-Sham Spin Density Functional Theory for the Electronic Ground State

In Kohn-Sham theory [2, 6] the density of electrons of spin $\sigma(=\uparrow$ or $\downarrow)$ is found by summing the squares of occupied orbitals,

$$n_\sigma(\mathbf{r}) = \sum_k |\psi_{k\sigma}(\mathbf{r})|^2 \theta(\mu - \varepsilon_{k\sigma}), \tag{1}$$

$$n(\mathbf{r}) = n_\uparrow(\mathbf{r}) + n_\downarrow(\mathbf{r}). \tag{2}$$

The Kohn-Sham orbitals are solutions of a selfconsistent one-electron equation

$$[-\frac{1}{2}\nabla^2 + v_\sigma(\mathbf{r}) + \int d^3r' \, \frac{n(\mathbf{r}')}{|\mathbf{r}' - \mathbf{r}|} + v_{xc}^\sigma(\mathbf{r})] \, \psi_{k\sigma}(\mathbf{r}) = \varepsilon_{k\sigma}\psi_{k\sigma}(\mathbf{r}), \tag{3}$$

where $v_\sigma(\mathbf{r})$ is a spin-dependent external potential (e.g., the electron-nucleus interaction). Aside from the nucleus-nucleus repulsion energy, the total energy is

$$E = \sum_{k\sigma} < \psi_{k\sigma}| - \frac{1}{2}\nabla^2|\psi_{k\sigma} > \theta\,(\mu - \varepsilon_{k\sigma}) + \sum_\sigma \int d^3r \, v_\sigma(\mathbf{r}) \, n_\sigma(\mathbf{r}) + U[n] + E_{xc}[n_\uparrow, n_\downarrow], \tag{4}$$

where

$$U[n] = \frac{1}{2}\int d^3r \int d^3r' \, \frac{n(\mathbf{r})n(\mathbf{r}')}{|\mathbf{r}' - \mathbf{r}|} \tag{5}$$

is the Hartree self-repulsion of the electron density, and

$$v_{xc}^\sigma(\mathbf{r}) = \delta E_{xc}/\delta n_\sigma(\mathbf{r}) \tag{6}$$

is a spin-dependent exchange-correlation potential.

Without the exchange-correlation energy $E_{xc}[n_\uparrow, n_\downarrow]$ and potential $v_{xc}^\sigma(\mathbf{r})$, these are the Hartree equations [7](without self-interaction correction). With the exact $E_{xc}$ and $v_{xc}^\sigma$, these equations deliver the exact ground-state spin densities and energy.

The exchange-correlation energy is the sum of exchange and correlation terms:

$$E_{xc}[n_\uparrow, n_\downarrow] = E_x[n_\uparrow, n_\downarrow] + E_c[n_\uparrow, n_\downarrow]. \tag{7}$$

$E_x$ corrects the self-interaction error in $U[n]$, and also accounts for the effects of the Pauli exclusion principle,which keeps electrons of the same spin apart. $E_c$ accounts for the effect on the wavefunction of Coulomb repulsion, which keeps all electrons apart. In practice, we must approximate the spin density functional for $E_{xc}$.

## III. Definition of Exchange and Correlation

The Hamiltonian for an N-electron system is

$$\hat{H} = \hat{T} + \sum_{i=1}^{N} v_{\sigma_i}(\mathbf{r}_i) + \hat{V}_{ee}, \tag{8}$$

where

$$\hat{T} = \sum_{i=1}^{N} -\frac{1}{2}\nabla_i^2 \tag{9}$$

is the kinetic energy operator and

$$\hat{V}_{ee} = \frac{1}{2}\sum_i \sum_{j\neq i} \frac{1}{|\mathbf{r}_i - \mathbf{r}_j|} \tag{10}$$

is the electron-electron repulsion operator.

Define[8] $\Psi^{\lambda}_{n_\uparrow,n_\downarrow}$ as that normalized antisymmetric N-electron wavefunction which yields the given spin densities $n_\uparrow(\mathbf{r}), n_\downarrow(\mathbf{r})$ and minimizes the expectation value $< \hat{T} + \lambda\hat{V}_{ee} >$. Then the exchange-correlation energy is

$$E_{xc}[n_\uparrow, n_\downarrow] \equiv < \Psi^{\lambda=1}_{n_\uparrow,n_\downarrow}|\hat{T} + \hat{V}_{ee}|\Psi^{\lambda=1}_{n_\uparrow,n_\downarrow} > - < \Psi^{\lambda=0}_{n_\uparrow,n_\downarrow}|\hat{T}|\Psi^{\lambda=0}_{n_\uparrow,n_\downarrow} > -U[n], \tag{11}$$

and the exchange energy is

$$\begin{aligned} E_x[n_\uparrow, n_\downarrow] &\equiv < \Psi^{\lambda=0}_{n_\uparrow,n_\downarrow}|\hat{V}_{ee}|\Psi^{\lambda=0}_{n_\uparrow,n_\downarrow} > -U[n] \\ &= -\frac{1}{4}\sum_\sigma \int d^3r \int d^3r' \frac{|\sum_k \psi^*_{k\sigma}(\mathbf{r})\psi_{k\sigma}(\mathbf{r}')|^2}{|\mathbf{r}' - \mathbf{r}|} \theta(\mu - \varepsilon_{k\sigma}) \leq 0, \end{aligned} \tag{12}$$

a Fock integral of Kohn-Sham orbitals. The correlation energy is then

$$E_c = < \Psi^{\lambda=1}_{n_\uparrow,n_\downarrow}|\hat{T} + \hat{V}_{ee}|\Psi^{\lambda=1}_{n_\uparrow,n_\downarrow} > - < \Psi^{\lambda=0}_{n_\uparrow,n_\downarrow}|\hat{T} + \hat{V}_{ee}|\Psi^{\lambda=0}_{n_\uparrow,n_\downarrow} > . \tag{13}$$

By the definition of $\Psi^{\lambda}_{n_\uparrow,n_\downarrow}$, $E_c \leq 0$. $E_c$ clearly has both kinetic and potential energy contributions.

From Eq. (11),

$$\begin{aligned} E_{xc}[n_\uparrow, n_\downarrow] &= \int_0^1 d\lambda \frac{d}{d\lambda} < \Psi^{\lambda}_{n_\uparrow,n_\downarrow}|\hat{T} + \lambda\hat{V}_{ee}|\Psi^{\lambda}_{n_\uparrow,n_\downarrow} > -U[n] \\ &= \int_0^1 d\lambda \{< \Psi^{\lambda}_{n_\uparrow,n_\downarrow}|\hat{V}_{ee}|\Psi^{\lambda}_{n_\uparrow,n_\downarrow} > -U[n]\} \\ &= \int_0^1 d\lambda\, W_\lambda[n_\uparrow, n_\downarrow], \end{aligned} \tag{14}$$

where

$$W_\lambda[n_\uparrow, n_\downarrow] = \frac{1}{2}\int d^3r \int d^3r' \frac{n(\mathbf{r})\, n^{\lambda}_{xc}(\mathbf{r}, \mathbf{r}')}{|\mathbf{r}' - \mathbf{r}|}. \tag{15}$$

In Eq. (15), $n^{\lambda}_{xc}(\mathbf{r}, \mathbf{r}')$ is the density at $\mathbf{r}'$ of the exchange-correlation hole around an electron at $\mathbf{r}$, in a system with ground-state spin densities $n_\uparrow(\mathbf{r}), n_\downarrow(\mathbf{r})$, electron-electron interaction $\lambda\hat{V}_{ee}$, and a $\lambda$-dependent external potential $v^{\lambda}_{\sigma}(\mathbf{r})$ which holds the spin densities fixed and independent of $\lambda$.

The coupling constant integral[9, 10, 11] of Eq. (14) makes $E_{xc}$ look like a pure potential energy, but averaged over a series of systems with coupling constants $\lambda$ between 0 and 1. $\lambda = 1$ is the real physical system of interest, while $\lambda = 0$ is the Kohn-Sham non-interacting system in which the wavefunction $\Psi^{\lambda=0}_{n_\uparrow,n_\downarrow}$ is a single Slater determinant and $W_{\lambda=0} = E_x$.

The exchange-correlation hole density $n^\lambda_{xc}(\mathbf{r},\mathbf{r}')$ can be found from the electron pair density $\rho^\lambda_2(\mathbf{r},\mathbf{r}')$ for coupling constant $\lambda$:

$$\begin{aligned}\rho^\lambda_2(\mathbf{r},\mathbf{r}') &\equiv N(N-1)\sum_{\sigma_1\cdots\sigma_N}\int d^3r_3\cdots d^3r_N\,|\Psi^\lambda_{n_\uparrow,n_\downarrow}(\mathbf{r}\sigma_1,\mathbf{r}'\sigma_2,\cdots\mathbf{r}_N\sigma_N)|^2\\ &= n(\mathbf{r})[n(\mathbf{r}') + n^\lambda_{xc}(\mathbf{r},\mathbf{r}')].\end{aligned} \tag{16}$$

The non-negative character of $\rho^\lambda_2$ implies that

$$n^\lambda_{xc}(\mathbf{r},\mathbf{r}') > -n(\mathbf{r}'). \tag{17}$$

The importance of the coupling constant integration of Eqs. (14) and (15) is that it establishes key properties [11, 12] of the exact $E_{xc}$ which approximations should (and usually do) have. For example, consider the exchange hole density

$$n_x(\mathbf{r},\mathbf{r}') \equiv n^{\lambda=0}_{xc}(\mathbf{r},\mathbf{r}'). \tag{18}$$

From Eq. (12),

$$n_x(\mathbf{r},\mathbf{r}') \leq 0. \tag{19}$$

From Eq. (16),

$$\int d^3r'\, n_x(\mathbf{r},\mathbf{r}') = -1. \tag{20}$$

The correlation hole density $n^\lambda_c(\mathbf{r},\mathbf{r}') \equiv n^\lambda_{xc}(\mathbf{r},\mathbf{r}') - n_x(\mathbf{r},\mathbf{r}')$ obeys

$$\int d^3r'\, n^\lambda_c(\mathbf{r},\mathbf{r}') = 0. \tag{21}$$

Eq. (19) says that the exchange hole density is non-positive, while Eqs. (20) and (21) assert that an electron at $\mathbf{r}$ is missing from all other positions $\mathbf{r}'$. Eqs. (19) - (21) will be referred to as "hole sum rules." Approximations to $E_{xc}$ of Eq. (15) which satisfy these sum rules are constrained to fairly realistic energies, at least when they have reasonably accurate on-top[13, 14, 15, 16] ($\mathbf{r}' = \mathbf{r}$) hole densities $n^\lambda_{xc}(\mathbf{r},\mathbf{r})$.

## IV. Local Spin Density and Generalized Gradient Approximations: Why Do They Work for Moderately-Correlated Systems?

The local spin density approximation[2, 6]

$$E^{LSD}_{xc}[n_\uparrow,n_\downarrow] = \int d^3r\, n(\mathbf{r})\,\varepsilon^{unif}_{xc}(n_\uparrow(\mathbf{r}),n_\downarrow(\mathbf{r})) \tag{22}$$

has been the workhorse of solid state physics for almost thirty years. In Eq. (22), $\varepsilon^{unif}_{xc}(n_\uparrow,n_\downarrow)$ is the exchange-correlation energy per electron in an electron gas of uniform spin densities $n_\uparrow, n_\downarrow$, a quantity that is accurately known[17, 18, 19] from Quantum Monte Carlo simulations[20].

LSD is exact for an electron gas of uniform or slowly-varying density, but that is *not* why it works so well for real systems. The correct explanation [10, 11, 12, 13, 14, 15, 16] is found in the exchange-correlation hole

$$n_{xc}^{\lambda,LSD}(\mathbf{r},\mathbf{r}') = n_{xc}^{\lambda,unif}(n_\uparrow(\mathbf{r}), n_\downarrow(\mathbf{r}); |\mathbf{r}'-\mathbf{r}|), \tag{23}$$

where $n_{xc}^{\lambda,unif}(n_\uparrow, n_\downarrow; |\mathbf{r}'-\mathbf{r}|)$is the hole in an electron gas with uniform spin densities $n_\uparrow, n_\downarrow$. Since that is a possible physical system, the LSD hole of Eq. (23) obeys [10, 11, 12] the sum rules (19)-(21).

Moreover, the LSD on-top exchange hole $n_x(\mathbf{r},\mathbf{r})$ is exact[13], since the on-top exchange hole of the single Slater determinant $\Psi_{n_\uparrow,n_\downarrow}^{\lambda=0}$ is determined by the local spin densities $n_\uparrow(\mathbf{r})$ and $n_\downarrow(\mathbf{r})$; this follows from the fact that *all* electrons of spin-$\sigma$ are excluded from the point $\mathbf{r}$ where an electron of spin $\sigma$ is located. The LSD on-top correlation hole $n_c^\lambda(\mathbf{r},\mathbf{r})$ is not exact, but it is correctly bounded [14] between $-n(\mathbf{r})/2$ and 0, and is often very accurate [15, 16]. Indeed, if the local approximation works at any $\mathbf{r}'$, it should work at $\mathbf{r}'=\mathbf{r}$. The accuracy of the LSD on-top hole explains[16] why spin density functionals typically work better than total density functionals,[21] even when the external potential is spin-independent and the Hohenberg-Kohn theorem [1] shows that only the total density is formally needed.

If LSD worked because real electron densities were slowly-varying over space, then the second-order gradient expansion approximation[1, 2, 22, 23]

$$E_{xc}[n_\uparrow, n_\downarrow] = E_{xc}^{LSD}[n_\uparrow, n_\downarrow] + \sum_{\sigma,\sigma'} \int d^3r \, \frac{C_{xc}^{\sigma\sigma'}(n_\uparrow, n_\downarrow)}{n_\sigma^{2/3} n_{\sigma'}^{2/3}} \nabla n_\sigma \cdot \nabla n_{\sigma'} \tag{24}$$

should provide a useful systematic correction to LSD for atoms, molecules and solids. In fact, the gradient terms of Eq. (24) usually worsen the energies of real systems.[22] While the gradient expansion of the hole barely changes[24] the on-top hole density $n_{xc}^\lambda(\mathbf{r},\mathbf{r})$, it violates[12, 25] the sum rules (19)-(21). These serious deficiencies of the second-order gradient expansion can be eliminated through generalization thereof to all orders in $\nabla$.[12, 22, 24, 26, 27, 28, 29, 30, 31, 32, 33, 34, 35]

The Perdew-Wang 1991 generalized gradient approximation[24, 32, 33, 34].

$$E_{xc}^{GGA}[n_\uparrow, n_\downarrow] = \int d^3r \, f(n_\uparrow, n_\downarrow, \nabla n_\uparrow, \nabla n_\downarrow) \tag{25}$$

starts from the second-order gradient expansion of the exchange-correlation hole density, then cuts off the spurious long-range ($|\mathbf{r}'-\mathbf{r}| \to \infty$) parts to restore the sum rules (19)-(21). An alternative non-empirical derivation[35] yields essentially the same GGA. This GGA is typically more accurate than LSD, especially for the rapidly-varying densities of atoms and molecules. In recent years, GGA's have made DFT popular in quantum chemistry.

For atoms, the errors in the exchange energy are about 10% in LSD and 1% in GGA, while the errors in the correlation energy are about 100% in LSD and about 5% in GGA[34]. These errors in exchange and correlation show some tendency to cancel one another, especially in LSD.

For twenty molecules, the mean absolute errors of the atomization energies are [35] 70 kcal/mole in the Hartree-Fock approximation (which neglects correlation), 30 kcal/mole in LSD, and 8 kcal/mole in GGA (where 1 ev = 23 kcal/mole). While the Hartree-Fock approximation underbinds and LSD overbinds, GGA provides a more realistic description even for weak (hydrogen[36] or van der Waals [37, 38, 39]) bonds. In other words, the Hartree-Fock brand of glue is too weak, and the LSD brand is too strong.

For the valence electrons in solid silicon, where nearly-exact Quantum Monte Carlo results are available, [40] the GGA error is about 0.1% in the exchange energy and 4% in the correlation energy.

LSD and GGA generally work well for moderately-correlated electrons. In some systems, however, the correlations among the electrons are stronger than might be expected from the local spin densities. Examples are the $H_2$ molecule at stretched bond length, the $Cr_2$ molecule at normal bond length, and certain transition metal oxides, in which the electrons tend to localize on lattice sites where fluctuations of electron number are suppressed.[41] In these systems, the on-top correlation hole density is typically *not* determined by the local spin densities. Then selfconsistent LSD and GGA calculations can produce false spin moments, although in many cases these calculations still yield accurate atomization energies and bond lengths.[42] Effectively, spurious spin polarization can mimic strong correlation, since both deepen the on-top exchange-correlation hole density.[15]

## V. SIMPLE LIMITS AND PHYSICAL EFFECTS OF EXCHANGE AND CORRELATION

Much can be learned from the study of simple limits, which are known either exactly or in the LSD and GGA approximations to the exchange-correlation energy. In particular, the characteristic physical effects of exchange and correlation can be extracted from such a study.

### Va. High and Low Density Limits

Given any density $n(\mathbf{r})$ and any scale parameter $\gamma > 0$, Levy and Perdew [43] defined the uniformly-scaled density as

$$n_\gamma(\mathbf{r}) = \gamma^3 n(\gamma \mathbf{r}). \tag{26}$$

Both $n_\uparrow(\mathbf{r})$ and $n_\downarrow(\mathbf{r})$ are scaled in the same way, and the electron number N is conserved:

$$\int d^3r\, n_\gamma(\mathbf{r}) = \int d^3r\, n(\mathbf{r}) = N. \tag{27}$$

$\gamma > 1$ is a compression, and $\gamma < 1$ an expansion. The high and low density limits are $\gamma \to \infty$ and $\gamma \to 0$, respectively.

The exchange energy of Eq. (12) scales homogeneously [43]:

$$E_x[n_\gamma] = \gamma E_x[n]. \tag{28}$$

Since the exchange energy of a uniform, spin-unpolarized density is

$$E_x^{LDA}[n] = \int d^3r\, n\, \varepsilon_x^{unif}(n), \tag{29}$$

Eq. (28) shows that

$$\varepsilon_x^{unif}(n) = A_x n^{1/3} < 0 \tag{30}$$

is the exchange energy per electron of the uniform gas.

The correlation energy does not scale homogeneously, but its high-density limit [44] is

$$\lim_{\gamma\to\infty} E_c[n_\gamma] = O(\gamma^o). \tag{31}$$

For a finite system, this limit is a constant, but for the uniform electron gas [45]

$$\lim_{n\to\infty} \varepsilon_c^{unif}(n) = c \ln n^{1/3} < 0. \tag{32}$$

The low-density limit[46] of correlation scales like exchange, i.e.,

$$\lim_{\gamma\to 0} E_c[n_\gamma] = O(\gamma), \tag{33}$$

whence

$$\lim_{n\to 0} \varepsilon_c^{unif}(n) = b n^{1/3} < 0. \tag{34}$$

As a density scales to the high density limit, both $E_x$ and $E_c$ become more negative, but $E_x$ dominates $E_c$. As a density scales to the low density limit, $E_x$ and $E_c$ approach zero from below; they become comparable in size, but always bounded by [47]

$$E_{xc}[n_\uparrow, n_\downarrow] \geq 2.3\, E_x^{LDA}[n]. \tag{35}$$

Clearly exchange and correlation favor processes in which the electron density increases, making $E_{xc}$ more negative.[48] Such processes include: (1) Transfer of an electron from a less to a more localized orbital, e.g., from 4s to 3d. (2) Bonding or cohesion of atoms in molecules and solids. (3) Bond shortening and stiffening. (4) Fusion of surfaces (i.e., exchange and correlation increase surface energies). (5) Filling of vacancies (i.e., exchange and correlation increase vacancy formation energies). (6) Symmetry breaking (e.g., exchange and correlation favor the formation of charge density waves).

In summary, exchange and correlation make interatomic interactions more attractive, and act to hold electrons together (e.g., by increasing the work function of a metal). More generally, exchange and correlation try to reduce the local Seitz radius $r_s$, defined by

$$n = 3/4\pi r_s^3. \tag{36}$$

## Vb. Spin Unpolarized and Fully Polarized Limits

The local relative spin polarization $\zeta$ is defined as

$$\zeta = (n_\uparrow - n_\downarrow)/n. \tag{37}$$

$\zeta = 1$ means all spins are up, $\zeta = -1$ means all spins are down, and $\zeta = 0$ means that the system is locally spin unpolarized.

Increasing $|\zeta|$ makes the exchange energy $E_x$ (which arises only between parallel-spin electrons) more negative, e.g., [6]

$$\varepsilon_x^{unif}(n_\uparrow, n_\downarrow) = A_x n^{1/3} 2^{1/3}[(\frac{1+\zeta}{2})^{4/3} + (\frac{1-\zeta}{2})^{4/3}]. \tag{38}$$

While the correlation energy $E_c$ becomes less negative as $|\zeta|$ increases, the exchange dominates within LSD and GGA so that $E_{xc}$ becomes more negative as $|\zeta| \to 1$. Thus exchange and correlation, taken together, favor magnetism in solids.

At least within LSD and GGA, $E_{xc}[n_\uparrow, n_\downarrow]$ always favors *ferromagnetic* spin alignment. The other possibilities (paramagnetism, antiferromagnetism, noncolinear magnetism) also arise, but are driven by the kinetic energy of the Kohn-Sham orbitals [49] (which must be generalized as spinors to get noncolinear magnetism).

### Vc. Small and Large Density-Gradient Limits

The local reduced density gradient is defined[28] as

$$s = |\nabla n|/2(3\pi^2)^{1/3} n^{4/3} = 0.7816\,|\nabla r_s|. \tag{39}$$

To a good approximation, the GGA of Eq. (25) may be written as

$$E_{xc}^{GGA}[n_\uparrow, n_\downarrow] \approx \int d^3r\, n\, \varepsilon_x^{unif}(n) F_{xc}(r_s, \zeta, s). \tag{40}$$

$F_{xc}$, the enhancement factor over local exchange,[33, 35] is a function of the three local variables $r_s, \zeta, s$. Fig. 1 of Ref. 35 shows the enhancement factor for the nonempirical GGA of Perdew, Burke, and Ernzerhof.

When the local reduced gradient vanishes, the GGA enhancement factor reduces to that of LSD, i.e.,

$$F_{xc}(r_s, \zeta, s = 0) = \varepsilon_{xc}^{unif}(n_\uparrow, n_\downarrow)/\varepsilon_x^{unif}(n). \tag{41}$$

In the high-density limit, where exchange dominates correlation, the GGA enhancement factor reduces to that for exchange alone:

$$F_{xc}(r_s = 0, \zeta, s) = F_x(\zeta, s). \tag{42}$$

Fig. 1 of Ref. 35 shows that $F_x$ and $F_{xc}$ increase with s, while $F_c$ decreases. The nonlocalities or s-dependences of exchange and correlation are opposite, and they almost cancel for valence electron densities in solids (where $0 \lesssim s \lesssim 1$ and $1 \lesssim r_s \lesssim 6$).

For typical densities of atoms, molecules and solids, the gradient corrections to LSD favor higher and more inhomogeneous densities, i.e., processes in which [50, 51]

$$\frac{d < s >}{< s >} \gtrsim \frac{d < r_s >}{2 < r_s >}, \tag{43}$$

where $< s >$ and $< r_s >$ are appropriately-defined averages. Typically, a process in which $< s >$ increases is also one in which $< r_s >$ increases, so both sides of the inequality (43) are positive.

For example, consider the atomization of a molecule or solid, in which the density becomes more inhomogeneous ($d < s > > 0$) and lower ($d < r_s > > 0$). In most cases, the inequality (43) is satisfied; gradient corrections to LSD lower the atomization energy. Similar processes are the fragmentation of a highly-bonded transition state [52](where gradient corrections raise the energy barrier between reactants and products), and the phase transition of a solid from a more close-packed to a more open structure.[53]

Bond stretching is another process in which $d < s > > 0$ and $d < r_s > > 0$. Usually the inequality (43) is satisfied, and gradient corrections to LSD increase equilibrium bond lengths and lattice constants. However, bonds between two hydrogen atoms behave differently: Because the inequality (43) is not satisfied under stretching of these bonds, the gradient corrections actually and correctly shrink the bond length.[42, 52].

### Vd. Summary of the Effects of Density, Spin Polarization, and Density Gradient

Effects[51]which increase the noninteracting kinetic energy functional $T_s[n_\uparrow, n_\downarrow]$ (decreasing $r_s$, increasing $|\zeta|$, and increasing $s$, in the language of sections V a-c) make the exchange energy $E_x[n_\uparrow, n_\downarrow]$ more negative and also reduce the ratio $E_c/E_x$. In the limits where $r_s \to 0$ or $s \to \infty$ everywhere, the ratio $E_c^{GGA}/E_x^{GGA}$ tends to zero. While this

GGA behavior is often correct, there are situations where it is not, e.g., the collapse of a three-dimensional electron gas to two dimensions,[54] in which the exact $E_c/E_x$ tends to a positive limit and not to zero.

### Ve. Short and Long Range Limits of the Exchange-Correlation Hole

To evaluate the exchange-correlation energy of Eqs. (14) and (15), only the spherical [11]and system[16]average of the exchange-correlation hole density is needed:

$$< n_{xc}^{\lambda}(u) > \equiv \frac{1}{N}\int d^3r\, n(\mathbf{r}) \int \frac{d\Omega_{\mathbf{u}}}{4\pi}\, n_{xc}^{\lambda}(\mathbf{r}, \mathbf{r}+\mathbf{u}), \tag{44}$$

$$E_{xc} = N\int_0^1 d\lambda \int_0^{\infty} du\, 4\pi u^2 \frac{< n_{xc}^{\lambda}(u) >}{u}, \tag{45}$$

where $u$ is the interelectronic separation. $u = 0$ is the on-top position, where the exchange hole is exact and the correlation hole is often surprisingly accurate in LSD and GGA (Section IV).

The exchange hole in the uniform electron gas [55] has a single length scale $\sim r_s$, as does the correlation hole in the low-density limit. But, in the high-density limit $r_s \to 0$, the correlation hole [55] has a second length scale $\sim \sqrt{r_s}$.

The LSD approximation is the zero-th order term of a gradient expansion for $E_{xc}[n_\uparrow, n_\downarrow]$ and $n_{xc}^{\lambda}(\mathbf{r}, \mathbf{r}+\mathbf{u})$. At the exchange-only level, this expansion has been carried to fourth order, and its structure has been identified:[56] Addition of each successive power of $\nabla$ improves $n_{xc}^{\lambda}(\mathbf{r}, \mathbf{r}+\mathbf{u})$ at small $u$, but worsens it at large $u$. This observation justifies the real-space cutoff construction of the GGA, as presented in Section IV.

The most long-ranged ($u \to \infty$) parts of the exact exchange and correlation holes cancel one another exactly in the uniform electron gas [55] and at a metal surface [57], and also show some tendency to cancel in a molecule.[58] It is only in rare situations (e.g., in stretched $H_2^+$)[59] that the exact exchange-correlation hole has an important long-range tail. Thus the sharp long-range cutoffs used to construct the GGA are usually justified for exchange and correlation together, although they are less justified for exchange and correlation separately. This is the reason why the atomizaton energies of molecules are better described by GGA exchange and correlation than by the combination of full exact exchange with GGA correlation.[60] In fact, the hybrid[61]

$$E_{xc}^{hybrid} = a\, E_x^{exact} + (1-a)E_x^{GGA} + E_c^{GGA} \tag{46}$$

gives optimum atomization energies for $a \approx 1/4$, for reasons that are now fully understood.[62, 63, 64] (Note that GGA exchange-correlation is more accurate at coupling-constant $\lambda = 1$ than at $\lambda = 0$.)

### Vf. Weakly and Strongly Correlated Limits

Weak correlation ($E_c/E_x << 1$) is found in the high-density (Section 4) or small coupling-constant ($\lambda \to 0$ in Eqs. (14) and (15)) limits, while strong correlation ($E_c/E_x \approx 1$) arises in the opposite limits of low density or strong coupling.

Most atoms, molecules and solids are weakly or moderately correlated, but even for these systems an understanding of the strong-coupling limit [65] is useful for construction of the $\lambda$-dependent integrand $W_\lambda[n_\uparrow, n_\downarrow]$ of Eqs. (14) and (15). (This integrand is pictured in Fig. 1 of Ref. 66). The new construction [65] invokes an analytic formula that interpotates between the $\lambda \to 0$ and $\lambda \to \infty$ limits. $W_{\lambda=0}$ and $dW_\lambda/d\lambda|_{\lambda=0}$ are taken from the exact exchange energy and second-order Goerling-Levy [67]correlation

taken from the exact exchange energy and second-order Goerling-Levy [67] correlation energy, respectively, and $W_{\lambda=\infty}$ is evaluated exactly for a two-electron system, or via a gradient expansion for $N > 2$, using the concept of "strict correlation".

The LSD and GGA approximations can to some extent describe strong correlation that arises due to low density, but tend to fail (at least in part) for strong correlation that arises from site localization, as described in Section IV. Site localization does not occur in the electron gas of uniform or slowly-varying density.

## VI. What's Beyond the Generalized Gradient Approximation?

Although GGA is an improvement over LSD, some errors persist in GGA: (1) sp and especially sd transfer energies in atoms are systematically in error in both LSD and GGA. [33, 68] (2) Lattice constants of some solids are too small in LSD; GGA sometimes corrects and sometimes overcorrects this error.[69, 70] (3) The atomization energies of multiply-bonded molecules, which are far too high in LSD, are still somewhat too high in GGA. [35, 42]

These errors of LSD and GGA appear to arise largely from the exchange interaction between different orbitals. While errors (1) and (2) probably arise from an incorrect LSD or GGA description of core-valence exchange,[3, 70] error (3) is more likely due to exchange between overlapped valence orbitals.[58]

The ingredients of the GGA exchange energy are essentially the local values of $r_s$, $\zeta$, and $s$, as defined in Eqs. (36), (37), and (39). This list can be expanded to include the local non-interacting kinetic energy density

$$\tau(\mathbf{r}) = \frac{1}{2}\sum_{\sigma}\sum_{k} |\nabla\psi_{k\sigma}(\mathbf{r})|^2\,\theta(\mu - \varepsilon_{k\sigma}), \tag{47}$$

as suggested by the density matrix expansion.[71, 72] The $\tau$-dependence can be chosen[73] to recover the known fourth-order gradient expansion of the exchange energy in the slowly-varying limit.[74]

Alternatively, the exact exchange energy of Eq. (12) can be employed, but only with an approximation for correlation which is "compatible", i.e., which can cancel the most long-ranged part of the exact exchange hole. A numerically-challenging but potentially accurate way to do this is to write [75, 76]

$$E_{xc}^{RPA+} = E_x^{exact} + E_c^{RPA} + (E_c^{GGA} - E_c^{GGA,RPA}). \tag{48}$$

In Eq. (48), the exact exchange energy is constructed from the occupied Kohn-Sham orbitals, the correlation energy is constructed in the random phase approximation from the occupied and unoccupied Kohn-Sham orbitals, and the short-range correction to RPA is constructed via GGA.

## ACKNOWLEDGMENTS

This article is based upon an invited talk at the Workshop on Electron Correlations and Material Properties, Heraklion, Crete, June 28 - July 3, 1998. The work described was supported in part by the U.S. National Science Foundation under Grant Nos. DMR95-21353 and DMR98-10620. The article was written during a visit to the Institute for Nuclear Theory at the University of Washington, sponsored by the U.S. Department of Energy. The hospitality of the Workshop and of the Institute are gratefully acknowledged.

## REFERENCES


1. P. Hohenberg and W. Kohn, Phys. Rev. **136**, B864 (1964).
2. W. Kohn and L.J. Sham. Phys. Rev. **140**, A1133 (1965).
3. R.O. Jones and O. Gunnarsson, Rev. Mod. Phys. **61**, 689 (1989).
4. R.M. Dreizler and E.K.U. Gross, *Density Functional Theory* (Springer, Berlin, 1990).
5. R.G. Parr and W. Yang, *Density Functional Theory of Atoms and Molecules* (Oxford University Press, New York, 1989).
6. U. von Barth and L. Hedin, J. Phys. C **5**, 1629 (1972).
7. D.R. Hartree, Proc. Camb. Phil. Soc. **24**, 111 (1927).
8. M. Levy, Proc. Natl. Acad. Sci. USA **76**, 6062 (1979).
9. J. Harris and R.O. Jones, J. Phys. F **4**, 1170 (1974).
10. D.C. Langreth and J.P. Perdew, Solid State Commun. **17**, 1425 (1975).
11. O. Gunnarsson and B.I. Lundqvist, Phys. Rev. B **13**, 4274 (1976).
12. J.P. Perdew, Phys. Rev. Lett. **55**, 1665 (1985); **55**, 2370 (1985) (E).
13. T. Ziegler, A. Rauk, and E.J. Baerends, Theoret. Chim. Acta **43**, 261 (1977).
14. J. Harris, Phys. Rev. A **29**, 1648 (1984).
15. J.P. Perdew, A. Savin, and K. Burke, Phys. Rev. A **51**, 4531 (1995).
16. K. Burke, J.P. Perdew, and M. Ernzerhof, J. Chem. Phys. **109**, 3760 (1998).
17. S.H. Vosko, L. Wilk, and M. Nusair, Can. J. Phys. **58**, 1200 (1980).
18. J.P. Perdew and A. Zunger, Phys. Rev. B **23**, 5048 (1981).
19. J.P. Perdew and Y. Wang, Phys. Rev. B **45**, 13294 (1992).
20. D.M. Ceperley and B.J. Alder, Phys. Rev. Lett. **45**, 566 (1980).
21. O. Gunnarsson, B.I. Lundqvist, and J.W. Wilkins, Phys. Rev. B **10**, 1319 (1974).
22. S.-K. Ma and K.A. Brueckner, Phys. Rev. **165**, 18 (1968).
23. M. Rasolt and H.L. Davis, Phys. Lett. **86A**, 45 (1981).
24. J.P. Perdew, K. Burke, and Y. Wang, Phys. Rev. B **54**, 16533 (1996); ibid. **57**, 14999 (1998)(E).
25. D.C. Langreth and J.P. Perdew, Phys. Rev. B **21**, 5469 (1980).
26. D.C. Langreth and M.J. Mehl, Phys. Rev. B **28**, 1809 (1983).
27. A.D. Becke, J. Chem. Phys. **84**, 4524 (1986).
28. J.P. Perdew and Y. Wang, Phys. Rev. B **33**, 8800 (1986); ibid. **40**, 3399 (1989) (E).
29. J.P. Perdew, Phys. Rev. B **35**, 8822 (1986); ibid. **34**, 7406 (1986) (E).
30. A.D. Becke, Phys. Rev. A **38**, 3098 (1988).
31. C. Lee, W. Yang, and R.G. Parr, Phys. Rev. B **37**, 785 (1988).
32. J.P. Perdew, in *Electronic Structure of Solids '91*, edited by P. Ziesche and H. Eschrig (Akademie Verlag, Berlin, 1991), p.11.
33. J.P. Perdew, J.A. Chevary, S.H. Vosko, K.A. Jackson, M.R. Pederson, D.J. Singh, and C. Fiolhais, Phys. Rev. B **46**, 6671 (1992); ibid. **48**, 4978 (1993) (E).
34. K. Burke, J.P. Perdew, and Y. Wang, in *Electronic Density Functional Theory: Recent Progress and New Directions*, edited by J.F. Dobson, G. Vignale, and M.P. Das (Plenum, NY, 1998).
35. J.P. Perdew, K. Burke, and M. Ernzerhof, Phys. Rev. Lett. **77**, 3865 (1996); ibid. **78**, 1396 (1997) (E).
36. D.R. Hamann, Phys. Rev. B **55**, R10157 (1997).
37. Y. Zhang, W. Pan, and W. Yang, J. Chem. Phys. **107**, 7921 (1997).
38. D.C. Patton and M.R. Pederson, Phys. Rev. A **56**, R2495 (1997).
39. B. Montanari, P. Ballone, and R.O. Jones, J. Chem. Phys. **108**, 6947 (1998).
40. R.Q. Hood, M.Y. Chou, A.J. Williamson, G. Rajagopal, and R.J. Needs, Phys. Rev. B **57**, 8972 (1998).
41. P. Fulde, *Electron Correlations in Molecules and Solids*, 2nd edition (Springer Verlag, Berlin, 1993).
42. D.C. Patton, D.V. Porezag, and M.R. Pederson, Phys. Rev. B **55**, 7454 (1997).
43. M. Levy and J.P. Perdew, Phys. Rev. A **32**, 2010 (1985).
44. M. Levy, Phys. Rev. A **43**, 4637 (1991).
45. M. Gell-Mann and K.A. Brueckner, Phys. Rev. **106**, 364 (1957).
46. M. Levy and J.P. Perdew, Phys. Rev. B **48**, 11638 (1993); **55**, 13321 (1997) (E).
47. E.H. Lieb and S. Oxford, Int. J. Quantum Chem. **19**, 427 (1981).

48. J.P. Perdew and S. Kurth, in *Density Functionals: Theory and Applications*, edited by D.P. Joubert (Vol. 500, Springer Lecture Notes in Physics, 1998).
49. L. Kleinman, preprint.
50. A. Zupan, K. Burke, M. Ernzerhof, and J.P. Perdew, J. Chem. Phys. **106**, 10184 (1997).
51. J.P. Perdew, M. Ernzerhof, K. Burke, and A. Zupan, J. Chem. Phys. **108**, 1522 (1998).
52. J.P. Perdew, M. Ernzerhof, A. Zupan, and K. Burke, in *Advances in Quantum Chemistry*, special issue edited by Jorge Seminario (Academic Press, 1998).
53. A. Zupan, P. Blaha, K. Schwarz, and J.P. Perdew, Rev. B (to appear).
54. L. Pollack and J.P. Perdew, in preparation.
55. Y. Wang and J.P. Perdew, Phys. Rev. B **44**, 13298 (1991).
56. Y. Wang, J.P. Perdew, J.A. Chevary, L.D. Macdonald, and S.H.Vosko, Phys. Rev. A **41**, 78 (1990).
57. D.C. Langreth and J.P. Perdew, Phys. Rev. B **15**, 2884 (1977).
58. M. Ernzerhof, J.P. Perdew, and K. Burke, Int. J. Quantum Chem. **64**, 285 (1997).
59. J.P. Perdew and M. Ernzerhof, in *Electronic Density Functional Theory: Recent Progress and New Directions*, edited by J.F. Dobson, G. Vignale, and M.P. Das (Plenum, 1998).
60. E. Clementi and S.J. Chakravorty, J. Chem. Phys. **93**, 2591 (1990).
61. A.D. Becke, J. Chem. Phys. **98**, 5648 (1993).
62. J.P. Perdew, M. Ernzerhof, and K. Burke, J. Chem. Phys. **105**, 9982 (1996).
63. M. Ernzerhof, Chem. Phys. Lett. **263**, 499 (1996).
64. K. Burke, M. Ernzerhof, and J.P. Perdew, Chem. Phys. Lett. **265**, 115 (1997).
65. M. Seidl, J.P. Perdew, and M. Levy, submitted to Phys. Rev. A.
66. K. Burke, J.P. Perdew, and M. Ernzerhof, in *Electronic Density Functional Theory: Recent Progress and New Directions*, edited by J.F. Dobson, G. Vignale, and M.P. Das (Plenum, 1998).
67. A. Goerling and M. Levy, Phys. Rev. B **47**, 13 105 (1993).
68. F.W.Kutzler and G.S. Painter, Phys. Rev. B **43**, 6865 (1991).
69. V. Ozolins and M. Koerling, Phys. Rev. B **48**, 18304 (1993).
70. M. Fuchs, M. Bockstedte, E. Pehlke, and M. Scheffler, Phys. Rev. B **57**, 2134 (1998).
71. J.W. Negele and D. Vautherin, Phys. Rev. C **5**, 1472 (1972).
72. T. Van Voorhis and G.E. Scuseria, J. Chem. Phys. **109**, 400 (1998).
73. J.P. Perdew, S. Kurth and A. Zupan, in preparation.
74. M. Springer, P.S. Svendsen, and U. von Barth, Phys. Rev. B **54**, 17392 (1996).
75. J.P. Perdew, Int. J. Quantum Chem. S **27**, 93 (1993).
76. S. Kurth and J.P. Perdew, submitted to Phys. Rev. B.

## Discussion

**Sawatzky:** Are the effects of quantum fluctuations like those in a singlet liquid or a RVB-like ground state in a 1D Heisenberg insulator properly included in the ground state energy and charge distribution in DFT?

**Perdew:** In principal, yes. If the exact density functional were known then the ground state energy and density will be exactly determined including the effects of quantum fluctuations.

# ON TIME-INDEPENDENT DENSITY-FUNCTIONAL THEORIES FOR EXCITED STATES

Mel Levy

Department of Chemistry and

Quantum Theory Group
Tulane University
New Orleans, Louisiana 70118

## INTRODUCTION

Ground-state density-functional theory (DFT) is now routinely used and generally provides the most powerful and efficient method today for electronic structure calculations [1-9]. My purpose here is to briefly review aspects of several excited-state formulations that are closely related to the ground-state time-independent Hohenberg-Kohn theorem. The appealing time-dependent theory for excited states [10] was described at this conference by Gross.

After a very short summary of the ground-state formulation, a variational functional for an individual excited state is defined and several of its properties are elucidated for the purpose of approximating the functional. Then, it is reviewed [11-15] that certain exact excitation energies may be obtained by subtracting the highest-occupied Kohn-Sham (KS) orbital energy of the ground-state calculation from the highest-occupied KS orbital energy of the appropriate excited-state calculation. This fact about orbital energies is then combined with a recently derived Koopmans theorem for ensembles [16] to obtain a new exact expression, for an excitation energy, that involves Kohn-Sham determinants from a ground-state calculation and an ensemble correlation potential for the situation where the excited state weight approaches zero. This expression is then utilized to give insight into the recent encouraging numerical results from of the adiabatic connection perturbation theory, ACPT, through first order, for approximating excitation energies from the valence shell [18]. The ACPT Hamiltonian was formulated by Görling and Levy [17] for a ground state perturbation theory and utilized by Görling for an excited-state perturbation theory [18].

## THE GROUND-STATE PROBLEM

Assume that $\hat{H}_v$ is our Hamiltonian of interest:

*Electron Correlations and Materials Properties*, edited by Gonis *et al.*
Kluwer Academic/Plenum Publishers, 1999.

$$\hat{H}_v = \hat{T} + \hat{V}_{ee} + \sum_{i=1}^{N} v(\vec{r}_i) \ , \tag{1}$$

where $\hat{T}$ is the kinetic energy operator, $\hat{V}_{ee}$ is the electron-electron repulsion operator, and $\hat{v}(\vec{r})$ is a local-multiplicative attractive potential. Then, by the Hohenberg-Kohn theorem, the ground-state energy is obtained from

$$E_o = \min_n \left\{ \int v(\vec{r}) n(\vec{r}) d^3r + F[n] \right\} = \int v(r) n_o d^3r + F[n_o] \ , \tag{2}$$

where the formal constrained-search identification of F[n] is [19, 20]

$$F[n] = \min_{\Psi \to n} <\Psi | \hat{T} + \hat{V}_{ee} | \Psi> \ , \tag{3}$$

or

$$F[n] = <\Psi_n^{\min} | \hat{T} + \hat{V}_{ee} | \Psi_n^{\min}> \ , \tag{4}$$

where $\Psi_n^{\min}$ is that antisymmetric function that yields the trial electron density n and simultaneously minimizes $<\hat{T} + \hat{V}_{ee}>$.

The minimization in Eq. (2) results in the following Kohn-Sham (KS) equations [2]:

$$\left\{ -\frac{1}{2}\nabla^2 + v_s([n_o]; \vec{r}) \right\} \varphi_i(r) = \epsilon_i \varphi_i(\vec{r}) \ , \ i = 1, 2, \cdots, \frac{N}{2} \tag{5}$$

where the KS potential $v_s$ is

$$v_s([n_o]; \vec{r}) = v(\vec{r}) + u([n_o]; \vec{r}) + v_{xc}(n_o]; \vec{r}) \ . \tag{6}$$

In Eq. (6), u is the Hartree potential given by $u([n]; \vec{r}) = \int d^3r' |\vec{r} - \vec{r}'|^{-1} \rho(\vec{r}')$, and $v_{xc}$ is the exchange-correlation potential. That is, $v_{xc}([n]; \vec{r}) = \dfrac{\delta E_{xc}[n]}{\delta n(\vec{r})}$, where $E_{xc}$ is the exchange-correlation functional. For simplicity of presentation, it shall be assumed here that the system of interest is such that each KS space orbital $\varphi_i$ is doubly occupied.

## VARIATIONAL FUNCTIONALS FOR INDIVIDUAL EXCITED STATES

Assume one is interested in the energy and density of the kth state of $\hat{H}_v$. Then extending the work of Levy and Perdew [21] and Lieb [22], Levy and Nagy [23] have observed very recently that there exists the following formal functional $F_k[n, v]$ for this purpose:

$$F_k[n, v] = <\Psi_k[n, v] | \hat{T} + \hat{V}_{ee} | \Psi_k[n, v]> \ , \tag{7}$$

where, in the spirit of Eq. (3), $\Psi_k[n, v]$ is that antisymmetric wavefunction that yields the trial density n, is orthogonal to the states 0, 1, 2, ⋯k-1 of $\hat{H}_v$, and simultaneously minimizes $<\hat{T} + \hat{V}_{ee}>$. With $F_k[n, v]$ thus defined, it follows that

$$E_k = \min_n <\Psi_k[n,v] \,|\, \hat{H}_v \,|\, \Psi_k[n,v]> \tag{8}$$

so that

$$E_k = \min_n \left\{ \int v(\vec{r})\, n(\vec{r}) d^3r \;+\; F_k[n,v] \right\}, \tag{9}$$

where $E_k$ is the energy of the kth state of $\hat{H}_v$.

Instead of the v in $F_k[n, v]$, we could use any entity that implies v. For instance, we could replace v in F[n, v] by $n_o$ or by any one of the occupied (or unoccupied) $\varphi_i$ in Eq. (5), because $\varphi_i$ identifies $v_s$ which identifies $n_o$ which identifies v.

**Properties of $F_k$[n, v]**

In order to approximate $F_k[n, v]$ one must know some of its key properties [23]. With this in mind, their definitions directly imply

$$F_{k+1}[n,v] \;\ge\; F_k[n,v] \;\ge\; F[n] \,. \tag{10}$$

Further,

$$\min_n <\Psi_k[n,v'] \,|\, \hat{H}_v \,|\, \Psi_k[n,v']> \;\le\; E_k\,, \tag{11}$$

or

$$\min_n \left\{ \int v(\vec{r}) n(\vec{r}) d^3r \;+\; F_k[n,v'] \right\} \;\le\; E_k\,, \tag{12}$$

where the inequality applies when v′ is any one-body potential that differs from v by more than an additive constant. Eq. (12) is a very severe constraint and thus quite useful. This inequality arises by the following argument: Observe that [23]

$$\min_n <\Psi_k[n,v'] \,|\, \hat{H}_v \,|\, \Psi_k[n,v']> \;\le\; <G|\hat{H}_v|G>\,, \tag{13}$$

where G is defined by

$$G \;=\; \sum_{\ell=o}^{k} C_\ell G_\ell\,, \tag{14}$$

where $G_\ell$ is the $\ell$-th eigenstate of $\hat{H}_v$, where the coefficients $C_\ell$ are determined such that the normalized function G is orthogonal to states 0 to k-1 of $\hat{H}_{v'}$, and where $\hat{H}_{v'}$, is $\hat{H}_v$ with v replaced by v′.

Next, use

$$<G|\hat{H}_v|G> \;=\; \sum_{\ell=0}^{k} |C_\ell|^2 E_\ell \;\le\; E_k\,. \tag{15}$$

The combination of equations (13) and (15) yields the inequality in expression (11). Finally, since Eq. (12) is an equality when v′ = v, it follows that

$$E_k \;=\; \max_{v'} \min_n \left\{ \int v(\vec{r}) n(\vec{r}) d^3r \;+\; F_k[n,v'] \right\}. \tag{16}$$

Recently, we have formulated [23] a Kohn-Sham theory that corresponds to the use of $F_k[n, v]$. With this KS theory, a constraint arises from the fact that the highest-occupied orbital energy equals the negative of the ionization energy with respect to the removal of an electron from state k to form the ground-state of the (N-1)-electron system [11-15].

## EXACT EXCITATION ENERGIES FROM ORBITAL ENERGIES

The excitation energy $E_k - E_o$ may be usefully expressed as

$$E_k - E_o = I_o - I_k , \tag{17}$$

where

$$I_o = E_o^{N-1} - E_o \tag{18}$$

and

$$I_k = E_o^{N-1} - E_k . \tag{19}$$

In equations (18) and (19), $E_o^{N-1}$ is the ground-state energy of $\hat{H}_v$ with one electron removed.

It is now becoming well known that [11-13, 15]

$$I_o = -\epsilon_{N/2} , \tag{20}$$

where $\epsilon_{N/2}$ is the highest-occupied orbital energy in Eq. (5). So, an exact Kohn-Sham ground-state calculation gives half what is needed in Eq. (17). Similarly, unless prevented by symmetry, $I_k$ is given by the negative of the highest-occupied orbital energy in an isolated excited-state calculation, as described in the previous section, or by the negative of the highest-occupied orbital energy in an ensemble calculation [14].

## APPROXIMATE EXCITATION ENERGIES FROM GROUND-STATE KOHN-SHAM CALCULATIONS

For simplicity of presentation, only the smallest excitation energy shall be considered, $E_1 - E_0$. I shall investigate three simple formulas to estimate this excitation. Each formula employs KS orbitals that are obtained from a ground-state calculation. Following are the three simple methods: (A) Use of the highest-occupied KS orbital energy to obtain $I_o$ exactly and the lowest-unoccupied KS orbital energy to approximate $I_1$. (B) Use of first-order standard perturbation theory (SPT) to approximate both $I_o$ and $I_1$. (C) Use of [17, 18] first-order adiabatic connection perturbation theory (ACPT) to obtain $I_o$ exactly and to approximate $I_1$.

All three formulas employ the same KS wavefunctions but utilize different operators. Define the non-interacting Hamiltonian $\hat{H}_s$ by

$$\hat{H}_s = \sum_{i=1}^{N} h_s(\vec{r}_i) , \tag{21}$$

where

$$h_s(\vec{r}) = -\frac{1}{2}\nabla^2 + v_s(\vec{r}) . \tag{22}$$

Then, Formula (A) approximates $E_1$ - $E_o$ by

$$E_1 - E_o \simeq <\Phi_1|\hat{H}_s|\Phi_1> - <\Phi_0|\hat{H}_s|\Phi_0> = \epsilon_{m+1} - \epsilon_m \tag{23}$$

where $\Phi_0$ is the ground-state of $\hat{H}_s$ and $\Phi_1$ is the first-excited state of $\hat{H}_s$. If the first-excited energy level of $\hat{H}_s$ is degenerate, then $\Phi_1$ will be defined as the function in the degenerate set that minimizes $<\hat{H}>$. In any case, $\Phi_1$ utilizes $\{\varphi_0, \varphi_1, \cdots, \varphi_{(N/2)+1}\}$ while $\Phi_0$ utilizes $\{\varphi_0, \varphi_1, \cdots, \varphi_{N/2}\}$. In $\Phi_0$, the first N/2 orbitals are doubly occupied, and the rest are unoccupied, while $\varphi_{N/2}$ and $\varphi_{(N/2)+1}$ are singly-occupied in $\Phi_1$.

Formula (B) approximates $E_1$ - $E_0$ by

$$E_1 - E_0 \simeq <\Phi_1|\hat{H}_v|\Phi_1> - <\Phi_0|\hat{H}_v|\Phi_0>, \tag{24}$$

and formula (C) approximates $E_1$ - $E_0$ by

$$E_1 - E_o \simeq <\Phi_1|\hat{H}_v + \sum_{i=1}^{N} v_c([n_o];\vec{r}_i)|\Phi_1> - <\Phi_o|\hat{H}_v + \sum_{i=1}^{N} v_c([n_o];\vec{r}_i)|\Phi_o>, \tag{25}$$

where $v_c$ is the correlation potential associated with the generation of the ground-state energy and density. That is,

$$v_c\,(n_o];\vec{r}) = \left.\frac{\delta E_c[n]}{\delta n(\vec{r})}\right|_{n=n_o}, \tag{26}$$

where $E_c[n]$ is the familiar ground-state correlation energy functional, defined by

$$E_c[n] = <\Psi_n^{min}|\hat{T} + \hat{V}_{ee}|\Psi_n^{min}> - <\Phi_n^{min}|\hat{T} + \hat{V}_{ee}|\Phi_n^{min}>, \tag{27}$$

where $\Phi_n^{min}$ yields n and minimizes just $<\hat{T}>$, while, of course, $\Psi_n^{min}$ yields n and minimizes $<\hat{T} + \hat{V}_{ee}>$.

Surprisingly, the best results, thus far, have been obtained by formula (C), Eq. (25), as shall be explained later. I use the expression "surprisingly" because it probably is not at all obvious to most readers why the inclusion of the $v_c$ in Eq. (25) should yield improved results in general. After all, v is the attractive potential for the system of interest, not v + $v_c$. With this in mind, in the next section I will appeal to ensemble density-functional theory to analyze the error in each formula by employing an exact formula for $I_1$ from ensemble theory.

## EXACT EXCITATION ENERGY EXPRESSIONS FROM ENSEMBLE THEORY

Define $\Psi_o$ as the ground state of $\hat{H}_v$ and $\Psi_1$ as its first excited state (Assume for discussion purposes that the ground level is non-degenerate). Then, by Gross, Oliviera, and Kohn [24],

$$\begin{aligned} &\min_{\{\Psi_A,\Psi_B\}} [(1-w)<\Psi_A|\hat{H}_v|\Psi_A> + w<\Psi_B|\hat{H}_v|\Psi_B>] \\ &= (1-w)<\Psi_o|\hat{H}_v|\Psi_o> + w<\Psi_1|\hat{H}_v|\Psi_1> \end{aligned} \tag{28}$$

where $\Psi_A$ and $\Psi_B$ are constrained to form an orthohromal set and where $0 < w \leq < ½$. Equation (28) becomes the earlier one of Theophilou [25] when $w = \frac{1}{2}$.

Next define the N-electron density $n_w$ by

$$n_w(\vec{r}) = (1-w)n_o(\vec{r}) + wn_1(\vec{r}), \tag{29}$$

where $n_o$ and $n_1$ are the densities of $\Psi_o$ and $\Psi_1$, the ground and first-excited states of $\hat{H}_v$. Let's also express $n_w$ as

$$n_w(\vec{r}) = (1-w)n_o^w(\vec{r}) + wn_1^w(\vec{r}), \tag{30}$$

where $n_0^w$ and $n_1^w$ are the ground-state and first-excited state densities of some non-interacting Hamiltonian with a local potential. These densities arise as the solutions of

$$\min_{\{\Phi_A,\Phi_B\}} [(1-w)<\Phi_A|\hat{H}_w|\Phi_A> + w<\Phi_B|\hat{H}_w|\Phi_B>]$$
$$= (1-w)<\Phi_o^w|\hat{H}_w|\Phi_o^w> + w<\Phi_1^w|\hat{H}_w|\Phi_1^w> \tag{31}$$

where

$$H_w = \hat{T} + \hat{V}_{ee} + \sum_{i=1}^{N} w(\vec{r}_i), \tag{32}$$

and where $\Phi_A$ and $\Phi_B$ form an orthonormal set and are simultaneously eigenstates of some non-interacting Hamiltonian with a local potential. (Observe that the local potential $w(\vec{r})$ depends upon the weight w in Eq. (28)).

Now let $w \to 0$. Then for the purpose of taking expectation values that are relevant to the present discussion, it is true that $\Phi_o^w \to \Phi_o$ and $\Phi_1^w \to \Phi_1$. However, as $w \to 0$, it is important to note that the exponential asymptotic decay of $n_w$ goes as the decay of $n_1$ and not as the decay of $n_0$, as $|\vec{r}| \to \infty$.

Next, define v′ by

$$v'(\vec{r}) = \lim_{w \to o} w(\vec{r}). \tag{33}$$

Then it can be shown that utilization of a recently derived ensemble Koopmans theorem gives the following exact relation [16]

$$-I_1 = <\Phi_1|\hat{H}_v + \sum_{i=1}^{N} v_c'(\vec{r}_i)|\Phi> - <\Phi_o^{N-1}|\hat{H}_v^{N-1} + \sum_{i=1}^{N-1} v_c'(\vec{r}_i)|\Phi_o^{N-1}>, \tag{34}$$

where $\hat{H}_v^{N-1}$ is $\hat{H}_v$ with one electron removed, where $\Phi_o^{N-1}$ is formed from $\Phi_o$ by removing an electron from $\varphi_{N/2}$, and where the ensemble correlation potential, $v_c'$, is

$$v_c'(\vec{r}) = v'(\vec{r}) - v(\vec{r}). \tag{35}$$

Equation (34) should be compared with the analogous one for $I_o$, which I observe to be

$$-I_0 = <\Phi_0|\hat{H}_v + \sum_{i=1}^{N} v_c(\vec{r}_i)|\Phi_0> - <\Phi_o^{N-1}|\hat{H}_v^{N-1} + \sum_{i=1}^{N-1} v_c(\vec{r}_i)|\Phi_o^{N-1}>. \tag{36}$$

Equation (36) results from manipulating

$$\int d^3r |\varphi_m(\vec{r})|^2 \frac{\delta V_{ee}[n]}{\delta n(\vec{r})} = <\Phi_0|\hat{V}_{ee}|\Phi_0> - <\Phi_0^{N-1}|\hat{V}_{ee}|\Phi_o^{N-1}> \tag{37}$$

In the above expression $m = \frac{N}{2}$, and $V_{ee}[n]$ is the electron-electron repulsion energy of that ground-state Kohn-Sham single determinant that yields n. Equation (37) was arrived at for both finite [26, 27] and infinite [28] systems. Expression (34) follows from an expression analogous to Eq. (37), but it involves, instead, the non-interacting ensemble electron repulsion functional [16]. To obtain Eq. (36) from Eq. (37), add

$$<\Phi_0|\sum_{i=1}^{N} -\frac{1}{2}\nabla_i^2 + v(\vec{r}_i) + v_c(\vec{r}_i)|\Phi_0>$$

to both sides of Eq. (37) and substract

$$<\Phi_0^{N-1}|\sum_{i=1}^{N-1} -\frac{1}{2}\nabla_i^2 + v(\vec{r}_i) + v_c(\vec{r}_i)|\Phi_0^{N-1}>$$

from both sides of this equation. Then observe that the left-hand-side of the modified Eq. (37) is $\epsilon_m$, which equals $-I_0$, and the right-hand-side of the modified Eq. (37) equals the right-hand-side of Eq. (36).

The combination of relations (34) and (36) yields the following exact expression for the excitation energy:

$$\begin{aligned} E_1 - E_0 &= <\Phi_1|\hat{H}_v + \sum_{i=1}^{N} v_c(\vec{r}_i)|\Phi_1> - <\Phi_0|\hat{H}_v + \sum_{i=1}^{N} v_c(\vec{r}_i)|\Phi_0> \\ &+ \int d^3r\, [v_c'(\vec{r}) - v_c(\vec{r})]\, |\varphi_{m+1}(\vec{r})|^2 , \end{aligned} \tag{38}$$

where, or course,

$$|\varphi_{m+1}(\vec{r})|^2 = <\Phi_1|\hat{n}|\Phi_1> - <\Phi_0|\hat{n}|\Phi_0> , \tag{39}$$

with $\hat{n}$ as the density operator. Our other exact expression for the excitation energy is

$$\begin{aligned} E_1 - E_0 &= <\varphi_{m+1}(\vec{r})| -\frac{1}{2}\nabla^2 + v(\vec{r}) + u(\vec{r}) + v_x'(\vec{r}) + v_c'(\vec{r})|\varphi_{m+1}(\vec{r})> \\ &- <\varphi_m(\vec{r})| -\frac{1}{2}\nabla^2 + v(\vec{r}) + u(\vec{r}) + v_x(\vec{r}) + v_c(\vec{r})|\varphi_m(\vec{r})> , \end{aligned} \tag{40}$$

or

$$\begin{aligned} E_1 - E_0 &= <\varphi_{m+1}| -\frac{1}{2}\nabla^2 + v_s|\varphi_{m+1}> - <\varphi_m| -\frac{1}{2}\nabla^2 + v_s|\varphi_m> \\ &+ \int d^3r [v_x' - v_x]|\varphi_{m+1}|^2 + \int d^3r [v_c' - v_c]\, |\varphi_{m+1}|^2 . \end{aligned} \tag{41}$$

where $[v_x' - v_x]$ is equal to the functional derivative of the non-interacting ensemble repulsion energy associated with the right-hand-side of Eq. (31), evaluated at $n_w$ as $w \to 0$, minus $\frac{\delta V_{ee}[n]}{\delta n}|_{n=n_o}$. The Hartree potential, $u(\vec{r})$, cancels out because the subtraction is made as $w \to 0$.

Expression (40) results from [14]

$$-I_1 = <\varphi_{m+1}| -\frac{1}{2}\nabla^2 + v + u + v'_x + v'_c|\varphi_{m+1}> \tag{42}$$

in combination with Eq. (20). An equation corresponding to Eq. (34) holds for each higher excitation energy. To obtain the desired equation, let all the weights approach zero, except the ground-state one, and make the highest excited-state weight the smallest.

## ERRORS OF APPROXIMATE FORMULAS

Comparison of expressions (23) and (41) gives the following relation for ΔA, the error in method A, defined as the exact expression minus the approximate expression:

$$\Delta A = \int d^3r [v'_{xc}(\vec{r}) - v_{xc}(\vec{r})] \, |\varphi_{m+1}(\vec{r})|^2 \,. \tag{43}$$

Comparison of Eq. (24) with Eq. (38) gives the error ΔB as

$$\Delta B = \int d^3r [v'_c(\vec{r})\varphi_{m+1}(\vec{r}) - v_c(\vec{r})\varphi_m(\vec{r})] \,, \tag{44}$$

and comparison of Eq. (25) with Eq. (38) gives the error ΔC as

$$\Delta C = \int d^3r [v'_c(\vec{r}) - v_c(\vec{r})] \, |\varphi_{m+1}(\vec{r})|^2 \,. \tag{45}$$

## NUMERICAL RESULTS AND ANALYSIS

Filipe, Umrigar, and Gonze [29] computed the excitation energies in the atoms He, $Li^+$, and Be. The first excitation energy was approximated by formulas (A), (B), and (C), as given here, and higher excitation energies were approximated by the obvious extensions of these formulas. The root mean square errors with the three formulas were obtained for He, $Li^+$, and Be, respectively. With formula (A), the errors, in Hartree atomic units, were 0.0067, 0.0167, and 0.0178. With formula (B), the errors were 0.0159, 0.0148, and 0.0358. And, with formula (C), the errors were 0.0039, 0.0056, and 0.0099. For two electrons, it has been ascertained [29] that formula (C) is equivalent to the single pole approximation in the time-dependent DFT for excited states [10].

For the systems tested, ΔC was smallest in magnitude. This seems reasonable by looking at expressions (43)-(45) for the errors ΔA, ΔB, and ΔC. The relation for ΔA containes $v'_{xc} - v_{xc}$, while the relation for ΔC contains only the correlation components of $v'_{xc}$ and $v_{xc}$; the relatively large exchange components are missing from ΔC. Further, although ΔB also contains only $v'_c$ and $v_c$, there is generally less chance of $v'_c - v_c$ cancellation because $v'_c$ and $v_c$ are multiplied by different orbital densities in Eq. (44).

## CLOSING REMARKS

The Hohenberg-Kohn ground-state functional is a special case of $F_k[n, v]$ with k = 0. The case of the lowest excited-state, of a symmetry different from the ground-state one, is a special case of $F_k[n, v]$ with $k = 1$. When approximating the F for this case, it is therefore important to keep in mind that $F_1[n, v] \geq F[n]$.

What makes the ground-state problem easier than the excited-state one is of course the fact that there are no orthogonality requirements in the ground-state situation. In spite of the orthogonality restrictions, it is hoped that variational constraints (12) and the ionization energy constraints will help make approximations to $F_k[n, v]$ feasible.

A non-interacting kinetic energy is needed to be defined [23] for a Kohn-Sham theory for an individual excited state. Let's, for simplicity of discussion, just look at k=1. Identify $\hat{H}_w$ as a non-interacting Hamiltonian with an excited-state density equal to the first-excited state density of $\hat{H}_v$. Moreover, of all such non-interacting Hamiltonians, $\hat{H}_w$ is the one whose ground-state density resembles closest the ground-state density of $\hat{H}_v$. This definition of $\hat{H}_w$ is not quite unique in that we could, for instance, substitute the word "wavefunction" for "density" in both places that it appears in this last sentence. In any case, define $T_s[n, v]$ by [23]

$$T_s[n,v] = \min_{\Phi \to n} <\Phi|\hat{T}|\Phi>,$$

where $\Phi$ is constrained to be orthogonal to the ground-state of $\hat{H}_w$. (For simplicity of discussion, the ground-states of both $\hat{H}_v$ and $\hat{H}_w$ are assumed to be non-degenerate here.) We [23] have found the KLI approximation [26] useful for the exchange component of the individual excited-state calculation. Also, Nagy has presented individual excited-state Kohn-Sham calculations at this conference [34].

In the second half of our presentation, simple formulas were analyzed that employ excited state determinants from a ground-state Kohn-Sham calculation. An explanation was given for the relative success of the first-order adiabatic connection perturbation theory (ACPT) formula, where the ground-state correlation potential is simply added to v. Utilization was made of ensemble theory with a vanishing excited-state weight. As mentioned earlier, for two electrons this perturbation theory has been observed [29] to be equivalent to the single pole approximation in the utilization of linear response in time-dependent DFT for excited states [10]. This time-dependent theory is attractive because it is able to provide an accurate correction to (A), the use of unoccupied KS orbital energies from a ground-state calculation. Ensemble theory should also be studied especially since key properties of the ensemble exchange-correlation functionals are now better known [14, 16 24, 25, 30-34].

## REFERENCES

[1] P. Hohenberg and W. Kohn, Phys. Rev. 136, B864 (1964).

[2] W. Kohn and L. J. Sham, Phys. Rev. 140, A1133 (1965).

[3] R. G. Parr and W. Yang, Density-Functional Theory of Atoms and Molecules (Oxford University Press, Oxford, 1989).

[4] R. M. Dreizler and E., K. U. Gross, Density Functional Theory (Springer-Verlag, Berlin, 1990).

[5] Density Functional Theory, edited by E. K. U. Gross and R. M. Dreizler (Plenum, New York, 1995).

[6] Recent Developments and Applications of Modern Density Functional Theory, edited by J. M. Seminario (Elsevier, Amsterdam, 1996).
[7] Á. Nagy, Phys. Rep. 398, 1 (1998).
[8] J. P. Perdew, K. Burke, and M. Ernzerhof, Phys. Rev. Lett. 77, 3865 (1996); 78, 1396 (1997)(E).
[9] J. P. Perdew and S. Kurth, in Density Functionals: Theory and Applications, edited by D. P. Joubert (Vol. 500, Springer Lecture Notes in Physics, 1997).
[10] M. Petersilka, U. J. Grossman, and E. K. U. Gross, Phys. Rev. Lett. 76, 1212 (1996); M. Petersilka and E. K. U. Gross, Int. J. Quantum Chem. S 30, 1393 (1996).
[11] J. P. Perdew, R. G. Parr, M. Levy, and J. L. Balduz, Jr., Phys. Rev. Lett. 49, 1691 (1982).
[12] M. Levy, J. P. Perdew, and V. Sahni, Phys. Rev. A 30, 2745 (1984).
[13] C. O. Almbladh and U. von Barth, Phys. Rev. B 31, 3231 (1985).
[14] M. Levy, Phys. Rev. A 52, R 4313 (1995).
[15] J. P. Perdew and M. Levy, Phys. Rev. B. 56, 16021 (1997).
[16] M. Levy and Á. Nagy, Phys. Rev. A, in press.
[17] A. Görling and M. Levy, Phys. Rev. B 47, 13105 (1993).
[18] A. Görling, Phys. Rev. A 54, 3912 (1996).
[19] M. Levy, Proc. Natl. Acad. Sci. USA 76, 6062 (1979).
[20] M. Levy, Phys. Rev. A 26, 1200 (1982).
[21] M. Levy and J. P. Perdew, in Density Functional Methods in Physics, edited by R. M. Dreizler and J. da Providencia (Plenum, New York, 1985).
[22] E. H. Lieb, in Ref. [21].
[23] M. Levy and Á. Nagy, unpublished.
[24] E. K. U. Gross, L. N. Oliveira, and W. Kohn, Phys. Rev. A 37, 2805, 2809 (1988).
[25] A. K. Theophilou, J. Phys. C 12, 5419 (1979).
[26] J. B. Kreiger, Y. Li, and G. J. Iafrate, Phys. Rev. A 45, 101 (1992); 46, 5453 (1992).
[27] M. Levy and A. Görling, Phys. Rev. A 53, 3140 (1996).
[28] J. P. Perdew in Ref. [21].
[29] C. Filippi, C. J. Umrigar, and X. Gonze, J. Chem. Phys. 107, 9994 (1997).
[30] W. Kohn, Phys. Rev. A 34, 737 (1986).
[31] Á. Nagy, Phys. Rev. A 42, 4388 (1990); 49, 3074 (1994).
[32] Á. Nagy, Int. J. Quantum Chem. 56, 225 (1995).
[33] Á. Nagy, Int. J. Quantum Chem. S 29, 297 (1995).
[34] Á. Nagy, Adv. Quantum Chem. 29, 159 (1997).

# QUASIPARTICLE AND OPTICAL EXCITATIONS IN SOLIDS AND CLUSTERS

Michael Rohlfing and Steven G. Louie

Department of Physics
University of California
Berkeley, California 94720-7300, USA, and
Materials Science Division
Lawrence Berkeley National Laboratory
Berkeley, CA 94720, USA

## INTRODUCTION

In the last decades much progress has been achieved in the description of the electronic structure of real materials from first principles. Density-functional theory (DFT) has proven to be a very powerful tool in investigating the electronic ground state and determining the atomic structure of condensed matter.[1,2] The evaluation of the DFT within the local-density approximation (LDA) or the generalized gradient approximation (GGA) for the exchange-correlation energy, and the use of powerful computers, allow for a highly efficient treatment of up to hundreds of atoms.

However, only limited information about spectroscopic properties is available from such ground-state calculations. Quasiparticle (QP) excitations, as they occur in photoemission and tunneling experiments, are not fully described by the Kohn-Sham eigenvalues. In fact, the band structures given by the LDA are often in distinct disagreement from experimental data, showing systematic deviations of band dispersions and band gaps. In addition to the failures in the band structure, two-particle excitations are also not ontained with satisfying accuracy. In particular, optical spectra, which correspond to electron-hole excitations in the electronic system, cannot be described by straight forward use of DFT or other ground-state theories. The shape of the spectra is often not correct, the peaks in the spectra are found at too low energies and do not have the correct strength. In addition to that, bound exciton states, formed from interacting electrons and holes, are not described within a band-structure picture, at all.

A rigorous approach to describe excited states of an electronic system is to investigate Green's functions of increasing order. One-particle excitations, i.e. the creation of quasi-electrons and quasi-holes, correspond to the spectral properties of the one-particle Green's function. The reliable determination of the one-particle Green's function, however, requires the accurate treatment of the electronic self energy, which is a complicated non-local, energy-dependent operator. A major achievement in the field was given by the *GW* approximation[3] (GWA) for the self energy, which allows for a very accurate evaluation of the self energy on the basis of results from a preceding DFT calculation.[4–6] Highly accurate band structures for real materials have been obtained by this method, including bulk semiconductors, insulators, metals, semiconductor surfaces, atoms, defects, and clusters. The band structures are typically in very good agreement with experimental data with remaining uncertainties in the order of 0.1 eV, thus making the GWA a standard tool in predicting the electronic quasiparticle spectrum of moderately correlated materials in various situations.

The QP band structure given by the GWA successfully describes the properties of individually excited electrons and holes of semiconductors. The calculation of optical spectra from a QP band structure yields better results than the LDA (simply because the gaps have been corrected), but still shows systematic deviations from experiment. The shape of the spectra is still incorrect, and bound excitons are not described. The reason for these shortcomings is that, in optical processes, electrons and holes are created simultaneously. The particles interact with each other and form bound excitons and unbound, correlated electron-hole pairs. This makes it impossible to treat electrons and holes as independent particles.[7–10] Instead, the investigation of coupled electron-hole pairs requires the calculation of the two-particle Green's function, fully including the electron-hole interaction, beyond a QP band-structure calculation. This is the goal of our present work.

The paper is organized as follows. We first introduce the theoretical framework of our approach. This is done in three steps: the calculation of the electronic ground state within DFT, the calculation of the one-particle Green's function and of the QP band structure, and finally the calculation of the two-particle Green's function and the evaluation of coupled electron-hole excitations and optical spectra. Thereafter, we present and discuss results for two types of systems. First we discuss the optical absorption spectrum and bound excitons of bulk crystals (GaAs and LiF). Then we study optical excitations in small semiconductor clusters (H-terminated Si clusters), in which the electronic spectrum is completely different from that of extended crystals due to the spatial confinement.

## THEORETICAL FRAMEWORK

The calculation of quasiparticle and optical excitations is done in three steps, corresponding to the increasing order of the Green's functions involved. As a first step, we carry out a DFT calculation of the system which yields the electronic ground state without any excitation. If necessary, the geometric structure can be addressed on this level. The second step consists of defining the one-particle Green's function and solving its equation of motion. This leads to QP excitations and their band structure. As a third step, we investigate the two-particle Green's function and solve the corresponding equation of motion. From the two-particle Green's function, we obtain the electron-hole excitations which allow for the evaluation of optical spectra.

### Quasiparticle Excitations

The methods of density-functional theory and of quasiparticle calculations within the *GW* approximation have been discussed extensively in the literature.[1–6] Here we give

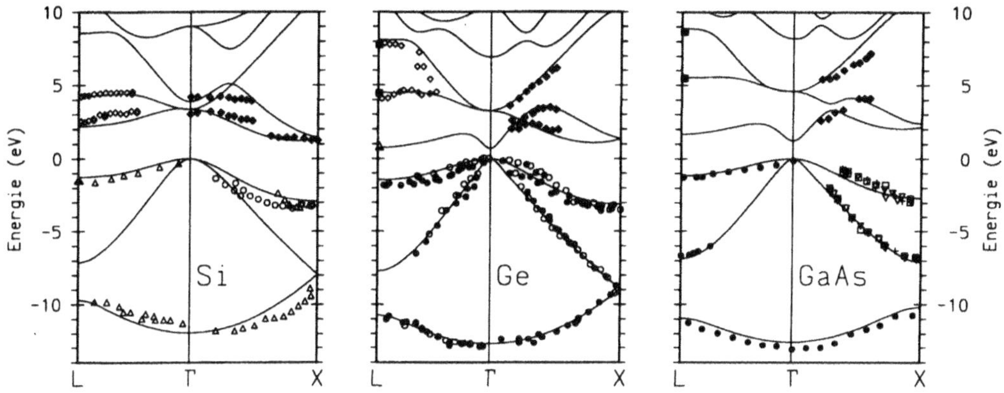


Figure 1. Calculated quasiparticle band structures of Si, Ge, and GaAs. The single symbols denote experimental data from angle-resolved direct and inverse photoemission experiments.[12]

only a very short overview.

The starting point of our approach is a DFT calculation which we perform within the local-density approximation (LDA) for the exchange-correlation potential. From this calculation we obtain the ground-state properties of the real system (geometric structure and charge density), as well as, LDA wave functions $\psi_n^{\rm LDA}(\mathbf{r})$ and energies $E_n^{\rm LDA}$ of individual electron and hole states of the associated Kohn-Sham non-interacting system. They result from the Kohn-Sham equations[1]

$$\left\{-\frac{\hbar}{2m}\nabla^2 + V_{\rm ion}(\mathbf{r}) + V_{\rm Coul}(\mathbf{r}) + V_{\rm xc}(\mathbf{r})\right\}\psi_n^{\rm LDA}(\mathbf{r}) = E_n^{\rm LDA}\psi_n^{\rm LDA}(\mathbf{r}). \tag{1}$$

For the external potential $V_{\rm ion}$ we employ norm-conserving, non-local *ab initio* pseudopotentials.[11] The LDA energies $E_n^{\rm LDA}$ should, in principle, not be interpreted as QP band-structure energies of individual electrons and holes. In fact, LDA band structures show well-known systematic deviations from experimental band-structure data. Band widths, band dispersions, and effective masses are often insufficiently described. In particular, the fundamental band gap of semiconductors and insulators is severely underestimated by LDA. Since our aim is studying optical excitations that go beyond the band-structure concept but are, nevertheless, based on it, we first have to calculate more accurate band-structure energies.

In order to obtain reliable QP band structures, one has to define the corresponding one-particle Green's function $G_1(\mathbf{r},\mathbf{r}',\omega)$ and solve the equation-of-motion for it.[3] This procedure can be understood as a perturbation on top of the electronic ground state, which is successfully described by LDA. One arrives at the following equation-of-motion for the quasiparticles:[4]

$$\left\{-\frac{\hbar}{2m}\nabla^2 + V_{\rm ion}(\mathbf{r}) + V_{\rm Coul}(\mathbf{r})\right\}\psi_n^{\rm QP}(\mathbf{r}) + \int d^3r'\Sigma(\mathbf{r},\mathbf{r}',E_n^{\rm QP})\psi_n^{\rm QP}(\mathbf{r}') = E_n^{\rm QP}\psi_n^{\rm QP}(\mathbf{r}) \tag{2}$$

The central quantity in this equation is the electron self-energy operator $\Sigma(\mathbf{r},\mathbf{r}',\omega)$, which appears as a non-local, energy-dependent generalization of the exchange-correlation potential $V_{\rm xc}(\mathbf{r})$ of the LDA. It can be calculated very accurately within the *GW* approximation (GWA):[3]

$$\Sigma = iG_1W \quad , \tag{3}$$

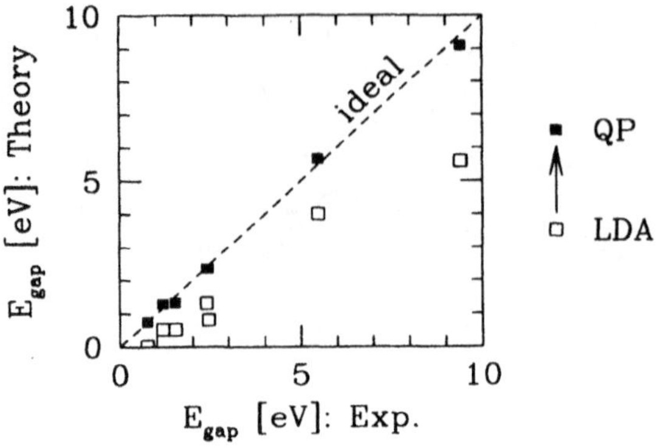

Figure 2. Gap energies of Ge, Si, GaAs, SiC, CdS, and LiCl (from left to right), calculated within LDA (open symbols) and GWA (full symbols).

i.e. as a product of the one-particle Green's function $G_1$ and the screened Coulomb interaction $W$. We calculate $W$ within the random-phase approximation from LDA results. The LDA wave functions $\psi_n^{\text{LDA}}$ are in most cases in excellent agreement with the QP amplitudes $\psi_n^{\text{QP}}$ of the one-particle Green's function. Therefore, Eq. (2) can often be solved perturbatively, leading to

$$E_n^{\text{QP}} = E_n^{\text{DFT}} + \langle \psi_n | \Sigma(E_n^{\text{QP}}) - V_{\text{xc}} | \psi_n \rangle \quad . \tag{4}$$

The expectation value of the difference $(\Sigma - V_{\text{xc}})$ defines QP corrections to the LDA band structure. Note that Eq. (4) has to be solved self-consistently since $\Sigma$ depends on $E_n^{\text{QP}}$. The resulting energies $E_n^{\text{QP}}$ form the QP band structure.

As illustrations, we show in Fig. 1 the QP band structures of Si, Ge, and GaAs (solid lines), in comparison with experimental data from angle-resolved direct and inverse photoemission spectroscopy (symbols).[12] The calculated energies are in very good agreement with the measured data for the entire energy range (−14 eV to +10 eV). In Fig. 2, we present the fundamental gap energies of several semiconductors and insulators. Ideally, the data should be on the dashed line (i.e., the calculated gap should coincide with the measured one). The LDA gap energies are too low for all materials. When the QP corrections are included, we arrive at QP gap energies in excellent agreement with experiment. These results demonstrate the accuracy of the *GW* approach. The QP energies will provide a reliable basis for the following discussion of electron-hole excitations.

## Optical Excitations

The QP excitations discussed in the last section yield very reliable band structures in excellent agreement with experimental data for single-particle excitations. When optical properties are computed from these QP band structures within an independent-particle picture, however, one encounters some systematic shortcomings. On one hand, the overall shape of the absorption spectrum of a semiconductor is not correctly described by free electron-hole transitions. The calculations typically underestimate the absorption strength at low energies and overestimate it at high energies (see, e.g., the dashed curve in Fig. 4 for the case of GaAs). On the other hand, bound exciton states are completely missing from the spectra. Both shortcomings have long been identified to originate from the missing electron-hole interaction,[7,8] but attempts to overcome these failures have been limited to relatively simple situations. The conventional approach is the effective-mass approximation (EMA) to bound excitons. The EMA relies on empirical parameters such as the effective mass tensor and the dielectric constant. It is extremely difficult to extend the EMA to higher-energy excitations above the fundamental gap. One early approach to calculate the continuum part of the absorption spectrum of a semiconductor has been presented by Hanke and Sham[9], using an empirical tight-binding approach. The theoretical study of optical excitations in electronic systems so far is thus limited and incomplete. It would be highly desirable to have a unified *ab-initio* approach that, in addition to the quasiparticle self-energy, can also handle all aspects of electron-hole interaction and optical excitations on equal footing. This is the goal of our present work. In particular, such an *ab-initio* method would be most useful for predicting the optical properties of new materials or materials under novel conditions, such as in reduced dimensions, confined geometries, or under pressure.

The main reason for the above systematic failures in describing optical excitation is that the band structure yields, by construction, information about individually excited electrons and holes. This corresponds to excitations and processes in which the total number of electrons of the system is increased (creation of an electron) or reduced (creation of a hole). In optical excitations, on the other hand, an electron and a hole are created simultaneously, and the total number of electrons remains unchanged. Such a process goes beyond the free-quasiparticle picture of a band structure, and one has to study electron-hole excitations including their interaction. To this end, we solve the Bethe-Salpeter equation (BSE) for the two-particle Green's function of electron-hole pairs. The BSE is expressed in terms of the wave functions and energies that are obtained from an *ab-initio* QP calculation within the *GW* approximation for the electron self energy as described above.[4,6] This concept has been discussed very carefully by Strinati.[10] Applications to real materials have recently been presented by Onida *et al.*[13] for the alkali metal cluster $Na_4$, by us[14] for Si clusters, by Albrecht *et al.*[15] for $Li_2O$ and for Si, as well as, by Benedict *et al.*[16] and by us[17] for various semiconductors and insulators.

Similar to the one-particle Green's function $G_1$, the two-particle Green's function $G_2$ is defined as[10]

$$G_2(12;1'2') = -\langle 0|T[\hat{\psi}(1)\hat{\psi}(2)\hat{\psi}^\dagger(2')\hat{\psi}^\dagger(1')]|0\rangle \quad . \tag{5}$$

In this equation, $\hat{\psi}$ and $\hat{\psi}^\dagger$ are electron annihilation and creation operators (field operators) in the Heisenberg picture. The set of variables (1) comprises position, spin, and time coordinates: $(1)=(\mathbf{x}_1,t_1)=(\mathbf{r}_1,\sigma_1,t_1)$. $T$ is Wick's time ordering operator, and $|0\rangle$ is the ground state of the $N$ particle system. Depending on the order of the times $t_1$, $t_2$, $t_1'$ und $t_2'$, $G_2$ describes the propagation of two electrons, two holes, or one electron and one hole. With the specifications $t_1' = t_1 + 0^+$ and $t_2' = t_2 + 0^+$, $G_2$ is restricted to the simultaneous creation of an electron-hole pair and its simultaneous annihilation at later time.

If there was no interaction between the quasi-electron and the quasi-hole, $G_2$ would be given by products of $G_1$. Due to the interaction, $G_2$ has a more complicated structure. In order to describe the correlation between electron and hole, one introduces the correlation function[10]

$$L(12;1'2') := -G_2(12;1'2') + G_1(1,1')G_1(2,2') \quad . \tag{6}$$

$L$ obeys the Bethe-Salpeter equation (BSE)

$$L(12;1'2') = G_1(1,2')G_1(2,1') + \int d(3456)G_1(1,3)G_1(4,1')K(35;46)L(62;52') \quad . \tag{7}$$

The interaction $K$ is given by the functional derivative

$$K(35;46) = \frac{\delta[V_{\rm Coul}(3)\delta(3,4) + \Sigma(3,4)]}{\delta G_1(6,5)} \tag{8}$$

Making the assumption that the self-energy operator $\Sigma$ is given by the GWA and that the derivative of the screened interaction $W$ with respect to $G_1$ can be neglected, we obtain:

$$K(35;46) = -i\delta(3,4)\delta(5^-,6)v(3,6) + i\delta(3,6)\delta(4,5)W(3^+,4) \quad . \tag{9}$$

The easiest way to solutions of the BSE is to carry out a Fourier transform with respect to the time difference $t_2 - t_1$. The electron-hole excitation energies are then given by the positions of frequency poles in $G_2$. Furthermore, we transform all equations from the twelve-dimensional space representation onto the QP states of $G_1$ which form a natural basis for the present problem. In this basis, the BSE is given as a non-Hermitean eigenvalue problem. In order to obtain a clear physical interpretation of the solutions of the BSE, we further assume that the excitations $|S\rangle$ of the $N$ electron system have the form

$$|S\rangle = \sum_v^{\rm hole}\sum_c^{\rm elec} A_{vc}^S \hat{a}_v^\dagger \hat{b}_c^\dagger |0\rangle \quad . \tag{10}$$

The operator $\hat{b}_c^\dagger$ creates a quasi-electron in the unoccupied conduction state $c$, and the operator $\hat{a}_v^\dagger$ creates a quasi-hole in the occupied valence state $v$. Four-particle processes and higher processes are neglected. We note that this form of the excitation is a restriction which may not be applicable to all systems. For the systems studied in this paper, however, we obtain results in excellent agreement with experiment, indicating that the above assumption is valid. If the electron-hole interaction was not present, the free electron-hole pair $\hat{a}_v^\dagger \hat{b}_c^\dagger|0\rangle$ would be an excited eigenstate with the excitation energy $E_c^{\rm QP} - E_v^{\rm QP}$. Due to the interaction, however, the pairs are coupled. From the BSE we obtain the following eigenvalue problem for the coefficients $A_{vc}^S$ and the excitation energies $\Omega_S$:[10]

$$(E_c^{\rm QP} - E_v^{\rm QP})A_{vc}^S + \sum_{v',c'} \langle vc|K^{eh}(\Omega_S)|v'c'\rangle A_{v'c'}^S = \Omega_S A_{vc}^S \quad . \tag{11}$$

The matrix elements of the interaction consist of a screened, direct term $K^{eh,d}$ (which corresponds to the functional derivative of the self-energy $\Sigma$ in Eq. (8)) and an exchange term $K^{eh,x}$ (which corresponds to the functional derivative of the Coulomb term $V_{\rm Coul}$ in Eq. (8)):[10]

$$\begin{aligned}\langle vc|K^{eh,d}(\Omega_S)|v'c'\rangle &= \int d^3x d^3x' \psi_c^*(\mathbf{x})\psi_{c'}(\mathbf{x})\psi_v(\mathbf{x}')\psi_{v'}^*(\mathbf{x}')\frac{i}{2\pi}\int d\omega e^{-i\omega 0^+} W(\mathbf{r},\mathbf{r}',\omega) \\ &\times \left[\frac{1}{\Omega_S - \omega - (E_{c'}^{\rm QP} - E_v^{\rm QP}) + i0+} + \frac{1}{\Omega_S + \omega - (E_c^{\rm QP} - E_{v'}^{\rm QP}) + i0+}\right]\end{aligned} \tag{12}$$

$$\langle vc|K^{eh,x}|v'c'\rangle = \int d^3x d^3x' \psi_c^*(\mathbf{x})\psi_v(\mathbf{x})v(\mathbf{r},\mathbf{r}')\psi_{c'}(\mathbf{x}')\psi_{v'}^*(\mathbf{x}') \quad . \tag{13}$$

The direct term $K^{eh,d}$ is frequency dependent; this results from the dynamical properties of the screened Coulomb interaction $W$. If the excitation energy $\Omega_S$ was set to be the difference of the QP energies in the denominator of Eq. (12), the frequency integral would result in static screening (this is often assumed in the literature). Due to the electron-hole interaction, this is no longer the case, and the frequency integration has to be performed explicitly. In most cases, however, the difference between $\Omega_S$ and the QP energy difference is much smaller than the excitation energies and plasmon frequencies of the systems. Therefore, the frequency integration can be performed analytically using a plasmon-pole model for $W(\omega)$,[18] and the effects of the screening on the excitation energies $\Omega_S$ can be treated perturbatively.[14]

In Eq. (11), the spin configuration deserves special attention. In systems without spin polarisation, the wave functions for positive and negative spin are the same and the spin structure is usually not explicitly marked. For electron-hole excitations, however, the spins play a crucial role. Without spin-orbit interaction, the BSE decouples into separate equations for spin-singlet and spin-triplet excitations. In the following we will mainly discuss spin-singlet excitations since they are observable in optical experiments whereas triplet excitations do not couple to light. On the other hand, in the presence of spin-orbit interaction, as in the case of excitons in GaAs discussed below, spin-triplet and singlet excitations are coupled, and the spin coordinates have to be included in solving the BSE.

For discussing optical spectra, the optical oscillator strengths of the electron-hole transitions are crucial. To first order in the external field, the interaction between the light and the electron system is given by $\vec{A}\cdot\vec{V}$ where $\vec{A} = i\vec{E}/\omega$ is the vector potential of the external electromagnetic wave with frequency $\omega$ and $\vec{V}$ is the velocity operator of the electronic system. $\vec{A}\cdot\langle v|\vec{V}|c\rangle$ is the corresponding matrix element for a free transition $v \to c$. The transition matrix element of the coupled excitation $|0\rangle \to |S\rangle$ is then given by

$$\vec{A}\cdot\langle 0|\vec{V}|S\rangle = \sum_{v}^{\text{hole}}\sum_{c}^{\text{elec}} A_{vc}^{S}(\vec{A}\cdot\langle v|\vec{V}|c\rangle) \quad . \tag{14}$$

From Eq. (14) and the excitation energies $\Omega_S$, we may then calculate the imaginary part $\epsilon_2(\omega)$ of the macroscopic dielectric function (neglecting the off-diagonal elements of the dielectric matrix):

$$\epsilon_2(\omega) = \frac{4\pi e^2}{\omega^2}\sum_S |\vec{\lambda}\cdot\langle 0|\vec{V}|S\rangle|^2\delta(\omega - \Omega_S) \tag{15}$$

where $\vec{\lambda} = \vec{A}/|\vec{A}|$ is the polarization vector of the light. From $\epsilon_2(\omega)$ we obtain all other optical properties, like the real part $\epsilon_1(\omega)$ of the dielectric function, the refraction index, the absorption coefficient, the reflectivity, etc. It is important to note that the dielectric function given by Eq. (15) is now different from the dielectric function which we had employed to calculate the screened Coulomb interaction.

### Solving the BSE for Periodic Crystals

In the case of isolated clusters, the electronic spectrum is discrete, and the BSE can directly be solved following the procedure presented in the last section. For periodic systems, on the other hand, the spectrum is continuous, and the wave vectors of the states require special attention.

We first have to specify the optical absorption process. Different from isolated clusters, the momentum $\mathbf{Q}$ of the absorbed photon cannot be neglected, and the excited electron-hole state now has the form

$$|S\rangle = \sum_{\mathbf{k}}\sum_{v}^{\text{hole}}\sum_{c}^{\text{elec}} A_{vc\mathbf{k}}^{S}\hat{a}_{v\mathbf{k}}^{\dagger}\hat{b}_{c,\mathbf{k}+\mathbf{Q}}^{\dagger}|0\rangle \quad . \tag{16}$$

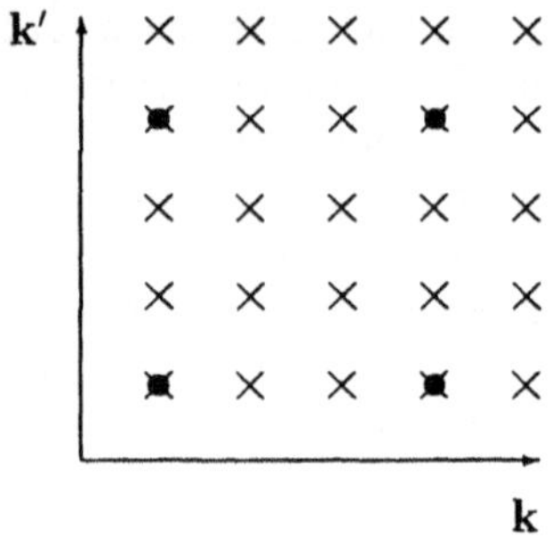


Figure 3. Illustration of the interpolation scheme for the electron-hole interaction. The matrix elements of the interaction are calculated for (**k**,**k**′) points on a coarse grid (•) and are then interpolated onto a much finer grid (×) on which the BSE is solved.

The magnitude of **Q** is unimportant due to the large wave length of the light compared to the atomic length scale; its *direction*, however, plays a crucial role for the exchange term of the electron-hole interaction which is not analytical for $|\mathbf{Q}|\to 0$. This leads, for example, to the splitting of excitons into transverse and longitudinal modes in some cubic crystals.

In extended systems, several occupied and unoccupied bands $v$ and $c$, as well as, several hundred wave vectors **k**, i.e. up to $10^4$ basis functions, are necessary to represent the transitions $|S\rangle$ relevant to the optical spectrum. The determination of the $10^8$ matrix elements of the electron-hole interaction, each of which requires a multi-dimensional integration given by Eq. (12), forms a major bottleneck of these calculations. In order to make the calculations feasible, we have developed a novel interpolation scheme[17] with respect to reciprocal-space vectors (see Fig. 3). The main idea is to calculate the interaction only for some **k** points on a coarse grid in reciprocal space (indicated as (•) in Fig. 3), and then to interpolate onto the much finer grid required for the BSE (indicated as (×)).

Before we can do this, the spatial nature of the screened Coulomb interaction $W$ has to be considered. The direct term of the electron-hole interaction is given by

$$\langle vc\mathbf{k}|K^{eh,d}|v'c'\mathbf{k}'\rangle = \int d^3r d^3r' \psi^*_{c,\mathbf{k}+\mathbf{Q}}(\mathbf{r})\psi_{c',\mathbf{k}'+\mathbf{Q}}(\mathbf{r})W(\mathbf{r},\mathbf{r}')\psi_{v\mathbf{k}}(\mathbf{r}')\psi^*_{v'\mathbf{k}'}(\mathbf{r}') \quad . \tag{17}$$

It is most convenient to write $W$ as a Fourier series:

$$W(\mathbf{r},\mathbf{r}') = \sum_{\mathbf{q}\mathbf{G}\mathbf{G}'} e^{-i(\mathbf{q}+\mathbf{G})\mathbf{r}} \frac{\epsilon^{-1}_{\mathbf{G},\mathbf{G}'}(\mathbf{q})}{|\mathbf{q}+\mathbf{G}||\mathbf{q}+\mathbf{G}'|} e^{i(\mathbf{q}+\mathbf{G}')\mathbf{r}'} \quad . \tag{18}$$

[For simplicity, we neglect the dynamical behavior of $W$ in this discussion.] Only the term $\mathbf{q}=\mathbf{k}'-\mathbf{k}$ contributes to Eq. (17). Eq. (18) contains terms behaving as $1/q^2$ und $1/q$. Therefore, the matrix elements of $K^d$ depend very sensitively on the reciprocal distance **q**, in particular when **q** is small. An interpolation scheme directly for $K^d$ would thus be very unstable and converge extremely slowly. Instead, we decompose $K^d$ into three contributions, motivated by the Fourier series for $W$. Corresponding to Eq. (18), the matrix elements of $K^d$ are given as double summations over **G** and **G**′. For small **q**, the "head"($\mathbf{G}=\mathbf{G}'=0$) and the "wings" ($\mathbf{G}=0$ or $\mathbf{G}'=0$) of $W$ behave as $1/\epsilon_0(q)q^2$ and $1/q$, respectively, whereas all other matrix elements of $W$ remain finite. Concomitantly the matrix elements of $K^d$ can be written as

$$\langle vc\mathbf{k}|K^{eh,d}|v'c'\mathbf{k}'\rangle = \frac{a_{vc\mathbf{k},v'c'\mathbf{k}'}}{\epsilon_0(q)q^2} + \frac{b_{vc\mathbf{k},v'c'\mathbf{k}'}}{q} + c_{vc\mathbf{k},v'c'\mathbf{k}'} \quad . \tag{19}$$

The coefficients $a$, $b$, and $c$ vary only weakly with respect to $\mathbf{k}$ and $\mathbf{k}'$.

We calculate $a$, $b$ and $c$ on the coarse grid $\{\tilde{\mathbf{k}}\}$ (indicated schematically as (•) in Fig. 3). From $(\tilde{\mathbf{k}}, \tilde{\mathbf{k}}')$, a matrix element of $a$ (or that of $b$ and $c$) for a point $(\mathbf{k}, \mathbf{k}')$ on the fine grid (×) is now approximately given by

$$a_{vc\mathbf{k},v'c'\mathbf{k}'} = \sum_{\tilde{v}\tilde{c}\tilde{v}'\tilde{c}'} d_{v\mathbf{k}}^{\tilde{v}\tilde{\mathbf{k}}} \left(d_{c\mathbf{k}}^{\tilde{c}\tilde{\mathbf{k}}}\right)^* \left(d_{v'\mathbf{k}'}^{\tilde{v}'\tilde{\mathbf{k}}'}\right)^* d_{c'\mathbf{k}'}^{\tilde{c}'\tilde{\mathbf{k}}'} a_{\tilde{v}\tilde{c}\tilde{\mathbf{k}},\tilde{v}'\tilde{c}'\tilde{\mathbf{k}}'} \tag{20}$$

where

$$d_{\nu\mathbf{k}}^{\tilde{\nu}\tilde{\mathbf{k}}} = \int \psi_{\tilde{\nu}\tilde{\mathbf{k}}}^*(\mathbf{r}) e^{i(\tilde{\mathbf{k}}-\mathbf{k})\mathbf{r}} \psi_{\nu\mathbf{k}}(\mathbf{r}) d^3r \quad . \tag{21}$$

The points $\tilde{\mathbf{k}}$ and $\tilde{\mathbf{k}}'$ should be chosen close to $\mathbf{k}$ and $\mathbf{k}'$. The integrals $d_{\nu\mathbf{k}}^{\tilde{\nu}\tilde{\mathbf{k}}}$ account for the changes in the wave functions when going from $\tilde{\mathbf{k}}$ to $\mathbf{k}$. Compared to the direct evaluation of all matrix elements of the interaction using Eqs. (12) and (17), the interpolation scheme allows for a radical reduction of the computational demand by more than two orders of magnitude.

Another problem in solving the BSE for large basis sets is the diagonalization of large Hamilton matrices of dimension $10^4$. Here we use the following approach. An electron-hole transition at $\Omega_S$ consists mainly of free electron-hole pairs whose energies $E_c^{\mathrm{QP}} - E_v^{\mathrm{QP}}$ are only a few excitonic binding energies lower or higher than $\Omega_S$. To determine the transitions with energies $\Omega_S$ close to a given reference energy $E_0$, the Hamiltonian can be restricted to a correspondingly small energy interval around $E_0$. The effect of electron-hole pairs outside this interval can later be included perturbatively. This procedure yields the excited states at and near $E_0$ with high accuracy. To obtain the entire spectrum, the calculation has to be repeated with different $E_0$, finally covering the entire desired energy range.

## ABSORPTION SPECTRUM AND EXCITONS OF BULK CRYSTALS

### The Spectrum of a Semiconductor: GaAs

In this section we discuss the optical absorption spectrum of GaAs as a prototype semiconductor (see Ref. 17). The dielectric screening of GaAs is strong ($\epsilon_\infty$=10.9) and the conduction band effective mass is small; therefore the electron-hole interaction is weak, leading to exciton binding energies of just a few meV. Nevertheless, the absorption spectrum is significantly changed by the interaction. Fig. 4 shows the optical spectrum of GaAs, calculated with (solid lines) and without (dashed lines) electron-hole interaction. The calculations employed the three highest valence bands and the six lowest conduction bands. We used 500 $\mathbf{k}$ points in the entire Brillouin zone. These points were equidistantly distributed. The grid was offset from the $\Gamma$ point with a shift different from all high-symmetry directions. This allows for a converged spectrum and high energy resolution with only 500 $\mathbf{k}$ points.

The spectrum resulting from the free quasiparticle transitions shows systematic deviations from experiment. At low energies (2–5 eV) the absorption strength is much lower than in experiment while it is too high for energies above 5 eV. When we include the electron-hole interaction, the peaks at 3 eV and at 5 eV are strongly enhanced; in addition, the peak structure at 5 eV is effectively shifted to lower energies. The spectrum including the interaction is in much better agreement with the measured data. The same type of improvement has been observed by Benedict et al.[16]

The modifications of the spectrum do not result from a negative shift of the transition energies, as one might naively expect from the attractive nature of the interaction.

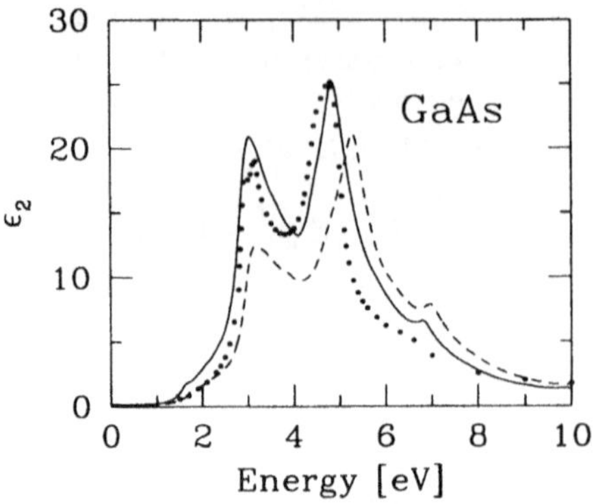


Figure 4. Calculated optical absorption spectrum of GaAs with (solid lines) and without (dashed lines) electron-hole interaction, using 3 valence bands, 6 conduction bands, 500 **k**-points in the BZ, and an artificial broadening of 0.15 eV. The dots denote experimental data.[20]

In fact, we find that the joint density-of-states of the electron-hole transitions remains nearly unchanged by the interaction (except for the formation of bound excitons; see below). The changes in the optical spectrum originate mainly from the *coupling* of the electron-hole configurations in the excited-state wave functions, which leads to a constructive coherent superposition of the oscillator strengths for transitions at lower energies and to a destructive superposition at energies above 5 eV. The apparent energy shift by 0.5 eV of the peak structure at 5 eV is, in fact, not an energy shift, but results from enhancement of oscillator strength at its left shoulder and suppression at its right shoulder. Such modifications of the spectrum occur also in other semiconductors.[9,15,16,19] We expect that this is a general effects in most group IV, III-V, and II-VI materials.

Fig. 4 does not exhibit bound exciton states because of the limited resolution (~ 0.15 eV) arising from the finite set of 500 **k**-points in the Brillouin zone (BZ) used. To describe the discrete exciton states, we have to increase the **k**-point grid density to 100 million **k**-points in the Brillouin zone. Fortunately, the use of about 1000 of these **k**-points near the position of the band extrema allows already for a converged representation of the $1s$, $2s$, and $2p$ excitons. Different from the calculation of the entire absorption spectrum (Fig. 4), the spin-orbit interaction must be included in this calculation. At the $\Gamma$ point, the valence-band maximum observes a spin-orbit splitting of 350 meV, whereas the exciton binding energies are no larger than 4 meV. We employ the two highest valence bands, which are degenerate at $\Gamma$, and the lowest conduction band in the calculation of the exciton spectrum. Since the excitons cannot be decoupled into pure spin-singlet and spin-triplet states, we include the full spin structure, which increases the basis to four valence bands and two conduction bands.

The exciton spectrum is given in Table 1 and in Fig. 5. The lower panel of Fig. 5 displays the exciton levels and their degeneracy from our *ab-initio* calculation. The first two peaks at −4.0 meV and −2.5 meV below $E_g$ are given by the $1s$ excitons. The states at −4.0 meV are composed of mixed (due to spin-orbit interaction) triplet and singlet excitonic states. The splitting of 1.5 meV between these states and the next one, which can be identified as a longitudinal exciton, is caused by the exchange part of the electron-hole interaction.[21] The $2s$ excitons at −0.9 meV and −0.7 meV form a

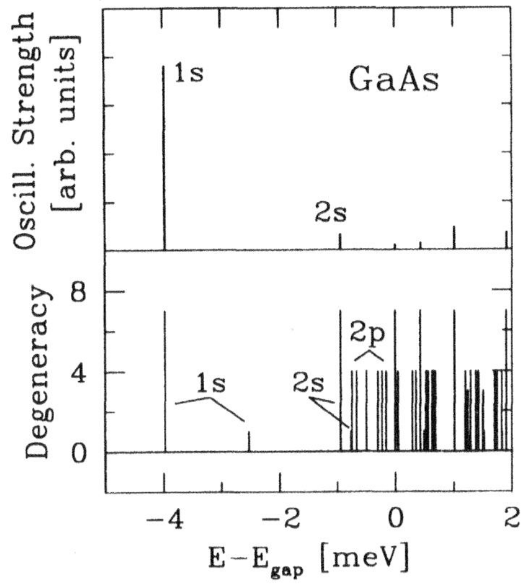


Figure 5. Calculated exciton spectrum of GaAs relative to the fundamental gap, showing the oscillator strength (upper panel) and the degeneracy (lower panel) of the exciton states. Two valence bands, one conduction band, and 1000 **k** points ($|\mathbf{k}| < 0.015$ a.u.) near the $\Gamma$ point were used. The band structure includes spin-orbit interaction.

similar multiplet. Between $-0.7$ meV and $-0.2$ meV we observe the $2p$ excitons. The splitting among these states and the energy difference to the $2s$ level are caused by the anisotropy of the valence bands.[22] Our calculated binding energies agree well with available experimental data (see Table 1). Due to the finite **k**-point sampling used, the higher transitions, which are more extended in real space, do not form accurately a hydrogen-like series in our spectrum. This problem may be overcome with a denser **k**-point grid or a non-uniform grid near the band minimum. In the upper panel of Fig. 5 we display the corresponding dipole oscillator strengths of the exciton states. Due to the selection rules, only the transverse $s$ states have a non-vanishing dipole matrix element and are observable in optical experiments.[21] In particular, the $1s$ excitons observe a coherent, constructive superposition of oscillator strength. The ratio $I_{2s}/I_{1s}$ of the strengths of the $2s$ and the $1s$ exciton is 0.1, which is close to the ideal ratio of 1/8 for hydrogen-like envelope functions. The oscillator strengths of the longitudinal $s$ excitons are zero.

Table 1. Calculated binding energies of excitons in GaAs (cf. Fig. 5), compared with data from optical absorption[23] and from two-photon absorption[24] measurements.

| | This work [meV] | Exp. [meV] |
|---|---|---|
| $E_{1s}$ | 4.0 | $4.2^{a}$ |
| $E_{2s}$ | 0.9 | $1.05^{a}$ |
| $E_{2p}$ | 0.2 – 0.7 | $\sim 0.1^{b}$ |

[a]Ref. 23 [b]Ref. 24

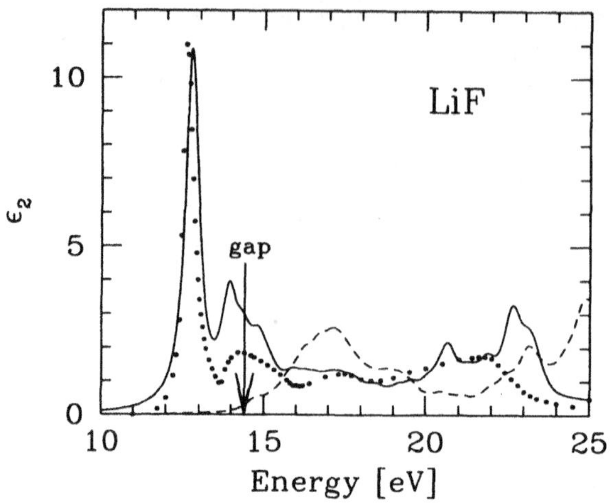


Figure 6. Same as Fig. 4, but for LiF. A broadening of 0.25 eV is included. The experimental data are from Ref. 25.

Of course, all these structures and effects are well-known for GaAs. An effective-mass approach with adjustable parameters can reproduce the experimental features. The novelty here is that the present approach yields the same results without any assumptions on the dispersion and wave functions of the bands and on the electron-hole interaction. Our scheme can thus directly be applied to situations in which simple empirical techniques do not hold.

## The Spectrum of an Insulator: LiF

Now we discuss a much different system, LiF (see Ref. 17). Compared to GaAs, LiF has a much larger fundamental gap (14.4 eV in our QP calculation) and the Coulomb interaction is only weakly screened ($\epsilon_\infty$=1.92). Therefore, the electron-hole interaction is much stronger, and the corresponding modifications of the optical absorption spectrum (shown in Fig. 6) are much more pronounced. In fact, the spectrum is completely altered from the non-interacting case. Again, the spectrum including the interaction is in much better agreement with experiment. The most pronounced feature in the calculated spectrum is the occurence of two strongly bound singlet excitons at 12.8 eV (transverse excitons). In addition, a longitudinal singlet exciton, which is not visible in the optical spectrum, is found at 13.3 eV. The energetic splitting of 0.5 eV between the transverse excitons and the longitudinal one is caused by the exchange part of the electron-hole interaction. In contrast to GaAs, the spin-orbit interaction can be neglected for LiF, and the transitions can be classified as pure spin-triplet and -singlet transitons. Here we focus on singlet transitions only. Concomitantly, the formation of transverse and longitudinal excitons is slightly different from the situation in GaAs. In the optical excitation process, the transverse excitons observe a coherent, constructive superposition of oscillator strength, which leads to the pronounced signal in the spectrum. The oscillator strength of the longitudinal exciton is zero.

Our approach further allows us to explicitly calculate the real-space electron-hole amplitude

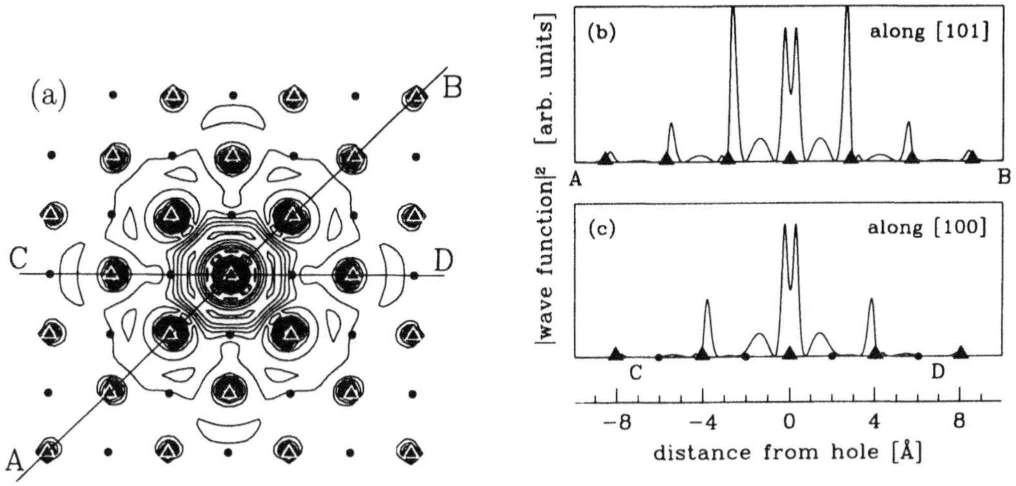


Figure 7. (a) Real-space distribution $|\chi_S(\mathbf{r}_h, \mathbf{r}_e)|^2$ of the electron ($\mathbf{r}_e$) with respect to the hole ($\mathbf{r}_h$) for the lowest-energy exciton in LiF (at 12.8 eV), in the (010) plane. Dots (triangles) denote Li (F) atoms. The hole ($\mathbf{r}_h$) is fixed at the central F atom. (b, c) The same quantity, along the lines $AB$ and $CD$ indicated in panel (a).

$$\chi_S(\mathbf{r}_h, \mathbf{r}_e) = \sum_{\mathbf{k}} \sum_{v}^{\text{hole}} \sum_{c}^{\text{elec}} A^S_{vc\mathbf{k}} \psi^*_{v\mathbf{k}}(\mathbf{r}_h) \psi_{c,\mathbf{k+Q}}(\mathbf{r}_e) \tag{22}$$

of each excitation. $\mathbf{r}_h$ ($\mathbf{r}_e$) refers to the coordinates of the hole (electron). $\chi_S$ is invariant to lattice-vector shifts when applied simultaniously to $\mathbf{r}_h$ and $\mathbf{r}_e$. To observe the real-space correlation between the electron and the hole, one has to fix the coordinates of one particle ($\mathbf{r}_h$ or $\mathbf{r}_e$) and discuss $\chi_S$ with respect to the coordinates of the other particle. As an example, we discuss the lowest exciton in LiF. In Fig. 7 (a) we show the distribution of the electron relative to the hole, which we fix at an F atom, in the (010) plane. The envelope function is $s$-like, with slight modifications due to the anisotropy of the cubic crystal, and has a mean radius of slightly over 4 Å or one lattice constant. Figs. 7 (b) and (c) show the same quantity as a line plot along the [101], as well as, along the [100] direction (indicated as lines $AB$ and $CD$, respectively, in panel (a)). The figures exhibit several interesting features. First, the charge density of the electron is low at the Li atoms. From an extreme Frenkel exciton-like picture, one might have expected that the electron hops from the central F atom to the nearest-neighbor Li atoms. Instead, it partly remains on the central F atom and partly hops to the first- and second-nearest F atoms and even farther away. Second, the electron distribution on these neighbor F atoms is not isotropic, but is highly polarized towards the central F atom. This results from the correlation of the quasielectron to the effective positive charge of the quasihole on the central F atom.

## SEMICONDUCTOR CLUSTERS

In this section we discuss optical excitations of hydrogen-terminated Si clusters (see Ref. 14). These systems have been investigated extensively both in experiment[26,27] and theory.[28,29] Theoretical studies of $Si_mH_n$ include many different approaches ranging from empirical tight-binding schemes to ab-initio LDA investigations and quantum

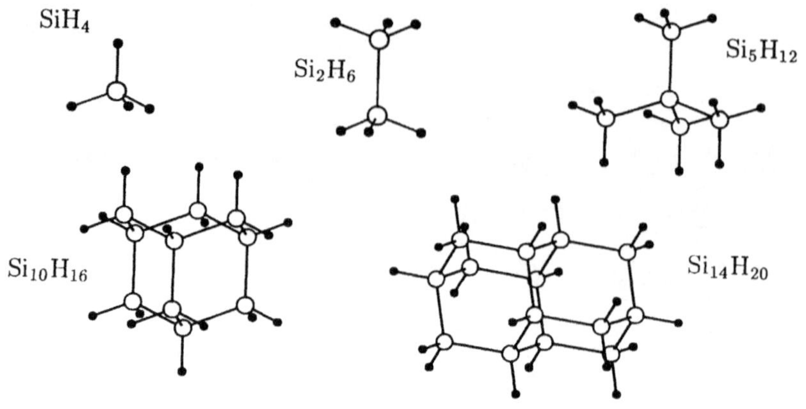


Figure 8. Five semiconductor clusters studied in this work, consisting of Si (○) and H atoms (•).

chemistry methods. Most of these approaches focus on the dependence of band edges and gap energies on the size and shape of the clusters. A calculation of the entire excitation spectrum of small semiconductor clusters including excitonic effects has not been done until now. Therefore, only limited information on the optical properties of such systems is available. Here we present a detailed study of the optical excitations of the electronic system of several small silicon clusters, fully including all relevant electronic correlation effects using the method described above. We have studied five different Si-H clusters (i.e. $SiH_4$, $Si_2H_6$, $Si_5H_{12}$, $Si_{10}H_{16}$, and $Si_{14}H_{20}$) with diameters ranging from about 3 Å to 9 Å, allowing for an investigation of quantum confinement effects as well as the effects of dynamical screening and mixing of electron-hole configurations on the exciton spectra as a function of cluster size. The five systems are shown in Fig. 8.

We have carefully checked the LDA energy spectrum, which is in excellent agreement with LDA results available in the literature,[30] as well as, the QP energies of single electrons and holes. As examples, the energies for the HOMO (highest occupied molecular orbital) are given in Fig. 9. As the cluster size decreases, the HOMO energies are lowered. This trend is much more pronounced for the QP energies than for the LDA levels. Because of the much lower polarizability and hence weaker dielectric screening in a semiconductor cluster as compared to the bulk crystal, the QP corrections are very large in the small systems studied here.[31] In fact, they are as large as 4 eV, which is one order of magnitude larger than in bulk Si. For the HOMO of $SiH_4$ ($Si_2H_6$), e.g., we obtain an LDA energy of −8.4 eV (−7.3 eV) and *GW* QP energy of −12.7 eV (−10.6 eV), respectively. The latter is in excellent agreement with the measured ionization energy of −12.6 eV (−10.7 eV).[27] Taking into account the excellent QP results for bulk Si[4] and for free and H-covered Si surfaces,[6,32] the GWA can thus be considered to be valid for the entire range of Si and Si-H systems from the molecular to the macroscopic scale.

Based on the QP results, we have calculated the electron-hole interaction in the clusters and solved the BSE for the electron-hole excitations. In Fig. 10 we display the calculated optical spectra of $SiH_4$, $Si_2H_6$, $Si_5H_{12}$, $Si_{10}H_{16}$, and $Si_{14}H_{20}$ (solid lines). The dotted lines show the respective absorption spectrum resulting from the QP energies only, i.e. completely neglecting the electron-hole interaction. The inclusion of the interaction modifies the shape of the spectrum significantly due to mixing of different electron-hole configurations. Most importantly, the entire spectrum is shifted to lower

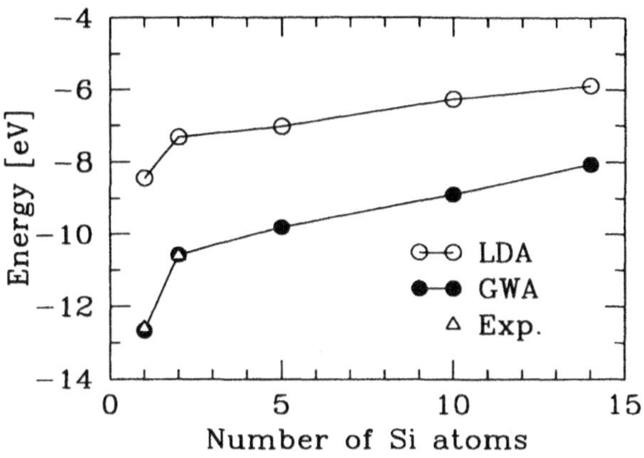


Figure 9. Energy of the highest occupied state (HOMO) of the five Si-H clusters, calculated in LDA and GWA. The experimental data (ionization energies) were measured by Itoh et al.[27]

energies by several eV. In Table 2 we compare our results for the pronounced transitions of $SiH_4$ and $Si_2H_6$ with measured absorption energies by Itoh et al.[27] The optical spectra of the two clusters consist of several well-defined peaks. For both systems, the lowest transition energy is in very good agreement with the measured data. The higher transitions also agree with experiment to within 0.5 eV, i.e. within an accuracy of 5%. For the larger systems, the optical spectra become more complex due to the increasing number of QP states and of electron-hole excitations. It should be noted that, since we are only interested in low-energy excitations, we have restricted ourselves to including no more than 30 unoccupied electron states $n$ in the solution of Eq. (11). This limits the number of resulting excitations. Fig. 10 does not show excitations above 11.9 eV (11.5 eV) for $Si_{10}H_{16}$ ($Si_{14}H_{20}$), as indicated by the vertical dashed lines.

The excitonic binding energy (i.e., the negative expectation value of $K$) amounts to several eV, up to 7.7 eV for the lowest spin-triplet excitation in $SiH_4$. It is thus more than two order of magnitudes larger than in bulk Si (15 meV). This large binding energy is caused by two effects. First, the spatial confinement of electrons and holes increases the overlap and thus the attractive interaction between them. Second, the dielectric screening in such small systems is much weaker than in extended bulk systems. Therefore, the electron-hole interaction is much less efficiently screened than in the bulk, increasing the exciton binding energy even further. In all cases, the binding energies for the triplet states are larger than for the singlet states by 0.5–1 eV since the triplet states do not observe the repulsive exchange interaction term $K^x$.

Table 2. Optical absorption energies of $SiH_4$ and $Si_2H_6$ (in eV). The experimental data are from Itoh et al.[27]

| $SiH_4$ | | $Si_2H_6$ | |
|---|---|---|---|
| This work | Exp. | This work | Exp. |
| 9.0 | 8.8 | 7.6 | 7.6 |
| 10.2 | 9.7 | 9.0 | 8.4 |
| 11.2 | 10.7 | 9.6–9.8 | 9.5, 9.9 |

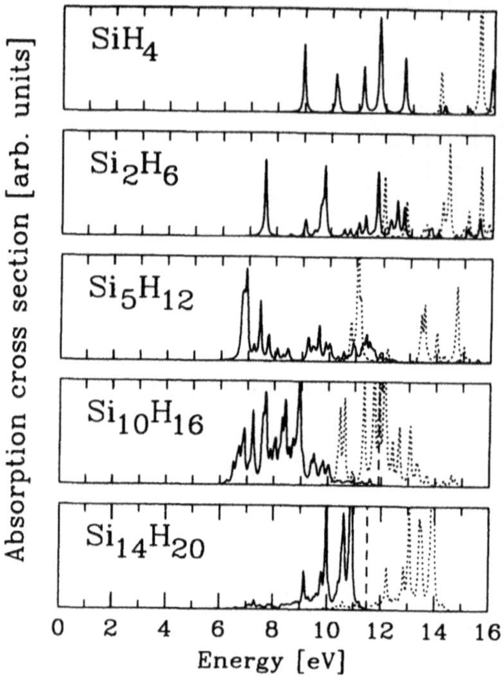


Figure 10. Calculated optical absorption spectra of the five Si-H clusters. The solid (dotted) lines show results with (without) electron-hole interaction. The spectra of $Si_{10}H_{16}$ and $Si_{14}H_{20}$ do not show transitions above 11.9 eV and 11.5 eV, respectively, since only a limited number of unoccupied electron states has been included in solving the BSE.

It has sometimes been presumed that at least the lowest-energy excitation of a semiconductor cluster might be given by a single transition from the HOMO to the LUMO state, i.e., it might be represented by a single "basis excitation" $\hat{a}_v^\dagger \hat{b}_c^\dagger |N, 0\rangle$ in Eq. (10). From our calculations, we observe that this is not the case. Even for the smallest clusters, the excited states involve a complicated mixing of a large number of electron-hole pairs $(v, c)$, as a result of the non-diagonal elements of the electron-hole interaction $K_{vc,v'c'}$. Neglecting these elements results in errors of the excitation energies of as much as 1 eV. In Fig. 11 we depict the contribution $|A_S(vc)|^2$ of different electron-hole pairs (denoted by arrows from occupied to empty levels) to the three transitions in $SiH_4$ discussed in Table 2. Note that the one-particle levels can consist up to three different states that are degenerate due to the symmetry of $SiH_4$, which further increases the complexity of the composition of the transitions. The contributions shown in Fig. 11 have been added up to account for these degeneracies. Similarly complex compositions are found for the spin-triplet transitions. As the size of the systems increases, the composition becomes even more complex due to the increasing number of states per energy interval. Roughly speaking, the energy interval involved in the formation of an exciton is of the same order of magnitude as its binding energy.

In Fig. 12 we compile the lowest excitation energies for spin-triplet and spin-singlet excitations for the Si clusters considered. As the size of the clusters increases, these excitation thresholds are reduced. The splitting between triplet and singlet excitations is reduced, as well, resulting from the decreasing overlap between the electron and hole, and the respectively decreasing exchange interaction $K^x$. In most cases, the

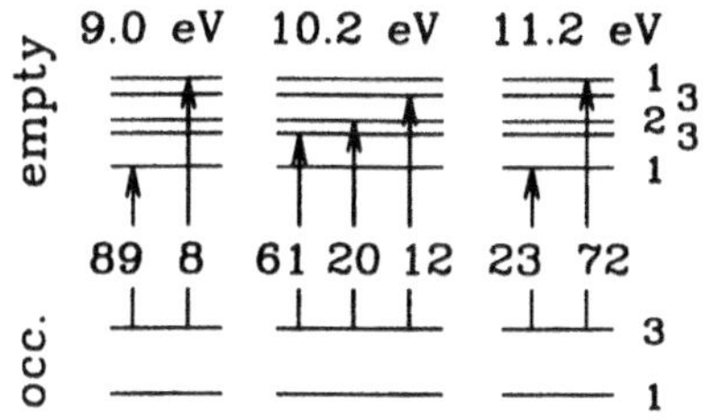


Figure 11. Contribution (in percent) of different electron-hole pair configurations to three spin-singlet transitions in $SiH_4$ (labelled by their transition energy $\Omega_S$ on the top). Note that the positions of the levels are not drawn to scale. The degeneracies of the one-particle levels are given on the right-hand side. Contributions smaller than 1 % are not shown.

calculated lowest spin-singlet excitation, although optically allowed in principle, is not observable in optical absorption due to a vanishing electric dipole oscillator strength. Therefore we include the lowest dipole-allowed spin-singlet excitations in Fig. 12, as well. Baierle et al.[29] have recently obtained the lowest excitation energy of $Si_5H_{12}$ and $Si_{10}H_{16}$ within configuration-interaction calculations with single-excitations (CIS) and within correlated Hartree-Fock calculations (included in Fig. 12). It is unclear from Ref. 29 whether the quoted transitions refer to spin-triplet or spin-singlet excitations. Our transition energies are higher than those results. We have also included the measured lowest optical absorption energy of $SiH_4$ and $Si_2H_6$ in Fig. 12.[27] Our results for the lowest dipole-allowed spin-singlet excitation are in excellent agreement with the data. This demonstrates that our approach of calculating electron-hole excitations in confined systems includes the important physical aspects, and the involved assumptions are valid.

## CONCLUSIONS

We have presented an approach that allows for the investigation of one-particle and two-particle excitations of electronic systems from first principles. The one-particle excitations are related to the quasiparticle band structure of the system. The investigation of two-particle excitations allows us to study coupled electron-hole excitations fully including the electron-hole interaction and the relevant aspects of electron-hole correlation. We obtain both the excitation energies and the wave functions of the electron-hole transitions. From these, we can evaluate optical properties. Very gratifying results are obtained both for the discrete, bound exciton states, as well as, for the continuum spectrum above the fundamental band gap of the material. We have investigated different systems, including semiconductors (GaAs), insulators (LiF), and semiconductor clusters (H-terminated Si clusters). For all systems considered, very good agreement with available experimental data is observed, confirming the general applicability of the approach to a range of materials under various conditions.

## ACKNOWLEDGMENTS

This work was supported by the Deutsche Forschungsgemeinschaft (Bonn, Germany) under Grant No. Ro-1318/1-1, by National Science Foundation Grant No. DMR-

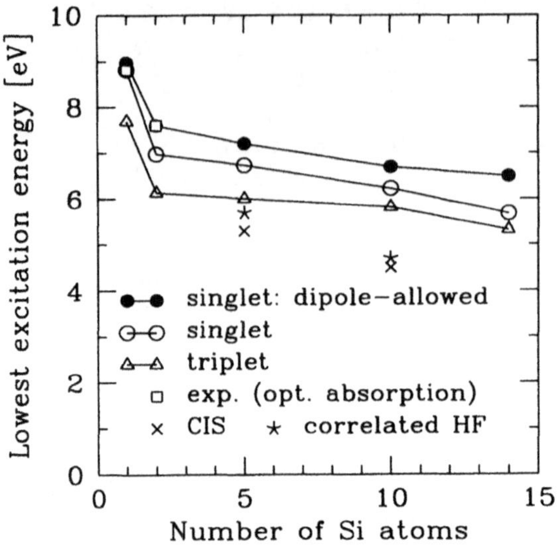


Figure 12. Excitation energies of the lowest spin-triplet, spin-singlet, and dipole-allowed spin-singlet excitations of the five $Si_mH_n$ clusters. We include theoretical results from CIS calculations (×) and from correlated Hartree-Fock calculations (⋆) by Baierle et al.,[29] as well as, experimental data (□) by Itoh et al.[27]

9520554 and by the Director, Office of Energy Research, Office of Basic Energy Sciences, Materials Sciences Division of the U.S. Department of Energy under Contract No. DE-AC03-76SF00098. Computational resources have been provided by the National Energy Research Scientific Computing Center (NERSC).

## REFERENCES


1. P. Hohenberg and W. Kohn, Phys. Rev. **136**, B864 (1964); W. Kohn and L. J. Sham, Phys. Rev. **140**, A1133 (1965).
2. G. P. Srivastava and D. Weaire, Advances in Physics **36**, 463 (1987).
3. L. Hedin, Phys. Rev. **139**, A796 (1965); L. Hedin and S. Lundqvist, in *Solid State Physics, Advances in Research and Application*, edited by F. Seitz, D. Turnbell, and H. Ehrenreich (Academic, New York, 1969), vol. 23, p. 1.
4. M. S. Hybertsen and S. G. Louie, Phys. Rev. Lett. **55**, 1418 (1985); Phys. Rev. B **34**, 5390 (1986); R. W. Godby, M. Schlüter, and L. J. Sham, Phys. Rev. B **37**, 10 159 (1988); M. Rohlfing, P. Krüger, and J. Pollmann, Phys. Rev. B **48**, 17 791 (1993); S. G. Louie, in *Quantum Theory of Real Materials*, ed. by J. R. Chelikowsky and S. G. Louie (Kluwer, Boston, 1996), p. 83.
5. F. Aryasetiawan and O. Gunnarsson, to appear in Reports on Progress in Physics (1998).
6. M. Rohlfing, P. Krüger, and J. Pollmann, Phys. Rev. Lett. **75**, 3489 (1995); Phys. Rev. B **52**, 1905 (1995).
7. P. Nozieres, *Theory of Interacting Fermi Systems* (Benjamin, New York, 1964).

8. L. J. Sham and T. M. Rice, Phys. Rev. **144**, 708 (1966).

9. W. Hanke and L. J. Sham, Phys. Rev. Lett. **43**, 387 (1979); Phys. Rev. B **21**, 4656 (1980).

10. G. Strinati, Phys. Rev. B **29**, 5718 (1984).

11. D. R. Hamann, M. Schlüter, and C. Chiang, Phys. Rev. Lett. **43**, 1494 (1979).

12. J. E. Ortega and F. J. Himpsel, Phys. Rev. B **47**, 2130 (1993).

13. G. Onida, L. Reining, R. W. Godby, R. DelSole, and W. Andreoni, Phys. Rev. Lett. **75**, 818 (1995).

14. M. Rohlfing and S. G. Louie, Phys. Rev. Lett. **80**, 3320 (1998).

15. S. Albrecht, G. Onida, and L. Reining, Phys. Rev. B **55**, 10 278 (1997); S. Albrecht, L. Reining, R. Del Sole, and G. Onida, Phys. Rev. Lett. **80**, 4510 (1998).

16. L. X. Benedict, E. L. Shirley, and R. B. Bohn, Phys. Rev. Lett. **80**, 4514 (1998); Phys. Rev. B **57**, R9385 (1998).

17. M. Rohlfing and S. G. Louie, Phys. Rev. Lett. **81** (1998), in print.

18. W. von der Linden and P. Horsch, Phys. Rev. B **37**, 8351 (1988).

19. M. Rohlfing and S. G. Louie, unpublished.

20. D. E. Aspnes and A. A. Sturge, Phys. Rev. B **27**, 985 (1983).

21. We believe that the longitudinal-transverse splitting observed in the calculation is a different effect than the occurence of longitudinal and transverse exciton-polaritons in optical measurements. From our calculated dipole oscillator strength of the $1s$ exciton, we estimate an LT-splitting in the order of 0.2 meV for the polaritons.

22. N. O. Lipari and A. Baldereschi, Phys. Rev. B **6**, 3764 (1972).

23. D. D. Sell, Phys. Rev. B **6**, 3750 (1972).

24. J. S. Michaelis, K. Unterrainer, E. Gornik, and E. Bauser, Phys. Rev. B **54**, 7917 (1996).

25. D. M. Roessler and W. C. Walker, J. Opt. Soc. Am. **57**, 835 (1967).

26. see, e.g., H. Takagi et al., Appl. Phys. Lett. **56**, 2379 (1990); K. A. Littau et al., J. Phys. Chem. **97**, 1224 (1993); W. A. Saunders et al., Mater. Res. Soc. Symp. **283**, 77 (1993).

27. U. Itoh, Y. Toyoshima, and H. Onuki, J. Chem. Phys. **85**, 4867 (1986).

28. see, e.g., S. Y. Ren and J. D. Dow, Phys. Rev. B **45**, 6492 (1992); C. Delerue, G. Allan, and M. Lannoo, Phys. Rev. B **48**, 11 204 (1993); B. Delley and E. F. Steigmeier, Phys. Rev. B **47**, 1397 (1993); L.-W. Wang and A. Zunger, J. Phys. Chem. **98**, 2158 (1994); M. Hirao and T. Uda, Surf. Sci. **306**, 87 (1994); N. A. Hill and K. B. Whaley, Phys. Rev. Lett. **75**, 1130 (1995); A. Francheschetti and A. Zunger, Phys. Rev. Lett. **78**, 915 (1997).

29. R. J. Baierle, M. J. Caldas, E. Molinari, and S. Ossicini, Solid State Comm. **102**, 545 (1997).

30. G. Onida and W. Andreoni, Chem. Phys. Lett. **243**, 183 (1995).

31. S. Öğüt, J. R. Chelikowsky, and S. G. Louie, Phys. Rev. Lett. **79**, 1770 (1997).

32. J. E. Northrup, M. S. Hybertsen, and S. G. Louie, Phys. Rev. Lett. **66**, 500 (1991); J. E. Northrup, Phys. Rev. B **47**, 10 032 (1993). X. Blase, X. Zhu, and S. G. Louie, Phys. Rev. B **49**, 4973 (1994).

## Discussion

**Perdew:** Can the GW method open up a fundamental gap when the LDA gap is zero?

**Louie**: Yes. Solid Ge is an example. However, if the topology of the GW bands is very different from that of the LDA bands, a problem could arise.

**Sawatzky**: Could you test the validity of the exciton calculations by doing this for a simple Hubbard model in 1 dimension where the solution also of the excitonic states are known?

**Louie**: Yes, that could be a good test.

**Sawatzky**: Are the excitonic states you obtain multi Slater determinate states involving real correlation, or are they still single Slater determinants states as those used to describe the ground state energy and density?

**Louie**: These are multi Slater determinant states that can be clearly seen in the singlet – triplet splittings.

**Sawatzky**: If that is the case, could you also calculate the d-d excitonic states of the 3d transition metal compounds like those of NiO I discussed in my talk?

**Louie**: Yes, indeed. I think we could and I would like to try this.

**Sawatzky**: In the case of the calculations of polyaceteline, did you take into account polaronic effects, which are thought by some to be very large?

**Louie**: No. In the present calculation we did not, and I agree these could be large. In fact, if we were not able to find an experimental value for the value for the band gap in polyaceteline.

**Zdetsis:** For the Si-cluster calculations have you optimized the geometry or you take some given structures. This could affect your predictions.

**Louie**: For each cluster we use LDA calculations within a given symmetry to optimize the geometry, so relaxation is included for this particular symmetry. However, this does not exclude the possibility of other lower lying structures of different symmetry.

# *Ab initio* STUDIES OF ELECTRONIC EXCITATIONS IN REAL SOLIDS

Adolfo G. Eguiluz and Wei Ku

Department of Physics and Astronomy, The University of Tennessee, Knoxville, TN 37996-1200

and Solid State Division, Oak Ridge National Laboratory Oak Ridge, TN 37831-6030

## ABSTRACT

The first part of this article centers on the fact that key features of the dynamical response of weakly-correlated materials (the alkalis, Al), have been found experimentally to differ qualitatively from simple-model behavior. In the absence of *ab initio* theory, the surprises embodied in the experimental data were imputed to effects of dynamical correlations. We summarize results of *ab initio* investigations of linear response, performed within time-dependent density-functional theory (TDDFT), in which the unexpected features of the observed spectra are shown to be due to band-structure effects. For example, the key process behind the anomalous dispersion of the plasmon linewidth in K turns out to be decay into electron-hole pairs involving flat *d*-bands; the much-discussed double peak in the dynamical structure factor of Al is traced to a final-state gap. These results entail a new way of thinking about the physics of the electronic response in these systems. Contrary to conventional wisdom, the response cannot be understood "universally," in terms of a simple scaling with the density, on going from metal to metal (e.g., through the alkali series) — even the *shape* of the dispersion curve for the plasmon energy is system-specific. The second part of this article starts out with the observation that a similar *ab initio* study of systems with more complex electronic structures would require the availability of a realistic approximation for the dynamical many-body kernel $f_{xc}$ entering the density-response function in TDDFT. In its absence, we outline a diagrammatic alternative, framed within the conserving-approximation method of Baym and Kadanoff. Using as a benchmark the band gap of Si obtained in the *GW* approximation, together with a diagrammatic (and conserving) solution of the ensuing Bethe-Salpeter equation (which includes dynamically-screened particle-hole ladders and other fluctuation processes), we discuss issues involving conservation laws, self-consistency, and sum rules. These conceptual issues are particularly important for the development of *ab initio* methods for the study of dynamical response and quasiparticle band structure of strongly-correlated materials. We argue that inclusion of short-range correlations absent in the *GW* approximation is a must, even in Si.

*Electron Correlations and Materials Properties*, edited by Gonis *et al.*
Kluwer Academic/Plenum Publishers, 1999.

## INTRODUCTION

The current theoretical understanding of the electronic correlations, and their impact on the properties of materials, is based, for the most part, on simple-model descriptions. In broad terms, the available models derive from two extreme pictures of a solid. On the one hand we have the homogeneous electron gas (or jellium) model, usually invoked in the study of simple metals.[1] In this model, the dominant excitation for small wave vectors is the plasmon, whose dispersion relation has traditionally been assumed to be well described by the long-range interactions contained in the random-phase approximation (RPA).[1] Because of the success of the local-density approximation (LDA) for the exchange-correlation energy functional introduced in Kohn-Sham density-functional theory,[2] the jellium model turns out to play an important "supporting role" in modern electronic structure calculations. On the other hand, we have the Hubbard model,[3,4] originally introduced for the study of itinerant magnetism in the 3*d* metals; this model emphasizes short-range aspects of the correlations, as it allows the electrons to interact only when they encounter at a given lattice site. In both extreme pictures the description of the band structure is reduced to a minimum: In jellium, it is ignored altogether (more precisely, there is one parabolic band); in the Hubbard model, it is represented by one (sometimes three, in the context of the copper-oxide superconductors) tight-binding bands.

The central theme of this article is the physics of the electronic excitations in real materials. This is a vast subject. The first part of our presentation is devoted explicitly to systems which, on account of their relative simplicity, have recently become amenable to *ab initio* theory. We start out with the observation that, even for jellium-like, weakly-correlated materials, such as the alkali metals and Al, high-resolution spectroscopic measurements[5-9] have uncovered significant "anomalies," whose explanation requires a new way of thinking about the physics of the excitations. For example, the dispersion of the plasmon *energy* in Cs turns out to be negative for small wave vectors, contrary to the case of, say, K, in which the dispersion is positive, as expected from textbook physics,[1] and general considerations of stability. Furthermore, the dispersion of the plasmon *linewidth* in K is positive,[5] contrary to the results of a simple, intuitive, model of the dielectric response which relies on the assumed weakness of the crystal potential (a negative plasmon linewidth dispersion is a "generic" feature for the alkalis series in this model).[10] In addition, the overall shape of the dynamical structure factor of Al (and other systems such as Be, Li, Si, and graphite)[7-9] is dominated, for large wave vectors, by a two-peak structure whose existence is beyond RPA theory for the homogeneous electron gas.

The above phenomenology is quite remarkable. The fact that experiment tells us that from the knowledge of the dispersion curve for the plasmon energy in, say, K, one cannot predict the *shape* of the dispersion curve in Cs, clearly calls for new theoretical insight into the physics of correlated electrons in these systems. A similar message follows from the qualitative deviations from simple-model predictions observed in the loss spectrum of Al and other systems. Such insight may be sought within two scenarios:

*(i)* We must abandon the long-cherished idea that the excitations in these materials can be understood "universally," i.e., in terms of a simple scaling, controlled solely by the density parameter $r_S$, on going from system to system. This viewpoint suggests that one must go beyond simple-model descriptions of the underlying band structure.

*(ii)* Scaling with the density may be restored if we assume that what breaks down is scaling à la RPA; in this case, then, for example, there must be some threshold value of $r_S$ in the range 4.8 –5.6 (corresponding to K and Cs, respectively) for which strong non-RPA correlations must emerge, and be responsible for switching the plasmon energy dispersion from positive to negative. This viewpoint is commonly pursued together with the premise that the effects of the band structure are not important.

In our view, to date there is no model of the correlations that is able to reproduce the experimental facts convincingly within scenario *(ii)*. In this article, whose scope is that of a pedagogical overview of some of our ideas about a rapidly-developing field, we show that scenario *(i)* leads to a compelling interpretation of the available experimental information on the basis of *ab initio* calculations of the excitation spectra.[11-20]

We consider first the plasmon in the alkali metals. Interestingly, in the absence of *ab initio* theory, it was suggested that the "anomalies" in the excitation spectra of K and Cs, observed via electron energy-loss spectroscopy (EELS) measurements,[5,6] must be due to unspecified dynamical short-range correlations —which falls in the realm of *(ii)*. We compare calculated and measured dispersion curves for the plasmon energy in Na, K, and Cs. Our results illustrate vividly the absence of conventional RPA scaling with the density. Band-structure effects account for the clear-cut breakdown of simple-model physics contained in the data; these effects are mainly due to the presence, in both K and Cs, of *d*-bands for energies comparable with that of the collective mode. Next, we show that the key mechanism behind the anomalous dispersion of the plasmon linewidth in K is decay into particle-hole pairs involving empty states of *d*-symmetry.[20] In all these calculations, quantitative agreement with the data is achieved within scenario *(i)*.

We also address the dynamical structure factor of Al, a notorious example of the duality of outlook implied by the above two scenarios. Under the simplifying assumption that "Al is jellium," most of the many papers published on this problem over the years sought to explain the two-peak loss structure originally observed via inelastic x-ray scattering spectroscopy (IXSS) measurements[21,22] in terms of dynamical correlations. For recent attempts within this approach, which is consistent with (*ii*), the reader is referred to papers by Platzman et al.[7] and Schülke et al.[8] Here we sketch results of *ab initio* calculations which lead to a "natural" explanation —along the lines of *(i)*— of the key features of the spectrum for large wave vectors.[12,15] The two-peak structure is shown to correlate strikingly with a wide quasi-gap in the density of states (DOS) of Kohn-Sham single-particle states.

The density-response calculations underlying the results just referred to are based on a linear-response implementation of time-dependent density functional theory (TDDFT).[23,24]

In the TDDFT linear-response scheme the many-body effects enter the loss spectrum in two ways: via a dynamical many-body kernel, $f_{xc}$ (loosely speaking, a vertex function), and through the effects of exchange and correlation included in the single-particle Kohn-Sham density-response function $\chi_s$. We note at the outset that the basic reason why we are able to elucidate the above-mentioned anomalies is that, upon switching off explicitly the dynamical correlations contained in $f_{xc}$, the calculated observables either agree nearly quantitatively with the data (e.g., the dispersion curve of the plasmon in the alkalis for small wave vectors), or reproduce the main qualitative features of the spectrum quite well (e.g., the loss function of Al for large wave vectors).

In this context, an interesting finding is the role played by the exchange-correlation effects built into the *unoccupied* Kohn-Sham states (which we treat in the LDA) in the explanation of the positive dispersion of the inverse lifetime of the plasmon in K. This novel result is traced to the fact that the damping process is dominated by a subtle interplay between flat *d*-bands and the collective mode.

Thus, for the relatively-simple systems alluded to above, the single-particle response $\chi_s$ provides a "starting point" which by itself contains a good deal of the physics of the electron-hole response as it manifests itself experimentally. However, for systems with more complex electronic structures (for example, transition metals, and their oxides) the dynamical correlations contained in $f_{xc}$ should play a significant role. Unfortunately, explicit representations for this kernel which would make possible the *ab initio* study of dynamical response in those systems are not yet available.[24] Thus, in the second part of this paper, we pursue a diagrammatic alternative within the Baym-Kadanoff method of conserving approximations.[25,26] In this approach, the counterpart of the TDDFT integral equation for the density-response function is the Bethe-Salpeter equation for the two-particle correlation function, whose kernel, the effective electron-hole interaction, depends on four coordinates, instead of just two, as $f_{xc}$ does. A diagrammatic solution of this equation within the shielded-interaction approximation (SIA) for the electron self-energy,[25] yields a formal result for the irreducible polarizability $P_{\rm SIA}$ which includes, to all orders, the effects of shielded particle-hole ladders and other fluctuation diagrams which arise naturally in the conserving, functional-differentiation approach. The formal relationship between $f_{xc}$ and the (inverse) polarizability is touched upon, as it may suggest ways of building into the former the dynamical processes contained in the latter.

Now the road to the two-particle problem goes through the one-particle problem, i.e., the Dyson equation, which we address first. In the important case of silicon, the SIA yields a value of the absolute band gap which is too large, compared to experiment.[27] This conclusion brings the "band-gap problem" back to the fore, as the SIA is nothing but Hedin's *GW* approximation,[28] which was supposed to have solved the problem —but only in the absence of self-consistency. (Recent, extensive reviews of work performed within the *GW* approximation are available.[29,30]) Thus, a more elaborate self-energy must be considered. Within the conserving scheme, this may be done by augmenting the Baym

$\Phi_{SIA}$ functional via the inclusion of short-range correlations —e.g., particle-hole and particle-particle ladders. In view of recent efforts to go beyond the *GW* approximation, and of related discussions of conservation laws and sum rules,[30-38] we address these issues in some detail, because of their significance for the development of meaningful approximations for the response function and quasiparticle band structure in correlated-electron systems.

## DYNAMICAL DENSITY RESPONSE IN TDDFT

A general formulation of the time evolution of an interacting electron system in an external potential $v_e(\vec{x},t)$ has been given by Runge and Gross.[23] These authors established the invertibility of the mapping $v_e(\vec{x},t) \to n(\vec{x},t)$, where $n(\vec{x},t)$ is the time-dependent density for the interacting system. This result constitutes a time-dependent generalization of the Hohenberg-Kohn theorem[39] of ground-state density-functional theory.† We then have that the time-evolved many-particle state is uniquely determined (for the given initial state) as a functional of $n(\vec{x},t)$ —up to a purely time-dependent phase. Such phase is eliminated when evaluating mean values of physical operators for the many-body system.

A particularly important example is the quantum mechanical action, $A[n]$, which has a stationary point at the exact solution of the many-particle Schrödinger equation. From the previous argument it then follows that

$$\frac{\delta A[n]}{\delta n(\vec{x},t)} = 0 \ , \tag{1}$$

i.e., the action is stationary at the exact density. Reasoning in a similar manner as done by Kohn and Sham[2] for the case of the ground state, from this variational principle one can prove[23] that $n(\vec{x},t)$ may be obtained within an equivalent single-particle picture, i.e., we can write

$$n(\vec{x},t) = \sum_{v}^{occupied} |\varphi_v(\vec{x},t)|^2 \ , \tag{2}$$

in terms of the solutions $\varphi_v(\vec{x},t)$ of the time-dependent Kohn-Sham equation

$$i\hbar\frac{\partial}{\partial t}\varphi_v(\vec{x},t) = \left(-\frac{\hbar^2}{2m}\nabla^2 + v_s[n](\vec{x},t)\right)\varphi_v(\vec{x},t) \ , \tag{3}$$

where the single-particle potential $v_s[n](\vec{x},t)$ contains, in addition to the external and Hartree potentials, the exchange-correlation potential $V_{xc}[n](\vec{x},t)$, whose *functional* dependence on the density is implied by our notation. (Note that this scheme assumes non-interacting *v*-representability for the time-dependent density.[40])

† There is a conceptual difference though, having to do with the dependence of the time-evolved state on the initial many-body state.[40]

The TDDFT formalism provides a convenient framework for the study of the *linear response* of a many-electron system to an external potential $\delta v_e(\vec{x},t)$.[24] In general, we define the density-response function $\chi$ by the equation $\delta n = \chi \delta v_e$. In TDDFT the linear change in density can also be calculated as $\delta n = \chi^{(s)} \delta v_s$, where $\chi^{(s)}$ is the *single-particle* response for the *unperturbed* Kohn-Sham system. From Eqs. (2) and (3) $\delta v_s$ can be related to $\delta v_e$, and this leads us to the following integral equation for the response function:

$$\chi = \chi^{(s)} + \chi^{(s)}(v + f_{xc})\chi , \tag{4}$$

where $v$ is the bare Coulomb interaction, and the exchange-correlation kernel $f_{xc}$ is defined by the equation

$$f_{xc}[n](\vec{x}t;\vec{x}'t') = \frac{\delta V_{xc}[n](\vec{x},t)}{\delta n(\vec{x}',t')} , \tag{5}$$

where the functional derivative is to be evaluated at the unperturbed density. We emphasize that Eqs. (4) and (5) are formally exact; all explicit effects of dynamical correlations are contained in $f_{xc}$.

The spectral representation for the single-particle response $\chi^{(s)}$ in terms of Kohn-Sham eigenfunctions and eigenvalues is of the usual form. For the case of a periodic crystal its Fourier transform with respect to the spatial coordinates takes on the form

$$\chi^{(s)}_{\vec{G},\vec{G}'}(\vec{k};\omega) = \frac{1}{V_N}\sum_{\vec{k}'}^{BZ}\sum_{j,j'} \frac{f_{\vec{k}',j} - f_{\vec{k}'+\vec{k},j'}}{E_{\vec{k}',j} - E_{\vec{k}'+\vec{k},j'} + \hbar(\omega + i\eta)} \left\langle \vec{k}',j \middle| e^{-i(\vec{k}+\vec{G})\cdot\hat{\vec{x}}} \middle| \vec{k}'+\vec{k},j' \right\rangle$$

$$\times \left\langle \vec{k}'+\vec{k},j' \middle| e^{i(\vec{k}+\vec{G}')\cdot\hat{\vec{x}}} \middle| \vec{k}',j \right\rangle , \tag{6}$$

where $V_N$ is the normalization volume, $\vec{G}$ is a vector of the reciprocal lattice, $j$ is a band index, and all wave vectors are in the first Brillouin zone (BZ). We evaluate Eq. (6) using Kohn-Sham states and eigenvalues obtained within the LDA for the functional $V_{xc}[n]$. In the Fourier representation of Eq. (6), Eq. (4) is turned into a matrix equation that we solve numerically.

The above outline of linear response within TDDFT was presented in order to emphasize the point that the Kohn-Sham response $\chi^{(s)}$ enters Eq. (4) *rigorously* —even if the individual states in the sum in Eq. (6) do not correspond to physical excitations (as are observed, e.g., in photoemission and inverse photoemission experiments). This issue is important for a proper interpretation of the results we discuss in the next two sections, where one of our aims is to sort out the effects of the one-particle band structure from those of dynamical many-body correlations; the latter are turned off by setting $f_{xc} = 0$.†

† For systems with more complex electronic structures we expect that, for $f_{xc} = 0$, theory would be sufficiently far off experiment that the success story of the next two sections would not be reproduced.

## PLASMONS IN THE ALKALI METALS; ENERGY AND LINEWIDTH DISPERSION

In Fig. 1 we compare our calculated dispersion curves for the plasmon energy in Na, K, and Cs, with the EELS data of vom Felde, Sprösser-Prou, and Fink.[5] The plasmon energy for a given wave vector $\vec{k}$ was read off the position of the main peak in the loss function $\mathrm{Im}\,\chi_{G=0,G=0}(\vec{k};\omega)$ (in the alkalis the plasmon dispersion curve is entirely confined to the first BZ). We have set $f_{xc}=0$ in Eq. (4). As discussed elsewhere,[20] the sampling of the BZ required in Eq. (6) is efficiently performed by carrying out the calculation on the imaginary frequency axis; the analytic continuation to the real axis is performed via Padé approximants. Since higher-lying core states turn out to affect the plasmon dispersion and lifetime of the alkali metals heavier than Na, we evaluate $\chi^{(s)}$ using the full-potential linearized augmented plane wave (LAPW) method.[41] Also shown in Fig. 1 are the dispersion curves for electrons in jellium with the density corresponding to each metal (solid lines); note that, in jellium, the response $\chi$ obtained from Eq. (4) for $f_{xc}=0$ is identical to the RPA response.

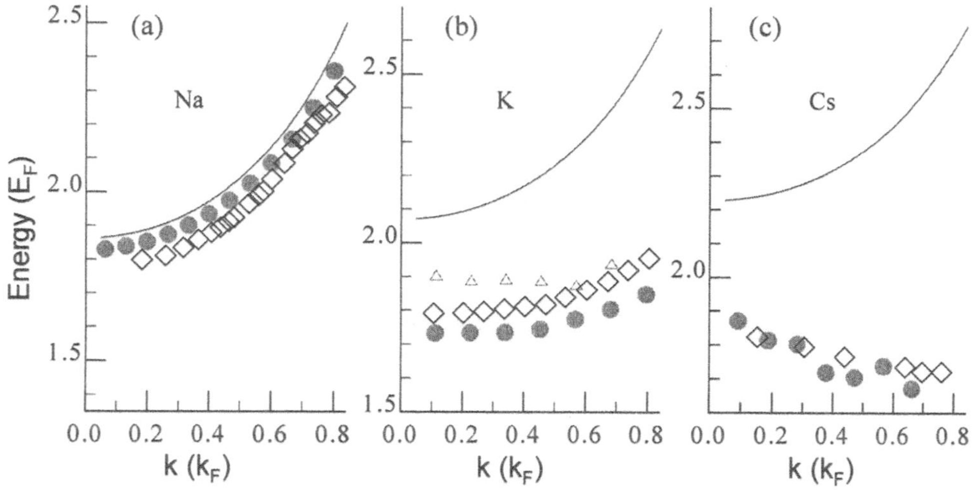


**Figure 1.** Absence of "universal" behavior, or RPA-like scaling, in the excitations in the alkali series. Shown are our calculated (solid circles) and measured[5] (empty diamonds) dispersion curves for the plasmon energy in Na, K, and Cs. Note: we have set $f_{xc}=0$ in Eq. (4). The solid lines are the corresponding results for electrons in jellium. The empty triangles in panel (b) correspond to the dispersion curve for the K plasmon obtained upon excluding the contribution to Eq. (6) from the core states. (Theory is for (1,1,0) propagation; the EELS data[5] are for polycrystalline samples.)

From a glance at Fig. 1 it is apparent that there is very good agreement between calculated (solid circles) and measured (empty diamonds) dispersion curves. This is a non-trivial test of *ab initio* theory, as this agreement includes the dramatic change in the shape of the dispersion curve which is realized on going from Na to Cs.[5,14] We emphasize that this change in the physics of screening taking place through the alkali series implies that the loss function $\mathrm{Im}\,\chi_{G=0,G=0}(\vec{k};\omega)$ (which, for small $k$'s is dominated by the plasmon) does not scale with, nor is controlled solely by, the density parameter $r_S$. We remind the reader that this conclusion is consistent with scenario *(i)* put forth in the Introduction.

The "anomalous" dispersion of the plasmon energy in Cs, and the consequent lack of universal behavior evident in Fig. 1, are traced to significant departures of the LDA band structure from nearly-free-electron behavior for energies of the order of the plasmon energy (whose value for zero wave vector we denote by $\omega_p(0)$). In Na the band structure is essentially free-electron like in that energy range; consequently, the plasmon dispersion curve is similar to the result for jellium, as seen in Fig. 1a. However, in K, as we illustrate below when we discuss plasmon damping, $\omega_p(0)$ falls in a spectral region where the LDA band structure is dominated by *d*-bands.¶ As a result, the plasmon dispersion curve (Fig. 1b) differs appreciably from the "canonical" jellium result. The extreme case is Cs (Fig. 1c), for which a qualitative breakdown of simple-model physics is realized: the plasmon dispersion is negative, for small wave vectors.†

Let us expand a little on the interplay between band-structure effects and the calculated plasmon dispersion curves for K and Cs.‡ Consider first the case of K, shown in Fig. 1b, where the empty triangles correspond to a calculation in which we have excluded the contribution to $\chi^{(s)}$ from the core states. From that result, and by comparison with the dispersion curve for jellium, we can visualize that the large "real-metal" effects at play recognize two sources: the "polarization effect" of the core states (mostly the $3p^6$-derived states), *and the polarization effect of d-bands lying just above* $\omega_p(0)$¶ —the latter effect being responsible for the major part of the difference between the solid line (jellium) and the empty triangles (crystal calculation without the effect of the core).$

As for the case of the plasmon in Cs, the dispersion curve given by the triangles in Fig. 2a corresponds to a calculation in which we have kept only the lowest three valence bands in the evaluation of $\chi^{(s)}$—one occupied, two empty bands. (The core states are also kept.) We stress the fact that the slope of the dispersion curve so obtained is *negative*; indeed, it differs little from the slope of the "all-band" dispersion curve (solid circles) —the shift between both dispersion curves is weakly $q$-dependent. This test establishes that *the*

¶ Cf. Fig. 4a shown below.

† Our results of Fig. 1 agree qualitatively with those of Aryasetiawan and Karlsson;[14] however, there appear to be differences in the analysis of the results. For example, in our calculations for K the *d*-bands are already degenerate with the plasmon; similarly, we find that the *d*-bands responsible for the negative dispersion of the Cs plasmon lie appreciably *below* the plasmon energy.

‡ For brevity, and since such discussion would take us away from the main theme of this paper, in Fig. 1 we have ignored the crystal local fields, i.e., the non-diagonal elements of the matrix defined in Eq. (6). Now, in Na these effects are very small. In K they are small but non-negligible; their inclusion shifts the calculated dispersion curve upward, relative to the solid circles in Fig. 1b, leading to an even better agreement with the data. In Cs these effects are larger, a reflection of the importance of the $5p^6$ semicore states, which are quite shallow —they lie at ~ 11 eV below the Fermi energy. As we have found out in earlier work,[18] in this case the many-body kernel $f_{xc}$ cannot be ignored, even for $k \to 0$.

$ Note that the core states affect not just the plasmon energy —which shifts upwards, when the core is excluded, as seen in Fig. 1b— but also the slope of the dispersion curve, which actually becomes slightly negative in the absence of these states.

*first two empty bands are responsible for the anomalous dispersion of the plasmon energy in Cs* —that is, for the key physical difference with a jellium-like description of the Cs plasmon. It is noteworthy that these bands are mostly of *d*-character (the partially-full lowest band is, of course, predominantly *s*-like); they are the origin of the sharp structure lying at ~ 1-2 eV above the Fermi surface in the DOS shown in Fig. 2b. Higher-lying valence bands, particularly the *d*-complex lying just above $\omega_p(0)$ (~ 3eV), account for the large polarization shift observed in Fig. 2a between the three-band dispersion curve and the all-band result.†

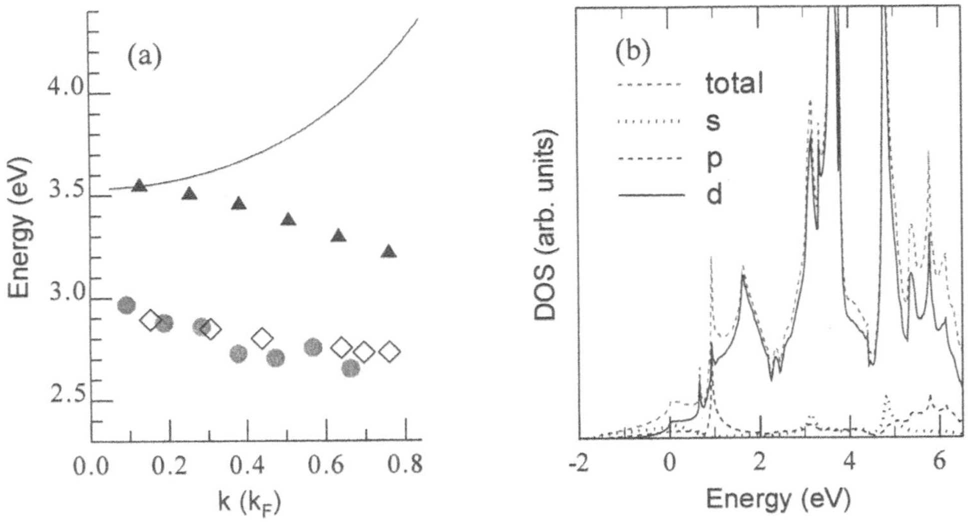


**Figure 2.** Physical explanation of the negative dispersion of the plasmon energy in Cs. (a): Plasmon dispersion curve (solid triangles) obtained upon keeping just the lowest three valence bands (plus the core states) in the evaluation of $\chi^{(s)}$. The solid circles, empty diamonds, and the solid line, are as in Fig. 1c. (b): LAPW-based DOS for Cs;[41] shown are the total DOS and contributions from states of *s*, *p*, and *d*, symmetry. The zero of energy is the Fermi energy. The spiky structure below 2eV corresponds to the first two empty bands kept in the calculation of (a). See text for details.

Finally, we note that the core states play an even more important role in the dynamical response of Cs than they do in the case of K. Indeed, in addition to being the source of a downward shift of the plasmon dispersion curve, the core states of Cs (mainly the $5p^6$-derived states, whose excitation threshold is at ~ 11 eV) have a profound impact on the *lineshape* of the plasmon loss. A related finding, which we will discuss elsewhere, is that the core states "screen out" the fine structure in the loss function due to interband transitions to states corresponding to the sharp peaks lying above $\omega_p(0)$ in the DOS shown in Fig. 2b.

† In previous work[18] we had concluded that the higher *d*-complex lying between 3eV and 4eV was responsible for the negative dispersion of the Cs plasmon. Figure 2a shows that, in agreement with the earlier analysis, this *d*-complex does bring about a large downward shift of the plasmon energy relative to the jellium result (this shift basically takes the triangles into the circles). However, the figure also shows that the effect of this complex on the *slope* of the dispersion curve is small.

Turning now to the physics of plasmon *damping*, Fig. 3a shows a well-converged calculation[20] of the dispersion of the plasmon *linewidth* in K (solid circles). The calculated full-width at half-maximum of the plasmon peak, $\Delta E_{1/2}(\vec{k})$, is compared with the value extracted by vom Felde et al.[5] from their EELS loss function (empty diamonds). (Note that Fig. 3 gives $\Delta E_{1/2}(\vec{k})$ relative to its value at zero wave vector.) Clearly, our result is in excellent agreement with experiment; since this agreement is obtained for $f_{xc} = 0$, we conclude that the plasmon linewidth *dispersion* of K is *not* controlled by a dynamical many-body mechanism.‡

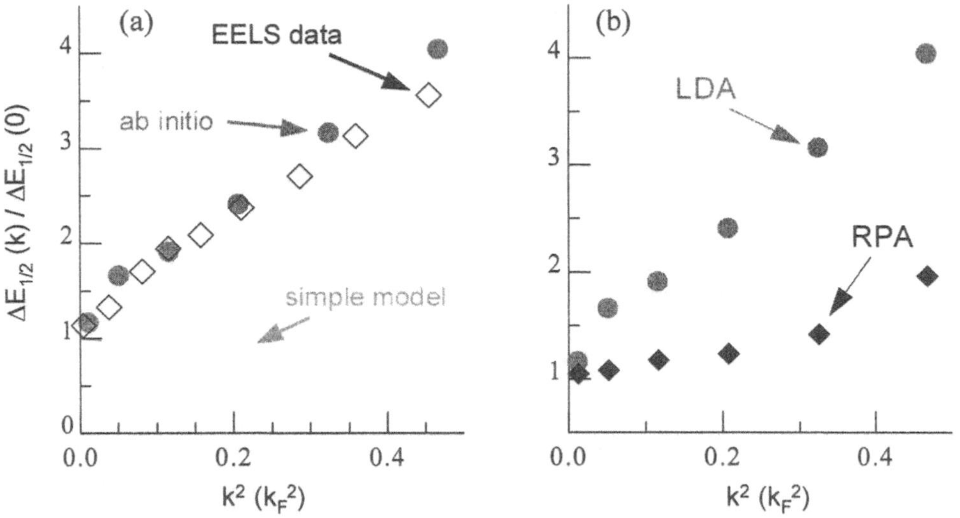


**Figure 3.** Physical explanation of the positive dispersion of the plasmon linewidth in K. (a): Comparison of calculated[20] (solid circles; labeled "*ab initio*") and measured[5] (empty diamonds; labeled "EELS data") linewidths. Note: we have set $f_{xc} = 0$ in Eq. (4). The solid line (labeled "simple model") corresponds to the calculations of Sturm and Oliveira.[10] (b): Plasmon linewidth dispersion obtained from Eq. (4) on the basis of RPA (solid diamonds) and LDA (solid circles) calculations of the single-particle response $\chi^{(s)}$. See text for details; cf. Fig. 4.

Our result is striking, as intuitive expectations based on the fact that the gap just above the Fermi surface at the *N*-point is small (cf. Fig. 4a) yield a $\Delta E_{1/2}(\vec{k})$ of the form shown by the solid line in Fig. 3a, which corresponds to an evaluation of the dielectric function to second order in the electron-lattice interaction[10] (modeled by a local, empirical pseudopotential). It is apparent that, although "physically reasonable," a simple-model description of plasmon damping in K breaks down, and badly so.

This breakdown can be understood by analyzing the impact of key "final state" bands, shown on Fig. 4a, in which the shaded strip is the energy interval representing all the single-particle states which may couple to the plasmon, as determined by the conservation laws of energy and crystal momentum. This analysis is done in detail elsewhere;[20] here we

‡ This conclusion does not imply that the dynamical correlations vanish identically, but, rather, that the effects of $f_{xc}$ have little impact on the *dispersion* of the linewidth.

simply summarize results of several tests we performed in order to open up the "black box." For example, if we keep just the first three valence bands (thin solid lines in Fig. 4a) in the evaluation of $\chi^{(s)}$ according to Eq. (6), the calculated $\Delta E_{1/2}(\vec{k})$ happens to be quite comparable with the simple-model result displayed in Fig. 3a; this is reasonable, as the dispersion of these bands is quite substantial, even if they are not strictly parabolic.¶ Now, when three additional bands (thick solid lines in Fig. 4a) are included in $\chi^{(s)}$, the plasmon linewidth dispersion curve changes *qualitatively*; the corresponding $\Delta E_{1/2}(\vec{k})$ turns out to agree well with the "all band" calculation of Fig. 3a (and thus, with the EELS data as well). Clearly, then, these three bands provide the key decay channels for the K plasmon. Note that the point here is not the addition of three extra bands in the evaluation of $\chi^{(s)}$. What is crucial is that these bands are flat, and overwhelmingly of *d*-character.[20] The phase space complexity which these bands introduce in the plasmon damping process cannot be approximated via nearly-free-electron states —thus the anomalous behavior of the linewidth dispersion.

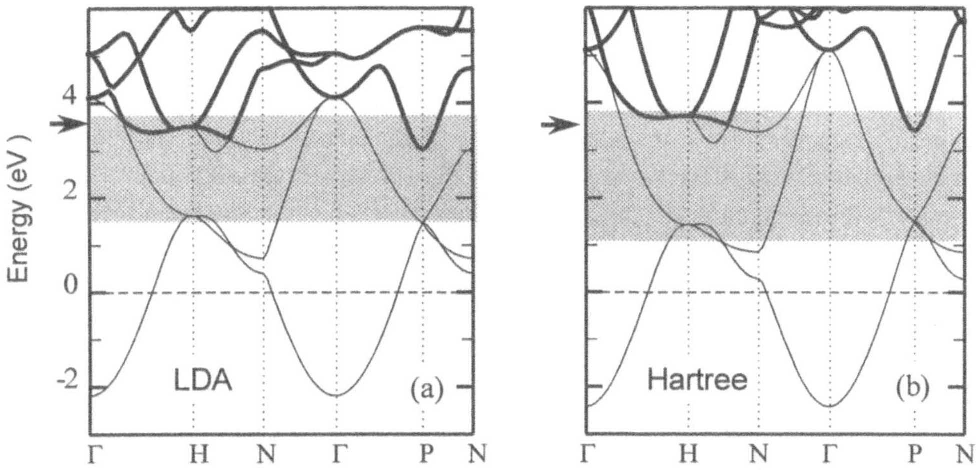


**Figure 4.** LAPW valence band structure of K for energies in the vicinity of the plasmon energy $\omega_p(0)$, whose value is indicated by an arrow. (a): LDA; (b) Hartree. The shaded strip illustrates, in each case, the energy location of all the possible empty states corresponding to electron-hole decay channels for the plasmon. See text for details; cf. Fig. 3.

We reiterate that the use of the Kohn-Sham states in Eq. (6) is "legal," and not a matter of expediency; the only approximation made in the evaluation of $\chi^{(s)}$ is the LDA. We pursue this issue further with reference to Fig. 3b, which shows results of additional calculations performed within the RPA, in which all exchange-correlation effects are left out, including those in the band structure, which corresponds to the Hartree approximation (i.e., we switch off $V_{xc}$ in the Kohn-Sham equation). The main physical change is that, as

¶ The lower edge of the shaded strip in Fig. 4 is given by (bottom of occupied band+$\hbar\omega_p(0)$); the upper edge is ($E_F + \hbar\omega_p(k_c)$), where $k_c$ is the wave vector for which the plasmon enters the continuum of particle-hole pair excitations; $\hbar\omega_p(0)$=3.7 eV.

seen in Fig. 4b, the key flat bands are shifted upwards in relation to the LDA *d*-bands of Fig. 4a. Although the magnitude of the shift is not large, it does have an important impact on the linewidth, as the Hartree *d*-bands lie almost entirely above the corresponding shaded strip —thus, the decay channels involving these bands become inaccessible to the plasmon. This effect is the root of the reason why, as seen in Fig. 3b, the $\Delta E_{1/2}(\vec{k})$ calculated in RPA has a much smaller slope than the TDDFT result (and the EELS data).

These results lead us to two interesting afterthoughts. First, we have shown that the dispersion of the K plasmon linewidth is controlled by decay into electron-hole pairs involving unoccupied Kohn-Sham states lying in a narrow energy window —states which are commonly said to "not to mean anything." At a formal level, our finding is on a sound basis because of the TDDFT framework. At a physics level, we can rationalize this result as follows. We have found (by comparison with experiment) that the effects of the many-body kernel $f_{xc}$ are not important in the present problem; this is akin to stating that the excitonic electron-hole attraction does not play a role, in the present case.† In this sense, it is not too surprising that a one-particle picture may provide a good description of the degrees of freedom of the many-particle system probed in the plasmon decay process [these degrees of freedom (electron-hole states) involve no change in the number of particles, similarly to the case of optical absorption]. What *is* striking, and this takes us to the second comment, is that the Kohn-Sham one-particle "picture" is so successful in the present problem. Put differently, it is intriguing that the exchange-correlation "dressing" of the empty *d*-bands highlighted in Fig. 4a manifests itself directly in the overall shape of the dispersion curve for the linewidth of the K plasmon (Fig. 3b).‡

## DYNAMICAL STRUCTURE FACTOR OF ALUMINUM

In the RPA, the dynamical density response for large wave vectors ($q \approx 2k_F$) corresponds to the excitation of particle-hole pairs which are nearly free (non-interacting). Furthermore, in the jellium model, $\mathrm{Im}\,\chi^{\mathrm{RPA}} \to \mathrm{Im}\,\chi^{(0)}$ for large $q$'s, where $\chi^{(0)}$ is the Lindhard function —which in this limit is a *smooth function of frequency*. Now seminal IXSS experiments reported many years ago by Platzman and collaborators[21,22] showed that the response of relatively-simple systems such as Be, Al, Li, Si, and graphite, deviates significantly from this "expected" behavior. Indeed, the IXSS spectra for large wave vectors are dominated by an anomalous "two peak structure," whose discovery triggered a vast amount of work, which has persisted over the years, aiming at an understanding of the dynamical correlations in these systems —and, by extension, in solids in general.

With the advent of synchrotron sources, and the consequent improvement in energy and wave vector resolution (and photon fluxes), major advances have been reported over

† Now $f_{xc}$ does not have the direct diagrammatic interpretation of a vertex correction for the irreducible polarizability; thus our argument is only meant to convey a flavor for the physics at play.

‡ It is not obvious whether this effect of exchange and correlation is dealt with without uncontrollable double-counting in many-body models of interacting electrons, which typically include empirical band structures in the hopping term in the Hamiltonian.

the last 10-15 years on the experimental front.[7-9,42,43] On the theoretical side,[44-50] until quite recently the thrust was directed almost exclusively towards the study of the effects of correlation for electrons in jellium.† It seems fair to state that this simple-model approach, which has featured treatments of the many-body problem of various degrees of sophistication, has not yielded spectra that one could consider to be faithful representations of the measured loss functions. This being the case, this approach remains inconclusive, in our view, since, in the absence of a small parameter for the interactions at play, comparison with the data plays a crucial role in the process of sorting out the relevance of proposed physical mechanisms. (For accounts of recent efforts within this approach, and differing views as to their relevance for an understanding of the data, see Refs. [7] and [9].)

It is only recently that *ab initio* calculations[11-20] have begun to change this state of affairs. The key reason behind this progress is the same "fortunate circumstance" we encountered above in the discussion of the collective mode in the alkali metals, where the Kohn-Sham single-particle response $\chi^{(s)}$ contained enough of the physics of the problem that we were close to the experimental data upon setting $f_{xc} = 0$ in Eq. (6). A similar situation is realized in the present case.

We recall that, in the first Born approximation, the double differential scattering cross section measured in IXSS is proportional to the dynamical structure factor $S(\vec{q};\omega)$, given by the equation

$$S(\vec{q};\omega) = -2\hbar V_N \operatorname{Im} \chi_{\vec{G},\vec{G}}(\vec{k};\omega)\,, \tag{7}$$

where $\vec{q} = \vec{k} + \vec{G}$. In Eq. (7) the wave vector *transfer* $\vec{q}$ is not restricted to the BZ. An appealing feature of IXSS is that it probes the electronic excitations over a wide range of wave vectors, from the small $q$'s for which the response is coherent (dominated by the plasmon) to the large $q$'s for which the response is incoherent (dominated by one-electron-like excitations), all the way to the Compton regime. For brevity, here we refer to a representative wave vector in the incoherent regime.‡ In Fig. 5a we discuss the $S(\vec{q};\omega)$ of Al for $|\vec{q}| = 1.7k_F$; this case has been discussed at length by Platzman et al.,[7] Schülke et al.,[8] and Larson et al.[9,43] The IXSS data,[7] labeled $S_{\text{exp}}$, are given by the empty circles; the most prominent feature of the measured loss function is the famous two-peak structure, whose minimum, centered at ~ 32 eV, is highlighted by a downward-directed arrow.

† The motivation for this widespread adoption of a simple-model approach was that the "double peak" (or one-peak, one-shoulder) loss structure was observed in systems with quite different band structures; furthermore, when the energy loss is measured in units of the Fermi energy, the double peak appears to suggest "universal" electron-gas behavior.

‡ It is noteworthy that the comparison which follows is made on the basis of theoretical and experimental spectra which are plotted in absolute units (by use of the *f*-sum rule); there is no lining up of peak heights, a procedure which would introduce a definite element of arbitrariness in the analysis. For example, if the comparison were based on lining up the height of the most prominent low-frequency peak, the overall agreement observed in Fig. 5a between theoretical and experimental spectra for frequencies above the two-peak structure would be destroyed. That agreement is an integral part of the reason why the conclusions we reach regarding the overall impact of the effects of the band structure are, in our view, quite compelling.[43]

Our analysis of the physics behind the data starts with a look at the loss function calculated using $\mathrm{Im}\,\chi^{(s)}$ in Eq. (7) in place of $\mathrm{Im}\,\chi$, i.e., the loss function $S_{KS}$ corresponding to the single-particle Kohn-Sham response (thick solid curve in gray tone.[15]) For reference, we also show the counterpart of $S_{KS}$ for electrons in jellium, $S_{Lindhard}$ (dots), which, when compared with $S_{exp}$, serves as a reminder of why the IXSS spectrum was originally described as being "anomalous."

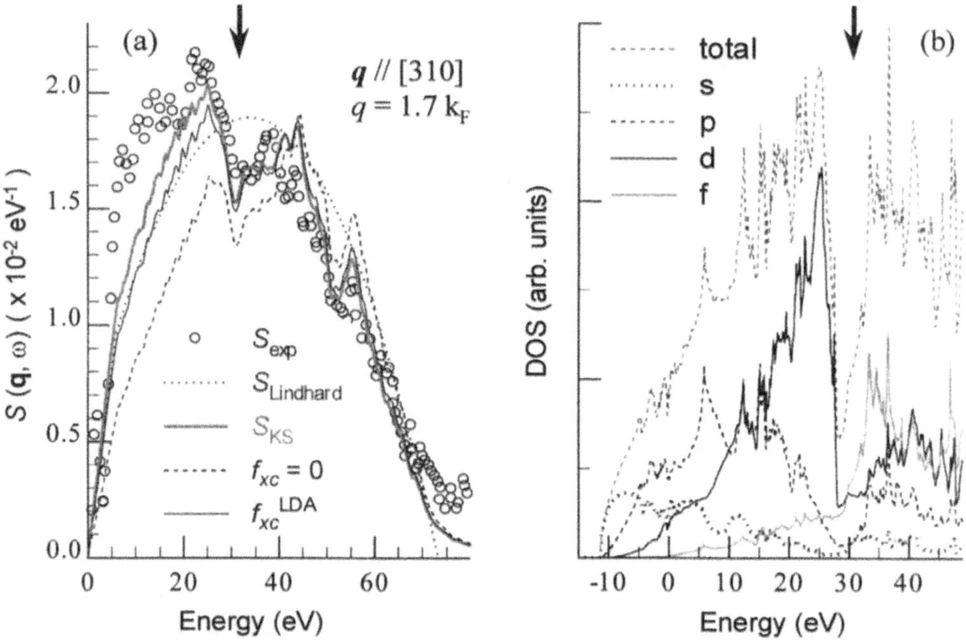


**Figure 5.** Explanation of the "double-peak" structure in the dynamical structure factor of Al. (a) The IXSS spectrum,[7] labeled $S_{exp}$ (empty circles), is compared with the loss function for Kohn-Sham electrons, labeled $S_{KS}$ (thick solid curve in gray tone), and with the loss function obtained from the solution of Eq. (4) for $f_{xc} = 0$ (dashes) and for the LDA value of this kernel, $f_{xc}^{LDA}$ (thin solid line), respectively. For reference, the loss spectrum for non-interacting electrons in jellium with $r_s = 2.07$ (dots) is also shown. (b): Calculated DOS for Al; total DOS and contributions from states of $s$, $p$, $d$, and $f$ symmetry, obtained in an LAPW band-structure calculation.[41] The zero of energy is the Fermi energy.

It is apparent that $S_{KS}$ reproduces the main features of $S_{exp}$ surprisingly well. In view of the exact nature of Eq. (4), this can only be true if, for the $(\vec{q},\omega)$ subset of phase space explored in Fig. 5a, the many-body kernel $f_{xc}$ largely compensates for (i.e., cancels) the effect of the bare Coulomb interaction in the TDDFT density-response function $\chi$. Note that, in the light of this statement, the result for $S_{KS}$ in Fig. 5a can be viewed as providing a strong hint as to what one would really like to *predict* via an appropriate treatment of the dynamical correlations contained in $f_{xc}$.¶

¶ The exaggerated fine structure present in the calculated spectra will presumable be suppressed by dynamical-correlation effects left out in this calculation. Note also that $S_{KS}$ in Fig. 5a was obtained from a pseudopotential-based band structure;[15] thus it does not include the core excitations observed in the IXSS data at ~ 72eV.

A functional form for $f_{xc}$ which can be evaluated self-consistently with the electronic structure corresponds to the *static* LDA expression

$$f_{xc}(\vec{x},\vec{x}';\omega) = \frac{dV_{xc}[n](\vec{x})}{dn(\vec{x})}\delta(\vec{x}-\vec{x}') \ , \tag{8}$$

where $V_{xc}[n](\vec{x})$ is the exchange-correlation potential entering the Kohn-Sham ground state problem within the LDA. That this approximate —and pragmatic— choice[51] for the many-body kernel leads to the type of cancellation effect suggested by the overall agreement between $S_{\mathrm{KS}}$ and $S_{\mathrm{exp}}$ is illustrated in Fig. 5a. First we note that, once we turn on the Coulomb interaction $v$ in Eq. (4), while still switching off the many-body kernel, the corresponding loss function, given by the dashes, and labeled $f_{xc} = 0$ in Fig. 5a, worsens considerably, relative to experiment —the intensity on the low-frequency side is far too low. Nonetheless, the fine structure built into the Kohn-Sham response is still present in that loss function. Next, we have that, upon including the many-body kernel $f_{xc}$ into Eq. (4) according to the LDA prescription given by Eq. (8) —curve labeled $f_{xc}{}^{\mathrm{LDA}}$ (thin solid line)— the quality of the low-frequency part of the calculated spectrum improves considerably. In fact, that result is quite close to the spectrum obtained for Kohn-Sham electrons, which is what we anticipated above.†

Of course, this near cancellation of the effects of the Coulomb interaction $v$ and the many-body vertex $f_{xc}$ in Eq. (4) does not hold "universally." For example, for wave vectors in the plasmon regime, $\mathrm{Im}\,\chi^{(s)}$ bears little resemblance to the IXSS data[52] —thus the cancellation is a non-issue; numerically, the effect of $f_{xc}$ is minor, in this case.

The remaining question is the origin of the two-peak structure in $S_{\mathrm{KS}}$.¶ This point can be elucidated particularly convincingly with reference to the calculated DOS for Al, shown in Fig. 5b. It is, in fact, remarkable that, on the energy scale relevant to the present argument, the DOS is dominated by a deep indentation ("quasi-gap"), located at about the same energy as the minimum in the two-peak structure in $S_{\mathrm{KS}}$ (note arrow in Fig. 5b). This wide quasi-gap in the DOS may seem to be a surprising feature of the band structure of Al, which is generally considered to be "jellium-like." However, there is no conflict. While, indeed, the bands near the Fermi surface are predominantly *sp*-like (see DOS in Fig. 5b), the excited-state band structure eventually differs significantly from simple parabolic bands. For example, there is the (2,0,0) zone-boundary gap, approximately 6 eV wide, centered at ~ 32 eV. Since the (310) direction which Fig. 5a refers to differs from the (100) direction by a relatively small angle, the final-state gap in the latter direction has a bearing on wave vector transfers along the former—thus the dip giving $\mathrm{Im}\,\chi_{\mathrm{KS}}$ its "two-peak" shape observed in Fig. 5a. But, why is the "gap" in the DOS (Fig. 5b) so large, if the crystal potential in Al is "weak?" The qualitative explanation for this apparently

† Note that the effect of $f_{xc}$ observed in Fig. 5 is qualitatively similar to the "continuum exciton" effect discussed by Hanke and Sham in their study of the optical response of Si.[52]

¶ See the more extensive discussion of this point presented recently by Larson et al.[43]

contradictory result is that, as shown in Fig. 5b, the angular-momentum content of the Bloch states is overwhelmingly *d*-like just below the gap at 32 eV, while above the gap both *d*-content and *f*-content dominate all other angular momentum components. Since the core states in Al contain neither $l$=2 nor $l$=3 components, the corresponding Bloch states near the gap are more exposed to the nuclear electrostatic potential than would be the case if these $l$-components contributed to the shielding between nucleus and valence states — thus the large gap at the (2,0,0) zone boundary, which is reflected in the pronounced dip in the DOS at 32 eV.‡ Naturally, this "final-state gap" depletes the available final states in the excitation process, and this gives rise to a dip in the loss function for Kohn-Sham electrons at about the same energy, as is clearly observed in Fig. 5a.

We summarize the above discussion by noting that the predominant two-peak structure in the IXSS spectrum of Al can be described pictorially as an indentation "carved out" from the Lindhard function (see Fig. 5a) because of the presence of a gap in the Kohn-Sham excited-state band structure. We emphasize again the sound basis which TDDFT gives to this argument. This framework has thus led us to the resolution of yet another "anomaly" in the experimental manifestation of the dynamical electronic response of a simple metal.

## EXCITATIONS IN REAL SOLIDS: CONSERVING DIAGRAMMATIC APPROACH

The above calculations, while successful, cannot be considered to represent a complete theory; to reach that goal we must incorporate dynamical correlations according to Eq. (5) for the same electronic structure from which $\chi^{(s)}$ is obtained. In the context of the $S(\vec{q};\omega)$ of Al discussed above, an appropriate inclusion of the effects of a frequency-dependent $f_{xc}$ is expected to smooth out the exaggerated fine structure observed in Fig. 5a, while preserving the indentation due to the band gap at ~ 32 eV. On a wider scope, the dynamical effects of $f_{xc}$ must be accounted for if we are to study at an *ab initio* level systems which are characterized by complex electronic structures and non-trivial correlations near the Fermi surface ("strongly-correlated" electron systems).

In the case of atoms, progress has been reported recently in the implementation of the TDDFT linear-response scheme for an exchange-only functional.[24] In addition, significant advances have been made in the elaboration of theories of $f_{xc}$ which include correlation effects beyond adiabatic and local approximations;[55] however, these schemes have yet to be implemented for realistic representations of the band structure. Thus, for completeness, in the remainder of this article we sketch a diagrammatic alternative.

A rigorous formulation of the problem of one-particle-like excitations in a many-body system starts out from the Dyson equation, which we represent diagrammatically as

‡ It is interesting to note that this gap shows up clearly in the photoemission data of Levinson, Greuter and Plummer;[54] furthermore, their analysis of the angular momentum content of the states around the gap is in good agreement with the LAPW results of Fig. 5b.

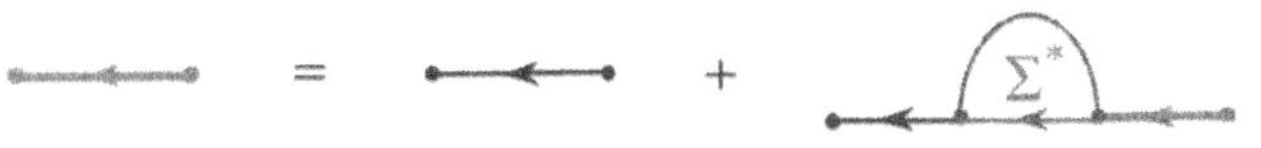


, (9)

where the thick line represents the actual one-particle Green's function $G(11')$, and the thin line represents the Green's function for "free" propagation in the LDA band structure, $G_{\text{LDA}}(1,1')$. All correlations beyond the LDA are formally built into the self-energy $\Sigma^*[G](1,1') = \Sigma[G](1,1') - (V_H(1) + V_{xc}(1))\delta(1-1')$, where $V_H(1)$ and $V_{xc}(1)$ are, respectively, the Hartree and exchange-correlation potentials entering the Kohn-Sham equation in LDA, and $\Sigma[G](1,1')$ is the self-energy for electron propagation in the "bare" external periodic crystal potential. Note that the cohesive effect of correlation (crystal binding) is built from the outset into the "unperturbed" Green's function $G_{LDA}(1,1')$.†

We proceed within the framework of the Baym-Kadanoff "conserving-approximation" scheme.[25] In the paper by Baym[26] it is shown that if an *approximate* model for the electron-electron interactions built into the self-energy is "Φ-derivable," i.e., if there exists a functional Φ, given by a set of Feynman diagrams composed of the bare interaction $v$ and the Green's function $G$, such that

$$\Sigma[G](11') = \frac{\delta\Phi}{\delta G(1'1)},$$

(10)

the ensuing Green's function is "conserving" —the self-consistent solution of Eqs. (9) and (10) fulfills all conservation laws (particle number, energy, momentum, angular momentum) *exactly*.‡ In addition, "thermodynamic consistency" is ensured, which means, e.g., that different schemes for calculating the partition function give the same result.¶

A choice for Φ which incorporates the physics of screening due to the long-range Coulomb interactions is the shielded-interaction approximation (SIA),[25] which consists of the Hartree and Fock diagrams plus an infinite series of electron-hole bubbles made out of "exact" $G$'s —i.e., of the self-consistent solution of the Dyson equation for the chosen Φ-functional— according to

---

† Comment on notation: The labels $1,1'$ denote space-time points; the time variables are Matsubara times $0 \le \tau, \tau' \le \beta\hbar$. The arrow in all Green's functions indicates propagation from the space-time point on the right to the one the left; i.e., from $1'$ to 1, in the case of $G(11')$. The arrow in the self-energy "cartoon" in Eq. (10) is simply a mnemonic device that helps decide the direction of the arrow for the propagators into which the former is to be inserted, as in Eq. (9). This convention about our cartoons (our nomenclature distinguishes cartoons from actual diagrams, such as the self-energy diagram in Eq. (12), etc.) is adhered to below in the case of other quantities which do not represent "propagation" of particles, such as the electron-hole interaction I, and the T-matrices $T^{ph}$ and $T^{pp}$.

‡ We note that the Φ-functional enters the exact expression for the thermodynamic potential $\Omega(T,\mu)$ obtained by Luttinger and Ward[56] from an analysis of the formal structure of many-body perturbation theory. In other words, Eq. (10) follows from the structure of the exact theory. The essential additional step taken by Baym was the demonstration that Eq. (10) ensures the fulfillment of the criteria for a conserving theory of $G$ put forth in the Baym-Kadanoff paper.[25]

¶ This conclusion is non-trivial, since it holds for *approximate* (conserving) calculations, and errors propagate differently in different schemes for evaluating the same physical observable.[26,31]

$\Phi_{\mathrm{SIA}}$ :

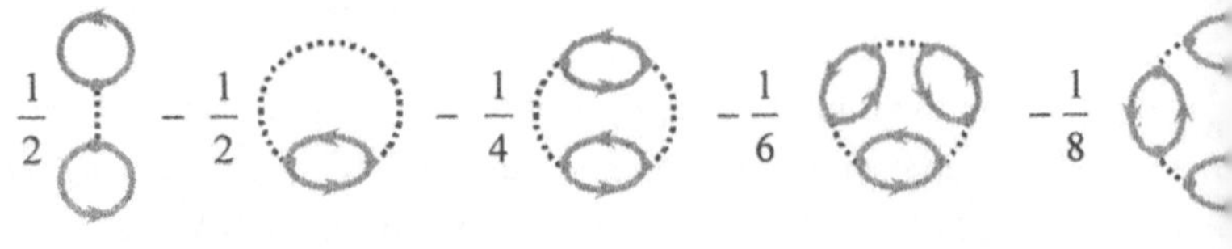

From $\Phi_{\mathrm{SIA}}$ we obtain, via Eq. (10), the following skeleton expansion for the self-energy:

$\Sigma_{\mathrm{SIA}}$ :

, (12)

which we have summed formally to all orders in $v$ via the introduction of the shielded interaction $W_{\mathrm{SIA}}(11')$; this effective, or screened, interaction obeys the integral equation

$W_{\mathrm{SIA}}$ :

. (13)

We note that the label "SIA" attached to $W$ in Eq. (13) is meant to serve as a reminder of the fact that in the present case the shielded interaction is to be evaluated for the Green's function defined by this very set of equations.† If we were to change $\Phi$, $W$ would change. Of course, having started by choosing a subset of the exact $\Phi$-functional to generate $\Sigma_{\mathrm{SIA}}[G_{\mathrm{SIA}}]$, this approximation is automatically conserving. ‡

We illustrate the physics contained in this scheme with reference to a recent calculation of the absolute band gap in Si.[27] In this context, it is relevant to note that the SIA is equivalent to Hedin's *GW* approximation,[28] which has proved extremely popular in *ab initio* calculations of quasiparticle band structures in metals and semiconductors, and more exotic forms of matter, such as the fullerines.[29,30,57] We work in the Bloch basis of the Kohn-Sham states, and turn Eqs. (9), (12), and (13) into matrix equations in the band

† One of the appealing features of the conserving method is the fact that self-consistency is built into it from the start —the self-energy arises naturally as a functional of the "exact" $G$. Of course, this adds to the computational complexity of the method, and helps explain why it has only been implemented recently (at the computational level),[27,31] starting with models of correlated electrons in two-dimensions, in the context of the high-$T_C$ superconductors.[58-61] The driving motivation for the resurgence of this method was the need to treat competing ordering mechanisms on an equal footing, self-consistency being crucial for even a qualitative description of the possibility of a change of phase.[62]

‡ To be precise, the scheme assumes the presence of an external, fictitious, potential $U(11')$, which enters the equation of motion of $G(11')$; thus, $\Sigma$ depends on $U$ via its functional dependence on $G$. This technical point is important in the formulation of the method, as it ensures that the two-particle correlation function $L$, defined below, also fulfills the conservation laws.

index.† In Fig. 6 we show the spectral function $A_{jj}(k;\omega) \equiv -\pi^{-1}\,\mathrm{Im}\,G_{jj}(k;\omega)$ for the $\vec{k}j$ pairs defining the absolute band gap (we use the notation introduced in Eq. (6)).‡ First we have the LDA; since this is a single-particle theory, the corresponding spectral function consists of delta functions, which, for ease of visualization, have been broadened numerically. The LDA band gap read off Fig. 6 (0.53 eV) differs from the experimental value (1.17 eV), by a factor of two. This was the original "band-gap problem."

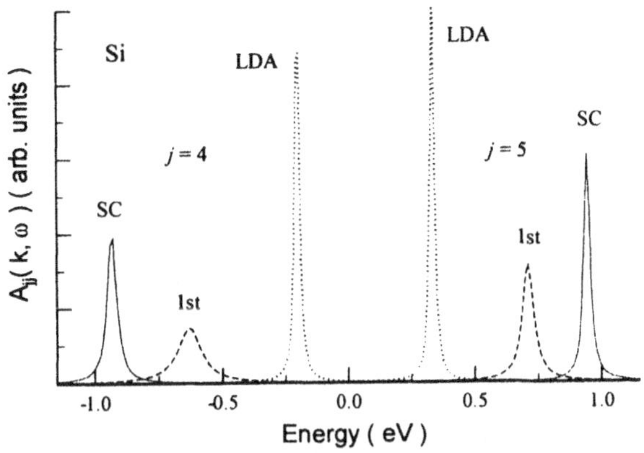


**Figure 6.** Impact of self-consistency on the absolute band gap in Si.[27] Shown is the spectral function $A_{jj}(k,\omega)$ for the states at the bottom ($j$=4) and the top ($j$=5) of the gap. Three cases are displayed, corresponding to the LDA, and to the first ("1st") and self-consistent ("SC") evaluations of the self energy. The zero of energy is the renormalized chemical potential for each case.

Next we have the gap corresponding to the quasiparticle band structure obtained from the self-energy $\Sigma_{\mathrm{SIA}}[G_{\mathrm{LDA}}]$, i.e., the solution of Eqs. (12) and (13) obtained with use of the LDA Green's function $G_{\mathrm{LDA}}$. This calculation, which is not conserving, corresponds to the usual level of implementation of the *GW* approximation. The numerical value of the gap (1.34 eV) is now in good agreement with experiment (1.17eV). This is the usual success story — the band-gap problem appears to have been solved.

Finally, we perform the calculation at the conserving level; the Dyson equation is actually solved, and the new Green's function is used to recompute the self-energy iteratively until convergence. The gap obtained from $\Sigma_{\mathrm{SIA}}[G_{\mathrm{SIA}}]$ —the true *GW* band gap of Si— turns out to be too large (~ 1.9 eV), as can be surmised from Fig. 6. In fact, the experimental value of the gap is overestimated by about the same amount that the LDA underestimates it. Thus, *we have a band-gap problem once again.*¶

---

† Some details of the procedure we have implemented so far for the evaluation of all the ingredients of this system of equations are given elsewhere.[27] These include the evaluation of frequency convolutions which are insensitive to the cutoff which must necessarily be introduced, BZ sampling, number of bands kept in the evaluation of $G_{LDA}$, effects of finite temperature, analytic continuation to the real axis, etc.

‡ The bottom of the gap corresponds to the fourth valence band, $j = 4$, at the $\Gamma$-point; the top of the gap (bottom of the conduction band) corresponds to the fifth band, $j = 5$, for $k \sim 0.8\Gamma X$.

¶ A similar situation is realized with regard to the calculated values of occupied bandwidths, which in the SIA are wider than the corresponding LDA values,[27,31] while experiment shows a narrowing, relative to the LDA.

It is clear that the treatment of correlation must be improved; the question is how to go about doing so. In the recent literature a number of suggestions related to this problem have been made.[30-38,63-68] It is important to bring these issues into the focus of the present discussion, since they have an impact on the development of a satisfactory description of quasiparticle band structure and density response.

For example, the issue has been brought up that in the SIA (that is, in the *GW* approximation, when implemented self-consistently) the electron-hole bubble entering Eq. (13) and the shielded interaction $W_{SIA}$ as well, "violate" the *f*-sum rule.[31,32] Here we stress the fact that both functions are simply building blocks for the evaluation of a conserving Green's function $G_{SIA}$, and of a conserving correlation function $L_{SIA}$ (see below); contact with experiment is made through *the latter* two functions. Thus, the fact that $W_{SIA}$ does not contain a plasmon pole, and, related to this, that it does not fulfill the *f*-sum rule,‡ is not to be viewed as a "failure" of self-consistency; rather, in the context of the conserving scheme, this is a non-issue.¶ We will return to this point below, when we introduce the irreducible polarizability for the conserving theory.

Related questions have been raised with regard to the satellite structure in $A_{jj}(\vec{k};\omega)$, whose spectral weight is strongly suppressed by the use of renormalized propagators in the SIA.[27,31] These issues, together with the fact that occupied bandwidths do not agree with experiment,[27,31] have prompted the suggestion that the effort invested in self-consistency may be "futile," since it leads to "inferior" results.[31] Similarly, it has been discussed whether "non-self-consistent *GW* is to be preferred over self-consistent *GW*.[63]"

With regard to the preceding question, we paraphrase it by asking whether it may be acceptable to ignore the fulfillment of basic conservation laws if this happens to lead to apparent agreement with experiment in some situations. Our viewpoint is that, if we are interested in understanding the electronic interactions at a fundamental level, what needs to be done is better many-body theory. And, in retrospect, why should one aspire to obtain, e.g., a "perfect" quasiparticle band gap while neglecting the contribution to the self-energy from the electron-hole attraction?[52,64-68]

It has also been suggested that the inclusion of vertex corrections should be pursued *instead* of adherence to self-consistency, when faced with the additional difficulty of carrying out these calculations for realistic models of the electronic structure.[31] This brings the problem into the realm of a rather large body of literature which focuses on the evaluation of the polarizability and dielectric function. Now, it is, of course, well-known that the effect of vertex corrections in the polarizability tends to compensate the effect of self-energy insertions in the same function. This "balancing" of both effects of correlation has been framed as a necessary condition for particle-number conservation.[33] In this light,

‡ We have checked numerically that this putative "violation" of the *f*-sum rule corresponds to a factor of 2 or more, depending on the wave vector; J. M. Sullivan, private communication.

¶ Incidentally, note that the screening in the SIA is not the same as in the RPA —*W* is evaluated for different Green's functions in both cases. This distinction is essential in the conserving scheme.[25]

the above SIA calculation (which includes a bubble which does not obey the *f*-sum rule) may seem to be flawed.[31] However, this is not the case: The solution $G(11')$ of the system of Eqs. (9), (12) and (13) is fully conserving, as we explained in detail above, even though the SIA self-energy functional does not include renormalized vertices. This point illustrates clearly why it is so important to agree on the language and framework that is being used.

We turn next to the two-particle problem, which is the proper context for *f*-sum rule arguments. As noted below Eq. (13), the Baym-Kadanoff scheme requires the presence of an external, fictitious, non-local potential $U(11')$, which is introduced in order to "polarize" the system virtually. This artifice allows us to define the correlation function

$$L(12;1'2') = \frac{\delta G(11')}{\delta U(2'2)}, \qquad (14)$$

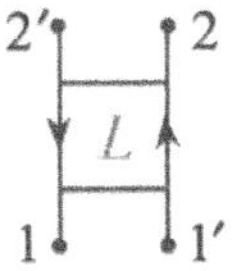


a subset of which,

$$\chi(12) = L(12;1^+2^+), \qquad (15)$$

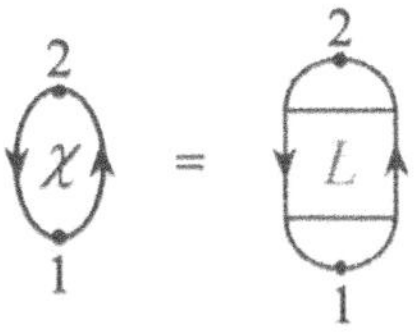


yields (upon setting $U = 0$) the dynamical structure factor $S(\vec{q};\omega)$ according to Eq. (7). Now the direct evaluation of Eq. (14) is not very convenient, as it would require the explicit solution of the one-particle problem out of equilibrium. A better approach is to obtain $L$ from the Bethe-Salpeter equation it satisfies, namely,[25]

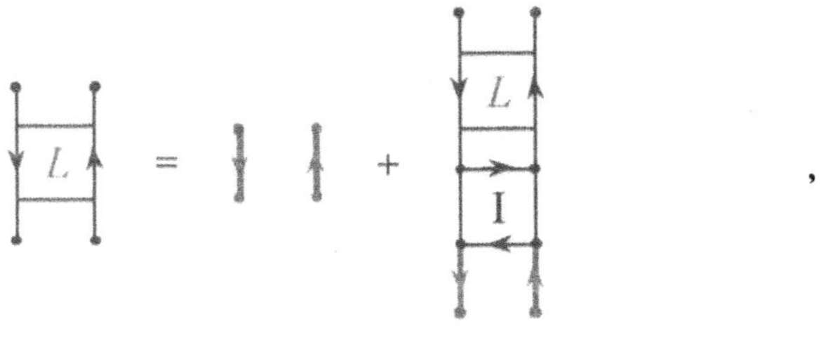


, (16)

where now $U = 0$, and the effective "electron-hole" interaction $\mathrm{I}(12;1'2')$ is given by

$$\mathrm{I}(12;1'2') = \frac{\delta \Sigma(11')}{\delta G(2'2)}. \qquad (17)$$

In principle, the physical content of the Bethe-Salpeter equation (16) is the same as that of the TDDFT response equation (4). If both schemes could be implemented exactly, they would yield the same loss spectrum $S(\vec{q};\omega)$. Of course, solving Eq. (16) is numerically much more demanding than solving Eq. (4), since the kernel $I(12;1'2')$ is a four-point function, while the kernel $f_{xc}(11')$ is a two-point function. The great conceptual merit of the conserving scheme is that, if the kernel $I(12;1'2')$ is generated according to Eq. (17), *on the basis of a conserving solution of the one-particle problem*, then $L(12;1'2')$ is also conserving —it fulfills all conservation laws in both $11'$ and $22'$ pairs of arguments; thus, in particular, it fulfills the *f*-sum rule. Note, again, that this exact fulfillment of the conservation laws for the two-particle response holds for an *approximate* functional $\Phi$ .

It is useful to note that the Bethe-Salpeter equation can be iterated, and cast in the form of an integral equation for the two-point density-response function $\chi$ defined according to Eq. (15), namely

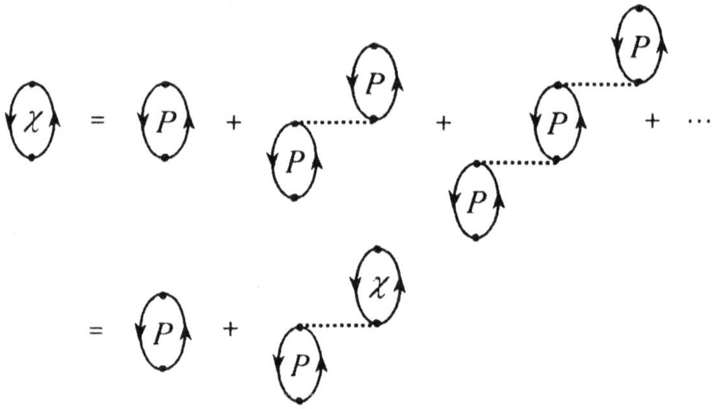


(18)

Equation (18), which is exact, is of the same *form* as Eq. (4).† Of course, all the complications of the many-particle processes contained in $I(12;1'2')$ are now hidden in the irreducible polarizability, $P(11')$. In fact, since both Eq. (18) and Eq. (4) yield the same response, we can readily write down an equation embodying a formal interrelation between $f_{xc}$ and $P$,

$$f_{xc}(11') = \chi_s^{-1}(11') - P^{-1}(11'), \tag{19}$$

to which we will refer (briefly) below.

We illustrate the physics contained in the above scheme with reference to the SIA. Taking the functional derivative of $\Sigma_{SIA}$ required in Eq. (17) yields the following result for $I_{SIA}$:

(20)

† Equation (18) can be derived by the usual many-body perturbation theoretic techniques.[68] The present approach has the added advantage that it yields a response $\chi$ which is guaranteed to be conserving.

The *structure* of the first two contributions to this effective electron-hole interaction corresponds to the so-called time-dependent screened Hartree-Fock approximation; that approximation is on a different footing —it is *not* conserving. The last two terms in Eq. (20) have traditionally been neglected in *ab initio* work for realistic models of solids; however, their inclusion is necessary if we seek a conserving response, $\chi_{\text{SIA}}$. Substituting Eq. (20) into Eq. (16), we obtain the counterpart of Eq. (18) for the specific case of the SIA, whose solution is given, in symbolic notation, by $\chi_{\text{SIA}} = (1 - vP_{\text{SIA}})^{-1} P_{\text{SIA}}$, in terms of the irreducible polarizability $P_{\text{SIA}}$, defined by the equation

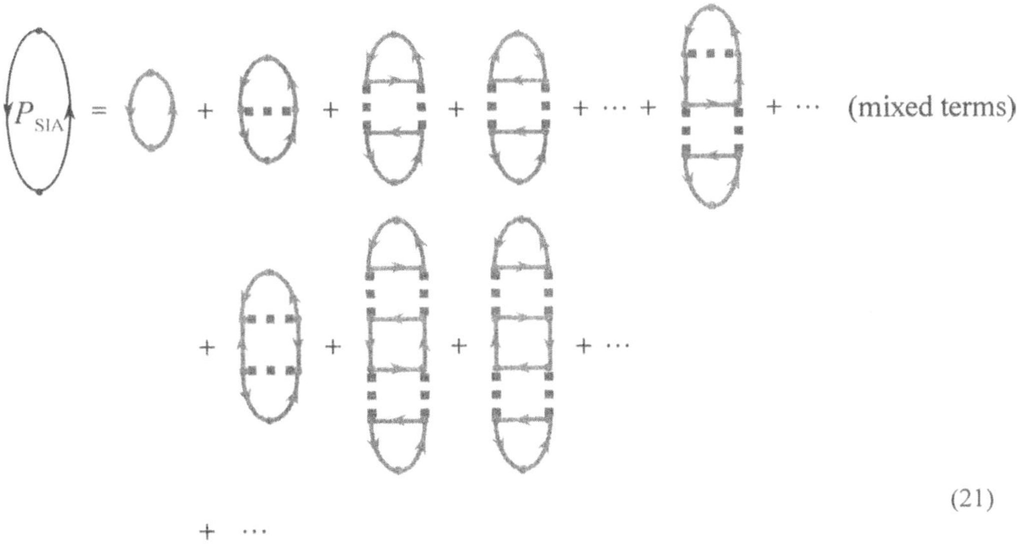


(21)

We emphasize that the first term on the right-hand-side of Eq. (21) is precisely the dressed electron-hole bubble which enters Eq. (13) for the self-energy $\Sigma_{\text{SIA}}$. It should now be apparent why we stated above that it was a non-issue for this bubble to "fail" to fulfill physical constraints, such as the *f*-sum rule.

In order to visualize the complex interactions built into $P_{\text{SIA}}$, we have grouped Eq. (21) in columns, each column beyond the first one having the structure of a ladder whose rungs are given by the respective terms in Eq. (20). The first such shielded ladder, which is traced to the second term in Eq. (20), has the structure of the ladders originally approximated by Hubbard via the introduction of a "local-field factor" (although in that case the ladders were not dynamically screened). The third and fourth columns in Eq. (21) correspond, respectively, to the last two terms in Eq. (20); these ladders are "fluctuation diagrams."[70] The remaining columns correspond to more complicated fluctuations containing a mixture of the processes which originate in the last three terms in $I_{\text{SIA}}$.[†]

[†] Note that, in order for $\chi_{\text{SIA}}$ to be strictly conserving, *all* the processes contained in Eq. (21) must be incorporated. Thus, in spite of the apparent simplification introduced on going from Eq. (16) to Eq. (18), we advocate the (very demanding) direct numerical solution of the Bethe-Salpeter equation.

We note that the first four diagrams in the first row of Eq. (21) correspond to the polarization diagrams whose contribution to $f_{xc}$, to the lowest order in $W_{\text{RPA}}$, was elucidated by Langreth and Vosko.[70†] In the light of Eq. (19), that work provides the beginnings of an explicit interrelation between the two approaches to correlated-electron response we have outlined here. We hope to explore this issue further in future work.

In addition, we note that the diagrams highlighted in the previous paragraph are responsible for the leading correction to the RPA density-response function for the high-density electron gas.[71] Furthermore, each of the these diagrams has non-integrable Fermi-surface-related singularities which become integrable upon summing them all up;[70] the weight of the integrable singularity increases with $r_s$.‡ Since the last two diagrams in this set originate from the last two terms in Eq. (18), it seems justified to assert that the common practice of neglecting these electron-hole interactions is unwarranted —of course, their inclusion is also required from the point of view of the conservation laws.

Finally, we stress the fact that the structure of Eq. (21) is consistent with the rules which have been discussed by Mahan for balancing vertex corrections and self-energy insertions in the irreducible polarizability, in compliance with a Ward identity.[33] However, it is extremely important to note that in the conserving method the polarization function is "predetermined" by the choice of the $\Phi$-functional, and by the definition of the electron-hole interaction according to Eq. (17), *which is to be implemented following the conserving solution of the one-particle problem* —as we have illustrated with our analysis of the polarization $P_{\text{SIA}}$, which follows from our choice of $\Phi$-functional, $\Phi_{\text{SIA}}$ (Eq. (11)). No such balancing of vertex corrections and self-energy insertions is imposed on the solution of the one-particle problem in order for $G(11')$ to be conserving —which is why this "balance" has no bearing on the conserving nature of $G_{\text{SIA}}$, or on the result for the Si band gap presented in Fig. 6, as we argued above.

We proceed to summarize the above presentation. First, we have discussed the implementation of the conserving scheme at the level of the one-particle problem. For the specific choice of $\Phi$-functional given by Eq. (11) —the SIA— we have noted that the self-consistent solution of Eqs. (9) and (12) yields a value of the absolute band gap in Si which disagrees with experiment. *There is a band gap problem, once again.*

One possible scenario is that there is some conceptual flaw with the SIA, which, as we have noted repeatedly, is the same as the *GW* approximation —when implemented self-consistently, which was the original intent in the Hedin scheme (although his method was not concerned with conserving approximations per se). We have considered this scenario at some length, and we hope that our discussion of the conserving scheme, including the crucial interrelation between the one-particle and two-particle problems, has made a clear case for the fact that the SIA self-energy functional is free of a variety of objections which

† $W_{\text{RPA}}$ has the structure of Eq. (13), but the bubble is evaluated for Hartree Green's functions.

‡ However, note that in the analysis of Langreth and Vosko the diagrams are evaluated with bare propagators. Nonetheless, their results are very suggestive of the importance of these diagrams for finite values of $r_s$.

have been raised in the literature —in the related context of the effect of self-consistency— with regard to sum rules, balance of vertex corrections and self-energy insertions, etc.

Another scenario —the one we are advocating herein— is that a better many-body calculation of the gap is needed. The apparent agreement with experiment obtained at the non-conserving level results from a cancellation of errors: lack of self-consistency (violation of the conservation laws), and absence of short-range correlation effects. In retrospect, this "new band gap problem" is not too surprising. In effect, it has been shown that the excitonic electron-hole attraction plays a significant role in the optical absorption process in Si.[52,64-66] In view of the results of Fig. 6, we expect that the electron-hole interaction must contribute to the value of the *quasiparticle* band gap.

The effects of the particle-hole interaction may be incorporated into the one-particle problem in terms of a $\Phi$-functional (displayed explicitly in the Appendix, where it is labeled $\Phi_T^{ph}$) which, via Eq. (10), generates a self-energy $\Sigma_T^{ph}$ of the form

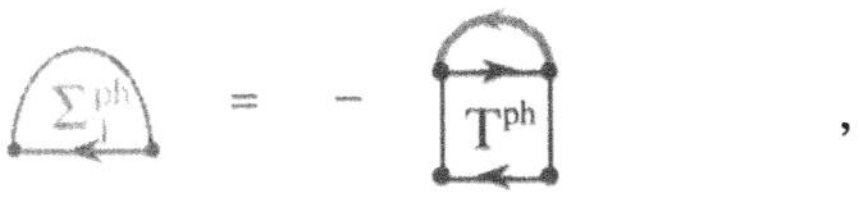

(22)

where the particle-hole T-matrix, $T^{ph}$, is the solution of the integral equation

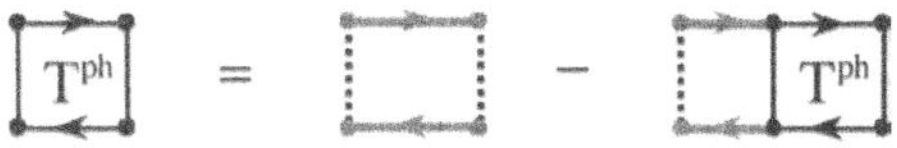

(23)

We note that, when adding the self-energy processes contained in Eq. (22) to those contained in Eq. (12), care must be exercised in order to avoid double counting the second-order diagram.

Another type of short-range correlation process left out in the SIA involves short-range particle-particle interactions (including exchange processes), which are known to be essential in the low-density limit.[72] These processes can be incorporated into a $\Phi$-functional (again, this functional is displayed explicitly in the Appendix, where it is labeled $\Phi_T^{pp}$) which, via Eq. (10), generates a self-energy $\Sigma_T^{pp}$ of the form

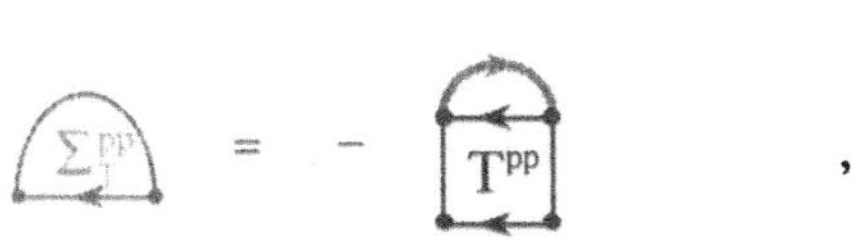

(24)

where the particle-particle T-matrix, $T^{pp}$, is the solution of the integral equation

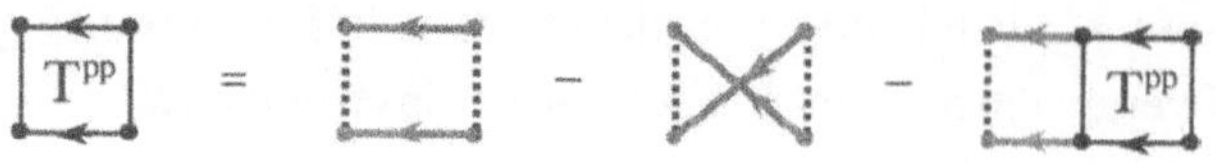


(25)

Again, double counting of the second-order term must be avoided, when adding this new contribution to the self-energy. Equations (12) and (25) provide a good interpolation between short-range and long-range correlations (or between low-density and high-density physics) in an electron gas.[46,48,73] Furthermore, the inclusion of the correlations contained in Eqs. (22) and (25) is expected to play a role in the description of more strongly-correlated materials, such as the transition metals, and their oxides.[67]

## APPENDIX

An important set of particle-hole interactions, absent in the SIA, is represented by the $\Phi$-functional

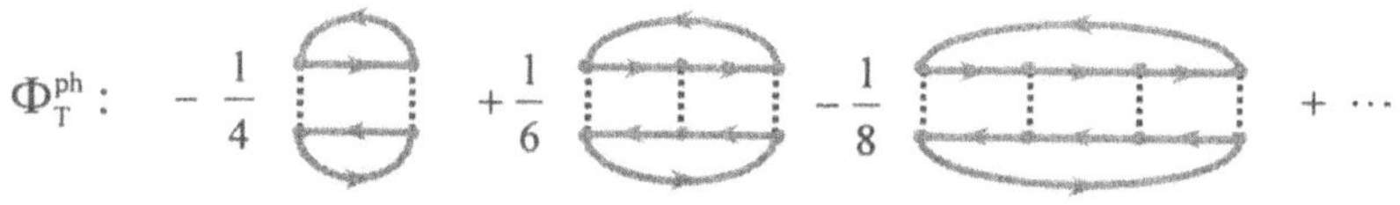


which, upon taking the functional derivative required in Eq. (10), generates a series of "particle-hole" self-energy ladders,

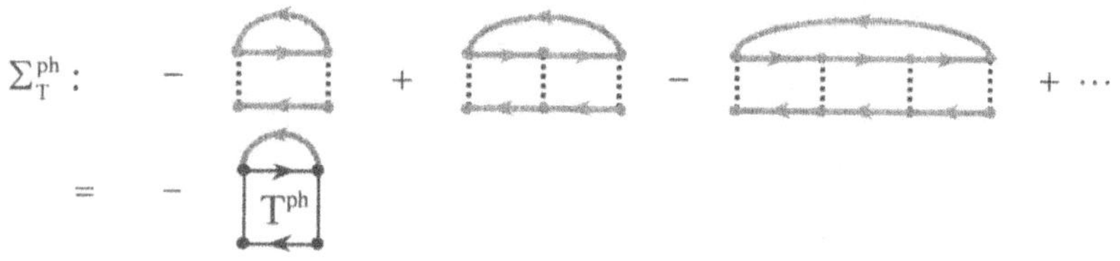


which we have formally summed to all orders via the definition of the particle-hole T-matrix, $T^{ph}$, which is to be obtained by solving Eq. (23).

A family of particle-particle interactions (including exchange proceses), also absent in the SIA, is represented by the $\Phi$-functional

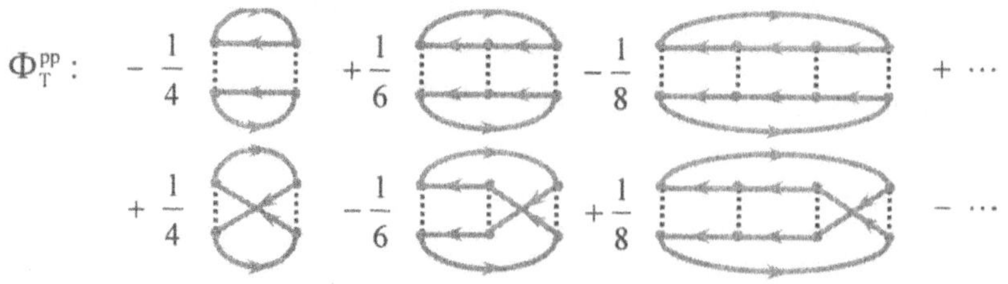


which, upon taking the functional derivative required in Eq. (10), generates a series of "particle-particle" self-energy ladders,

$\Sigma_T^{pp}$ :

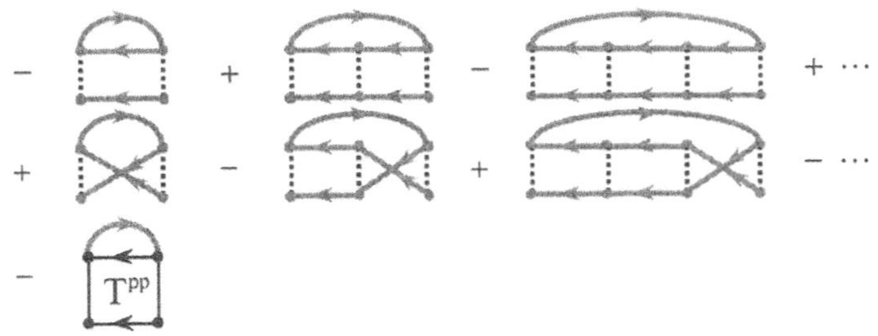

which we have formally summed to all orders via the definition of the particle-particle T-matrix, $T^{pp}$, which is to be obtained by solving Eq. (25).


## ACKNOWLEDGEMENTS

This work was supported by National Science Foundation Grant No. DMR-9634502 and the National Energy Research Supercomputer Center. Oak Ridge National Laboratory is managed by Lockheed Martin Energy Research Corp. for the Division of Materials Sciences, U.S. DOE under contract DE-AC05-96OR2464.


## REFERENCES


1. D. Pines and P. Nozières, *The Theory of Quantum Liquids, Vol. I: Normal Fermi Liquids,* Benjamin, New York (1966).
2. W. Kohn and L. J. Sham, Self-consistent equations including exchange and correlation effects, *Phys. Rev.* 140:A1133 (1965).
3. J. Hubbard, Electron correlations in narrow energy bands III. An improved solution, *Proc. Roy. Soc. A.* 281:401 (1964).
4. D. J. Scalapino, Does the Hubbard model have the right stuff?, in: *Perspectives in Many Particle Physics,* R. A. Broglia and J. R. Schrieffer, eds., North-Holland, Amsterdam (1994), p. 95.
5. A. vom Felde, J. Sprösser-Prou, and J. Fink, Valence-electron excitations in the alkali metals, *Phys. Rev. B* 40:10181 (1989).
6. J. Fink, Transmission electron energy-loss spectroscopy, in: *Unoccupied Electronic States,* J. C. Fuggle and J. E. Inglesfield, eds., Springer Verlag, Berlin (1992), p. 203.
7. P. M. Platzman, E. D. Isaacs, H. Williams, P. Zschack, and G. E. Ice, X-ray scattering determination of the dynamical structure factor of Al metal, *Phys. Rev. B* 46:12943 (1992).
8. W. Schülke, H. Schulte-Schrepping, and J. R. Schmitz, Dynamic structure of electrons in Al metal studied by inelastic x-ray scattering, *Phys. Rev. B* 47:12426 (1993); W. Schülke, J. R. Schmitz, H. Schulte-Schrepping, and A. Kaprolat, Dynamic and static structure factor of electrons in Si: Inelastic x-rays scattering results, *ibid.,* 52:11721 (1995).
9. B. C. Larson, J. Z. Tischler, E. D. Isaacs, P. Zschack, A. Fleszar, and A. G. Eguiluz, Inelastic x-ray scattering as a probe of the many-body local-field factor in metals, *Phys. Rev. Lett.* 77:1346 (1996).
10. K. Sturm and L. E. Oliveira, Wave-vector dependent plasmon linewidth in the alkali metals, *Phys. Rev. B* 24:3054 (1981).
11. A. A. Quong and A. G. Eguiluz, First-principles evaluation of dynamical response and plasmon dispersion in metals, *Phys. Rev. Lett.* 70:3955 (1993).

12. N. E. Maddocks, R. W. Godby, and R. J. Needs, Band structure effects in the dynamical response of aluminum, *Europhys. Lett.* 27:681 (1994).

13. N. E. Maddocks, R. W. Godby, and R. J. Needs, *Ab initio* calculations of the dynamical response of beryllium, *Phys. Rev. B* 49:R8502 (1994).

14. F. Aryasetiawan and K. Karlsson, Energy loss spectra and plasmon dispersion in alkali metals, *Phys. Rev. Lett.* 73:1679 (1994).

15. A. Fleszar, A. A. Quong, and A. G. Eguiluz, Band-structure and many-body effects in the dynamical response of aluminum metal, *Phys. Rev. Lett.* 74:590 (1995).

16. K. Karlsson and F. Aryasetiawan, Plasmon lifetime, zone-boundary collective states, and energy-loss spectra of lithium, *Phys. Rev. B* 52:4823 (1995).

17. K. –H. Lee and K. J. Chang, Analytic continuation of the response function using an *N*-point Padé approximant, *Phys. Rev. B* 54:R8285 (1996).

18. A. Fleszar, R. Stumpf, and A. G. Eguiluz, One-electron excitations, correlation effects, and the plasmon in cesium metal, *Phys. Rev. B* 55:2068 (1997).

19. M. Ehrnsperger and H. Bross, Calculation of the dielectric matrix of Si, *J. Phys.: Cond. Matt.* 9:1225 (1997).

20. W. Ku and A. G. Eguiluz, Plasmon lifetime in K: A case study of correlated electrons in solids amenable to *ab initio* theory, *cond-mat*/9811281.

21. P. Eisenberger, P. M. Platzman, and K. C. Pandy, Investigation of x-ray plasmon scattering in single-crystal beryllium, *Phys. Rev. Lett.* 31:311 (1973).

22. P. M. Platzman and P. Eisenberger, Presence of an incipient Wigner electron lattice in solid-state electron gases, *Phys. Rev. Lett.* 33:152 (1974).

23. E. Runge and E. K. U. Gross, *Phys. Rev. Lett.* 52:2850 (1984); E. K. U. Gross and W. Kohn, Time-dependent density-functional theory, in: *Density Functional Theory of Many-Fermion Systems,* S. B. Trickey, ed., Academic, New York (1990).

24. M. Petersilka, U. J. Gossmann, and E. K. U. Gross, Excitation energies from time-dependent density-functional theory, *Phys. Rev. Lett.* 76:1212 (1996).

25. G. Baym and L. P. Kadanoff, Conservation laws and correlation functions, *Phys. Rev.* 124:287 (1961).

26. G. Baym, Self-consistent approximations in many-body systems, *Phys. Rev.* 127:1391 (1962).

27. A. G. Eguiluz and W. -D. Schöne, Electronic excitations and correlation effects in metals, *Mol. Phys.* 94:87 (1998); W. -D. Schöne and A. G. Eguiluz, Self-consistent calculations of quasiparticle states in metals and semiconductors, *Phys. Rev. Lett.* 81:1662 (1998).

28. L. Hedin, New method for calculating the one-particle Green's function with application to the electron-gas problem, *Phys. Rev.* 139: A796 (1965).

29. W. G. Aulbur, L. Jönsson, and J. W. Wilkins, Quasiparticle calculations in solids, in: *Solid State Physics, Vol. 55,* F. Seitz, D. Turnbull, and H. Ehrenreich, eds., (Academic, New York, 1999).

30. F. Aryasetiawan and O. Gunnarsson, The *GW* method, *Rep. Prog. Phys.* 61:237 (1998).

31. B. Holm and U. von Barth, Fully self-consistent *GW* self-energy of the electron gas, *Phys. Rev. B* 57:2108 (1998).

32. M. Hindgren and C. –O. Almbladh, Improved local-field corrections to the $G_0W$ approximation in jellium: Importance of consistency, *Phys. Rev. B* 56:12832 (1997).

33. G. D. Mahan, *GW* approximations, *Comments Cond. Mat. Phys.* 16:333 (1994); G. D. Mahan, Electron-electron interactions: Ward identities, *Int. J. Mod. Phys. B* 6:3381 (1992).

34. E. L. Shirley, Self-consistent *GW* and higher order calculations of electron states in metals, *Phys. Rev. B* 54:7758 (1996).

35. A. Schindlmayr, Violation of particle-number conservation in the *GW* approximation, *Phys. Rev. B* 56:3528 (1997).

36. B. Farid, Self-consistent density-functional approach to the correlated ground states and an unrestricted many-body perturbation theory, *Phil. Mag.* 76:145 (1997).
37. A. Schindlmayr and R. W. Godby, Systematic vertex corrections through iterative solution of Hedin's equations beyond the *GW* approximation, *Phys. Rev. Lett.* 80:1702 (1998).
38. R. T. M. Ummels, P. A. Bobbert, and W. van Haeringen, First-order corrections to random-phase approximation *GW* calculations in silicon and diamond, *Phys. Rev. B* 57:11962 (1998).
39. P. Hohenberg, and W. Kohn, Inhomogeneous electron gas, *Phys. Rev. B* 136:B864 (1964).
40. E. K. U. Gross, C. A. Ullrich, and U. J. Gossmann, Density functional theory of time-dependent systems, in: *Density Functional Theory*, E. K. U Gross and R. Dreizler, eds., Plenum, New York (1994).
41. D. J. Singh, *Plane Waves, Pseudopotentials, and the LAPW Method*, Kluwer, Boston (1994); P. Blaha, K. Schwarz, P. Dufek, and R. Augustyn, WIEN95, Technical University of Vienna (1995).
42. J. P. Hill, C. –C. Kao, W. A. C. Caliebe, D. Gibbs, and J. P. Hastings, Inelastic x-ray scattering study of solid and liquid Li and Na, *Phys. Rev. Lett.* 77:3665 (1996).
43. B. C. Larson, J. Z. Tischler, A. Fleszar, and A. G. Eguiluz, Correlation and band structure effects in the electronic energy loss spectra of Al: An experimental perspective, *J. Phys. Chem. Solids* (1998), submitted.
44. G. Mukhopadyay, R. K. Kalia and K. S. Singwi, Dynamic structure factor of an electron liquid, *Phys. Rev. Lett.* 34:950 (1975); P. K. Aravind, A. Holas, and K. S. Singwi, Dynamical correlations in an electron gas. Kinetic-equation approach, *Phys. Rev. B* 25:561 (1982).
45. K. Utsumi, and S. Ichimaru, Dielectric formulation of strongly-coupled electron liquid at metallic densities. III. Dynamic structure factor, *Phys. Rev. B* 23:3291 (1981).
46. K. Awa, H. Yasuhara, and T. Asahi, Dynamical structure factor of an electron liquid. II. *Phys. Rev. B* 25:3687 (1982).
47. S. Rahman and G. Vignale, Fine structure in the dynamic form factor of an electron liquid, *Phys. Rev. B* 30: 6951 (1984); T. K. Ng and B. Dabrowski, Off-shell self-energy effects and the dynamic structure factor of an electron liquid, *Phys. Rev. B* 33: 5358 (1986).
48. F. Green, D. Neilson, and J. Szymanski, Conserving dynamic theory for the electron gas, *Phys. Rev. B* 31:2779 (1985).
49. F. Brosens and J. T. Devreese, Frequency-dependent solution of the time-dependent Hartree-Fock equation in the electron gas, *Phys. Status Solidi B* 147:173 (1988), and references therein.
50. S. Moroni, D. M. Ceperley, and G. Senatore, Static response and local-field factor of the electron gas, *Phys. Rev. Lett.* 75:689 (1995).
51. A. Zangwill and P. Soven, Density-functional approach to local-field effects in finite systems: Photoabsorption in the rare gases, *Phys. Rev. A* 21: 1561 (1980).
52. W. Hanke and L. J. Sham, Many-particle effects in the optical spectrum of a semiconductor, *Phys. Rev. B* 21:4656 (1980)
53. A. G. Eguiluz, W. Ku, and J. M. Sullivan, Dynamical response of correlated electrons in solids probed by inelastic scattering experiments: An *ab initio* theoretical perspective, *J. Phys. Chem. Solids* (1998), submitted.
54. H. J. Levinson, F. Greuter, and E. W. Plummer, Experimental band structure of Al, *Phys. Rev. B* 27:727 (1983).
55. G. Vignale and W. Kohn, Current-dependent exchange-correlation potential for dynamical linear response theory, *Phys. Rev. Lett.* 77: 2037 (1996); J. F. Dobson, M. J. Bünner, and E. K. U. Gross, Time-dependent density-functional theory beyond linear-response: An exchange-correlation potential with memory, *Phys. Rev. Lett.* 79: 1905 (1997); G. Vignale, C. A. Ullrich, and S. Conti, Time-dependent density functional theory beyond the adiabatic local density approximation, *Phys. Rev. Lett.* 79: 4878 (1997); R. Nifosì, S. Conti, and M. P. Tosi, Dynamic exchange-correlation potentials for the electron gas in dimensionality $D$=3 and $D$=2, *Phys. Rev. B* 58:12758 (1998).

56. J. M. Luttinger and J. C. Ward, Ground state energy of a many-fermion system. II. *Phys. Rev.* 118:1417 (1960).

57. M. S. Hybertsen and S. G. Louie, Electron correlations in semiconductors and insulators: Band gaps and quasiparticle energies, *Phys. Rev. B* 34:5390 (1986); R. W. Godby, M. Schlüter, and L. J. Sham, Self-energy operators and exchange-correlation potentials in semiconductors, *Phys. Rev B* 37:10159 (1988); M. Rohlfing, P. Krüger, and J. Pollmann, Quasiparticle band structure calculations for C, Si, Ge, GaAs, and SiC using Gaussian orbital basis sets, *Phys. Rev. B* 48:17791 (1993); E. L. Shirley and S. G. Louie, Electron excitations in solid $C_{60}$: Energy gap, band dispersions, and effects of orientational disorder, *Phys. Rev. Lett.* 71:133 (1993); S. G. Louie, Quasiparticle theory of surface electronic excitation energies, *Surf. Sci.* 299/300:346 (1994); A. Fleszar and W. Hanke, Spectral properties of quasiparticles in a semiconductor, *Phys. Rev. B* 56:10228 (1997).

58. N. E. Bickers, D. J. Scalapino, and S. R. White, Conserving approximations for strongly-correlated electron systems: Bethe-Salpeter equation and dynamics for the two-dimensional Hubbard model, *Phys. Rev. Lett.* 62:961 (1989); G. Esirgen and N. E. Bickers, Fluctuation-exchange theory for general lattice systems, *Phys. Rev. B* 55:2122 (1997).

59. J. W. Serene and D. W. Hess, Quasiparticle properties of the two-dimensional Hubbard model in a propagator-renormalized fluctuation-exchange approximation, *Phys. Rev. B* 44:R3391 (1991); J. J. Deisz, D. W. Hess, and J. W. Serene, *Phys. Rev. B* 55:2089 (1997).

60. R. Putz, R. Preuss, A. Muramatsu, and W. Hanke, Conserving approximation for the three-band Hubbard model: Flat quasiparticle dispersion, *Phys. Rev. B* 53:5133 (1996).

61. M. Langer, J. Schmalian, S. Grabowski, and K. H. Bennemann, Theory of the excitation spectrum of high-$T_C$ superconductors: Quasiparticle dispersion and shadows of the Fermi surface, *Phys. Rev. Lett.* 75:4508 (1995).

62. R. D. Mattuck, *A Guide to Feynman Diagrams in the Many-Body Problem*, Dover, New York (1992), Ch. 17.

63. H. J. de Groot, R. T. M. Ummels, P. A. Bobbert, and W. van Haeringen, Lowest-order corrections to the RPA polarizability and *GW* self-energy of a semiconducting quantum wire," *Phys. Rev. B* 54:2374 (1996).

64. M. Rohlfing and S. G. Louie, Electron-hole excitations in semiconductors and insulators, *Phys. Rev. Lett.* 81:2312 (1998).

65. F. Bechstedt, K. Tenelsen, B. Adolph, and R. Del Sole, Compensation of dynamical quasiparticle and vertex corrections in optical spectra, *Phys. Rev. Lett.* 78:1528 (1997).

66. S. Albrecht, G. Onida, and L. Reining, Ab initio calculation of the quasiparticle spectrum and excitonic effects in $Li_2O$, *Phys. Rev. B* 55:10278 (1997).

67. M. Springer, F. Aryasetiawan, and K. Karlsson, First-principles *T*-matrix theory with application to the 6 eV satellite in Ni, *Phys. Rev. Lett.* 80:2389 (1998).

68. R. Schepe, T. Schmielau, D. Tamme, and K. Henneberger, Damping and T-matrix in dense e-h plasmas, *Phys. Stat. Sol.* (b) 206:273 (1998).

69. G. D. Mahan, *Many Particle Physics*, 2nd ed. (Plenum, New York, 1990), p.436.

70. D. C. Langreth and S. H. Vosko, Exact electron-gas response functions at high density, *Phys. Rev. Lett.* 59:497 (1987).

71. D. J. W. Geldart and S. H. Vosko, The screening function of an interacting electron gas, *Can. J. Phys.*44:2137 (1966).

72. A. L. Fetter and J. D. Walecka, *Quantum Theory of Many-Particle Systems*, (Mc Graw-Hill, New York, 1971), p. 128.

73. H. Yasuhara, H. Suehiro, and Y. Ousaka, A new interpolation between long- and short-range correlations in an electron liquid: III, *J. Phys. C: Solid State Phys.* 21:4045 (1988).

## Discussion

**Gross**: When you solve the response equation

$$\chi = \chi_0 + \chi_0 \, ( W_{clb} + f_{xc} ) \, \chi$$

you have employed the Hartree response function for $\chi_0$ . Shouldn't $\chi_0$ be the Kohn--Sham response function?

**Eguiluz**: In principle yes. However, in the cases for which we did this calculation, the Hartree $\chi_0$ - and the KS $\chi_0$ were practically identical.

**Sawatzky**: In the calculation of the quasi particle dispersion of Na metal (or was it Potassium) you found that the quasi particle dispersion was larger for the calculation including electron correlation effects than for the standard LDA-like calculation. This seems strange to me because a dressed particle, as it will be in the correlated case, should always have a quasi particle weight less than 1, with a part of the weight in the incoherent part of the one electron greens function, and as a consequence, the dispersion should always be less than that of the undressed particle. Can you explain why this seems to be different in your calculation?

**Eguiluz**: I agree that this is indeed strange, but it is difficult to extract from the calculation exactly why this happens.

# PAIR DENSITIES, PARTICLE NUMBER FLUCTUATIONS, AND A GENERALIZED DENSITY FUNCTIONAL THEORY

P. Ziesche

Max Planck Institute for the
Physics of Complex Systems
Nöthnitzer Straße 38
D-01187 Dresden, Germany

## 1 INTRODUCTION

In the theory of atoms, molecules, clusters, and solids the quantum-mechanical many-electron problem is treated with a variety of well-studied and well-tried methods[1]. Methods applicable to finite as well as extended systems are
– the wave-function based Hartree-Fock (HF) method with its exchange (X) hole for electrons with parallel spins (and – following from this – its non-local X-potential), but with its complete neglect of the correlation (C) hole for electrons with antiparallel spins,
– the density-based density-functional theory (DFT)[2] with its local density approximation (LDA) and generalized gradient approximation (GGA)[3] of the local XC potential or with its recent optimized effective potential (OEP) method version[4], where the XC-energy depends implicitly on the density via orbitals,
– the Green function based quasi-particle theory (QPT) with its GW approximation for the non-local and energy dependent self-energy operator,
– the wave function based quantum Monte Carlo (QMC) method.
The wave-function based quantum-chemical correlation methods are applicable only to finite systems (one-dimensional extended polymers are an exception):
– the variational and non-size extensive configuration interaction (CI) method,
– the projective and size-extensive coupled cluster (CC) method,
– the Møller-Plesset (MP) many-body perturbation method.
The recently developed incremental method allows to apply the accurate quantum-chemical methods also to extended (non-metallic) systems. This is effected by treating

*Electron Correlations and Materials Properties*, edited by Gonis *et al.*
Kluwer Academic/Plenum Publishers, 1999.

the electron correlation with a size-extensive wave-function based correlation method (e. g. the CC method) in a small (but from one computational step to the next increasing) crystal region embedded in an otherwise HF only treated environment; from step to step the non-additive contributions to the correlation energy are calculated until saturation is reached[5]. – In Ref. 6 the electron correlation problem is treated non-variationally within a BCS-type theory by means of reduced density matrices (RDMs).

There are recent attempts to try and add further methods by generalizing the effective one-particle descriptions HF and DFT: in Refs. 7 and 8 a generalized DFT in terms on multi-particle densities is considered, in Ref. 9 a generalized HF method in terms of the two-particle density is proposed, and in Refs. 10 and 11 the conventional DFT is generalized towards effective two-particle descriptions. In DFT one-particle orbitals $\psi_\kappa(x)$, where $x = (\underline{r}, \sigma)$, are obtained from an effective one-particle equation, the Kohn-Sham (KS) equation, in such a way that the one-particle density (or one-density for short) follows as $\rho(x) = \sum_\kappa^{\rm occ} \mid \psi_\kappa(x) \mid^2$ . In Refs. 10, 11 the question is posed whether an effective two-particle equation may give two-particle orbitals or geminals $\psi_K(1,2)$, where $i =$ (1 or 2) stands for $x_i$, in such a way that the two-particle density or pair density (PD) follows as $n(1;2) = \sum_K^{\rm occ} \mid \psi_K(1,2) \mid^2$. In such a pair density functional theory (PDFT) the PD $n(1;2)$ would be the key quantity, whereas in DFT the one-density $\rho(1)$ is the key quantity. If such a PDFT or a similar scheme for the cumulant PD $w(1;2) = \rho(1)\rho(2) - n(1;2)$ can be really developed, then one may say 'a white spot is filled'. The existence of such a white spot is suggested by comparing the pair of one-particle schemes HF (describing the self-consistent motion of each particle in the mean field of all the others) and DFT (replacing the non-local X potential of HF by a local XC potential) with the Bethe-Goldstone equation (describing the self-consistent motion of each particle pair in the mean field of all the other particles). Similarly one may compare the QPT (related to the one-particle Green function) and DFT with the Bethe-Salpeter equation (related to the two-particle Green function) and one may feel again an appropriate PDFT is missing.

We consider systems described by

$$\hat{H} = \hat{T} + \hat{V}_{\rm ext} + \hat{V}_{\rm int}. \tag{1.1}$$

The first two terms are the one-particle part

$$\hat{H}_0 = \sum_i^N \hat{h}_0(i), \qquad \hat{h}_0(i) = -\frac{\hbar^2}{2m}\Delta_i + v_{\rm ext}(i), \tag{1.2}$$

the last term of $\hat{H}$ is the two-particle interaction

$$\hat{V}_{\rm int} = \frac{1}{2}\sum_{i \neq j}^N v_{\rm int}(i,j). \tag{1.3}$$

With $v_{\rm int}(i,j) = \epsilon^2/r_{ij}$ and appropriate $v_{\rm ext}(i)$ the Hamiltonian $\hat{H}$ describes the motion of electrons in such (finite) systems as atoms, molecules or clusters. The starting point is the Pauli principle and the Schrödinger equation $\hat{H}\Psi = E\Psi$ and symmetry constraints. Together they guarantee the existence of an antisymmetric wave function (WF) $\Psi(1,\ldots,N)$ for the ground state (GS).

This GS WF again guarantees the hierarchy of RDMs to exist via contraction procedures, which successively destroy information. This hierarchy starts with the one-DM $\gamma_1(1,1')$, the two-DM $\gamma_2(1,1';2,2')$ and so on. These RDMs are mutually related by a hierarchy of again successively information destroying contractions, the simplest one is

$$\gamma_1(1,1') = \int \frac{d2}{N-1}\gamma_2(1,1';2,2), \tag{1.4}$$

and by a hierarchy of equations of motion (EOMs), the simplest one is

$$[\hat{h}_0(1) - \hat{h}_0(1')]\gamma_1(1,1') = -\int d2\,[v_{\rm int}(1,2) - v^*_{\rm int}(1',2)]\gamma_2(1,1';2,2). \tag{1.5}$$

The normalizations of their diagonals are fixed by $\int d1\gamma_1(1,1) = N$. They possess symmetry properties such as e. g. $\gamma_1^*(1',1) = \gamma_1(1,1')$ and $\gamma_2(2,1';1,2') = -\gamma_2(1,1';2,2')$. These hierarchies, normalizations and symmetry properties are on the one hand equivalent to the original Pauli principle and Schrödinger equation, but they have on the other hand the advantage to survive in the thermodynamic limit (TDL). So they can be used to describe also extended systems such as crystalline solids; in this case the existence of a correlated many-body WF is replaced by the existence of a generating functional, which contains all the RDMs as Taylor-Volterra 'coefficients'. Another advantage is the following: While $\Psi(\ldots)$, which depends on $N$ spatial vector coordinates $\underline{r}_i$ and $N$ spin variables $\sigma_i$, has (for $N > 2$ or worse $\gg 2$) a giant, non-recordable and therefore not directly accessible content of information of all the aspects of the electron motion[12], the one-DM $\gamma(1,1')$ [we delete its index in the following] and its diagonal, the one-density $\rho(1) = \gamma(1,1)$ which is normalized as

$$\int d1\,\rho(1) = N, \tag{1.6}$$

have a more or less easily accessible and surveyable content of information. This holds also for the PD $n(1;2) = \gamma_2(1,1;2,2)$ from which follows once more the one-density

$$\rho(1) = \int \frac{d2}{N-1}n(1;2) \tag{1.7}$$

in agreement with Eq. (1.4). The charm and power of the one-density $\rho(1)$ is exploited in the DFT. The PD $n(1;2)$ is of course more complicated being the price for more information. Its charm and power is the driving force to study the possibility of a PDFT as a generalized DFT. As the KS equation of DFT survives in the TDL, also the effective two-particle equation of the hopefully existing PDFT should have the same property. Then the PD would be available (in addition to the one-density of the DFT) also for extended systems (for finite systems the WF based quantum-chemical methods provide the PD and also the one-DM).

In Sections 2 and 3 it is summarized which information is available with $\rho(1)$, $\gamma(1,1')$, and $n(1;2)$. Motivations are given to have in addition to the $\rho(1)$ of the DFT also $n(1;2)$ of a PDFT available, especially for extended systems. It is explained to what extent the PD $n(1;2)$ contains information on particle number fluctuations in a fragment (a spatio-spin region) and correlation between the particle number fluctuations in different fragments. For such discussions the concept and existence of the particle number distribution, i. e. the probability of finding a certain number of particles in a fragment is important (Appendix A). Section 4 presents Euler equations resulting from the minimization of the total energy as a functional of the PD. Section 5 describes the idea of an effective two-particle equation, its self-consistent solution would give geminals from which the PD follows.

# 2 THE ONE-DENSITY AND AVERAGE PARTICLE NUMBERS

Which information is contained in the one-density $\rho(1)$ ?

1. From the spin-dependent densities $\rho^{\uparrow}(\underline{r})$ and $\rho^{\downarrow}(\underline{r})$ follows the spin-traced density $\rho(\underline{r}) = \rho^{\uparrow}(\underline{r}) + \rho^{\downarrow}(\underline{r})$.

2. We consider an arbitrary region (a fragment) $\Omega$ of the whole three-dimensional space, e. g. a sphere or the Wigner-Seitz cell around an atom or the bond region between two atoms, together with a 'fragment' $\Sigma$ of the 'spin space' (with $\Sigma =$ " $\uparrow$ " or " $\downarrow$ " or " $\uparrow$ and $\downarrow$ ") and we ask for the average number of particles in this spatio-spin fragment $X = (\Omega, \Sigma)$. The answer is

$$N_X = \int_X d1\, \rho(1), \tag{2.1}$$

which means

$$N_\Omega^{\uparrow} = \int_\Omega d^3r\, \rho^{\uparrow}(\underline{r}) \quad \text{or} \quad N_\Omega^{\downarrow} = \int_\Omega d^3r\, \rho^{\downarrow}(\underline{r}) \quad \text{or} \quad N_\Omega = \int_\Omega d^3r\, \rho(\underline{r}). \tag{2.1$a,b,c$}$$

For $\Omega \to \infty$ of course these quantities approach $N^{\uparrow}(< N)$, the total number of spin-up particles, respectively $N^{\downarrow}(< N)$, the total number of spin-down particles, respectively $N = N^{\uparrow} + N^{\downarrow}$, the total number of particles. Here it is assumed that no spin-flip processes take place, so there are no fluctuations of the total number of spin-up or spin-down particles. $N_\Omega = N_\Omega^{\uparrow} + N_\Omega^{\downarrow}$ is the average particle number and $\zeta_\Omega = (N_\Omega^{\uparrow} - N_\Omega^{\downarrow})/N_\Omega$ is the average spin polarization in $\Omega$.

3. Eqs. (2.1) mean that there exists an unknown particle number distribution $P_X(M)$ with

$$\sum_{M=0}^{N} P_X(M) = 1 \tag{2.2}$$

and

$$\sum_{M=0}^{N} P_X(M)M = N_X. \tag{2.3}$$

The existence and representation of such particle number distributions are discussed in Refs. 13 - 15 and in Appendix A. $P_\Omega^\uparrow(M)$ is the probability of finding $M(=0,\ldots,N^\uparrow)$ spin-up particles in $\Omega$ and simultaneously an arbitrary number of spin-down particles. $P_\Omega^\downarrow(M)$ has the complementary meaning. The probability of finding $M$ particles with spin up or spin down (then $\Sigma =\uparrow$ and $\downarrow$) is denoted with $P_\Omega(M)$. $P_X(M)$ describes with a maximum near $M \approx N_X$ the particle number fluctuations in $X$ via the variance

$$(\Delta N)_X^2 = \sum_{M=0}^{N} P_X(M)(M-N_X)^2 = (N^2)_X - (N_X)^2. \tag{2.4}$$

While $N_X$ follows from the one-density, the first term on the r.h.s. needs to know the PD. This is true also for the quantity

$$\begin{aligned} C_{X_1,X_2} &= -\sum_{M_1,M_2} P\begin{pmatrix} M_1 & M_2 \\ X_1 & X_2 \end{pmatrix} (M_1 - N_{X_1})(M_2 - N_{X_2}) \\ &= -\sum_{M_1,M_2} P\begin{pmatrix} M_1 & M_2 \\ X_1 & X_2 \end{pmatrix} M_1 M_2 + N_{X_1}N_{X_2}, \end{aligned} \tag{2.5}$$

which describes the correlation between the particles in two different fragments $X_1$ and $X_2$. Here $P\begin{pmatrix} M_1 & M_2 \\ X_1 & X_2 \end{pmatrix}$ is the probability of finding $M_1$ particles in $X_1$ and simultaneously $M_2$ particles in $X_2(\neq X_1)$. Examples for such different fragments are: (i) $X_1 = (\Omega,\uparrow)$ and $X_2 = (\Omega,\downarrow)$ or (ii) $X_1 = (\Omega_1,\Sigma)$ and $X_2 = (\Omega_2,\Sigma)$, where $\Omega_1$ and $\Omega_2$ are non-overlapping spatial regions. In the first case the quantity $C_{X_1,X_2}$ describes the correlation between particles of different spin directions within a certain spatial region $\Omega$. In the second case the quantity $C_{X_1,X_2}$ describes the correlation of particles in different spatial regions $\Omega_1$ and $\Omega_2$ (where $\Sigma$ may be $\uparrow$ or $\downarrow$ or $\uparrow$ and $\downarrow$). So Eqs. (2.4) and (2.5) show, that for the simplest fluctuation and correlation analysis the PD (beyond the one-density) is needed.

4. From $\rho(1)$ follows also

$$V_{\text{ext}}[\rho] = \int d1\, \rho(1) v_{\text{ext}}(1), \tag{2.6}$$

the expectation value of the external potential, as well as the Hartree term of $V_{\text{int}}$,

$$V_{\text{H}}[\rho] = \frac{1}{2}\int d1\, d2\, \rho(1)\rho(2) v_{\text{int}}(1,2), \tag{2.7}$$

and

$$v_{\text{H}}(1) = \frac{\delta V_{\text{H}}(\rho)}{\delta\rho(1)} = \int d2\, \rho(2) v_{\text{int}}(1,2), \tag{2.8}$$

the Hartree potential.

A bit more information is contained in the one-DM. What does $\gamma(1,1')$ tell us ?

1. The one-density follows as $\rho(1) = \gamma(1,1)$.

2. The spin-traced one-DM is $\gamma(\underline{r}_1, \underline{r}_1') = \sum_{\sigma_1} \gamma(1,1') \mid_{\sigma_1'=\sigma_1}$.

3. The diagonalization of $\gamma(1,1')$ yields the spin-dependent natural orbitals $\psi_\kappa(1)$ and the spin-dependent natural occupation numbers $n_\kappa$ with $\sum_\kappa n_\kappa = N$ and $n_\kappa = 1$ or 0 if $\Psi$ is a single Slater determinant only and $0 < n_\kappa < 1$ if $\Psi$ is a sum of several (or many) Slater determinants (MRSCF or CI expansion). $\gamma(1,1')$ is in the first case idempotent, in the second case non-idempotent. The $q^{\text{th}}$-order non-idempotency

$$c(q) = 1 - \frac{1}{N} \sum_\kappa (n_\kappa)^q \tag{2.9}$$

and the correlation entropy $s = c'(1)$ or

$$s = -\frac{1}{N} \sum_\kappa n_\kappa \ln n_\kappa \tag{2.10}$$

vanish for idempotent one-DMs and increase with increasing correlation (where a single Slater determinant is the reference state for 'no correlation')[16]. So $c(q)$ and $s$ are - in addition to the usual correlation energy - purely quantum-kinematic measures of the correlation strength. Note the necessity to distinguish between non-dynamic correlations (with a few Slater determinants which take into account the symmetry of the problem or near degeneracies and determine the long-range behavior of the PD with Friedel oscillations in the case of the uniform electron gas) and dynamic correlations (the 'tail' of the CI expansion being responsible for the short-range behavior of the PD with curvature and coalescing cusp properties). Another point is to distinguish between intra- and interatomic, intra- and interbond correlations[1].

4. From $\gamma(1,1')$ follows with $\underline{p}_1 = \frac{\hbar}{i}\nabla_1$

$$T[\gamma] = \int d1 \, \frac{1}{2m} \underline{p}_1^2 \gamma(1,1') \mid_{1'=1} = \sum_\kappa n_\kappa \langle \kappa \mid \frac{1}{2m} \underline{p}_1^2 \mid \kappa \rangle, \tag{2.11}$$

the expectation value of the kinetic energy, and

5. also

$$\underline{j}(1) = \text{Re} \frac{1}{m} \underline{p}_1 \gamma(1,1') \mid_{1'=1} = \sum_\kappa \psi_\kappa^*(1) \frac{1}{m} \underline{p}_1 \psi_\kappa(1), \tag{2.12}$$

the one-current density.

6. The Fourier transform of $\gamma(1,1')$ yields the one-DM in the reciprocal or $\underline{k}$ space, its diagonal is the spin-dependent momentum distribution $n(\underline{k}, s)$.

So from $\gamma(1,1')$ follow in addition to $\rho(1)$ quantities like $n(\underline{k}, s)$ and $T$.

# 3 THE PAIR DENSITY AND PARTICLE NUMBER FLUCTUATIONS

What do we know with the PD $n(1;2)$?

1. The contraction formula (1.7) yields the one-density.

2. Spin summations yield $n^p(\underline{r}_1;\underline{r}_2)$, the PD of particles with parallel spins, and $n^a(\underline{r}_1;\underline{r}_2)$, the PD of particles with antiparallel spins, and the spin-traced PD $n(\underline{r}_1;\underline{r}_2) = n^p(\underline{r}_1;\underline{r}_2) + n^a(\underline{r}_1;\underline{r}_2)$.

3. From $n(1;2)$ follows

$$\binom{N}{2}_X = \int_X \frac{d1\,d2}{2!}\, n(1;2), \tag{3.1}$$

the average number of particle pairs in $X$. With Eq. (3.1) or

$$(N^2)_X - N_X = \int_X d1 \int_X d2\, n(1;2) \tag{3.2}$$

and with the definition (2.4) of the variance this latter quantity results as

$$(\Delta N)^2_X = \int_X d1 \int_X d2\, n(1;2) - N_X(N_X - 1). \tag{3.3}$$

With the spin-dependent cumulant PD or XC-hole $w(1;2)$, defined by

$$n(1;2) = \rho(1)\,\rho(2) - w(1;2) \tag{3.4}$$

and with the abbreviation (for its physical meaning see Eq. (A.11))

$$W_{X_1,X_2} = \int_{X_1} d1 \int_{X_2} d2\, w(1;2) \tag{3.5}$$

Eq. (3.3) takes the explicit size-extensive form

$$(\Delta N)^2_X = N_X - W_{X,X}. \tag{3.6}$$

Note the contraction property

$$\rho(1) = \int d2\, w(1;2) \tag{3.7}$$

and consequently the (size-extensive) normalization of the XC-hole

$$\int d1\, d2\, w(1,2) = N. \tag{3.8}$$

Note the (non-size consistent) normalization of the PD $n(1;2)$ with $N(N-1)$. With the spin-summed XC-hole $w(\underline{r}_1;\underline{r}_2)$ and with the notation

$$W_{\Omega_1,\Omega_2} = \int_{\Omega_1} d^3r_1 \int_{\Omega_2} d^3r_2\, w(\underline{r}_1;\underline{r}_2) \tag{3.9}$$

from Eq. (3.6) follows (for $\Sigma = \uparrow$ and $\downarrow$) the particle number fluctuation in $\Omega$ as

$$(\Delta N)^2_\Omega = N_\Omega - W_{\Omega,\Omega} = W_{\Omega,\infty-\Omega}. \tag{3.10}$$

Note $N_\Omega = W_{\Omega,\infty}$. For $\Omega \to \infty$ it is $N_\Omega$, $W_{\Omega,\Omega} \to N$, thus $W_{\Omega,\infty-\Omega}$, $(\Delta N)^2_\Omega \to 0$. (This limiting case needs a more sophisticated treatment for extended systems.) If $\Omega$ consists of two non-overlapping spatial fragments ($\Omega = \Omega_1 + \Omega_2$), then

$$(\Delta N)^2_{\Omega_1+\Omega_2} = (\Delta N)^2_{\Omega_1} + (\Delta N)^2_{\Omega_2} - 2W_{\Omega_1,\Omega_2}. \tag{3.11}$$

Thus the XC-hole reduces the fluctuations in $\Omega_1 + \Omega_2$ compared with the individual fluctuations in $\Omega_1$ and in $\Omega_2$ alone. So $C_{\Omega_1,\Omega_2}$ describes to what extent the fluctuations in $\Omega_1$ are correlated with the fluctuations in $\Omega_2$, see also Eq. (A.11).

Because $w(\underline{r}_1;\underline{r}_2)$ decreases with increasing separation $r_{12}$ between the two particles also $C_{\Omega_1,\Omega_2}$ becomes smaller and smaller the more the regions $\Omega_1$ and $\Omega_2$ are separated from each other. Thus the particle number fluctuations in largely separated fragments becomes additive. This suggests that $w(\underline{r}_1;\underline{r}_2)$ and $C_{\Omega_1,\Omega_2}$ should be also useful tools for giving the recently discussed nearsightedness of (normal) electron systems[17, 18] the appropriate mathematical description.

The XC-hole is the sum of the X-hole (resulting from particles with parallel spins) and the C-hole (resulting from particles with antiparallel spins):

$$w(\underline{r}_1;\underline{r}_2) = w^p(\underline{r}_1;\underline{r}_2) + w^a(\underline{r}_1;\underline{r}_2). \tag{3.12}$$

The quantities (partial normalizations)

$$X_\Omega = \int_\Omega d^3r_1 \int_\Omega d^3r_2\, w^p(\underline{r}_1;\underline{r}_2) \tag{3.13}$$

and

$$C_\Omega = \int_\Omega d^3r_1 \int_\Omega d^3r_2\, w^a(\underline{r}_1;\underline{r}_2) \tag{3.14}$$

describe, how particles in $\Omega$ with parallel and antiparallel spins, respectively, influence each other. The relation

$$X_\Omega + C_\Omega = W_{\Omega,\Omega} \tag{3.15}$$

compared with Eq. (3.10) shows how exchange, correlation, and particle number fluctuations are mutually related. For $\Omega \to \infty$ the quantities $X = X_\infty$ and $C = C_\infty$ are the (total) normalizations of the X-hole and C-hole, respectively. Note

$$X + C = N. \tag{3.16}$$

No correlation (HF or no Coulomb repulsion) means $C = 0$, inclusion of correlation narrows the X-hole and creates the C-hole. So the C-hole normalization is a further (extensive) measure of the correlation strength [19]. The on-top C-hole $w^a(\underline{r};\underline{r})$ is a local measure.

The partial C-hole normalization $C_\Omega$ appears also when the fluctuations of the spin-up particles in $\Omega$

$$\left[(\Delta N)^\uparrow_\Omega\right]^2 = N^\uparrow_\Omega - \int_\Omega d^3r_1 \int_\Omega d^3r_2\, w(\underline{r}_1,\uparrow;\underline{r}_2,\uparrow) \tag{3.17}$$

and the analog expression for the fluctuations of the spin-down particles are compared with $(\Delta N)^2_\Omega$ of Eq. (3.10):

$$(\Delta N)^2_\Omega = \left[(\Delta N)^\uparrow_\Omega\right]^2 + \left[(\Delta N)^\downarrow_\Omega\right]^2 - C_\Omega. \tag{3.18}$$

Thus the C-hole reduces the fluctuations in $\Omega$ in comparison with the sum of the independent fluctuations of the spin-up and spin-down particles alone. So these fluctuations are additive for 'no correlation'. $C_\Omega$ describes to what extent the fluctuations of the spin-up particles are correlated with the fluctuations of the spin-down particles, see also Eq. (A.11). Note the difference to Eq. (3.11), where the XC-hole appears.

4. From $N^\uparrow_\Omega$ and $\Delta N^\uparrow_\Omega$ follows an approximate particle number distribution

$$P^\uparrow_\Omega(M) \sim \mathrm{e}^{-\alpha^\uparrow(M-\beta^\uparrow)^2} \tag{3.19}$$

of the spin-up particles with $\alpha^\uparrow \approx 1/\Delta N^\uparrow_\Omega$ and $\beta^\uparrow \approx 1/N^\uparrow_\Omega$. A similar expression holds for the spin-down particles. An approximate probability of finding $M_1$ spin-up particles in $\Omega$ and simultaneously $M_2$ spin-down particles is

$$P_\Omega(M_1, M_2) \sim \mathrm{e}^{-\alpha_1(M_1-\beta_1)^2-\alpha_2(M_2-\beta_2)^2+\gamma(M_1-\beta_1)(M_2-\beta_2)} \tag{3.20}$$

with the six coefficients $\alpha_{1,2}$, $\beta_{1,2}$, $\gamma$, and the prefactor to be determinated from $N^\uparrow_\Omega$, $\Delta N^\uparrow_\Omega$, $N^\downarrow_\Omega$, $\Delta N^\downarrow_\Omega$, $C_\Omega$, and from the normalization condition. It is $C_\Omega \approx \gamma$. No correlation between the spin-up and the spin-down particles (no Coulomb repulsion) means $\gamma = 0$ and $C_\Omega = 0$. The exact expressions instead of the approximate Eqs. (3.19) and (3.20) are Eqs. (A.4) and (A.8). Trivial relations between $P_\Omega(M_1, M_2)$, $P^\uparrow_\Omega(M_1)$, and $P^\downarrow_\Omega(M_2)$ are

$$\sum_{M_2} P_\Omega(M_1, M_2) = P^\uparrow_\Omega(M_1), \qquad \sum_{M_1} P_\Omega(M_1, M_2) = P^\downarrow_\Omega(M_2). \tag{3.21a,b}$$

With $M = M_1 + M_2$, $m = M_1 - M_2$ and

$$\tilde{P}_\Omega(M, m) = P_\Omega(\frac{M+m}{2}, \frac{M-m}{2}) \tag{3.22}$$

the probability of finding in $\Omega$ just $M$ particles and simultaneously a spin polarization $\zeta = m/N_\Omega$ is introduced. From it follows the already above mentioned probability of finding $M$ particles in $\Omega$

$$P_\Omega(M) = \sum_m \tilde{P}_\Omega(M, m), \tag{3.23}$$

which gives $N_\Omega$, the average number of particles in $\Omega$, and the probability of finding in $\Omega$ a spin polarization $m/N_\Omega$

$$p_\Omega(m) = \sum_M \tilde{P}_\Omega(M, m), \tag{3.24}$$

which gives the average spin polarization $\zeta_\Omega = (N^\uparrow_\Omega - N^\downarrow_\Omega)/N_\Omega$ in $\Omega$.

5. With the cumulant expansion (3.4) the expectation value of the Coulomb repulsion is given by $V_{\rm int}[n] = V_{\rm H} + V_{\rm XC}[w]$ with the Hartree contribution

$$V_{\rm H}[\rho] = \frac{1}{2}\int d1\, d2\, \rho(1)\, \rho(2)\, v_{\rm int}(1,2) \tag{3.25}$$

and the XC-contribution

$$V_{\rm XC}[w] = -\frac{1}{2}\int d1\, d2\, w(1;2)\, v_{\rm int}(1,2) \tag{3.26}$$

with its X and C 'components'

$$V_{\rm X}[w] = -\frac{1}{2}\int d1\, d2\, w^p(1;2)\, v_{\rm int}(1,2)$$

and

$$V_{\rm C}[w] = -\frac{1}{2}\int d1\, d2\, w^a(1;2)\, v_{\rm int}(1,2).$$

6. With the extracule (or center-of-mass) and intracule (or relative) coordinates

$$\underline{R} = \frac{1}{2}(\underline{r}_1 + \underline{r}_2) \quad \text{and} \quad \underline{r} = \underline{r}_1 - \underline{r}_2 \tag{3.27}$$

the spin-traced PD $n(\underline{r}_1;\underline{r}_2)$ can be written as

$$\tilde{n}(\underline{R},\underline{r}) = n(\underline{R} + \frac{1}{2}\underline{r}; \underline{R} - \frac{1}{2}\underline{r}). \tag{3.28}$$

Further reductions or projections yield the extracule PD

$$E(\underline{R}) = \int d^3r\, \tilde{n}(\underline{R},\underline{r}) \tag{3.29}$$

and the intracule PD

$$I(\underline{r}) = \int d^3R\, \tilde{n}(\underline{R},\underline{r}). \tag{3.30}$$

Spherically averaged PD's and radial PD's are

$$e(R) = \langle E(\underline{R})\rangle, \quad i(r) = \langle I(\underline{r})\rangle \quad \text{and} \quad 4\pi R^2 e(R), \quad 4\pi r^2 i(r).$$

Important properties of the intracule PD are: It is an even function, i.e. $I(-\underline{r}) = I(\underline{r})$ because of $n(1;2) = n(2;1)$. It has the coalescing cusp property

$$\frac{\hbar^2}{m}\langle \nabla I(0)\rangle = \epsilon^2 I(0) \tag{3.31}$$

with the spherically averaged on-top gradient on the l.h.s.[20, 21] and from $I(\underline{r})$ follows

$$V_{\rm int} = \frac{1}{2}\int d^3r\, I(\underline{r})\, \frac{\epsilon^2}{r}, \tag{3.32}$$

the expectation value of the Coulomb repulsion (note that only $i(r)$ is needed). $I(\underline{r})$ proved to be an important tool of the PD analysis: calculations show how sensitive it is on electron correlation and on basis sets. Different bonding schemes can be seen from the change when going from the promolecules (independent atom model, independent ion model) to the molecules. It can be visualized how electrons from different nuclei form pairs. Total scattering of X-rays or high-energy electrons yields information on the PD $n(1;2)$ via the intracule PD $I(\underline{r})$: elastic scattering gives the Fourier transform of the one-density,

$$F(\underline{q}) = \int d^3r\, e^{i\underline{q}\underline{r}} \rho(\underline{r}), \tag{3.33}$$

which appears in the relation between $I(\underline{r})$ and the static structure factor

$$\int_0^\infty dE\, S(\underline{q}, E) = \int d^3r\, e^{i\underline{q}\underline{r}}\, I(\underline{r}) - |F(\underline{q})|^2 + N, \tag{3.34}$$

where $S(\underline{q}, E)$ is the dynamic structure factor measured from inelastic scattering. The static structure factor proves to be a sensitive probe for correlated wave functions.

A recent example where organic molecules ($C_6H_6$ etc.) have been studied both experimentally and theoretically (fully ab initio with the use of large CI spaces) is Ref. 26. Also the case of $H_2O$ shows good agreement between theory[27] and experiment[28]. For $N_2$ discrepancies between measurements[29] and theory with highly correlated MR-SDCI wave functions[30] were found. Atoms[31, 32] and molecules[27, 33] were studied theoretically. Crystals were also studied: Si experimentally and theoretically (with calculations based upon a jellium model and Hartree-Fock and random phase approximations) in Ref. 34 and LiF experimentally in Ref. 35 and theoretically (ab initio) in Ref. 36. For the future it is hoped that better experiments provide reduced error bars.

7. The parallel-spin PD $n^p(\underline{r}_1;\underline{r}_2)$ vanishes for $\underline{r}_1 = \underline{r}_2$ (on top) because of the Pauli principle. So for small $\underline{u} = \underline{r}_2 - \underline{r}_1$ the Fermi or X-hole exists. How to display it? One way is to consider around a given reference point $\underline{r}$ the parallel PD $n^p(\underline{r}, \underline{r} + \underline{u})$ as a function of $\underline{u}$, vanishing for $\underline{u} = 0$. If spherically averaged a function of $u$ results with $\underline{r}$ as a parameter, for small $u$ it behaves like $D(\underline{r})\, u^2$, where $D(\underline{r})$ is the X-hole curvature, which has been studied (including its anisotropy) in Ref. 37. In the HF-method this curvature (in comparison with the X-hole curvature of the uniform electron gas with density $\rho(\underline{r})$) is used to define $ELF(\underline{r})$, the electron localization function[38]. The Coulomb or C-hole present in the antiparallel PD $n^a(\underline{r};\underline{r} + \underline{u})$ can be discussed similarly, but one has to take into account, that it does not vanish for $\underline{u} = 0$ and exhibits a coalescing cusp property.

8. So the spin-dependent PD $n(1;2)$ resulting from a (fermionic) ground state $\Psi$ is full of informations about the internal kinematics of this quantum state: fluctuations and polarizations in a fragment, correlations between the particles in different fragments. The close relation between fluctuations and correlation has been stressed by P. Fulde[1]. Therein the molecule $C_2H_2$ is considered as an example, and the probability of finding in the ground state 0, 1, ..., or 8 valence electrons at a C-atom is considered. The average value is 4, but the variance depends on whether or not correlations are taken into account. A correlation-induced narrowing is found accompanied by a suppression

of fluctuations: The probability for configurations with large deviations from the average value, i.e., 0 or 1, respectively, 7 or 8 valence electrons is reduced considerably (to almost zero). Similarly in Ref. 39 a detailed fluctuation analysis for $H_2$ and $N_2$ is performed. With all this in mind one may perhaps say: if available (especially for extended systems), $n(1;2)$ would have even more charm and power than $\rho(1)$.

# 4 PAIR DENSITY FUNCTIONALS AND EULER EQUATIONS

In Refs. 10 and 11 the kinetic energy $T$ as a functional of the PD has been introduced via the constrained-search method of M. Levy[40]. Under the assumption that an approximate explicit expression for $T[n(1;2)]$ is available, the minimization of

$$E[n] = T[n] + V_{\rm ext}[\rho[n]] + V_{\rm int}[n] \tag{4.1}$$

under the constraint of the normalization (1.7) yields the Euler equation

$$\frac{\delta T[n]}{\delta n(1;2)} + \frac{1}{N-1}\sum_{i=1}^{2} v_{\rm ext}(i) + v_{\rm int}(1,2) = \mu. \tag{4.2}$$

Note the contraction property (1.7) and

$$\frac{\delta n(1';2')}{\delta n(1;2)} = \delta(1'-1)\delta(2'-2) + \delta(1'-2)\delta(2'-1). \tag{4.3}$$

For given $T[n]$ the solution of Eq. (4.2) yields the PD $n(1;2)$ parametrically depending on $\mu$ which is then fixed by the normalization (1.7).

A slightly different version of the above sketched scheme avoids the explicit appearance of the total particle number (which does not allow to perform the TDL) by defining $T$ as a functional of the cumulant PD $w(1;2) = \rho(1)\rho(2) - n(1;2)$. Note the contraction property (3.7). Then minimization of

$$E[w] = T[w] + V_{\rm ext}[\rho[w]] + V_{\rm H}[\rho[w]] + V_{\rm XC}[w] \tag{4.4}$$

under the constraint of the normalization (3.8) yields the Euler equation

$$\frac{\delta T[w]}{\delta w(1;2)} + \sum_{i=1}^{2}[v_{\rm ext}(i) + v_{\rm H}(i)] - v_{\rm int}(1,2) = \lambda. \tag{4.5}$$

For the definition of $V_{\rm H}, v_{\rm H}(i), {\rm and} V_{\rm XC}$ see Eqs. (2.7), (2.8), and (3.26), respectively. For given $T[w]$ the solution of Eq. (4.5) yields the cumulant PD $w(1;2)$ parametrically depending on $\lambda$ which is then fixed by the normalization (3.8).

Eqs. (4.2) and (4.5) may be considered as generalized Thomas-Fermi-Dirac equations[41]. Note that Eq. (4.2) contains $N$ and the 'bare' external potential $v_{\rm ext}(i)$, whereas Eq. (4.5) does not contain $N$ explicitly and the external potential $v_{\text{ext}}(i)$ is 'screened' by the Hartree potential $v_{\text{H}}(i)$. And note the different signs in front of $v_{\text{int}}(1,2)$: the repulsive $v_{\text{int}}$ deepens the on-top hole of $n(1;2)$, which means for $w(1;2)$ an enhanced maximum, as if it would 'feel' an attractive particle-interaction.

# 5 AN EFFECTIVE TWO-PARTICLE EQUATION

With a geminal ansatz directly for the PD, namely $n(1;2) = \sum_K^{\text{occ}} \mid \psi_\kappa(1,2) \mid^2$ an effective two-particle equation results, which explicitly contains the total particle number $N$ similarly as in Eq. (4.2) (version I in Ref. 11). This does not allow to perform the TDL. Two other attempts, which avoid this drawback, are: ansatz for the cumulant PD $w(1;2)$, where not only antisymetric but also symmetric geminals appear (version II) and ansatz for $u(1;2) = w(1;2) - \mid \gamma(1,2) \mid^2$ with the complication to determine simultaneously the one-DM (version III). In the following version II is summarized.

The starting point is – as motivated by the considerations of Appendix B – the decomposition of the cumulant two-DM

$$w(1,1';2) = \frac{1}{2}\sum_{K_+}^{\text{occ}} \psi_{K_+}^{+}(1,2)[\psi_{K_+}^{+}(1',2)]^* - \frac{1}{2}\sum_{K_-}^{\text{occ}} \psi_{K_-}^{-}(1,2)[\psi_{K_-}^{-}(1',2)]^* + \delta w(1,1';2) \tag{5.1}$$

into two geminal parts and an unknown remainder which describes XC beyond the first terms and which has the property $\delta w(1,1;2) = 0$ leaving

$$w(1;2) = \frac{1}{2}\sum_{K_+}^{\text{occ}} \mid \psi_{K_+}^{+}(1,2) \mid^2 - \frac{1}{2}\sum_{K_-}^{\text{occ}} \mid \psi_{K_-}^{-}(1,2) \mid^2 \tag{5.2}$$

as a variational ansatz for the cumulant PD $w(1;2)$ in terms of symmetric and antisymmetric geminals. With the contraction (B.3) the kinetic energy

$$T[\gamma[w]] = \frac{1}{2}\sum_{K_+}^{\text{occ}} \langle K_+ \mid \hat{t}(1) + \hat{t}(2) \mid K_+ \rangle - \frac{1}{2}\sum_{K_-}^{\text{occ}} \langle K_- \mid \hat{t}(1) + \hat{t}(2) \mid K_- \rangle + T_{\text{XC}}[w] \tag{5.3}$$

consists then of a part, explicitly given in terms of geminals [which are functionals of $w(1;2)$] and an unknown remainder. So all the 'components' of the total energy

$$E = T[\gamma] + V_{\text{ext}}[\rho] + V_{\text{H}}[\rho] + V_{\text{XC}}[w] \tag{5.4}$$

are functionals of $w$. Note the contraction (3.7).

Minimization under the constraint (3.8) yields two coupled effective two-particle equations

$$\left[\hat{h}_{\text{H}}(1) + \hat{h}_{\text{H}}(2) - v_{\text{int}}(1,2) + \frac{\delta T_{\text{XC}}[w]}{\delta w(1;2)}\right] \psi_K(1,2) = \varepsilon_K \psi_K(1,2), \tag{5.5}$$

where $\hat{h}_H(1) = \hat{h}_0(1) + v_H(1)$ is the Hartree Hamiltonian. The bare interaction potential is made effective by an additional XC term. For further details and discussions see Ref. 11.

The main problem is to find reasonable approximations for $T_{XC}[w]$, respectively for an effective interaction possibly in terms of a dielectric function. Two ways may be helpful of finding approximate effective interactions: (i) scaling properties and (ii) the integral relation between the two-particle integral kernel of the Bethe-Salpeter equation and the effective interaction. (A similar relation exists for the one-particle descriptions QPT and DFT between the self-interaction of QPT and the effective potential of the KS equation). Another problem is a 'mental barrier', because "scientists have not yet learned to think in terms of $\binom{N}{2}$ geminals, rather than in $N$ (one- particle) orbitals" (Davidson[42]). But in view of the PD's information content it is worth to overcome this barrier.

# A Reduced distribution functions and their meaning

From the existence of the antisymmetric (ground state) wave function $\Psi(1,...,N)$ follows the existence of all the reduced distribution functions

$$n_s(1;...;s) = \int \frac{d(s+1)...dN}{(N-s)!} \mid \Psi(1,...,s,s+1,...,N) \mid^2, \tag{A.1}$$

where $n_1(1) = \rho(1)$ and $n_2(1;2) = n(1;2)$. With the operator $\hat{N}_X$, which counts the number of particles in a spatio-spin fragment $X = (\Omega, \Sigma)^{43}$, the average number of $s$-tuples of particles in $X$ is

$$\begin{aligned} \binom{N}{s}_X &\equiv \frac{1}{s!} < \hat{N}_X(\hat{N}_X - 1)\cdots(\hat{N}_X - s + 1) > \\ &= \frac{1}{s!}\int_X d1 ... \int_X ds\; n_s(1;...;s) \equiv \frac{1}{s!}\Theta_X^s n_s. \end{aligned} \tag{A.2}$$

In the last step a short hand notation is introduced. Eq. (A.2) shows that $n_s(1;...;s)$ is the density of the average number of $s$-tuples or for short it is the $s$-particle density. For $X \to \infty$ the normalization

$$\int d1...ds\; n_s(1;...;s) = N(N-1)\cdots(N-s+1) \tag{A.3}$$

results.

If one asks for the probability of finding $M$ particles in $X$, i.e., for $< \delta_{\hat{N}_X,M} >$, then the answer is[13]

$$P_X(M) = \frac{1}{M!}\left[\Theta_X^M n_M - \frac{1}{1!}\Theta_X^{M+1} n_{M+1} + \frac{1}{2!}\Theta_X^{M+2} n_{M+2} - \cdots\right], \tag{A.4}$$

which means, not only the $M$-particle density $n_M(...)$ is needed, but also all the higher-order densities $n_{M+1}(...), n_{M+2}(...), \cdots$. Note, that for $M = 0$ the first term in the brackets equals 1. The summation of $P_X(M)$ for all the possible alternatives $M = 0$ or 1 or ... or $N$ gives of course 1, see Eq. (2.2). For $X \to (\infty, \uparrow$ and $\downarrow)$ it is $P_X(M) \to \delta_{M,N}$. The probability of finding in particular one particle in $X$ is

$$P_X(1) = \Theta_X n_1 - \frac{1}{1!}\Theta_X^2 n_2 + \frac{1}{2!}\Theta_X^3 n_3 - \cdots . \qquad (A.5)$$

This means for an infinitisimally small spatial fragment $d^3r$ at $\underline{r}$ (and $\Sigma = \uparrow$ and $\downarrow$)

$$\frac{P_{d^3r}(1)}{d^3r} = \rho(\underline{r}) + 0(d^3r). \qquad (A.6)$$

So $\rho(\underline{r})$ has not only the meaning of the one-particle density with Eq. (2.1) or

$$N_\Omega \equiv \sum_M P_\Omega(M)M = \int_\Omega d^3r \rho(\underline{r}) < N \qquad (A.7)$$

giving the average number of particles in the spatial fragment $\Omega$, but $\rho(\underline{r})$ has according to (A.6) simultaneously the meaning of the probability (per unit volume) of finding one particle at $\underline{r}$. For $X \to \infty$ (and $N > 1$) it is $P_X(1) \to 0$, whereas $N_X \to N$.

If one asks for the probability of finding $M_1$ particles in a fragment $X_1$ and simultaneously $M_2$ particles in another fragment $X_2$, i.e. for $< \delta_{\hat{N}_{X_1},M_1} \delta_{\hat{N}_{X_2},M_2} >$, then the answer is[13]

$$P\begin{pmatrix} M_1 & M_2 \\ X_1 & X_2 \end{pmatrix} = \frac{1}{M_1!M_2!} \Theta_{X_1}^{M_1}\Theta_{X_2}^{M_2} \Big[ n_{M_1+M_2} - \frac{1}{1!}(\Theta_{X_1} + \Theta_{X_2}) n_{M_1+M_2+1} + \frac{1}{2!}(\Theta_{X_1} + \Theta_{X_2})^2 n_{M_1+M_2+2} - \cdots \Big]. \qquad (A.8)$$

The probability of finding one particle in $d^3r_1$ at $\underline{r}_1$ and one particle in $d^3r_2$ at $\underline{r}_2$ (with $\Sigma_{1,2} = \uparrow$ and $\downarrow$) is related to the spin-traced PD $n(\underline{r}_1; \underline{r}_2)$ according to

$$\frac{1}{d^3r_1 d^3r_2} P\begin{pmatrix} 1 & 1 \\ d^3r_1 & d^3r_2 \end{pmatrix} = n(\underline{r}_1; \underline{r}_2) + 0\left(d^3r_1\right) + 0\left(d^3r_2\right). \qquad (A.9)$$

So $n(\underline{r}_1; \underline{r}_2)$ is not only the pair density with Eq. (3.1) or

$$\binom{N}{2}_\Omega \equiv \sum_M P_\Omega(M)\binom{M}{2} = \frac{1}{2!}\int_\Omega d^3r_1 \int_\Omega d^3r_2\, n(\underline{r}_1; \underline{r}_2) < \binom{N}{2} \qquad (A.10)$$

giving the average number of particle pairs in the spatial fragment $\Omega$, but $n(\underline{r}_1; \underline{r}_2)$ has according to (A.9) simultaneously the meaning of the probability (per unit volume squared) of finding one particle at $\underline{r}_1$ and one particle at $\underline{r}_2$.

Finally the meaning of $W_{X_1,X_2}$, defined in Eq. (3.5), in terms of particle number operators in the fragments $X_{1,2}$ is

$$W_{X_1,X_2} = < \hat{N}_{X_1} >< \hat{N}_{X_2} > - < \hat{N}_{X_1}\hat{N}_{X_2} > \tag{A.11}$$

in agreement with Eq. (3.11) and (3.18), respectively with a corresponding equation for spatio-spin fragments $X_1$ and $X_2$.

# B Cumulant expansion of the two-DM

The two-DM has the following cumulant expansion in terms of the one-DM and a non-reducible remainder

$$\gamma(1,1';2,2') = \gamma(2,2') - w(1,1';2,2'), \tag{B.1}$$

$$w(1,1';2,2') = \gamma(1,2')\gamma(2,1') + u(1,1';2,2'). \tag{B.2}$$

From Eq. (1.4) follows the contraction

$$\gamma(1,1') = \int d2\; w(1,1';2,2), \tag{B.3}$$

which generalizes Eq. (3.7). Note that thc remainder $u$ (like the two-DM itself) changes sign, when 1 and 2 or $1'$ and $2'$ are interchanged. The first term of $w$ does not have this symmetry property, but it can be decomposed into a symmetric and an antisymmetric part. So $w$ has the structure $w = w^+ + w^-$ with

$$w^+(1,1';2,2') = \frac{1}{2}\left[\gamma(1,1')\gamma(2,2') + \gamma(1,2')\gamma(2,1')\right], \tag{B.4}$$

$$w^-(1,1';2,2') = -\frac{1}{2}\left[\gamma(1,1')\gamma(2,2') - \gamma(1,2')\gamma(2,1')\right] + u(1,1';2,2'). \tag{B.5}$$

Diagonalizing $w^\pm$ yields

$$w^\pm(1,1';2,2') = \pm\frac{1}{2}\sum_{K_\pm} n_{K_\pm}\; \psi^\pm_{K_\pm}(1,1)\; [\psi^\pm_{K_\pm}(1',2')]^* \tag{B.6}$$

with symmetric geminals $\psi^+_{K_+}(1,2)$ and antisymmetric geminals $\psi^-_{K_-}(1,2)$.

In some cases (Be, C and its isoelectronic series, BH) the natural occupation numbers of the two-DM have been studied and so-called supergeminals were found, where these numbers are larger than one[44]. It should be of interest to study, what this finding means for the natural occupation numbers of $w^\pm$ and of $u$.

**ACKNOWLEDGMENTS**

The author thanks A. J. Coleman, D. Edwards, A. Gonis, M. Levy, J. P. Perdew, J. P. Piris, A. Shukla, and V. H. Smith for stimulating discussions, R. Heid and S. Kalvoda for critically reading the manuscript, P. Fulde for support, and Mrs. Schuppe, B. Schmidt, and H. Wonn for technical help.

# References

[1] P. Fulde. *Electron Correlations in Molecules and Solids*, Springer, Berlin (1991, 3rd enlarged ed.1995).

[2] Recent books on DFT are D.P. Chong (Ed.). *Recent Advances in Density Functional Theory*, World Scientific, Singapore (Part I, 1995; Part II, 1997);
J.M. Seminario (Ed.). *Recent Developments and Applications of Modern Density Functional Theory*, Elsevier, Amsterdam (1996).
H. Eschrig. *The Fundamentals of Density Functional Theory*, Teubner, Stuttgart (1996).

[3] P. Ziesche, S. Kurth, and J.P. Perdew, *Computational Material Science* 11:122(1998) and references therein.

[4] T. Grabo, T. Kreibich, and E.K.U. Gross, *Molecular Engineering* 7:27(1997) and references therein.

[5] M. Yu, S. Kalvoda, and M. Dolg, *Chem. Phys.* 224:121(1997);
S. Kalvoda, M. Dolg, H.-J. Flad, and P. Fulde, *Phys. Rev.* B 57:2127(1998) and references therein.

[6] M. Piris and R. Cruz, *Int. J. Quantum Chem.* 53:353(1995);
M. Piris, A. Montero, and N. Cruz, *J. Chem. Phys.* 107:180(1997);
M. Piris, *J. Math. Chem.*, in press;
M. Piris, *J. Chem. Phys.*, submitted.

[7] A. Gonis, T.C. Schulthess, J. van Ek, and P.E.A. Turchi, *Phys. Rev. Lett.* 77:2981(1996).

[8] A. Gonis, T.C. Schulthess, and P.E.A. Turchi, *Phys. Rev.* B 56:9335(1997).

[9] A. Gonis and T.C. Schulthess *J. Phys.: Condens. Matter* 10:3535(1998).

[10] P. Ziesche, *Phys. Lett.* A 190:201(1994).

[11] P. Ziesche, *Int. J. Quantum Chem.* 60 *(Quantum Chem. Symp.* 30*)*:1361[or 149 in some libraries] (1996).

[12] In terms of RDMs the situation is slightly different: of course more and more secrets of the electron motion are hidden in the higher-order irreducible parts of the RDMs, the lowest one of which is $u_2(...)$ in

$$\gamma_2(1,1';2,2') = \gamma_1(1,1')\gamma_1(2,2') - \gamma_1(1,2')\gamma_1(2,1') - u_2(1,1';2,2'),$$

but for the physical understanding and discussion it is sufficient to know the PD $n(1;2)$ and possibly also the one-DM $\gamma_1(1,1')$ and to resign information contained in the off-diagonal elements of $\gamma_2(...)$ or $u_2(...)$ and in all the higher-order RDMs.

[13] P. Ziesche, *Habilitation Thesis*, TU Dresden(1967);
P. Ziesche and V.A. Sagrebnov, *JINR* P4-8220, Dubna(1974).

[14] R. Lenk, *Ann. Physik (Leipzig)* 19:88(1967).

[15] R.F. Bader, *Atoms in Molecules*, Clarendon, Oxford (1990), p. 332.

[16] P. Ziesche, *Int. J. Quantum Chem.* 56:363(1995);
P. Gersdorf, W. John, J.P. Perdew, and P. Ziesche, *Int. J. Quantum Chem.* 61:935(1997);
P. Ziesche, O. Gunnarsson, W. John, and H. Beck, *Phys. Rev.* B 55:10270(1997);
P. Ziesche, V.H. Smith, M. Hô, S.P. Rudin, P. Gersdorf, and M. Taut, submitted.

[17] W. Kohn, *Phys. Rev. Lett.* 76:3168(1996).

[18] It should be of interest, to have a better understanding of the relations between nearsightedness, incremental method and Bethe-Goldstone equation.

[19] It should be of interest, to compare the different correlation measures: the correlation entropy $s$ of Eq. (2.8), which results from the one-DM, and $c = C/N$ of Eq. (3.14), which results from the PD, and the correlation strength proposed in Ref. 1, which is based on the particle number variance.

[20] A.J. Thakkar and V.H. Smith, *Chem. Phys. Lett.* 42:476(1976).

[21] For the uniform electron gas (see e.g. Ref. 22) with $n(1;2) = n^2 g(r_{12})$ the cusp theorem (3.31) means $g'(0) = g(0)/a_B$, $a_B = \hbar^2/m\epsilon^2$[23]. In Ref. 24 it is shown, that $g(0)$ not only determines $g'(0)$ but also the large-$k$ behavior of the momentum distribution $n(k)$. Relations between $n(k)$ and $g(r)$ on the basis of the virial theorem are discussed in Ref. 25.

[22] P. Ziesche and G. Paasch, *Das Elektronengas*, in: *Ergebnisse in der Elektronentheorie der Metalle*, P. Ziesche and G. Lehmann, ed., Akademie- and Springer-Verlag, Berlin (1983), p. 14.

[23] J.C. Kimball, *Phys. Rev.* A 7:1648(1973).

[24] J.C. Kimball, *J. Phys.* A 8:1513(1975).

[25] W. Macke and P. Ziesche, *Ann. Physik (Leipzig)* 13:26(1964).

[26] N. Watanabe, H. Hayashi, and Y. Udagawa, *J. Chem. Phys.* 108:4545(1998).

[27] J. Wang, A.N. Tripathi, and V.H. Smith, *J. Chem. Phys.* 101:4842(1994).

[28] N. Watanabe, H. Hayashi, and Y. Udagawa, *Bull. Chem. Soc. Japan* 70:719(1997).

[29] R.S. Barbieri and R.A. Bonham, *Phys. Rev.* A 44:7361(1991).

[30] M. Meyer, T. Müller, and A. Schweig, *Chem. Phys. Lett.* 236:497(1995).

[31] J. Wang, R.O. Esquivel, V.H. Smith, and A.V. Bunge, *Phys. Rev.* A 51:3812(1995).

[32] J. Cioslowski, B.B. Stefanov, A. Tan, and C.J. Umrigar, *J. Chem. Phys.* 103:6093(1995).

[33] J. Wang and V.H. Smith, *Int. J. Quantum Chem.* 49:147(1994).

[34] W. Schülke, J.R. Schmitz, H. Schulte-Schrepping, and A. Kaprolat, *Phys. Rev.* B 52:11721(1995), see also references therein.

[35] G. Calzuola, C. Petrillo, and F. Sacchetti, *Phys. Rev.* B, submitted.

[36] A. Shukla, *Phys. Rev.* B, submitted.

[37] J. F. Dobson, *J. Chem. Phys.* 94:4328(1991).

[38] A. Savin, R. Nesper, S. Wengert, and T.F. Fässler, *Angew. Chem. Int. Ed. Engl.* 36:1808(1997) and references therein.

[39] M. Mödl, M. Dolg, and H. Stoll, *J. Chem. Phys.* 105:2353(1996).

[40] M. Levy, *Proc. Math. Acad. Sci. USA* 76:6062(1979);
see also S.M. Valone, *J. Chem. Phys.* 73:1344(1980).

[41] The idea to derive from $T[n]$ or $T[w]$ Euler equations is due to M. Levy and D. Edwards.

[42] E.R. Davidson. *Reduced Density Matrices in Quantum Chemistry*, Academic, New York (1976), p. 97.

[43] In the wave or second quantization the particle number operator in a fragment $X = (\Omega, \Sigma)$ is

$$\hat{N}_X = \int_X d1\, \psi^\dagger(1)\psi(1) \equiv \int_{\underline{r}\in\Omega} d^3r \sum_{\sigma\in\Sigma} \psi^\dagger(\underline{r},\sigma)\psi(\underline{r},\sigma),$$

where $\psi^\dagger(1)$ and $\psi(1)$ are creation and annihilation operators.

[44] Jiahu Wang, *PhD Thesis*, Queen's University Kingston (1994) and references therein.

# THE TWO-PARTICLE PICTURE AND ELECTRONIC STRUCTURE CALCULATIONS

A. Gonis[1], T. C. Schulthess[2], and P. E. A. Turchi[1]

[1]Lawrence Livermore National Laboratory (L-353) P.O.Box 808
Livermore CA 94551
[2]Metals and Ceramics Division, Oak Ridge National Laboratory
Oak Ridge, TN 37831

## INTRODUCTION

The consept of self-energy figures prominently in the studies of interacting quantum many-particle systems. This quantity is generally defined in terms of the energy change that accompanies the addition of a single particle to the interacting system and incorporates exactly the effects of correlation and the statistics, Fermi-Dirac or Bose-Einstein. The single-particle self-energy is non-local, generally complex, and also energy-dependent. Its evaluation is ususally associated with various approximate, perturbative or decoupling, schemes as described[1,2] in the literature.

In this paper, we introduce a point of view which differs from canonical treatments of interacting quantum systems. Instead of building the self-energy from individual contributions to the interaction of a particle with the rest of the system, e.g., summations of selected diagrams, we treat the entire system of $N$ interacting three-dimensional particles as a *single particle* in a configurational space of $3N$ dimensions. In the non-relativistic limit considered here, the external potentials acting on the particles as well as their mutual interactions appear as a time-independent externally applied field. Consequently, the Green function of the system can be expressed in the familiar form obtained in the traditional single-particle picture. It is shown that from a knowledge of the Green function describing $n$ interacting particles (in an $N$-particle system, with $N$ being either finite or infinite) Green functions for $m$ interacting particles, $m < n$, can be obtained. Inverting the resulting $m$-particle Green function leads directly to the self-enrgy of the $m$-particle unit, extending the concept of self-energy to an $m$-particle quantity. Furthermore, the formulation in terms of t-matrices given below provides a particularly flexible framework for the development of approximate computational schemes for evaluating Green functions and the corresponding self-energies.

Another aim of the discussion is to investigate any possible formal connections between canonical many-body theory and band theory, the methodology used in the study of the electronic structure of matter, and properties related to it. Band theory is currently being implemented within the context of the local (spin) density approximation [L(S)DA] of density functional theory (DFT)[3]. The DFT-LDA approach has been very successful in the treatment of metallic solids[4], but is often criticized for its approximate treatment of correlation effects. It is known to provide a poor description of the gap in the single-particle spectrum of semiconductors and insulators, either understimating the gap, or missing it altogether predicting metallic behavior. It also fails to describe the volume collapse that accompanies increased pressure in many elemental solids of the Lanthanide and Actinide series. Possibly most important is the lack of a well-defined procedure for extending the DFT-LDA approach to alleviate these shortcomings.

It is shown below that it is possible to view DFT-LDA and canonical many-body theory from a single perspective as two parts of a coherent whole. As such, it affords a unique, formally exact procedure for improvement, and provides a single parameter, namely the number of particles treated explicitly, as a gauge for convergence.

The following discussion is confined to the non-relativistic limit so that the interaction between two particles is taken to be instantaneous, $U(\mathbf{r}t, \mathbf{r}'t') = U(\mathbf{r}, \mathbf{r}')\delta(t-t')$. It is, however, general enough to accommodate a diverse class of potentials that include the nuclear forces in an atom or a molecule, as well as in extended systems such as a solid.

## GREEN FUNCTION AND SELF-ENERGY

We confine our attention to an interacting, $N$-particle system described by the Hamiltonian,

$$H^{(N)} = \sum_{i=1}^{N} h(x_i) + \frac{1}{2}\sum_{i\neq j} U(x_i - x_j), \tag{1}$$

where $h(x_i)$ is the Hamiltonian describing a single particle and includes the kinetic energy and any external potential acting on the system, while $U(x_i - x_j)$ denotes the interparticle interaction. Here, $x_n = (\mathbf{r}_n t_n)$ is a combined space-time coordinate for the $n$th particle, and $d^4x_n = d^3r_n dt_n$. (This coordinate can also incorporate the spin variable but this will not be made explicit in the discussion.)

We begin with the familiar[5,6] equation of motion for the many-particle Green function. Treating explicitly the case of fermions, and denoting the $n$-particle Green function by the symbol $G^{(n)}$, we have

$$\begin{aligned}
&\left[i\hbar\frac{\partial}{\partial t_1} + \frac{\hbar^2}{2m}\nabla_1^2\right] G^{(n)}(x_1, x_2, \cdots, x_n; x_1', x_2', \cdots, x_n') \\
&+ i\hbar \int d^4x_{n+1} U(x - x_{n+1}) \\
&\times G^{(n+1)}(x_1, \cdots, x_{n+1}; x_1', \cdots, x_{n+1}') \\
&= \sum_{i=1}^{n} \delta^{(4)}(x_1 - x_i')(-1)^{i-1} \\
&\times G^{(n-1)}(x_2, \cdots, x_n; x_1', \cdots, x_{i-1}', x_{i+1}', \cdots, x_n').
\end{aligned} \tag{2}$$

Also, any external "single-particle" potential, such as that due to the nuclei in a molecule or a solid, can be incorporated explicitly inside the brackets in the last expression. For the sake of clarity of presentation, we continue to treat explicitly the uniform case, $h(x) = \nabla_{\mathbf{r}}^2$.

It is seen that in general the equation for the $n-$particle Green function contains both the $n+1$- and the $n-1$-particle Green functions. Defining the free-particle propagator, $G_0 = \left[i\hbar\frac{\partial}{\partial t} + \frac{\hbar^2}{2m}\nabla^2\right]^{-1}$, setting all $t_i = t$ and $t_i' = t'$, (describing the probability amplitude of creating *all* particles at time $t'$ and annihilating them at time $t$), and Fourier transforming with respect to $t'-t$ into the energy domain, we obtain,

$$\begin{aligned}
&G^{(N)}(r_1, r_2, \cdots, r_N; r_1', r_2', \cdots, r_N'; E) \\
&= \sum_{i=1}^{N}(-1)^{i-1}\int d\omega G_0(r_1, r_i'; \omega) \\
&\times G^{(N-1)}(r_2, \cdots, r_N; r_1', \cdots, r_{i-1}', r_{i+1}', \cdots, r_N'; E-\omega).
\end{aligned} \tag{3}$$

This expression is properly antisymmetric with respect to interchange of fermion coordinates. It describes the probability amplitude that the system of $N$ particles propagates from a state with position $\mathbf{r}_1', \cdots, \mathbf{r}_N'$ to the state with positions $\mathbf{r}_1, \cdots, \mathbf{r}_N$.

Let us now return to Eq.(3), (still for the case of distinguishable particles), and sum (or integrate) over the coordinate of a given particle, say particle 1. Assuming that $h^1$ is translationally invariant (or uniform), this summation yields the $\mathbf{k}=0$ element of the single-particle Green function, $G_0$. Since this is given by the general expression $G_0(\mathbf{k}=0; z) = [z - h(\mathbf{k}=0)]^{-1}$, the remaining integration yields the exact $N-1$-order Green function which can be written in the form,

$$G^{(N-1)}(z) = \left[z - H_0^{(N-1)} - \Sigma^{(N-1)}\right]^{-1}. \tag{4}$$

where $H_0^{(N-1)}$ denotes the Hamiltonian describing $N-1$ non-interacting particles. Here, the self-energy $\Sigma^{(N-1)}$ is non-local in the coordinates of the $N-1$ particles, complex, and energy dependent. As defined here, it also contains a shift by the energy at $\mathbf{k}=0$ of a uniform (or periodic) single-particle system, but that shift will not be shown explicitly. *Thus, summation over the coordinates of the Green function elements associated with a single particle in $G^{(N)}$ yields the exact Green function $G^{(N-1)}$ for the $N$-particle system.* Let us denote this reduced-order quantity by a hat symbol, (because it contains a shifted self-energy), and refer as *downfolding* to the process of summing over the coordinates (or Green function elements) associated with given particles in the system. In operator space, the previous discussion can be summarized in the form,

$$\begin{aligned}
&\hat{G}^{(N-1)}(i_2, \cdots, i_N; i_2', \cdots, i_N'; z) \\
&= \frac{1}{\Omega}\sum_{i_1, i_1'} G^{(N)}(i_1, i_2, \cdots, i_N; i_i', i_2', \cdots, i_N'; z),
\end{aligned} \tag{5}$$

where in the coordinate representation the summations are to be interpreted as integrals over the volume, $\Omega$, of the system. More generally, a reduced Green function for $n$ particles can be obtained by downfolding the exact $N$-particle Green function,

$$\hat{G}^{(n)}(z) = \frac{1}{\Omega^{N-n}} \times \sum_{\substack{i_1,i_2,\cdots,i_{N-n}\\ i'_1,i'_2,\cdots,i'_{N-n}}} G^{(N)}(i_1,i_2,\cdots,i_N;i'_1,i'_2,\cdots,i'_N;z), \tag{6}$$

over the coordinates of $N-n$ particles. Finally, we see that the *self-energy* for a system of $n$ particles is given by the general expression,

$$\Sigma^{(n)}(i_1,i_2,\cdots,i_n;i'_1,i'_2,\cdots,i'_n;z) = z - H_0^{(n)} - [\hat{G}^{(n)}]^{-1}. \tag{7}$$

The single-particle Green function and self-energy can now be obtained by setting $n=1$, and downfolding $G^{(N)}$ over the coordinates of $N-1$ particles.

Although not as easy to visualize, the treatment of *indistinguishable* particles proceeds from Eq.(3) along the lines just described. In retrospect, the downfolding procedure introduced here is a direct consequence of the definition of the Green function as a probability amplitude for the creation and annihilation of particles. Thus, to obtain the Green function for $n$ particles, one sums over all possible initial and final states for the remaining $N-n$ particles.

The procedure introduced above is exact. At the same time it suggests approximate ways for evaluating the system Green function and the single-particle self-energy, $\Sigma_{ij}$. For example, the Green function for an interacting $N$-particle system can be obtained as the average of all possible $n$-particle Green functions $(N < n)$ for each configuration of the remaining $N-n$ particles (thus neglecting configuration-configuration interactions between the $n$-particle subsystems). From this Green function, a single-particle Green function and self-energy follow upon summing over the coordinates (space) of $n-1$ particles.

## T-MATRIX

It is starightforward to cast the formal developments presented above in terms of an effective t-matrix. Here, the t-matrix of the exact $N$-particle system is defined by the expression,

$$G^{(N)} = G_0^{(N)}[1 + T^{(N)}G_0^{(N)}]. \tag{8}$$

The downfolding procedure, Eq.(6), can be applied to $T^{(N)}$ in order to obtain an effective t-matrix in $n$-particle space,

$$\hat{T}^{(n)}(z) = \frac{1}{\Omega^{N-n}} \times \sum_{\substack{i_1,i_2,\cdots,i_{N-n}\\ i'_1,i'_2,\cdots,i'_{N-n}}} T^{(N)}(i_1,i_2,\cdots,i_N;i'_1,i'_2,\cdots,i'_N;z), \tag{9}$$

where we have summed over the coordinates of $N-n$ particles. Also, the self-energy in $n$-particle space is given by the well-known relation,

$$\Sigma^{(n)} = \hat{T}^{(n)}[1 + \hat{\tilde{G}}_0^{(n)} \hat{T}^{(n)}]^{-1}. \tag{10}$$

It follows from the derivation that the self-energy obtained through this procedure is identical to that obtained through the Green function, Eq.(7). However, the evaluation of the t-matrix is considerably easier than that of the Green function because the t-matrix is non vanishing only when all of its arguments lie in the potential region. Thus, the downfolding process extends only over a finite region, namely where the interparticle potential is acting, rather than over all space as required when dealing with the Green function itself.

## NUMERICAL RESULTS

The previous discussion can be illustrated by means of a "small" one-dimensional system described by a Hubbard-like model Hamiltonian,

$$H = \sum_i \epsilon_i |i\rangle\langle i| + \sum_{i \neq j} t_{ij} |i\rangle\langle j| + \sum_i U_i n_{i\uparrow} n_{i\downarrow}. \tag{11}$$

This Hamiltonian is taken to describe a system of four electrons with total spin zero arranged on the four sites of a closed linear ring characterized by site energies $\epsilon_i = 0.0$ (uniform potential). The quantities $|i\rangle$ denote the state associated with site $i$, $t_{ij}$ is a transfer integral between sites $i$ and $j$, and $U_i$ is an on-site Coulomb repulsion term that is associated with two electrons of opposite spin occupying the same site. The quantities $n_{i\sigma}$ ($\sigma =\uparrow$ or $\downarrow$) are the number operators for an electron on site $i$ with spin $\sigma$. The exact four-particle Green function for this system can be obtained[7] by inverting the four-dimensional structure formed by the orthogonal combination of four four-sited rings. This space contains 256 configurations describing all possible arrangements of four electrons, two with spin up and two with spin down, on the four sites of the ring. These configurations are consistent with the Pauli exclusion principle, in that an "infinite" ($\approx 10^7$) potential was associated with any configuration in which two or more electrons of the same spin occupy the same site. In the calculations reported here, we restrict the range of electron hopping, $t$, to nearest neighbors, and set $U/t = 10.0$. Also, for ease of illustration, the spectra are exhibited for values of the energy that contain an imaginary part, $\Im E = t/4$.

An approximate evaluation of the Green function and the self-energy can be made by considering separately the sixteen different two-particle Green functions obtained for each configuration of the other two particles in the system. There are sixteen such configurations whose mutual interactions are neglected. Thus, each two-particle unit is described by an effective Hamiltonian which corresponds to each possible configuration of the remaining two particles in the system.

The solid line in the upper panel of Fig. 1 shows the exact single-particle spectrum obtained from downfolding the exact four-particle Green function. The dotted line corresponds to the approximate evaluation described in the previous paragraph. The lower two panels of the figure shows the real and imaginary parts of the corresponding site-diagonal element of the self-energy. It is seen that the approximate spectra reproduce rather well the main features of the exact ones, in particular the impurity-like peak at about $\Re E = U$. The exact spectra lead to a narrower distribution of eigenvalues (band in solids) since they take proper account of the configuration interactions. Because of the Pauli principle, which prevents two particles of the same spin from occupying the same site, the available configuration space of the system is reduced leading to a narrowing of the "band" at the four-particle level as well as at the single-particle level. The difference in structure inside the main part of the spectrum between the exact and approximate spectra can be attributed to the interactions among different two-particle configurations which are ignored in the approximate treatment. We also note that the approximate treatment misses some of the many-body contributions to the spectra at the bottom of the band. Finally, we note that the self-energies agree considerably better than the Green functions. This is an important point to keep in mind when basing the treatment of interacting quantum systems on the self-energy. It is well known that even when the self-energy is approximated to be strictly local, the corresponding Green function maintains its global character. It is seen that basing an approximate treatment of interacting quantum systems on the Green function rather than the self-energy places much more demanding criteria on agreement with exact results but can also be expected to lead to a more accurate procedure.

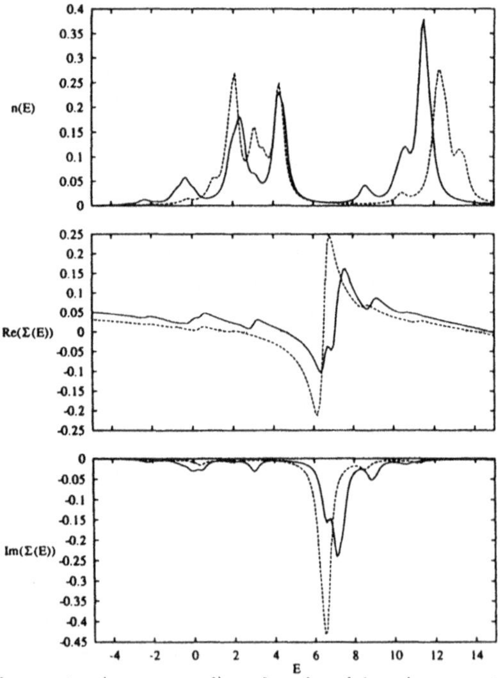


FIG. 1. Single-particle spectra (upper panel) and real and imaginary parts (central and lower panels) and of self-energies obtained from the exact four-particle Green function (solid lines) and from and effective two-particle Green function (dashed lines). The energy is given in arbitrary units.

The last comment provides a strong motivating factor for developing the formalism introduced in this letter. In the case of non-relativistic particles, the Green function of the entire system can be formulated within essentially a single-particle formalism and, since the time-independent Hamiltonian acting on the system as a whole is known, (consisting of interparticle interactions), this Green function can be trivially written down. As has been shown, an integration over the coordinates of $n$ particles yields the exact $N-n$-particle Green function, but with its poles shifted by the energies of $n$ single, non-ineracting particles at the $\mathbf{k}=0$ point[8]. The corresponding self-energy follows upon inversion of this quantity. This approach also illustrates in a natural way the possibility of extending the self-energy concept to describe the interaction of $n$ particles with the remaining particles in the system.

## DFT FOR N-PARTICLE STATES

In order to account for the effects of a finite electron density, especially in infinite systems, we generalize the concept of DFT to the space of $N$ particles.

We consider the Hamiltonian of a fully interacting N-particle system,

$$\hat{H}=\sum_i -\nabla_i^2+\sum_i v_i+\frac{1}{2}\sum_{ij} v_{ij}, \tag{12}$$

where $\nabla_i^2$ is the Laplacian operator for particle $i$, $v_i = v(\mathbf{r}_i)$ is a single-particle potential for particle $i$, and $v_{ij}=v(\mathbf{r}_i,\mathbf{r}_j)$ is the interparticle potential, with $v_{ii}=0$. Ordinarily, we are interested in the solutions of the many-particle wave equation,

$$H\Psi = E\Psi, \tag{13}$$

where $\Psi$ denotes the many-body wave function for the interacting system.

We consider the density $n(\mathbf{r})$ for the (non-degenerate) ground state of the system. The usual proof of the uniqueness of the external potential as a functional of the density proceeds by *reductio ad absurdum.* Let there be two external potentials, $v(\mathbf{r})$ and $v'(\mathbf{r})$, differing by more than a constant and each giving the same ground-state density $n(\mathbf{r})$. Let also $\hat{H}$ and $\hat{H}'$ be the corresponding many-body Hamiltonians, and $\Psi$ and $\Psi'$ the associated wave functions. Both $\hat{H}$ and $\hat{H}'$ would correspond to the *same* ground-state density, but the wave functions $\Psi$ and $\Psi'$ would be different. Take $\Psi'$ to be a trial wave function for the system described by $\hat{H}$ (which includes $v(\mathbf{r})$) and use the fact that the expectation value of the energy is a minimum for the correct wave function to obtain the result

$$\begin{aligned} E_0 < \langle\Psi'|\hat{H}|\Psi'\rangle &= \langle\Psi'|\hat{H}'|\Psi'\rangle + \langle\Psi'|\hat{H}-\hat{H}'|\Psi'\rangle \\ &= E_0' + \int n(\mathbf{r})\,[v(\mathbf{r})-v'(\mathbf{r})]\,\mathrm{d}^3r, \end{aligned} \tag{14}$$

where $E_0$ and $E_0'$ denote, respectively, the energy of the $N$-electron system (excluding nuclear-nuclear repulsion), under the influence of $v$ and $v'$. The integral $\int n(\mathbf{r})v(\mathbf{r})\mathrm{d}^3r$ describes the interaction of the charge distribution with the external field. Similarly, taking $\Psi$ as a trial wave function for the system described by $\hat{H}'$, we have

$$\begin{aligned} E_0' < \langle\Psi|\hat{H}'|\Psi\rangle &= \langle\Psi|\hat{H}|\Psi\rangle + \langle\Psi|\hat{H}'-\hat{H}|\Psi\rangle \\ &= E_0 - \int n(\mathbf{r})\,[v(\mathbf{r})-v'(\mathbf{r})]\,\mathrm{d}^3r, \end{aligned} \tag{15}$$

Adding the last two equations, we obtain

$$E_0+E_0' < E_0'+E_0. \tag{16}$$

These inequalities constitute a contradiction to the premise that there can be two different $v$'s, differing by more than a constant, that give the same density $n(\mathbf{r})$ for the ground state of the system. This proves the theorem (the second theorem of Hohenberg and Kohn[16]). We now show that this theorem carries through virtually intact in a hyperspace in which $n$ particles are considered as a single particle. We consider explicitly the case $n = 2$.

Let us consider again the Hamiltonian of Eq.(12) but as a sum of distinct, non-overlapping pairs of particles (so that a given particle belongs to only one pair), labeled by $I$, and write the Hamiltonian in the form,

$$H = \sum_I -\nabla_I^2 + \sum_I V_I + \frac{1}{2}\sum_{IJ} \frac{1}{2} V_{IJ}. \tag{17}$$

This Hamiltonian is the same as that above, Eq.(12), except that it is expressed in terms of pairs of particles. We consider explicitly systems with infinite numbers of particles (or with finite but even numbers) so that the partition into pairs can be effected exactly. Also, we consider singlet pairs only so that exchange effects on the wave function arise only from the interchange of particles across pairs. This partition does not constitute a restriction for zero spin systems. Now, we note that for each such pair we can write,

$$\begin{aligned} \nabla_i^2 + \nabla_j^2 &= \frac{\partial^2}{\partial x_i^2} + \frac{\partial^2}{\partial y_i^2} + \frac{\partial^2}{\partial z_i^2} + \frac{\partial^2}{\partial x_j^2} + \frac{\partial^2}{\partial y_j^2} + \frac{\partial^2}{\partial z_j^2} \\ &= \sum_{\alpha=1}^{6} \frac{\partial^2}{\partial x_{I\alpha}^2} = \nabla_I^2 = \nabla_{\mathbf{x}}^2, \end{aligned} \tag{18}$$

where the coordinates of the two particles have been combined in a single coordinate in six dimensions

$$\begin{aligned} \mathbf{x} &= (x_i, y_i, z_i, x_j, y_j, z_j) \\ &= (x_1, x_2, x_3, x_4, x_5, x_6). \end{aligned} \tag{19}$$

This six-dimensional coordinate forms the direct part of the phase-space (hyperspace) of the two-particles, in which the pair appears as a single particle. The other terms that appear in Eq.(17) are uniquely defined in terms of the original Hamiltonian. Thus, with $\mathbf{x} = (\mathbf{r}_i, \mathbf{r}_j)$, we have,

$$\begin{aligned} V_I \equiv V(\mathbf{x}) &= V(\mathbf{r}_i, \mathbf{r}_j) \\ &= v(\mathbf{r}_i) + v(\mathbf{r}_j) + \frac{1}{|\mathbf{r}_i - \mathbf{r}_j|}, \end{aligned} \tag{20}$$

and

$$\begin{aligned} V_{IJ} \equiv V(\mathbf{x}, \mathbf{x}') &= V(\mathbf{r}_i, \mathbf{r}_j; \mathbf{r}'_i, \mathbf{r}'_j) \\ &= \frac{1}{|\mathbf{r}_i - \mathbf{r}'_i|} + \frac{1}{|\mathbf{r}_i - \mathbf{r}'_j|} + \frac{1}{|\mathbf{r}_j - \mathbf{r}'_i|} + \frac{1}{|\mathbf{r}_j - \mathbf{r}'_j|}. \end{aligned} \tag{21}$$

This completely defines the Hamiltonian in Eq.(17).

The important feature to notice about this Hamiltonian is that it has a form *identical* to that in Eq.(12). Therefore, it suggests a treatment using a single-particle framework, where the particles are in six-dimensional (rather than three-dimensional) coordinate space. This form of the Hamiltonian allows one to carry through essentially unchanged the proof of the Hohenberg-Kohn theorem so that it holds in the phase-space of two-particles (or $n$ particles). Thus the potential, and hence the energy, is a unique functional of the density, so that we can write for the energy of the electron gas,

$$E[n(\mathbf{x})] = T[n(\mathbf{x})] + U[n(\mathbf{x})], \tag{22}$$

where the various terms are assigned their familiar meaning from ordinary DFT, but must be interpreted as quantities in an appropriate phase space. For example, for $n = 2$ the term $U[n(\mathbf{x})]$ contains all electrostatic interactions among and between particles in six-dimensional space, as well as their interaction with the nuclei.

This generalization of the Hohenberg-Kohn theorem is a straightforward consequence of the observation that the original proof is not dependent on dimensionality. In other words, the theorem holds intact in a configurational space of 3N dimensions for systems described by the Hamiltonian of Eq.(17). As is the case with respect to its canonical form, nDFT yields an energy functional which is stationary with respect to variations in the density and assumes its lowest value for the correct density of the ground state.

## Kohn-Sham equations

In order to make DFT a usefull tool, explicit expressions for the energy functional $E[n]$ are needed. The Kohn-Sham (KS) formalism[11] provides such an expression for ordinary DFT. The KS formalism is reviewed in a number of texts and review articles[12,3] and only some of its more salient features will be mentioned here.

We introduce single-particle orbitals, or generalized KS functions, such that

$$n(\mathbf{x}) = \sum_I |\Psi_I(\mathbf{x})|^2. \tag{23}$$

The meaning of these functions as well as the extent of the sum over $I$ is commented upon below. For the moment, we consider these functions as the solutions of an effective "single-particle" Schrödinger equation obtained as follows.

We define a single-body potential by the relation,

$$\begin{aligned} V(\mathbf{x}) = {} & \frac{1}{|\mathbf{r}_i - \mathbf{r}_j|} + \int d^6x' V(\mathbf{x},\mathbf{x}')n(\mathbf{x}') \\ & - \sum_n Z_n \left[\frac{1}{|\mathbf{r}_i - \mathbf{R}_n|} + \frac{1}{|\mathbf{r}_j - \mathbf{R}_n|}\right], \end{aligned} \tag{24}$$

where $\mathbf{x} = (\mathbf{r}_i, \mathbf{r}_j)$, and $V(\mathbf{x},\mathbf{x}')$ is given by Eq.(21). For a given density, the energy is evaluated in the usual manner,

$$\begin{aligned} E[n(\mathbf{x})] = {} & T_s[n(\mathbf{x})] + \int d^6x V(\mathbf{x})n(\mathbf{x}) \\ & + J[n(\mathbf{x})] + E_{xc}[n(\mathbf{x})], \end{aligned} \tag{25}$$

where

$$T_s[n(\mathbf{x})] = \sum_I \int d^6x \Psi_I^*(\mathbf{x}) \nabla_{\mathbf{x}} \Psi_I(\mathbf{x}), \tag{26}$$

represents the kinetic energy functional of a non-interacting gas,

$$J[n(\mathbf{x})] = \frac{1}{2} \int d^6x \int d^6x' n(\mathbf{x}) n(\mathbf{x}') V(\mathbf{x}, \mathbf{x}'), \tag{27}$$

and $E_{xc}[n(\mathbf{x})]$ contains the difference between the exact kinetic energy $T[n(\mathbf{x})]$ and that represented by $T_s[n(\mathbf{x})]$, as well as the difference between the exact interparticle interaction and its "classical" approximation[14] given by the expression

$$E_{xc}[n(\mathbf{x})] = T[n(\mathbf{x})] - T_s[n(\mathbf{x})] + \int d^6x_1 \int d^6x_2 \\ \times V(\mathbf{x}_1, \mathbf{x}_2) \left[ n(\mathbf{x}_1, \mathbf{x}_2) - n(\mathbf{x}_1) n(\mathbf{x}_2) \right]. \tag{28}$$

Now, the single-particle Schrödinger equation yielding the solutions $\Psi_I(\mathbf{x})$ takes the form,

$$\left[ -\nabla_{\mathbf{x}}^2 + V(\mathbf{x}) + \mu_{xc}(\mathbf{x}) \right] \Psi_I(\mathbf{x}) = E_I \Psi_I(\mathbf{x}), \tag{29}$$

where the *exchange-correlation* potential is given by the expression,

$$\mu_{xc}(\mathbf{x}) = \frac{\delta E_{xc}[n]}{\delta n(\mathbf{x})}. \tag{30}$$

Up to this point, the KS scheme is formally identical to that originally proposed for the case of three-dimensional systems. Most importantly, the wave functions (orbitals) $\Psi_I(\mathbf{x})$ are to be interpreted only as "aid to calculation", their function being to reproduce the density by means of Eq.(23).

## APPLICATION TO SOLIDS

The determination of the two-particle Green function and t-matrix requires the solution of Eq.(29). This is a particularly taxing task in the case of an infinite system, such as a solid, and depends crucially on the prior approximate knowledge of the exchange-correlation energy and potential. Once an exchange-correlation potential has been determined, (for possible ways of accomplishing this, see below), methods well-established in the treatment of the electronic structure of solids within band theory can be used to obtain the solution. One possible way is to generalize the well-known Green function method of Korringa, Kohn, and Rostoker (KKR) to six-dimensional space, as discussed in previous work[9]. Such an approach would lead to a two-particle-state representation of the electronic structure. The corresponding single-particle quantities, self-energy and Green function can be obtained from the procedures outlined above (see also discussion in final section).

The determination of the exchange-correlation potential could in principle be carried out in analogy with the determination of that potential in the canonical version of the Kohn-Sham scheme. Here, one would treat numerically a collection of six-dimensional particles which interact via the potential of Eq.(21). In this determination, the Coulomb repulsion among the members of any one pair must be turned off because this potential appears explicitly in the Schrödinger equation. A numerical representation of the exchange-correlation energy as function of two-particle density would then be applicable within the local density approximation. However, the details of this determination are far from simple. It is now required to obtain a parametrization of the energy of a collection of pairs of particle in terms of the pair density that unlike its single-particle counterpart is not constant.

## DISCUSSION

The formalism presented above provides one way for attaining a fairly wide and systematic treatment of correlation effects as they manifest themselves in the calculation of the electronic structure of matter. Its basis is not dissimilar from that of recent works[10,17] in the literature in which the LDA or the GW approximation are used as the basis for corrective treatments determining the two-particle t-matrix and hence the single-particle self-energy. The most important difference between the present proposal and those works is the suggestion that DFT (and the LDA) can be extended and applied in two-particle space, endowing the entire process with a self-consistent character. In this regard, the methodology is very similar to that proposed by Ziesche[18] and discussed elsewhere in this volume.

The use of DFT-LDA in two-particle space depends crucially on the (approximate) knowledge of an exchange-correlation functional. One possible method for determining this functional was outlined above, although the details of the calculation and the required parametrizations are still to be worked out. Given such a functional, methods used traditionaly in the determination of the electronic structure, such as KKR, can be generalized[9] to obtain the two-particle Green function and, hence, the single-particle Green function and self-energy.

The present formalism affords the possibility of treating a large spectrum of electron-structure related properties, such as ground-state properties, gaps and satellites, and transport within a unified and systematic framework. Although the method relies on a static potential in two-particle space, it does yield energy-dependent and non-local self-energies, which are in general complex quantities. Whether these self-energies provide an accurate treatment of materials properties could be determined through application.

## ACKNOWLEDGEMENTS

This work was suported by the U.S. Department of Energy through contract No. W-7405-ENG-48 with Lawrence Livermore National Laboratory. Research at Oak Ridge was sponsored by the Division of Materials Science, U.S. Department of Energy under contract DE-AC05-96OR22464 with Lockheed Martin Energy Research Corporation.

## REFERENCES

1. Alexander L. Fetter and John Dirk Wallecka, *Quantum Theory of Many-Particle Systems*, McGraw-Hill, New York (1971).
2. N. H. March, W. H. Young, and S. Sampanthar, *The Many-Body Problem in Quantum Mechanic*, Cambridge University Press, Cambridge (1967).
3. R. M. Dreitzler and E. K. U. Gross, *Density Functional Theory*, Springer-Verlag, Berlin, New York (1990).

4. V. L. Moruzzi, J. F. Janak, and A. R. Williams, *Calculated Electronic Properties of Metals,* Pergamon, New York (1978).
5. E. K. U. Gross and E. Runge, *Vielteilchentheorie*, Teubner, Stuttgart, (1986).
6. Tomokazu Kato, Tetsuro Kobayashi, and Mikio Namiki, *Supplement of the Progress of Theoretical Physics*, (Japan), No. **15**, (1960).
7. The DOS shown in the upper panel of the figure was reported earlier in connection with the generalization of DFT to two-particle states in A. Gonis, T. C. Schulthess, J. van Ek, and P. E. A. Turchi, *Phys. Rev. Lett.* **77**, 2981 (1996).
8. The case of a non-uniform but periodic system can be formulated through the extension of scalar quantities to matrices.
9. A. Gonis, T. C. Schulthess, P. E. A. Turchi, and J. van Ek, *Phys. Rev. B* **56**, 9335 (1997-I).
10. M. Springer, F. Aryasetiawan, and K. Karlsson, *Phys. Rev. Lett.* **80**, 2389 (1998).
11. W. Kohn and L. J. Sham, *Phys. Rev.* **140**, A1133 (1965).
12. R. G. Par and Yang, *Density Functional Theory of Atoms and Molecules*, Oxford University Press, Oxford (1989).
13. The exact interparticle interaction in a many-electron system is given by the correlated density, n(**r**,**r'**), rather than the product of two-single-particle densities.
14. Thomas Muir, *Treatise on the Theory of Determinants*, Dover, New York (1960).
15. P. Hohenberg and W. Kohn, *Phys. Rev.* **136**, B864 (1964).
16. C. Calandra and F. Mangi, *Phys. Rev. B* **50**, 2061 (1994).
17. Paul Ziesche, *Physics Letters* A **195**, 213 (1994), and this volume.

# ORBITAL FUNCTIONALS IN STATIC AND TIME-DEPENDENT DENSITY FUNCTIONAL THEORY

E.K.U. Gross, T. Kreibich, M. Lein, and M. Petersilka

Institut für Theoretische Physik
Universität Würzburg
Am Hubland
D-97074 Würzburg
Germany

## INTRODUCTION

Density functional theory (DFT) is among the most powerful quantum mechanical methods for calculating the electronic structure of atoms, molecules and solids [1, 2, 3]. In this introduction we present a brief overview of DFT, putting particular emphasis on the appearance of orbital functionals. To describe the electronic many-body system, we have to deal with the Hamiltonian

$$\hat{H} = \hat{T} + \hat{W}_{\mathrm{Clb}} + \hat{V} \tag{1}$$

where

$$\hat{T} = \sum_{i=1}^{N}\left(-\frac{\hbar^2}{2m}\nabla_i^2\right) \tag{2}$$

denotes the kinetic-energy operator,

$$\hat{W}_{\mathrm{Clb}} = \frac{1}{2}\sum_{\substack{i,j=1\\ i\neq j}}^{N}\frac{e^2}{|\mathbf{r}_i - \mathbf{r}_j|} \tag{3}$$

represents the Coulomb interaction between the electrons, and

$$\hat{V} = \sum_{i=1}^{N} v(\mathbf{r}_i) \tag{4}$$

contains all external potentials acting on the electrons, typically the Coulomb potentials of the nuclei.

Modern DFT is based on the celebrated theorem of Hohenberg and Kohn (HK) [4] which, for systems with nondegenerate ground states, may be summarized by the following three statements:

*Electron Correlations and Materials Properties*, edited by Gonis *et al.*
Kluwer Academic/Plenum Publishers, 1999.

1. The ground-state density $n$ uniquely determines the ground-state wave function $\Psi[n]$ as well as the external potential $v = v[n]$ (up to within an irrelevant additive constant). As a consequence, any observable of a static many-particle system is a functional of its ground-state density.

2. The total-energy functional

$$E_{v_0}[n] := \langle \Psi[n] | \hat{T} + \hat{W}_{\text{Clb}} + \hat{V}_0 | \Psi[n] \rangle \tag{5}$$

of a particular physical system characterized by the external potential $v_0$ is equal to the exact ground-state energy $E_0$ if and only if the exact ground-state density $n_0$ is inserted. For all other densities $n \neq n_0$, the inequality

$$E_0 < E_{v_0}[n] \tag{6}$$

holds. Consequently, the exact ground-state density $n_0$ and the exact ground-state energy $E_0$ can be determined by solving the Euler-Lagrange equation

$$\frac{\delta}{\delta n(\mathbf{r})} E_{v_0}[n] = 0. \tag{7}$$

3. The functional

$$F[n] := \langle \Psi[n] | \hat{T} + \hat{W}_{\text{Clb}} | \Psi[n] \rangle \tag{8}$$

is universal in the sense that it is independent of the external potential $v_0$ of the particular system considered, i.e. it has the same functional form for all systems with a fixed particle-particle interaction ($\hat{W}_{\text{Clb}}$ in our case).

The Hohenberg-Kohn theorem can be viewed as an *exact* reformulation of quantum mechanics, only using the one-particle electronic density instead of the much more complicated many-body wave function. In practice, however, one has to resort to approximations for the functional $F[n]$. A systematic way for constructing such approximations arises from an expansion of $F[n]$ in powers of the Coulomb-coupling constant $e^2$:

$$F[n] = F^{(0)}[n] + e^2 F^{(1)}[n] + e^4 F^{(2)}[n] + \ldots \tag{9}$$

Here, the zeroth-order term is given by the kinetic energy of non-interacting particles,

$$F^{(0)}[n] = T_s[n] \, . \tag{10}$$

In first order one obtains the Hartree- and exchange-energy functionals,

$$e^2 F^{(1)}[n] = \frac{e^2}{2} \int d^3r \int d^3r' \, \frac{n(\mathbf{r})n(\mathbf{r}')}{|\mathbf{r} - \mathbf{r}'|} + E_x[n] \, , \tag{11}$$

and the remaining terms define the correlation-energy functional,

$$\sum_{i=2}^{\infty} (e^2)^i F^{(i)}[n] =: E_c[n] \, . \tag{12}$$

Using this decomposition, the HK total-energy functional (5) can be written as

$$E_{v_0}[n] = T_s[n] + \int d^3r \, n(\mathbf{r}) v_0(\mathbf{r}) + \frac{e^2}{2} \int d^3r \int d^3r' \, \frac{n(\mathbf{r})n(\mathbf{r}')}{|\mathbf{r} - \mathbf{r}'|} + E_{xc}[n] \, , \tag{13}$$

where $E_x$ and $E_c$ have been merged, yielding the so-called exchange-correlation functional $E_{xc}$.

Eq. (13) contains two non-trivial density functionals: the non-interacting kinetic-energy functional $T_s[n]$ and the exchange-correlation-energy functional $E_{xc}[n]$. In historical retrospective we may identify three generations of density functional schemes which may be classified according to the level of approximations used for the universal functionals $T_s[n]$ and $E_{xc}[n]$.

In the *first generation* of DFT, *explicitly* density-dependent functionals were used to approximate both $T_s[n]$ and $E_{xc}[n]$. The simplest approximation of this kind is the Thomas-Fermi model, where $E_{xc}[n]$ is neglected completely and $T_s[n]$ is approximated by

$$T_s^{\mathrm{TF}}[n] = \frac{3}{10}\left(3\pi^2\right)^{2/3}\frac{\hbar^2}{m}\int d^3r\, n(\mathbf{r})^{5/3}. \tag{14}$$

For functionals of this type the HK variational principle (7) can be used directly, leading to equations of the Thomas-Fermi type. As these equations only contain one basic variable, namely the density $n(\mathbf{r})$ of the system, they are readily solved numerically. The results obtained in this way, however, are generally of only moderate accuracy.

The major advance of the *second generation* of DFT which was introduced by Kohn and Sham [5] lies in the use of the *exact* functional for the kinetic energy of non-interacting particles:

$$T_s^{\mathrm{exact}}[n] = \sum_{\substack{i=1\\ \text{lowest } \varepsilon_i}}^{N} \int d^3r\, \varphi_i^*[n](\mathbf{r})\left(-\frac{\hbar^2}{2m}\nabla^2\right)\varphi_i[n](\mathbf{r}), \tag{15}$$

where $\{\varphi_i\}$ are suitably chosen single-particle orbitals (see below).

Although the expression (15) depends explicitly on single-particle orbitals, it is indeed a functional of the density. This can be seen by applying the HK theorem to a system of non-interacting particles: Since the proof of the HK theorem does not depend on the particular form of the particle-particle interaction, it is valid for *any* given interaction $\hat{W}$, in particular also for $\hat{W} \equiv 0$, i.e. for non-interacting systems described by Hamiltonians of the form

$$\hat{H}_s = \hat{T} + \hat{V}_s. \tag{16}$$

Hence the potential $v_s(\mathbf{r})$ is uniquely determined by the ground-state density:

$$v_s(\mathbf{r}) = v_s[n](\mathbf{r}). \tag{17}$$

As a consequence, all single-particle orbitals satisfying the Schrödinger equation

$$\left(-\frac{\hbar^2}{2m}\nabla^2 + v_s[n](\mathbf{r})\right)\varphi_j(\mathbf{r}) = \varepsilon_j\varphi_j(\mathbf{r}) \tag{18}$$

are functionals of the density as well:

$$\varphi_j(\mathbf{r}) = \varphi_j[n](\mathbf{r}). \tag{19}$$

Therefore, *any* functional depending explicitly on the set of single-particle orbitals – like the expression (15) for the kinetic energy – is an (implicit) functional of the density, provided the orbitals come from a local (i.e. multiplicative) potential.

Treating the kinetic-energy functional exactly, the total-energy functional employed in the second generation of DFT becomes:

$$E_{v_0}^{\rm KS}[n] = T_{\rm s}^{\rm exact}[n] + \int d^3r\, v_0(\mathbf{r})n(\mathbf{r}) + \frac{e^2}{2}\int d^3r \int d^3r'\, \frac{n(\mathbf{r})n(\mathbf{r}')}{|\mathbf{r}-\mathbf{r}'|} + E_{\rm xc}[n]\,, \tag{20}$$

where only the xc energy has to be approximated.

In order to calculate the minimizing density of the energy functional (20), we now determine the potential $v_{\rm s}$ of the non-interacting system (16) in such a way that it yields the correct ground-state density of the interacting many-body system:

$$n(\mathbf{r}) = \sum_{\substack{i=1 \\ \text{lowest } \varepsilon_i}}^{N} |\varphi_i(\mathbf{r})|^2\,, \tag{21}$$

where the orbitals satisfy the single-particle Schrödinger equation (18). If such a system exists, the local effective potential $v_{\rm s}[n]$ is uniquely determined and can be obtained from the HK variational principle: Plugging Eq. (20) into Euler equation (7) and using Eq. (18), we obtain

$$v_{\rm s}[n](\mathbf{r}) = v_0(\mathbf{r}) + e^2\int d^3r'\, \frac{n(\mathbf{r}')}{|\mathbf{r}-\mathbf{r}'|} + v_{\rm xc}[n](\mathbf{r}) \tag{22}$$

where we have defined the xc potential as

$$v_{\rm xc}[n](\mathbf{r}) := \frac{\delta E_{\rm xc}[n]}{\delta n(\mathbf{r})}. \tag{23}$$

The single-particle Schrödinger equation (18), together with the effective potential (22) and the prescription for calculating the density (21) are known as Kohn-Sham equations. In practice, these equations have to be solved self-consistently employing approximate but explicitly density-dependent functionals for $E_{\rm xc}[n]$. The simplest and surprisingly successful approximation is the well-known local-density approximation (LDA):

$$E_{\rm xc}^{\rm LDA}[n] = \int d^3r\, n(\mathbf{r})\epsilon_{\rm xc}(n(\mathbf{r}))\,, \tag{24}$$

where $\epsilon_{\rm xc}(n)$ is the known xc energy per particle of the homogeneous electron gas of constant density $n$. More recently, density-gradient-dependent approximations – the so-called generalized gradient approximations (GGAs),

$$E_{\rm xc}^{\rm GGA}[n] = \int d^3r\, n(\mathbf{r})\epsilon_{\rm xc}(n(\mathbf{r}), \nabla n(\mathbf{r}), \ldots)\,, \tag{25}$$

have become increasingly popular and give excellent results for a wide range of atomic, molecular and solid-state systems.

Finally, in the *third generation of DFT,* one employs, in addition to the *exact* expression for $T_{\rm s}$, also the *exact* expression for the exchange energy, given by

$$E_{\rm x}^{\rm exact}[n] = -\frac{e^2}{2}\sum_{\sigma=\uparrow,\downarrow}\sum_{j,k=1}^{N_\sigma}\int d^3r \int d^3r'\, \frac{\varphi_{j\sigma}^*(\mathbf{r})\varphi_{k\sigma}^*(\mathbf{r}')\varphi_{k\sigma}(\mathbf{r})\varphi_{j\sigma}(\mathbf{r}')}{|\mathbf{r}-\mathbf{r}'|}. \tag{26}$$

Only the correlation part of $E_{\rm xc}[n]$ needs to be approximated in this approach. In contrast to the conventional second-generation KS scheme, the third generation allows for the treatment of explicitly orbital-dependent functionals for $E_{\rm c}$ as well, giving more flexibility in the construction of such approximations. Therefore, this scheme naturally lends itself for the use of correlation-energy functionals derived from many-body-perturbation theory. There the correlation energy is given as the sum of all higher-order diagrams in terms of the non-interacting Greens function, which itself is a simple functional of single-particle orbitals.

The central equation in the third generation of DFT is still the KS equation (18). The difference between the second and the third generation lies in the level of approximation to the xc energy. As a consequence of the orbital dependence of $E_{\rm xc}$ in the third generation of DFT the calculation of $v_{\rm xc}[n](\mathbf{r})$ from Eq. (23) is somewhat more complicated. It has to be determined from an integral equation, known as the optimized effective potential (OEP) equation [6, 7]. The OEP integral equation as well as a semi-analytical method of approximately solving it will be discussed in the following section, along with a selection of numerical results.

## THE OEP METHOD

### Derivation of the OEP integral equation

We are going to derive the OEP equations for the spin-dependent version of DFT [8, 9], where the basic variables are the spin-up and spin-down densities $n_\uparrow(\mathbf{r})$ and $n_\downarrow(\mathbf{r})$, respectively. They are obtained by self-consistently solving the single-particle Schrödinger equations (atomic units are used from now on)

$$\left(-\frac{\nabla^2}{2} + v_{s\sigma}[n_\uparrow, n_\downarrow](\mathbf{r})\right) \varphi_{j\sigma}(\mathbf{r}) = \varepsilon_{j\sigma}\varphi_{j\sigma}(\mathbf{r}) , \qquad j = 1, \ldots N_\sigma , \qquad \sigma = \uparrow, \downarrow \tag{27}$$

where

$$n_\sigma(\mathbf{r}) = \sum_{i=1}^{N_\sigma} |\varphi_{i\sigma}(\mathbf{r})|^2. \tag{28}$$

The Kohn-Sham potentials $v_{s\sigma}(\mathbf{r})$ may be written in the usual way as

$$v_{s\sigma}(\mathbf{r}) = v_0(\mathbf{r}) + \int d^3r' \, \frac{n(\mathbf{r}')}{|\mathbf{r} - \mathbf{r}'|} + v_{\rm xc\,\sigma}(\mathbf{r}), \tag{29}$$

where

$$n(\mathbf{r}) = \sum_{\sigma=\uparrow,\downarrow} n_\sigma(\mathbf{r}) , \tag{30}$$

and

$$v_{\rm xc\,\sigma}(\mathbf{r}) := \frac{\delta E_{\rm xc}[n_\uparrow, n_\downarrow]}{\delta n_\sigma(\mathbf{r})}. \tag{31}$$

If the xc-energy functional is now given as an explicit (approximate) functional of spin orbitals and therefore only an *implicit* functional of the spin densities $n_\uparrow$ and $n_\downarrow$,

$$E_{\rm xc}^{\rm OEP} = E_{\rm xc}^{\rm OEP}[\{\varphi_{j\tau}\}] , \tag{32}$$

the evaluation of the functional derivative in Eq. (31) is somewhat more complicated. In order to calculate the xc potentials, we use the chain rule for functional derivatives to obtain

$$v_{\mathrm{xc}\,\sigma}^{\mathrm{OEP}}(\mathbf{r}) = \frac{\delta E_{\mathrm{xc}}^{\mathrm{OEP}}[\{\varphi_{j\tau}\}]}{\delta n_\sigma(\mathbf{r})} \tag{33}$$

$$= \sum_{\alpha=\uparrow,\downarrow}\sum_{\beta=\uparrow,\downarrow}\sum_{i=1}^{N_\alpha}\int d^3r'\int d^3r''\left(\frac{\delta E_{\mathrm{xc}}^{\mathrm{OEP}}[\{\varphi_{j\tau}\}]}{\delta\varphi_{i\alpha}(\mathbf{r}')}\frac{\delta\varphi_{i\alpha}(\mathbf{r}')}{\delta v_{\mathrm{s}\,\beta}(\mathbf{r}'')}+\mathrm{c.c.}\right)\frac{\delta v_{\mathrm{s}\,\beta}(\mathbf{r}'')}{\delta n_\sigma(\mathbf{r})}\,.$$

The last term on the right-hand side is the inverse $\chi_{\mathrm{s}}^{-1}$ of the static density response function of a system of non-interacting particles

$$\chi_{\mathrm{s}\,\alpha\beta}(\mathbf{r},\mathbf{r}') := \frac{\delta n_\alpha(\mathbf{r})}{\delta v_{\mathrm{s}\,\beta}(\mathbf{r}')}. \tag{34}$$

This quantity is diagonal with respect to the spin variables so that Eq. (33) reduces to

$$v_{\mathrm{xc}\,\sigma}^{\mathrm{OEP}}(\mathbf{r}) = \sum_{\alpha=\uparrow,\downarrow}\sum_{i=1}^{N_\alpha}\int d^3r'\int d^3r''\left(\frac{\delta E_{\mathrm{xc}}^{\mathrm{OEP}}[\{\varphi_{j\tau}\}]}{\delta\varphi_{i\alpha}(\mathbf{r}')}\frac{\delta\varphi_{i\alpha}(\mathbf{r}')}{\delta v_{\mathrm{s}\,\sigma}(\mathbf{r}'')}+\mathrm{c.c.}\right)\chi_{\mathrm{s}\,\sigma}^{-1}(\mathbf{r}'',\mathbf{r})\,. \tag{35}$$

Acting with the response operator (34) on both sides of Eq. (35) one obtains

$$\int d^3r'\; v_{\mathrm{xc}\,\sigma}^{\mathrm{OEP}}(\mathbf{r}')\chi_{\mathrm{s}\,\sigma}(\mathbf{r}',\mathbf{r}) = \sum_{\alpha=\uparrow,\downarrow}\sum_{i=1}^{N_\alpha}\int d^3r'\;\frac{\delta E_{\mathrm{xc}}^{\mathrm{OEP}}[\{\varphi_{j\tau}\}]}{\delta\varphi_{i\alpha}(\mathbf{r}')}\frac{\delta\varphi_{i\alpha}(\mathbf{r}')}{\delta v_{\mathrm{s}\,\sigma}(\mathbf{r})}+\mathrm{c.c.} \tag{36}$$

To further evaluate this equation, we note that the first functional derivative on the right-hand side of Eq. (36) is readily computed once an explicit expression for $E_{\mathrm{xc}}^{\mathrm{OEP}}$ in terms of single-particle orbitals is given. The remaining functional derivative on the right-hand side of Eq. (36) can be exactly calculated by first-order perturbation theory. This yields

$$\frac{\delta\varphi_{i\alpha}(\mathbf{r}')}{\delta v_{\mathrm{s}\,\sigma}(\mathbf{r})} = \delta_{\alpha,\sigma}\sum_{\substack{k=1\\k\neq i}}^{\infty}\frac{\varphi_{k\sigma}(\mathbf{r}')\varphi_{k\sigma}^*(\mathbf{r})}{\varepsilon_{i\sigma}-\varepsilon_{k\sigma}}\varphi_{i\sigma}(\mathbf{r}). \tag{37}$$

Using this equation, the response function

$$\chi_{\mathrm{s}\,\alpha\beta}(\mathbf{r},\mathbf{r}') = \frac{\delta}{\delta v_{\mathrm{s}\,\beta}(\mathbf{r}')}\left(\sum_{i=1}^{N_\alpha}\varphi_{i\alpha}^*(\mathbf{r})\varphi_{i\alpha}(\mathbf{r})\right) \tag{38}$$

is readily expressed in terms of the orbitals as

$$\chi_{\mathrm{s}\,\sigma}(\mathbf{r},\mathbf{r}') = \sum_{i=1}^{N_\sigma}\sum_{\substack{k=1\\k\neq i}}^{\infty}\frac{\varphi_{i\sigma}^*(\mathbf{r})\varphi_{k\sigma}(\mathbf{r})\varphi_{k\sigma}^*(\mathbf{r}')\varphi_{i\sigma}(\mathbf{r}')}{\varepsilon_{i\sigma}-\varepsilon_{k\sigma}}+\mathrm{c.c.} \tag{39}$$

Inserting (37) and (39) in Eq. (36), we obtain the standard form of the OEP integral equation:

$$\sum_{i=1}^{N_\sigma}\int d^3r'\varphi_{i\sigma}^*(\mathbf{r}')\left(v_{\mathrm{xc}\,\sigma}^{\mathrm{OEP}}(\mathbf{r}')-u_{\mathrm{xc}i\sigma}(\mathbf{r}')\right)G_{\mathrm{s}i\sigma}(\mathbf{r}',\mathbf{r})\,\varphi_{i\sigma}(\mathbf{r})+\mathrm{c.c.}=0 \tag{40}$$

where

$$u_{\mathrm{xc}i\sigma}(\mathbf{r}) := \frac{1}{\varphi_{i\sigma}^*(\mathbf{r})}\frac{\delta E_{\mathrm{xc}}^{\mathrm{OEP}}[\{\varphi_{j\tau}\}]}{\delta\varphi_{i\sigma}(\mathbf{r})} \tag{41}$$

and

$$G_{\mathrm{s}i\sigma}(\mathbf{r},\mathbf{r}') := \sum_{\substack{k=1\\k\neq i}}^{\infty} \frac{\varphi_{k\sigma}(\mathbf{r})\varphi^*_{k\sigma}(\mathbf{r}')}{\varepsilon_{i\sigma}-\varepsilon_{k\sigma}}. \tag{42}$$

This integral equation has to be solved simultaneously with the KS equation (27).

The derivation of the OEP integral equation (40) described here was given by Shaginyan [10] and by Görling and Levy [11]. It is important to note that the same expression results [6, 7, 12, 13, 14, 15] if one demands that the local single-particle potential appearing in Eq. (27) be the *optimized* one yielding orbitals minimizing the total-energy functional, i.e., that

$$\left.\frac{\delta E^{\mathrm{OEP}}_{v_0}}{\delta v_{\mathrm{s}\,\sigma}(\mathbf{r})}\right|_{v_{\mathrm{s}\,\sigma}=v^{\mathrm{OEP}}} = 0. \tag{43}$$

This equation is the historical origin [6] of the name *optimized effective potential.* The name might be somewhat misleading, since *no additional* optimization for the potential is used. Indeed, as was first pointed out by Perdew and co-workers [16, 17], Eq. (43) is equivalent to the HK variational principle. This is most easily seen by applying the functional chain rule to Eq. (43) yielding

$$0 = \frac{\delta E^{\mathrm{OEP}}_{v_0}}{\delta v_{\mathrm{s}\,\sigma}(\mathbf{r})} = \sum_{\alpha}\int d^3r'\, \frac{\delta E^{\mathrm{OEP}}_{v_0}}{\delta n_\alpha(\mathbf{r}')}\frac{\delta n_\alpha(\mathbf{r}')}{\delta v_{\mathrm{s}\,\sigma}(\mathbf{r})}. \tag{44}$$

Once again, the last term on the right-hand side of Eq. (44) can be identified with the static KS response function (34). Hence, acting with the inverse response operator on Eq. (44) leads to the HK variational principle

$$0 = \frac{\delta E^{\mathrm{OEP}}_{v_0}}{\delta n_\sigma(\mathbf{r})}. \tag{45}$$

**Approximation of Krieger, Li, and Iafrate**

The OEP method discussed in the preceding paragraph opens up a way for using orbital-dependent functionals within the KS scheme. However, as a price to pay, one has to solve an integral equation self-consistently with the KS equation. Due to the rather large computational effort involved in this scheme, it has not been used extensively. Indeed, full solutions of the OEP integral equation have been achieved so far only for systems of high symmetry such as spherical atoms [13, 7, 18, 19, 20]. For solids, further numerical simplifications have been employed, such as the atomic sphere approximation [21, 22, 23, 24] or the use of pseudopotentials for the core electrons [25]. Clearly, practical applications of the OEP scheme to a greater variety of systems require some simplification.

Krieger, Li, and Iafrate (KLI) have suggested an approximation leading to a highly accurate but numerically tractable scheme preserving many important properties of the exact OEP method [13, 26, 21, 27, 28, 18, 19, 12, 29]. It is most easily derived by replacing the energy denominator of the Greens function (42) by a single constant, i.e.

$$G_{\mathrm{s}i\sigma}(\mathbf{r}',\mathbf{r}) \approx \frac{1}{\Delta\varepsilon}\left(\delta(\mathbf{r}'-\mathbf{r}) - \varphi_{i\sigma}(\mathbf{r})'\varphi^*_{i\sigma}(\mathbf{r})\right). \tag{46}$$

Substituting this into the OEP integral equation (40) leads to an approximate equation, known as the KLI approximation:

$$v^{\mathrm{KLI}}_{\mathrm{xc}\,\sigma}(\mathbf{r}) = \frac{1}{2n_\sigma(\mathbf{r})}\sum_{i=1}^{N_\sigma} |\varphi_{i\sigma}(\mathbf{r})|^2 \left[u_{\mathrm{xc}i\sigma}(\mathbf{r}) + (\bar{v}^{\mathrm{KLI}}_{\mathrm{xc}i\sigma} - \bar{u}_{\mathrm{xc}i\sigma})\right] + c.c. \tag{47}$$

where $\bar{v}^{\rm KLI}_{{\rm xc}i\sigma}$ is defined as

$$\bar{v}^{\rm KLI}_{{\rm xc}i\sigma} := \int d^3r \varphi^*_{i\sigma}(\mathbf{r}) v^{\rm KLI}_{{\rm xc}\,\sigma}(\mathbf{r}) \varphi_{i\sigma}(\mathbf{r}) \,. \tag{48}$$

In contrast to the full OEP equation (40), the KLI equation, still being an integral equation, can be solved explicitly in terms of the orbitals $\{\varphi_{i\sigma}\}$: Multiplying Eq. (47) by $|\varphi_{j\sigma}(\mathbf{r})|^2$ and integrating over space yields

$$\bar{v}^{\rm KLI}_{{\rm xc}j\sigma} = \bar{v}^{\rm S}_{{\rm xc}j\sigma} + \sum_{i=1}^{N_\sigma - 1} N_{ji\sigma} \left( \bar{v}^{\rm KLI}_{{\rm xc}i\sigma} - \frac{1}{2} \left( \bar{u}_{{\rm xc}i\sigma} + \bar{u}^*_{{\rm xc}i\sigma} \right) \right), \tag{49}$$

where

$$\bar{v}^{\rm S}_{{\rm xc}j\sigma} := \int d^3r \frac{|\varphi_{j\sigma}(\mathbf{r})|^2}{n_\sigma(\mathbf{r})} \sum_{i=1}^{N_\sigma} |\varphi_{i\sigma}(\mathbf{r})|^2 \frac{1}{2} \left( u_{{\rm xc}i\sigma}(\mathbf{r}) + u^*_{{\rm xc}i\sigma}(\mathbf{r}) \right) \tag{50}$$

and

$$N_{ji\sigma} := \int d^3r \frac{|\varphi_{j\sigma}(\mathbf{r})|^2 |\varphi_{i\sigma}(\mathbf{r})|^2}{n_\sigma(\mathbf{r})}. \tag{51}$$

The term corresponding to the highest occupied orbital $\varphi_{N_\sigma \sigma}$ has been excluded from the sum in Eq. (49) because $\bar{v}^{\rm KLI}_{{\rm xc}N_\sigma\sigma} = \bar{u}_{{\rm xc}N_\sigma\sigma}$ [13]. The remaining unknown constants $(\bar{v}^{\rm KLI}_{{\rm xc}i\sigma} - \bar{u}_{{\rm xc}i\sigma})$ are determined by the linear equation

$$\sum_{i=1}^{N_\sigma - 1} (\delta_{ji} - N_{ji\sigma}) \left( \bar{v}^{\rm KLI}_{{\rm xc}i\sigma} - \frac{1}{2} \left( \bar{u}_{{\rm xc}i\sigma} + \bar{u}^*_{{\rm xc}i\sigma} \right) \right) = \left( \bar{v}^{\rm S}_{{\rm xc}j\sigma} - \frac{1}{2} \left( \bar{u}_{{\rm xc}j\sigma} + \bar{u}^*_{{\rm xc}j\sigma} \right) \right) \tag{52}$$

with $j = 1, \ldots N_\sigma - 1$. Solving Eq. (52) and substituting the result into Eq. (47), we obtain an explicitly orbital dependent functional. The approximation (46) might appear rather crude. However, it can be justified by a much more rigorous derivation, showing that the KLI equation can be interpreted as a mean-field type approximation [13, 12].

We finally mention that the OEP and KLI methods have recently been generalized to relativistic situations [30, 31, 32], which arise e.g. in the treatment of heavier systems.

### Rigorous properties of the OEP and KLI methods

To conclude this section, we summarize some exact properties of the OEP and KLI methods.

1. No spurious self-interactions

   Since both the OEP and the KLI method are constructed to deal with orbital-dependent functionals, one can employ xc-energy functionals – like the expression for the exact exchange-energy functional – which exactly cancel the spurious self-interaction contained in the Hartree term.

2. Asymptotics

   If the exact expression for the exchange energy is used, both the OEP and the KLI potential fall of as $-1/r$ as $r \to \infty$ (for finite systems) [13, 33], thus reflecting the proper cancelation of self-interaction errors. One has to emphasize that this holds true for all orbitals (i.e. for the occupied as well as the unoccupied orbitals)

**Table 1.** Various self-consistently calculated x-only results for the Ar atom. All values in atomic units.

| | OEP | KLI | B88 | xPW91 | xLDA |
|---|---|---|---|---|---|
| $E_{tot}$ | -526.8122 | -526.8105 | -526.7998 | -526.7710 | -524.5174 |
| $\varepsilon_{1s}$ | -114.4524 | -114.4279 | -114.1890 | -114.1887 | -113.7159 |
| $\varepsilon_{2s}$ | -11.1534 | -11.1820 | -10.7911 | -10.7932 | -10.7299 |
| $\varepsilon_{2p}$ | -8.7339 | -8.7911 | -8.4107 | -8.4141 | -8.3782 |
| $\varepsilon_{3s}$ | -1.0993 | -1.0942 | -0.8459 | -0.8481 | -0.8328 |
| $\varepsilon_{3p}$ | -0.5908 | -0.5893 | -0.3418 | -0.3441 | -0.3338 |
| $\varepsilon_{4s}$ | -0.1607 | -0.1616 | -0.0102 | -0.0122 | -0.0014 |
| $< r^2 >$ | 1.4465 | 1.4467 | 1.4791 | 1.4876 | 1.4889 |
| $< r^{-1} >$ | 3.8736 | 3.8738 | 3.8731 | 3.8729 | 3.8648 |
| $n(0)$ | 3839.7 | 3832.6 | 3847.3 | 3847.0 | 3818.7 |

because the KS potential is the same for all orbitals. In contrast, the Hartree-Fock (HF) potential falls off like $-1/r$ only for the occupied, but *not* for the unoccupied orbitals, leading to much too weakly bound virtual HF orbitals. As a consequence, the OEP orbitals and orbital energies can serve as a much better starting point for the calculation of excited state properties.

3. Derivative discontinuities

   If one considers the exact xc potential as a function of the particle number $N$, it exhibits discontinuities at integer values of $N$ [34]. This has important consequences for the values of band gaps in insulators of semi-conductors (for a detailed description see e.g. [2]). Neither the LDA nor the GGAs can reproduce these discontinuities. To date, the OEP and the KLI potential are the only known approximations of $v_{xc}(\mathbf{r})$ reproducing this feature.

## Selected results

We begin with a comparison of x-only results. The x-only limit of the xc-energy functional is given by the exact Fock term, Eq. (26). As explained in the preceding section, the OEP method then provides the corresponding exchange potential $v_x^{exact}(\mathbf{r})$ and therefore represents the exact implementation of x-only DFT. Consequently it provides a benchmark for testing approximate exchange-energy functionals employed within the Kohn-Sham scheme. In Table 1 we show, as a typical example, various results for the argon atom obtained with different x-only methods. Besides the OEP and KLI methods employing the exact exchange-energy functional (26) described above, we also list results from conventional KS-DFT obtained with the approximations due to Becke (B88) [35], Perdew and Wang (PW91) [36], and from the well-known x-only LDA approximation. The KLI results given in the second column of Table 1 clearly demonstrate the high quality of the KLI approximation as all results differ only slightly from the exact OEP ones. For all standard DFT methods, the disagreement is much more pronounced, especially for the highest occupied orbital energies and even more so for the unoccupied ones. As already mentioned, the reason for this lies in the asymptotic behavior of the corresponding exchange potentials: While both the OEP and the KLI potential yield the correct $-1/r$ decay for large $r$ and therefore become indistinguishable in this region, the potentials derived from the conventional density-dependent approximations fall off much too rapidly as can be seen from Figure 1.

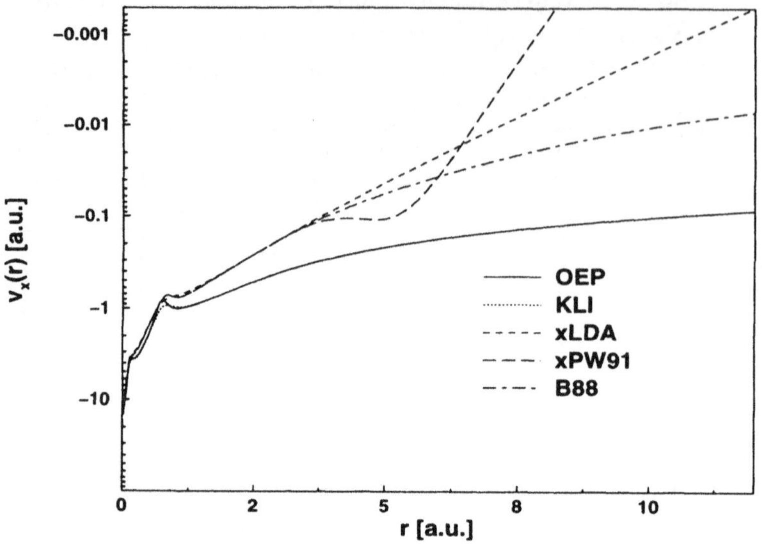


**Figure 1.** Exchange potentials $v_x(r)$ for Ar from various self-consistent x-only calculations in the valence region.

If one aims at the description of heavier elements one should take relativistic effects into account. As shown in Ref. [37, 31, 32], the derivation based on relativistic DFT (for the case of vanishing external magnetic fields) leads to the same expressions for the OEP integral equation (40) and the KLI approximation (47), only with the KS spin orbitals replaced by relativistic KS spinors. Fully relativistic calculations then yield results similar to the non-relativistic case: The data obtained from the relativistic x-only OEP and KLI methods agree closely with each other, while the relativistic LDA [38, 39] and GGA [40] approximations yield larger errors. This becomes evident from Table 2, where we list the relativistic contribution to the total energy, i.e. the difference between the total energy obtained from fully relativistic and non-relativistic calculations: The second column of Table 2 shows that the relativistic version of the KLI method introduces almost no additional errors, thus maintaining the quality of the non-relativistic KLI approximation. In contrast, the conventional density-dependent approximations – in addition to the deviations already known from the non-relativistic case – are not able to reproduce the effects induced by relativity with the same degree of accuracy.

The inclusion of correlation effects into the OEP scheme is straightforward, as already indicated in the introduction. As we will demonstrate in this section, the correlation-energy functional developed by Colle and Salvetti (CS) [41, 42] is well suited for atomic systems. This functional has been obtained through a series of approximations starting from a Jastrow-type ansatz for the correlated total wave function . It may be written in spin-polarized form as [43]

$$E_c^{CS}[\{\varphi_{j\tau}\}] = -ab\int d^3r\ \gamma(\mathbf{r})\xi(\mathbf{r})\Big[\sum_\sigma n_\sigma(\mathbf{r})\sum_i |\nabla\varphi_{i\sigma}(\mathbf{r})|^2 - \frac{1}{4}|\nabla n(\mathbf{r})|^2 - \frac{1}{4}\sum_\sigma n_\sigma(\mathbf{r})\Delta n_\sigma(\mathbf{r}) + \frac{1}{4}n(\mathbf{r})\Delta n(\mathbf{r})\Big] - a\int d^3r\ \gamma(\mathbf{r})\frac{n(\mathbf{r})}{\eta(\mathbf{r})}, \tag{53}$$

**Table 2.** Relativistic Contribution to the total energy calculated from various self-consistent x-only approximations. $\bar{\Delta}$ denotes the mean absolute deviation and $\bar{\delta}$ the average relative deviation (in 0.1 percent) from the exact ROEP values. All numbers in atomic units. Taken from [31].

| | ROEP | RKLI | RB88 | xRLDA |
|---|---|---|---|---|
| He | 0.000 | 0.000 | 0.000 | 0.000 |
| Be | 0.003 | 0.003 | 0.003 | 0.002 |
| Ne | 0.145 | 0.145 | 0.145 | 0.138 |
| Mg | 0.320 | 0.320 | 0.321 | 0.308 |
| Ar | 1.867 | 1.867 | 1.867 | 1.821 |
| Ca | 2.953 | 2.953 | 2.952 | 2.888 |
| Zn | 16.770 | 16.770 | 16.779 | 16.555 |
| Kr | 36.820 | 36.820 | 36.822 | 36.432 |
| Sr | 46.553 | 46.553 | 46.552 | 46.092 |
| Pd | 106.526 | 106.526 | 106.526 | 105.715 |
| Cd | 128.243 | 128.243 | 128.243 | 127.323 |
| Xe | 214.858 | 214.858 | 214.825 | 213.522 |
| Ba | 252.222 | 252.221 | 252.176 | 250.725 |
| Yb | 676.551 | 676.549 | 676.590 | 673.785 |
| Hg | 1240.513 | 1240.511 | 1240.543 | 1236.349 |
| Rn | 1736.144 | 1736.142 | 1736.151 | 1730.890 |
| Ra | 1934.770 | 1934.768 | 1934.781 | 1929.116 |
| No | 3953.155 | 3953.151 | 3953.979 | 3944.569 |
| $\bar{\Delta}$ | | 0.001 | 0.056 | 1.788 |
| $\bar{\delta}$ | | 0.009 | 1.14 | 33.7 |

**Table 3.** Total absolute ground-state energies for first-row atoms from various self-consistent calculations. Quantum chemistry (QC) values from [45]. $\bar{\Delta}$ denotes the mean absolute deviation from the exact non-relativistic values [46]. All numbers in Hartrees. Taken from [15] and modified.

| | KLICS | xcLDA | BLYP | PW91 | QC | exact |
|---|---|---|---|---|---|---|
| He | 2.9033 | 2.8346 | 2.9071 | 2.9000 | 2.9049 | 2.9037 |
| Li | 7.4829 | 7.3433 | 7.4827 | 7.4742 | 7.4743 | 7.4781 |
| Be | 14.6651 | 14.4465 | 14.6615 | 14.6479 | 14.6657 | 14.6674 |
| B | 24.6564 | 24.3525 | 24.6458 | 24.6299 | 24.6515 | 24.6539 |
| C | 37.8490 | 37.4683 | 37.8430 | 37.8265 | 37.8421 | 37.8450 |
| N | 54.5905 | 54.1344 | 54.5932 | 54.5787 | 54.5854 | 54.5893 |
| O | 75.0717 | 74.5248 | 75.0786 | 75.0543 | 75.0613 | 75.067 |
| F | 99.7302 | 99.1112 | 99.7581 | 99.7316 | 99.7268 | 99.734 |
| Ne | 128.9202 | 128.2299 | 128.9730 | 128.9466 | 128.9277 | 128.939 |
| $\bar{\Delta}$ | 0.0047 | 0.3813 | 0.0108 | 0.0114 | 0.0045 | |

where

$$\gamma(\mathbf{r}) = 4\,\frac{n_\uparrow(\mathbf{r})n_\downarrow(\mathbf{r})}{n(\mathbf{r})^2}, \tag{54}$$

$$\eta(\mathbf{r}) = 1 + d\,n(\mathbf{r})^{-\frac{1}{3}}, \tag{55}$$

$$\xi(\mathbf{r}) = \frac{n(\mathbf{r})^{-\frac{5}{3}} e^{-cn(\mathbf{r})^{-\frac{1}{3}}}}{\eta(\mathbf{r})}. \tag{56}$$

The constants $a$, $b$, $c$, and $d$ are given by

$$a = 0.04918, \quad b = 0.132,$$
$$c = 0.2533, \quad d = 0.349.$$

In Table 3 we compare total ground state energies of first-row atoms calculated self-consistently with various approximations. The first column, headed KLICS, shows the results from the KLI method employing the exact exchange-energy functional (26) plus the CS correlation-energy functional, while the next columns show conventional KS results. The latter were obtained with an LDA functional using the parametrization of the correlation energy of a homogeneous electron gas by Perdew and Wang [44]; the x-energy functional due to Becke [35] combined with the c-energy functional of Lee, Yang, and Parr [43], denoted as BLYP; and the generalized gradient approximation due to Perdew and Wang [36], referred to as PW91. The quantum chemistry values, headed QC, are from a highly accurate complete basis set quadratic CI calculation using atomic pair natural orbitals [45]. The exact non-relativistic energies, i.e. the experimental values with relativistic effects subtracted, have been taken from [46]. The mean absolute deviations of the calculated from the exact values, denoted by $\bar{\Delta}$, clearly show that the KLICS approach is significantly more accurate than the conventional KS methods and almost as accurate as the traditional QC scheme. We emphasize that the numerical effort involved in the KLICS scheme for atoms is only slightly higher than in the LDA and GGA schemes.

Apart from total energies, the highest occupied orbital energies, which should be equal to the exact ionization potential in an exact implementation of DFT, are much closer to the experimental ionization potentials in the KLICS scheme than in the conventional KS approaches. This is shown in Table 4: All the numbers of the conventional KS calculations are off by about 100%, which can be traced back to the wrong asymptotic behavior of the xc-potentials in these approximations. Again, only the KLICS approximation provides the correct $-1/r$ decay for $r \to \infty$, which results in much more reliable predictions for quantities depending strongly on this region of space. As shown here, this is of crucial importance for the ionization potential, but it also becomes important, when excitation energies are calculated from DFT, as we will show later.

The feasibility of the KLI approximation for more complex systems has been shown recently [47, 48, 32] by x-only and KLICS calculations for diatomic molecules. To this end, we have implemented these approaches in our fully numerical basis-set-free code for diatomic molecules. For comparison, we have also performed calculations employing the conventional LDA [44] and the PW91 functional for $E_{\mathrm{xc}}$.

In Table 5 we display results for the bond lengths. It is apparent that the KLICS scheme leads to equilibrium distances which are generally too short, an effect present

**Table 4.** Ionization potentials from the highest occupied orbital energy of neutral atoms. $\bar{\Delta}$ denotes the mean absolute deviation from the experimental values, taken from [49]. All values in Hartrees. Taken from [15] and modified.

| | KLICS | xcLDA | BLYP | PW91 | experiment |
|---|---|---|---|---|---|
| He | 0.945 | 0.570 | 0.585 | 0.583 | 0.903 |
| Li | 0.200 | 0.116 | 0.111 | 0.119 | 0.198 |
| Be | 0.329 | 0.206 | 0.201 | 0.207 | 0.343 |
| B | 0.328 | 0.151 | 0.143 | 0.149 | 0.305 |
| C | 0.448 | 0.228 | 0.218 | 0.226 | 0.414 |
| N | 0.579 | 0.309 | 0.297 | 0.308 | 0.534 |
| O | 0.559 | 0.272 | 0.266 | 0.267 | 0.500 |
| F | 0.714 | 0.384 | 0.376 | 0.379 | 0.640 |
| Ne | 0.884 | 0.498 | 0.491 | 0.494 | 0.792 |
| Na | 0.189 | 0.113 | 0.106 | 0.113 | 0.189 |
| Mg | 0.273 | 0.175 | 0.168 | 0.174 | 0.281 |
| Al | 0.222 | 0.111 | 0.102 | 0.112 | 0.220 |
| Si | 0.306 | 0.170 | 0.160 | 0.171 | 0.300 |
| P | 0.399 | 0.231 | 0.219 | 0.233 | 0.385 |
| S | 0.404 | 0.228 | 0.219 | 0.222 | 0.381 |
| Cl | 0.506 | 0.305 | 0.295 | 0.301 | 0.477 |
| Ar | 0.619 | 0.382 | 0.373 | 0.380 | 0.579 |
| $\bar{\Delta}$ | 0.030 | 0.176 | 0.183 | 0.177 | |

**Table 5.** Calculated bond lengths of the closed-shell-first-row dimers and hydrides. HF values taken from [50]. Experimental values from [51] except where noted. All values in atomic units. Taken from [48].

| | KLICS | PW91 | xcLDA | HF | experiment |
|---|---|---|---|---|---|
| $H_2$ | 1.378 | 1.414 | 1.446 | 1.379 | 1.401[a] |
| $Li_2$ | 5.086 | 5.153 | 5.120 | 5.304 | 5.051 |
| $Be_2$ | -[b] | 4.588 | 4.522 | - | 4.63[c] |
| $C_2$ | 2.306 | 2.367 | 2.354 | - | 2.3481 |
| $N_2$ | 1.998 | 2.079 | 2.068 | 2.037 | 2.074 |
| $F_2$ | 2.465 | 2.669 | 2.615 | 2.542 | 2.6682 |
| LiH | 2.971 | 3.030 | 3.030 | 3.092 | 3.0154 |
| BH | 2.274 | 2.356 | 2.373 | - | 2.3289 |
| FH | 1.684 | 1.756 | 1.761 | 1.722 | 1.7325 |

[a] Exact value from [52].
[b] There is no local minimum in the electronic potential curve.
[c] From [53].

**Table 6.** Absolute total ground-state energies of the closed-shell-first-row dimers and hydrides calculated at the bond lengths given in Table 5. Estimates for exact values calculated using dissociation energies from Table 7 and nonrelativistic, infinite nuclear mass atomic ground-state energies from [54]. All numbers in atomic units. Taken from [48].

| | KLICS | PW91 | xcLDA | exact |
|---|---|---|---|---|
| $H_2$ | 1.171444 | 1.170693 | 1.137692 | 1.174448[a] |
| $Li_2$ | 14.9982 | 14.9819 | 14.7245 | 14.9954 |
| $Be_2$ | 29.3197[b] | 29.3118 | 28.9136 | 29.3385 |
| $C_2$ | 75.7736 | 75.8922 | 75.2041 | 75.922 |
| $N_2$ | 109.4683 | 109.5449 | 108.6959 | 109.5424 |
| $F_2$ | 199.4377 | 199.5699 | 198.3486 | 199.5299 |
| LiH | 8.0723 | 8.0625 | 7.9189 | 8.0705 |
| BH | 25.2857 | 25.2688 | 24.9770 | 25.29 |
| FH | 100.4241 | 100.4715 | 99.8490 | 100.4596 |

[a] Exact value from [52].
[b] Calculated at the experimental bond length 4.63 a.u.

in the HF approximation as well. Most notable is the fact that the potential energy curve of $Be_2$ displays no local minimum in this approximation. Good agreement with experiment is found only for $Li_2$. Except for this molecule, the PW91 values are clearly superior.

Total absolute ground-state energies calculated at the bond lengths given in Table 5 are shown in Table 6. For the lighter molecules $H_2$, $Li_2$, $Be_2$, LiH, and BH the KLICS and PW91 results are of the same good quality, yielding errors of a few mHartrees. For the heavier molecules, however, the KLICS results are worse. Being the simplest approximation, it is not surprising that the LDA gives values for the total energies which show the largest errors.

Apart from $H_2$ and LiH, the dissociation energies obtained within the KLICS approach are disappointing, as may be seen in Table 7. In most cases, the magnitude is underestimated considerably and for $Be_2$ and $F_2$ even the wrong sign is obtained. Since the corresponding *atomic* ground-state energies given in the previous section are of excellent quality, the error must be due to correlation effects present in molecules only. In particular, the left-right correlation error [56] well-known in HF theory also occurs in DFT when the exact Fock expression (26) for $E_x$ is employed. Apparently, this error is not sufficiently corrected for by the CS functional. The LDA and PW91 results are clearly much better, the latter reducing the over-binding tendencies of the former.

In the asymptotic region, however, the KLICS xc potential is of better quality than the conventional xc potentials: In Table 8 we list the absolute values of the highest occupied molecular orbital energies, which, in *exact* DFT, should be identical with the ionization potentials of the systems under consideration. It is evident that the conventional KS approaches represented by the LDA and PW91 functionals yield results which are typically 40 percent too high, while the KLICS values are much closer to the experimental results. As for atomic systems, this fact may be traced back to the correct asymptotic behavior of the KLICS xc potential for large $r$.

**Table 7.** Dissociation energies of the closed-shell-first-row dimers and hydrides calculated at the bond lengths given in Table 5. HF values taken from [50]. All numbers in mHartrees. Taken from [48].

| | KLICS | PW91 | xcLDA | HF | experiment |
|---|---|---|---|---|---|
| $H_2$ | 171.444 | 167.665 | 180.270 | 121.0 | 174.475[b] |
| $Li_2$ | 32.4 | 33.5 | 37.9 | 3.5 | 39.3[c] |
| $Be_2$ | -10.5[a] | 15.9 | 20.6 | - | 3.8[c] |
| $C_2$ | 75.6 | 239.2 | 267.5 | - | 232[d] |
| $N_2$ | 287.3 | 387.5 | 427.1 | 167.5 | 364.0[d] |
| $F_2$ | -22.7 | 106.7 | 126.2 | -54.7 | 62.1[d] |
| LiH | 89.4 | 86.8 | 96.9 | 48.4 | 92.4[d] |
| BH | 129.3 | 137.4 | 145.8 | - | 135[e] |
| FH | 193.9 | 238.4 | 259.1 | 130.8 | 225.7[e] |

[a] Calculated at the experimental bond length 4.63 a.u.
[b] Exact value from [52].
[c] From [53].
[d] From [51].
[e] From [55].

**Table 8.** Absolute values for the highest occupied orbital energies of the closed-shell-first-row dimers and hydrides calculated at the bond lengths given in Table 5. Experimental values are the ionization potentials taken from [51]. All numbers in atomic units. Taken from [48].

| | KLICS | PW91 | xcLDA | experiment |
|---|---|---|---|---|
| $H_2$ | 0.621563 | 0.382656 | 0.373092 | 0.5669 |
| $Li_2$ | 0.1974 | 0.1187 | 0.1187 | 0.18 |
| $Be_2$ | 0.2560[a] | 0.1678 | 0.1660 | - |
| $C_2$ | 0.4844 | 0.2942 | 0.2987 | 0.4465 |
| $N_2$ | 0.6643 | 0.3804 | 0.3826 | 0.5726 |
| $F_2$ | 0.6790 | 0.3512 | 0.3497 | 0.5764 |
| LiH | 0.3237 | 0.1621 | 0.1612 | 0.283[b] |
| BH | 0.3692 | 0.2058 | 0.2041 | 0.359 |
| FH | 0.6803 | 0.3567 | 0.3594 | 0.5894 |

[a] Calculated at the experimental bond length 4.63 a.u.
[b] From [57].

## TIME-DEPENDENT SYSTEMS

Density functional theory can be generalized to allow for the description of time-dependent many-particle systems [58, 59, 60, 61]. An atom being placed in the focus of a laser field constitutes an example of such a time-dependent system. Owing to rapid experimental progress in the field of laser physics, femto-second laser pulses of very high intensity have become available in recent years. The electric field produced in such pulses can reach or even exceed the strength of the static nuclear Coulomb field as experienced by an electron on the lowest Bohr orbit. If an atomic system is placed in the focus of such a pulse, one observes a wealth of new phenomena [62, 63, 64] which cannot be explained by traditional perturbation theory. In principle, the solution of the fully interacting time-dependent Schrödinger equation is required. The aim of a density-functional description in this intensity regime is to "replace" the solution of the full time-dependent Schrödinger equation by the solution of time-dependent Kohn-Sham equations.

Traditional spectroscopy, on the other hand, mainly aims at the interaction of low-intensity laser fields with matter, in order to probe properties of the undisturbed system. Theoretical descriptions in this intensity regime are typically based on the calculation of the linear density response $n_1(\mathbf{r}, t)$ to a weak external perturbation

$$v_1(\mathbf{r}, t) = \mathcal{E} z \cos(\omega t) \tag{57}$$

describing an electric dipole field of amplitude $\mathcal{E}$ and frequency $\omega$, linearly polarized along the $z$-direction. The dynamical polarizability then follows from

$$\alpha(\omega) = -\frac{1}{\mathcal{E}} \int d^3r \, z \, n_1(\mathbf{r}, \omega) \,. \tag{58}$$

This quantity is directly related to the photo-absorption cross section $\sigma(\omega)$ via

$$\sigma(\omega) = -\frac{4\pi\omega}{c} \Im \alpha(\omega) \,. \tag{59}$$

### Time-dependent Hohenberg-Kohn-Sham formalism

In contrast to the ground-state formalism [4], where the existence proof relies on the Rayleigh-Ritz principle for the energy, the proof for time-dependent systems [58] is based directly on the Schrödinger equation

$$i\frac{\partial}{\partial t}\Psi(t) = \hat{H}(t)\Psi(t) \tag{60}$$

where

$$\hat{H}(t) = \hat{T} + \hat{W}_{\mathrm{Clb}} + \hat{V}(t) \,. \tag{61}$$

We shall investigate the densities $n(\mathbf{r}, t)$ of electronic systems evolving from an *arbitrary but fixed* initial (many-particle) state

$$\Psi(t_0) = \Psi_0 \tag{62}$$

under the influence of different external potentials of the form

$$\hat{V}(t) = \sum_{i=1}^{N} v(\mathbf{r}_i, t) \,. \tag{63}$$

Following the proof first given by Runge and Gross [58], one can formulate the following Hohenberg-Kohn-type theorem: The densities $n(\mathbf{r},t)$ and $n'(\mathbf{r},t)$ evolving from a common initial state $\Psi_0 = \Psi(t_0)$ under the influence of two potentials $v(\mathbf{r},t)$ and $v'(\mathbf{r},t)$ are always different provided that the potentials differ by more than a purely time-dependent ($\mathbf{r}$-independent) function:

$$v(\mathbf{r},t) \neq v'(\mathbf{r},t) + c(t) \, .^{*} \tag{64}$$

To prove this theorem, the potentials $v$ and $v'$ are required to be expandable in Taylor series about the initial time $t_0$. The condition (64) then guarantees that there exists some (lowest-order) $k$ where the Taylor coefficients $v_k(\mathbf{r}) = \partial^k/\partial t^k v(\mathbf{r},t)\big|_{t_0}$ and $v'_k(\mathbf{r}) = \partial^k/\partial t^k v'(\mathbf{r},t)\big|_{t_0}$ differ by more than a constant. Using the equation of motion for the expectation value of the paramagnetic current density operator and the continuity equation, one arrives at the central equation of the proof:

$$\left(\frac{\partial}{\partial t}\right)^{k+2} (n(\mathbf{r},t) - n'(\mathbf{r}t))\bigg|_{t=t_0} = \nabla \cdot (n(\mathbf{r},t_0)\nabla(v_k(\mathbf{r}) - v'_k(\mathbf{r}))\, . \tag{65}$$

It can be shown that the right-hand side of (65) cannot vanish identically. Hence the densities $n(\mathbf{r},t)$ and $n'(\mathbf{r},t)$ will become different infinitesimally later than $t_0$. The detailed proof and the formulation of precise conditions for the external potentials can be found in [59]. An extension of the Runge-Gross proof to the spin-dependent situation was formulated by Liu and Vosko [65].

Note that the right-hand side of Eq. (65) is linear in $v_k(\mathbf{r}) - v'_k(\mathbf{r})$. Consequently, the difference between $n(\mathbf{r},t)$ and $n'(\mathbf{r},t)$ is non-vanishing already in first order of $v(\mathbf{r},t) - v'(\mathbf{r},t)$. This ensures the invertibility of time-dependent linear response operators.

The uniqueness of the external potential reproducing a $v$-representable density (i.e. a density evolving out of a given initial state under the influence of an external potential $v$) can be established for *any fixed* particle-particle interaction, in particular also for $\hat{W}_{\text{Clb}} \equiv 0$ (this can be seen from Equation (65), in which the the particle-particle interaction does not occur). Hence, given a non-interacting $v$-representable density $n(\mathbf{r},t)$, the potential $v_s[n](\mathbf{r},t)$ leading to this density in a non-interacting system is uniquely determined. Assuming that the density of the interacting system of interest is simultaneously non-interacting $v$-representable, it can be obtained from

$$n(\mathbf{r},t) = \sum_{j=1}^{N} |\phi_j(\mathbf{r},t)|^2 \tag{66}$$

with orbitals $\phi_{j\sigma}(\mathbf{r},t)$ satisfying the time-dependent Kohn-Sham equation

$$i\frac{\partial}{\partial t}\phi_j(\mathbf{r},t) = \left(-\frac{\nabla^2}{2} + v_s[n](\mathbf{r},t)\right)\phi_j(\mathbf{r},t) \quad . \tag{67}$$

Usually, the single-particle potential $v_s$ is written as

$$v_s[n](\mathbf{r},t) = v(\mathbf{r},t) + \int d^3r' \, \frac{n(\mathbf{r}',t)}{|\mathbf{r}-\mathbf{r}'|} + v_{\text{xc}}[n](\mathbf{r},t)\, , \tag{68}$$

where $v(\mathbf{r},t)$ is the external time-dependent field. Equation (68) *defines* the time-dependent xc potential. In practice, this quantity has to be approximated.

*In analogy to the ground-state formalism, where the potentials are required to differ by more than a constant, the condition (64) ensures that the resulting wave functions $\Psi(t)$ and $\Psi'(t)$ differ by more than just a purely time-dependent phase factor.

One has to emphasize that the time-dependent Kohn-Sham (TDKS) scheme does *not* follow from a variational principle, but rather from the basic 1-1 mapping (applied to non-interacting particles) and the assumption of non-interacting $v$-representability.

Like in static DFT, the great advantage of the time-dependent Kohn Sham scheme lies in its computational simplicity compared to other methods such as time-dependent Hartree-Fock or time-dependent configuration interaction [66–73]. In contrast to time-dependent Hartree-Fock, the effective single-particle potential $v_s$ is a *local* potential, i.e., a multiplicative operator in configuration space.

It is important to keep in mind that in the time-dependent case the 1–1 correspondence between potentials and densities can be established only for a *fixed* initial many-body state $\Psi_0$. Consequently, $v[n]$ and $v_{xc}[n]$ implicitly depend on the initial many-body wave function and on the initial Kohn-Sham Slater determinant. However, for a large class of systems, namely those where both $\Psi_0$ and the initial KS Slater determinant are non-degenerate ground states, the Kohn-Sham potential is indeed a functional of the density *alone*. This is because any non-degenerate ground state $\Psi_0$ is a unique functional of its density $n_0(\mathbf{r})$ by virtue of the traditional HK theorem. In particular, the initial KS orbitals are *uniquely* determined as well in this case.

### Approximations of the time-dependent exchange-correlation potential

**Adiabatic Local Density Approximation** The simplest possible approximation of the time-dependent xc potential is the so-called *time-dependent* or *"adiabatic" local density approximation* (ALDA) [59]. It was first introduced in semiconductor physics by T. Ando [74] and later, for finite systems, by Zangwill and Soven [75]. The ALDA employs the functional form of the static LDA with a time-dependent density:

$$v_{\mathrm{xc}}^{\mathrm{ALDA}}[n](\mathbf{r},t) = v_{\mathrm{xc}}^{\mathrm{hom}}(n(\mathbf{r},t)) = \frac{d}{d\rho}\left(\rho\epsilon_{\mathrm{xc}}^{\mathrm{hom}}(\rho)\right)\Big|_{\rho=n(\mathbf{r},t)} . \tag{69}$$

Here $\epsilon_{\mathrm{xc}}^{\mathrm{hom}}$ is the xc energy per particle of the homogeneous electron gas. Parameterizations of this quantity can be found in [76, 44]. By its very definition, the ALDA can be expected to be a good approximation only for nearly homogeneous densities, i.e., for functions $n(\mathbf{r},t)$ that are slowly varying both spatially and temporally. It will turn out, however, that the ALDA gives rather accurate results even for rapidly varying densities.

**Time-Dependent Optimized Effective Potential** Like in the stationary case, the optimized effective potential method offers a viable route to go beyond the local density approximation. In particular, it offers a clear cut way to construct self-interaction-free functionals. In order to derive a time-dependent generalization [77] of the OEP we consider an $N$-electron system subject to a potential of the form

$$v_{\mathrm{ext}\,\sigma}(\mathbf{r},t) = \begin{cases} v_{0\sigma}(\mathbf{r}) & ; \quad t < t_0 \\ v_{0\sigma}(\mathbf{r}) + v_{1\sigma}(\mathbf{r},t) & ; \quad t \geq t_0 \end{cases} \tag{70}$$

For all times until $t_0$, the system has been in the ground state associated with the static external potential $v_{0\sigma}(\mathbf{r})$. We assume that the corresponding stationary OEP problem has been solved for that system, i. e. a local effective potential for each spin orientation and a set of $N$ spin orbitals $\{\varphi_{j\sigma}\}$ (with energy eigenvalues $\varepsilon_{j\sigma}$) minimizing a given energy functional $E[\varphi_{j\sigma}]$ are assumed to be known. Our goal is to determine the time evolution of the system under the influence of the total external potential

$v_{\text{ext}\,\sigma}(\mathbf{r}t) = v_{0\sigma}(\mathbf{r}) + v_{1\sigma}(\mathbf{r}t)$ from $t_0$ until an arbitrary later time $t_1$. The construction of the optimized local effective potential starts with the quantum mechanical action

$$A[\phi_{j\sigma}] = \sum_\sigma \sum_j^{N_\sigma} \int_{-\infty}^{t_1} dt \int d^3r\, \phi_{j\sigma}^*(\mathbf{r}t) \left(i\,\partial/\partial t + \nabla^2/2\right) \phi_{j\sigma}(\mathbf{r}t)$$
$$- \sum_\sigma \int_{-\infty}^{t_1} dt \int d^3r\, n_\sigma(\mathbf{r}t) v_{\text{ext}\,\sigma}(\mathbf{r}t) - \frac{1}{2} \int_{-\infty}^{t_1} dt \int d^3r \int d^3r'\, \frac{n(\mathbf{r}t)n(\mathbf{r}'t)}{|\mathbf{r}-\mathbf{r}'|} - A_{\text{xc}}[\phi_{j\sigma}] \quad (71)$$

written as a functional of $N = \sum_\sigma N_\sigma$ time-dependent spin orbitals $\{\phi_{j\sigma}(\mathbf{r}t)\}$. In the following we will not refer to any specific approximation for the exchange-correlation functional $A_{\text{xc}}[\phi_{j\sigma}]$. We mention however, that in an exchange-only theory $A_{\text{xc}}$ would be replaced by the time-dependent Hartree-Fock expression [78]

$$A_{\text{x}} = -\frac{1}{2} \sum_\sigma \sum_{i,j}^{N_\sigma} \int_{-\infty}^{t_1} dt \int d^3r \int d^3r'\, \phi_{i\sigma}^*(\mathbf{r}'t)\phi_{j\sigma}(\mathbf{r}'t)\phi_{i\sigma}(\mathbf{r}t)\phi_{j\sigma}^*(\mathbf{r}t) \,/\, |\mathbf{r}-\mathbf{r}'| \quad . \quad (72)$$

The orbitals are solutions of the spin-dependent TDKS equation

$$i\frac{\partial}{\partial t}\phi_{j\sigma}(\mathbf{r},t) = \left(-\frac{\nabla^2}{2} + v_{\text{s}\sigma}[n](\mathbf{r},t)\right) \phi_{j\sigma}(\mathbf{r},t) \quad . \quad (73)$$

with the initial condition $\phi_{j\sigma}(\mathbf{r}t) = \varphi_{j\sigma}(\mathbf{r}) \exp[-i\varepsilon_{j\sigma}(t-t_0)]$ for $-\infty < t \leq t_0$. The local effective potential $v_{\text{s}\,\sigma}(\mathbf{r}t)$ has to be determined in such a way that the $\{\phi_{j\sigma}(\mathbf{r}t)\}$, resulting from Eq. (67), render the total action functional $A[\phi_{j\sigma}]$ stationary. Therefore, we have to solve the following variational problem:

$$\frac{\delta A[\phi_{j\sigma}]}{\delta v_{\text{s}\,\sigma}(\mathbf{r}t)} = \sum_{\sigma'} \int_{-\infty}^{+\infty} dt' \int d^3r' \sum_j^{N_{\sigma'}} \left( \frac{\delta A[\phi_{j\sigma}]}{\delta \phi_{j\sigma'}(\mathbf{r}'t')} \frac{\delta \phi_{j\sigma'}(\mathbf{r}'t')}{\delta v_{\text{s}\,\sigma}(\mathbf{r}t)} + \frac{\delta A[\phi_{j\sigma}]}{\delta \phi_{j\sigma'}^*(\mathbf{r}'t')} \frac{\delta \phi_{j\sigma'}^*(\mathbf{r}'t')}{\delta v_{\text{s}\,\sigma}(\mathbf{r}t)} \right) = 0 \,. \quad (74)$$

The functional derivative of $A$ with respect to the orbitals is given by

$$\frac{\delta A[\phi_{j\sigma}]}{\delta \phi_{j\sigma}(\mathbf{r}'t')} = \left[v_{\text{xc}\,\sigma}(\mathbf{r}'t') - u_{\text{xc}\,j\sigma}(\mathbf{r}'t')\right] \phi_{j\sigma}^*(\mathbf{r}'t')\, \theta(t_1 - t') \,, \quad (75)$$

with the abbreviation

$$u_{\text{xc}\,j\sigma}(\mathbf{r}t) = \frac{1}{\phi_{j\sigma}^*(\mathbf{r}t)} \frac{\delta A_{\text{xc}}[\phi_{j\sigma}]}{\delta \phi_{j\sigma}(\mathbf{r}t)} \,. \quad (76)$$

$\theta(x)$ denotes the usual step function (1 for $x > 0$, 0 for $x < 0$). The functional derivative $\delta A/\delta\phi_{j\sigma}^*$ is just the complex conjugate of expression (75). From first-order perturbation theory, we can read off the functional derivative

$$\frac{\delta \phi_{j\sigma'}(\mathbf{r}'t')}{\delta v_{\text{s}\,\sigma}(\mathbf{r}t)} = i\delta_{\sigma\sigma'} \sum_{k=1}^{\infty} \phi_{k\sigma}^*(\mathbf{r}t)\phi_{j\sigma}(\mathbf{r}t)\phi_{k\sigma}(\mathbf{r}'t')\, \theta(t_1 - t)\, \theta(t - t') \quad (77)$$

and the complex conjugate expression for $\delta\phi_{j\sigma'}^*/\delta v_{\text{s}\,\sigma}$. Insertion of Eqs. (75) and (77) in the variational equation (74) results in the time-dependent OEP (TDOEP) integral equation for the local exchange-correlation potential $v_{\text{xc}\,\sigma}(\mathbf{r}t)$:

$$\int_{-\infty}^{t_1} dt' \int d^3r' \left( iK_\sigma(\mathbf{r}t, \mathbf{r}'t') \sum_j^{N_\sigma} \phi_{j\sigma}(\mathbf{r}t)\phi_{j\sigma}^*(\mathbf{r}'t') \left[v_{\text{xc}\,\sigma}(\mathbf{r}'t') - u_{\text{xc}\,j\sigma}(\mathbf{r}'t')\right] + \text{c.c.} \right) = 0 \,. \quad (78)$$

The kernel $K_\sigma(\mathbf{r}t, \mathbf{r}'t') = \sum_{k=1}^{\infty} \phi^*_{k\sigma}(\mathbf{r}t)\phi_{k\sigma}(\mathbf{r}'t')\,\theta(t-t')$ is identical with the Green's function of the system.

The TDOEP scheme is now complete: the integral equation (78) has to be solved for $v_{\mathrm{xc}\,\sigma}(\mathbf{r}t)$ simultaneously with the Kohn-Sham equation (73).

Clearly, the full TDOEP scheme is computationally extremely demanding. It is therefore highly desirable to construct approximations of $v_{\mathrm{xc}\,\sigma}(\mathbf{r}t)$ which are *explicit* functionals of the orbitals $\{\phi_{j\sigma}\}$. Similar manipulations to those developed by KLI for the stationary case [79, 12], lead to the so-called TDKLI expression [77]

$$\begin{aligned}\tilde{v}_{\mathrm{xc}\,\sigma}(\mathbf{r}t) &= \frac{1}{n_\sigma(\mathbf{r}t)} \sum_j^{N_\sigma} n_{j\sigma}(\mathbf{r}t)\,\frac{1}{2}\left(u_{\mathrm{xc}\,j\sigma}(\mathbf{r}t) + u^*_{\mathrm{xc}\,j\sigma}(\mathbf{r}t)\right) \\ &+ \frac{1}{n_\sigma(\mathbf{r}t)} \sum_j^{N_\sigma} n_{j\sigma}(\mathbf{r}t)\left[\bar{\tilde{v}}_{\mathrm{xc}\,j\sigma}(t) - \frac{1}{2}\left(\bar{u}_{\mathrm{xc}\,j\sigma}(t) + \bar{u}^*_{\mathrm{xc}\,j\sigma}(t)\right)\right] \\ &+ \frac{i}{4n_\sigma(\mathbf{r}t)} \sum_j^{N_\sigma} \nabla^2 n_{j\sigma}(\mathbf{r}t) \int_{-\infty}^{t} dt' \left(\bar{u}_{\mathrm{xc}\,j\sigma}(t') - \bar{u}^*_{\mathrm{xc}\,j\sigma}(t')\right) . \end{aligned} \tag{79}$$

The tilde indicates that this is only an approximate solution of the TDOEP equation (78) and the over-bar is a shorthand notation for the orbital average, i.e. $\bar{\tilde{v}}_{\mathrm{xc}\,j\sigma}(t) = \int d^3r\, n_{j\sigma}(\mathbf{r},t)\tilde{v}_{\mathrm{xc}\,\sigma}(\mathbf{r},t)$. The last term of Eq. (79) vanishes identically for a large class of exchange-correlation functionals $A_{\mathrm{xc}}$. This includes all functionals which depend on $\{\phi_{j\sigma}\}$ only through the combinations $\phi_{j\sigma}(\mathbf{r},t)\phi^*_{j\sigma}(\mathbf{r}',t)$ (such as the exact x-only functional, Eq. (72)). Eq. (79) is still an integral equation for $\tilde{v}_{\mathrm{xc}\,\sigma}$. In contrast to the full TDOEP equation it can be solved analytically [80]: Multiplying Eq. (79) by $n_{k\sigma}(\mathbf{r}t)$ and integrating over all space yields

$$\bar{\tilde{v}}_{\mathrm{xc}\,k\sigma}(t) = \bar{w}_{\mathrm{xc}\,k\sigma}(t) + \sum_j^{N_\sigma} N_{kj\sigma}(t)\bar{\tilde{v}}_{\mathrm{xc}\,j\sigma}(t)\,, \tag{80}$$

where

$$\begin{aligned} w_{\mathrm{xc}\,\sigma}(\mathbf{r}t) &= \frac{1}{n_\sigma(\mathbf{r}t)} \sum_j^{N_\sigma} n_{j\sigma}(\mathbf{r}t)\,\frac{1}{2}\left(u_{\mathrm{xc}\,j\sigma}(\mathbf{r}t) + u^*_{\mathrm{xc}\,j\sigma}(\mathbf{r}t)\right) \\ &- \frac{1}{n_\sigma(\mathbf{r}t)} \sum_j^{N_\sigma} n_{j\sigma}(\mathbf{r}t)\left[\frac{1}{2}\left(\bar{u}_{\mathrm{xc}\,j\sigma}(t) + \bar{u}^*_{\mathrm{xc}\,j\sigma}(t)\right)\right] \\ &+ \frac{i}{4n_\sigma(\mathbf{r}t)} \sum_j^{N_\sigma} \nabla^2 n_{j\sigma}(\mathbf{r}t) \int_{-\infty}^{t} dt' \left(\bar{u}_{\mathrm{xc}\,j\sigma}(t') - \bar{u}^*_{\mathrm{xc}\,j\sigma}(t')\right) \end{aligned} \tag{81}$$

and

$$N_{kj\sigma}(t) = \int d^3r\, \frac{n_{k\sigma}(\mathbf{r}t)n_{j\sigma}(\mathbf{r}t)}{n_\sigma(\mathbf{r}t)} \quad . \tag{82}$$

The solution of Eq. (80) for $\bar{\tilde{v}}_{\mathrm{xc}\,j\sigma}(t)$ requires inversion of the $N_\sigma \times N_\sigma$ matrix $A_{kj\sigma}(t) = \delta_{kj} - N_{kj\sigma}(t)$ and leads to

$$\bar{\tilde{v}}_{\mathrm{xc}\,j\sigma}(t) = \sum_k^{N_\sigma} \left(A_\sigma^{-1}(t)\right)_{jk} \bar{w}_{\mathrm{xc}\,k\sigma}(t) \quad . \tag{83}$$

By substitution of Eq. (83) into Eq. (79), $\tilde{v}_{\mathrm{xc}\,\sigma}(\mathbf{r}t)$ is obtained as an explicit functional of the orbitals $\{\phi_{j\sigma}(\mathbf{r}t)\}$.

The TDOEP scheme has been successfully applied to the non-perturbative description of atoms [81, 82, 83] and clusters [84, 85, 86] in strong laser pulses. In Fig. 2 we show the harmonic generation spectrum of the He atom in a two-color laser pulse [87].

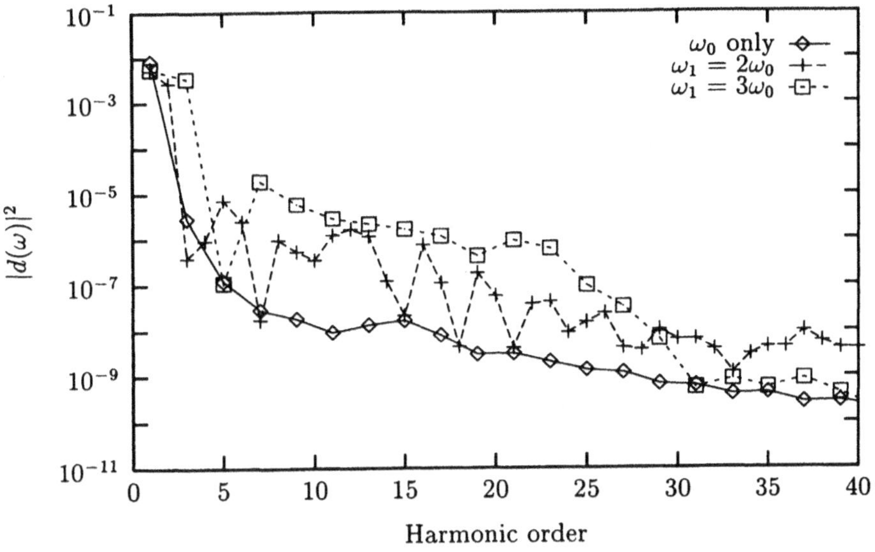


**Figure 2.** Harmonic spectra of helium resulting from a one–color calculation and from two two–color calculations. One of the latter includes the second and the other includes the third harmonic in addition to the fundamental frequency. The laser parameters used are $\lambda = 616$ nm and $I = 7.0 \times 10^{14}$ W/cm$^2$.

## Time-dependent Kohn-Sham equations in the linear-response regime

If the external field intensity is low, it is sufficient to consider only changes of the time-dependent density which are *linear* in the applied field. Within the framework of time-dependent density-functional theory, a formally exact representation of the linear spin-density response $n_{1\sigma}(\mathbf{r}, t)$ of interacting electrons can be derived [88] in terms of the response function $\chi_s$ of the corresponding (non-interacting) Kohn-Sham system:

$$n_{1\sigma}(\mathbf{r}, t) = \sum_{\sigma'} \int dt' \int d^3r' \, \chi_{s\,\sigma\sigma'}(\mathbf{r}, t, \mathbf{r}', t') \, v_{s,1\,\sigma'}(\mathbf{r}', t') \,. \tag{84}$$

$v_{s,1}$ is the linearized time-dependent KS potential

$$\begin{aligned} v_{s,1\,\sigma}(\mathbf{r}, t) &= v_{1\sigma}(\mathbf{r}, t) + \sum_{\sigma'} \int d^3r' \, \frac{n_{1\sigma'}(\mathbf{r}', t)}{|\mathbf{r} - \mathbf{r}'|} \\ &+ \sum_{\sigma'} \int d^3r' \int dt' \, f_{\mathrm{xc}\,\sigma\sigma'}(\mathbf{r}, t, \mathbf{r}', t') \, n_{1\sigma'}(\mathbf{r}', t') \,. \end{aligned} \tag{85}$$

It contains the external perturbation $v_{1\sigma}$ and the Hartree- and exchange-correlation contributions to first order in the perturbing potential $v_{1\sigma}$. The so-called time-dependent xc kernel

$$f_{\mathrm{xc}\,\sigma\sigma'}(\mathbf{r}, t, \mathbf{r}', t') := \left. \frac{\delta v_{\mathrm{xc}\,\sigma}[n_\uparrow, n_\downarrow](\mathbf{r}, t)}{\delta n_{\sigma'}(\mathbf{r}', t')} \right|_{n_{0\uparrow}, n_{0\downarrow}} \tag{86}$$

is a functional of the initial ground-state densities. Combining Eqs. (84) and (85) and taking the Fourier transform with respect to time, the *exact* frequency-dependent linear density response is seen to be

$$n_{1\sigma}(\mathbf{r},\omega) = \sum_{\nu} \int d^3y\, \chi_{s\,\sigma\nu}(\mathbf{r},\mathbf{y};\omega) v_{1\nu}(\mathbf{y},\omega) \tag{87}$$
$$+ \sum_{\nu\nu'} \int d^3y \int d^3y' \chi_{s\,\sigma\nu}(\mathbf{r},\mathbf{y};\omega) \left( \frac{1}{|\mathbf{y}-\mathbf{y}'|} + f_{\mathrm{xc}\,\nu\nu'}[n_0](\mathbf{y},\mathbf{y}';\omega) \right) n_{1\,\nu'}(\mathbf{y}',\omega) .$$

The Kohn-Sham response function $\chi_s$ is readily expressed in terms of the static unperturbed Kohn-Sham orbitals $\varphi_{k\sigma}$:

$$\chi_{s\,\sigma\sigma'}(\mathbf{r},\mathbf{r}';\omega) = \delta_{\sigma\sigma'} \sum_{j,k} (f_{k\sigma} - f_{j\sigma}) \frac{\varphi_{j\sigma}(\mathbf{r})\varphi^*_{k\sigma}(\mathbf{r})\varphi^*_{j\sigma}(\mathbf{r}')\varphi_{k\sigma}(\mathbf{r}')}{\omega - (\epsilon_{j\sigma} - \epsilon_{k\sigma}) + i\eta} . \tag{88}$$

Here, $(f_{k\sigma}, f_{j\sigma})$ are the occupation numbers (0 or 1) of the KS orbitals. The summation in (88) ranges over both occupied and unoccupied orbitals, including the continuum states.

The exchange-correlation kernel $f_{\mathrm{xc}}$, given by Eq. (86), comprises all dynamical exchange and correlation effects to first order in the perturbing potential. For spin-unpolarized ground states, the spin-dependent exchange-correlation kernel has the form

$$f_{\mathrm{xc}\uparrow\uparrow} = f_{\mathrm{xc}\downarrow\downarrow} = f_{\mathrm{xc}} + \mu_0^2 G_{\mathrm{xc}} \tag{89}$$
$$f_{\mathrm{xc}\uparrow\downarrow} = f_{\mathrm{xc}\downarrow\uparrow} = f_{\mathrm{xc}} - \mu_0^2 G_{\mathrm{xc}} \tag{90}$$

($\mu_0$ denotes the Bohr magneton). In practice, $f_{\mathrm{xc}\sigma\sigma'}$ is of course only approximately known. Again, the simplest possible approximation is the spin-dependent version of the adiabatic local density approximation (69). For spin-unpolarized ground states, this leads to

$$f_{\mathrm{xc}}^{\mathrm{ALDA}}(\mathbf{r},\mathbf{r}';\omega) = \delta(\mathbf{r}-\mathbf{r}') \frac{d^2}{d\rho} \left( \rho\epsilon_{\mathrm{xc}}^{\mathrm{hom}}(\rho,\zeta=0) \right)\Big|_{\rho=n_0(\mathbf{r})} \tag{91}$$
$$= \delta(\mathbf{r}-\mathbf{r}')\, f_{\mathrm{xc}}^{\mathrm{hom}}(\rho,\zeta=0,q=0,\omega=0)\Big|_{\rho=n_0(\mathbf{r})}$$

and

$$G_{\mathrm{xc}}^{\mathrm{ALDA}}(\mathbf{r},\mathbf{r}';\omega) = \delta(\mathbf{r}-\mathbf{r}') \frac{\alpha_{\mathrm{xc}}(\rho)}{\mu_0^2\rho}\Bigg|_{\rho=n_0(\mathbf{r})} , \tag{92}$$

where

$$\alpha_{\mathrm{xc}} = \frac{\delta^2}{\delta\zeta^2} \left( \epsilon_{\mathrm{xc}}^{\mathrm{hom}}(\rho,\zeta) \right)\Bigg|_{\zeta=0} \tag{93}$$

is the exchange-correlation contribution to the spin-stiffness ($\zeta$ stands for the relative spin polarization). Most notably, the Fourier transforms (91) and (92) of the time-dependent response kernels in the ALDA have no frequency dependence at all. Beyond this adiabatic limit, frequency-dependent approximations of the xc kernel have been obtained [89, 90, 91, 92]. We finally mention an explicitly orbital-dependent approximation of the xc kernel,

$$f_{\mathrm{xc}\,\sigma\sigma'}^{\mathrm{PGG}}(\mathbf{r},\mathbf{r}') = -\delta_{\sigma\sigma'} \frac{1}{|\mathbf{r}-\mathbf{r}'|} \frac{\left| \sum_k f_{k\sigma}\, \varphi_{k\sigma}(\mathbf{r})\varphi^*_{k\sigma}(\mathbf{r}') \right|^2}{n_\sigma(\mathbf{r}) n_\sigma(\mathbf{r}')} . \tag{94}$$

which Petersilka, Gossmann, and Gross (PGG) have obtained from a linearization of the TDOEP equation [88, 93] using the x-only functional (72).

## Calculation of excitation energies

The traditional Hohenberg-Kohn theorem guarantees that every observable of a stationary physical system can be expressed in terms of its ground-state density. In principle, this is also true for the set of excited-state energies: Knowledge of the external potential, which is a unique functional of the ground-state density, allows the calculation of any excited state and its energy. In practice, however, the description of excited states within stationary density-functional theory is a notoriously difficult subject [94–105]. In particular, the Kohn-Sham eigenvalues, introduced as purely mathematical constructs into density-functional theory, cannot simply be interpreted as excited-state energies.

In this chapter a different approach [88, 106] to the calculation of excitation energies is presented which is based on time-dependent density functional theory in the linear response regime. The formalism has been successfully applied to the calculation of molecular excitation spectra [107–114].

In the standard linear response formalism, the full response function

$$\chi_{\sigma\sigma'}(\mathbf{r},\mathbf{r}';\omega) = \lim_{\delta\to 0^+}\sum_m \left(\frac{\langle\psi_0|\hat{n}_\sigma(\mathbf{r})|\psi_m\rangle\langle\psi_m|\hat{n}_{\sigma'}(\mathbf{r}')|\psi_0\rangle}{\omega-(E_m-E_0)+i\delta} - \frac{\langle\psi_0|\hat{n}_{\sigma'}(\mathbf{r}')|\psi_m\rangle\langle\psi_m|\hat{n}_\sigma(\mathbf{r})|\psi_0\rangle}{\omega+(E_m-E_0)+i\delta}\right). \tag{95}$$

is formally constructed from the exact many-body eigenfunctions and energies

$$H(t_0)|\Psi_m\rangle = E_m|\Psi_m\rangle \tag{96}$$

of the full static many-body Hamiltonian $H(t_0)$ at $t_0$. It is obvious from Eq. (95) that the frequency-dependent linear density response has poles at the exact excitation energies $\Omega = E_m - E_0$.

The key idea of a density functional calculation of excitation energies is to start from a particular KS excitation energy $\epsilon_{j\sigma} - \epsilon_{k\sigma}$ (which are the poles of the Kohn-Sham response function (88)) and to use the formally exact representation (87) of the linear density response to calculate the shifts of the Kohn-Sham excitation energies towards the true excitation energies $\Omega$.

To calculate these shifts, we rewrite Eq. (87) as a linear integral equation for the frequency dependent linear density response:

$$\sum_{\nu'}\int d^3y' \left[\delta_{\sigma\nu'}\delta(\mathbf{r}-\mathbf{y}') - \sum_\nu\int d^3y \chi_{s\,\sigma\nu}(\mathbf{r},\mathbf{y};\omega)\left(\frac{1}{|\mathbf{y}-\mathbf{y}'|} + f_{xc\,\nu\nu'}(\mathbf{y},\mathbf{y}';\omega)\right)\right] n_{1\nu'}(\mathbf{y}',\omega)$$
$$= \sum_\nu \int d^3y\, \chi_{s\,\sigma\nu}(\mathbf{r},\mathbf{y};\omega) v_{1\nu}(\mathbf{y},\omega). \tag{97}$$

In general, the true excitation energies $\Omega$ are not identical with the Kohn-Sham excitation energies $\epsilon_{j\sigma} - \epsilon_{k\sigma}$. Therefore, the right-hand side of Eq. (97) remains finite for $\omega \to \Omega$. Since, on the other hand, the exact spin-density response $n_{1\sigma}$, has poles at the true excitation energies $\Omega$, the integral operator acting on $n_{1\sigma}$ on the left-hand side of Eq. (97) cannot be invertible for $\omega \to \Omega$.

Consequently, the true excitation energies $\Omega$ are characterized as those frequencies where the integral operator acting on the spin-density vector in Eq. (97) becomes non-invertible, i.e. has vanishing eigenvalues. Hence, the true excitation energies $\Omega$ are those frequencies where the eigenvalues $\lambda(\omega)$ of

$$\sum_{\nu'}\int d^3y' \sum_\nu\int d^3y\, \chi_{s\,\sigma\nu}(\mathbf{r},\mathbf{y};\omega)\left(\frac{1}{|\mathbf{y}-\mathbf{y}'|} + f_{xc\,\nu\nu'}(\mathbf{y},\mathbf{y}';\omega)\right)\gamma_{\nu'}(\mathbf{y}',\omega) =$$
$$= \lambda(\omega)\gamma_\sigma(\mathbf{r},\omega) \tag{98}$$

satisfy

$$\lambda(\Omega) = 1\,. \tag{99}$$

This condition rigorously determines the true excitation spectrum of the interacting system at hand.

To facilitate the notation, we introduce double indices $q \equiv (j,k)$ so that $\omega_{q\sigma} \equiv \epsilon_{j\sigma} - \epsilon_{k\sigma}$ denotes the excitation energy of the single-particle transition $(k\sigma \to j\sigma)$. Moreover, we define $\alpha_{q\sigma} := f_{k\sigma} - f_{j\sigma}$ and

$$\Phi_{q\sigma}(\mathbf{r}) := \varphi^*_{k\sigma}(\mathbf{r})\varphi_{j\sigma}(\mathbf{r}) \tag{100}$$

and set

$$\xi_{q\sigma}(\omega) := \sum_{\nu'} \int d^3y' \int d^3y\, \Phi_{q\sigma}(\mathbf{y})^* \left( \frac{1}{|\mathbf{y}-\mathbf{y}'|} + f_{\mathrm{xc}\,\sigma\nu'}(\mathbf{y},\mathbf{y}';\omega) \right) \gamma_{\nu'}(\mathbf{y}',\omega)\,. \tag{101}$$

Using these definitions, we arrive at [106]

$$\sum_{\sigma'}\sum_{q'} \frac{M_{q\sigma\,q'\sigma'}(\omega)}{\omega - \omega_{q'\sigma'} + i\eta} \xi_{q'\sigma'}(\omega) = \lambda(\omega)\xi_{q\sigma}(\omega)\,, \tag{102}$$

where we have introduced the matrix elements

$$M_{q\sigma\,q'\sigma'}(\omega) = \alpha_{q'\sigma'} \int d^3r \int d^3r'\, \Phi^*_{q\sigma}(\mathbf{r}) \left( \frac{1}{|\mathbf{r}-\mathbf{r}'|} + f_{\mathrm{xc}\,\sigma\sigma'}(\mathbf{r},\mathbf{r}';\omega) \right) \Phi_{q'\sigma'}(\mathbf{r}')\,. \tag{103}$$

So far, no approximations have been made.

In order to actually calculate the excitation spectrum, the eigenvalue problem (102) has to be truncated in one way or another. For finite systems, where we encounter well separated poles of the linear density response due to the discrete level structure, one possibility is to expand all quantities in Eq. (102) about one particular KS-orbital energy difference $\omega_{p\tau}$ [88, 106]. The true excitation energies $\Omega$ are then determined by the solution of

$$\lambda(\Omega) = \frac{A(\omega_{p\tau})}{\Omega - \omega_{p\tau}} + B(\omega_{p\tau}) + \ldots = 1 \tag{104}$$

For non-degenerate single-particle poles $\omega_{p\tau}$, the coefficients in Eq. (104) are given by

$$A(\omega_{p\tau}) = M_{p\tau\,p\tau}(\omega_{p\tau}) \tag{105}$$

and

$$B(\omega_{p\tau}) = \left.\frac{dM_{p\tau p\tau}}{d\omega}\right|_{\omega_{p\tau}} + \frac{1}{M_{p\tau p\tau}(\omega_{p\tau})} \sum_{q'\sigma' \neq p\tau} \frac{M_{p\tau\,q'\sigma'}(\omega_{p\tau}) M_{q'\sigma'\,p\tau}(\omega_{p\tau})}{\omega_{p\tau} - \omega_{q'\sigma'} + i\eta}\,. \tag{106}$$

If the pole $\omega_{p\tau}$ is $\wp$-fold degenerate, $\omega_{p_1\tau_1} = \omega_{p_2\tau_2} = \ldots = \omega_{p_\wp\tau_\wp} \equiv \omega_0$, the lowest-order coefficient $A$ in Eq. (104) is determined by a $\wp$-dimensional matrix equation

$$\sum_{k=1}^{\wp} M_{p_i\tau_i\,p_k\tau_k}(\omega_0)\xi^{(n)}_{p_k\tau_k} = A_n(\omega_0)\xi^{(n)}_{p_i\tau_i}\,, \quad i = 1\ldots\wp\,, \tag{107}$$

leading to $\wp$ different coefficients $A_1 \ldots A_\wp$. For excitation energies $\Omega$ close to $\omega_0$, the lowest-order term of the above Laurent expansion will dominate the series. In this single-pole approximation (SPA), Eq. (104) reduces to

$$\lambda_n(\Omega) \approx \frac{A_n(\omega_0)}{\Omega - \omega_0} = 1\,. \tag{108}$$

The condition (99) and its complex conjugate, $\lambda^*(\Omega) = 1$, finally lead to to a simple (approximate) expression for the excitation energies.

$$\Omega_n \approx \omega_0 + \Re A_n(\omega_0)\,. \tag{109}$$

For closed-shell systems, every Kohn-Sham orbital eigenvalue is degenerate with respect to spin, i.e. the spin multiplet structure is absent in the bare Kohn-Sham eigenvalue spectrum. Within the SPA, the dominant terms in the corrections to the Kohn-Sham eigenvalues towards the true multiplet energies naturally emerge from the solution of the $(2 \times 2)$ eigenvalue problem

$$\sum_{\sigma'=\uparrow,\downarrow} M_{p\sigma p\sigma'}(\omega_0)\xi_{p\sigma'}(\omega_0) = A\xi_{p\sigma}(\omega_0)\,. \tag{110}$$

Then, the resulting excitation energies are:

$$\Omega_{1,2} = \omega_0 + \Re\left\{M_{p\uparrow p\uparrow} \pm M_{p\uparrow p\downarrow}\right\}\,. \tag{111}$$

Using the explicit form of the matrix elements (103) one finds[†]

$$\Omega_1 = \omega_0 + 2\Re \int d^3r \int d^3r'\, \Phi_p^*(\mathbf{r}) \left(\frac{1}{|\mathbf{r} - \mathbf{r}'|} + f_{\mathrm{xc}}(\mathbf{r}, \mathbf{r}'; \omega_0)\right) \Phi_p(\mathbf{r}') \tag{112}$$

$$\Omega_2 = \omega_0 + 2\Re \int d^3r \int d^3r'\, \Phi_p^*(\mathbf{r}) \mu_0^2 G_{\mathrm{xc}}(\mathbf{r}, \mathbf{r}'; \omega_0) \Phi_p(\mathbf{r}')\,. \tag{113}$$

The latter kernel, $G_{\mathrm{xc}}$, is responsible for the exchange and correlation effects in the Kohn-Sham equation for the linear response of the frequency-dependent magnetization density $m(\mathbf{r}, \omega)$ [65]. The fact that the magnetization density response naturally involves spin-flip processes suggests that $\Omega_2$ can be attributed to spin triplet excitation energies of many-electron systems. The corresponding spin singlet excitation energies, on the other hand, are given by $\Omega_1$. Hence the spin-multiplet structure, originally absent in the bare Kohn-Sham eigenvalue spectrum, is recovered.

Apart from the explicit calculation of the dominant contributions to the shifts of the Kohn-Sham eigenvalue differences towards the true excitation energies, one can also solve Eq. (102) directly. At the frequencies $\omega = \Omega$, Eq. (102) can be written as

$$\sum_{q'\sigma'} \left(M_{q\sigma\, q'\sigma'}(\Omega) + \delta_{q\sigma\, q'\sigma'}\omega_{q\sigma}\right) \beta_{q'\sigma'}(\Omega) = \Omega \beta_{q\sigma}(\Omega)\,, \tag{114}$$

where we have introduced

$$\beta_{q\sigma}(\Omega) := \xi_{q\sigma}(\Omega)/(\Omega - \omega_{q\sigma})\,. \tag{115}$$

Hence, the exact excitation energies $\Omega$ are the exact solutions of the nonlinear matrix-equation (114). Truncation of the infinite-dimensional matrix in Eq. (114) amounts to the approximation of $\chi^{(0)}$ by a finite sum

$$\chi^{(0)}(\mathbf{r}, \mathbf{r}', \omega) \approx \sum_{\sigma=\uparrow\downarrow} \sum_{q}^{Q} \alpha_q \frac{\Phi_q(\mathbf{r})\Phi_q(\mathbf{r}')}{\omega - \omega_{q\sigma}}\,. \tag{116}$$

[†]Since we are dealing with spin saturated systems, we have dropped the spin-index of $\Phi_{p\sigma}$ for simplicity.

**Table 9.** The lowest S→P excitation energies of the alkaline earths, calculated from the x-only KLI potential. The experimental values (first column) [115] are compared with results calculated in SPA from Eq. (112) for the singlet and from Eq. (113) for the triplet using the ALDA (91, 92) for the xc kernel. For comparison, the ordinary LDA-$\Delta_{SCF}$ values are given in the third column. The Kohn-Sham orbital-energy differences $\omega_{KS}$ are shown in the last column (All values in atomic units).

| Atom | State | Expt. | SPA | $\Delta_{SCF}$ | $\omega_{KS}$ | Atom | State | Expt. | SPA | $\Delta_{SCF}$ | $\omega_{KS}$ |
|---|---|---|---|---|---|---|---|---|---|---|---|
| Be | $^1P_1$ | 0.194 | 0.199 | 0.166 | 0.130 | Zn | $^1P_1$ | 0.213 | 0.209 | 0.202 | 0.157 |
| | $^3P_0$ | 0.100 | | | | | $^3P_0$ | 0.147 | | | |
| | $^3P_1$ | 0.100 | 0.098 | 0.091 | 0.130 | | $^3P_1$ | 0.148 | 0.140 | 0.158 | 0.157 |
| | $^3P_2$ | 0.100 | | | | | $^3P_2$ | 0.150 | | | |
| Mg | $^1P_1$ | 0.160 | 0.165 | 0.150 | 0.117 | Sr | $^1P_1$ | 0.099 | 0.106 | 0.096 | 0.071 |
| | $^3P_0$ | 0.100 | | | | | $^3P_0$ | 0.065 | | | |
| | $^3P_1$ | 0.100 | 0.098 | 0.103 | 0.117 | | $^3P_1$ | 0.066 | 0.059 | 0.068 | 0.071 |
| | $^3P_2$ | 0.100 | | | | | $^3P_2$ | 0.068 | | | |
| Ca | $^1P_1$ | 0.108 | 0.118 | 0.106 | 0.079 | Cd | $^1P_1$ | 0.199 | 0.185 | 0.173 | 0.135 |
| | $^3P_0$ | 0.069 | | | | | $^3P_0$ | 0.137 | | | |
| | $^3P_1$ | 0.070 | 0.065 | 0.072 | 0.079 | | $^3P_1$ | 0.140 | 0.120 | 0.136 | 0.135 |
| | $^3P_2$ | 0.070 | | | | | $^3P_2$ | 0.145 | | | |

Thus, instead of expanding about a single pole, we explicitly take into account several poles of the noninteracting response function. Using the ALDA for the xc kernels, the matrix elements $M_{q\sigma\, q'\sigma'}$ become real and frequency independent, and the excitation energies $\Omega$ are then given as the eigenvalues of the $(Q \times Q)$ matrix $M_{q\sigma\, q'\sigma'}(\Omega = 0) + \delta_{q\sigma, q'\sigma'}\omega_{q\sigma}$.

The scheme developed above involves three different types of approximations:

1. In the calculation of the Kohn-Sham orbitals $\varphi_k(\mathbf{r})$ and their eigenvalues $\epsilon_k$, one employs some approximation of the static xc potential $v_{xc}$.

2. Given the stationary Kohn-Sham orbitals and the ground state density, the functional form of the xc kernel $f_{xc\,\sigma\sigma'}$ needs to be approximated in order to calculate the matrix elements defined in Eq. (103).

3. Once the matrix elements are obtained, the infinite-dimensional eigenvalue problem (102) (or, equivalently, (114)) must be truncated in one way or another.

**Results for atoms and molecules** As a first application of the method, we have calculated the lowest excitation energies of the alkaline earth elements and the zinc series. The values listed in Table 9 are from a calculation using the KLI-x-only approximation for the static xc potential and the adiabatic local density approximation for the xc kernels. Apparently, the KS-eigenvalue differences alone do not resolve singlet-triplet splittings, but are always located in between the experimentally measured singlet and triplet excitation energies. The multiplet structure, however, is already obtained within the SPA, i.e. by adding the lowest order correction terms given by Eqs. (112) and (113). The experimentally measured excitations are quite well reproduced.

In addition to spin-multiplet splittings, large splittings of spatial multiplets are observed in molecules. In Table 10, the excitation energies of the CO molecule are given. The stationary KS orbitals and eigenvalues were calculated using the LDA for exchange and correlation. For the functional form of the xc kernels the ALDA was used. Again, the single-pole approximation gives the dominant corrections of the Kohn-Sham

**Table 10.** Excitation energies for CO from an xcLDA–calculation at $R = 2.1322$ a.u. The LDA was employed for $v_{xc}$ and the ALDA for the xc kernels. $\omega_{KS}$ denotes the KS orbital energy difference. All numbers in mHartrees.

| State | | $\omega_{KS}$ | SPA | full[a] | exp[b] |
|---|---|---|---|---|---|
| A $^1\Pi$ | $5\sigma \to 2\pi$ | 252.3 | 326.8 | 310.2 | 312.7 |
| a $^3\Pi$ | | | 223.8 | 221.4 | 232.3 |
| B $^1\Sigma^+$ | $5\sigma \to 6\sigma$ | 333.2 | 338.9 | 338.0 | 396.2 |
| b $^3\Sigma^+$ | | | 331.5 | 331.6 | 382.2 |
| I $^1\Sigma^-$ | $1\pi \to 2\pi$ | 362.6 | 362.6 | 362.6 | 363.1 |
| e $^3\Sigma^-$ | | | 362.6 | 362.6 | 363.1 |
| a′ $^3\Sigma^+$ | | | 318.1 | 314.9 | 312.7 |
| D $^1\Delta$ | | | 381.2 | 380.7 | 375.9 |
| d $^3\Delta$ | | | 340.4 | 339.6 | 344.0 |
| c $^3\Pi$ | $4\sigma \to 2\pi$ | 438.8 | 420.4 | 420.2 | 424.5 |
| E $^1\Pi$ | $1\pi \to 6\sigma$ | 443.6 | 443.5 | 443.5 | 423.7 |

[a] Neglecting continuum states.
[b] From [116].

eigenvalue differences towards the experimentally measured energies. The results from a solution of Eq. (114) are listed in the column headed "full". Here we have coupled all occupied and the lowest 9 unoccupied molecular orbitals. In general, this leads to a more refined approximation of the spectrum, since more configurations are taken into account. However, the differences with respect to the SPA are usually small, especially for the triplet states.

Considering the computational simplicity of the approach, the quality of the results is very promising. This opens up the road towards a reliable calculation of spectra for larger systems, like clusters and large molecules. Since the present scheme is mainly concerned with the calculation of corrections to the Kohn-Sham eigenvalue spectrum, it has to be stressed, however, that a quantitative calculation of excitation energies substantially depends on the accuracy of the ground state xc potential employed. In the LDA and in the popular GGAs for instance, which are not free of spurious self-interaction effects, the highest occupied orbital eigenvalues are in error by about a factor of two. Cancellation of self-interaction errors has been observed for lower excited states [111, 112, 114], but, in general, a resulting excitation spectrum which is accurate for all states can only be expected if self-interaction free xc potentials are used. The development of orbital functionals should offer a viable route in this direction.

## VAN DER WAALS INTERACTIONS

Van der Waals (vdW) forces are caused by correlations between distant electronic fluctuations. In an intuitive picture, the origin of the vdW energy can be understood as follows: Consider a system consisting of two neutral subsystems $A$ and $B$ at a distance $R$ and suppose an electric dipole moment is spontaneously created in one of the subsystems. As a consequence, a second dipole is induced in the other subsystem. The interaction between the two dipole moments creates the vdW energy. According to this argument, the $R^{-3}$ decay of dipole fields enters twice so that the asymptotic behavior of the vdW energy is

$$E_{\mathrm{vdW}} \simeq -\frac{C_6}{R^6} \quad \text{for} \quad R \to \infty \tag{117}$$

with the vdW coefficient $C_6$. (The $R^{-6}$ behavior only applies to finite subsystems; for extended subsystems such as metal surfaces one obtains different power laws.)

The subject can be treated rigorously by applying second-order perturbation theory. This leads to the Zaremba-Kohn formula [117]

$$E_{\mathrm{vdW}} = -\frac{1}{2\pi} \int\limits_A d^3x\, d^3x' \int\limits_B d^3r\, d^3r' \int\limits_0^\infty du\, \frac{\chi^A(\mathbf{x},\mathbf{x}';iu)\,\chi^B(\mathbf{r},\mathbf{r}';iu)}{|\mathbf{x}-\mathbf{r}|\ \ |\mathbf{x}'-\mathbf{r}'|}\,, \tag{118}$$

which expresses the vdW energy in terms of the imaginary-frequency response functions $\chi^A$ and $\chi^B$ of the isolated and unperturbed subsystems $A$ and $B$. Through a multipole expansion of Eq. (118) one recovers the asymptotic behavior (117) with the vdW coefficient now given by

$$C_6 = \frac{1}{2\pi} \sum_{i,j=1}^{3} (1-\delta_{3i})(1-3\delta_{3j}) \int\limits_0^\infty du\, \alpha_{ij}^A(iu)\, \alpha_{ij}^B(iu)\,, \tag{119}$$

where the 3-axis is chosen to point from $A$ to $B$ and, for each subsystem, $\alpha_{ij}$ is the polarizability tensor (i.e. the tensorial generalization of Eq. (58)):

$$\alpha_{ij}(\omega) = -\int d^3r\, d^3r'\, r_i r_j'\, \chi(\mathbf{r},\mathbf{r}';\omega) = -\frac{1}{\mathcal{E}} \int d^3r\, r_i\, n_1(\mathbf{r},\omega)_j\,. \tag{120}$$

Here $n_1(\mathbf{r},\omega)_j$ is the linear density response to the perturbation $v_1(\mathbf{r}t) = \mathcal{E} r_j \cos(\omega t)$.

TDDFT provides a very efficient way of calculating the $C_6$ coefficients: Given an approximation for the xc kernel, Eqs. (87) and (88) can be used to determine the density responses $n_1(\mathbf{r},\omega)_j$ which, by Eqs. (119) and (120) give the vdW coefficients. With this method, van Gisbergen et al [118, 119] have obtained vdW coefficients for a large variety of molecules in rather good agreement with experiment.

In the present section, however, we want to go beyond the asymptotic limit: Our goal is to construct a functional for the xc energy that yields accurate total energies *and* has the correct asymptotic behavior (117). In other words, we want a "seamless" scheme which gives reliable results at any subsystem separation and is still applicable in the case of overlapping subsystems.

Static DFT, as a matter of principle, yields the exact ground-state energy, including long-range vdW contributions. The commonly used approximate functionals such as the LDA and the GGAs, however, are not able to reproduce vdW forces, at least not the asymptotic $R^{-6}$ behavior.

Several suggestions have been made for the treatment of vdW interactions in DFT [120, 121, 122, 123, 124]. Yet, there has been only one seamless calculation so far, performed by Dobson and Wang [125]. It follows the concept of [123] and deals with a jellium slab situation. In the case of atoms and molecules there is no functional that has been shown to provide both vdW coefficients and total correlation energies accurately. The scheme proposed by Kohn, Meir, and Makarov [124] should in principle be capable of this, but its computational implementation appears to be rather demanding.

Here, we suggest an approximation for the correlation-energy functional which will prove to describe the vdW limit correctly. The derivation is based on the adiabatic connection formula and on TDDFT. Numerical results will be presented for total correlation energies as well as vdW coefficients.

**Construction of a correlation-energy functional containing the vdW energy**

Since the HK theorem is valid for any particle-particle interaction, we can apply it to a system with arbitrary Coulomb coupling constant $\lambda$. Thus for each value of $\lambda$ between 0 and 1 there exists a uniquely determined external potential that reproduces the spin densities of the fully interacting system. (In the special case $\lambda = 0$ this potential is equal to the KS potential.) The adiabatic connection formula expresses the xc energy as an integral over the coupling constant (see e.g. [122]):

$$E_{xc} = -\frac{1}{2}\int_0^1 d\lambda \int d^3r\, d^3r'\, \frac{1}{|\mathbf{r}-\mathbf{r}'|}\left\{\left[\frac{1}{\pi}\int_0^\infty du \sum_{\sigma\sigma'} \chi^{(\lambda)}_{\sigma\sigma'}(\mathbf{r},\mathbf{r}';iu)\right] + n(\mathbf{r})\delta(\mathbf{r}-\mathbf{r}')\right\} . \tag{121}$$

Here, $\chi^{(\lambda)}_{\sigma\sigma'}(\mathbf{r},\mathbf{r}';iu)$ denotes the spin-density response function at coupling strength $\lambda$ and imaginary frequency $iu$. The exact exchange energy is obtained from Eq. (121) if $\chi^{(\lambda)}_{\sigma\sigma'}$ is replaced by the KS reponse function $\chi_{s\,\sigma\sigma'}$. Hence, the correlation energy $E_c = E_{xc} - E_x$ is given by

$$E_c = -\frac{1}{2\pi}\int_0^1 d\lambda \int d^3r\, d^3r'\, \frac{1}{|\mathbf{r}-\mathbf{r}'|}\int_0^\infty du \sum_{\sigma\sigma'}\left[\chi^{(\lambda)}_{\sigma\sigma'}(\mathbf{r},\mathbf{r}';iu) - \chi_{s\,\sigma\sigma'}(\mathbf{r},\mathbf{r}';iu)\right] . \tag{122}$$

$\chi^{(\lambda)}_{\sigma\sigma'}$ is related to the KS reponse function $\chi_{s\,\sigma\sigma'}$ by a Dyson-type equation involving the $\lambda$-dependent xc kernel $f^{(\lambda)}_{xc}$:

$$\begin{aligned}\chi^{(\lambda)}_{\sigma\sigma'}(\mathbf{r},\mathbf{r}';\omega) &= \chi_{s\,\sigma\sigma'}(\mathbf{r},\mathbf{r}';\omega) + \\ &+ \sum_{\nu\nu'}\int d^3x\, d^3x'\, \chi_{s\,\sigma\nu}(\mathbf{r},\mathbf{x};\omega)\left[\frac{\lambda}{|\mathbf{x}-\mathbf{x}'|} + f^{(\lambda)}_{xc\,\nu\nu'}(\mathbf{x},\mathbf{x}';\omega)\right]\chi^{(\lambda)}_{\nu'\sigma'}(\mathbf{x}',\mathbf{r}';\omega) .\end{aligned} \tag{123}$$

Substituting (123) into (122) leads to

$$\begin{aligned}E_c = &-\frac{1}{2\pi}\int_0^1 d\lambda \int d^3r\, d^3r'\, d^3x\, d^3x'\, \frac{1}{|\mathbf{r}-\mathbf{r}'|}\int_0^\infty du \\ &\sum_{\nu\nu'\sigma\sigma'} \chi_{s\,\sigma\nu}(\mathbf{r},\mathbf{x};iu)\left[\frac{\lambda}{|\mathbf{x}-\mathbf{x}'|} + f^{(\lambda)}_{xc\,\nu\nu'}(\mathbf{x},\mathbf{x}';iu)\right]\chi^{(\lambda)}_{\nu'\sigma'}(\mathbf{x}',\mathbf{r}';iu) .\end{aligned} \tag{124}$$

Eq.(124) is the starting point for our construction of an approximate correlation-energy functional. To this end, the xc kernel is approximated by the simple orbital-dependent formula suggested by Petersilka, Gossmann, and Gross [88, 93], given in Eq. (94). The $\lambda$-dependent version of it is linear in $\lambda$. Further, we approximate the interacting response function on the right-hand side of Eq. (124) by the non-interacting response function:

$$\chi^{(\lambda)}_{\sigma\sigma'}(\mathbf{r},\mathbf{r}';\omega) \simeq \chi_{s\,\sigma\sigma'}(\mathbf{r},\mathbf{r}';\omega) . \tag{125}$$

Then the integrations over $\lambda$ and $u$ in Eq. (124) can be done analytically. The resulting orbital functional reads as follows:

$$\begin{aligned}E_c^{\chi_s} = &-\frac{1}{2}\sum_{\substack{jkmn\\ \sigma\sigma'}} f_{jk\sigma} f_{mn\sigma'} \int d^3r\, d^3r'\, d^3x\, d^3x' \\ &\frac{1}{|\mathbf{r}-\mathbf{r}'|} f_{hxc\,\sigma\sigma'}(\mathbf{x},\mathbf{x}')\frac{Z_{jk\sigma}(\mathbf{r},\mathbf{x})Z_{mn\sigma'}(\mathbf{x}',\mathbf{r}')}{\omega_{j\sigma,k\sigma} + \omega_{m\sigma',n\sigma'}} .\end{aligned} \tag{126}$$

Here

$$Z_{jk\sigma}(\mathbf{r},\mathbf{r}') = \varphi^*_{k\sigma}(\mathbf{r})\varphi_{j\sigma}(\mathbf{r})\varphi_{k\sigma}(\mathbf{r}')\varphi^*_{j\sigma}(\mathbf{r}')\,, \tag{127}$$

$$\omega_{j\sigma,k\sigma} = \epsilon_{j\sigma} - \epsilon_{k\sigma}\,, \tag{128}$$

$$f_{jk\sigma} = f_{k\sigma}(1 - f_{j\sigma})\,, \tag{129}$$

and

$$f_{\mathrm{hxc}\,\sigma\sigma'}(\mathbf{x},\mathbf{x}') = \frac{1}{|\mathbf{x}-\mathbf{x}'|} + f^{\mathrm{PGG}}_{\mathrm{xc}\,\sigma\sigma'}(\mathbf{x},\mathbf{x}')\,. \tag{130}$$

Apparently, Eq. (126) is similar to second-order perturbation theory. This means that we can expect reasonable results for finite systems. On the other hand, the functional will diverge if it is applied to the uniform electron gas; i.e., it will fail for bulk metals.

In the case of two non-overlapping subsystems $A$ and $B$ at distance $R \to \infty$, the correlation energy (126) consists of a constant part (equal to the correlation energies of the isolated subsystems) and a vdW part. A multipole expansion of the latter shows that it exhibits the correct $R^{-6}$ dependence with the vdW coefficient

$$C_6^{X_s} = \sum_{i=1}^{3}(1+3\delta_{3i}) \sum_{\substack{jkmn\\ \sigma\sigma'}} f_{jk\sigma} f_{mn\sigma'} \frac{\left|\langle k\sigma|r_i|j\sigma\rangle^{(A)}\right|^2 \left|\langle n\sigma'|r_i|m\sigma'\rangle^{(B)}\right|^2}{\omega_{j\sigma,k\sigma}+\omega_{m\sigma',n\sigma'}}\,, \tag{131}$$

which, in the numerator, involves dipole-operator matrix elements of the isolated subsystems. We note that the xc kernel has dropped out. This is due to the fact that $f_{\mathrm{xc}\,\sigma\sigma'}(\mathbf{r},\mathbf{r}';\omega)$ goes to zero much faster than the Coulomb interaction for $|\mathbf{r}'-\mathbf{r}| \to \infty$.

Görling and Levy have derived Eq. (131) by applying perturbation theory to the vdW energy directly [126]. This shows that, as far as the $C_6$ coefficients are concerned, our approximation is exactly equivalent to second-order perturbation theory.

**Results**

Table 11 shows the resulting vdW coefficients for combinations of rare-gas atoms, alkali atoms, and hydrogen. The hydrogen wave functions were calculated from the exact potential. For the rare-gas and alkali atoms we used x-only-KLI, Colle-Salvetti-KLI (KLICS), and LDA-SIC-KLI orbitals, as indicated. (LDA-SIC is LDA with self-interaction correction [129].) We conclude that the agreement with the empirical values is very good for light atoms (H, He, Li, Ne, Na), but becomes worse for heavier elements. Compared to the rare gases, the alkali atoms are more sensitive to what kind of KS orbitals we insert.

The response of a hydrogen atom is given by the non-interacting response function without approximation. Therefore, Eq. (131) is *exact* in the case of two vdW-interacting hydrogen atoms. Indeed, the calculated $C_6$(H-H) of 6.499 a.u. (not shown in Table 11) equals the empirical value [128] of $(6.49 \pm 0.02)$ a.u.

In Table 12 we present correlation energies for He, Be, Ne, and Ar. Here, we have evaluated Eq. (126) with x-only-KLI wave functions. We compare our results with self-consistent DFT correlation energies following from well-known functionals (LDA, LDA-SIC, KLICS, PW91). Further we compare with the exact DFT correlation energies of He, Be, and Ne [130, 131] and with a semi-empirical value [54] for the quantum-chemical

**Table 11.** VdW coefficients $C_6$.[a]

| Atoms | x-KLI | KLICS | LDA-SIC-KLI | empirical |
|---|---|---|---|---|
| He-He | 1.664 | 1.639 | 1.594 | 1.458 |
| He-Ne | 3.490 | 3.424 | 3.495 | 3.029 |
| Ne-Ne | 7.447 | 7.284 | 7.761 | 6.383 |
| Ar-Ar | 128.5 | 124.4 | 133.2 | 64.30 |
| Kr-Kr | 282.4 | 271.3 | 290.0 | 129.6 |
| Xe-Xe | 730.7 | 697.1 | 734.6 | 285.9 |
| | | | | |
| Li-Li | 1460 | 1313 | 1423 | 1390 |
| Li-Na | 1689 | 1453 | 1485 | 1450 |
| Na-Na | 1957 | 1614 | 1560 | 1510 |
| K-K | 6665 | 5265 | 4662 | 3890 ± 20 |
| Rb-Rb | 9624 | 7505 | 6347 | 4870 ± 70 |
| | | | | |
| H-He | 3.022 | 2.995 | 2.948 | 2.82 ± 0.02 |
| H-Ne | 6.060 | 5.976 | 6.227 | 5.71 ± 0.07 |
| H-Li | 67.99 | 64.96 | 67.22 | 66.4 ± 0.5 |
| H-Na | 81.14 | 75.43 | 74.50 | 71.8 ± 0.3 |

[a]Coefficients calculated from Eq. (131), first three rows, are compared with empirical numbers. Hydrogen wave functions were calculated from the exact potential, all other wave functions from x-only-KLI, KLICS, or LDA-SIC-KLI potentials as indicated. Empirical $C_6$ values from [127] and [128]. Empirical values of the alkali atoms are accurate to 3 digits if no uncertainty is indicated. All numbers in atomic units.

**Table 12.** Atomic correlation energies from various approximations.[a]

| Atom | $E_c^{\chi_s}$ | LDA | LDA-SIC-KLI | KLICS | PW91 | exact/empirical |
|---|---|---|---|---|---|---|
| He | -0.048 | -0.111 | -0.0582 | -0.0416 | -0.0450 | -0.0421 |
| Be | -0.13 | -0.224 | -0.1169 | -0.0934 | -0.0942 | -0.0962 |
| Ne | -0.41 | -0.739 | -0.4283 | -0.3757 | -0.3784 | -0.394 |
| Ar | -0.67 | -1.423 | -0.8330 | -0.7435 | -0.7687 | -0.72 |

[a]The $\chi_s$ approximation (126) using the PGG xc kernel (94), first row, is compared with self-consistent DFT results and with exact DFT correlation energies for He, Be, Ne [130, 131] and the conventional quantum-chemical correlation energy for Ar [54]. All numbers in Hartrees.

**Table 13.** Total atomic ground-state energies from various approximations.[a]

| Atom | $E^{\chi_s}$ | LDA | LDA-SIC-KLI | KLICS | PW91 | exact/empirical |
|---|---|---|---|---|---|---|
| He | -2.910 | -2.835 | -2.920 | -2.903 | -2.900 | -2.9037 |
| Be | -14.71 | -14.446 | -14.695 | -14.665 | -14.648 | -14.6674 |
| Ne | -128.96 | -128.230 | -129.287 | -128.920 | -128.947 | -128.939 |
| Ar | -527.48 | -525.940 | -528.432 | -527.553 | -527.539 | -527.604 |
| $\bar{\delta}(\%)$ | 0.14 | 1.19 | 0.29 | 0.016 | 0.070 | |

[a]Total energies obtained from the $\chi_s$ approximation using the PGG xc kernel are compared with self-consistent DFT results and exact/empirical values [46, 2]. $\bar{\delta}$ denotes the mean relative deviation from the exact/empirical values. All energies in Hartrees.

correlation energy of Ar. (This quantity is defined by $E_c^{QC} = E - E^{HF}$ and should be very close to the DFT correlation energy; cf. [132].)

By adding the "$\chi_s$ correlation energy" to the x-only-KLI energy we obtain the total ground-state energies shown in Table 13. Again we compare with self-consistent DFT calculations and with exact/empirical reference values.

From Tables 12 and 13 we infer that our approximation is performing better than LDA and LDA-SIC. Especially the correlation energies of the rare-gas atoms are described accurately. However, the energies produced by the Colle-Salvetti and the PW91 functionals are still closer to the reference values.

To summarize, we have introduced a correlation-energy functional which reproduces the correct asymptotic $R^{-6}$ behavior of the vdW interaction. This functional was then used to calculate atomic correlation energies and vdW coefficients. The numerical results for the vdW coefficients of light atoms and the energies of noble-gas atoms are quite satisfactory. In view of this, the present approach appears to be a promising starting point for further refinement.

## REFERENCES


1. *Density Functional Theory,* Vol. 337 of *NATO ASI Series B,* edited by E.K.U. Gross and R.M. Dreizler (Plenum Press, New York, 1995).
2. R.M. Dreizler and E.K.U. Gross, *Density Functional Theory* (Springer, Berlin, 1990).
3. R.G. Parr and W. Yang, *Density-Functional Theory of Atoms and Molecules* (Oxford University Press, New York, 1989).
4. P. Hohenberg and W. Kohn, Phys. Rev. **136**, B864 (1964).
5. W. Kohn and L.J. Sham, Phys. Rev. **140**, A1133 (1965).
6. R.T. Sharp and G.K. Horton, Phys. Rev. **90**, 317 (1953).
7. J.D. Talman and W.F. Shadwick, Phys. Rev. A **14**, 36 (1976).
8. U. von Barth and L. Hedin, J. Phys. C **5**, 1629 (1972).
9. M.M. Pant and A.K. Rajagopal, Sol. State Commun. **10**, 1157 (1972).
10. V.R. Shaginyan, Phys. Rev. A **47**, 1507 (1993).
11. A. Görling and M. Levy, Phys. Rev. A **50**, 196 (1994).
12. J.B. Krieger, Y. Li, and G.J. Iafrate, in *Density Functional Theory,* edited by R.M. Dreizler and E.K.U. Gross (Plenum Press, New York, 1995), p. 191.
13. J.B. Krieger, Y. Li, and G.J. Iafrate, Phys. Rev. A **45**, 101 (1992).
14. M.R. Norman and D.D. Koelling, Phys. Rev. B **30**, 5530 (1984).
15. T. Grabo and E.K.U. Gross, Chem. Phys. Lett. **240**, 141 (1995).
16. V. Sahni, J. Gruenebaum, and J.P. Perdew, Phys. Rev. B **26**, 4371 (1982).
17. J.P. Perdew and M.R. Norman, Phys. Rev. B **26**, 5445 (1982).
18. Y. Li, J.B. Krieger, and G.J. Iafrate, Chem. Phys. Lett. **191**, 38 (1992).
19. Y. Li, J.B. Krieger, and G.J. Iafrate, Phys. Rev. A **47**, 165 (1993).
20. E. Engel and S.H. Vosko, Phys. Rev. A **47**, 2800 (1993).
21. J.B. Krieger, Y. Li, M.R. Norman, and G.J. Iafrate, Phys. Rev. B **44**, 10437 (1991).
22. T. Kotani, Phys. Rev. B **50**, 14816 (1994), and Phys. Rev. B **51**, 13903 (1995)(E).
23. T. Kotani, Phys. Rev. Lett. **74**, 2989 (1995).
24. T. Kotani and H. Akai, Phys. Rev. B **52**, 17153 (1995).
25. M. Städele, J.A. Majewski, P. Vogl, and A. Görling, Phys. Rev. Lett. **79**, 2089 (1997).
26. J.B. Krieger, Y. Li, and G.J. Iafrate, Phys. Lett. A **146**, 256 (1990).
27. J.B. Krieger, Y. Li, and G.J. Iafrate, Phys. Rev. A **46**, 5453 (1992).
28. J.B. Krieger, Y. Li, and G.J. Iafrate, Int. J. Quantum Chem. **41**, 489 (1992).
29. J.B. Krieger, J. Chen, Y. Li, and G.J. Iafrate, Int. J. Quantum Chem. Symp. **29**, 79 (1995).
30. E. Engel, S. Keller, A. Facco Bonetti, H. Müller, and R.M. Dreizler, Phys. Rev. A **52**, 2750 (1995).
31. T. Kreibich, E.K.U. Gross, and E. Engel, Phys. Rev. A **57**, 138 (1998).
32. T. Grabo, T. Kreibich, S. Kurth, and E.K.U. Gross, in *Strong Coulomb Correlations in Electronic Structure: Beyond the Local Density Approximation,* edited by V.I. Anisimov (Gordon&Breach, Tokyo, 1998).

33. T. Kreibich, S. Kurth, T. Grabo and E.K.U. Gross, Adv. Quantum Chem. (1998), in press.
34. J.P. Perdew, R.G. Parr, M. Levy and J.L. Balduz, Phys. Rev. Lett. **49**, 1691 (1982).
35. A.D. Becke, Phys. Rev. A **38**, 3098 (1988).
36. J.P. Perdew, K. Burke, and Y. Wang, Phys. Rev. B **54**, 16533 (1996).
37. E. Engel, H. Müller, C. Speicher, and R.M. Dreizler, in *Density Functional Theory*, edited by E.K.U. Gross and R.M. Dreizler (Plenum Press, New York, 1995), p. 65.
38. A. K. Rajagopal, J. Phys. C **11**, L943 (1978).
39. A.H. MacDonald and S.H. Vosko, J. Phys. C **12**, 2977 (1979).
40. E. Engel, S. Keller, and R.M. Dreizler, Phys. Rev. A **53**, 1367 (1996).
41. R. Colle and D. Salvetti, Theoret. Chim. Acta **37**, 329 (1975).
42. R. Colle and D. Salvetti, Theoret. Chim. Acta **53**, 55 (1979).
43. C. Lee, W. Yang, and R.G. Parr, Phys. Rev. B **37**, 785 (1988).
44. J.P. Perdew and Y. Wang, Phys. Rev. B **45**, 13244 (1992).
45. J.A. Montgomery, J.W. Ochterski, and G.A. Petersson, J. Chem. Phys. **101**, 5900 (1994).
46. E.R. Davidson, S.A. Hagstrom, S.J. Chakravorty, V.M. Umar, and C. Froese Fischer, Phys. Rev. A **44**, 7071 (1991).
47. T. Grabo and E.K.U. Gross, Int. J. Quantum Chem. **64**, 95 (1997).
48. T. Grabo, T. Kreibich, and E.K.U. Gross, Mol. Engineering **7**, 27 (1997).
49. A.A. Radzig and B.M. Smirnov, *Reference Data on Atoms and Molecules* (Springer, Berlin, 1985).
50. B.G. Johnson, P.M.W. Gill, and J.A. Pople, J. Chem. Phys. **98**, 5612 (1993).
51. K.P. Huber and G. Herzberg, *Molecular Spectra and Molecular Structure: IV. Constants of Diatomic Molecules* (Van Nostrand Reinhold, New York, 1979).
52. W. Kolos and L. Wolniewicz, J. Chem. Phys. **49**, 404 (1968).
53. C. Filippi and C.J. Umrigar, J. Chem. Phys. **105**, 213 (1996).
54. S.J. Chakravorty, S.R. Gwaltney, E.R. Davidson, F.A. Parpia, and C. Froese Fischer, Phys. Rev. A **47**, 3649 (1993).
55. K. A. Peterson, R. A. Kendall, and T. H. Dunning, J. Chem. Phys. **99**, 1930 (1993).
56. M.A. Buijse and E.J. Baerends, in *Density Functional Theory of Molecules, Clusters and Solids*, edited by D.E. Ellis (Kluwer Academic Publishers, Amsterdam, 1995), p. 1.
57. O.V. Gritsenko, R. van Leeuwen, and E.J. Baerends, J. Chem. Phys. **104**, 8535 (1996).
58. E. Runge and E.K.U. Gross, Phys. Rev. Lett. **52**, 997 (1984).
59. E.K.U. Gross and W. Kohn, Adv. Quant. Chem. **21**, 255 (1990).
60. E.K.U. Gross, J.F. Dobson, and M. Petersilka, in *Density Funcional Theory II*, Vol. 181 of *Topics in Current Chemistry*, edited by R.F. Nalewajski (Springer, Berlin, 1996), p. 81.
61. K. Burke and E.K.U. Gross, in *Density Funcionals: Theory and Applications*, Vol. 500 of *Lecture Notes in Physics*, edited by D. Joubert (Springer, Berlin, 1998), p. 116.
62. *Atoms in Intense Laser Fields*, edited by M. Gavrila (Academic Press, Boston, 1992).
63. in *Super Intense Laser-Atom Physics IV*, Vol. 13 of *NATO ASI Series*, edited by H.G. Muller and M.V. Fedorov (Kluwer, Dordrecht, 1996).
64. in *Multiphoton Processes 1996*, No. 154 in *Institute of Physics Conference Series*, Institute of Physics, edited by P. Lambropoulos and H. Walther (Institute of Physics Publishing, Bristol, 1997).
65. K.L. Liu and S.H. Vosko, Can. J. Phys. **67**, 1015 (1989).
66. D. Tiszauer and K.C. Kulander, Phys. Rev. A **29**, 2909 (1984).
67. L.F. Errea, L. Méndez, A. Riera, M. Yáñez, J. Hanssen, C. Harel, and A. Salin, J. Physique **46**, 719 (1985).
68. I.L. Cooper, A.S. Dickinson, S.K. Sur, and C.T. Ta, J. Phys. B **20**, 2005 (1987).
69. A. Henne, H.-J. Lüdde, A. Toepfer, and R.M. Dreizler, Phys. Lett. A **124**, 508 (1987).
70. W. Fritsch and C.D. Lin, Phys. Lett. A **123**, 128 (1987).
71. J.F. Reading and A.L. Ford, Phys. Rev. Lett. **58**, 543 (1987).
72. J.F. Reading and A.L. Ford, J. Phys. B **20**, 3747 (1987).
73. X. Tang, H. Rudolph, and P. Lambropoulos, Phys. Rev. A **44**, R6994 (1991).
74. T. Ando, Z. Phys. B **26**, 263 (1977).
75. A. Zangwill and P. Soven, Phys. Rev. A **21**, 1561 (1980).
76. S.H. Vosko, L. Wilk, and M. Nusair, Can. J. Phys. **58**, 1200 (1980).
77. C.A. Ullrich, U.J. Gossmann, and E.K.U. Gross, Phys. Rev. Lett. **74**, 872 (1995).
78. P.A.M. Dirac, Proc. Cam. Phil. Soc. **26**, 376 (1930).
79. J.B. Krieger, Y. Li, and G.J. Iafrate, Phys. Rev. A **46**, 5453 (1992).
80. J.B. Krieger, Y. Li, and G.J. Iafrate, Phys. Rev. A **45**, 101 (1992).

81. C.A. Ullrich, U.J. Gossmann, and E.K.U. Gross, Ber. Bunsenges. Phys. Chem. **99**, 488 (1995).
82. C.A. Ullrich, S. Erhard, and E.K.U. Gross, in *Super Intense Laser Atom Physics IV*, edited by H.G. Muller and M.V. Fedorov (Kluwer, Dordrecht, 1996), p. 267.
83. C.A. Ullrich and E.K.U. Gross, Comments At. Mol. Phys. **33**, 211 (1997).
84. C.A. Ullrich, P.G. Reinhard, and E. Suraud, J. Phys. B: At. Mol. Opt. Phys. **30**, 5043 (1997).
85. C.A. Ullrich, P.G. Reinhard, and E. Suraud, J. Phys. B: At. Mol. Opt. Phys. **31**, 1871 (1998).
86. C.A. Ullrich, P.G. Reinhard, and E. Suraud, Phys. Rev. A **57**, 1938 (1998).
87. S. Erhard and E.K.U. Gross, in *Multiphoton Processes*, Vol. 154 of *Institute of Physics Conference Series*, edited by P. Lambropoulos and H. Walther (Institute of Physics, Bristol, 1997), p. 37.
88. M. Petersilka, U.J. Gossmann, and E.K.U Gross, Phys. Rev. Lett. **76**, 1212 (1996).
89. E.K.U. Gross and W. Kohn, Phys. Rev. Lett. **55**, 2850 (1985), (E) ibid. **57**, 932 (1986).
90. K.L. Liu, Can. J. Phys. **69**, 573 (1991).
91. J.F. Dobson, Phys. Rev. Lett. **73**, 2244 (1994).
92. G. Vignale, C.A. Ullrich and S. Conti, Phys. Rev. Lett. **79**, 4878 (1997).
93. M. Petersilka, U. Gossmann, and E.K.U. Gross, in *Electronic Density Functional Theory: Recent Progress and New Directions*, edited by G. Vignale J.F. Dobson and M.P. Das (Plenum, New York, 1998), p. 177.
94. O. Gunnarsson and B.I. Lundqvist, Phys. Rev. B **13**, 4274 (1976).
95. T. Ziegler, A. Rauk, and E.J. Baerends, Theoret. Chim. Acta **43**, 261 (1977).
96. U. von Barth, Phys. Rev. A **20**, 1693 (1979).
97. A. Theophilou, J. Phys. C **12**, 5419 (1979).
98. W. Kohn, Phys. Rev. A **34**, 5419 (1986).
99. E.K.U. Gross, L.N. Oliveira, and W. Kohn, Phys. Rev. A **37**, 2805 (1988).
100. E.K.U. Gross, L.N. Oliveira, and W. Kohn, Phys. Rev. A **37**, 2809 (1988).
101. L.N. Oliveira, E.K.U. Gross, and W. Kohn, Phys. Rev. A **37**, 2821 (1988).
102. A. Nagy, Phys. Rev. A **42**, 4388 (1990).
103. A. Nagy, Int. J. Quantum Chem. Symp. **29**, 297 (1995).
104. M. Levy, Phys. Rev. A **52**, R4313 (1995).
105. A. Görling, Phys. Rev. A **54**, 3912 (1996).
106. M. Petersilka and E.K.U. Gross, Int. J. Quantum Chem. **60**, 181 (1996).
107. Ch. Jamorski, M.E. Casida, and D.R. Salahub, J. Chem. Phys. **104**, 5134 (1996).
108. R. Bauernschmitt and R. Ahlrichs, Chem. Phys. Lett. **256**, 454 (1996).
109. R. Bauernschmitt, M. Hauser, O. Treutler, and R. Ahlrichs, Chem. Phys. Lett. **264**, 573 (1997).
110. T. Grabo, Ph.D. thesis, Universität Würzburg, 1997.
111. S.J.A. van Gisbergen, F. Kootstra, P.R.T. Schipper, O.V. Gritsenko, J.G. Snijders, and E.J. Baerends, Phys. Rev. A **57**, 2556 (1998).
112. M.E. Casida, C. Jamorski, K.C. Casida, and D.R. Salahub, J. Chem. Phys. **108**, 4439 (1998).
113. S.J.A. van Gisbergen, Ph.D. thesis, Vrije Universiteit, Amsterdam, 1998.
114. T. Grabo, M. Petersilka, and E.K.U. Gross, Phys. Rev. A (1999), in press.
115. C.E. Moore, Nat. Stand. Ref. Data Ser. (U.S. Nat. Bur. Stand) **35**, Vol. I,II, and III (1971).
116. E.S. Nielsen, P. Jørgensen, and J. Oddershede, J. Chem. Phys. **73**, 6238 (1980).
117. E. Zaremba and W. Kohn, Phys. Rev. B **13**, 2270 (1976).
118. S.J.A. van Gisbergen, J.G. Snijders, and E.J. Baerends, J. Chem. Phys. **103**, 9347 (1995).
119. S.J.A. van Gisbergen, V.P. Osinga, O.V. Gritsenko, R. van Leeuwen, J.G. Snijders, and E.J. Baerends, J. Chem. Phys. **105**, 3142 (1996).
120. K. Rapcewicz and N.W. Ashcroft, Phys. Rev. B **44**, 4032 (1991).
121. Y. Andersson, D.C. Langreth, and B.I. Lundqvist, Phys. Rev. Lett. **76**, 102 (1996).
122. J.F. Dobson and B.P. Dinte, Phys. Rev. Lett. **76**, 1780 (1996).
123. J.F. Dobson, B.P. Dinte, and J. Wang, in *Electronic Density Functional Theory: Recent Progress and New Directions*, edited by G. Vignale J.F. Dobson and M.P. Das (Plenum, New York, 1998), p. 261.
124. W. Kohn, Y. Meir, and D.E. Makarov, Phys. Rev. Lett. **80**, 4153 (1998).
125. J.F. Dobson and J. Wang, to be published.
126. A. Görling and M. Levy, to be published.
127. A. Kumar and W.J. Meath, Mol. Phys. **54**, 823 (1985).
128. K.T. Tang, J.M. Norbeck, and P.R. Certain, J. Chem. Phys. **64**, 3063 (1976).
129. J.P. Perdew and A. Zunger, Phys. Rev. B **23**, 5048 (1981).
130. C.J. Umrigar and X. Gonze, Phys. Rev. A **50**, 3827 (1994).
131. C.J. Umrigar and X. Gonze, in *High Performance Computing and its Application to the Physical Sciences*, edited by D.A. Browne et al (World Scientific, Singapore, 1993).

132. E.K.U. Gross, M. Petersilka, and T. Grabo, in *Chemical Applications of Density Functional Theory, ACS Symposium Series 629*, edited by B.B Laird, R.B. Ross, and T. Ziegler (American Chemical Society, Washington, DC, 1996), p. 42.

# Discussion

**Perdew**: You calculated accurate Van Der Waals coefficients for the light atoms, and less accurate coefficients for the heavier atoms. Can you explain the difference?

**Gross**: No

**Sawatzky:** You mentioned that the off diagonal Spin Orbit coupling terms would be related to the gradient in the superconducting order parameter, and that a large gradient could be obtained experimentally using artificial multilayer systems. Would it not also be possible to use simply a type 2 superconductor in which at finite magnetic fields above Hc1 the Vortex lattice would provide a natural inhomogeneous superconducting order parameter?

**Gross:** In principle, yes, but according to our estimates the gradients would be much smaller than for the superlattice I mentioned.

# UNDERSTANDING ELECTRONIC WAVE FUNCTIONS

D. M. Ceperley

Department of Physics and NCSA
University of Illinois,
Urbana-Champaign, IL, 61801, USA

## INTRODUCTION

In this article I discuss some aspects of what is known about correlated electronic wavefunctions, in particular what has been learned from quantum Monte Carlo studies of the homogeneous electron gas. First, I motivate the discussion, then briefly give references to the quantum Monte Carlo methods, discuss the pair product wavefunction and its generalization to backflow and three body wavefunctions. Finally, I conclude with some current difficulties in extending these ideas to electrons in an external potential.

Before we dive in, let me open this short review with some philosophy, appropriate to the conference setting. What does it mean to say one understands something about electron correlation? We will take the somewhat controversial position that an important aspect is to understand the many-body wavefunction. The conventional picture, dating back to Bohr and Heisenberg, is that all we should strive to do is to calculate measurable quantities, the response functions, densities and so forth. More recently, density functional theory takes the point of view that the density functional is the main object to understand. The debate goes back to the founders of quantum mechanics with the Copenhagen interpretation arguing that one should just deal with measurable objects. However, many physicists felt that quantum theory become more definite when Schrödinger wrote down his equation and introduced the idea of a wavefunction, leading materialists to hope that the wavefunction was the reality behind quantum mechanics.

If we can, by some means, specify a many-body wavefunction with sufficient accuracy then all matrix elements are determined. The wavefunction contains complete information but it is a complicated $3N$ dimensional function. We will not necessarily be able to compute any properties analytically but they can be calculated straightforwardly using Monte Carlo techniques. The understanding of quantum systems will have been reduced to understanding of classical systems .This is one justification for seeking to determine the many-body wavefunction. It is by no means obvious to me that the goal of determining the wave function is any more difficult that determining the density functional. Whether or not we can achieve this goal, a challenging intellectual goal is to determine the many body wave function, and is a puzzle of some practical importance.

The many-body wavefunction for N electrons is a 3N dimensional antisymmetric function, even if we forget about electron spin. Feynman argued[1] that a completely

*Electron Correlations and Materials Properties*, edited by Gonis *et al.*
Kluwer Academic/Plenum Publishers, 1999.

general understanding is ruled out, simply because the initial conditions require too much information. In quantum mechanics, one is allowed to take any initial conditions on the wavefunction. If each dimension requires a mesh or basis of say 100 functions, the entire function will require $100^{3N}$ complex numbers. To seek understanding in the general case is hopeless; we have to restrict ourselves to simpler situations: either low-lying eigenstates, or, a possibility more common but not considered here, systems in thermodynamic equilibrium.

Many problems will require the density matrix because, except for isolated non-degenerate systems, the density matrix is more fundamental. For example typical highly excited states are chaotic, and it is expected that the wave function is effectively random. However, the thermal density matrix, at a high or a low temperature, is a smooth function. I will not worry about the difference in this talk, but assume that much of what is written about the wavefunction can be generalized to the many-body density matrices, though that is not always true.

Even given these caveats what does it mean to understand the wavefunction? There are at least two aspects of understanding:

*A. Modeling.* By this I mean to find an analytically solvable model whose wavefunction is close to the exact wavefunction. One of the earliest successful models in science was the two sphere model of astronomy. This model allows one to correlate various facts with a simple geometrical picture, but fails in calculating precise trajectories of the planets. That was supplanted by the Ptolemaic models, which are more accurate but less understandable. If one has a simple fairly accurate model (*e.g.* Fermi liquid theory or BCS theory) but which is not able to calculate important properties, there is a question of whether we can really claim full understanding. We must beware of models that manage to account for most but not all of the features, since they may turn out to be completely wrong.

*B. Practical calculations* of electronic systems are essential to many areas of science and technology. We note in a similar vein, that modern science resulted from the need for more accurate astronomical calculations. Copernicus took pains to make clear that he was only introducing a simpler numerical scheme. This argues that the struggle to compute things may itself lead to understanding.

Let us consider to what extent we understand something as simple as the square root of a number? Of course we can determine that $\sqrt{2} = 1.4142135623730950488\ldots$ One knows that the digits do not repeat but that is negative information. There are connections to all sorts of mathematical properties of functions (e.g. in the complex plane) that are clearly very useful. We have similar understanding of general properties of wavefunctions. Clearly tabulating values of the square root, useful activity a century ago, has little to do with understanding. Now we compute it as we need it. So is understanding wavefunctions simply akin to understanding what the square root button does on a calculator?

One sort of understanding most of us have is the algorithm for calculating $\sqrt(x)$ using pencil and paper. A general procedure involves Newton iteration:

**Procedure for square root of y**

```
initialize x
repeat
   error = x^2-y
   x = x - error/(2x)
while | error | > tolerance
```

We find an initial approximation, a procedure for estimating the error and for using this error to make a better approximation. For the square root function, it is not hard to prove that such a procedure must converge if the initial approximation is good enough. In fact, you can easily determine how many iterations you will need for a given number of decimal places. If we come up with a procedure like that for the wavefunction, would that constitute an understanding?

Now, let us return to the problem of understanding the correlated electronic wavefunction. In many cases, the role of the good approximation or model (A) is given by a single determinant wave function, and it accounts for most of the energy. What is difficult in chemistry is that so much accuracy is required that even if the Slater determinant is very good, one must do much better in practice. So, what new terms come in when electrons are correlated?

As mentioned above, one cannot simply make a big real-space table of the wavefunction as that would be too verbose. One systematic improvement for practical calculations (B) is to express the wavefunction as a sum over a complete set of single determinants:

$$\Psi(R) = \sum_{k=1}^{L} c_k Det_k(R) \tag{1}$$

with the $c_k$ solved for using a variational procedure. This is known as the Configuration Interaction(CI) method in quantum chemistry. This is clearly an improvement over simply making a big table of wavefunction values since the first term (a mean field description) accounts for much of the physics and the corrections are single particle states, chosen to be low in energy. Perturbation theory allows us in advance to select which $k$'s will have large coefficients. But does a long list of $c_k$'s constitute an understanding? The values of $c_k$ are analogous to the digits in the expansion of the square root. The main practical problem with this method is that it is still too verbose as the number of electrons increase because electron correlation is not succinctly expressed in terms of determinants. As $N \rightarrow \infty$ then $L \propto \exp(\kappa N)$. Even with today's computers, 30 fully correlated electrons are difficult to treat to "chemical accuracy". Another problem is size-consistency, namely there is no simple relation between the coefficients for the super- molecule $AB$ to those of the molecules $A$ and $B$.

I propose the following two guidelines for rating the understandability of wavefunctions:

A. Other things being equal, for example the same error in the energy, a more "understandable" wavefunction has fewer adjustable parameters. It is not clear if we have not really solved anything fundamental with this definition since we now have to define "adjustable parameters". Adjustable parameters require lengthy calculations to determine their values and are not universal.

B. A purely algorithmic guideline is related the cost of determining a wavefunction value or property to a given accuracy; in other words the speed and memory of a calculator that can compute the value of $\Psi$. We do not have to write down the wave function in a big table; we just have to be able to deliver it on demand quickly. For a more understandable wavefunction, this cost is lower.

## QUANTUM MONTE CARLO METHODS

There are several Quantum Monte Carlo methods that can be used either A) to assess the accuracy of a given model or B) to provide a computational tool to determine the wavefunction or one of its expectations.

i) Variational Monte Carlo (VMC), where one samples the square of a trial function as if it were a classical Boltzmann distribution, is a way to determine the properties of a given wavefunction[4]. By optimizing the energy or variance defined as:

$$E_V = \langle\psi|\mathcal{H}|\psi\rangle \tag{2}$$

$$\sigma^2 = \langle\psi|[\mathcal{H} - E_V]^2|\psi\rangle \tag{3}$$

we can choose the best wave function from a given class. In some cases we can determine how accurate a functional form is. In contrast to methods such as SCF and CI of quantum chemistry, any analytic form can be used as a trial function, the only limitation being the cost and memory of computation.

ii) Diffusion Monte Carlo (DMC) is a way to optimize automatically the wavefunction. It is an exact method for bosonic systems. Thus, it can be used to check on the optimization of VMC and put in all kinds of complicated but positive pieces into the wavefunction. For fermion systems, one typically uses the fixed-node approximation, which gives an upper bound to the ground state energy, the best upper bound consistent with the assumed nodes of the trial function. Exact fermion methods (transient estimate or release node) are useful for small fermion systems but they become exponentially slow as the size of the system grows, because of the fermion sign problem.

iii) Path Integral Monte Carlo (PIMC)is conceptually similar to DMC but determines finite temperature properties. Because there are as yet few numerical applications to low temperature electronic systems I will not discuss this further.

**THE PAIR PRODUCT OR SLATER-JASTROW WAVEFUNCTION** In the

following sections, I will take the simplest continuum example of a many body system: non- relativistic particles interacting with a pair potential:

$$\mathcal{H} = -\hbar^2/2m\sum_{i=1}^{N}\nabla_i^2 + \sum_{i<j} v(r_{ij}). \tag{4}$$

In the electron gas, the particles interacts with the Coulomb potential $v(r) = e^2/r$ where $r_s = a/a_0$ is the only dimensionless parameter $\propto n_e^{-1/3}$. For $r_s > 100$ the stable ground state is a quantum crystal. In the small $r_s$ limit, the electrons become uncorrelated and methods based on perturbing around the free Fermi gas are successful. Solid state systems are in the range of $1 < r_s < 5$.

The almost universal choice for a correlated wavefunction is the pair product or Jastrow function (other examples are the Gutzwiller wavefunction for lattice spin systems and the Laughlin wave function for quantum Hall liquids):

$$\Psi(R) = Det(\phi_k(\mathbf{r}_i))\exp(-\sum_{i<j} u(r_{ij})) \tag{5}$$

where $u(r)$ is the *correlation factor* and the *orbitals* determined by translational symmetry are plane waves $\phi_k(\mathbf{r}) = e^{i\mathbf{kr}}$. $\mathbf{k}$ lies beneath the Fermi surfaces. We know analytically the behavior of the correlation function at both large and small $r$: in the large limit the elementary excitations are plasmons (Bohm and Pines) and in the small $r$ limit one has the cusp condition on the correlation factor.

Remarkably, for the electron gas there is an analytic form wavefunction having no adjustable parameters, the Gaskell-RPA trial function[2] which has these short and long-range properties. It has a simple form in Fourier space:

$$2u(k) = -1/S_{NI}(k) + \sqrt{1/S_{NI}(k)^2 + 12r_s/k^4}. \tag{6}$$

Here, $S_{NI}(k)$ is the structure function for the Slater determinant. This form gets about 95% of the correlation energy. No better form, even those having lots of adjustable parameters[15], has been found. Other properties such as correlation functions are good. Hence, at this level of accuracy, we understand the wavefunction of the electron gas.

Do we understand why the form is so accurate for the electron gas? It turns out that this form is not so accurate on systems which have more than one length scale. Probably it is so accurate for the electron gas because the large and small r behaviors more or less determine everything.

One of the predictions from VMC and DMC calculations with this trial function is that the low density liquid is likely to be a partially spin-polarized liquid between $20 \leq r_s \leq 100$[3, 5, 6]. Recent experiments have confirmed this ferromagnetism in a solid state system[16]. The ground state prefers to break local spin symmetry and put correlation in a Slater determinant, rather than building up more two body correlation. This is a general rule for highly correlated systems, related to Hund's rules in atoms.

### The Wigner crystal

The liquid wavefunction breaks down at very low densities when the system crystallizes. The pair product form using plane wave orbitals predicts crystallization at a much too low a density. One needs to explicitly break translational symmetry by introducing the lattice symmetry into the wavefunction. Shadow wavefunctions[8] probably would spontaneously give the crystal symmetry without having to put it in, but they are more closely related to the path integral description and are not easy to use for fermion systems. We can break translational symmetry by having the orbitals be Wannier functions localized around the lattice sites, $\{\mathbf{z}_k\}$:

$$\phi_k(r) = f(\mathbf{r} - \mathbf{z}_k). \tag{7}$$

The simplest choice of the orbital is a Gaussian $f(r) = \exp(-cr^2)$, which introduces one additional parameter, $c$ controlling the electron localization. This wavefunction is also very accurate. Using it, VMC and DMC calculations have predicted that Wigner crystallization will occur at $r_s \approx 100$ in 3d and $r_3 \approx 37$ in 2d. Experiments[17] on a clean GaAs/AlGa/As hetero-structure have recently confirmed this prediction in 2d.

Questions remain such as how the system "knows" to break translational symmetry. What would happen in a finite system such as a quantum dot or around a defect to the wavefunction? Path integrals provide a natural way to answer some of these questions computationally but they do not necessarily lead to an understandable wavefunction.

### Higher Order Wavefunctions

Why go beyond the pair level of a wavefunction? First of all some properties are completely wrong at the pair level, such as the spin dependent Fermi liquid parameters[13]. Also, we need better nodes for DMC. Feenberg proposed an expansion for the wave function for bosonic systems. Can we do something analogous be done for fermion systems?

In calculating the square root, we iterated an initial guess using the error of the guessed solution to make an improved solution. Let us try to do the same for a function,

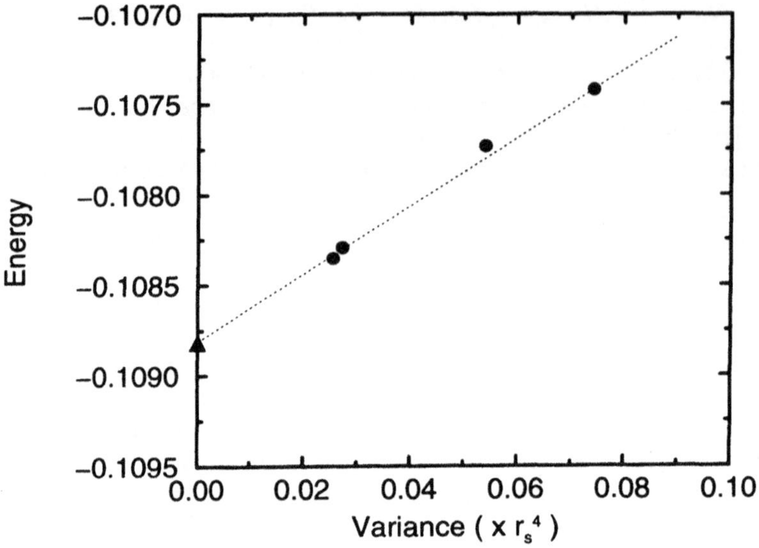


Figure 1. Variational energy versus the variance of the local energy for 3d electron fluid at a density $r_s = 10$. Each point represents one variational calculation from higher to lower energies: pair product, three-body, backflow,and backflow+three body. The filled triangle is the backflow fixed-node result and the dotted line a linear fit through the points.

not just a number. This is the basis for iterative eigenvalue solvers such as the Lanczos method. The problem with directly applying those iterative methods is that one quickly ends up having to represent the complete many-body wavefunction, which is infeasible in many-dimensions and also not understandable. We now describe a method which does not have these drawbacks. Brief accounts of this have appeared elsewhere[13].

The error of a trial function $\phi_n(R)$ is its local or residual energy:

$$E_n(R) = \phi_n^{-1}(R)\mathcal{H}\phi_n(R). \tag{8}$$

If the trial function were exact, the local energy would be constant. The method of using the local energy to correct the trial function is based on the generalized Feynman-Kacs formula[19]: the exact wave function at a point R, is related to the average of the local energy over random walks starting at R and generated using a drift term proportional to $\nabla_i \ln[\phi_n(R)]$.

Let us take as our initial approximation a mean field solution, the Hartree state:

$$\phi_1(R) = \prod_i \chi_i(\mathbf{r}_i). \tag{9}$$

Then the basic iteration is a correction of the form:

$$\phi_{n+1}(R) = \phi_n(R)\exp(-\tilde{E}_n(R)) \tag{10}$$

where $\tilde{E}$ means to smooth the local energy and should really done by path integrals. In practice, we take the same many-body functional form, but optimize with VMC within that form. Finally, we use the permutation operator to make a proper antisymmetric function:

$$\hat{\phi}_n(R) = \frac{1}{N!}\sum_P (-1)^P \phi_n(PR) \tag{11}$$

Numerically, we have carried out this procedure to third order. For homogeneous systems, at second order one gets the pair product function or Jastrow function. At third order, one generates backflow and polarization terms. Figure (1) shows how the energy and variance of the trial function improve order by order. Third order seems to get about 80% of the energy missing at second order.

Hence this smoothed-local-energy method produces a sequence of increasingly accurate correlated wavefunctions for homogeneous systems. The other system on which it has been tried out is liquid $^3$He[22] There is also seems to be quite accurate. At each order, one introduces new one particle functions. At second order, one introduces the correlation factor $u(r)$ determined primarily by the long and short-range properties. At third order one introduces a backflow and polarization functions. Polarization is a bosonic correlation while the backflow dresses the particles and changes the fermion nodes.

The sequence of functions so introduced is understandable in the sense I discussed in the introduction: only a few parameters are needed to specify a $3N$ dimensional function, where $N$ itself is arbitrary $N \gg 1$. We also get computationally feasible functions since computer time needed to estimate some property of the wavefunction is roughly proportional to $N^3$. As yet we do not know if this procedure can be continued to higher order and thereby make arbitrarily accurate functions. Also the rôle of the initial state is not known and how can we generate other states such as superconductors, spin and charge density waves.

## WAVEFUNCTIONS FOR MOLECULES

The bad news is the accuracy of this type of wavefunction when there is an external potential such as would be created by an array of fixed nuclei. Over the last decade there have been a number of attempts to determine accurate energies for very simple molecules using correlated wavefunctions. We discuss low-Z molecules because the experiments and theory are easier than is possible with extended solids since there are no finite size effects or pseudo-potentials to worry about.

Let us try to generalize the Jastrow pair wavefunction of Eq.(5) to a molecule. The first problem is that the orbitals are not determined by translation symmetry as they are in the uniform systems. In computational quantum chemistry this problem has been extensively worked on; the result has been an established method using an expansion of "contracted" Gaussians. This is a practical solution, but the answer is not very understandable. Such orbitals are normally used as the starting points for correlated calculations with VMC or DMC.

When we go to the next order, the correlation factor, $u(\mathbf{r}_i, \mathbf{r}_j)$ becomes a six dimensional function instead of the one dimensional function it is in homogeneous system. The technical difficulty in representing the correlation function becomes very important. The basis set problem of quantum chemistry (what basis set is best for a given problem, which is most compact, fastest etc.) gets correspondingly worse. Eventually this problem will be solved, it is just a difficult technical problem. But the correlation factor is now less understandable than it was for the uniform electron gas because of the number of coefficients that come in.

Even after extensive optimization at the second order, the results are not very good; see Fig.(2). Shown are results with single determinants and multiple determinants, with VMC and DMC calculations. Typical errors for the first row atoms and diatomics are often greater than 1 eV, equivalent to 10,000 K. This is not nearly accurate enough.

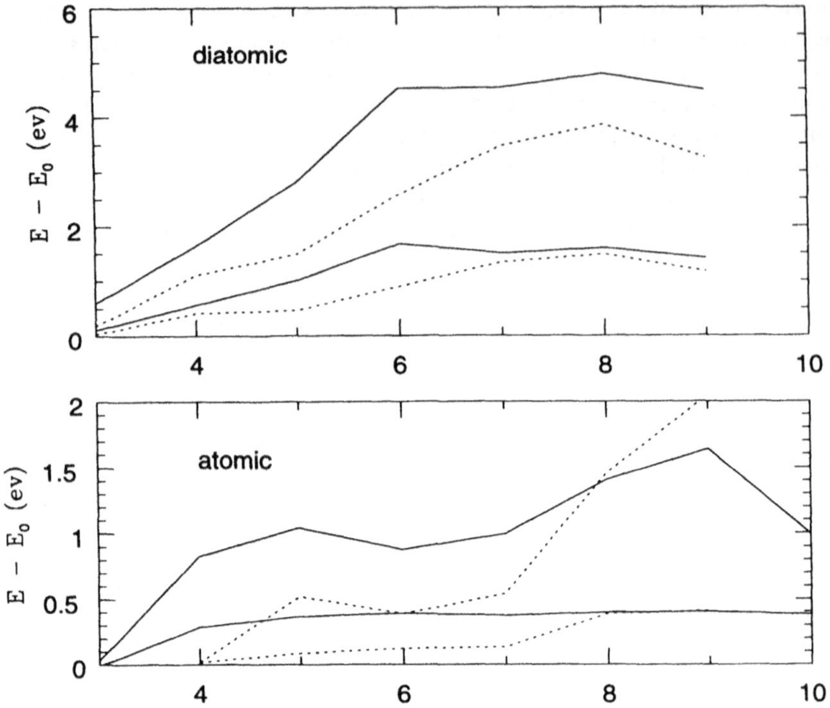


Figure 2. The accuracy of quantum Monte Carlo calculations for the total energy of first row atoms and homo-nuclear diatoms. The horizontal axis is thc atomic number, the vertical axis is the error in the total energy (relative to experiment). Solid lines are single determinant wave functions, dashed lines are multiple determinants. The upper two curves in each plot are VMC calculations, the lower are DMC calculations[20, 21]

We have not learned to put in correlation in a wavefunction for something as simple as an atom in an understandable way, if the electron orbitals involve more than one symmetry, *i. e.* if s and p orbitals are both occupied. One does get higher accuracy if one replaces inner electrons with a pseudo-potentials. Hence, the problem seems to be in an inefficient treatment of electron correlation of electrons in different shells. This is a key problem in understanding electron correlation.

## CONCLUSIONS

What good have Quantum Monte Carlo methods been in learning about electron correlation? As mentioned earlier, QMC is useful for assessing the quality of a trial function and for providing essentially exact results for simple systems. The impact of QMC has been small in materials science except for homogeneous systems such as the electron gas, very light atoms such as solid hydrogen where no pseudo potentials are needed. Recent calculations with atoms and molecules using pseudo-potentials are starting to have a major impact in calculations for difficult problems in chemistry. Even in simple systems, one is not certain that the right phase (e.g. a paired superconductor) has been found. The following list summarizes some of the correlated properties that have been determined for the e- gas:

- Energy versus density and polarization [5, 7]

- Momentum distribution of the electron gas [15]
- Phase diagram of electron gas [5, 12]
- Static linear response of 2 and 3D electron gas [10]
- Pair correlation and structure functions[15]
- Electron gas surface [9, 11]
- Fermi liquid parameters [18]

In recent years GPS devices have revolutionized the way we locate our position relative to the earth. The complicated theory and technology to make it work was the result of many years of scientific endeavor, but unimportant to the end user, who is only concerned with its cost and accuracy. When will the day arrive when understanding of electron wavefunctions and the ability to perform routine calculations allow science to make a gadget with the wizardry of a GPS device for quantum physics?

I have been supported by the NSF grant DMR-98-02373 and the Department of Physics at the University of Illinois at Urbana-Champaign. More references are found on http://www.ncsa.uiuc.edu/Apps/CMP.

## References


[1] Feynman, R. P., Int. J. Theo. Phys. **21**, 467 (1985).

[2] Gaskell, T., Proc. Phys. Soc. **77**, 1182 (1961); **80**, 1091 (1962).

[3] Ceperley, D., Phys. Rev. **18**, 3126 (1978).

[4] Ceperley, D., Chester, G. V. and Kalos, M. H., Phys. Rev. **B16**, 3081 (1977).

[5] Ceperley, D. M. and B. J. Alder, Phys. Rev. Letts. **45**, 566 (1980).

[6] Alder, B. J., Ceperley, D. M. and E. L. Pollock, Int. Journ. Quant. Chem. **16**, 49 (1982).

[7] Tanatar, B. and D. M. Ceperley, Phys. Rev. **B39**, 5005 (1989).

[8] Vitiello, S. A., K. J. Runge and M. H. Kalos, Phys. Rev. Lett. **60**, 1970 (1988).

[9] Li. X.-P., R. J. Needs, R. M. Martin and D. M. Ceperley, Phys. Rev. **B45**, 6124 (1992).

[10] Moroni, S., D. M. Ceperley, and G. Senatore, Phys. Rev. Letts. **75**, 689 (1995).

[11] Acioli, P. H. and D. M. Ceperley, Phys. Rev. **B54**, 17199 (1996).

[12] Jones, M. D. and D. M. Ceperley, Phys. Rev. Letts. **76**, 4572 (1996).

[13] Kwon, Y., D. M. Ceperley, and R. M. Martin, Phys. Rev. **B48**,12037 (1993).

[14] Kwon, Y., D. M. Ceperley, and R. M. Martin, Phys. Rev. **B58**,6800 (1998).

[15] Ortiz, G. and P. Ballone, Phys. Rev. **B50**, 1391 (1994); **56**, 9970 (1997).

[16] D. P. Young et al., Nature, Jan. (1999).

[17] Yoon, J., C. C. Li, D. Shar, D. C. Tsui and M. Shayegan, preprint xxx.lanl.gov. cond-mat/9807235.

[18] Kwon, Y., D. M. Ceperley, and R. M. Martin, Phys. Rev. **B50**,1694 (1994).

[19] Pollock, E. L. and D. M. Ceperley, Phys. Rev. **B30**, 2555 (1984).

[20] Filippi, C. and C. J. Umrigar, J. Chem. Phys. **105**, 213 (1996).

[21] Flad, H. J., M. Caffarel and A. Savin, in *Recent Advances in Quantum Monte Carlo Methods*, ed. W. A. Lester Jr., World Scientfic, (1997).

[22] Moroni, S., S. Fantoni and G. Senatore, Phys. Rev. **B 52**, 13547 (1995).

## Discussion

**Gross**: Is there a way of calculating the frequency-dependent density-density response function of the uniform electron gas with Monte-Carlo techniques?

**Ceperley**: Presently we do not know how to do it. Probably one could calculate frequency moments of the response function. But the full frequency dependence is currently beyond scope.

## DENSITY FUNCTIONAL THEORY FOR THE STUDY OF SINGLE-MOLECULE ELECTRONIC SYSTEMS

Jorge M. Seminario and James Tour

Department of Chemistry and Biochemistry
University of South Carolina
Columbia, South Carolina 29208

## INTRODUCTION

A top-priority objective in the field of design of new materials for nanotechnology is the study of small metal clusters in contact through *alligator clips* with small organic molecules so that their properties can be know precisely and used as extrapolations for the study of larger systems. Several properties using metal atoms can not be determined experimentally with the same precision, as it is possible with first and second row atoms. Achieving this goal clearly requires a good understanding of the factors that determine single interactions between metal clusters and molecules. This will allow us to predict interactions for proposed target molecules prior to undertaking what are often lengthy and costly synthetic efforts. The use of ab initio techniques help us tremendously to learn how structural, electronic and external factors influence the chemistry of metal and transition metal clusters, and to quantitatively predict the characteristics of their interactions with other elements in the periodic table.

We briefly describe in this review the ab initio methods available for the precise study of the macroscopic/microscopic interface. This interface reduces, for most of the cases, into the metal contact or tip making connection through an atom or molecule to a single molecule with specific features similar to those of a logical gate. Presently, the main goal is to understand the electrical properties of the interface metal-molecule so we could be able to use such a system as a logical gate, an ambitious substitution of mesoscopic semiconductor devices by single-molecule devices, practically at the reach of present technology, and ready to revolutionize all fields of present knowledge.

Molecules are quantum mechanical entities whose behavior can not be explained by classical methods, which embody our macroscopic world. A molecular system with N electrons can be successfully approximated (in zero order) as a group of one-electron orbitals $\varphi_i$ where every electron, with a determined energy $\varepsilon_i$, moves under the field of the nuclei and of the other N-1 electrons. This is the Hartree-Fock (HF) approximation. The inherent errors introduced by this approximation can be systematically corrected to any degree of precision limited only by the available computational resources. Methods like

Møller-Plesset (MP), configuration interaction (CI), coupled cluster (CC) and combinations of them are typical ways to upgrade results obtained by the HF approximation. Even more encouraging nowadays, the modern Hohenberg-Kohn-Sham-Levy density functional theory (DFT) allows an exact description of the molecular system using auxiliary orbitals of noninteracting electrons [1]. Therefore, even within the quantum mechanical picture, one-electron orbitals can be successfully used to study any molecular system. This does not imply that electrons are assigned to a specific orbital but that there is one or two indistinguishable electron per orbital.

## SINGLE MOLECULAR ELECTRONICS

Distinguishing between two types of gates in molecular circuits is also in order [2]. There are those gates at the interface of the logical circuits or CPU. The gates at the input interface would have to be able to take signals from standard electronic circuits, lasers, scanning probe microscopy (SPM) tips, or in general, from well-localized electrical fields. Molecular gates at the output would have to drive or inject a current onto a minuscule tip or metal cluster, or be able to excite the input of standard bulk devices. The vast majority of molecular gates, however, would be within the molecular CPU and will thus perform the information processing by controllably affecting the electrostatic potential of neighboring molecules. The ideas described here belong to the large group of gates that would allow the processing of information and the interconnection between them. The bulk/molecular interfacing structures are not being considered here.

Present technologies are also beset with problems from working at very high frequencies [3]. The use of electrostatics also addresses this difficulty. Small spurious capacitances (C) usually appear between portions of devices and circuits that otherwise should be totally electrically disconnected, or at infinite impedance (Z). If the operational frequency of the device increased, the impedance would decrease proportionally with the frequency (f) of operation ($Z = 1/2\pi fC$) rendering the device useless after a characteristic cutoff frequency. This cutoff frequency constitutes an upper limit to the operational frequency, which, in present technology, approaches 1 GHz (which corresponds to a 1 ns clock cycle time). Since the capacitance is also inversely dependent on the distance, making smaller circuits yields larger spurious capacitances, and therefore smaller impedances, further exacerbating the move toward smaller devices. Since electrostatic potentials are only limited by the time in which the electron density rearranges in a molecule (<1 fs) and there are no electron currents, the problems of low impedance between supposedly isolated circuits are eliminated. This point deserves a lot of attention since it is sometimes believed that smaller circuits are faster; however at microscopic sizes, in the range of present integrated circuits, it is well known that the faster circuits are relatively bigger than the slower ones. This is because the cut off frequency is larger in larger circuits.

**Table 1.** Main characteristics of bulk and molecular CPU circuits

| Feature | Bulk Electronics | Molecular Electronics |
|---|---|---|
| Time cycle | 1 ns | 1 fs |
| Electrons needed for bit of information | 16,000 electrons | Much less than one electron |
| Power sources to operate | Need external power supply | Molecules are always ready to operate $10^{13}$ gates/cm$^2$ |
| Integration scale | $10^8$ gates/cm$^2$ | 2 Å |
| Gate length of a transistor | 300 Å | |

The advantages that can be obtained with the use of single molecular devices are indicated in Table 1, which shows the characteristics of present bulk electronics and the proposed single molecular electronics

The use of molecular electronic devices can be exploited to increase the possibility to have larger and more powerful computers. An analysis of the present state of the art in the computational hardware as compared to the possibilities given by the use of single molecules instead of transistors is shown in Table 2.

**Table 2.** The hardware-software duality as a CPU-memory compromise

| Characteristic | CPU | MEMORY |
|---|---|---|
| Basic units | Faster and bigger | Smaller and slower |
| Programming | Wired logic | Programmed logic |
| Size of problems | Very small: few bit adders and few logical operations | Very large: largest possible memory, programs data |
| Rate of usage | Practically all circuits at a time | Less than one in $10^6$ at a time |
| Power dissipation | Very high (design limit) | Very low |
| Real problem | Can not get below 1 ns | Limited by speed, transfer. |
| Present solution | Going parallel (several problems) | Except for cost, no problem. |

## COMPUTATIONAL TECHNIQUES

Matter, in the form we know it, follows the laws of quantum mechanics (QM). Although at the macroscopic level Newton's laws can be applied very precisely, at the atomistic level the only possible way to study accurately the behavior of matter is by using quantum mechanics. The laws of quantum mechanics regarding problems in chemistry have been well understood since the 1930s; however, there is still much controversy regarding their interpretation, nevertheless the predictions of quantum mechanics have never been in contradiction with the observed behavior of matter. So far the only limitation, and the constrain in predictions, is the mathematical complexity of QM equations. However, since the solutions of the quantum mechanical equations allow us to predict the behavior of matter, efforts to develop the most sophisticated techniques and algorithms are increasing widely and exponentially with so encouraging results that theory now has the ability to correct several of the conclusions drawn from experiments. For example, several techniques are very accurate for predicting bond energies with errors smaller than 1%; even more sophisticated methods have claimed recently tolerances of less than 0.2 kcal/mol [4], which are better than those from most of the standard experimental methods.

There are several levels of theory available to study matter such that it can be very confusing. In order to clarify this situation, let us classify the several techniques in two major categories. The quantum and the classical mechanics techniques, both can be called first principles, but since we are dealing in most of the cases at the atomistic level, the equivalent term *ab initio* has been reserved for the quantum mechanical techniques. In addition, both techniques, quantum and classical, can be used with empirical parameters obtained from experiments, or obtained by other ab initio calculations on smaller or prototype systems. This adds a problem to the nomenclature since the parameters now are not empirical but from first principles calculations on smaller systems. Terms like molecular modeling for instance, which have been extensively used to describe a molecular system using classical mechanics techniques with empirical parameters, sometimes are confused with present quantum mechanical procedures to solve the Schrödinger equation, which is an exact equation of nature, and do not require of experimental input.

In this section, we describe several quantum mechanical techniques and their extensions. Matter most of the time is in a stationary state that can be studied by the time independent Schrödinger equation. In addition, most of the systems of interest deal with situations where relativistic effects are not too important, and if they are, still they can be treated as corrections to the nonrelativistic effects. One further simplification is given by the fact that electrons move much faster than nuclei, thereby permitting us to consider the wave function for electrons separated from the wave function for nuclei. In any case, if corrections due to any of the above effects are necessary, they can always be treated by a perturbational approach. Time dependent cases are treated using the famous Fermi golden rule or modifications of it, where the time dependence is considered as a perturbation. For large systems where time dependent effects involve pressure and temperature, in most cases introducing classical mechanics corrections to the quantum results can treat these effects.

The simplest useful case corresponds to a two-body problem, the hydrogen like systems, one nucleus and one electron. For this system the Schrödinger equation can be solved exactly and the wave function and the energy can be expressed in closed form. The solutions of this system, hydrogenic wavefunctions (1s, 2s, $2p_m$,...), are the building blocks or basis to construct the solutions of multielectron systems; however, expressed in a functionally different form (Slater-type function, Gaussian-type functions, etc.). The next case corresponds to the helium-like atoms. The helium case is actually the proof of the validity and incredible precision of the Schrödinger equation. Several variational and perturbational techniques have been developed to solve the helium atom [5]. The precision obtained goes beyond the precision of any other theory or method in any area of knowledge. The present value is -2.9037243770341 19479 Ha (1 Ha = 627.509 kcal/mol) where at least 15 significant figures are correct! This amazing precision is equivalent to predicting a distance between two points: one in New York and the other in Paris with less than 10 Å of error. Notice that chemical accuracy (about 1 kcal/mol) is already surpassed with the fourth decimal figure.

The situation becomes more complicated when we try to solve the Schrödinger equation for molecules or larger atoms. Basis sets have to be of reduced complexity due to the large number of integrals needed to solve the quantum eigenvalue problem and to the practical convenience to perform these integrals analytically rather than numerically. There are two well-known routes to solve the electronic, non-relativistic, time independent Schrödinger equation,

$$\hat{H}\Psi = E\Psi,$$

with Hamiltonian partitioned as

$$\hat{H} = \hat{T} + \hat{V}_{ne} + \hat{V}_{ee},$$

n-electron wavefunction

$$\Psi = \Psi(\vec{r}_1, \vec{r}_2, \vec{r}_3, \cdots, \vec{r}_N),$$

and Hamiltonian explicitly written as

$$\hat{H} = -\sum_{i=1}^{N} \frac{1}{2}\nabla_i^2 - \sum_{i=1}^{N}\sum_{A=1}^{M} \frac{Z_A}{r_{Ai}} + \sum_{i=1}^{N-1}\sum_{j=i+1}^{N} \frac{1}{r_{ij}}:$$

## Density Functional Theory (DFT) Methods

Standard correlated ab initio methods are usually very precise. However, they require extremely large amount of computational resources. In order to obtain just chemical accuracy, methods that scale as $N^7$ or higher are required. In addition, the need for larger basis sets also increase with the level of theory. For instance, methods that scale as $N^7$ usually require polarization f- and g-functions or even of higher angular momentum. Therefore, standard systems of interest for chemistry are practically intractable with these levels of theory. Parallel to the development of the standard ab initio methods when HF was the only alternative, several routes for its simplification were developed. The bottleneck of HF was the calculation of the two-electron four center integrals that appear in the direct and exchange terms of the electron-electron repulsion energy. Therefore, methods were developed in order to avoid these terms obtaining in this way an approximation to the HF procedure. The direct terms were no major problem because they are quadratic in the density; therefore, by expressing the density in some basis sets, the four-center integrals are eliminated. The problem was with the exchange terms. There is not straightforward way to eliminate the four-center integrals, therefore functionals of the density were developed that were able to express the exchange energy as a functional of the density. This functional was calculated for a uniform electron gas (also called jellium) by Bloch who found that the exchange energy for this system was proportional to $\rho^{\frac{1}{3}}$ [6]. Later, this expression was also used by Dirac [7] in a well-known paper where exchange was included in the Thomas-Fermi atom [8,9]. This form of exchange functional with an empirical parameter $\alpha$ was later on used by Slater [10] in order to simplify the HF procedure.

The time independent, nonrelativistic, Born-Oppenheimer approximated Schrödinger equation,

$$\hat{H}\Psi = E\Psi,$$

$$\left(\hat{T} + \hat{\upsilon}_{ext} + \hat{V}_{ee}\right)\Psi = E\Psi$$

yields the energy $E$,

$$E = \int \upsilon_{ext}(\vec{r})N\int d\vec{r}_2 d\vec{r}_3 ... d\vec{r}_N \left|\Psi(\vec{r}_1, \vec{r}_2, \vec{r}_3, ..., \vec{r}_N\right|^2 d\vec{r} + \left\langle \Psi \middle| \hat{T} + \hat{V}_{ee} \middle| \Psi \right\rangle$$

where the nuclei-electron energy interaction is explicitly calculated using the many-electron wavefunction,

$$V_{ne} = \int \upsilon_{ext}(\vec{r})N\int d\vec{r}_2 d\vec{r}_3 ... d\vec{r}_N \left|\Psi(\vec{r}_1, \vec{r}_2, \vec{r}_3, ..., \vec{r}_N\right|^2 d\vec{r}.$$

Due to the fact that electrons are much faster than nuclei, and the classical energy interaction between a static charge distribution with a set of fixed charges or external potential $\upsilon_{ext}$ is

$$V_{ne}^{class} = \int \upsilon_{ext}(\vec{r})\rho(\vec{r})d\vec{r},$$

the function $N\int d\vec{r}_2 d\vec{r}_3 ... d\vec{r}_N |\Psi(\vec{r}_1, \vec{r}_2, \vec{r}_3, ..., \vec{r}_N|^2$ has exactly the same meaning as the classical function $\rho(\vec{r})$ and therefore, provided that the Born-Oppenheimer approximation is valid, the function $N\int d\vec{r}_2 d\vec{r}_3 ... d\vec{r}_N |\Psi(\vec{r}_1, \vec{r}_2, \vec{r}_3, ..., \vec{r}_N|^2$ can be regarded as a charge distribution, or more precisely, as the electron density. This is another statement of the Hellmann-Feynman electrostatic theorem [11,12]. Therefore using

$$E = \langle \Psi | \hat{H} | \Psi \rangle = \int \upsilon_{ext}(\vec{r})\rho(\vec{r})d\vec{r} + \left\langle \Psi \middle| \hat{T} + \hat{V}_{ee} \middle| \Psi \right\rangle,$$

and the variational principle, Hohenberg and Kohn [13], by a trivial *reductio ad absurdum* showed that there is a one to one relation between the density function ρ and the external potential $\upsilon_{ext}$. Therefore, implying that the energy E, and any expectation value are functionals of the electronic density of the ground state, i.e. for a given external potential $\upsilon$,

$$E = E_\upsilon[\rho]$$

Which can be rewritten as

$$E_\upsilon[\rho] = \int \upsilon(\vec{r})\rho(\vec{r})d\vec{r} + \left\langle \Psi \middle| \hat{T} + \hat{V}_{ee} \middle| \Psi \right\rangle = \int \upsilon(\vec{r})\rho(\vec{r})d\vec{r} + F[\rho]$$

where $F[\rho]$ is a universal functional (it does not depend on the external potential) identical for any system. Together with this theorem, Hohenberg and Kohn also derived a second theorem that provides a variational principle for the energy which established that for any trial density $\tilde{\rho}$, the energy is an upper bound to the exact energy of the ground state, i.e.,

$$E = \min_{\tilde{\rho}} \left\{ \int \upsilon(\vec{r})\tilde{\rho}(\vec{r})d\vec{r} + F[\tilde{\rho}] \right\}$$

for the exact density $\rho$, we have

$$E = \int \upsilon(\vec{r})\rho(\vec{r})d\vec{r} + F[\rho]$$

There are certain cases in which these theorems have problems associated with the so-called v- and N-representability. These problems were subsequently solved by Levy [14] and the correct ground state energy can be unambiguously expressed as

$$E = \min_{\rho} \left\{ \int \upsilon(\vec{r})\rho(\vec{r})d\vec{r} + \left\langle \Psi_\rho^{\min} \middle| \hat{T} + \hat{V}_{ee} \middle| \Psi_\rho^{\min} \right\rangle \right\}$$

This expression is what constitutes the Levy's constrained search, with the universal functional

$$F[\tilde{\rho}] = \left\langle \Psi_{\tilde{\rho}}^{\min} \middle| \hat{T} + \hat{V}_{ee} \middle| \Psi_{\tilde{\rho}}^{\min} \right\rangle.$$

Notice that $\Psi_{\tilde{\rho}}^{\min}$ is the wavefunction that yields $\tilde{\rho}$ and minimizes only $\hat{T} + \hat{V}_{ee}$ and not the total Hamiltonian. A similar universal functional was developed by Percus for the non-interacting system [15].

Let us now assume a system with Hamiltonian [16]

$$\hat{H}_\lambda = \hat{T} + \hat{\upsilon}_\lambda + \lambda \hat{V}_{ee}$$

where $\lambda$ can vary between 0 and 1. In addition, we constrain the $\hat{\upsilon}_\lambda$ such that it always yields the density $\rho(\bar{r})$of the real system [17]. The Schrödinger equation for this system is

$$\hat{H}_\lambda \Psi_\lambda = E_\lambda \Psi_\lambda$$

where $E_\lambda$and $\Psi_\lambda$ are the corresponding eigenvalue and eigenfunction of the system with the constraint on the density $\rho(\bar{r})$. The case $\lambda = 0$ corresponds to an ideal system of noninteracting electrons having the density of the real system under an effective potential $\upsilon_{eff}$ or $\upsilon_s$ and the case $\lambda=1$ corresponds to the real system with $\upsilon_{ext}$. The system of non-interacting electrons is trivially solved since the exact solution is, in most cases, a Slater-determinant.

Therefore, the energy $E_\lambda$ can be written as

$$E_\lambda = \langle \Psi_\lambda | \hat{H}_\lambda | \Psi_\lambda \rangle = \int \upsilon_\lambda(\bar{r})\rho(\bar{r})d\bar{r} + \left\langle \Psi_\lambda \middle| \hat{T} + \lambda \hat{V}_{ee} \middle| \Psi_\lambda \right\rangle$$

Derivation and integration of the above equation with respect to $\lambda$ yields the difference of the energies of the real and noninteracting systems ($E$-$E_S$) in the left hand side, and $E_S$ is exactly $V_s+T_s$, i.e., the sum of the potential energy plus the kinetic energy of a noninteractive system of electrons with a density identical to the one of the real system, and using the Hellmann-Feynman theorem and rearranging terms, we have

$$E(\rho) = \int \upsilon_{ext}(\bar{r})\rho(\bar{r})d\bar{r} + T_s + \int_0^1 \langle \Psi_\lambda | V_{ee} | \Psi_\lambda \rangle d\lambda$$

where the integral on $\lambda$ is still unsolvable. It is convenient to extract the classical electron-electron interaction energy, $V_{class}$, from the second integral and define the difference as the exchange-correlation energy, because of its close resemblance to their ab initio counterparts.

$$\int_0^1 \langle \Psi_\lambda | V_{ee} | \Psi_\lambda \rangle d\lambda = \frac{1}{2} \int\int d\bar{r}_1 d\bar{r}_2 \frac{\rho(\bar{r}_1)\rho(\bar{r}_2)}{|\bar{r}_1 - \bar{r}_2|} + E_{xc}(\rho)$$

However, it can be easily noted that $E_{xc}$ does not correspond to the $E_c + E_x$ of the standard HF based definitions. The total energy of the electronic system can be written as

$$E[\rho] = \int \upsilon_{ext}(r)\rho(r)dr + T_s + V_{class}[\rho] + E_{xc}[\rho]$$

This expression constitutes the basis of the famous Kohn-Sham (K-S) procedure [18], which gives a practical use of the Hohenberg-Kohn theorem. Up to this point, there are not approximations; not even the mean field approximation has been used, as is the case of the HF procedure. The K-S procedure uses a set of molecular orbitals with the sole purpose of representing the electron density of the system of non-interacting. All this development comes out directly from the Schrödinger equation. All functionals or quantities in the above equation are known, except the $E_{xc}$. $T_s$ is an unknown functional of the density, but it can be obtained exactly using the wavefunction of the noninteracting system. Therefore, it is not required to have a functional, either in the interactive or noninteractive systems, for its exact calculation.

The only unknown functional is the $E_{xc}[\rho]$. All the others are known or can be easily calculated from the wavefunction of the noninteracting system. Once $E_{xc}$ is known the total energy $E$ can be minimized with respect to the density ρ yielding the Kohn-Sham equations that can be solved self-consistently.

The wavefunction of the noninteractive system is exactly the determinant made using the orbitals $\varphi_i$,

$$\Psi_s(\vec{r}_1,\vec{r}_2,...,\vec{r}_N) = \det|\varphi_1\varphi_2...\varphi_N|$$

Keep in mind that $\Psi_s$ is not the wavefunction of the real system, neither are the φ's related to real electrons, except for fact that they yield the same density through the sum:

$$\rho(\vec{r}) = \sum_{i=1}^{N} |\varphi_i(\vec{r})|^2$$

The total energy for the real system will be evaluated using those orbitals that yield the minimum energy, which in most of the cases corresponds to those with the lowest energies. The first functionals related to the $E_{XC}$ were not made with DFT in mind, but were oriented to the description of exchange effects. Therefore, as a practical way, the partition,

$$E_{xc} = E_x + E_c$$

is invoked. The exchange energy using the formula for the uniform electron gas is the local density approximation (LDA) written in our notation is

$$E_x^{LDA}[\rho] = -\int d\vec{r}\, \tfrac{3}{4}\left(\tfrac{3}{\pi}\right)^{\frac{1}{3}} \rho^{\frac{4}{3}}$$

The corresponding potential is

$$\upsilon_x^{LDA}(\vec{r}) = -\left(\tfrac{3}{\pi}\rho(\vec{r})\right)^{\frac{1}{3}}$$

Similarly, this uniform gas of electrons bring us the first approximation for the $E_{XC}$,

$$E_{xc}^{LDA} = \int \varepsilon_{xc}^{LDA}[\rho(\vec{r})]\rho(\vec{r})d\vec{r}$$

where $\varepsilon_{xc}^{LDA}$ is the exchange-correlation energy distribution per unit of volume and it only depends on the density at the point where it is evaluated. The functional for the ideal gas of electrons of uniform density reduces to

$$E[\rho] = T_s + E_{xc}[\rho]$$

which can be solved numerically as precisely as it is needed for any situation using plane waves, and analytically for some limiting cases. Therefore, it will be possible to get the $E_{xc}$ and subtracting the $E_x$ already known, the $E_c$ can be calculated.

Perdew and Wang (PW92) [19] have proposed a local correlation functional that has become the standard local approximation of DFT

To correct for the nonuniformity of the electron density, gradients of the density are introduced into the exchange and correlation functionals, creating in this way, nonlocal or semilocal functionals. The first gradient approximation was not successful since it did not fulfill many of the requirements of the exchange-correlation functional; it was even worst than the local approximation. To solve this problem of the gradient approximation, generalized gradient approximations (GGA) were designed: $E_{xc}^{GGA}[\rho] = \int d\bar{r}\varepsilon_{xc}[\rho, |\nabla\rho|, \nabla^2\rho]$ As a result of a series of improvements to the gradient approximation, the Langreth-Mehl exchange-correlation functional [20,21] were developed in 1983. The Perdew-Wang (PW86) for exchange and the Perdew (P86) for correlation functionals were of wider application in the most important DFT programs. Finally the Perdew-Wang (PW91) exchange and correlation [19,22,23] functionals, the first GGA's designed from purely first principles.

## APPLICATIONS OF THE METHODS AND AB INITIO NATURE OF DFT

Specifically, we have used DFT techniques since these are the most powerful tools to deal with a relatively large number of atoms [1]. The use of DFT is fully justified due to the fact that it is an *ab initio* technique able to deal with a broad variety of systems. Several successful applications of DFT have been reported using the hybrid functionals where a portion of the exchange functional is calculated as a fully nonlocal functional of the wave function of an auxiliary noninteracting system of electrons. Since this resembles the exchange in the HF procedure (actually in any wave-function procedure), it is common to refer to this functional, or procedure, as a DFT-HF hybrid. However, we have to consider that the exchange is being calculated using a noninteractive wave function whose density, but not its wave function, corresponds to the real system. A detailed analysis and its theoretical rigor was recently reviewed [24]. The functional that we used is the B3PW91 with the triple split-valence basis set 6-311G** which includes polarization functions for all atoms [19,23,25].

A pure ab initio method is one that solves exactly or approximately the Schrödinger equation. This means using only the principles of quantum mechanics and not using empirical or other information extrapolated, interpolated, or transferred from other systems. Whether or not the method is a very rough approximation is not going to eliminate its ab initio character. In this sense, we have a spectrum of methods from the very precise to the very inaccurate ones, forcing us to a trade-off because the cost of the method is usually a direct function of its precision. In general, any practical computational method will be part ab initio and part empirical in character, and it is not much problem to determine if a given method can be classified as ab initio, empirical or semiempirical. The selection and use of a basis set is always based on experience. Whether, theoretical or experimental, one can always create or improve a basis set based on its previous performances. The basis set tells the method the spectral region to look for the solutions but it does not constitute the solution itself. A 100% analytical ab initio method would be limited to the solution of the hydrogen-like atoms. Another 100% ab initio, but nevertheless approximated method, would be any of the variational-perturbational methods that solves the helium-like atoms with about 15 figures of precision, and where the basis set has been chosen with the only criteria of the lowest possible energy. The standard ab initio or HF-based methods (HF, MP, CC, CI, MRHF,...) present several degrees of accuracy and also several degrees of small empirical (due to experiment or experience from earlier calculations) corrections and truncations. We call all of them ab initio, and there is an implicit understanding that this is where our definition of ab initio is settled. Small corrections to those methods, like 10% correction to the ZPE do not change the ab initio character of the methods. There exists the accepted Gaussian2 theory where corrections, not necessarily small, are made to the energies based on precise experimental results. Methods are being made faster but not more precise. MP4 was widely available in 1982, and MP5 just appeared in 1992. It will take more time until we have MP7 or MP8 widely available. This is why higher performances of computers are required.

Now to the question if DFT methods are ab initio: First of all, we derived DFT directly from the Schrödinger equation even without recurring, as in HF methods, to the mean field approximation. The fact that a well defined but nowadays unsolvable functional shows up can not exclude the ab initio character of DFT. As in any of the ab initio methods above, approximations are made. A classification could be done whether the functional was derived using entirely first principles information or if it was obtained by using experimental or other transferred information. Functionals like the local approximation, gradient approximation are exact functionals up to their order, and are they were obtained from purely first principles arguments. The GGA of Perdew and Wang for exchange-correlation, PW91, is strictly ab initio and there is evidence that it performs as well as QCI and MP4 methods. A predecessor to the PW91, the PW86, has been used with systems containing up to transition metals with great success. DFT methods can also be developed using fully numerical techniques, eliminating the use of basis sets. Using this method together with one of the ab initio functionals yields a 100% ab initio procedure, i.e., more ab initio than any of the standard ones is.

**Table 3**. Average errors (kcal/mol) of the theoretical atomization energies for the 32 molecules from the Gaussian-1 basis [26] and error per formal bond. Also shown are the formal scaling and the number of basis per heavy atom (Li, Be, B, C, N, O, and F). The errors are with respect to precise experimental atomization energies, which have been measured with tolerances smaller than 1 kcal/mol.

| Method | Error | Error/bond | Formal Scaling | Basis/heavy-atom |
|---|---|---|---|---|
| MP4SDTQ/6-311G(d,p) | 15.6 | 6.2 | $N^7$ | 18 |
| MP4SDTQ/6-311+G(d,p) | 14.6 | 5.8 | $N^7$ | 22 |
| MP4SDTQ/6-311G(2df,p) | 8.3 | 3.3 | $N^7$ | 30 |
| QCISD(T)/6-311G(d,p) | 16.7 | 6.7 | $N^7$ | 18 |
| PW86/DZP | 5.3 | 2.1 | $N^3$ | 15 |

The molecules in the Gaussian-1 basis are $H_2$, LiH, BeH, CH, $CH_2(^1A_1)$, $CH_2(^3B_1)$, $CH_3$, $CH_4$, NH, $NH_2$, $NH_3$, OH, $OH_2$, FH, $Li_2$, LiF, $C_2H_2$, $C_2H_4$, $C_2H_6$, CN, HCN, CO, HCO, $H_2CO$, $H_3COH$, $N_2$, $N_2H_4$, NO, $O_2$, $O_2H_2$, $F_2$, and $CO_2$. MP4 and QCI errors were calculated from the energies reported in Ref. [26]. PW86 errors were calculated from total energies reported in Ref. [27]. Experimental atomization energies are reported in Ref. [26] where pointers to the experimental reports are indicated. The DZP basis set is equivalent to a 6-31G(d,p) which also uses 15 basis function for the heavy atoms.

Results in Table 3 indicate that density functional theory, which is computationally less expensive, is generally more accurate than full fourth-order Møller-Plesset perturbation theory and quadratic configuration interaction. These are the convincing arguments for the use of DFT methods. The atomization energies are one of the most difficult energies to compute since it implies the rupture of all the bonds on the molecule; therefore, most of the error cancellations in standard calculations are not present. Modern functionals, like the PW91 and B3PW91 perform even better than the PW86 shown above with the additional advantage that the PW91 functional is a purely first principles functional; it has no fitted parameters. Table 4 shows these errors for the particular case of water. It can be noticed that the local density approximation is not suitable for precise calculations; however, any of the nonlocal density functional performs much well than any of the standard ab initio methods. The study of nanometer devices depends strongly on the precise determination of energies.

Preliminary studies have permitted us to analyze, using DFT, the effects on real experimental situations performed at our own and other's laboratories. For instance, the modeling of gold [28] or palladium tips [29] to be used as contacts for molecular devices, the study of the effects on conductivity of single molecules [30] and the rationale for molecular scales systems [2] have all been investigated. These works are the first of their kind. Several ab initio methods including density functional theory [1] will be used to determined the viability of several target molecules to be used as logical devices. In addition, these studies will aid in directing us to more viable target molecules for synthesis. This studies are the initial step toward the design of a molecular-CPU.

**Table 4.** Water atomization energy errors (kcal/mol). The column labeled error shows the errors with respect to the experimental value of 219.3 kcal/mol at 0 K. The error per bond is also shown.

| Method | Error | Error/bond |
|---|---|---|
| HF/6-311G(2df,p) | 79.9 | 40.0 |
| MP2/6-311G(2df,p) | 11.7 | 5.9 |
| MP4/6-311G** | 16.9 | 8.5 |
| MP4/6-311+G | 12.5 | 6.3 |
| MP4/6-311G(2df,p) | 13.2 | 6.6 |
| QCISD(T)/6-311G** | 17.3 | 8.5 |
| CCSD(T)/6-311G** | 17.2 | 8.6 |
| LSD(VWN)/DZVP | 32.9 | 16.5 |
| BLYP/6-31G** | 5.7 | 2.9 |
| BLYP/6-311G(3df,3pd) | 1.2 | 0.6 |
| PW86/DZVP | 4.3 | 2.2 |
| PW86/TZVP | 3.7 | 1.9 |
| PW91/TZVP | 1.9 | 0.9 |
| BP86/6-31G** | 1.9 | 0.9 |
| B3PW91/6-311G(2df,p) | 3.0 | 1.5 |

The experimental water atomization energy is from the heat of formation given in the JANNAF tables [31].

**Acknowledgments**
This work is supported by the Defense Advanced Research Projects Agency. We also thank NASA for supercomputer support.

## REFERENCES

[1] J. M. Seminario Eds, *Recent Developments and Applications of Modern Density Functional Theory* (Elsevier, Amsterdam, 1996).

[2] J. M. Tour, M. Kosaki, and J. M. Seminario, *Molecular Scale Electronics: A Synthetic/Computational Approach to Digital Computing*, J. Am. Chem. Soc. in press **1998**.

[3] W. P. Kirk and M. A. Reed Eds, *Nanostructures and Mesoscopic Systems* (Academic: New York, New York, 1992).

[4] J. M. L. Martin, *Benchmark ab initio calculations of the total atomization energies of the first-row hydrides $AH_n$ (A = Li–F)*, Chem. Phys. Lett. 273, 98-106 **1997**.

[5] F. C. Sanders, *An Introduction to High-precision Computational Methods for Simple Atomic and Molecular Systems* in *Recent Developments and Applications of Modern Density Functional Theory*, edited by J. M. Seminario (Elsevier, Amsterdam, 1996), p. 99-150.

[6] F. Bloch, *Bemerkung zur Elektronentheorie des Ferromagnetismus und der elektrischen Leitfähigkeit*, Z. Phys. 57, 545-555 **1929**.

[7] P. A. M. Dirac, *Note on exchange phenomena in the Thomas atom*, Proc. Camb, Phil. Soc. 26, 376-385 **1930**.

[8] L. H. Thomas, *The calculation of atomic fields*, Proc. Camb. Phil. Soc. 23, 542-548 **1927**.

[9] E. Fermi, *Eine statistische Methode zur Bestimmung einiger Eigenschaften des Atoms und ihre Anwendung auf die Theorie des periodischen Systems der Elemente*, Z. Phys. 48, 73-79 **1928**.

[10] J. Slater, *The Self-Consistent Field for Molecules and Solids: Quantum Theory of Molecules and Solids* (Mc Graw-Hill, New York, 1974).

[11] H. Hellmann, *Einführung in die Quantenchemie* (Deuticke, Leipzig, Vienna, 1937).

[12] R. P. Feynman, *Forces in Molecules*, Phys. Rev. 56, 340-343 **1939**.

[13] P. Hohenberg and W. Kohn, *Inhomogeneous electron gas*, Phys. Rev. B 136, 864-871 **1964**.

[14] M. Levy, *Universal variational functionals of electron densities, first-order density matrices, and natural spin-orbitals and solution of the v-representability problem*, Proc. Natl. Acad. Sci. USA 76, 6062-6065 **1979**.

[15] J. K. Percus, *The Role of Model Systems in the Few-Body Reduction of the N-Fermion Problem*, Int. J. Quantum Chem. 13, 89-124 **1978**.

[16] J. Harris and R. O. Jones, *The surface energy of a bounded electron gas*, J. Phys. F 4, 1170-1186 **1974**.

[17] D. C. Langreth and J. P. Perdew, *The exchange-correlation energy of a metallic surface*, Solid State Commun. 17, 1425-1429 **1975**.

[18] W. Kohn and L. J. Sham, *Self-consistent equations including exchange and correlation effects*, Phys. Rev. A 140, 1133-1138 **1965**.

[19] J. P. Perdew and Y. Wang, *Accurate and Simple Analytic Representation of the Electron-Gas Correlation Energy*, Phys. Rev. B 45, 13244-13249 **1992**.

[20] D. C. Langreth and M. J. Mehl, *Beyond the local-density approximation in calculations of ground-state electronic properties*, Phys. Rev. B 28, 1809-1834 **1983**.

[21] D. C. Langreth and M. J. Mehl, *Beyond the local-density approximation in calculations of ground-state electronic properties-Erratum*, Phys. Rev. B 29, 2310 **1984**.

[22] J. P. Perdew, *Unified Theory of Exchange and Correlation beyond the Local Density Approximation* in *Electronic Structure of Solids*, edited by P. Ziesche and H. Eschrig (Akademie Verlag, Berlin, 1991), p. 11-20.

[23] J. P. Perdew, J. A. Chevary, S. H. Vosko, K. A. Jackson, M. R. Pederson, D. J. Singh, and C. Fiolhais, *Atoms, Molecules, Solids, and Surfaces: Applications of the Generalized Gradient Approximation for Exchange and Correlation*, Phys. Rev. B 46, 6671-6687 **1992**.

[24] A. Görling and M. Levy, *Hybrid schemes combining the Hartree-Fock method and density-functional theory: Underlying formalism and properties of correlation functionals*, J. Chem. Phys. 106, 2675-2680 **1997**.

[25] A. D. Becke, *A new mixing of Hartree-Fock and local density-functional theories*, J. Chem. Phys. 98, 1372-1377 **1993**.

[26] J. A. Pople, M. Head-Gordon, D. J. Fox, K. Raghavachari, and L. A. Curtiss, *Gaussian-1 theory: A general procedure for prediction of molecular energies*, J. Chem. Phys. 90, 5622-5629 **1989**.

[27] J. M. Seminario, *Energetics using DFT: comparisons to precise ab initio and experiment*, Chem. Phys. Lett. 206, 547-554 **1993**.

[28] J. M. Seminario and J. M. Tour, *Systematic Study of the Lowest Energy States of $Au_n$ (n=1-4) Using DFT*, Int. J. Quantum Chem. 65, 749-758 **1997**.

[29] A. G. Zacarias, M. Castro, J. M. Seminario, and J. M. Tour, *Lowest Energy States of $Pd_4$ Using DFT: Nanotips for Single Molecule Electronics*, J. Am. Chem. Soc. Submitted **1998**.

[30] J. M. Seminario, A. G. Zacarias, and J. M. Tour, *Theoretical Interpretation of Conductivity Measurements of Thiotolane Sandwich. A Molecular Scale Electronic Controller*, J. Am. Chem. Soc. 120, 3970-3974 **1998**.

[31] M. W. Chase, C. A. Davies, J. R. DowneyJr, D. J. Frurip, R. A. McDonald, and A. N. Syverud, *JANAF Thermochemical Tables*, J. Phys. Chem. Ref. Data 14, 1-1856 **1985**.

# DENSITY FUNCTIONAL THEORY FOR A SINGLE EXCITED STATE

Á. Nagy

Institute of Theoretical Physics
Kossuth Lajos University
H-4010 Debrecen
Hungary

## INTRODUCTION

Density functional theory has originally been developed for the ground state [1]. It can be applied only for the lowest-energy state in each symmetry class [2, 3]. Rigorous generalization for excited states was given by Theophilou [4] and by Gross, Oliveira and Kohn [5]. Several calculations have been done with this method [6-13] Relativistic generalization of this formalism [13]. and the optimized potential method [14] have also been introduced. This approach has the disadvantage that one has to calculate all the ensemble energies lying under the given ensemble energy to obtain the desired excitation energy. It is especially inconvenient to use it if one is interested in highly excited states. Time-dependent density functional theory [15, 16] can also be applied to obtain excitation energies. The work formalism proposed by Sahni and coworkers [17]has also been used in excited-state density functional calculations [18]. Generalizing a recent perturbation theory [20], Görling[34] presented a new density functional formalism for excited states.

Recently, a new approach of treating a single excited state is proposed [21]. This approach is based on Kato's theorem [22] and is valid for Coulomb external potential (i.e. atoms, molecules and solids). It has the advantage that one can treat a single excited state. In this paper this new approach is discussed. It is shown that for a given set of occupation numbers the Kohn-Sham potential can be uniquely constructed. Then the excited-state exchange-correlation potential is compared with that of the ground state. Excitation energies calculated with the generalized KLI (Krieger-Li-Iafrate) method are presented for several atoms.

## THEORY FOR A SINGLE EXCITED STATE

We know that a knowledge of the ground state electron density is sufficient in principle to determine all molecular properties. For Coulomb systems this statement can be simply understood following Bright Wilson's [23] argument: A well-known

*Electron Correlations and Materials Properties*, edited by Gonis *et al.*
Kluwer Academic/Plenum Publishers, 1999.

theorem of quantum mechanics, Kato's theorem [22] states that

$$Z_\beta = -\frac{1}{2n(r)} \left.\frac{\partial n(r)}{\partial r}\right|_{r=R_\beta}, \tag{1}$$

where the partial derivatives are taken at the nuclei $\beta$. So the cusps of the density tell us where the nuclei are ($\mathbf{R}_\beta$) and what the atomic numbers $Z_\beta$ are. On the other hand, the integral of the density gives us the number of electrons:

$$N = \int n(\mathbf{r})d\mathbf{r} \,. \tag{2}$$

Thus from the density the Hamiltonian can be readily obtained:

$$\hat{H} = \hat{T} + \hat{V} + \hat{V}_{ee} \,, \tag{3}$$

where

$$\hat{T} = \sum_{i=1}^{N}(-\frac{1}{2}\nabla_i^2), \tag{4}$$

$$\hat{V} = \sum_{i=1}^{N} v(r_i), \tag{5}$$

and the electron-electron energy operators

$$\hat{V}_{ee} = \sum_{i=1}^{N-1}\sum_{j=i+1}^{N} \frac{1}{|\mathbf{r_i} - \mathbf{r_j}|}, \tag{6}$$

and

$$v(\mathbf{r}) = -\sum_{\beta=1}^{M} \frac{Z_\beta}{|\mathbf{r} - \mathbf{R}_\beta|}. \tag{7}$$

From the Hamiltonian, in principle every property can be determined. So, the density contains all information on the system. This argument is valid only for Coulomb systems (e.g. atoms, molecules, solids), while the density functional theory is valid for any external potential.

Kato's theorem is valid for the excited states, too. So, from the density $n_i$ of the $i$-th electron state the Hamiltonian $\hat{H}$ can in principle be obtained and the eigenvalue problem

$$\hat{H}\Psi_k = E_k\Psi_k \qquad (k = 1, ..., i, ...) \tag{8}$$

can be solved.

In the ground-state density functional theory the first Hohenherg-Kohn theorem [1] states that the external potential $v(\mathbf{r})$ is determined within a trivial additive constant by the knowledge of the electron density $n(\mathbf{r})$. We see that the external potential $v(\mathbf{r})$ can be constructed if the density $n_i(\mathbf{r})$ of the $i$th excited state is known.

In the ground-state density functional theory the problem of $v$-representability was discussed by several authors and solved by Levy and Lieb [24, 25]. A $v$-representable density is one that is associated with a ground-state wave function of some Hamiltonian with a local external potential. Similarly, we can define a $v$-representable excited-state density as a density that is associated with an excited-state wave function of a Hamiltonian having the form of Eq. (3). Obviously, not every density is $v$-representable. Densities without cusp(s) are not associated with an excited-state wave function arising from Couloumb-type external potential so not $v$-representable. Clearly, all excited state densities of real electronic systems are $v$-representable.

Using the concept [2, 26] of adiabatic connection Kohn-Sham-like equations can be derived. We suppose the existence of a continuous path between the interacting and the non-interacting systems. The density $n_i$ of the $i$-th electron state is the same along the path.

$$\hat{H}_i^\alpha \Psi_k^\alpha = E_k^\alpha \Psi_k^\alpha , \tag{9}$$

where

$$\hat{H}_i^\alpha = \hat{T} + \alpha \hat{V}_{ee} + \hat{V}_i^\alpha . \tag{10}$$

The subscript $i$ denotes that the density of the given excited state is supposed to be the same for any value of the coupling constant $\alpha$. $\alpha = 1$ corresponds to the fully interacting case, while $\alpha = 0$ gives the non-interacting system:

$$\hat{H}_i^0 \Psi_k^0 = E_k^0 \Psi_k^0 , \tag{11}$$

For $\alpha = 1$ the Hamiltonian $\hat{H}_i^\alpha$ is independent of $i$. For any other values of $\alpha$ the 'adiabatic' Hamiltonian depends on $i$, we have different Hamiltonian for different excited states. Thus the non-interacting Hamiltonian($\alpha = 0$) is different for different excited states.

## CONSTRUCTION OF EXCITED-STATE EFFECTIVE POTENTIAL

Recently, there is a growing interest in constructing the Kohn-Sham potential from the ground-state electron density [27-35] In line with these works, now, a generalization to excited states is studied.

It is shown that for a given electron-configuration, the non-interacting effective potential can be given uniquely from the density of the excited state $n_i$.

In the nondegenerate case the non-interacting wave-function can be given as a Slater determinant of orbitals satisfying the one-electron equations

$$\left[ -\frac{1}{2} \nabla^2 + V_i^0(\mathbf{r}) \right] \phi_j^i = \varepsilon_j^i \phi_j^i . \tag{12}$$

(The proof for degenerate case will be detailed elsewhere.)

Let us first consider a two-level independent-particle problem with the effective potential $V_i^0$ [27]. In a spherically symmetric system the radial equations take the form

$$(P_j^i)'' + 2(\epsilon_j^i - V_i^0) P_j^i = 0 \qquad j = 1, 2 , \tag{13}$$

where $P_j^i$ are the radial wave functions. Using the Dawson–March transformation [36]:

$$P_1^i = \frac{1}{\sqrt{\lambda_1^i}} \varrho_i^{1/2} \cos \theta_i \tag{14}$$

$$P_2^i = \frac{1}{\sqrt{\lambda_2^i}} \varrho_i^{1/2} \sin \theta_i \tag{15}$$

Eqs.(13) and (14) lead to the relation between the radial electron density $\varrho_i = 4r^2\pi n_i$ and the phase $\theta_i$:

$$\theta_i'' + \frac{\varrho_i'}{\varrho_i}\theta_i' = 2\xi_i \sin(2\theta_i) \tag{16}$$

and an expression for the effective potential

$$V_i^0 = \frac{1}{4}\frac{\varrho_i''}{\varrho_i} - \frac{1}{8}\left(\frac{\varrho_i'}{\varrho_i}\right)^2 + \zeta_i + \xi_i \cos(2\theta_i) - \frac{1}{2}(\theta_i')^2 \,, \tag{17}$$

where

$$\xi_i = \frac{1}{2}(\epsilon_1^i - \epsilon_2^i) \,, \tag{18}$$

and

$$\zeta_i = \frac{1}{2}(\epsilon_1^i + \epsilon_2^i) \,. \tag{19}$$

$\lambda_1^i$ and $\lambda_2^i$ are the occupation numbers. It was shown by several authors [28, 32, 37, 38] that Eq. (16) with the boundary conditions

$$\int_0^\infty dr\, \varrho_i \sin^2\theta_i = \int_0^\infty dr\, \varrho_i \cos^2\theta_i = 1 \,, \tag{20}$$

$$\theta_i \rightarrow -\frac{\pi}{2} \qquad as \quad r \rightarrow \infty \tag{21}$$

and

$\theta_i$ must have $k$ nodes

can be solved uniquely. For instance, for the first excited state of the He atom ($1s^1 2s^1$)

$$\lambda_1^1 = \lambda_2^1 = 1 \tag{22}$$

and

$$k = 1 \,. \tag{23}$$

Determining $\theta_i$ Eq. (17) gives the effective potential uniquely.

Aryasetiawan and Stott [31] studied the model problem of two spinless fermions moving in the simple harmonic potential $x^2/2$. They calculated the effective potential not only from the ground-state density. Then they determined the effective potential from the first excited state density. Fig. 2. in Ref.[31] shows that the effective potential for the first excited state differs from that of the ground state with the appearence of a peak which tend to raise the first excited state energy with respect to the ground state. In another paper Aryasetiawan and Stott [32] studied three spinless fermions moving in the square-well potential. They determined the effective potentials corresponding to the ground and the first excited state densities. In the last section the ground and the first excited state exchange potentials of the He atom calculated in the KLI approximation are compared.

Let us turn to a more general case. Following Stott et al. [32] the one-electron orbital with the lowest energy can be expressed as

$$\phi_1^i = \left( n_i - \sum_{j=2}^{M} \lambda_j^i (\phi_j^i)^2 \right)^{1/2} . \tag{24}$$

The occupation numbers $\lambda_j^i$ can be 0 or 1. M denotes the subscript for the highest occupied orbital. If we know $\lambda_1^i$, $\lambda_2^i$, ... $\lambda_M^i$, i. e. the electron configuration is known, one can obtain the effective potential

$$\begin{aligned} V_i^0 &= \frac{1}{2\phi_1^i} \nabla^2 u_1 + \epsilon_1^i \\ &= \frac{1}{2\left( n_i - \sum_{j=2}^{M} \lambda_j^i (\phi_j^i)^2 \right)^{1/2}} \nabla^2 \left( n_i - \sum_{j=2}^{M} \lambda_j^i (\phi_j^i)^2 \right)^{1/2} + \epsilon_1^i . \end{aligned} \tag{25}$$

Combinings Eqs. (12) and (25) a set of coupled nonlinear second-order differential equations

$$-\frac{1}{2}\nabla^2 \phi_j^i + \left[ \frac{1}{2\left( n_i - \sum_{j=2}^{M} (\phi_j^i)^2 \right)^{1/2}} \nabla^2 \left( n_i - \sum_{j=2}^{M} (\phi_j^i)^2 \right)^{1/2} \right] \phi_j^i = \bar{\varepsilon}_j^i \phi_j^i ,$$

$$j = 2, 3, ..., M \tag{26}$$

are obtained, where

$$\bar{\varepsilon}_j^i = \varepsilon_j^i - \varepsilon_1^i . \tag{27}$$

If the density $n_i$ and the occupation numbers $\lambda_j^i$ are known the orbitals $\phi_j^i$ can be determined from Eq. (26). Then Eq. (25) gives the effective potential. In the last section the ground- and the first excited state exchange potentials of several atoms calculated in the KLI approximation are compared.

## OPTIMIZED POTENTIAL METHOD FOR A SINGLE EXCITED STATE

In order to perform calculations one needs an explicit expression for the non-interacting effective potential. The optimized potential method [39] can be generalized for a single excited state.

In the optimized potential method one faces the following problem: find the potential such that when it is given a small variation, the energy of the system remains stationary. It can also be written as

$$\frac{\delta E}{\delta V} = 0 . \tag{28}$$

It was shown [21] that from the fact that the energy is stationary at the true wave function follows that the energy is stationary at the true potential.

It is well-known that considering the energy as a functional of the wave function $E[\Psi]$, the eigenvalues of the Hamiltonian are stationary points of $E$

$$\frac{\delta E}{\delta \Psi_k} = 0 \qquad (k = 1, ..., i, ...) \ , \tag{29}$$

and only the eigenvalues are stationary points.

From our arguments above, from the density of a given excited state $n_i$, one can obtain the Hamiltonian, the eigenvalues and eigenfunctions and through adiabatic connection the non-interacting effective potential $V_i^{\alpha=0}$ and certainly the solution of equations of the non-interacting system leads to the density $n_i$. Thus, we can consider the total energy as a functional of the non-interacting effective potential:

$$E[\Psi_i] = E[\Psi_i[V_i^0]] \ . \tag{30}$$

Making use of Eq. (29) we obtain

$$\frac{\delta E}{\delta V_i^0} = \int \frac{\delta E}{\delta \Psi_i}\frac{\delta \Psi_i}{\delta V_i^0} + c.c. = 0 \ . \tag{31}$$

So an optimized effective potential can be found for the given excited state.

Now, following the usual method [39] deriving the optimal potential the well-known integral equation for the effective exchange potential $V_x$ can be obtained. The integral equation for the effective exchange-correlation potential $v_{xc}^i$ has the form

$$\int L_j^i(\mathbf{r}, \mathbf{r}') v_{xc}^i(\mathbf{r}') d\mathbf{r}' = Q^i(\mathbf{r}) \ , \tag{32}$$

where

$$L^i(\mathbf{r}, \mathbf{r}') = \sum_j \phi_j^{i*}(\mathbf{r}) G_j^i(\mathbf{r}, \mathbf{r}') \phi_j^i(\mathbf{r}') \ , \tag{33}$$

$$Q^i(\mathbf{r}) = \sum_j \int d\mathbf{r}' \phi_j^{i*}(\mathbf{r}) G_j^i(\mathbf{r}, \mathbf{r}') v_{xc,j}^i(\mathbf{r}') \phi_j^i(\mathbf{r}') \tag{34}$$

and $G_j^i(\mathbf{r}, \mathbf{r}')$ is the Green function. The orbital dependent potential $v_{xc,j}^i$ is given by

$$v_{xc,j}^i(\mathbf{r}) = \frac{\delta E_{xc,i}[\phi_j^i]}{\phi_j^i(\mathbf{r}) \delta \phi_j^{i*}(\mathbf{r})} \ . \tag{35}$$

To find the optimized potential is very tedious even in the ground-state. However, Krieger, Li and Iafrate [40] introduced a very accurate approximation that can be used here, too. Recently, an alternative derivation of the KLI approximation to the optimized potential method has been presented. This method can be readily generalized to excited states that is detailed in [21]. The generalized KLI exchange potential for the $i$th excited state reads

$$v_x^i = v_S^i + \sum_j \frac{|\phi_j|^2}{n_i} (\bar{v}_{xj}^i - \bar{v}_{xj}^{i,HF}) \ , \tag{36}$$

where $\bar{v}_{xj}^i$ and $\bar{v}_{xj}^{i,HF}$ are the expectation values of the exchange potential $v_x^i$ (36) and the Hartree-Fock exchange potentials $v_{xj}^{i,HF}$ ( defined in the usual way) with respect to orbital $\phi_j$.

$$v_S^i(\mathbf{r}) = -\frac{1}{2} \int d\mathbf{r}' \frac{|\sum_j \phi_j^{i*}(\mathbf{r}') \phi_j^i(\mathbf{r})|^2}{n_i(\mathbf{r}') |\mathbf{r} - \mathbf{r}'|} \ . \tag{37}$$

is the Slater potential.

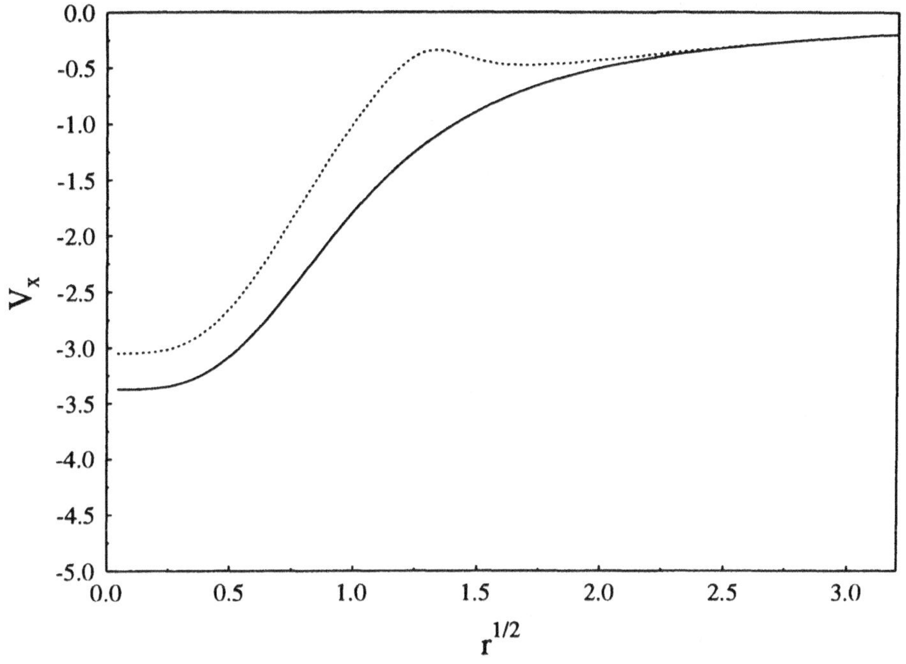


**Figure 1.** Ground(—) and first (...) excited-state exchange potentials for He atom.

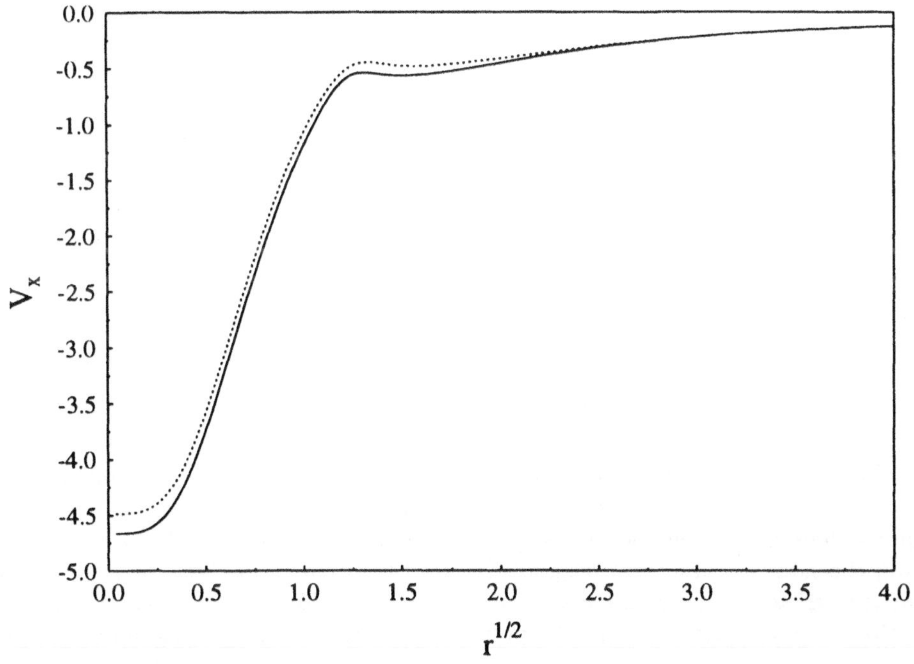


**Figure 2.** Ground (—) and first (...) excited-state exchange potentials for Li atom.

## COMPARISON OF GROUND AND EXCITED STATE EXCHANGE POTENTIALS

The fact, however, that the energy is only stationary and not minimum makes it more difficult to find adequate approximations. The non-interacting wave function should be orthogonal to the exact non-interacting wave function(s) of the lower state(s) corresponding to the same non-interacting Hamiltonian. Since the exact non-interacting wave functions are not known, one is satisfied if approximate orthogonality with respect to the approximate lower non-interacting wave function(s) is assured.

To illustrate the present method excitation energies for a couple of states are determined using the generalized KLI approximation. For simplicity, in this work only finite nondegenerate systems are considered. The treatment of degenerate states is slightly more complicated and was addressed elsewhere[21].

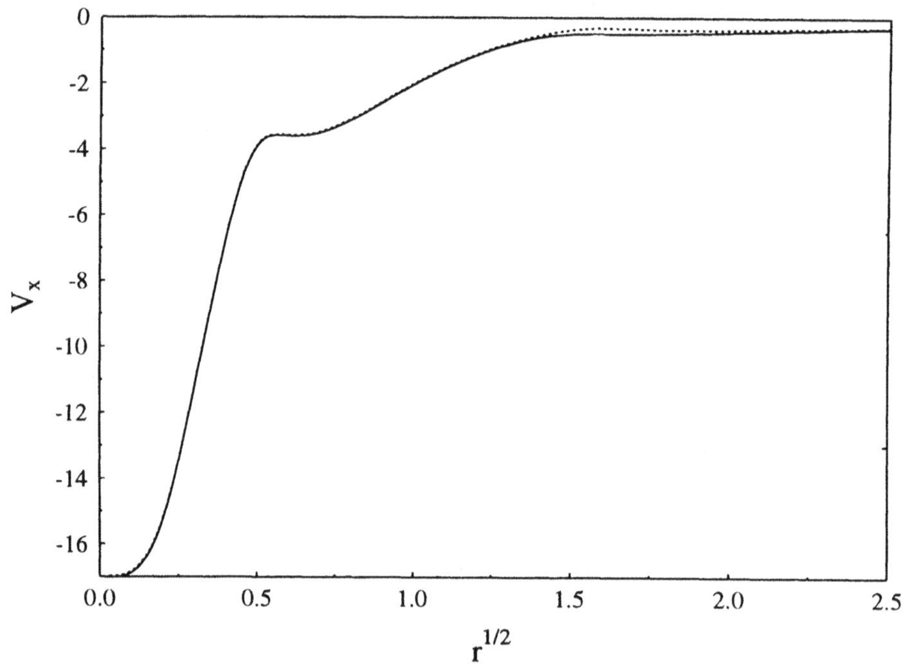


**Figure 3.** Ground(—) and first(...) excited-state exchange potentials for Na atom.

Table 1. presents the first excitation energy for the atoms $Li$, $Na$ and $K$. For the latter another excitation energy is also shown. Results are compared with accurate Hartree-Fock excitation energies [41]. The accuracy of the generalized KLI results is about the same as that of the original KLI for the ground state [40]. It has been shown that the HF total energy is lower than the KLI one [40, 34, 42]. It is, of course, true for the present results, too. In the present application exact orthogonality is assured by the fact that these excited states have different symmetries than the ground states.

Table 2. presents total energies for states that are not the lowest of a symmetry species. Here, only an approximate orthogonality is ensured. Table 2. presents total energies for the $^2P(1s^22p^5)$ states of the $N$ atom and the $O^+$ ion and for the $^1S(1s^22p^6)$ state of the $O$ atom. For comparison Hartree-Fock values [43] are also shown. Result of the KLI calculations are very close to the Hartree-Fock values.

In Figs.1.-4. the KLI exchange potentials are compared. Fig.1. presents the KLI exchange potentials for the ground- and first excited states of the He atom. (The electron configurations are $1s^2$ and $1s2s$, respectively.) They are obviously different. In Figs.2. and 3. the KLI exchange potentials are compared for the ground- and first excited states of the Li and Na atoms. (The electron configurations are $1s^22s$ and $1s^22p$, and $[Ne]3s$ and $[Ne]3p$, respectively.) Fig.4. presents the KLI exchange potentials for the ground-, first and second excited states of the K atom. (The electron configurations are $[Ar]4s$, $[Ar]4p$ and $[Ar]3d$, respectively.) It can be clearly seen that the difference between the ground and the excited state exchange potencials is less and less. It is not surprising at all. Having a glance at the KLI exchange potential expression (36), we note at once that while in the He atom the excited state orbital gives a large contribution to the exchange potential as the there are only two electrons and one of them is excited, in the K atom the ratio of the number of excited electrons (1) to the whole number of electrons (19) is much less and consequently, the contribution of the excited state orbital to the exchange potential is not so significant.

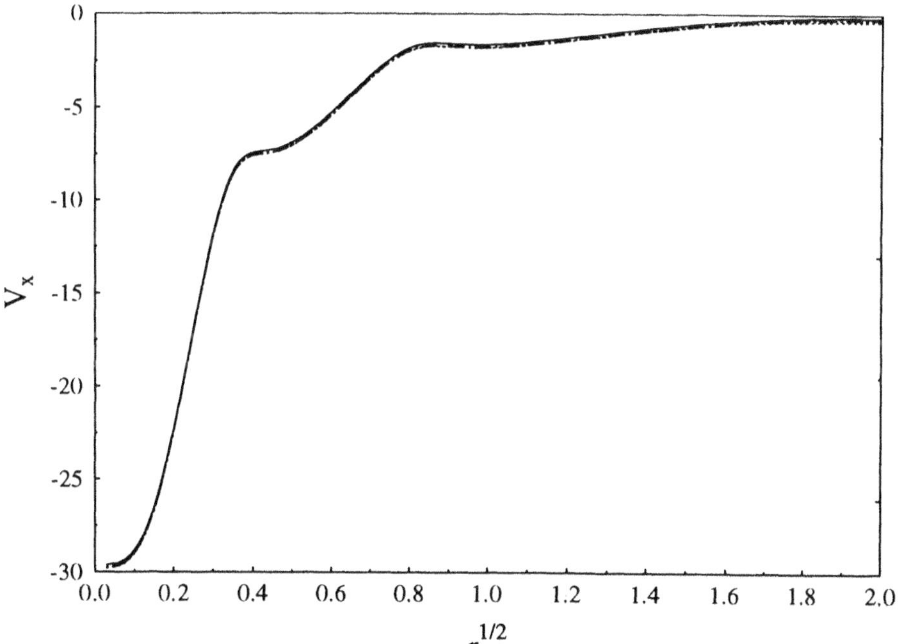


**Figure 4.** Ground(—), first(...) and second(-.-) excited-state exchange potentials for K atom.

In summary, a method was proposed to treat a single excited state using Kato's theorem and the adiabatic connection were applied to derive Kohn-Sham-like equations. It is emhasized that the energy functional is only stationary, there is no minimum. The optimized potential method is generalized and using the generalized KLI approximation quite accurate results are presented for some illustrative examples.

**Table 1.** Hartree-Fock and exchange-only density functional excitation energies (in Ry)

| | $HF$ | $DF$ |
|---|---|---|
| $Li(2s^1 \rightarrow 2p^1)$ | 0.1353 | 0.1354 |
| $Na(3s^1 \rightarrow 3p^1)$ | 0.1450 | 0.1440 |
| $K(4s^1 \rightarrow 4p^1)$ | 0.1033 | 0.1032 |
| $K(4s^1 \rightarrow 3d^1)$ | 0.1782 | 0.1816 |

**Table 2.** Total energies for a couple of states that are not the lowest of a symmetry species (in Ry)

| | $HF$ | $DF$ |
|---|---|---|
| $N(1s^2 2p^5)^2P$ | -106.4435 | -106.4419 |
| $O(1s^2 2p^6)^1S$ | -146.6116 | -146.6099 |
| $O^+(1s^2 2p^5)^2P$ | -145.6561 | -145.6543 |

## ACKNOWLEDGEMENTS

The grant 'Széchenyi' from the Hungarian Ministry of Culture and Education is gratefully acknowledged. This work was supported by the grants OTKA No. T16623 and FKFP 0314/1997.

## REFERENCES

[1] P. Hohenberg and W. Kohn, Phys. Rev. B **136**:864(1964).

[2] O. Gunnarsson and B.I. Lundqvist, Phys. Rev. B **13**:4274(1976); O. Gunnarsson, M. Jonson and B.I. Lundqvist, Phys. Rev. B **20**:3136(1979)

[3] U. von Barth, Phys. Rev. A **20**:1693(1979).

[4] A.K. Theophilou, J. Phys. C **12**:5419(1978).

[5] E.K.U. Gross, L.N. Oliveira and W. Kohn, Phys. Rev. A **37**: 2805, 2809, 2821 (1988).

[6] Á. Nagy, Phys. Rev. A **42**:4388(1990).

[7] Á. Nagy, J. Phys. **B 24**:4691(1991).

[8] Á. Nagy and I. Andrejkovics, J. Phys. B **27**:233(1994).

[9] Á. Nagy, Int. J. Quantum. Chem. **56**:225(1995).

[10] Á. Nagy , J. Phys. B **29**, 389 (1996).

[11] Á. Nagy , Int. J. Quantum. Chem. S. **29**:297 (1995).

[12] Á. Nagy , Adv. Quantum. Chem. (in press).

[13] Á. Nagy, Phys. Rev. A **49**:3074(1994).

[14] Á. Nagy , Int. J. Quantum. Chem. **29**:159(1997).

[15] E. U. K. Gross, J. F. Dobson and M. Petersilka, in Density Functional Theory, (Topics in Current Chemistry vol. 181. p 81.) R. Nalewajski, ed. Springer-Verlag, Heidelberg (1996).

[16] M. F. Casida, in Recent advances in the density functional methods, (Recent Advances in Computational Chemistry vol.1 p.155.) D. P. Chong, ed. World Scientific, Singapore (1996).

[17] M. K. Harbola and V. Sahni, Phys. Rev. Lett. **62**:489(1989); see also V. Sahni, in Density Functional Theory (Topics in Current Chemistry vol.182. p.1.) R. Nalewajski, ed. Springer-Verlag, Berlin (1996).

[18] K. D. Sen, Chem. Phys. Lett. **168**,510(1992); R. Singh and B. M. Deb, Proc. Indian Acad. Sci. **106**:1321(1994); J. Mol. Struc. Theochem **361**:33(1996); J. Chem. Phys. **104**:5892 (1996); A. K. Roy, R. Singh and B. M. Deb, J. Phys. B **30**:4763(1997); Int. J. Quantum. Chem. **65**:317(1997); A. K. Roy, and B. M. Deb, Phys. Lett. A **234**: 465(1997);

[19] A. Görling, Phys. Rev. A **54**:3912(1996).

[20] A. Görling and M. Levy, Phys. Rev. A **47**:13105(1993); Phys. Rev. A **47**:196(1994); Int. J. Quantum. Chem. S. **29**:93(1995).

[21] Á. Nagy , Int. J. Quantum. Chem. (in press).

[22] T. Kato, Commun. Pure Appl. Math. **10**:151(1957); E. Steiner, J. Chem. Phys. **39**:2365(1963); N. H. March, Self-consistent fields in atoms (Pergamon, Oxford, 1975)

[23] N. C. Handy, in Quantum Mechanical Simulation Methods for Studying Biological Systems Eds. D. Bicout and M. Field (Springer–Verlag, Heidelberg,1996) p.1.

[24] M. Levy, Proc. Natl. Acad. Sci. USA **76**: 6002 (1979).

[25] M. Lieb, Int. J. Quantum. Chem. **65**:317(1997).

[26] J. Harris and R. O. Jones, J. Phys. F **4**:1170(1974); J. Harris, Phys. Rev. A **29**:1648(1984).

[27] Á. Nagy, J. Phys. B **26**:43(1993) ; Phil. Mag. B **69**:779(1994).

[28] Á. Nagy and N. H. March, Phys. Rev. A **39**:5512(1989) ; **40**:5544(1989).

[29] Q. Zhao and R. G. Parr, J. Chem. Phys. **98**:543(1993) ; R. G. Parr, Phil. Mag. B **69**:737(1994); Q. Zhao, R. C. Morrison and R. G. Parr, Phys. Rev. A **50**:2138(1994). R. C. Morrison and Q. Zhao, Phys. Rev. A **51**:1980(1995). R. G. Parr and S. Liu, Phys. Rev. A **51**:3564(1995).

[30] C. O. Almbladh and A. P. Pedroza, Phys. Rev. A **29**:2322(1984).

[31] F. Arysetiawan and M. J. Stott, Phys. Rev. B **34**:4401(1986).

[32] F. Arysetiawan and M. J. Stott, Phys. Rev. B **38**:2974(1988).

[33] J. Chen and M. J. Stott, Phys. Rev. A **44**:2816(1991); J. Chen, R. O. Esquivel and M. J. Stott, Phil. Mag. B **69**: (1994).

[34] A. Görling, Phys. Rev. A **46**:3753(1992); A. Görling and M. Ernzerhof,**51**:4501(1995).

[35] R. van Leeuwen and E. J. Baerends, Phys. Rev. A **49**:2421(1994).

[36] K.A. Dawson and N.H. March, J. Chem. Phys. **81**:5850(1984).

[37] Q. Zhao and R. G. Parr, Phys. Rev. A **46**:2337(1992).

[38] A. Holas, N. H. March, Y. Takahashi and C Zhang, Phys. Rev. A **48**: 2708(1993).

[39] R. T. Sharp and G. K. Horton, Phys. Rev. A **30**:317(1953); K. Aashamar, T. M. Luke and J. D. Talman, At. Data Nucl. Data Tables **22**:443(1978).

[40] J. B. Krieger, Y. Li and G. J. Iafrate, Phys. Lett. A **146**:256 (1990); Int. J. Quantum. Chem. **41**:489(1992); Phys. Rev. A **45**:101(1992); Phys. Rev. A **46**:5453(1992) ; in Density Functional Theory, E. K. U. Gross and R. M. Dreizler eds., Plenum, New York(1995).

[41] H. Tatewaki, T. Koga, Y. Sakai and A. J. Thakkar, J. Chem. Phys. **101**:4945(1994); H. Tatewaki and T. Koga, Chem. Phys. Lett. **228**:562(1994).

[42] A. Holas and N.H. March, in Density Functional Theory, (Topics in Current Chemistry Vol. 180) R. Nalewajski ed., Springer-Verlag, Heidelberg (1996).

[43] C. C. J. Roothaan and P. S. Kelly, Phys. Rev. **131**:1177(1963).

# CONSTRUCTION OF AN ACCURATE SELF-INTERACTION-CORRECTED CORRELATION ENERGY FUNCTIONAL BASED ON AN ELECTRON GAS WITH A GAP

Joseph B. Krieger, Jiqiang Chen

Physics Department
Brooklyn College, CUNY
Brooklyn, NY 11210

Gerald J. Iafrate

Department of Electrical Engineering
University of Notre Dame
Notre Dame, IN 46556-5637

Andreas Savin

Laboratoire de Chimie Théorique (CNRS)
Université P. et M. Curie
4, place Jussieu, F-75252 Paris

## I. INTRODUCTION

The usual starting point of any first principles construction of an approximate electron correlation energy functional $E_c[\{n_\sigma\}]$ is the homogeneous electron gas weakly perturbed by an external potential. To lowest non-vanishing order in the gradient of the electron density, $\nabla n$, the correlation energy density, per electron, $\epsilon_c(r_s,\zeta)$, may be written[1,2] as the usual local spin density (LSD) expression, $\epsilon_c^{LSD}$, plus a term proportional to $|\nabla n|^2$. When perturbation theory is employed to calculate $\epsilon_c^{LSD}$, the result diverges in each order of the electron-electron coupling due to the zero energy gap

*Electron Correlations and Materials Properties*, edited by Gonis *et al.*
Kluwer Academic/Plenum Publishers, 1999. 

between states at the Fermi energy and the neighboring unoccupied states in the continuum. By employing many body perturbation theory techniques, this series may be summed in the high density limit to yield a finite result[3]. Other numerical calculations have been employed to calculate $\epsilon_c^{LSD}$ for arbitrary uniform density[4].

However, when the resulting $\epsilon_c^{LSD}(r_s, \zeta)$ is employed to calculate $E_c$ for atoms, the results are generally too large in magnitude by approximately a factor of two. Thus the gradient corrections to $E_c^{LSD}$ must be approximately equal to $-E_c$, suggesting that a simple gradient expansion of $E_c$ is inadequate to describe the contribution from the rapid variation in electron density in these physical systems. This has prompted the construction of generalized gradient approximations (GGA's) which reduce to the gradient expansion approximation (GEA) results in the slowly varying limit, but are constructed to satisfy certain important scaling properties[6] known to be satisfied by the exact $E_c[n]$ but not satisfied by a simple GEA gradient or by an expansion including higher order derivatives of the density. Perhaps the most interesting GGA is the recent one by Perdew, Burke and Enzerhof[7] (PBE), who employ only the $\epsilon_c^{LSD}(r_s, \zeta)$, the lowest order term in the gradient expansion and an ingenious ansatz to obtain an expression for $E_c[\{n_\sigma\}]$ that yields results for molecular dissociation energies which are nearly identical to those predicted by the much more complicated PWGGA.[5] However, although these results reduce the error in the LSD calculations by a factor of four, the PBE dissociation energies still exhibit an average error of $\approx$ 8 kcal/mole, which is eight times larger than the quantum-chemistry accuracy criterion. Moreover, from a more theoretical point of view, the PBE correlation energy functional may be criticized because, as with other pure spin density functionals, it is not self-interaction-free[8] i.e. for the hydrogen atom it predicts[9] $E_c \approx -6$ mhartree $\approx -4$ kcal/mole.

It has long been understood that the LSD overestimate of the magnitude of the correlation energy for localized electrons arises primarily from the fact that the ground state of such systems is separated from the higher lying states by a non-zero energy with the gap to the continuum given by the ionization energy. Thus, if one employs perturbation theory taking the zeroth order wavefunction to be either the Hartree-Fock (Moller-Plesset perturbation theory) or the exchange-only Kohn-Sham ground state, the calculation of $E_c$ for atoms converges after only a few orders to nearly the same result as that obtained by configuration interaction techniques. Moreover, early work by Tong[10] demonstrated that if electrons were confined to a box of atomic dimensions such that the density was equivalent to that found in atoms, the existence of discrete states with resulting energy gaps led to results for $E_c$ that were about a factor of two smaller than that given by the local density approximation. The foregoing observations therefore suggest that a more appropriate model to employ in performing density functional calculations of $E_c$ is one that includes an appropriately chosen energy gap.

Generally, it is not possible to define a local potential that yields a homogeneous electron gas with a non-zero energy gap at the Fermi level for an infinite system. However, Rey and Savin[11] have employed a non-local one body operator to shift the virtual (non-interacting) levels by an arbitrary energy gap $G$. They then performed calculations of the correlation energy density of a homogeneous unpolarized electron gas as a function of both $r_s$ and $G$ i.e. $\epsilon_c(r_s, G, \zeta = 0)$ which reduces to the usual local density approximation when $G = 0$. In addition, they have obtained an accurate analytic representation of $\epsilon_c(r_s, G, \zeta = 0)$ so that this functional may be employed to calculate

$E_c$ for any spin unpolarized system once a prescription is given to determine $G$ for the system of interest.

In section II we discuss their results and propose that, to a first approximation, $G[n]$ be taken as the simplest function of the gradient of the electron density that reduces to the ionization energy in the large $r$ limit, i.e. the energy gap to the continuum. It then follows that $\epsilon_c(r_s, G[n], \zeta = 0) \to 0$ as $|\nabla n| \to \infty$ and that $\epsilon_c(r_s, G[n], \zeta = 0)$ saturates under uniform scaling to the high density limit without the need to introduce any further ansatz to construct a GGA. The correlation energy density for any $\zeta$ is then obtained by employing the von Barth-Hedin[12] interpolation formula between the $\zeta = 0$ and $\zeta = 1$ results, the latter having recently been obtained by Savin[13]. When the resulting $\epsilon_c(r_s, G[n], \zeta)$ is employed to calculate $E_c$ for atoms and positive ions for $Z \leq 18$, we find that the discrepancy from highly accurate CI calculations[14] is less than half that obtained by employing the LSD with $G = 0$.

The $\epsilon_c(r_s, G[n], \zeta)$ functional is further improved by performing an orbital dependent self-interaction-correction similar to one first suggested by Becke.[15] The resulting $\epsilon_c^{SIC}$ is invariant under a unitary transformation among the occupied orbitals. We find that the corresponding $E_c^{SIC}$ for both atoms and positive ions with $Z \leq 18$ are significantly improved being approximately 10 to 20% too large in magnitude.

Finally, in order to ensure that $\epsilon_c(r_s, G[n], \zeta)$ for $\zeta = 0$ and $\zeta = 1$ yields the correct small gradient expansion in the high density limit, we construct the simplest possible $\epsilon_c^{GGA}(r_s, |\nabla n|, \zeta)$ to replace $\epsilon_c^{LSD}(r_s, \zeta)$ in the analytic representation of $\epsilon_c(r_s, G, \zeta)$. This GGA is constructed by requiring the correct small gradient expansion as well as satisfying the conditions that $\epsilon_c^{GGA} \to 0$ for $|\nabla n| \to \infty$ and $\epsilon_c^{GGA}$ saturates under uniform scaling in the high density limit (the three conditions required by PBE).

With this modification, the resulting $\epsilon_c^{GGASIC}(r_s, |\nabla n|, G[n], \zeta)$ yields results for $E_c/N$ the average energy per electron, in atoms and positive ions with $Z \leq 18$, that are on average in error by $\approx$ 1 mhartree compared to an average error of $\approx$ 40 mhartree when the LSD approximation is employed. Our results for $E_c$ generally lie between those given by the PWGGA and PBE functionals and are closer to the latter with the PWGGA calculations generally lying deeper and the PBE calculations always lying more shallow than the exact results for atoms and ions with more than three electrons.

## II. CONSTRUCTION OF THE CORRELATION ENERGY FUNCTIONAL

### A) Local Density Approximation for a Homogeneous Electron Gas with an Energy Gap

In order to account explicitly for the fact that the perturbation theory calculation of the correlation energy for systems containing localized electrons depends on the energy gap between the ground state and the higher lying states of the non-interacting (Kohn-Sham) system, Rey and Savin[11] have performed detailed calculations of $\epsilon_c(r_s, G)$, the correlation energy per particle for an unpolarized uniform electron gas with energy gap G separating the Fermi energy from the continuum. The results of their calculations have been tabulated for a wide range of densities $0.1 \leq r_s \leq 10$ relevant to atomic and molecular electron densities with $0 \leq G \leq 20\epsilon_F$ where $\epsilon_F$ is the Fermi energy of the homogeneous electron gas. In addition, they have provided an analytic fit to $\epsilon_c(r_s,G)$ which is accurate to a few tenths of a milli-hartree when compared to their numerical results.

They find that

$$\epsilon_c(r_s,G) = \frac{\epsilon_c^{LDA}(r_s) + C_1(r_s)G}{1 + C_2(r_s)G + C_3(r_s)G^2} \tag{1}$$

where $C_1, C_2, C_3$ are given as analytic functions of $r_s$. Obviously, in the limit $G = 0$, $\epsilon_c(r_s,G) \to \epsilon_c^{LDA}(r_s)$.

In order to apply eq. (1) to physical systems, in addition to the usual local density approximation for $r_s = r_s[n(r)]$, it is necessary to choose an approximate ansatz for $G$.

If in the zeroth approximation we neglect any contribution to the correlation energy calculation from unoccupied localized KS orbitals then the smallest excitation energy is given by the ionization energy to the continuum, $I$. However, it is clear that such a choice for $G$ is not consistent with the condition that the $E_c$ for two dissimilar atoms that are very far apart must be the sum of the two separate correlation energies. This follows from the fact that $E_c$ for each separate atom will depend on its own $I$, but the composite system with large interatomic distance will have only a single $I$, that of the atom with the lowest ionization energy. We therefore seek a definition of $G[n]$ which is local so that the $E_c$ of two sub-systems far apart will be the sum of the separate correlation energies and that for any isolated system, $G \to I$ as $r \to \infty$. In addition we require that $G \to 0$ when applied to a homogeneous electron gas so that $\epsilon_c(r_s,G) \to \epsilon_c^{LDA}$ in that limit. This suggests that $G$ depends on gradients of the density. Moreover, since $G \geq 0$, the latter suggests that $G$ is a function of $|\nabla n|^2$.

But for any KS system of non-interacting electrons for which the external potential vanishes as $r \to \infty$, $V_{KS} \to 0$ as $r \to \infty$[16] so the density for large $r$ is dominated by the highest lying occupied orbital corresponding to

$$n(r) \underset{r \to \infty}{\sim} \exp -2[(-2\epsilon_m)^{1/2} r] \tag{2}$$

Consequently if we define

$$G^0[n] \equiv \frac{1}{8} |\nabla \ln n|^2 \tag{3}$$

then it follows from eqs. (2) and (3) that

$$G^0[n] \underset{r \to \infty}{\to} -\epsilon_m \tag{4}$$

But in an exact KS theory[17], $-\epsilon_m = I$ so

$$G^0[n] \underset{r \to \infty}{\to} I \;. \tag{5}$$

We also note that since

$$n(r) \underset{r \to 0}{\sim} \exp(-2Zr) \tag{6}$$

near the nucleus of an atom of atomic number $Z$, then

$$G^0[n] \underset{r \to 0}{\to} Z^2/2 > I \quad \text{for } Z > 1 \ , \tag{7}$$

so that $G^0$ tends to increase in the atomic core and approximates the excitation energy necessary to move an electron to the continuum from these deeper levels i.e. approximates the energy denominators appearing in a perturbation theory treatment. In taking $G^0 = G^0[n]$ we are of course depending on the fundamental theorem of density functional theory[18] which guarantees that $E_c = E_c[n]$ so that in applying eq.(1) to real physical systems, it should be possible to write $G = G[n]$.

Moreover, if eq.(3) is employed to approximate $G$ in eq.(1), then it not only follows that

$$\epsilon_c(r_s, G^o[n]) \to \epsilon_c^{LDA}(r_s) \quad \text{for } |\nabla n| \to 0 \tag{8}$$

but in addition

$$\epsilon_c(r_s, G^o[n]) \to 0 \quad \text{for } |\nabla n| \to \infty \ . \tag{9}$$

Furthermore, under uniform scaling i.e. $n(r) \to n_\lambda(r) = \lambda^3 n(\lambda r)$ in the high density limit ($\lambda \gg 1$) $r_s \sim \lambda^{-1}$, $\epsilon_c^{LDA} \sim -\gamma \ln\lambda$, $G^0[n] \sim \lambda^2$, and employing the analytical behavior[11] of $C_1$, $C_2$, $C_3$ in eq.(1) we find

$$\epsilon_c(r_s, G^0[n]) \underset{\lambda \to \infty}{\to} \frac{C_1}{C_3 G^0} \sim \frac{1}{r_s^2 G^0} \sim \lambda^0 \tag{10}$$

i.e. the correlation energy saturates. Thus $\epsilon_c(r_s, G^o[n])$ automatically satisfies some of the important exact scaling properties often imposed in constructing approximate GGA's for the correlation energy[7].

**B) Extension to Spin Polarized System**

In order to perform calculations on spin polarized systems, we require $\epsilon_c(r_s, G, \zeta)$ which is not yet available for partial spin polarization i.e. $0 < \zeta < 1$. However, since the results for $\zeta = 0$[11] and $\zeta = 1$[13] have been calculated, we can employ the von Barth-Hedin[12] interpolation formula to obtain the results for intermediate $\zeta$ i.e.

$$\epsilon_c(r_s, G, \zeta) = \epsilon_c(r_s, G, \zeta = 0) + f(\zeta)[\epsilon_c(r_s, G, \zeta = 1) - \epsilon_c(r_s, G, \zeta = 0)] \tag{11}$$

where

$$f(\zeta) \equiv \frac{(1+\zeta)^{4/3} + (1-\zeta)^{4/3} - 2}{2(2^{1/3}-1)} \quad . \tag{12}$$

In order to employ eq. (11) for continuous values of $r_s$ and $G$, we require an analytic fit to Savin's numerical results for $\epsilon_c(r_s, G, \zeta = 1)$. Taking the same functional form given by eq. (1) we find $C_1(r_s, \zeta = 1) = 0.70\, C_1(r_s, \zeta = 0)$, $C_2(r_s, \zeta = 1) = 1.50\, C_2(r_s, \zeta = 0)$, and $C_3(r_s, \zeta = 1) = 2.59\, C_3(r_s, \zeta = 0)$ yields $\epsilon_c(r_s, G, \zeta = 1)$ which are within 1 mhartree for $0.1 \leq r_s < 10$ and $0 \leq G \leq 20\, \epsilon_F$. Although the accuracy of the von Barth-Hedin formula cannot be tested for arbitrary $G$, we have compared the results of employing this interpolation to those given by a more precise spin polarized fit to numerical calculations for $G = 0$. We find that when the two $\epsilon_c^{LSD}(r_s, \zeta)$ are employed to calculate the correlation energy for atoms using exact exchange-only densities[19], the two results for $E_c$ are within a few tenths of one percent for atoms with $Z \leq 18$ with the exception of nitrogen for which they differ by 1%.

When the local density approximation is made for $r_s$ and $G$ is given by eq.(3), we shall refer to eq.(10) as the LSDGAP approximation.

### C) Inclusion of Self-Interaction-Correction

The exact correlation energy functional must vanish indentically for any one-electron system independent of the form of the external potential. As in the case of the LSD approximation, the LSDGAP expression for $E_c[n]$ does not satisfy this condition and thus a self-interaction correction (SIC) must be introduced.

Early work by Perdew and Zunger[20] employed an ad hoc SIC for $E_c$ which consisted of subtracting the sum of the separate orbital contributions to $E_c$ of each electron. The resulting expression for $E_c^{SIC}$ was self-interaction free and significantly reduced the overestimate of $E_c^{LSD}$. However, this method is not invariant under a unitary transformation of the occupied orbitals and thus does not yield a unique $E_c$ for different orbital representations of the same density. Subsequently, Becke[15] showed that the correlation energy of any system of electrons may be written as the sum of the contribution from interactions between electrons with opposite spins and a contribution from electrons with parallel spin, the latter being proportional to an orbital dependent factor $D_\sigma \equiv \sum_i |\nabla\phi_{i\sigma}|^2 - |\nabla n_\sigma^{1/2}|^2$. $D_\sigma$ is invariant under a unitary transformation of the occupied orbitals and is identically zero if there is only one occupied orbital of a given spin.

Then, if $\epsilon_c[n_\alpha, n_\beta]$ is the uncorrected correlation energy density,

$$E_{copp} = \int d\tau\, \{n\epsilon_c[n_\alpha, n_\beta] - \sum_\sigma n_\sigma \epsilon_c[n_\sigma, 0]\} \tag{13}$$

and

$$E_{c\sigma\sigma} = \int d\tau \sum_\sigma \frac{\sum_i |\nabla\phi_{i\sigma}|^2 - |\nabla n_\sigma^{1/2}|^2}{\sum_i |\nabla\phi_{i\sigma}|^2}\, n_\sigma \epsilon_c[n_\sigma, 0] \quad . \tag{14}$$

The only difference between eqs.(13) and (14) and Becke's prescription is the choice of the denominator in eq. (14). In Becke's original treatment, the kinetic energy density in the LSD approximation was employed as a normalizing factor so the factor multiplying $n_\sigma \epsilon_c[n_\sigma,0]$ should reduce to unity in the homogeneous gas limit. However, he found that for atoms with large interatomic distances, this could lead to an enhancement of $E_{c\sigma\sigma}$ due to this SIC which was unphysical. The choice of the exact kinetic energy density as the normalizing factor obviously ensures that the fraction is always $\leq 1$[21].

Adding eqs.(13) and (14) yields

$$E_c^{SIC} = \int d\tau \{ n\epsilon_c[n_\alpha,n_\beta] - \sum_\sigma \frac{|\nabla n_\sigma^{1/2}|^2}{\sum_i |\nabla \phi_{i\sigma}|^2} n_\sigma \epsilon_c[n_\sigma,0] \} \quad . \tag{15}$$

When the LSDGAP expression given by eq.(11) is employed for $\epsilon_c$ in eq. (15), we shall refer to this approximation as LSDGAPSIC. As in the case of the LSD approximation for $E_c[\{n_\sigma\}]$, the correlation energy functional given by the LSDGAP and LSDGAPSIC approximations have no adjustable parameters, depending only on results of calculations on the homogeneous electron gas with a gap.

### D) Inclusion of Gradient Correction to $\epsilon_c^{LSD}$

In the limit of high but slowly varying density it has been shown[1,2] that

$$\epsilon_c^{LSD}(r_s,\zeta,t) = \epsilon_c^{LSD}(r_s) + \beta\phi^3 t^2 + \ldots \tag{16}$$

where

$$\beta = 0.066725 \frac{e^2}{a_0} \quad , \quad \phi(\zeta) \equiv \frac{[(1+\zeta)^{2/3} + (1-\zeta)^{2/3}]}{2},$$

$$t \equiv \frac{|\nabla n|}{2\phi k_s n} \quad , \quad k_s \equiv \left(\frac{4k_F}{\pi a_0}\right)^{1/2} \tag{17}$$

However, if we take the small $G$ expansion of the LSDGAP approximation discussed in Sec. IIB, employ eq. (3) and write the result in terms of $t^2$ instead of $G$, we find that the correction to $\epsilon^{LSD}(r_s)$ of order $t^2$ vanishes in the high density limit.

The simplest approach to ensure the correct small $t^2$ expansion of $\epsilon_c^{LSDGAP}$ is to modify the $\epsilon_c^{LDA}(r_s)$ term appearing in eq. (1) for $\zeta = 0$ and in the corresponding expression for $\zeta = 1$ so that $\epsilon_c(r_s,G,\zeta)$ given by eq.(11) exactly reduces to eq.(16) at least for $\zeta = 0$ and 1. As in the case of the LSD approximation, it is not possible to simply employ eq.(16) to describe the inhomogeneous electron gas for arbitrary $t^2$ because for sufficiently large $t^2$ in the high density limit the expansion for $\epsilon_c$ would become positive.

Alternatively, we consider the construction of $\epsilon_c(r_s, G, \zeta, t)$ for extended states such that in the $t = 0$ limit, the resulting $\epsilon_c(r_s, G, \zeta)$ is identical to $\epsilon_c^{LSDGAP}$ given by eq.(11), and in the $G = 0$ limit is identical to eq.(16) for $t^2 \ll 1$ (at least for $\zeta = 0$ and $\zeta = 1$), and in addition satisfies the three conditions required by PBE in constructing a suitable GGA[7].

In the simplest approximation this can be accomplished by replacing $\epsilon_c^{LDA}(r_s)$ in eq.(1) (and in the corresponding expression for $\zeta = 1$[13]) by an $\epsilon_c^{GGA}(r_s, \zeta, t)$ that satisfies the conditions given by eqs. (9,10) and reduces to $\epsilon_c(r_s, \zeta, t)$ given by eq.(16) for $t^2 \ll 1$.

We consider a GGA of the form

$$\epsilon_c^{GGA}(r_s, \zeta, t) = \frac{\epsilon_c^{LSD}(r_s, \zeta)}{1 + f(r_s, \zeta, t)} \tag{18}$$

Then, under uniform scaling in the high density limit[7]

$$\epsilon_c^{LSD} \underset{\lambda \to \infty}{\sim} -\gamma ln\lambda \quad \text{and} \quad t^2 \underset{\lambda \to \infty}{\sim} \lambda \tag{19}$$

so that if $\epsilon_c^{GGA}$ saturates in the limit $\lambda \to \infty$, then

$$f \underset{\lambda \to \infty}{\sim} ln\lambda \sim lnt^{2p} \tag{20}$$

where $p$ can be taken as a constant $> 0$. Furthermore, if we take

$$f = \beta ln \left[ 1 + \frac{\phi^3 t^2}{p\,|\epsilon_c^{LSD}(r_s, \zeta)|} \right]^p \tag{21}$$

with $p > 0$, then it follows from eq.(18) that $\epsilon_c^{GGA} \to 0$ as $t^2 \to \infty$, and for $t^2 \ll 1$, the expansion of $\epsilon_c^{GGA}$ satisfies eq.(16) independent of the choice of $p$. In addition, since $f \geq 0$, $\epsilon_c^{GGA} \leq 0$ for all $r_s$ and $t$.

We thus find that there is a family of $\epsilon_c^{GGA}$, each corresponding to a different choice of the parameter $p$, that satisfy the PBE conditions for an acceptable GGA. This result suggests that it may not be possible to derive a unique $\epsilon_c^{GGA}$ even if more scaling conditions are imposed and that the only way to obtain a highly accurate $\epsilon_c^{GGA}$ by employing only scaling properties and the small $t^2$ expansion of $\epsilon_c$ is to bias the choice of parameters that enter $\epsilon_c^{GGA}$. It is clear that the PBEGGA yields results close to those of PWGGA, not simply because it satisfies the conditions given by eq.(9,10,16) but because it was constructed to mimic the PWGGA which is not a unique theoretical fit but is somewhat biased toward atoms by the use of a sharp real-space cut-off of the gradient expansion for the correlation hole[9].

However, as long as we employ the exact exchange-only functional[22], we can in principle determine the best $p$ based on purely theoretical considerations by requiring the ionization theorem be satisfied i.e. the self consistently calculated $-\epsilon_m$ and $I \equiv E(N-1) - E(N)$ agree as closely as possible. Such a method has been successfully employed to calculate the best exchange-mixing coefficient[23]. For simplicity, the calculations reported here employ the simplest choice of $p$ in eq.(21) i.e. $p = 1$.

In the following we shall refer to the method in which $\epsilon_c^{LSD}(r_s,G,\zeta)$ is replaced by $\epsilon_c^{GGA}(r_s,G,\zeta,t)$ in the LSDGAPSIC approximation as the GGAGAPSIC approximation.

## III. DISCUSSION OF NUMERICAL RESULTS

It is interesting to study the effect on the calculated correlation energy density as the sequence of approximations described in Section II are implemented. All calculations reported here are performed by employing the KLI approximation[22] for the exact exchange-only single particle wavefunctions.

Figure 1 is a graph of the radial correlation energy density for helium versus the distance from the nucleus so that the area under the horizontal axis is the corresponding $E_c$ in each approximation. For helium, the exact[14] $E_c = -0.0420$ a.u. and $E_c^{LSD} = -0.1125$ a.u. using the PW[24] parametrization of $\epsilon_c^{LSD}(r_s)$ which will be employed throughout these calculations. We see that when the LSDGAP for $\epsilon_c$ is employed with $G$ given by eq.(3), the resulting radial correlation energy density is significantly smaller in magnitude with the resulting $E_c^{LSDGAP} = -0.0715$ a.u., thus reducing the error in the LSD result by $\approx$ 60%. In addition, when the self-interaction correction is included, the correlation energy density is further everywhere reduced with $E_c^{LSDGAPSIC} = -0.0470$ a.u. which corresponds to an error of $\approx$ 7% of the original LSD error. Finally, the implementation of the GGAGAPSIC approximation leads to a further, smaller, decrease in the magnitude of the correlation energy density yielding $E_c^{GGAGAPSIC} = -0.0408$ a.u., thus overshooting the exact result and leaving an error of only 1.2 mhartree which is less than 2% of the LSD error.

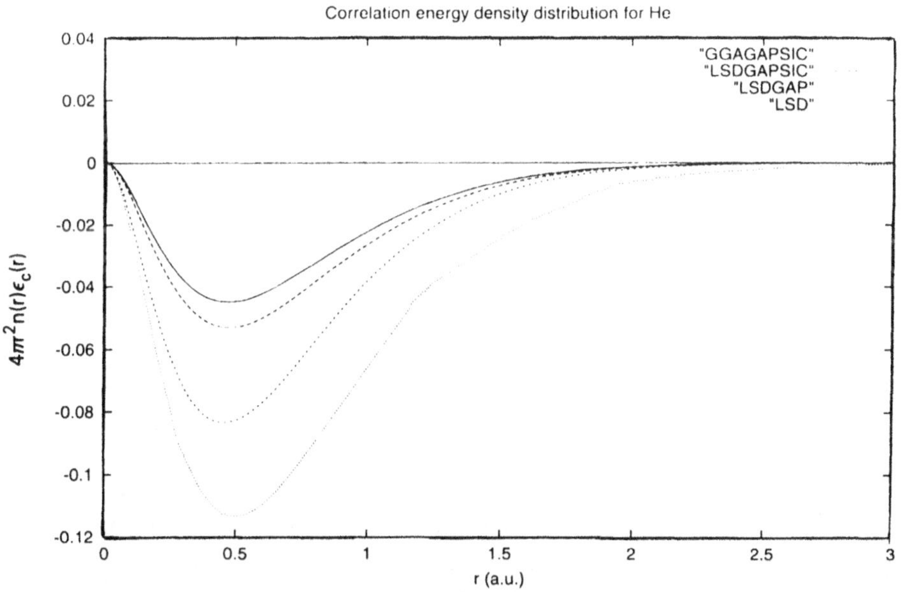


**Fig. 1** The radial correlation energy density for the helium atom in several approximations vs. the distance from the nucleus. Results are exhibited in the local spin density (LSD) approximation as well as approximations described in Section II.

Similarly, Figure 2 is a graph for the magnesium atom for which the shell structure is apparent due to the weighting of the correlation energy density by the radial density. Here the best available estimate of the correlation energy[14] (see also eq. (23)) is $E_c$ = -0.4424 a.u., and $E_c^{LSD}$ = -0.8876 a.u. We see that the application of the LSDGAP energy functional decreases the depth of $\epsilon_c$ in the vicinity of the $n = 1$ shell i.e. near $r \approx 0.1$ a.u., to about 60% of its LSD value whereas the corresponding decrease for $r$ near the $n = 2$ radial density maximum i.e. $r \approx 0.5$ a.u., is to approximately 85% of its LSD value. This is a general characteristic of the LSDGAP functional: the core electron contribution to $E_c$ is reduced by a greater factor than that from the higher lying electron shells and for atoms with larger $Z$, the radial correlation energy density near the $n = 2$ shell is deeper than that near the $n = 1$ shell. For magnesium, $E_c^{LSDGAP}$ = -0.6183 a.u. which corresponds to a reduction in the LSD error of $\approx$ 60%. As in the $Z = 2$ case, the implementation of the LSDGAPSIC approximation further reduces the magnitude of $\epsilon_c$ everywhere with the greatest depth of the radial correlation energy density now in the vicinity of the $n = 2$ subshell, with $E_c^{LSDGAPSIC}$ = -0.5029 a.u. corresponding to an error of $\approx$ 14% of the LSD error. Finally, the application of the GGAGAPSIC approximation results in a further decrease in the magnitude of $\epsilon_c$ everywhere and a value of $E_c^{GGAGAPSIC}$ = -0.4362 a.u. thus overshooting the needed correction by less than 2% of the LSD error.

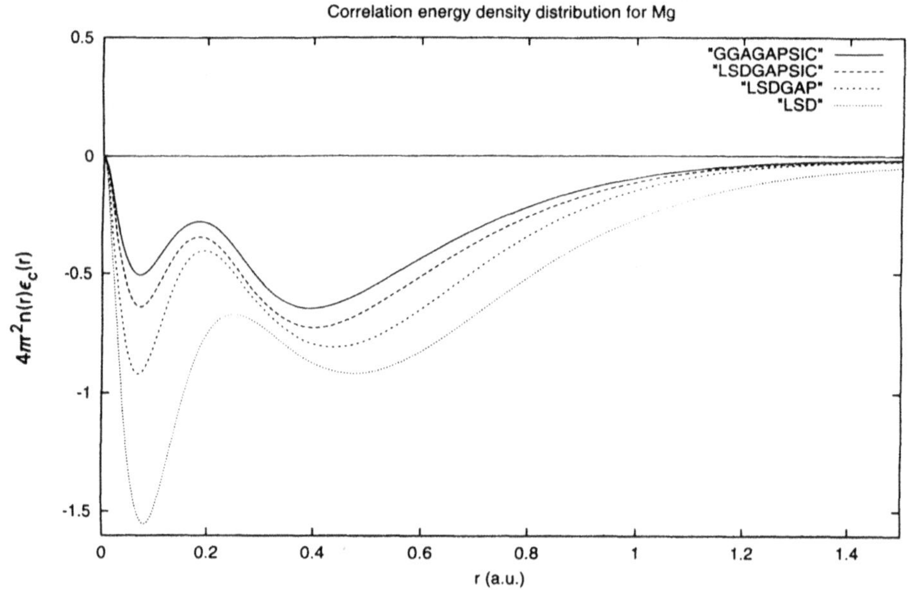


**Fig. 2** The radial correlation energy density for the magnesium atom vs. the distance from the nucleus. The approximations employed are the same as in Fig. 1.

Figures 3 and 4 exhibit the correlation energy per electron for atoms and positive ions versus the number of electrons in each system for which accurate CI calculations have been performed[14] i.e. $Z \leq 18$. We see that as each successive correction is made to the correlation energy functional, the calculated $E_c$ more closely approximate the most accurate available $E_c^{DFT}$ results with the GGAGAPSIC yielding a discrepancy of $\approx$ 1 mhartree on average for the correlation energy per electron compared to an average error of $\approx$ 40 mhartree for $E_c/N$ when the LSD approximation is employed.

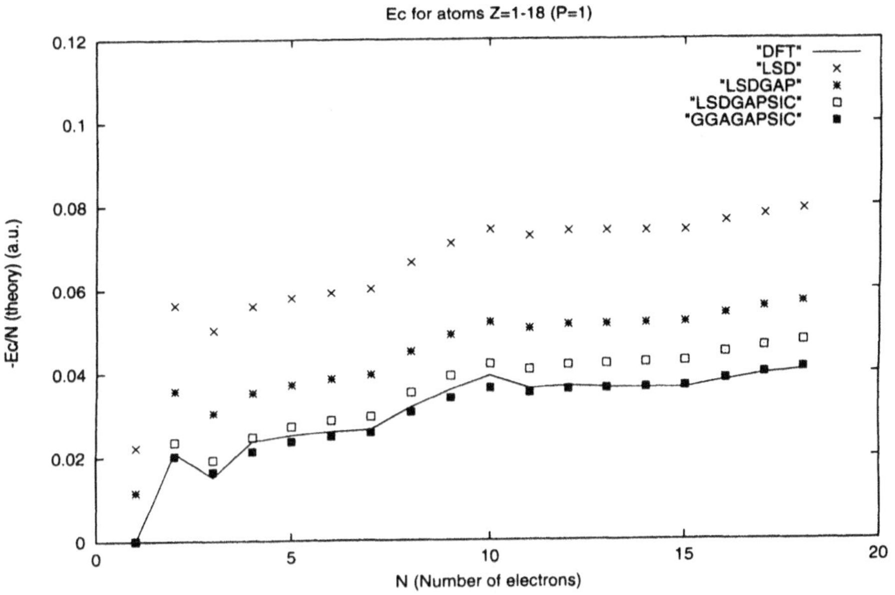


**Fig. 3** The negative of the correlation energy per electron vs. the number of electrons, N, for atoms with nuclear charge Z ≤ 18. The DFT results are obtained from eq. (23) and ref. 14. The straight lines connecting the DFT results for integer values of N have no physical significance. The approximations employed are the same as in Fig. 1.

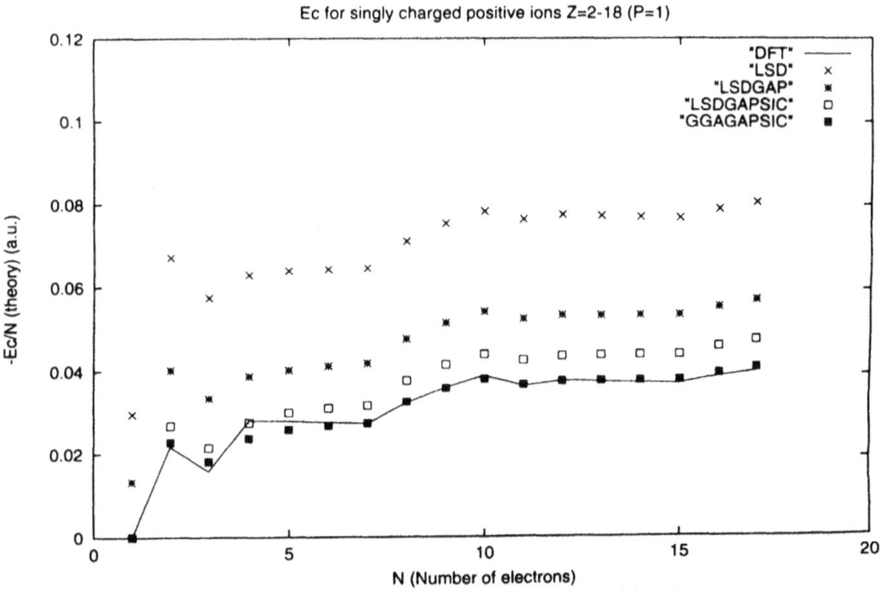


**Fig. 4** The negative of the correlation energy per electron vs. the number of electrons, N, for singly charged positive ions with nuclear charge Z ≤ 18. The methods employed are the same as in Fig. 3.

Table 1. Comparison of calculations of the correlation energy (-$E_c$ in hartrees) of neutral atoms for Z=1 to 18 with accurate DFT results

| Z | DFT | PWGGA | PBEGGA | GGAGAPSIC |
|---|---|---|---|---|
| 1 | 0.0000 | 0.0066 | 0.0060 | 0.0000 |
| 2 | 0.0420 | 0.0459 | 0.0420 | 0.0408 |
| 3 | 0.0455 | 0.0574 | 0.0514 | 0.0498 |
| 4 | 0.0950 | 0.0941 | 0.0854 | 0.0860 |
| 5 | 0.1255 | 0.1269 | 0.1158 | 0.1187 |
| 6 | 0.1561 | 0.1599 | 0.1473 | 0.1499 |
| 7 | 0.1858 | 0.1932 | 0.1797 | 0.1804 |
| 8 | 0.2553 | 0.2583 | 0.2383 | 0.2466 |
| 9 | 0.3249 | 0.3217 | 0.2956 | 0.3085 |
| 10 | 0.3929 | 0.3822 | 0.3510 | 0.3665 |
| 11 | 0.3988 | 0.4061 | 0.3711 | 0.3903 |
| 12 | 0.4424 | 0.4497 | 0.4104 | 0.4362 |
| 13 | 0.4740 | 0.4896 | 0.4466 | 0.4756 |
| 14 | 0.5094 | 0.5316 | 0.4856 | 0.5153 |
| 15 | 0.5447 | 0.5742 | 0.5258 | 0.5548 |
| 16 | 0.6095 | 0.6406 | 0.5865 | 0.6209 |
| 17 | 0.6728 | 0.7062 | 0.6467 | 0.6843 |
| 18 | 0.7314 | 0.7702 | 0.7060 | 0.7452 |

Table 2. Comparison of calculations of the correlation energy (-$E_c$ in hartrees) of singly charged positive ions for Z=2 to 18 with accurate DFT results

| Z | DFT | PWGGA | PBEGGA | GGAGAPSIC |
|---|---|---|---|---|
| 2 | 0.0000 | 0.0074 | 0.0062 | 0.0000 |
| 3 | 0.0435 | 0.0506 | 0.0448 | 0.0461 |
| 4 | 0.0477 | 0.0620 | 0.0541 | 0.0551 |
| 5 | 0.1119 | 0.1028 | 0.0919 | 0.0955 |
| 6 | 0.1394 | 0.1368 | 0.1236 | 0.1298 |
| 7 | 0.1658 | 0.1703 | 0.1556 | 0.1614 |
| 8 | 0.1909 | 0.2039 | 0.1885 | 0.1922 |
| 9 | 0.2590 | 0.2733 | 0.2506 | 0.2626 |
| 10 | 0.3257 | 0.3384 | 0.3091 | 0.3261 |
| 11 | 0.3906 | 0.3999 | 0.3654 | 0.3852 |
| 12 | 0.4028 | 0.4222 | 0.3838 | 0.4085 |
| 13 | 0.4548 | 0.4669 | 0.4238 | 0.4557 |
| 14 | 0.4897 | 0.5085 | 0.4619 | 0.4961 |
| 15 | 0.5257 | 0.5507 | 0.5014 | 0.5359 |
| 16 | 0.5596 | 0.5932 | 0.5417 | 0.5753 |
| 17 | 0.6266 | 0.6621 | 0.6045 | 0.6436 |
| 18 | 0.6884 | 0.7286 | 0.6655 | 0.7079 |

Tables 1 and 2 provide a comparison between our results and those of both the PWGGA and PBE approximations. The $E_c^{DFT}$ results are obtained from reference 14 adjusted for the fact that the spin-density functional theory definition of the correlation energy is slightly different from that employed in quantum chemistry (QC) i.e. we define $E_c^{DFT}$ by

$$E_{tot} = E^{RHF} + E_c^{QC} = E^{UOEPX} + E_c^{DFT} \quad . \tag{22}$$

Therefore

$$E_c^{DFT} = E_c^{QC} + (E^{RHF} - E^{UOEPX}) \tag{23}$$

i.e. the adjustment is the difference between the total energy in the restricted Hartree Fock (RHF) approximation and the corresponding exact unrestricted optimized effective potential[19] result for exchange-only (UOEPX) which corresponds to exact KS exchange-only theory. (This definition of $E_c^{DFT}$ is slightly different from the conventional $E_c^{KS}$ which would employ the expectation value of the Hamiltonian, H, with a determinant of the exact KS orbitals rather than $E^{UOEPX}$, which is the expectation value of H with a determinant of the exact exchange-only KS orbitals. $E_c^{DFT}$ given by eq. (23) is then easy to calculate because $E^{UOEPX}$ is available for any atom, whereas $<H>^{KS}$ is not because the exact KS orbitals are difficult to obtain. We expect that the different DFT definitions of $E_c$ lead to $|E_c^{KS}| > |E_c^{DFT}|$ by $\leq$ 1 mhartree.)

We find that with few exceptions the PWGGA overestimates the magnitude of $E_c$ and the PBE underestimates the same with the GGAGAPSIC results generally lying between the two and closely approximating the PBE results especially for first row atoms. For second row atoms, the GGAGAPSIC results always lie between the results of the other two approximations and generally are significantly closer to the DFT results.

One can attempt to improve the accuracy of the GGAGAPSIC approximation by treating $p$ as a parameter and studying the $p$ dependence of the calculated $E_c$. We find that in this approximation, $|E_c|$ is monotomically decreasing for increasing $p$ for each atom and positive ion which follows from the observation that $f$ increases with increasing $p$ so $\epsilon_c^{GGA}$ given by eq.(18) decreases in magnitude with increasing $p$. As a result, for a choice of $p = 5/4$, all calculated $E_c$ are reduced in magnitude by $\approx$ 2% so the GGAGAPSIC calculation for $Z = 15$ to 18 yields results within 1 mhartree of $E_c^{DFT}$ for atoms and within 6 mhartree for the positive ions. However, the decrease in $|E_c|$ for lighter atoms and ions generally yields results which are slightly further from the exact values. This negates the utility of simply increasing $p$ because the largest percent errors in our calculated $E_c$ for chemically active atoms and ions are for $N = 4,5$, and 6, all of which are already too small in magnitude.

From a more fundamental point of view, we recognize that in constructing the correlation energy functionals discussed in Section II, we have neglected the contributions from the unoccupied orbitals with single particle energy eigenvalues lying in the gap when we required that $G^0 \rightarrow I$ rather than to an energy gap corresponding to the energy difference to the lowest lying unoccupied orbital. However, we find that no matter how we modify our definition of $G^0$ to include this contribution, the resulting GGAGAPSIC correlation energy functional always saturates for fixed $N$ when $Z \rightarrow \infty$. On the other hand, it is well known[14] that $E_c$ diverges for the isoelectronic series with $N = 4,5$ or 6 electrons and converges for the other first row atoms. As can be seen from Moller- Plesset perturbation theory, the reason for this is clear: for these isoelectronic series there are unoccupied $2p$ levels in both spin channels that are nearly degenerate with the highest lying occupied states in the $n = 2$ shell. As a result, the energy denominator in second order perturbation theory corresponding to two electrons from the $n = 2$ shell excited into these $2p$ states is proportional to $Z$ in the large $Z$ limit, while the matrix element in the numerator (which is squared) is proportional to $Z$. Thus, the contribution to $E_c$ is $\sim Z^2/Z \sim Z$ instead of converging for other first row atom isoelectronic series for which the two particle excitation energy to the lowest lying states $\sim Z^2$. Although this behavior, as well as the saturation of the correlation energy under uniform scaling in the large $\lambda$ limit, are both in principle properties of the exact DFT $E_c$, it is difficult to imagine how both properties can be obtained from any GGA which saturates such as our GGAGAPSIC or the PBE functional without explicitly including the contributions from nearly degenerate states. Although this divergence is not immediately relevant to questions concerning the binding of atoms or molecules, we believe the omission of contributions to $E_c$ from low lying nearly degenerate electron states is the primary reason for the significant disparity between the calculated and exact $E_c$ for atoms and ions with $N = 4,5$ or 6 electrons.

We therefore plan to investigate whether accurate values of $E_c$ may be calculated by employing the GGAGAPSIC to approximate the contribution to $E_c$ from states in the continuum and to employ second order perturbation theory (or a more sophisticated wavefunction approach) to include the contributions to $E_c$ from the low lying KS states above the ground state which have been previously omitted for all atoms and ions. Preliminary calculations indicate that this latter contribution is generally small compared to the contributions to $E_c$ from the continuum and is the correct order of magnitude to eliminate most of the discrepancy of the GGAGAPSIC results from the exact values. We anticipate that the inclusion of such terms will be of particular importance in improving the accuracy of calculations of $E_c$ for molecules due to the frequent existence of low-lying unoccupied orbitals in such systems.

In concluding, we note that the inclusion of matrix elements involving unoccupied KS orbitals does not violate the Hohenberg-Kohn (HK) theorem that the correlation energy is a functional of the ground state electron density. This follows from the fact that the external potential and consequently all the orbitals, whether occupied or unoccupied, are in principle functionals of $n$. Thus any expression involving the single particle KS orbitals is an implicit functional of the density as in the case of exact KS exchange[22].

Finally, we observe that the suggestion that the correlation energy may be accurately calculated by making a density functional approximation for the large contribution from states that are relatively far from the highest occupied states and to include the contribution from nearby states by calculating matrix elements is similar to the work of Krieger *et al.*[25] on accurately calculating the ionization potential and electron affinity, $A$, in an exchange-only theory. They found that if the exchange energy for the interaction of the core electrons with each other, which is the greatest contribution to $E_x$, was approximated by a density functional and the interaction between the electrons in the two occupied subshells having the highest orbital energy with each other and with the core electrons was calculated by employing the exact exchange integrals, then $I$ and $A$ could be calculated with errors of only 0.1 mhartree compared with an exact KS exchange-only calculation. Moreover, they found that the KS exchange potential could be accurately calculated in the region occupied by the valence electrons with the correct asymptotic behavior as $r \to \infty$ leading to accurate highest occupied orbital energy eigenvalues as well.

## ACKNOWLEDGMENTS

This research was supported in part by the U.S. Army Research Office. One of us (J.B.K.) is indebted to Prof. Walter Kohn for pointing out the importance of explicitly accounting for energy gaps in DFT calculations of the correlation energy for localized electrons, and to Profs. John P. Perdew and Mel Levy for several stimulating discussions.

## REFERENCES

1. S.-K. Ma and K.A. Brueckner, Phys. Rev. **165**, 18 (1968).
2. M. Rasolt and D.J.W. Geldart, Phys. Rev. Lett. **35**, 1234 (1975); D.J.W. Geldart and M. Rasolt, Phys. Rev. **B13**, 1477 (1976); D.C. Langreth and J.P. Perdew, Phys. Rev. **B21**, 5469 (1980); Y. Wang and J.P. Perdew Phys. Rev. **B43**, 8911 (1991).
3. M. Gell-Mann and K.A. Brueckner, Phys. Rev. **106**, 364 (1957).
4. D.M. Ceperley and B.J. Alder, Phys. Rev. Lett. **45**, 566 (1980).
5. J.P. Perdew and Y. Wang, Phys. Rev. **B33**, 8800, (1986); J.P. Perdew, in Electronic Structure of Solids '91, edited by P. Ziesche and H. Eschrig (Akademie Verlag, Berlin, 1991), pg. 11.
6. M. Levy and J. P. Perdew, Phys. Rev. **B48**, 11,638 (1993).
7. J.P. Perdew, K. Burke, and M. Enzerhof, Phys. Rev. Lett. **77**, 3865 (1996).
8. J.P. Perdew and A. Zunger, Phys. Rev. **B23**, 5048 (1991).
9. J.P. Perdew (private communication).
10. B.Y. Tong, Phys. Rev. **A4**, 1375 (1971).
11. J. Rey and A. Savin, (Duke Symposium on Density Functional Theory, 1997), Int. J. Quantum Chem. **69**, 581 (1998).
12. U. von Barth and L. Hedin, J. Phys. **C5**, 1629 (1972).
13. A. Savin (unpublished).
14. S.J. Chakravorty, S.R. Gwaltney, E.R. Davidson, F.A. Parpia, and C.F. Fischer, Phys. Rev. **A47**, 3649 (1993); E.R. Davidson, S.A. Hagstrom, S.J. Chakravorty, V.M. Umar, and C.F. Fischer, **44**, 7071 (1991).
15. A.D. Becke, J. Chem. Phys. **88**, 1053 (1988), **104**, 1040 (1996).
16. M. Levy, J.P. Perdew, and V. Sahni, Phys. Rev. **A30**, 2745 (1984).
17. J.P. Perdew, M. Levy, R.G. Parr, and J.L. Balduz, Phys. Rev. Lett. **49**, 1691 (1982).
18. P. Hohenberg and W. Kohn, Phys. Rev. **136B**, 864 (1964).
19. Y.Li, J.B. Krieger, and G.J. Iafrate, Phys. Rev. **A47**, 165 (1993).
20. J.P. Perdew and A. Zunger, Phys. Rev. **B23**, 5048 (1981).
21. A.D. Becke (private communication).
22. J.B. Krieger, Y. Li, and G.J. Iafrate in *Density Functional Theory*, edited by E.K.U. Gross and R.M. Dreizler (Plenum, New York, 1995).
23. J.B. Krieger, J. Chen, and G.J. Iafrate, (Duke Symposium on Density Functional Theory, 1997) Int. J. Quantum Chem., **69**, 255 (1998).
24. J.P. Perdew and Y. Wang, Phys. Rev. **B45**, 13,244 (1992).
25. J.B. Krieger, Y. Li, Y. Liu, and G.J. Iafrate, Int. J. Quantum Chem. **61**, 273 (1997).

# TOWARDS NEW APPROXIMATIONS FOR THE EXCHANGE-CORRELATION FUNCTIONAL USING MANY-BODY PERTURBATION THEORY

Stefan Kurth

Department of Physics and Quantum Theory Group
Tulane University
New Orleans, LA 70118
USA

An exact expression for the exchange-correlation (xc) energy of density functional theory in terms of the one-particle Green's function and the xc part of the self-energy is derived. An exact diagrammatic representation of the xc self-energy is given. Employing this representation, an approximate scheme to calculate the xc energy and the xc potential is proposed. This scheme can be viewed as an extension of the exchange-only optimized-effective-potential method to treat correlation in a systematic manner.

Density functional theory (DFT) as based on the fundamental theorems of Hohenberg, Kohn and Sham [1,2] has become one of the main tools for calculating the electronic structure of atoms, molecules and solids [3,4,5]. This is mainly due to the availability of increasingly accurate approximations of the exchange-correlation (xc) energy functional $E_{xc}$. The most widely used approximations of $E_{xc}$, the local density approximation (LDA) [2] and the generalized gradient approximations (GGA's) [6,7,8,9] have in common that the xc energy is expressed as an explicit functional of the density.

Recently, however, the so-called optimized-effective-potential (OEP) method [10,11] has enjoyed increasing interest [12,13,14,15,16,17,18]. In this approach $E_{xc}$ is approximated as an explicit functional of the Kohn-Sham (KS) orbitals, which are themselves functionals of the density. In other words, the xc functional is given only as an *implicit* functional of the density.

Since the exchange energy functional is defined in terms of single-particle orbitals (atomic units are used throughout),

$$E_x = -\frac{1}{2}\sum_\sigma \int d^3r \int d^3r' \frac{|\gamma_\sigma(\mathbf{r},\mathbf{r}')|^2}{|\mathbf{r}-\mathbf{r}'|} \tag{1}$$

with

$$\gamma_\sigma(\mathbf{r},\mathbf{r}') = \sum_{i,occ} \varphi_{i\sigma}(\mathbf{r})\varphi^*_{i\sigma}(\mathbf{r}') \ , \tag{2}$$

*Electron Correlations and Materials Properties*, edited by Gonis *et al.*
Kluwer Academic/Plenum Publishers, 1999.

the OEP method allows for the exact treatment of exchange and only the correlation energy functional needs to be approximated. Since the OEP method is capable of dealing with explicitly orbital-dependent functionals for the correlation energy as well, this allows for more flexibility in the construction of such approximate functionals.

As in conventional DFT, the OEP method itself gives no hint to the construction of improved orbital-dependent approximations of the correlation energy functional. To find such approximations in a systematic manner, two approaches have been discussed in the literature: Görling and Levy [19, 20] have used a coupling-constant perturbation theory to derive expressions for the correlation energy, while Casida [21] used an (approximate) expression for the total energy as a functional of a single-particle Green's function to generalize the exchange-only (x-only) OEP method.

In this paper we develop a systematic approximation scheme to obtain the xc potential and the xc energy by using the powerful techniques of many-body perturbation theory. This scheme includes the x-only OEP method as well as Casida's approximation as special cases. It is based on the work of Sham and Schlüter [22, 23] who established an exact connection between DFT and many-body perturbation theory.

We consider systems of interacting electrons described by the Hamiltonian:

$$\hat{H} = \hat{T} + \hat{V}_0 + \hat{U} \tag{3}$$

where, in second quantized notation,

$$\hat{T} = \sum_\sigma \int d^3r \; \hat{\psi}^\dagger_\sigma(\mathbf{r}) \left(-\frac{\nabla^2}{2}\right) \hat{\psi}_\sigma(\mathbf{r}) \tag{4}$$

is the kinetic energy operator,

$$\hat{V}_0 = \sum_\sigma \int d^3r \; \hat{\psi}^\dagger_\sigma(\mathbf{r}) v_0(\mathbf{r}) \hat{\psi}_\sigma(\mathbf{r}) \tag{5}$$

is the operator of the external (nuclear) potential and

$$\hat{U} = \frac{1}{2}\sum_{\sigma,\sigma'} \int d^3r \int d^3r' \; \hat{\psi}^\dagger_\sigma(\mathbf{r}) \hat{\psi}^\dagger_{\sigma'}(\mathbf{r}') \frac{1}{|\mathbf{r}-\mathbf{r}'|} \hat{\psi}_{\sigma'}(\mathbf{r}') \hat{\psi}_\sigma(\mathbf{r}) \tag{6}$$

is the mutual Coulomb repulsion of the electrons. For the purpose of deriving some exact equations we assume, for the time being, that the exact xc potential $v_{xc\sigma}(\mathbf{r}) = \delta E_{xc}/\delta\rho_\sigma(\mathbf{r})$ of the system under consideration is known. Then we rewrite the Hamiltonian (3) by addition and subtraction as

$$\begin{aligned} \hat{H} &= \hat{T} + \hat{V}_0 + \hat{V}_H + \hat{V}_{xc} + (\hat{U} - \hat{V}_H - \hat{V}_{xc}) \\ &=: \hat{H}_{KS} + \hat{H}_1 \end{aligned} \tag{7}$$

where $\hat{V}_H$ and $\hat{V}_{xc}$ are the operators of the Hartree and xc potentials defined in analogy to Eq. (5) with $v_0(\mathbf{r})$ replaced by

$$v_H(\mathbf{r}) = \int d^3r' \; \frac{\rho(\mathbf{r}')}{|\mathbf{r}-\mathbf{r}'|} \tag{8}$$

and $v_{xc\sigma}(\mathbf{r})$, respectively.

$$\hat{H}_{KS} = \hat{T} + \hat{V}_0 + \hat{V}_H + \hat{V}_{xc} \tag{9}$$

is the Hamiltonian of non-interacting electrons moving in the spin-dependent KS potential

$$v_{s\sigma}(\mathbf{r}) = v_0(\mathbf{r}) + v_H(\mathbf{r}) + v_{xc\sigma}(\mathbf{r}) \;, \tag{10}$$

and will serve as the unperturbed reference Hamiltonian, while

$$\hat{H}_1 = \hat{U} - \hat{V}_H - \hat{V}_{xc} \tag{11}$$

will be treated as a perturbation on $\hat{H}_{KS}$. Application of the techniques of many-body perturbation theory then leads to the Dyson equation

$$G_\sigma(\mathbf{r}, \mathbf{r}', \omega) = G_\sigma^{KS}(\mathbf{r}, \mathbf{r}', \omega) + \int d^3y \int d^3y' \; G_\sigma^{KS}(\mathbf{r}, \mathbf{y}, \omega)\Sigma_\sigma(\mathbf{y}, \mathbf{y}', \omega)G_\sigma(\mathbf{y}', \mathbf{r}', \omega) \tag{12}$$

where $G_\sigma(\mathbf{r}, \mathbf{r}', \omega)$ is the Green's function of the interacting system and $\Sigma_\sigma(\mathbf{r}, \mathbf{r}', \omega)$ is the irreducible self-energy corresponding to the perturbation $\hat{H}_1$. The KS Green's function $G_\sigma^{KS}(\mathbf{r}, \mathbf{r}', \omega)$ can easily be expressed in terms of the KS orbitals and energy eigenvalues by

$$G_\sigma^{KS}(\mathbf{r}, \mathbf{r}', \omega) = \lim_{\eta \to 0^+} \sum_j \varphi_{j\sigma}(\mathbf{r})\varphi_{j\sigma}^*(\mathbf{r}') \left( \frac{\Theta(\varepsilon_{j\sigma} - \varepsilon_F)}{\omega - \varepsilon_{j\sigma} + i\eta} + \frac{\Theta(\varepsilon_F - \varepsilon_{j\sigma})}{\omega - \varepsilon_{j\sigma} - i\eta} \right) \quad . \tag{13}$$

Due to the KS theorem [2], the spin-density of the interacting system can be obtained both via the Green's function $G_\sigma$ and the KS Green's function $G_\sigma^{KS}$ through

$$\rho_\sigma(\mathbf{r}) = -i \int \frac{d\omega}{2\pi} \; G_\sigma(\mathbf{r}, \mathbf{r}, \omega) = -i \int \frac{d\omega}{2\pi} \; G_\sigma^{KS}(\mathbf{r}, \mathbf{r}, \omega) \tag{14}$$

where the path of the frequency integration follows the real $\omega$-axis and is closed through a semi-circle in the upper complex half-plane. Writing the self-energy as

$$\Sigma_\sigma(\mathbf{r}, \mathbf{r}', \omega) = \Sigma_{xc\sigma}(\mathbf{r}, \mathbf{r}', \omega) - \delta(\mathbf{r} - \mathbf{r}')v_{xc\sigma}(\mathbf{r}) \quad , \tag{15}$$

Eqs. (12) and (14) finally lead to

$$\begin{aligned} &\int d^3y \; v_{xc\sigma}(\mathbf{y}) \int \frac{d\omega}{2\pi} \; G_\sigma^{KS}(\mathbf{r}, \mathbf{y}, \omega)G_\sigma(\mathbf{y}, \mathbf{r}, \omega) \\ &= \int d^3y \int d^3y' \int \frac{d\omega}{2\pi} \; G_\sigma^{KS}(\mathbf{r}, \mathbf{y}, \omega)\Sigma_{xc\sigma}(\mathbf{y}, \mathbf{y}', \omega)G_\sigma(\mathbf{y}', \mathbf{r}, \omega) \quad . \end{aligned} \tag{16}$$

This is the integral equation for the xc potential in the form given by Sham and Schlüter [22, 23]. Sham [23] also derived an expression for the xc energy in terms of the Green's function and self-energy by employing the techniques of Luttinger and Ward [24]. The resulting expression for $E_{xc}$, however, is rather difficult to handle in practice since it requires the evaluation of a power series of an integral operator. In the following we will derive a much simpler expression for the xc energy from the following formula for the total ground state energy [25]:

$$\begin{aligned} E_{GS} &= -\frac{i}{2} \sum_\sigma \int \frac{d\omega}{2\pi} \int d^3r \lim_{\mathbf{r}' \to \mathbf{r}} \left( \omega - \frac{\nabla_\mathbf{r}^2}{2} + v_0(\mathbf{r}) \right) G_\sigma(\mathbf{r}, \mathbf{r}', \omega) \\ &= -\frac{i}{2} \sum_\sigma \int \frac{d\omega}{2\pi} \int d^3r \lim_{\mathbf{r}' \to \mathbf{r}} \left( \omega - h_{s\sigma}(\mathbf{r}) + 2h_{s\sigma}(\mathbf{r}) \right) G_\sigma(\mathbf{r}, \mathbf{r}', \omega) \\ &\quad -\frac{1}{2} \sum_\sigma \int d^3r \left( v_H(\mathbf{r}) + v_{xc\sigma}(\mathbf{r}) \right) \rho_\sigma(\mathbf{r}) \\ &= -i \sum_\sigma \int \frac{d\omega}{2\pi} \int d^3r \lim_{\mathbf{r}' \to \mathbf{r}} \left( -\frac{\nabla_\mathbf{r}^2}{2} \right) G_\sigma(\mathbf{r}, \mathbf{r}', \omega) + \int d^3r v_0(\mathbf{r})\rho(\mathbf{r}) + \frac{1}{2} \int d^3r \int d^3r' \frac{\rho(\mathbf{r})\rho(\mathbf{r}')}{|\mathbf{r} - \mathbf{r}'|} \\ &\quad -\frac{i}{2} \sum_\sigma \int \frac{d\omega}{2\pi} \int d^3r \int d^3r' \Sigma_{xc\sigma}(\mathbf{r}, \mathbf{r}', \omega)G_\sigma(\mathbf{r}', \mathbf{r}, \omega) \end{aligned} \tag{17}$$

where

$$h_{s\sigma}(\mathbf{r}) = -\frac{\nabla_{\mathbf{r}}^2}{2} + v_{s\sigma}(\mathbf{r}) \ , \tag{18}$$

is the KS single-particle Hamiltonian and the last line of Eq. (17) follows from Dyson's equation (12) and the equation of motion of the KS Green's function,

$$(\omega - h_{s\sigma}(\mathbf{r}))G_\sigma^{KS}(\mathbf{r},\mathbf{r}',\omega) = \delta(\mathbf{r}-\mathbf{r}') \ . \tag{19}$$

Comparison of Eq. (17) with the defining equation for $E_{xc}$,

$$E_{GS}[\rho_\uparrow,\rho_\downarrow] = T_s[\rho_\uparrow,\rho_\downarrow] + \int d^3r v_0(\mathbf{r})\rho(\mathbf{r}) + \frac{1}{2}\int d^3r \int d^3r' \frac{\rho(\mathbf{r})\rho(\mathbf{r}')}{|\mathbf{r}-\mathbf{r}'|} + E_{xc}[\rho_\uparrow,\rho_\downarrow] \ , \tag{20}$$

leads to the simple expression

$$E_{xc}[\rho_\uparrow,\rho_\downarrow] = -\frac{i}{2}\sum_\sigma \int \frac{d\omega}{2\pi}\int d^3r \int d^3r' \Sigma_{xc\sigma}(\mathbf{r},\mathbf{r}',\omega)G_\sigma(\mathbf{r}',\mathbf{r},\omega) + T[\rho_\uparrow,\rho_\downarrow] - T_s[\rho_\uparrow,\rho_\downarrow] \ , \tag{21}$$

where

$$T[\rho_\uparrow,\rho_\downarrow] = -i\sum_\sigma \int \frac{d\omega}{2\pi}\int d^3r \lim_{\mathbf{r}'\to\mathbf{r}}\left(-\frac{\nabla_{\mathbf{r}}^2}{2}\right) G_\sigma(\mathbf{r},\mathbf{r}',\omega) \tag{22}$$

is the kinetic energy of the interacting system while, correspondingly,

$$T_s[\rho_\uparrow,\rho_\downarrow] = -i\sum_\sigma \int \frac{d\omega}{2\pi}\int d^3r \lim_{\mathbf{r}'\to\mathbf{r}}\left(-\frac{\nabla_{\mathbf{r}}^2}{2}\right) G_\sigma^{KS}(\mathbf{r},\mathbf{r}',\omega) \tag{23}$$

is the kinetic energy of the non-interacting KS system.

The derivation of the above formulas did not involve any approximations. It required, however, the use of the exact xc potential $v_{xc\sigma}(\mathbf{r})$ which, of course, is not known in practice. A possible strategy to obtain approximations of $v_{xc\sigma}(\mathbf{r})$ is based on the observation that Eq. (16) reduces to the x-only OEP equation [10, 11] if one replaces $\Sigma_{xc\sigma}$ by the Fock self-energy

$$\Sigma_{xc\sigma}(\mathbf{r},\mathbf{r}',\omega) \approx \Sigma_\sigma^F(\mathbf{r},\mathbf{r}') = -\frac{\gamma_\sigma(\mathbf{r},\mathbf{r}')}{|\mathbf{r}-\mathbf{r}'|} \tag{24}$$

and at the same time replaces $G_\sigma$ by the KS Green's function $G_\sigma^{KS}$. Within our perturbative approach this simply amounts to the evaluation of Eq. (16) to first order in the perturbation $\hat{H}_1$. Likewise, the evaluation of the xc energy Eq. (21) to first order in $\hat{H}_1$ leads to the exchange energy functional of Eq. (1). Therefore the x-only OEP method is equivalent to the described perturbative approach in first order in the sense that one solves the first-order approximation to Eq. (16) for the unknown $v_{xc\sigma}$ self-consistently with the KS equation

$$h_{s\sigma}(\mathbf{r})\varphi_{j\sigma}(\mathbf{r}) = \varepsilon_{j\sigma}\varphi_{j\sigma}(\mathbf{r}) \tag{25}$$

which, via Eq. (13), determines the KS Green's function.

This observation suggests that correlation can be treated in a straightforward manner by including higher orders in the perturbation $\hat{H}_1$. For that purpose we propose a scheme which makes use of an approximation to the xc part $\Sigma_{xc\sigma}$ of the self-energy as a functional of the full Green's function. Then Dyson's equation (12) for $G_\sigma$ has to be solved together with the integral equation (16) for $v_{xc\sigma}$ in a self-consistent fashion.

To derive an exact diagrammatic representation of $\Sigma_{xc\sigma}$ in terms of the full Green's function $G_\sigma$, we recall that in the conventional formulation of many-body perturbation

a) b)

**Figure 1.** The two diagrams which constitute the self-energy $\tilde{\Sigma}_\sigma$. The thick arrowed lines represent the Green's function of the interacting system, the thin wavy line is the bare Coulomb interaction, the thick wavy line is the screened interaction, and $\Gamma_\sigma$ is the irreducible vertex function. The xc self-energy can be exactly represented solely by diagram b).

theory the interaction $\hat{U}$ is treated as a perturbation on the Hamiltonian $\hat{H}_0 = \hat{T} + \hat{V}_0$. Of course, Dyson's equation then reads

$$G_\sigma(\mathbf{r},\mathbf{r}',\omega) = G_\sigma^{(0)}(\mathbf{r},\mathbf{r}',\omega) + \int d^3y \int d^3y' \; G_\sigma^{(0)}(\mathbf{r},\mathbf{y},\omega)\tilde{\Sigma}_\sigma(\mathbf{y},\mathbf{y}',\omega)G_\sigma(\mathbf{y}',\mathbf{r}',\omega) \quad (26)$$

with the Green's function $G_\sigma^{(0)}$ corresponding to $\hat{H}_0$ and the irreducible self-energy $\tilde{\Sigma}_\sigma$ corresponding to the perturbation $\hat{U}$. Using the equation of motion for $G_\sigma^{(0)}$,

$$(\omega - h^{(0)}(\mathbf{r}))G_\sigma^{(0)}(\mathbf{r},\mathbf{r}',\omega) = \delta(\mathbf{r}-\mathbf{r}') \quad (27)$$

with

$$h^{(0)}(\mathbf{r}) = -\frac{\nabla_\mathbf{r}^2}{2} + v_0(\mathbf{r}) \; , \quad (28)$$

one obtains from Eq. (26)

$$(\omega - h^{(0)}(\mathbf{r}))G_\sigma(\mathbf{r},\mathbf{r}',\omega) = \delta(\mathbf{r}-\mathbf{r}') + \int d^3y' \; \tilde{\Sigma}_\sigma(\mathbf{r},\mathbf{y}',\omega)G_\sigma(\mathbf{y}',\mathbf{r}',\omega) \; . \quad (29)$$

Similarly, by employing the equation of motion of the KS Green's function, Eq. (19), one obtains from Eqs. (12) and (15)

$$\begin{aligned}(\omega - h_{s\sigma}(\mathbf{r}))G_\sigma(\mathbf{r},\mathbf{r}',\omega) &= (\omega - h_0(\mathbf{r}) - v_H(\mathbf{r}) - v_{xc\sigma}(\mathbf{r}))G_\sigma(\mathbf{r},\mathbf{r}',\omega) \\ &= \delta(\mathbf{r}-\mathbf{r}') - v_{xc\sigma}(\mathbf{r})G_\sigma(\mathbf{r},\mathbf{r}',\omega) + \int d^3y' \; \Sigma_{xc\sigma}(\mathbf{r},\mathbf{y}',\omega)G_\sigma(\mathbf{y}',\mathbf{r}',\omega) \; . \end{aligned} \quad (30)$$

By comparison with Eq. (29), one can identify

$$\tilde{\Sigma}_\sigma(\mathbf{r},\mathbf{r}',\omega) = v_H(\mathbf{r})\delta(\mathbf{r}-\mathbf{r}') + \Sigma_{xc\sigma}(\mathbf{r},\mathbf{r}',\omega) \; . \quad (31)$$

As it is known from the conventional formulation of many-body perturbation theory [25], $\tilde{\Sigma}_\sigma$ can be expressed as the sum of the two diagrams shown in Fig. 1 employing the irreducible vertex function $\Gamma_\sigma$ and the effective screened interaction. Alternatively, it can be represented by the sum of all dressed skeleton diagrams [25]. Since the diagram of Fig. 1 a) yields $v_H(\mathbf{r})\delta(\mathbf{r}-\mathbf{r}')$, the xc-part of the self-energy $\Sigma_{xc\sigma}$ can be represented solely by the diagram of Fig. 1 b) or, alternatively, as the sum of all dressed skeleton diagrams except the diagram of Fig. 1 a). This diagrammatic representation of the xc-part of the self-energy is well suited to approximate $\Sigma_{xc\sigma}$ as a functional of the full Green's function $G_\sigma$. One could, e.g., approximate $\Sigma_{xc\sigma}$ by a sum of all dressed skeleton diagrams up to second order, i.e.,

$$\Sigma_{xc\sigma}(\mathbf{r},\mathbf{r}',\omega) \approx \Sigma^{(2)}_{xc\sigma}[G_\sigma](\mathbf{r},\mathbf{r}',\omega) = i\int \frac{d\omega}{2\pi}\frac{G_\sigma(\mathbf{r},\mathbf{r}',\omega)}{|\mathbf{r}-\mathbf{r}'|}$$
$$+\sum_{\sigma'}\int\frac{d\omega_1}{2\pi}\int\frac{d\omega_2}{2\pi}\int d^3y\int d^3y' G_\sigma(\mathbf{r},\mathbf{r}',\omega-\omega_1+\omega_2)\frac{G_{\sigma'}(\mathbf{y},\mathbf{y}',\omega_1)}{|\mathbf{r}-\mathbf{y}|}\frac{G_{\sigma'}(\mathbf{y}',\mathbf{y},\omega_2)}{|\mathbf{r}'-\mathbf{y}'|}$$
$$-\int\frac{d\omega_1}{2\pi}\int\frac{d\omega_2}{2\pi}\int d^3y\int d^3y' G_\sigma(\mathbf{y},\mathbf{y}',\omega+\omega_1-\omega_2)\frac{G_\sigma(\mathbf{r}',\mathbf{y},\omega_1)}{|\mathbf{r}-\mathbf{y}|}\frac{G_\sigma(\mathbf{y}',\mathbf{r},\omega_2)}{|\mathbf{r}'-\mathbf{y}'|}\ . \quad (32)$$

This leads us to propose the following iterative scheme for the approximate treatment of electron correlation:

1. Specify an approximation to $\Sigma_{xc\sigma}$ as a functional of $G_\sigma$ :

$$\Sigma_{xc\sigma}(\mathbf{r},\mathbf{r}',\omega) \approx \Sigma^{approx}_{xc\sigma}[G_\sigma](\mathbf{r},\mathbf{r}',\omega)\ , \quad (33)$$

   and a starting approximation for the xc potential $v_{xc\sigma}$ (e.g., the x-only $v^{OEP}_{x-only}$ ).

2. Calculate the KS Green's function $G^{KS}_\sigma$ corresponding to $v_{xc\sigma}$.

3. Solve Dyson's equation (12) for the Green's function $G_\sigma$, with $\Sigma_\sigma$ approximated by

$$\Sigma_\sigma(\mathbf{r},\mathbf{r}',\omega) \approx \Sigma^{approx}_{xc\sigma}[G_\sigma](\mathbf{r},\mathbf{r}',\omega) - \delta(\mathbf{r}-\mathbf{r}')v_{xc\sigma}(\mathbf{r})\ . \quad (34)$$

4. With the solution $G_\sigma$ of Dyson's equation and the corresponding xc self-energy, solve Sham's integral equation (16) for $v_{xc\sigma}$.

5. Return to point 2 and iterate until self-consistency is achieved.

6. Calculate the ground-state energy via Eqs. (20) and (21) using the self-consistently obtained $G^{KS}_\sigma$, $G_\sigma$ and $\Sigma^{approx}_{xc\sigma}$.

The only approximation involved in the above scheme is the approximation to the functional $\Sigma_{xc\sigma}[G_\sigma]$. Even for the simplest approximation to the xc self-energy, namely using only the first order dressed skeleton diagram, the converged solution goes - in diagrammatic terms - far beyond the results of the x-only OEP method, since the latter is equivalent to a purely first-order evaluation of Sham's equation (16), while any scheme involving the solution of Dyson's equation (12) already includes an infinite number of diagrams. It should also be noted that the KS Green's function of the non-interacting reference system through the xc potential already incorporates some correlation effects. Therefore the hope is that the described scheme might have the merit of including the important part of electron correlation already through low order approximations to the xc self-energy.

Financial support through the Deutsche Forschungsgemeinschaft is gratefully acknowledged.

## REFERENCES

1. P. Hohenberg and W. Kohn, Phys. Rev. **136**, B864 (1964).
2. W. Kohn and L. Sham, Phys. Rev. **140**, A1133 (1965).
3. R.G. Parr and W. Yang, *Density-Functional Theory of Atoms and Molecules* (Oxford University Press, New York, 1989).

4. R.M. Dreizler and E.K.U. Gross, *Density Functional Theory* (Springer, Berlin, 1990).
5. *Density Functional Theory*, Vol. B337 of *NATO ASI Series*, edited by E.K.U. Gross and R.M. Dreizler (Plenum Press, New York, 1995).
6. A. Becke, Phys. Rev. A **38**, 3098 (1988).
7. C. Lee, W. Yang, and R. Parr, Phys. Rev. B **37**, 785 (1988).
8. J. Perdew, in *Electronic Structure of Solids '91*, edited by P. Ziesche and H. Eschrig (Akademie Verlag, Berlin, 1991), p. 11.
9. J. Perdew, K. Burke, and M. Ernzerhof, Phys. Rev. Lett. **77**, 3865 (1996), ibid, **78**, 1396 (1997)(E).
10. R. Sharp and G. Horton, Phys. Rev. **90**, 317 (1953).
11. J. Talman and W. Shadwick, Phys. Rev. A **14**, 36 (1976).
12. E. Engel, J. Chevary, L. Macdonald, and S. Vosko, Z. Phys. D **23**, 7 (1992).
13. E. Engel and S. Vosko, Phys. Rev. A **47**, 2800 (1993).
14. M. Norman and D. Koelling, Phys. Rev. B **30**, 5530 (1984).
15. J. Krieger, Y. Li, and G. Iafrate, Phys. Rev. A **45**, 101 (1992).
16. J. Krieger, Y. Li, and G. Iafrate, Phys. Rev. A **46**, 5453 (1992).
17. T. Grabo and E.K.U. Gross, Chem. Phys. Lett. **240**, 141 (1995), ibid, **241**, 635 (1995)(E).
18. T. Grabo, T. Kreibich, S. Kurth, and E.K.U. Gross, in *The Strong Coulomb Correlations and Electronic Structure Calculations: Beyond Local Density Approximations*, edited by V. Anisimov (Gordon and Breach, Amsterdam), to appear.
19. A. Görling and M. Levy, Phys. Rev. B **47**, 13105 (1993).
20. A. Görling and M. Levy, Phys. Rev. A **50**, 196 (1994).
21. M. Casida, Phys. Rev. A **51**, 2005 (1995).
22. L. Sham and M. Schlüter, Phys. Rev. Lett. **51**, 1888 (1983).
23. L. Sham, Phys. Rev. B **32**, 3876 (1985).
24. J. Luttinger and J. Ward, Phys. Rev. **118**, 1417 (1960).
25. E.K.U. Gross, E. Runge, and O. Heinonen, *Many-Particle Theory* (Adam Hilger, Bristol, 1991).

# ELECTRONIC STRUCTURE AND MAGNETISM OF ITINERANT *5f* FERROMAGNETS URhSi AND URhGe

Martin Diviš,[1,2] Peter Mohn,[1] Karlheinz Schwarz,[1] Peter Blaha,[1] and Pavel Novák[3]

[1]Institute of Technical Elektrochemistry, TU Wien, Getreidemarkt 9/158 Austria
[2]Department of Metal Physics, Charles University, Ke Karlovu 5, 121 16 Prague 2, Czech Republic
[3]Institute of Physics, Czech Academy of Sciences, Cukrovarnická 10 162 00 Praha 6, Czech Republic

## INTRODUCTION

The electronic structure of actinide intermetallic compounds is mainly determined by partially occupied *f*-electron states. The magnetic and other electronic properties have, however, only little in common with the behavior of their lanthanide counterparts, and in many cases they resemble *3d*, *4d*, or *5d* transition metal intermetallics. This is particularly visible when comparing physical properties of light lanthanide with light actinide intermetallics. In the former case the *4f*-states are usually atomic-like, in contrast to the delocalized Bloch-like *5f*-states in the latter case. The nature of the *5f*-states can be also remarkably affected by external variables like pressure and magnetic field as well as by alloying with other elements. Thus a wide range of properties of light actinide compounds is connected with the degree of localization of the *5f*-electrons including the strongly correlation regime, which gives rise to exotic phenomena like heavy fermion behavior and superconductivity. Another important ingredients of *5f*-magnetism are the relativistic effects especially the spin-orbit (SO) interaction, which provides significant orbital polarization of the *5f* band and leads to a huge magnetocrystalline anisotropy frequently observed e.g. in uranium intermetallics.[1]

So far a general theory does not exist for the description and understanding of all physical properties of uranium intermetallics.[2] A partial description at the "first principle" level can be obtained from density functional theory (DFT), which treats exchange and correlation effects for the ground state in a effective one-electron approach.[3] The actual calculations are usually performed using the local spin density approximation (LSDA).[4]

There are many uranium intermetallic compounds which have a reduced magnetic

*Electron Correlations and Materials Properties*, edited by Gonis *et al.*
Kluwer Academic/Plenum Publishers, 1999.

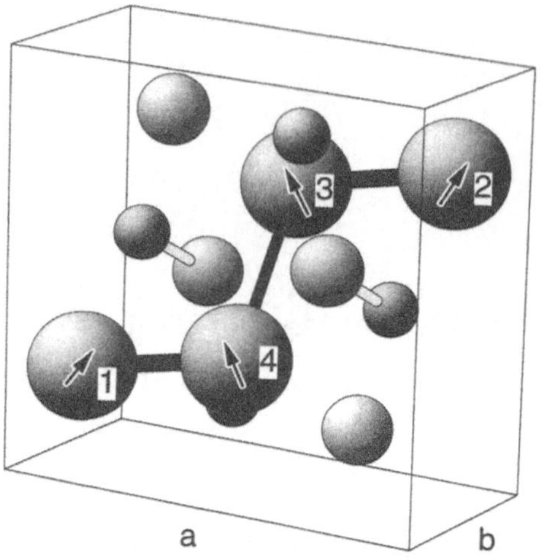


**Figure 1.** The crystal structure of URhGe. The largest and smallest circle denotes U and Ge atom, respectively. The noncollinear ground state magnetic structure[6] is schematically shown.

moments and show spin-fluctuation effects.[2] To obtain a quantitative microscopic picture of the influence of the spin fluctuations on the magnetic ordering in the *5f*-systems we performed electronic structure calculations on a selected group of URhX (X = Si, Ge) ternary compounds, which exhibit ferromagnetic ordering with a strong magnetocrystalline anisotropy of the magnetic moment.[2,5,6] In the present work we are going beyond LSDA and also used the generalized gradient approximation (GGA), a recent improvement in DFT.

It was confirmed by means of neutron powder diffraction[5,6] that the URhX compounds crystallize in the orthorhombic TiNiSi-type of structure (see Fig. 1) with space group Pnma. URhSi[5,6] and URhGe[6] undergo a magnetic phase transition into a ferromagnetic structure at 9.5 and 9.0 K, respectively. In the case of URhSi the magnetic structure is collinear with the magnetic moments parallel to the c-axis[5,6] while for URhGe the magnetic structure is non-collinear, where the magnetic moments of uranium lie inside the ac-plane but are tilted by an angle of about 31°.[6] The values of the uranium magnetic moments derived from the powder neutron diffraction data are rather small ($M_U \leq 0.5\ \mu_B$). All these experimental findings suggest that the URhX compounds can be classified as itinerant *5f* ferromagnets.[2] We decided to study theoretically especially these ferromagnets since they differ in their magnetic ground state from the antiferromagnetic order often observed in related *5f*-systems.[1,2]

## THEORETICAL METHOD AND COMPUTATIONAL DETAILS

The compounds under investigation crystallize in the orthorhombic TiNiSi-type structure (space group Pnma, No. 62) shown in Fig.1. The U, Rh and X atoms all occupy the 4c positions with free structural parameters x and z. In our calculations the literature values for the lattice parameters and the corresponding six structural parameters of URhSi[5] and URhGe[6] were used. In the case of URhSi all structural data of Prokeš et al[5] agree well with those reported in.[6] For URhGe structural data are only available in paper of Tran at al.[6]

The main part of calculations were performed with the highly accurate full-potential linearized augmented plane wave method (LAPW) using the optimized version WIEN97 of the originally published WIEN code[7] were used. The atomic (AS) sphere radii of 0.148, 0.116 and 0.116 pm were taken for U, Rh, and X, respectively. We used about 1200 plane wave basis functions ($R_{AS}\ K_{max} = 8$) , and a maximum $l = 12$ in the expansion of the radial wave functions inside the AS spheres. Both, the potential and the charge density, were expanded inside the spheres into crystal harmonics up to $L = 6$, but in the interstitial region in the Fourier series with about 4 000 **K** stars. For the Brillouin (BZ) zone integration a modified tetrahedron method[7] with 36 special **k** points in the irreducible wedge of BZ (IBZ) was used. We have checked that with these parameters the calculations are converged. Tests with larger basis set ($R_{MT}\ K_{max} = 9.5$, up to 2000 plane waves), changed sphere radii, a longer Fourier series (up to 10 000 **K** stars) and a finer BZ **k**-point sampling (up to 112 **k**-points in the IBZ) yielded only very small changes in our results (e.g. the spin and orbital magnetic moments changed by less than one percent).

The relativistic effects were treated in the standard scalar relativistic approximation[8] to which spin-orbit coupling was added that was treated self-consistently via the second order variational scheme.[9] For the LSDA studies Perdew and Wang's parametrization[10] of Ceperley and Alder's results were used for the DFT functional. For the GGA the functional "PBE"[11] was employed. Local orbitals were used to treat the U *6s*, and *6p*, Rh *4p* and Ge *3d* states with the valence states in a single energy window. In all present calculations the uranium *5f*-states were treated as band states and thus are part of the envelope functions used in variational procedure.

Different orientations of the magnetic moments with respect to the orthorhombic crystal axes yield different eigenvalues in the presence of spin-orbit coupling. Therefore magnetocrystalline anisotropy energy (MAE) can be calculated as the difference of the corresponding total energies.

In order to explore the possibility of non-collinear magnetic order or "spin-spiral states" we employed the augmented spherical wave (ASW) method[12] in the scalar relativistic version. This scheme has been implemented into the ASW code by formulating the density functional theory of non-collinear magnetism.[13] A spin-spiral is characterized by a combination of a lattice translation and a spin rotation which leads to a generalized Bloch theorem. Since both the spin rotation and the original lattice translation commute with the scalar relativistic crystal hamiltonian the relevant **k**-vectors for the Brillouin zone integration still lie in the first Brillouin zone. However, the respective eigenfunctions are bi-spinor wave functions which whose eigenvalues simultaneously carry spin-up and spin-down character.

Our ASW calculations in the atomic sphere approximation (ASA) were performed on a uniform **k**-mesh with at least 125 **k**-points in the IBZ, whose symmetry also depends on the direction of the spin spiral. The exchange and correlation effects were treated using LSDA according to with von Barth and Hedin form.[14] The calculation of the total energy as a function of the overlapping ASA sphere radii allows us to find the optimum radii of the spheres, which minimizes the total energy and the charge transfer between the spheres. We have found the optimum radius $R_U = 0.182$ and $R_U = 0.183$ pm for URhSi and URhGe, respectively, where the ratio $R_U/R_{Rh} = R_U/R_X = 1.2$ was kept constant for all ASW calculations. Tests with a larger number of **k**-points in the IBZ (maximum 1750) were performed to check the stability our numerical results. To ensure good convergence in the total energy, the self-consistency-cycle was carried out until the inter band charge transfer was less then $10^{-6}$ electrons.

## RESULTS AND DISCUSSION

In the first part of this work we performed both LAPW and ASW DFT calculations assuming a non-spinpolarized ground state and the scalar relativistic form of the Kohn-Sham Hamiltonian. A small charge transfer was found between the different constituents of the URhX compounds and thus the bonding is based on the hybridization of different states.

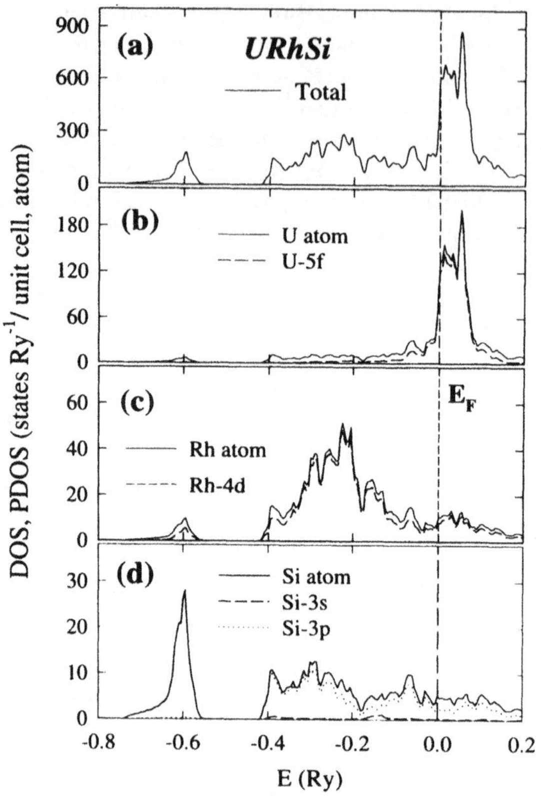


**Figure 2.** Calculated DOS of URhSi **(a)** and the site projected *l*-decomposed DOS curves for U **(b)**, Rh **(c)** and Si **(d)** atoms.

In Fig. 2 and 3 the total density of states (DOS) is shown for URhSi and URhGe. In both cases there is a separated DOS below –0.5 Ry originating from *3s*-states (*4s*-states) of Si (Ge), which are subjected to a stronger attractive nuclear potential compared to other valence states. The site projected *l*-decomposed DOS curves (see Fig. 2, 3) show that the uranium electronic states are mainly of *5f* character, the rhodium states are dominated by *4d* character and the X states stem from *3p* (*4p*) states. From the similarity and the relative intensities of the projected DOS a strong mixing between rhodium and silicon (germanium) states can be seen in the energy range –0.4 Ry < E < –0.2 Ry. The hybridization of uranium *5f*-states, rhodium *4d*-states and silicon (germanium) *3s*-states (*4s*-states) is significant in the region –0.1 Ry < E < +0.1 Ry around the Fermi level ($E_F$). The hybridization seems to be more pronounced for the case of URhSi as the occupied total DOS and unoccupied *5f*-DOS are broader (compare Fig. 2a - 2d and Fig. 3a - 3d). In both compounds the $E_F$ is situated in a the region of high DOS which originate from uranium *5f*-states. The U *6p*-states (not shown in the Figs. 2, 3) are situated around the –1.2 Ry (*6p-3/2*) and –1.7 Ry (*6p-1/2*) below $E_F$. We note that especially the U *6p-3/2* electronic states hybridize with the actual valence states and therefore contribute to the bonding energy. The DOS results of LAPW and ASW calculations are very similar and also agree with results obtained previously[5] using the optimized linear combination of atomic orbital (OLCAO) method.

Itinerant magnetism arises from the competition of two contributions to the energy. If the spin-up and spin-down DOS are shifted in energy against each other, the kinetic energy

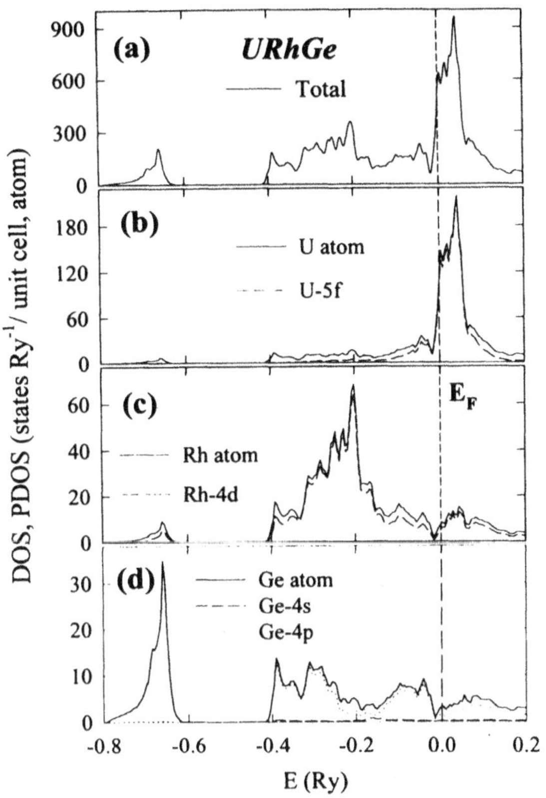


**Figure 3.** Calculated DOS of URhGe **(a)** and the site projected *l*-decomposed DOS curves for U **(b)**, Rh **(c)** and Ge **(d)** atoms.

is increased but the exchange energy is decreased. If there is a net energy gain the system is unstable with respect to the formation of a magnetic moment. This situation may be expressed more generally in the case of a monoatomic solid, like for e.g. Fe, Co, Ni, by writing the enhanced magnetic susceptibility $\chi$ as

$$\chi = \frac{\chi_0}{1 - IN(E_F)} \tag{1}$$

where $I$ is the exchange integral and $N(E_F)$ is the DOS at the Fermi level. In our case we found that the Stoner criterion $I\ N(E_F) > 1$ is fulfilled at the uranium site, since the Stoner product is 1.67 and 1.31 for URhSi and URhGe, respectively. Therefore uranium states are spin-polarized and induce a magnetic moment on the other sites via hybridization.

In the second part spin-polarized LSDA (LAPW and ASW) calculations were performed for both URhX compounds. We carried out two sets of such calculations assuming a ferromagnetic spin arrangement

- without SO (FSP) and
- with SO coupling (FSP-SO) assuming a ferromagnetic arrangements of the spins along the c-axis.

Using the ASW method and optimum spheres we obtained in the FSP calculations the spin magnetic moments 1.53 $\mu_B$ (URhSi) and 1.22 $\mu_B$ (URhGe) which originate mainly from the spin polarization of the uranium *5f*-states. We found only a small moment (< 0.1 $\mu_B$ ) belonging to the rhodium, silicon and germanium electronic states. These moments are induced by hybridization of the U *5f*-states with Rh *4d*-states and X *p*-states and therefore have a direction opposite to the uranium spin moment (ferrimagnetic). The stabilization energy $\Delta E = E_{FSP} - E_{NSP}$ was found – 5.5 mRy/f.u. and – 3.4 mRy/f.u. for URhSi and URhGe, respectively.

The calculations performed with WIEN97 confirmed these results which are given in Table 1. The reason, that the LAPW moments are systematically smaller that the ASW moments is due to the fact that the atomic sphere radii are smaller in the LAPW. In order to allow a detailed comparison between the ASW and LAPW magnetic moments the decomposition must be done with same sphere radii. For that purpose we used the spin-up

**Table 1.** Spin magnetic moments obtained from FSP calculations using ASW and LAPW methods. The symbol $M_{IS}^S$ means spin magnetic moment in the interstitial region.

| Compound | Method | $M_U^S$ ($\mu_B$) | $M_{Rh}^S$ ($\mu_B$) | $M_X^S$ ($\mu_B$) | $M_{IS}^S$ ($\mu_B$) |
|---|---|---|---|---|---|
| *URhSi* | ASW | 1.53 | -0.04 | <0.001 | - |
| *URhSi* | LAPW | 1.47 | -0.06 | -0.02 | 0.58 |
| *URhGe* | ASW | 1.22 | -0.09 | -0.04 | - |
| *URhGe* | LAPW | 0.95 | -0.04 | -0.02 | 0.39 |

and spin-down density from ASW and integrated them up to the same radii yielding magnetic moments which are almost exactly as in LAPW.

In the case of FSP-SO calculations we used the LAPW method and the results are summarized in Table 2. In the formation of the *5f* bands its width results from an interplay between the delocalization (in the periodic crystal potential) and the SO strength. Table 2 shows a large stabilization due to SO close to –100 mRy in the case of URhGe, so that the inclusion of SO effects is essential. In the comprehensive electronic structure study[15]

**Table 2.** Spin, orbital magnetic moments inside uranium MT sphere, stabilization energy and MAE obtained from FSP-SO calculations using LAPW method.

| | $M_U^S$ ($\mu_B$) | $M_U^L$ ($\mu_B$) | $\left|M_U^S - M_U^L\right|$ ($\mu_B$) | $\Delta E_{FSP-SO}$ (mRy) | $\Delta_{MAE}$ (mRy) |
|---|---|---|---|---|---|
| *URhSi* | 1.19 | -1.51 | 0.32 | -48 | -3.6 |
| *URhGe* | 1.00 | -1.31 | 0.31 | -94 | -2.0 |

of $UT_2Ge_2$ (T = *3d* transition metal) a similar SO stabilization energy has been found. In comparison to the FSP calculation the spin moment decreased by 20 % for URhSi but increased by 5 % for URhGe. We found on U a large orbital moment of –1.51 $\mu_B$ and –1.31 $\mu_B$ for URhSi and URhGe, respectively. Its orientation is opposite to the spin moment according to third Hund's rule for a less than half-filled *f*-subshell. The orbital contribution to the Rh magnetic moment is very small ($\left|M_{Rh}^L\right| \le 0.01\mu_B$) and is oriented parallel to the spin moment. For Si and Ge the orbital magnetic moments are negligibly small. The magnitude of the resulting total uranium magnetic moments $\approx 0.3$ $\mu_B$ is in fairly good agreement with measured data, namely of 0.11 $\mu_B$ and 0.5 $\mu_B$ for URhSi[5,6] and 0.5 $\mu_B$ for URhGe.[6]

We have estimated the magnetic anisotropy energy (MAE) by performing self-consistent FSP-SO calculations with the magnetic moment aligned along in the c-axis (001) or the a-axis (100). The difference between these two total energies defines the MAE given in Tab.2. In URhSi the c-axis should be the "easy magnetization axis" as was observed experimentally.[5,6] On the other hand for URhGe such calculations have only limited meaning since the actual magnetic ground state is non-collinear.[6] In a previous work on URhSi[5] the MAE has also been calculated but with "force theorem", i.e. taking the difference of the sum of eigenvalues of the occupied valence states corresponding to the two moment orientations[16] leading to a value of $\Delta = -4.3$ mRy. Our present LAPW result led to a total energy difference of $\Delta = -3.6$ mRy, which is in close agreement considering the difference in these two schemes (total energy difference vs. force theorem, and LAPW vs. OLCAO). The similar result found from these two independent calculations strongly desires a quantitative experimental determination of the MAE for the URhSi compound.

It is well known that GGA enhances magnetism in *3d* transition metals and leads to larger magnetic moments.[17] Therefore we investigated whether or not this trend holds for the present systems as well. We used in our FSP-SO calculations for URhSi the recent form of GGA[10] and really found an increase of the spin $M_U^S = 1.44\mu_B$ and orbital moment $M_U^L = -1.70\mu_B$, but due to the opposite orientation the total moment slightly decreased to 0.26 $\mu_B$.

We have also tested the influence of the orbital polarization scheme,[18] which is widely use in case of uranium intermetallics[3,13,18] and in which a term $-1/2\,E^3 L_z^2$ is added to the DFT functional to improve the description of the intra-atomic interaction energies. Here, $E^3$ is the Racah parameter and $L_z$ is the z-component of the total angular momentum. This term lowers the energy of Kohn-Sham states of a given spin and a high quantum number $m_l$ and causes larger spin and magnetic moments than a regular spin-polarized calculation with spin-orbit interaction. Such a development has recently been done within the OLCAO method.[19] For U a Racah parameter $E^3$ was calculated self-consistently from Slater type integrals[20] and we have found the value $E^3 \cong 4.2$ mRy for both compounds. Our preliminary results found a U total magnetic moment of $-1.68$ $\mu_B$ and $-1.62$ $\mu_B$ for URhSi and URhGe, respectively. Therefore in the case of URhX the ab-initio orbital polarization scheme largely overestimates the magnetic moment of U in comparison to available experimental data.[5,6]

All calculations described above were performed for the experimental lattice constants and structural parameters x and z. In figures 4a and 5a we show the calculated total energy (based on FSP-SO in the LSDA approximation) as a function of volume for URhSi and URhGe, respectively. The data were fitted to the Murnaghan equation of state[21] from which the equilibrium volume $V_0$, the bulk modulus $B_0$ and the pressure derivative of bulk modulus $B_0'$ was obtained by analytic differentiation (Table 3). It should be mentioned that we did not minimize the structural parameters x, z for each volume, but we found that the atomic forces calculated at the experimental positions were quite small and thus must be close to the theoretical equilibrium structure parameters. The computed equilibrium volumes are 4.4 % and 1.9 % smaller than the experimental for URhSi and URhGe, respectively. The deviations may come from the limitations of LSDA (at T = 0) and the thermal expansion. It would be quite interesting to compare the predicted values of bulk moduli with experimental measurements which are presently not available.

In figures 4b and 5b we show the calculated values of the U spin, orbital and total magnetic moments as a function of pressure, which is defined as $P = -\partial E/\partial V$, the required pressure to keep the unit cell at the volume $V$. We found quite a pronounced dependence of the magnetic moments on pressure. The orbital moment is more sensitive to pressure than

**Table 3.** Equilibrium constants for URhX. The equilibrium volume $V_0$ is related to experimental volume $V_{exp} = 0.2159\ nm^3$ and $V_{exp} = 0.2203\ nm^3$ for URhSi and URhGe, respectively.

| | $V_0/V_{exp}$ | $B_0$ (GPa) | $B'_o$ |
|---|---|---|---|
| *URhSi* | 0.956 | 173 | 3.9 |
| *URhGe* | 0.981 | 189 | 0.3 |

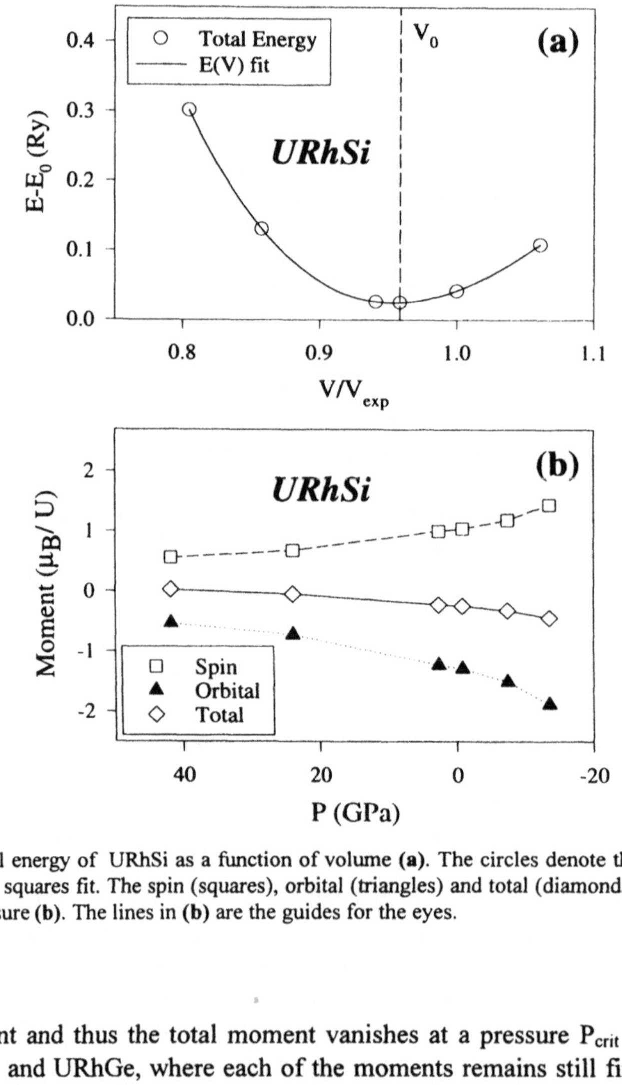


**Figure 4.** The total energy of URhSi as a function of volume **(a)**. The circles denote the calculated data and the line is the least squares fit. The spin (squares), orbital (triangles) and total (diamonds) uranium moment as a function of pressure **(b)**. The lines in **(b)** are the guides for the eyes.

the spin moment and thus the total moment vanishes at a pressure $P_{crit}$ = 36 GPa and 33 GPa for URhSi and URhGe, where each of the moments remains still finite ($M_S = -M_L$ = 0.55 $\mu_B$ for URhSi and $M_S = -M_L$ = 0.72 $\mu_B$ for URhGe). These pressures $P_{crit}$ are in principle experimentally available and therefore could be compared with our theoretical prediction.

In order to investigate the origin of the non-collinear magnetic arrangements in URhGe we performed ASW calculations assuming the magnetic structure found by Tran et al.[6]

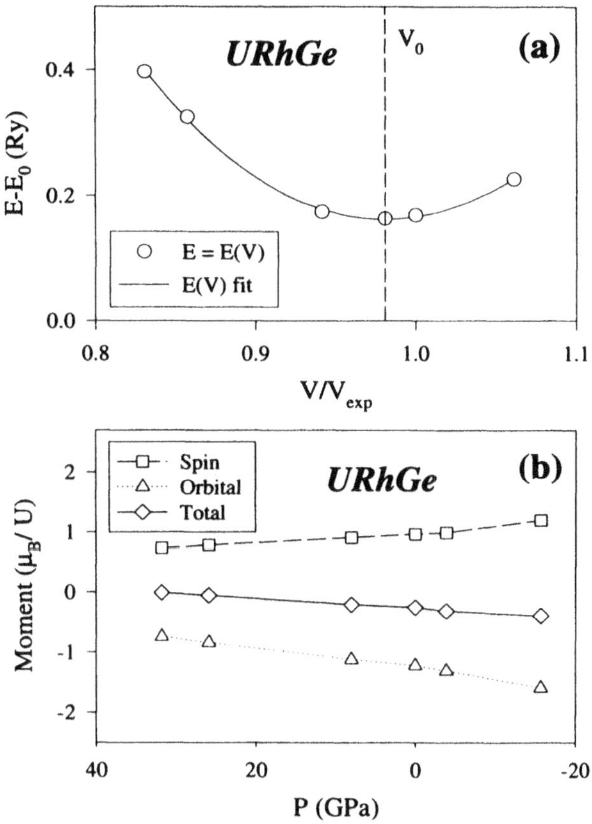


**Figure 5.** The total energy of URhGe as a function of volume **(a)**. The circles denote the calculated data and the line is the least squares fit. The spin (squares), orbital (triangles) and total (diamonds) uranium moment as a function of pressure **(b)**. The lines in **(b)** are the guides for the eyes.

In this compound the uranium magnetic moments lie in the ac-plane and are ferromagnetically ordered, but tilted with respect to the c-axis by an angle of about 31° (Fig.1). This tilting is such that there is a pairwise compensation of the U moment with respect to the a-axis ($M_a(1) = - M_a(3)$, $M_a(2) = -M_a(4)$) and this magnetic structure can be unambiguously described by the tilting angle $\theta$ and two polar angles $\varphi_1 = \varphi_2 = 0°$ and $\varphi_3 = \varphi_4 = 180°$. The total energy was calculated by ASW as the function of $\theta$ (Fig. 6a, 6b). It should be noted that the energy difference between the ferromagnet ($\theta = 0°$) and the antiferromagnet ($\theta = 90°$) is 5.6 mRy and 13.1 mRy for URhSi and URhGe, respectively. Therefore the effective magnetic field, which would switch the uranium moment of 0.5 $\mu_B$ from the ferromagnet to the antiferromagnet can be roughly estimated to be 660 T and 1540 T for the URhSi and URhGe, respectively. In ASW the collinear magnetic structure ($\theta = 0$) is the ground state for both systems, which agrees with experiment for URhSi[5,6] whereas for URhGe the tilted state ($\theta = 31°$) is found experimentally.[6] According to our ASW calculation this state lies only about 2 mRy above the collinear ferromagnetic state. It should be pointed out that this result was obtained for a spin only calculation where no SO-coupling was taken into account. It has been shown[13] that SO-coupling in these heavy systems very often breaks the symmetry towards a spin canting ground state. The missing

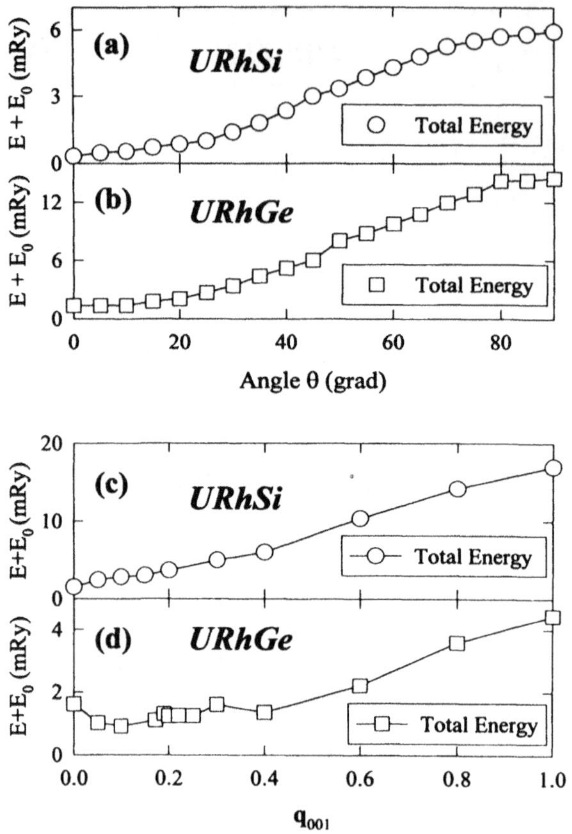


**Figure 6.** The total energy as a function of tilting angle θ for URhSi **(a)** and URhGe **(b)** and as a fuction of spin-spiral vector **q** in the (001) direction for URhSi **(c)** and URhGe **(d)**. The energy $E_0 = -264268.012$ and $E_0 = -278723.845$ Ry for URhSi **(a)** and URhGe, respectively. The lines are the guides for the eyes.

stabilization energy for the found spin canting state[6] is thus likely to stem from the spin-orbit interaction which was not taken into account in our present ASW calculations.

The knowledge of the spin-projected density of states at the Fermi energy allows us to calculate the Stoner temperature $T_S$ and the spin fluctuation temperature $T_{SF}$ according to Mohn and Wohlfarth's theory.[22] We have found the values for the $T_S = 2012$ K, $T_{SF} = 776$ K and $T_S = 2989$ K, $T_{SF} = 585$ K for URhSi and URhGe, respectively. The large value of the $T_S$ indicates that the single-particle excitations play a minor role for the ground state magnetism of the URhX compounds. Using the well-known equation for the Curie temperature

$$\frac{T_C^2}{T_S^2} + \frac{T_C}{T_{SF}} - 1 = 0 \tag{2}$$

we would find rather high values for the calculated Curie temperature $T_C \cong T_{SF}$ in contrast to the experimental values of 9.5 and 9.0 K for URhSi[5,6] and URhGe,[6] respectively. If, however we use the total moment instead of the spin magnetic moment only (i.e. 0.32 and 0.31 $\mu_B$) one would find much the smaller values of $T_{SF} = 33$ K and $T_{SF} = 38$ K for URhSi and URhGe, respectively. Therefore by the simple assumption, that not the spin moment but the total magnetic moment fluctuates, one can obtain a consistent qualitative picture, namely that both Curie temperatures are similar (despite their quite different spin moments) and are small as in experiment.

To investigate the nature of the itinerant magnetism of URhX we also performed spin-spiral calculations using the ASW method but again neglecting SO effects. Our spin-spiral state is represented by a propagation vector **q** in (001) direction. The total energy as a function of **q** was obtained. This function steadily increases for URhSi and provides another proof that the collinear ferromagnetic arrangement is the true ground state (see Fig. 6c). For URhGe the situation is much more complicated (see Fig. 6d). There are a few (meta-) stable states within an energy range of 1mRy which are spin-spiral states. At one minimum corresponding to q = (0, 0, 0.17) the relative angles of the four uranium spins are $\varphi_1$ = 12.63°, $\varphi_2$ = 49.36°, $\varphi_3$ = 43.63° and $\varphi_4$ = 18.36°. The difference in the z-coordinate for the two pairs of U atoms (1,3) and (2,3) are almost half of the c-lattice constant and $\Delta\varphi_I = \varphi_3 - \varphi_1 = 31°$, $\Delta\varphi_{II} = \varphi_2 - \varphi_4 = 31°$. Therefore without spin-orbit coupling there is a strong tendency to create a spin-spiral state with a tilt angle of 31°, whereas with SO coupling a canted spin structure is stabilized. This is in line with the general statement made by Sandratskii,[13] that SO coupling does not allow a separate transformation of the spin and space variables and thus it is destructive for the spin-spiral states. Since SO coupling is strong in uranium intermetallics in URhGe a canted instead of a spin-spiral states is observed. A collinear spin structure as in URhSi is always possible even with SO coupling.

## CONCLUSIONS

We performed DFT electronic structure calculations for URhSi and URhGe employing the LAPW and the ASW method. These URhX intermetallic compounds were chosen because of there unique "ferromagnetic" ground state. Both compounds exhibit an almost complete cancellation of the spin and orbital magnetic moment, which explains the small respective values found by experiment. As expected, the spin magnetic moment are due to a polarization of the U-*5f* states. The large density of states at $E_F$ causes the Stoner criterion to be fulfilled for these itinerant electron ferromagnets. Assuming fluctuations of the total (small) magnetic moment leads to Curie temperatures of about 38 K for both systems which should be compared with the experimental result of about 9 K. While the ground state for URhSi is truly ferromagnetic, URhGe was found experimentally to show a spin canted state. Our ASW calculations for this state did not give such spin canted ground state, a deficiency which may be due to the neglect of SO-coupling during this calculation. However allowing for a spin spiral state also confirmed the ferromagnetic ground state for URhSi but led to a spin spiral ground state for URhGe. Since SO-coupling always destroys spin spiral states, we conjecture that also in this case the remaining interaction could lead to a spin canted ground state as found in experiment.

**Acknowledgements**

M.D. wishes to thank to Prof. H. Eschrig and Doc. M. Richter for allowing him to use the OLCAO relativistic code for the present project. The work of M. D. was supported by Austrian Science Foundation (Project Nr. M00413-PHY). The work of P. N. was supported by the Grant Agency of the AVCR (Project A1010715) and partially by Grant Agency of the Czech Republic (Project 202/96/0207) and Grant Agency of Charles University (Project 40/97).

## REFERENCES


1. V. Sechovský and L. Havela, Intermetallic compounds of actinides, Vol. 4 in: *Handbook of Magnetic Materials*, E. P. Wohlfarth and K. H. J. Buschow, eds., Elsevier Science B.V., Amsterdam (1988).
2. V. Sechovský and L. Havela, Magnetism of ternary intermetallic compounds of uranium, Vol. 11 in: *The Handbook of Magnetic Materials*, K H. J. Buschow, ed., Elsevier Science B.V., Amsterdam (1998).
3. M.S.S. Brooks and B. Johansson, Density functional theory of the ground state magnetic properties of rare earths and actinides, Vol. 7 in: *Handbook of Magnetic Materials*, K. H. J. Buschow, ed., Elsevier Science B.V., Amsterdam (1993).
4. M.R. Norman and D.D. Koelling, Electronic structure, Fermi surfaces, and superconductivity an f-electron metals, Vol. 17 in: *Handbook on the Physics and Chemistry of Rare Earths*, edited by K. A. Gschneidner,Jr., L. Eyring, G. R. Choppin, and G. H. Lander, Elsevier Science B.V., Amsterdam (1993).
5. K. Prokeš, M. Diviš, V. Sechovský, P. Svoboda, M. Richter, K.H.J. Buschow and F.R. de Boer, Electronic structure of URhSi, *Solid State Communications* 104:597 (1997).
6. V.H. Tran, R. Troc, G. Andre, Magnetic ordering in URhSi and URhGe, *Journal of Magnetism and Magnetic Materials*, in press.
7. P. Blaha, K. Schwarz, and J. Luitz, WIEN97, Vienna University of Technology 1997. (Improved and updated Unix version of the original copyrighted WIEN-code, which was published by P. Blaha, K. Schwartz, P. Sorantin, and S.B. Trickey, Full-potential, linearized augmented plane wave programs for crystalline systems, *Computer Physics Communications* 59:399 (1990).
8. D.D. Koelling and B.N. Harmon, A techique for relativistic spin-polarised calculations, *J. Phys. C: Solid State Phys.* 10:3107 (1977).
9. A.H. MacDonald, W.E. Pickett, and D.D. Koelling, A linearised relativistic augmented-plane-wave method utilising approximate pure spin basis functions, *J. Phys. C: Solid State Phys.* 13:2675 (1980).
10. J.P. Perdew and Y. Wang, Accurate and simple analytic representation of the electron-gas correlation energy, *Physical Review B* 45:13244 (1992).
11. J.P. Perdew, S. Burke, and M. Ernzerhof, Generalized gradient approximation made simple, *Physical Review Letters* 77:3865 (1996).
12. A.R. Williams, J. Kübler, and C.D. Jr. Gelatt, Cohesive properties of metallic compounds: augmented spherical wave calculations, *Physical Review B* 19:6094 (1979).
13. L.M. Sandratskii, Noncollinear magnetism in itinerant-electron systems: theory and applications, *Advances in Physics* 47:91 (1998).
14. U. von barth and L. Hedin, A local exchange-correlation potential for the spin-polarized case: I, *J. Phys. C: Solid State Phys.* 4:1629 (1972).
15. S.F. Matar, V. Eyert, A. Mavromaras, S. Najm, B. Chevalier, and J. Etourneau, Chemical bonding and magnetism in the ternary germanides $UT_2Ge_2$ (T = *3d* transition metal) from local spin density functional calculations, *Journal of Magnetism and Magnetic Materials* 174:219 (1997).
16. G.H.O. Daalderop, P.J. Kelly, M.F.H. Schuurmans, Magnetocrystalline anisotropy of $YCo_5$ nad related $RECo_5$ compounds, *Physical Review B* 53:14415 (1996).
17. M. Battocletti, H. Ebert , and H. Akai, Influence of gradient corrections to the local-density-approximation on the calculation of hyperfine fields in ferromagnetic Fe, Co, and Ni, *Physical Review B* 53:9776 (1996).
18. O. Eriksson, M.S.S. Brooks, and B. Johansson, Orbital polarization in narrow-band systems: Application to volume collapses in light lanthanides, *Physical Review B* 41:7311 (1990).
19. M. Diviš, M. Richter, and H. Eschrig, Magnetism and magnetocrystalline anisotropy of $Np_2T_2X$ (T = Co, Ni, Ru, Rh, Pd, Ir, Pt; X = In, Sn) compounds derived from density functional calculations, *Journal of Alloys and Compounds* 255:11 (1997).
20. A. Svane, J. Trygg, B. Johansson, and O. Eriksson, Electronic-structure calculations of praseodymium metal by means of modified density-functional theory, *Physical Review B* 56:7143 (1997).
21. F.D. Murnaghan, *Proc. Natl. Acad. Sci. U.S.A.* 30:244 (1944).
22. P. Mohn and E.P. Wohlfarth, The Curie temperature of the ferromagnetic transition metals and their compounds, *J. Phys. F: Metal Physics* 17:2421 (1987).


# Discussion

**Perdew**: In your GGA calculation for URhSi and URhGe, what equilibrium volumes did you find?

**Divis**: We used the experimental volumes.

# PRESSURE-INDUCED PHASE TRANSITIONS IN ALKALI HALIDES: HF AND DFT STUDY

Tomaž Sušnik and Aleš Zupan

Department of Physical and Organic Chemistry
Jožef Stefan Institute
Jamova cesta 39
SI-1000 Ljubljana
SLOVENIA

## INTRODUCTION

The alkali halides have been, due to their simplicity, a well visited playground in which new theories and experimental techniques have been tested. Their structural, elastic, and thermochemical properties are well known. They generally crystallize under ambient conditions in B1 (fcc) or B2 (bcc) phase and undergo pressure- and temperature-induced first order transitions (see Ref. 1 for an original review). The pressure-induced phase transition between B1 and B2 phase is particularly well documented[2] and this process is a subject of our *ab-initio* Hartree-Fock (HF) and Density Functional (DFT) Kohn-Sham (KS)[3] study.

The early theoretical efforts in the study of alkali halides have been made by Madelung[4], Ewald[5], and Born and Huang[6], who studied energy related properties such as the elastic tensor, band structure, and phase transitions. In the seventies a more fundamental approach, still partially empirical and based on the electron-gas theory[7-9] was applied systematically to alkali halides. In the past decade the implementation of reasonably reliable fully *ab-initio* computational schemes allowed the non empirical evaluation of the structural properties of ionic systems both with reference to Density Functional[10], perturbed ion[11], and Hartree-Fock[12,13] hamiltonians. In addition, some quantum chemistry methods have been used, like coupled cluster approach, to treat the correlation effects *a-posteriori*[14].

Results of these different computational methods are getting closer and closer to the experimental values, so it is time to study quantities of higher complexity that should show the limits and the advantages of the adopted hamiltonians and computational methods. In the present paper we report results obtained with CRYSTAL 95[15], the *ab-*

*Electron Correlations and Materials Properties*, edited by Gonis *et al.*
Kluwer Academic/Plenum Publishers, 1999. 

*initio* Hartree-Fock (HF) and Kohn-Sham (KS) Linear Combination of Atomic Orbital (LCAO) Self Consistent Field (SCF) computer program. General information on the method can be found in Ref. 16. In the present study we combine four cations ($Li^+$, $Na^+$, $K^+$, and $Rb^+$) with three anions ($F^-$, $Cl^-$, and $Br^-$) to get 12 different crystals. For all the ions all-electron Gaussian basis sets are used.

We focus mainly on the comparison of the HF and DFT results. We try to explain why GGA approximation to $E_{XC}$ expands LSD equilibrium lattice parameters and increases LSD transition pressures by using the reduced density-gradient analysis[17].

## COMPUTATIONAL METHOD

In our study we used both HF and KS computational approaches. In the KS method one has a liberty to choose from a welter of approximations for the exchange-correlation energy $E_{XC}$. Two main (non-empirical) groups of functionals are used contemporarily: (i) local spin-density (LSD) functionals, which use only electronic density as an input, and (ii) generalized gradient approximation (GGA) functionals, which use electronic density and its gradient to model the exchange-correlation energy $E_{XC}$ of a slowly varying electron gas. We chose the exact result for the exchange energy of a homogeneous electron gas by Dirac[18] and the parameterization of the correlation energy of the homogeneous gas by Perdew and Wang[19] for the local description. For the semilocal gradient-enhanced description of exchange-correlation energy we used generalized gradient approximation by Perdew, Burke, and Ernzerhof (PBE)[20].

For each system we calculated the total energy of the unit cell (either fcc or bcc) for a number of different volumes around the equilibrium. Then we fitted the Murnaghan equation of state[21] to these energy points to obtain the equilibrium volume $V_0$, the bulk modulus $B_0$ at equilibrium, its pressure derivative $B_0'$, and the total energy $E_0$ at equilibrium. Pressure is then easily calculated as a derivative of the energy with respect to the volume (at constant temperature) $p = -(dE/dV)_T$. The free energy at zero temperature becomes then:

$$G = E + pV = E - \frac{dE}{dV}V\,. \tag{1}$$

The transition pressure is determined at the point where the free energies for the two crystal phases equal: $G^{\text{fcc}}(p_t) = G^{\text{bcc}}(p_t)$. The result was checked with a geometrical Gibbs common tangent construction and the same transition pressures were obtained. In this paper we report only results for the equilibrium lattice parameters $a_0$ and the transition pressures $p_t$.

Experimental values for the equilibrium lattice parameters at room temperature are taken from Ref. 22. They are extrapolated from room temperature data to the static limit, i.e., to 0 K temperature without zero point motion using thermal expansion coefficients from Ref. 23. Experimental transition pressures are taken from a number of available papers[24-27]. No extrapolation to zero temperature is made, since not many experimental studies are available for this data. Probably the most elaborated study is the one by Li and Jeanloz[26], who measured the transition pressure for NaCl $p_t = 26.8$ GPa at room temperature. They extrapolated the value to 0 K and got a value of $p_t = 28.2$ GPa.

## RESULTS AND DISCUSSION

### Equilibrium Lattice Parameters

In Tab. 1 we report the equilibrium lattice parameters for all the alkali halides in the studied set. For the Hartree-Fock results we observe an uniform trend, all the equilibrium lattice parameters are too long, and the error increases with the increasing size of the anion or the cation. The smallest relative error is obtained for LiF (0.7%) and the biggest for RbBr (8.1%). This trend is a consequence of a too strong non-locality of the exchange energy (see for example curve $r_s = 0$ in Fig. 1 of Ref. 20); the structure that is more expanded and thus more inhomogeneous is over-favored.

Surprisingly, within the KS method, simple local LSD approximation gives the most accurate results. Here the equilibrium lattice parameters are usually slightly underestimated (1.4 % for LiF), but with the increasing size of the ions the error gets less negative, is equal nearly to zero (in our case for KCl) and then becomes positive (1.0 % for RbBr).

**Table 1.** Calculated equilibrium lattice parameters in Å. $\Delta = \frac{a_0 - a_0^{expt.}}{a_0^{expt.}} \cdot 100$ denotes the deviation in % from the experimental values.

| $a_0$ [Å] | | expt. | HF | Δ | LSD | Δ | PBE | Δ |
|---|---|---|---|---|---|---|---|---|
| LiF | fcc | 3.99 | 4.02 | 0.7 | 3.93 | -1.4 | 4.09 | 2.6 |
| | bcc | | 2.54 | | 2.48 | | 2.57 | |
| NaF | fcc | 4.58 | 4.64 | 1.2 | 4.52 | -1.3 | 4.71 | 2.9 |
| | bcc | | 2.82 | | 2.74 | | 2.87 | |
| KF | fcc | 5.30 | 5.49 | 3.5 | 5.27 | -0.6 | 5.53 | 4.3 |
| | bcc | | 3.29 | | 3.14 | | 3.33 | |
| RbF | fcc | 5.59 | 5.84 | 4.4 | 5.55 | -0.8 | 5.82 | 4.2 |
| | bcc | | 3.51 | | 3.32 | | 3.50 | |
| LiCl | fcc | 5.06 | 5.28 | 4.4 | 5.01 | -1.0 | 5.19 | 2.5 |
| | bcc | | 3.37 | | 3.20 | | 3.29 | |
| NaCl | fcc | 5.58 | 5.79 | 3.8 | 5.50 | -1.5 | 5.72 | 2.5 |
| | bcc | | 3.58 | | 3.38 | | 3.52 | |
| KCl | fcc | 6.22 | 6.59 | 6.0 | 6.22 | 0.0 | 6.51 | 4.6 |
| | bcc | | 3.97 | | 3.72 | | 3.91 | |
| RbCl | fcc | 6.51 | 6.94 | 6.6 | 6.52 | 0.1 | 6.81 | 4.6 |
| | bcc | | 4.15 | | 3.86 | | 4.07 | |
| LiBr | fcc | 5.42 | 5.71 | 5.4 | 5.32 | -1.8 | 5.53 | 2.0 |
| | bcc | | 3.67 | | 3.44 | | 3.55 | |
| NaBr | fcc | 5.90 | 6.20 | 5.0 | 5.81 | -1.5 | 6.06 | 2.8 |
| | bcc | | 3.84 | | 3.60 | | 3.74 | |
| KBr | fcc | 6.53 | 6.99 | 7.0 | 6.55 | 0.4 | 6.87 | 5.1 |
| | bcc | | 4.21 | | 3.90 | | 4.11 | |
| RbBr | fcc | 6.78 | 7.33 | 8.1 | 6.85 | 1.0 | 7.18 | 5.9 |
| | bcc | | 4.39 | | 4.05 | | 4.27 | |

GGA correction to LSD exchange-correlation energy greatly expands the lattice parameters for all the crystals studied and overestimates them on the average by 3.7%. We discuss this typical behavior later. Here we only comment that this change goes in the right direction; a proper treatment of the long-range attractive dispersion interaction, which is included neither in LSD nor in PBE approximation, would shrink the lattice parameters and compensate the PBE expansion. If one evaluates the dispersion energy empirically by summing the terms in the form of $1/r^6$ and $1/r^8$ over the infinite crystal lattice, one finds the contraction of the lattice parameters on the average of 0.04 Å for fluorides and 0.06 Å for chlorides[13].

### Transition Pressure

Evaluation of the transition pressure requires a much higher overall accuracy of the calculation than for the equilibrium lattice parameter. Not only the total energies of the unit cell over a wide range of different volumes are needed, also the pressure for each volume (i.e., the derivative of the total energy with respect to the volume) is required to obtain the transition pressure. In order to minimize the impact of the computational method on the calculated results a very high cutoff criteria for the truncation of the Coulomb series (and of the exchange series in the HF method) were adopted. Beside that, a "frozen index" option was used in the program. This means that the region (i.e., the indices of the unit cells) in which the bielectronic integrals are to be calculated according to truncation criteria was determined for the smallest, most compressed volume of the unit cell. For larger volumes, the same index table of integrals was used, thus reducing the numerical noise in the $E(V)$ curve. Additionally, the stability of the Murnaghan fit was tested by changing the fitting region of the volume and the density of energy points in the fitting set. We estimate that the fitting error in each calculated transition pressure is not more than ± 3 %.

**Table 2.** Calculated transition pressures $p_t$ in GPa. For missing numbers see comments in the text.

| $p_t$ [GPa] | expt. | Ref. | HF | LSD | PBE |
|---|---|---|---|---|---|
| LiF | | | | | |
| NaF | 23.0 | 24 | 32.9 | 19.9 | 25.7 |
| KF | 1.7 | 25 | 7.1 | 3.2 | 7.0 |
| RbF | 0.9 | 25 | 6.1 | 3.0 | 5.7 |
| LiCl | | | | | |
| NaCl | 28.2 | 26 | 38.9 | 23.6 | 26.5 |
| KCl | 2.0 | 27 | 4.0 | 1.1 | 3.1 |
| RbCl | 0.5 | 27 | 2.2 | 0.2 | 2.0 |
| LiBr | | | | | |
| NaBr | 33.0 | 24 | 31.1 | 33.8 | |
| KBr | 1.9 | 27 | 3.4 | 0.1 | 2.2 |
| RbBr | 0.5 | 27 | 2.1 | | 1.4 |

We report the calculated values of the transition pressure $p_t$ in Tab. 2. First a comment is needed for experimental values. Pressure-induced phase transitions in alkali halides show a considerable hysteresis (for example in NaCl B1 and B2 phases coexist metastably over a 10-GPa pressure, with the transition pressure determined at ≈ 30 GPa[26]), so the uncertainty of the experimental data is relatively high and a caution needs to be exercised when comparing them to the calculated zero temperature transition pressures. In agreement with experiment, no transitions were detected for lithium halides for the pressures of up to 110 GPa, regardless of the computational method used. We were unable to calculate the transition pressures for two further cases. For NaBr in the PBE approximation the fcc and bcc equilibrium volumes are almost the same and the fcc energy is always more negative than bcc energy, disallowing a common tangent to be found. On the other hand, LSD predicts a stable bcc RbBr phase, so no transition pressure can be calculated.

**Table 3.** Calculated $V_t/V_0^{expt.}$ ratios. No experimental data is available for the transition pressure $V_t$ at zero temperature, so we used room temperature data instead to calculate the experimental ratios in the first two columns of the table. The experimental values are taken from the same papers as in Tab. 2, but for NaCl crystal, where the value is taken from Ref. 24.

| $V_t/V_0^{expt.}$ | expt. | | HF | | LSD | | PBE | |
|---|---|---|---|---|---|---|---|---|
| | fcc | bcc | fcc | bcc | fcc | bcc | fcc | bcc |
| LiF | | | | | | | | |
| NaF | 0.770 | 0.690 | 0.759 | 0.696 | 0.784 | 0.711 | 0.813 | 0.745 |
| KF | 0.916 | 0.843 | 0.945 | 0.830 | 0.917 | 0.788 | 0.954 | 0.841 |
| RbF | 0.955 | 0.942 | 0.966 | 0.856 | 0.910 | 0.792 | 0.963 | 0.850 |
| LiCl | | | | | | | | |
| NaCl | 0.657 | 0.619 | 0.668 | 0.639 | 0.685 | 0.649 | 0.709 | 0.673 |
| KCl | 0.915 | 0.803 | 1.001 | 0.886 | 0.960 | 0.823 | 0.995 | 0.874 |
| RbCl | 0.970 | 0.830 | 1.069 | 0.926 | 0.996 | 0.829 | 1.029 | 0.888 |
| LiBr | | | | | | | | |
| NaBr | | | 0.715 | 0.679 | 0.613 | 0.605 | | |
| KBr | 0.912 | 0.807 | 1.017 | 0.904 | 1.004 | 0.848 | 1.022 | 0.893 |
| RbBr | 0.967 | 0.834 | 1.094 | 0.955 | | | 1.075 | 0.922 |

The transition pressures for all other crystals come too high in the HF method. Again, this is a consequence of the strong nonlocality of the exchange in the exchange-only (HF) calculation. Hartree-Fock overly favors more open structures (fcc in our case) relative to close-packed ones (bcc). Energy difference between the two phases becomes too big and thus the transition pressure is overestimated. LSD values are usually smaller than HF ones and usually underestimate the transition pressure. This is a well known behavior, LSD does not, speaking in terms of the total energy, distinguish enough between different phases (or different coordination numbers of atoms) due to the local description of the exchange-correlation energy. The energy difference between the two phases is too small and hence also the transition pressure. Sometimes LSD even predicts a wrong ground state and no transition pressure at all (as in our case for RbBr). GGA improves the behavior of the LSD, increases the transition pressures and sometimes brings them in a better agreement with the experiment. The relation between the LSD and the GGA results is discussed in the next section.

## INTERPRETATION OF DFT RESULTS

We base our interpretation of LSD and GGA results and particularly of their relation on the energy difference between LSD and GGA exchange-correlation energy.

Commonly, GGA exchange-correlation energy can be written in the following form:

$$E_{\rm XC}^{\rm GGA}[n_\uparrow, n_\downarrow] = \int d^3r\, n\, \varepsilon_{\rm X}^{\rm LDA}(r_s)\, F_{\rm XC}^{\rm GGA}(r_s, \zeta, s)\,, \tag{2}$$

where the electronic density is a sum of spin-up and spin-down contribution $n = n_\uparrow + n_\downarrow$, the Seitz radius of the electron $r_s = (4\pi n/3)^{-1/3}$, the relative spin polarization $\zeta = (n_\uparrow - n_\downarrow)/n$, and the reduced density-gradient $s = |\nabla n|^2/(2k_F n)$. Term $\varepsilon_{\rm X}^{\rm LDA}(r_s) = -3k_F/4\pi$ denotes the exchange energy per electron of the uniform electron gas and $k_F = (3\pi^2 n)^{1/3}$ is the local Fermi wavevector. $F_{\rm XC}^{\rm GGA}$ is GGA exchange-correlation enhancement factor, which reduces to LSD form in the uniform limit ($F_{\rm XC}^{\rm GGA}(r_s, \zeta, 0) \to F_{\rm XC}^{\rm LSD}(r_s, \zeta)$). For the plot of PBE enhancement factor $F_{\rm XC}$ for different values of $r_s$, $\zeta$, and $s$, see Fig. 1 in Ref. 20.

To simplify the interpretation of the role of gradients in DFT we define average values of the density parameters $\langle r_s\rangle$, $\langle|\zeta|\rangle$, and $\langle s\rangle$ (for details see Ref. 17), that fix the exchange-correlation energy of the system:

$$E_{\rm XC} = N\varepsilon_{\rm X}(\langle r_s\rangle)F_{\rm XC}(\langle r_s\rangle, \langle|\zeta|\rangle, \langle s\rangle)\,, \tag{3}$$

where $N$ denotes the number of the electrons in the system. By neglecting small differences in GGA and LSD electronic densities, the difference of GGA and LSD total energy of the system becomes the difference of GGA and LSD exchange-correlation energies: $\Delta = (E_{\rm XC}^{\rm GGA} - E_{\rm XC}^{\rm LSD})/N = \varepsilon_{\rm X}(\langle r_s\rangle)\,(F_{\rm XC}(\langle r_s\rangle, \langle|\zeta|\rangle, \langle s\rangle) - F_{\rm XC}(\langle r_s\rangle, \langle|\zeta|\rangle, 0))$. Now let's consider a small change in the volume of the unit cell. The density decreases ($\langle r_s\rangle$ increases) and at the same time the inhomogeneity of the system increases ($\langle s\rangle$ increases). The gradient correction will favor such a process only if the change in the energy is smaller than zero ($d\Delta < 0$). By making a total derivative of $\Delta$ we arrive to an inequality, which for non spin-polarized systems ($\zeta = 0$) reads:

$$\frac{d\langle s\rangle}{\langle s\rangle} > P\frac{d\langle r_s\rangle}{2\langle r_s\rangle}\,, \qquad P \approx 1\,. \tag{4}$$

This inequality shows that a common belief that "GGA favors density inhomogeneity" does not completely describe the effect. A competition between a bigger inhomogeneity (bigger $\langle s\rangle$) and a density contraction (smaller $\langle r_s\rangle$) is always present, though in most cases the inhomogeneity effect prevails. Note that also situations exist where the density effect prevails, one example is a bond between two hydrogen atoms, a bond at transition state for the reaction $H_2$+H $\to$ H + $H_2$, or a bond between a hydrogen and a heavier atom[17,30]. All these bonds are shorter in PBE than in LSD approximation.

We have tested the inequality (4) for a number of systems in other papers (see Refs. 17, 28, and 29) and in the present paper we did this for the lattice parameter expansion of NaCl in the fcc structure. "Infinitesimal" change of density parameters is obtained by expanding and shrinking the lattice parameter of NaCl by ±1% ($d\langle r_s\rangle = 1.23\cdot 10^{-4}$, $d\langle s\rangle = 5.77\cdot 10^{-4}$). Average values at equilibrium are: $\langle r_s\rangle = 0.335227$, $\langle s\rangle = 0.641558$, and $P = 1.118850$. The inequality (4) with these parameters is obeyed, $8.99\cdot 10^{-4} > 2.05\cdot 10^{-4}$, which shows that the inhomogeneity effect wins. So the GGA correction expands the lattice parameter of LSD; this is confirmed in Tab. 1 for all the systems and both phases.

The inequality (4) can also be implemented for small finite changes by substituting the infinitesimal changes of density parameters with finite ones. One such process, where small finite changes of density parameters occur, is an inverse phase transition from bcc to fcc structure in alkali halides. We calculated the value of $\langle r_s \rangle$ and $\langle s \rangle$ for each alkali halide at equilibrium volume for each phase. Small finite changes $d\langle r_s \rangle$ and $d\langle s \rangle$ are the differences between the average values for bcc and fcc phases and $\langle r_s \rangle$ and $\langle s \rangle$ are arithmetic averages of the corresponding values for bcc and fcc phases.

To do a consistent analysis we first have to eliminate the numbers for Li halides from Tab. 4. No transition occurs for these compounds, all three computational methods predict bigger bcc than fcc unit cells. This is seen in Tab. 4, $\langle r_s \rangle$ is bigger for bcc than for fcc phase. For all other systems, where the transition occurs, one can observe that what really changes between the two phases is the inhomogeneity (fcc phase is more inhomogeneous; bigger $\langle s \rangle$), the average density (inverse $\langle r_s \rangle$) remains almost the same. Even without a strict numerical evaluation of the inequality (4) one sees that gradient effect prevails, the inequality is fulfilled, and GGA favors more inhomogeneous fcc structure.

**Table 4.** Calculated average values of the density parameters $\langle r_s \rangle$ in $\langle s \rangle$. They are evaluated using the LSD electronic densities at LSD equilibrium lattice parameters. These numbers change only slightly when calculated for PBE densities and PBE equilibrium lattice parameters; the qualitative predictive power of the inequality (4) remains unchanged.

| $\langle r_s \rangle$ | fcc | bcc |
|---|---|---|
| LiF | 0.507456 | 0.507913 |
| NaF | 0.415744 | 0.415675 |
| KF | 0.325000 | 0.324929 |
| RbF | 0.207169 | 0.207156 |
| LiCl | 0.338653 | 0.338750 |
| NaCl | 0.335227 | 0.335204 |
| KCl | 0.296084 | 0.296036 |
| RbCl | 0.209730 | 0.209710 |
| LiBr | 0.201864 | 0.201905 |
| NaBr | 0.216282 | 0.216251 |
| KBr | 0.215339 | 0.215315 |
| RbBr | 0.186211 | 0.186198 |

| $\langle s \rangle$ | fcc | bcc |
|---|---|---|
| LiF | 0.719139 | 0.718944 |
| NaF | 0.687957 | 0.686789 |
| KF | 0.633993 | 0.632840 |
| RbF | 0.529693 | 0.529213 |
| LiCl | 0.634792 | 0.634756 |
| NaCl | 0.641558 | 0.640979 |
| KCl | 0.617407 | 0.616624 |
| RbCl | 0.536333 | 0.535854 |
| LiBr | 0.522504 | 0.522628 |
| NaBr | 0.540285 | 0.539023 |
| KBr | 0.542913 | 0.542409 |
| RbBr | 0.513488 | 0.513118 |

Therefore, the gradient correction to the LSD energy will be bigger for fcc than for bcc phase, yielding a bigger energy difference between the two phases. Hence, the transition pressure $p_t = -\Delta E_t/\Delta V_t$, where $\Delta E_t$ is the difference of the total energies of the fcc and bcc unit cell at the transition point and $\Delta V_t$ the difference of volumes between the two phases at the same point, becomes larger. This is confirmed for all the transitions studied (see Tab. 2). Note that the value of $p_t$ is also influenced by the change of volume $\Delta V_t = V_t^{fcc} - V_t^{bcc}$. Though, as seen in Tab. 3, the difference between $\Delta V_t$ for LSD and GGA approximations is small and does not contribute significantly to the difference between calculated LSD and GGA transition pressure $p_t$.

## CONCLUSIONS

We have performed HF and DFT study of the structural properties of alkali halides and their pressure-induced phase transition from fcc to bcc structure. We have found that DFT method with either LSD or GGA approximation for the exchange-correlation energy on the overall performs better than HF method.

LSD slightly underestimates the lattice parameters or predicts them right, whereas GGA greatly over-corrects. Inclusion of the attractive van der Waals interactions in the calculation would shorten the lattice parameters, move the LSD values away form and bring the GGA values closer to the experimental results. One the other hand, GGA corrects or slightly over-corrects the transition pressures between the two phases. It is not easy to say which of the two approximations (LSD or GGA) predicts transition pressures $p_t$ better because a direct comparison of the experimental and theoretical values is sometimes difficult due to experimental uncertainty. Nevertheless, we have shown that these results are the consequence of the tendency of gradient corrections to stabilize structures with more inhomogeneous electron structures.

## REFERENCES


1. J. C. Slater, Phys. Rev. **23**, 488 (1924).
2. P. W. Bridgman, *The Physics of High Pressure* (Dover, New York, 1970).
3. W. Kohn and L. J. Sham, Phys. Rev. **140**, A1133 (1965).
4. E. Madelung, Phys. Z. **19**, 524 (1918).
5. P. P. Ewald, Ann. Phys. (Leipzig) **64**, 253 (1921).
6. M. Born, and K. Huang, *Dynamical Theory of Crystal lattices* (Oxford University Press, London, 1954).
7. R. G. Gordon and Y. S. Kim, J. Chem. Phys. **56**, 3122 (1972).
8. C. Muhlhauser and R. Gordon, Phys. Rev. B **23**, 900 (1981); **24**, 2147 (1981).
9. N. V. K. Prabhakar, R. K. Singh, N. K. Gaur, and N. N. Sharma, J. Phys. Condens. Matter **2**, 3445 (1990).
10. A. Zunger and A. J. Freeman, Phys. Rev. B **16**, 2901 (1977); J. S. Melvin and D. C. Hendry, J. Phys. C: **15**, 2093 (1982); S. C. Erwin and C. C. Lin, J. Phys. C: Solid State Phys. **21**, 4285 (1988); P. Cortona, Phys. Rev. B **46**, 2008 (1992); P. Cortona, Nuovo Cimento **15 D**, 243 (1993).
11. V. Luaña and L. Pueyo, Phys. Rev. B **41**, 3800 (1990); A. Martín Pendás, V. Luaña, J. M. Recio, M. Flórez, E. Francisco, M. A. Blanco, and L. N. Kantorovich, *ibid.* **49**, 3066 (1994).
12. E. Aprà, M. Causà, M. Prencipe, R. Dovesi, and V. R. Saunders, J. Phys. C: Condensed Matter **5**, 2969 (1993).
13. M. Prencipe, A. Zupan, R. Dovesi, E. Aprà, and V. R. Saunders, Phys. Rev. B **51**, 3391 (1995).
14. K. Doll and H. Stoll, Phys. Rev. B **56**, 10121 (1997); **57**, 4327 (1998).
15. R. Dovesi, V. R. Saunders, C. Roetti, M. Causà, N. M. Harrison, R. Orlando, and E. Aprà, *CRYSTAL 95 User's Manual*, University of Torino, Torino (1996).
16. C. Pisani, R. Dovesi, and C. Roetti, *Hartree-Fock Ab Initio Treatment of Crystalline Solids*, Lecture Notes in Chemistry Series, (Springer Verlag, Berlin, 1988), Vol. 48.
17. A. Zupan, K. Burke, M. Ernzerhof, and J. P. Perdew, J. Chem. Phys. **106**, 10184 (1997).
18. P. A. M. Dirac, Proc. Cambridge Phil. Soc. **26**, 376 (1930).
19. J. P. Perdew and Y. Wang, Phys. Rev. B **45**, 13244 (1992).
20. J. P. Perdew, K. Burke, and M. Ernzerhof, Phys. Rev. Lett. **77**, 3865 (1996); **78**, 1396(E) (1997).
21. F. D. Murnaghan, Proc. Nat. Acad. Sci. U.S.A. **30**, 244 (1944).
22. R. W. G. Wyckoff, *Crystal structures*, (Interscience Publishers, 1963), Vol. 1, 2$^{nd}$ edition.

23. J. E. Rapp and H. D. Merchant, J. Appl. Phys. **44**, 3919 (1973).
24. Y. Sato-Sorensen, J. Geophys. Res. **88**, 3543 (1983).
25. S. N. Vaidya and G. C. Kennedy, J. Phys. Chem. Solids **32**, 951 (1971).
26. X. Li and R. Jeanloz, Phys. Rev. B **36**, 474 (1987).
27. A. J. Cohen and R. G. Gordon, Phys. Rev. B **12**, 3228 (1975).
28. A. Zupan, J. P. Perdew, K. Burke, and M. Causà, Int. J. Quantum Chem. **61**, 835 (1997).
29. A. Zupan, P. Blaha, K. Schwarz, and J. P. Perdew, submitted to Phys. Rev. B (1998).
30. E. I. Proyonov, E. Ruiz, A. Vela, and D. R. Salahub, Int. J. Quantum Chem. **S29**, 61 (1995).

# A Quantum Monte Carlo Study of the Exchange-Correlation Hole in Silicon Atom and System-Averaged Correlation Holes of Second Row Atoms

Antonio C. Cancio*, C. Y. Fong
Department of Physics
University of California
Davis, Ca 95616-8677, USA

and

J. S. Nelson
Semiconductor Physics Division, Sandia National Laboratories
Albuquerque, NM 87185-5800, USA

**Abstract**

The variational Monte Carlo method using the correlated estimates technique and the correlated (Slater-Jastrow) wave function to study the exchange-correlation hole, $n_{xc}$, in open shell atoms are discussed in detail. The spin decomposed $n_{xc}$ for the silicon atom and the system-averaged correlation holes for the second row atoms are presented. We find important nonlocal contributions to the opposite spin $n_{xc}$. The system-averaged correlation holes show approximate scaling behavior with respect to the number of electron pairs and a correlation radius.

## I. Introduction

Much progress using accurate numerical techniques [1,2] and analytical models [3] has been made in the study of the electron correlation for the system of homogenous electron gas. In a many-electron system with inhomogeneous charge distribution, the understanding of the pair correlation function and related quantities, such as the exchange-correlation hole and exchange-correlation energy remains a challenging task. Recently, efforts have been made to calculate the correlation properties in filled shell atoms [4] and periodic solids [5,6]. There has not been much study of the correlation effects on atoms with open valence shells. They are, however, important physical systems to test the understanding of correlation effects and theoretical approaches.

The variational Monte Carlo (VMC) method was used in Refs. 5 and 6. The trial wave function is in the form of Slater-Jastrow wave function [5,6]. Typically, with a few parameters determined variationally, this kind scheme can recover 85% of the correlation energy in all electron calculations of atoms and the dissociation energy of molecules. Unlike the configuration interaction (CI) approach [7], this kind scheme can describe with equal ease both the nondynamic or orbital-dependent correlations and dynamic correlations such as the cusp condition [8].

*Electron Correlations and Materials Properties*, edited by Gonis *et al.*
Kluwer Academic/Plenum Publishers, 1999.

In this paper, we apply the VMC method with the correlated estimates technique and use an accurate trial wave function to study the spin decomposed exchange-correlation hole, spin decomposed $n_{xc}$ or $n_{xc}$, in a silicon (Si) atom and the system-averaged correlation (opposite spin) holes, system averaged $n_{xc}$ or $<n_{xc}>$, of the second row atoms. In section II, we shall present in detail our approach of the VMC method and the trial wave function. In section III, the results of the $n_{xc}$ of Si and the $<n_{xc}>$ of the second row atoms will be described. Finally, a summary will be given.

## II. Calculational method

We divide this section into four parts: A. quantities of interest, B. general comments about the approach, C. correlated estimates technique within the VMC method specifically for calculating the quantities of interest, and D. the form of the trial wave function.

### A. Quantities of interest

Two common but physically revealing quantities of interest for studying electron correlation in atoms are the spin decomposed $n_{xc}$ and the system-averaged $n_{xc}$. The former is defined as

$$n_{xc}(r_o,\sigma_o;\, r,\sigma) = \frac{n^{(2)}(r_o,\sigma_o;\, r,\sigma)}{n(r_o,\sigma_o)} - n(r,\sigma) \qquad (1)$$

where $n(r,\sigma)$ is the mean spin (particle- not charge-) density of a many-electron system at position **r** with spin $\sigma$. $n^{(2)}(r_o,\sigma_o;\, r,\sigma)$ is the pair spin (particle- not charge-) density of one particle at **r** with spin $\sigma$ given another particle having spin $\sigma_o$ and staying at $r_o$ and it is related to the standard pair correlation function:

$$g(r_o,\sigma_o;\, r,\sigma) = \frac{n^{(2)}(r_o,\sigma_o;\, r,\sigma)}{n(r,\sigma)\, n(r_o,\sigma_o)}$$

$g(r_o;\, r) = g(r_o,\uparrow;\, r,\uparrow) + g(r_o,\uparrow;\, r,\downarrow)$ is the exchange-correlation function. It is often useful to compare the $n_{xc}$ to the exchange hole, $n_x$, which is the corresponding quantity measured with respect to the Hartree-Fock wave function that reproduces the ground state single particle density. The difference between the two often called the Coulomb or correlation hole measures the correlations that are generated by the Coulomb interaction on top of purely statistical correlations of the Hartree-Fock wave function. In this paper, we shall study the explicit spin-decomposition of $n_{xc}$ given in Eq. (1).

The system-averaged $n_{xc}$ is defined as

$$<n_{xc}(u;\sigma_o,\sigma)> = \int d^3r \int d\Omega_u [n^{(2)}(r,\sigma_o;\, r+u,\sigma) - n(r,\sigma_o)*n(r+u,\sigma)] \qquad (2)$$

where **u** is the displacement between two particles. The integral is over the position of one of the particles and the angle of the displacement **u** to obtain a function of the interparticle distance, u, only. As the Coulomb interaction itself is only a function of interparticle distance and is further independent of spin, the spin-summed and system-averaged $n_{xc}$ contains the correlations directly necessary for determining the exchange-correlation potential energy. The same- and opposite-spin system-averaged $n_{xc}$ determine spin decomposed potential energy terms that are similar (but not equal) to exchange and correlation energies, respectively.

In this paper, we are studying the explicit spin-decomposition of $<n_{xc}>$. The spin decomposition of the total $<n_{xc}>$ is very convenient for VMC measurements, and provides a decomposition similar to that of correlation and exchange. Note that the exchange hole has contributions only in the same-spin component – opposite spin particles do not experience Pauli correlations in their spatial distribution. On the other hand, the correlation hole is predominantly observed in the opposite spin channel, with only small corrections to the same-spin $n_{xc}$.

B. General comments on the approach

We are interested mainly in the correlation between valence electrons. To this end, we study the following Hamiltonian:

$$H = \Sigma_v p_v^2/2m + \Sigma_v V_p(r_v) + \Sigma_{v<v'} e^2/|r_v - r_{v'}| \tag{3}$$

The sum over v is over the valence electrons. $V_p$ is the ab-initio norm conserved pseudopotential [9] to characterize the interaction between the valence electrons and the nucleus plus the core electrons. No core electron is treated in this paper. The last term in H is the electron-electron interaction. The quantities of interest are determined as expectation values of the ground state of this Hamiltonian.

In principle, the VMC method is our general approach to obtain an optimized variational wave function to model the ground state of the Hamiltonian, and from it to calculate the quantities of interest. For this approach, a trial wave function, $\Psi(\{r,\sigma\})$, in Slater-Jastrow form is used, where $\{r,\sigma\}$ is the spatial and spin configurations of the valence electrons. In the Slater determinants, functions in a single-particle orthonormal set obtained from self-consistent Hartree-Fock method or density functional theory can be used. The Jastrow wave function characterizing the correlation will be discussed in detail in II.D. The parameters in the Jastrow factor were determined by minimizing the variance of the energy:

$$\varepsilon_E^2 = < |\Psi(\{r,\sigma\})^{-1} H \Psi(\{r,\sigma\}) - <E>|^2 > \tag{4}$$

where H is Hamiltonian of the system (Eq. (3)) and $<E>$ is the ground state energy. When $\Psi(\{r,\sigma\})$ approaches an eigenvector of H, the variance vanishes. Otherwise, it is positive definite. These features make the variance estimation to be a robust method for finding the optimal wave function. The minimization was done by using the Levenberg-Marquardt algorithm [10] which is a method to minimize the least squares fit of a smooth function to a noisy data set. As we shall discuss in the next section, the straightforward MC to calculate the quantities of interest needs to be modified in order to reduce the statistical noise.

C. Correlated estimates technique within the variational Monte Carlo method

The Coulomb correlation hole, though nonperturbatve, is a fairly small correction compared to the contribution of the exchange hole. A straightforward MC calculation of this quantity can suffer from statistical noise. We adapted the correlated estimates technique [11,12], developed for calculating arbitrarily small differences in expectation values resulting from small changes in system parameters or variational parameters.

To apply the correlated estimates technique for the quantities given in Eqs. (1) and (2), we express the exchange-correlation hole, $n_{xc}(r_o,\sigma_o;r,\sigma)$, and the exchange hole $n_x(r_o,\sigma_o;r,\sigma)$ in terms of the density expectations of the optimized trial wave function, $n_t(r,\sigma)$, and the Hartree-Fock wave function, $n_{HF}(r,\sigma)$, respectively. That is:

$$n_{xc}(r_o,\sigma_o;r,\sigma) = n_t(r_o,\sigma_o|r,\sigma)_t - n_t(r,\sigma) \qquad (5)$$

$$n_x(r_o,\sigma_o;r,\sigma) = n_{HF}(r_o,\sigma_o|r,\sigma) - n_{HF}(r,\sigma) \qquad (6)$$

where $n_t(r,\sigma)$ is the mean density for the trial wave function and $n_t(r_o,\sigma_o|r,\sigma)$ is the corresponding conditional density of $N_v - 1$ particles at r with spin $\sigma$ given one particle located at $r_o$ with spin $\sigma_o$. $N_v$ is the total number of the valence electrons. Corresponding quantities for the Hartree-Fock wave function are given with the HF subscript. We calculate $n_x$ analytically and reexpress the full $n_{xc}$ in terms of the difference expression:

$$n_{xc}(r_o,\sigma_o;r,\sigma) = \Delta n(r_o,\sigma_o|r,\sigma) - \Delta n(r,\sigma) + n_x(r_o,\sigma_o;r,\sigma) \qquad (7)$$

where $\Delta n(r,\sigma)$ is the difference in mean density between the fully correlated and Hartree-Fock wave functions. $\Delta n(r_o,\sigma_o|r,\sigma)$ is the difference in condition density. Then we use the Monte Carlo technique to sample the density difference rather than the density directly for each of these two cases:

$$\Delta n(r,\sigma) = \left(\frac{\langle(|\Psi_t(\{r,\sigma\})|^2/P(\{r,\sigma\}))\Sigma_v\delta(r-r_v)\rangle_P}{\langle|\Psi_t(\{r,\sigma\})|^2/P(\{r,\sigma\})\rangle_P}\right) - \left(\frac{\langle(|\Psi_{HF}(\{r,\sigma\})|^2/P(\{r,\sigma\})\Sigma_v\delta(r-r_v)\rangle_P}{\langle|\Psi_{HF}(\{r,\sigma\})|^2/P(\{r,\sigma\})\rangle_P}\right) \qquad (8)$$

where $\Psi_t(\{r,\sigma\})$ and $\Psi_{HF}(\{r,\sigma\})$ are the correlated trial and the Hartree-Fock ground state wave functions, respectively. Then the notation $\langle\ \rangle_P$ represents the average over all configurations $\{r,\sigma\}$ with a probability distribution $P(\{r,\sigma\})$; for the conditional density, one electron is further assumed to be located at the position $r_o$ with spin $\sigma$. Averages are statistical estimated over a random set of configurations sampled from the distribution P using the standard Metropolis algorithm [13]. The key step is that statistical estimates for both the expectations using $\Psi_t(\{r,\sigma\})$ and $\Psi_{HF}(\{r,\sigma\})$ are taken from the same random set of configurations. If the two wave functions are sufficiently similar the statistical fluctuations in each measurement will then be strongly correlated and thus by taking the difference much of the noise will cancel out.

The choice of the sampling function, P, in this technique has been guided first of all by the consideration of conditions in which the correlated estimates method may fail to improve results or worse, make the calculation unfeasible. The standard failure occurs when either of the weighting factors

$$w_{t\ \text{or}\ HF} = (|\Psi_{t\ \text{or}\ HF}(\{r,\sigma\})|^2 / P(\{r,\sigma\})) / \langle|\Psi_{t\ \text{or}\ HF}(\{r,\sigma\})|^2 / P(\{r,\sigma\})\rangle_P \qquad (9)$$

in Eq. (9) becomes too large (>10). In this case, bias errors in the statistical estimates of the ratio of two averages may become important. More importantly, it signals the undersampling of the region of configuration space in which the large reweighting factors occur with possibly fatal results. A typical example of undersampling occurs in atoms when P is chosen to decay exponentially with a faster decay rate than that of $|\Psi_t(\{r,\sigma\})|^2$ (or $|\Psi_{HF}(\{r,\sigma\})|^2$). In this case, one of the w in Eq. (9) increases exponentially as the difference in rates as an electron moves away from the nucleus. The longer a random walk is taken, the more divergent become the contributions to the average as walkers stray away from the atom, so that statistical estimates of the energy and other expectations using this weight factor do not converge.

In the present calculations, we choose a probability distribution, P, as:

$$P(\{r,\sigma\}) = (1-\varepsilon)||\Psi_t(\{r,\sigma\})|^2 - \alpha|\Psi_{HF}(\{r,\sigma\})|^2| + \varepsilon(|\Psi_t(\{r,\sigma\})|^2 + \alpha|\Psi_{HF}(\{r,\sigma\})|^2) \quad (10)$$

where $\alpha$ is a parameter chosen to be roughly the ratio of the normalization of the two wave functions. $\varepsilon$ is a small parameter chosen to avoid accidental nodes where $\Psi_t$ and $\Psi_{HF}$ have the same value. The first term at the right hand side of Eq. (10) weighs most favorably the region of configuration space in which the two wave functions are most different (e.g., in the center of the correlation hole) and minimizing the sampling of configurations where they are the same. To see the role of the small parameter, $\varepsilon$, we consider (a) $\varepsilon = 0$, and (b) $\varepsilon = 1$ with the optimal $\alpha = 1$. For (a), it is possible to have $P = 0$ when two wave functions cross. This causes the w in Eq. (9) to diverge. For (b), it would result in a purely additive mixture of the two wave functions and would limit the variation of the weighting ratios to a finite upper limit of 2. Such a choice, however, would not have the advantage of weighting the random walk where the wave functions differed but rather to where they are the same. By tuning $\varepsilon$, we found that a choice of $\varepsilon = 0.05$ limited upper bound of the reweighting factors to be about 5 to 10, while obtaining a noticeable reduction in noise from the case (b). Therefore, the feature of this reweighting scheme is that statistical estimates for both expectations in Eq. (9) are taken from the same random set of configurations and the undersampling can be avoided. It can give quite good results in about $10^5$ sampling points as compared to $10^7$ or more by conventional sampling methods.

We consider the specific spin dependence of the system-averaged $n_{xc}$, $\langle n_{xc}(u;\uparrow,\downarrow)\rangle$, given in Eq. (2):

$$\langle n_{xc}(u;\uparrow,\downarrow)\rangle = \langle n^{(2)}(\mathbf{r},\uparrow;\mathbf{r}+\mathbf{u},\downarrow)\rangle - \int d^3r \int d\Omega_u\, n(\mathbf{r};\uparrow)\, n(\mathbf{r}+\mathbf{u};\downarrow) \quad (11)$$

All the expectations can be calculated in VMC, again using correlated estimates to obtain a difference of trial and Hartree-Fock estimates of this quantity, as in Eqs. (5) and (6). $\langle\rangle$ in Eq. (11) means integration over spatial coordinates and angular integration of $\mathbf{u}$. $d\Omega_u$ stands for integral over the solid angle of the vector $\mathbf{u}$.

D. Trial wave function

The basic type of wave function used for VMC calculations starts with a Slater determinant or combination of Slater determinants of single particle orbitals obtained from Hartree-Fock or density functional method. To account for the correlation effect, the Jastrow [14] type of wave function is often used. The Slater-Jastrow wave function is a product of a correlated factor and a Slater determinant.

$$\Psi_t(\{r,\sigma\}) = \exp(-\Sigma_v \Sigma_{v'<v} f(\mathbf{r}_v,\mathbf{r}_{v'},\mathbf{r}_{vv'}))\Pi_\sigma \det_\sigma \quad (12)$$

where f depends on the coordinates of v-th, and v'-th electrons, $\mathbf{r}_v$ and $\mathbf{r}_{v'}$, which are measured from the center of the atom. $\mathbf{r}_{vv'}$ is the displacement between electron v and electron v', the relative displacement between the two particles. $Det_\sigma$ is a Slater determinant for states with spin $\sigma$. Since we are interested in the pair correlation as a function of the separation between an electron pair, the vector nature of the coordinates are neglected in our calculations.

The optimal form for f in an inhomogeneous system is difficult to predict in an a priori fashion. In order to obtain a simple and systematically improvable form, we expand f in the form of Boys and Handy [15]:

$$f(r_v, r_{v'}, r_{vv'}) = \Sigma_{lmn, l+m+n<N} C_{lmn} R_b(r_v)^l R_b(r_{v'})^m R_b(r_{vv'})^n \quad (13)$$

where $R_b$ is given by

$$R_b(r) = br/(1+br) \quad (14)$$

1/b defines the range of the correlation function. $R_b$ approaches a constant when the interparticle distance becomes large to take into account the less correlated physical situation when the two particles are far apart. The number of terms in the expansion is limited by imposing the cusp conditions. In particular, the coefficients of all terms with n = 1 are:

$$C_{001} = -0.5, \text{ for antiparallel spins} \quad (15)$$

$$C_{001} = -0.25, \text{ for parallel spins} \quad (16)$$

and

$$C_{lml} = 0, \text{ for all nonvanishing l and m} \quad (17)$$

Particle exchange symmetry requires that $C_{lmn} = C_{mln}$. With these restrictions and neglecting explicit spin dependence except for the cusp conditions, there are eight terms survive for the N = 4 order correlation function. These are: $C_{00n}$, n = 2, 3, and 4 which give a refinement of the basic repulsive correlation form, and $C_{l00}$, l = 2, 3, and 4, which are the single particle terms that act to restore the single-particle density in the presence of the repulsive correlations. In addition, $C_{220}$ and $C_{202}$ provide density dependent refinements to the correlation factor.

### III. Results and discussion

We shall present first the results of the $n_{xc}$ for the Si atom and then the $<n_{xc}(u;\uparrow,\downarrow)>$ for the second row atoms as functions of the separation between a pair of electrons.

A. The spin decomposed exchange-correlation holes of the Si atom

The exchange-correlation hole given in Eq. (7) of the Si atom is calculated by using the VMC method with the correlated estimates technique. It is useful to review the Hartree-Fock picture for this open shell atom in order to discuss the resulting correlation holes. The Hartree-Fock ground state of the valence shell of the Si atom consists of four electrons in the $3s^2 3p^2$ configuration. Here the 3s subshell forms a spin singlet. Hund's rules require that the two p-electrons form a symmetric spin triplet state with total spin S=1 and an antisymmetric orbital angular momentum state with L=1. The ground state configuration is $^3P$. There are nine configurations of the p-electrons with these values of L and S, each with a different projection of spin and orbital angular momentum, $M_S$ and $M_L$. The most important distinction for spatial correlations is the different cases for orbital angular momentum. For $M_L=0$, the two occupied p orbitals are the $p_x$ and $p_y$ states, with the $p_z$ state left unoccupied; the single particle density of this configuration is roughly that of a "pancake". For $M_L=\pm1$, the $p_z$ and $p_+$ or $p_-$ states are occupied, so that there is a "cigar'-like distribution of charge oriented along the z-axis with at least some p-eletron density along each axis. The pancake projection offers a distinction between correlations in a direction perpendicular to and parallel to the occupied p orbitals and therefore a clear picture of the effects of orbital occupation on electron correlation. In this paper, we focus on this $M_L=0$ configuration. The corresponding Slater determinant for the $^3P$ ground state is the product of a spin up determinant with the single particle states to be one 3s and two 3p orbitals and a down spin determinant consisting of one 3s orbital.

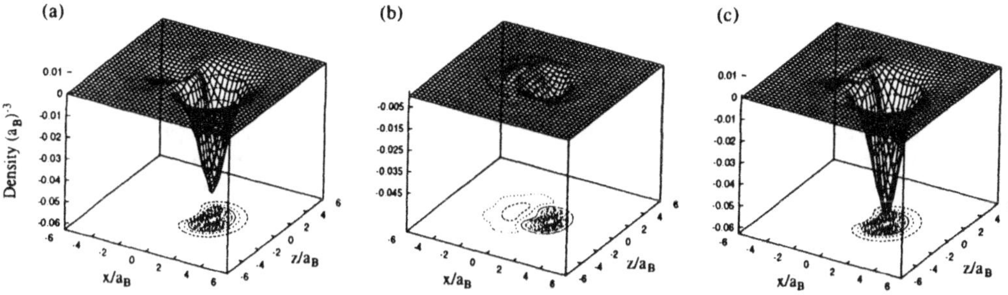


Fig. 1 The same spin (a), opposite spin (b) and total (c) exchange-correlation hole about a spin-up particle located at $1.4a_B$ from the atom center on the x-axis. The surface plot shows the change in density along a plane cutting through the origin along the x- and z-axes.

In Fig. 1, we show the $n_{xc}$ for an electron with spin-up located at $x = 1.4a_B$ and $z = 0$ away from the nucleus (at the origin), where $a_B$ is the Bohr radius. In the Hartree-Fock picture, the fixed electron could occupy possibly either a 3s or $3p_x$ orbital of the valence shell. The exchange hole is, therefore, a hybrid of the two, such as a $sp^2$ orbital and is the dominant contribution to the exchange hole around the fixed electron (Fig. 1(a)). The remaining two spin-up electrons must be in other hybridized orbitals. When the Coulomb correlation is added, there is little change in the same spin channel. The opposite spin channel shows a relatively small reduction of charge near the fixed electron (14% of the total strength at the minimum of the density) with some polarization at farther distance from the hole center (Fig. 1(b)). Fig. 1(c) shows the total exchange-correlation hole for this case. The effect of exchange and correlation is primarily around the fixed electron. This indicates the $sp^2$ hybrid hole screens the electron efficiently.

Next, consider that an electron with spin-up is located at $x = 0$ and $z = 1.4a_B$ away from the nucleus. In the Hartree-Fock picture, the fixed electron must be occupying the 3s orbital of the valence shell because the $p_x$ and $p_y$ orbitals are zero on the z-axis. In this case, the exchange hole which describes the density difference between the density of the two remaining spin up electrons given one fixed, and the average density, is therefore formed by removing the electron in the 3s state. With the addition of correlation, the same-spin density (Fig. 2(a)) loses the rotational symmetry of the s-state. The polarization effect of the two 3p-states creates a double valley on either side of the s-electron. Fig. 2(b) shows the polarization due to the correlation for opposite spin case. A well centered about the fixed electron and a strong dipole response at larger range are obtained. This polarization effect is nonlocal. The total exchange-correlation hole for a fixed particle at the z-axis is shown in Fig. 2(c) which indicates a smooth interpolation of the two-spin contributions leading to a large crescent-shaped area near the fixed electron from which charge is repelled. The correlation hole contributes significantly to the total exchange-correlation hole. It is up to 40% of the total correlation.

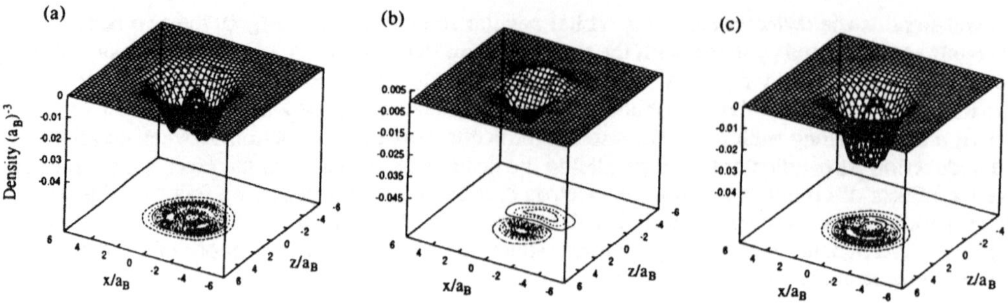


Fig. 2 The same spin (a), opposite spin (b) and total (c) exchange-correlation hole about a spin-up particle located at $1.4a_B$ from the atom center on the z-axis. The surface plot shows the change in density along the same plane as in Fig. 1.

Comparing the two cases for the fixed electron at different axes, the opposite spin $n_{xc}$ exhibits more anisotropy in the case when the electron is at a point along the z-axis than the case when the electron is at the x-axis. The reason is that in the former case the occupied orbitals do not screen effectively for the fixed electron. Consequently, the down spin 3s electron can easily be pushed from one side of the atom along the z-axis to the other side. On the other hand, when the up spin electron is at $x=1.4a_B$, the two remaining spin up electrons occupying hybridized orbitals in the x-y plane because of the Pauli principle, can block the polarization of the down spin electron occupying the 3s orbital.

B. The system-averaged correlation holes of the second row atoms

We calculated the $\langle n_{xc}(u;\sigma_o,\sigma)\rangle$ for the second row atoms in order to understand how the basic features of the correlation hole, such as the length scale and the correlation energy of the hole, varies with particle number and system size. In Fig. 3, we plotted the opposite spin component weighted by $4\pi u/Z_p$, $4\pi u\langle n_{xc}(u;\uparrow,\downarrow)\rangle/Z_P$ for the seven atoms (Magnesium (Mg) to Argon (Ar)) as functions of u in units of $r^*$. $r^*$ is the distance where the pair density changes from

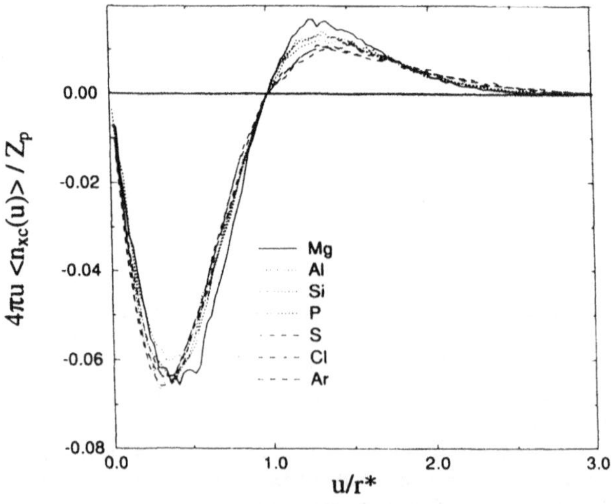


Fig. 3 The system-averaged exchange-correlation holes for the second row atoms.

a negative to a positive value when it is plotted against u in Bohr units. $Z_P$ is the number of opposite spin electron pairs. The density is weighted by u, taking into account a phase space factor of $u^2$ and a Coulomb interaction of 1/u, so as to measure the contribution at each u to the exchange-correlation potential energy. The curves for the different atoms are plotted with different symbols as indicated in the legend. It is interesting to note that the $\langle n_{xc}(u;\uparrow,\downarrow)\rangle$ weighted by $4\pi u/Z_P$ are very similar as functions of a normalized separation between the electrons, u/r*. The total peak height and shape of the correlated density for each case is roughly proportional to the number of opposite spin pairs, indicating that the quantity that determines the nature of the correlation is the correlation between each individual pair. Notice a charge depletion region for small u and charge piling region at a farther distance. In addition to some small shell effects, r*, or the correlation length, is the main physical variable that distinguishes each case. It varies from about 1.5 times the atom radius for Ar to 2.0 times the atomic radius for Mg. This change in r* along with the 'sharper' profile for the scaled Mg curve vs. that of Ar suggests that the lighter the atom the more the polarization effect.

## IV. Summary

We have used the variational Monte Carlo method with correlated estimates technique to calculate the spin decomposed exchange-correlation holes for the Si atom the system-averaged correlation holes for the second row atoms. One of the features of the spin decomposed exchange-correlation holes is the nonlocal response of the down-spin electron to an up-spin electron. Our results suggest that the correlation effects depend on the orbitals occupied by the other electrons. This kind of orbital-dependent correlation is not included in the local density approximation for electron correlation. The system-averaged correlation holes for the second row atoms show a very similar qualitative feature if we plot the correlated holes weighted by $4\pi u/Z_P^2$ against u/r* where r* is a crossover radius which separates regions of charge depletion and charge accumulation. This crossover radius decreases as Z, the atomic number, increases and decreases in comparison to atomic radius as Z increases.

## Acknowledgments

This work was supported in part by Sandia National Laboratories under contract AP-7094 and a Campus Laboratories Collaboration grant of the University of California. A grant from San Diego Supercomputer Center is greatly appreciated.

## References

[1] D. M. Ceperley and B. J. Alder, Phys. Rev. Lett. 45, 566 (1980).
[2] Warren E. Pickett and Jeremy Q. Broughton, Phys. Rev. B48, 14859 (1993-II).
[3] R. O. Jones and O. Gunnarson, Rev. Mod. Phys. 61, 689, (1989).
[4] M. Ernzerhof, J. P. Perdew and K. Burke, in "Density Functional Theory", ed. by R. Nalewajski. Springer-Verlag, Berlin, (1996).
[5] S. Fahy, X. W. Wang and Steven G. Louie, Phys. Rev. Lett. 61, 1631 (1988).
[6] Randolph Q. Hood et al., Phys. Rev. Lett. 78, 3350 (1997).
[7] See for example, Peter Fulde, "Electron Correlation in Molecules and Solids" Springer-Verlag, Berlin, (1991).
[8] J. C. Kimball, Phys. Rev. A7, 1648 (1973).
[9] G. B. Bachelet, D. R. Hamann and M. Schluter, Phys. Rev. B26, 4199 (1982).
[10] See W. H. Press et al., in "Numerical Recipes (Fortran Version)", Cambridge University Press, Cambridge, 523 (1989).
[11] D. M. Ceperley and M. H. Kalos in "Monte Carlo Methods in Statistical Physics", ed. by K. Binder, Springer-Verlag, Berlin, (1979).

*Present address: School of Physics, Georgia Tech. Atlanta, GA30332, USA

[12] M. Kalos and P. Whitlock, in "Monte Carlo Method", vol. 1, Wiley and Sons, New York, 107 (1986).
[13] N. Metropolis, et al. J. Chem. Phys. 21, 1087 (1953).
[14] R. Jastrow, Phys. Rev. 98, 1479 (1955).
[15] N. C. Handy, J. Chem. Phys. 58, 279 (1973).

# STRONGLY CORRELATED ELECTRONS: DYNAMICAL VERTEX RENORMALIZATION

Václav Janiš

Institute of Physics, Academy of Sciences of the Czech Republic
Na Slovance 2, CZ-180 40 Praha, Czech Republic

## INTRODUCTION

Strongly correlated electrons and heavy fermions display a number of phenomena not understandable within simple Fermi-liquid theory of metals. These unusual effects include dynamical formation of magnetic local moments and magnetic orders, mixed-valence fluctuations, Kondo and heavy-fermion behavior, non-Fermi-liquid phases and metal-semiconductor transitions. At present there are two principal approaches for treating electron correlations. Ab-initio investigation within the density-functional theory and the local-density approximation takes explicitly into account the Coulomb repulsion to equilibrium states within elementary cells (muffin-tin sphere) spread over a few interatomic distances. Short-range effects of the electron correlations in the ground state are quite well captured. Much less reliable are the results of the local-density approximations for long-range effects and low-energy excitations above the ground state. This failure can be attributed to the underestimation of the Pauli exclusion principle and long-range fluctuations spreading over thousands of lattice constants. The other approach to the treatment of the electron correlations is based on tight-binding model calculations. The microscopic structure of the electron lattice gas is significantly simplified by taking into account only a few atomic (Wannier) orbitals. On the other hand, large-scale effects and low-energy excitations are well described. Presently the two approaches are more or less complementary.

A realistic description of the electron correlations has to join the two existing approaches to enable a sufficiently accurate treatment of both short as well as long distances. However, to succeed in a more reliable description of the electron correlations on small and large scales, it is mandatory to pick up and maintain the relevant aspects of each of the construction. The former supply us with realistic static properties of the ground-state. The latter offers qualitatively accurate estimates of dynamical fluctuations and collective responses to low-energy excitations of strongly correlated systems.

*Electron Correlations and Materials Properties*, edited by Gonis *et al.*
Kluwer Academic/Plenum Publishers, 1999.

It appears that most of the physics of strongly correlated electrons is covered by the archetypal Anderson impurity and Hubbard lattice models with only a local electron interaction. The weak-coupling regime is governed by a Hartree-Fock mean field and Fermi-liquid theories. Extended systems at low temperatures are Pauli paramagnets with smeared out local magnetic moments. For bipartite lattices antiferromagnetic long-range order sets in at half filling at arbitrarily small interaction. In the strong-coupling regime the Hubbard model at half filling maps onto a Heisenberg antiferromagnet with pronounced local magnetic moments and the Curie-Weiss law for the staggered magnetic susceptibility. The spectral structure is dominated by separated lower and upper Hubbard bands and the strongly correlated system seems insulating even in the paramagnetic phase. The most difficult task of the model calculations is to address the transition region between the weak and strong coupling limits.

We present in this paper a theoretical approach based on many-body perturbation theory with sophisticated renormalizations of one as well as two-particle Green functions (propagators). In particular we show how the complicated algebra of two-particle multiple scatterings can be simplified to a manageable form. The presented single-band theory can be extended to the multi-band case and can hence be used for a more realistic electron-structure calculations.

## TRANSITION FROM WEAK TO STRONG COUPLING IN THE HUBBARD MODEL

We have at our disposal relatively reliable techniques to describe the two extreme limits of weak and strong couplings of the Hubbard model. We can use perturbation expansions in each of the limits. In the former case it is the expansion in the interaction strength while in the latter in the kinetic energy (hopping matrix). The problem is that the two expansions do not match each other and break down before the transition regime is reached. The effective Coulomb repulsion becomes comparable with the kinetic energy in the transition region and there is no apparent small parameter at intermediate coupling.

Dynamical fluctuations control the low-temperature physics of interacting electrons at intermediate coupling and neither weak-coupling nor atomic-like perturbation theories are adequate. Weak and strong-coupling perturbation expansions with the bare interaction can no longer reflect the relevant physics with creation and annihilation of long-living electron-hole pairs. The electron-hole pairs start to condense and lead to metal-insulator and magnetic transitions generating poles in appropriate two-particle Green functions.

The failure of the existing analytic-numerical methods to describe properly the transition between weak and strong coupling lies in an inadequate treatment of the two-particle criticality. While the one-particle functions are usually renormalized due to a self-consistency in the self-energy, the coupling constant and two-particle vertices remain unrenormalized. However, the two-particle criticality demands a theory with a dynamical renormalization of the relevant two-particle Green functions.

To succeed in a quantitative description of this complex situation, it is necessary to reformulate perturbation theory in such a way that the bare interaction be systematically replaced with fully renormalized two-particle functions determined from multiple two-particle scattering processes. It means that we have to use a perturbation theory at the level of two-particle Green functions including dynamical vertex renormalizations.

## PARQUET DIAGRAMS

The most advanced approximation for two-particle functions are the so-called parquet diagrams. Parquet diagrams were introduced to describe interaction of mesons more effectively [1]. Since then a number of attempts have been made to utilize the nontrivial renormalization scheme of the parquet algebra in condensed matter. Kondo effect [2], x-ray edge problem [3], formation of the local magnetic moment [4] are among the most well known applications. Inability to solve the parquet equations effectively has impeded broader application of the method.

Parquet diagrams represent a systematic way of a summation and renormalization of Feynman graphs for two-particle Green functions. Instead of concentrating on the one-particle irreducible diagrams and the Dyson equation, the parquet approach sums two-particle diagrams contributing to vertex functions for which Bethe-Salpeter equations are constructed. The resulting algebra is much more complicated than that of the one-particle approximations. There are namely three topologically inequivalent definitions of a two-particle irreducibility. It may be defined according to cutting an electron-hole or electron-electron pair propagation, or according to cutting a polarization bubble shielding the electron-electron interaction. Each possibility defines a two-particle channel of multiple pair scatterings characterized by a different binding of independent variables in the vertex functions.

A general two-particle quantity will be denoted in this paper as in Fig. 1. Each

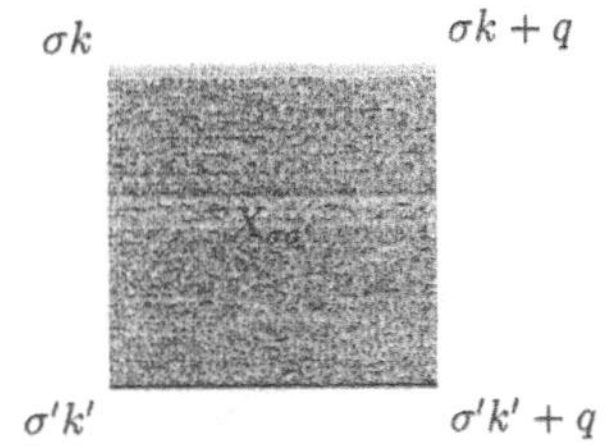


Figure 1. Generic two-particle function with three independent four-momenta and a defined order of incoming and outgoing fermions.

two-particle function carries three independent four-momenta and two spin indices. We generally denote the fermionic four-momenta $k = (\mathbf{k}, i\omega_n)$ and the transferred bosonic ones as $q = (\mathbf{q}, i\nu_m)$, where $\omega_n = (2n+1)\pi T$ and $\nu_m = 2m\pi T$ are the respective Matsubara frequencies at temperature $T = \beta^{-1}$.

It is convenient to introduce a matrix notation in the spin indices to distinguish different two-particle channels in the parquet diagrams. We define a $2 \times 2$ matrix for the generic two-particle function $X_{\sigma\sigma'}$:

$$\widehat{X} = \begin{pmatrix} X_{\uparrow\uparrow} & X_{\uparrow\downarrow} \\ X_{\downarrow\uparrow} & X_{\downarrow\downarrow} \end{pmatrix} . \tag{1}$$

We speak about singlet and triplet contributions to a two-particle function if the spins of the involved fermions are antiparallel or parallel, respectively.

We define three matrix multiplication schemes for the two-particle quantities

$$\left[\widehat{X} \bullet \widehat{Y}\right]_{\sigma\sigma'}(k, k'; q) = \frac{1}{\beta N} \sum_{q''} X_{\sigma\sigma'}(k, k'; q'') Y_{\sigma\sigma'}(k + q'', k' + q''; q - q''), \tag{2a}$$

$$\left[\widehat{X} \circ \widehat{Y}\right]_{\sigma\sigma'}(k, k'; q) = \frac{1}{\beta\mathcal{N}} \sum_{q''} X_{\sigma\sigma'}(k, k' + q''; q - q'') Y_{\sigma\sigma'}(k + q - q'', k'; q''), \quad (2b)$$

$$\left[\widehat{X} \star \widehat{Y}\right]_{\sigma\sigma'}(k, k'; q) = \frac{1}{\beta\mathcal{N}} \sum_{\sigma'' k''} X_{\sigma\sigma''}(k, k''; q) Y_{\sigma''\sigma'}(k'', k'; q) \quad (2c)$$

representing summations over the intermediate states in the three inequivalent two-particle channels, electron-hole, $(eh)$, electron-electron, $(ee)$, and interaction, $(U)$, respectively. We see that the variables of the two-particle functions are convoluted differently in inequivalent channels. Note that only the interaction channel mixes the singlet and triplet contributions.

We decompose the full two-particle Green function into a sum of always reducible and irreducible projections onto each inequivalent channel

$$\mathcal{K}_{\sigma\sigma'}(k, k'; q) = \mathcal{K}^{\alpha}_{\sigma\sigma'}(k, k'; q) + I^{\alpha}_{\sigma\sigma'}(k, k'; q) \quad (3)$$

where $\alpha = eh, ee, U$ refers to a two-particle channel.

The parquet diagrams can at best be represented graphically. Having in mind the above introduced notation and a general convention that double primed variables are summed over, we can write

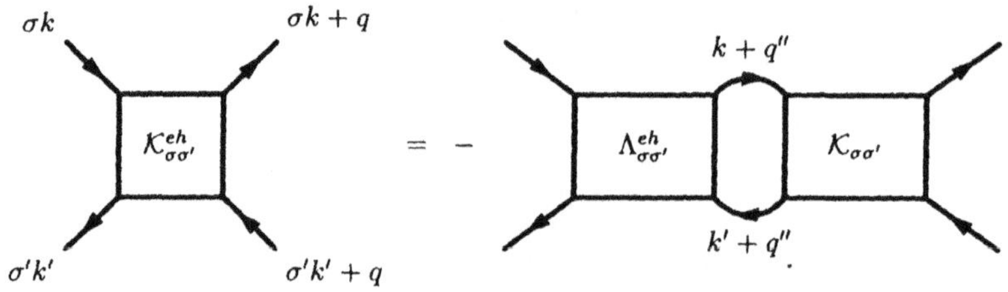


(4a)

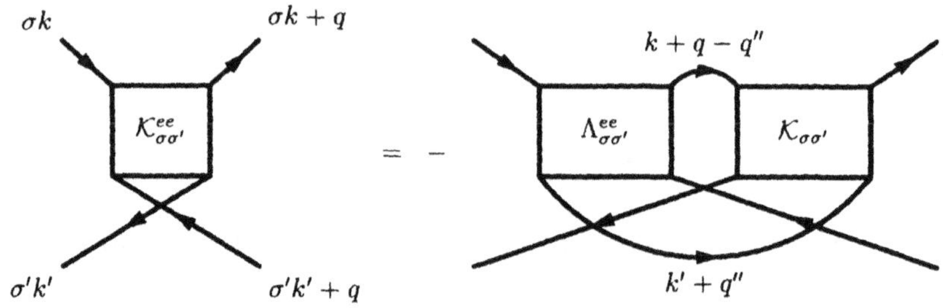


(4b)

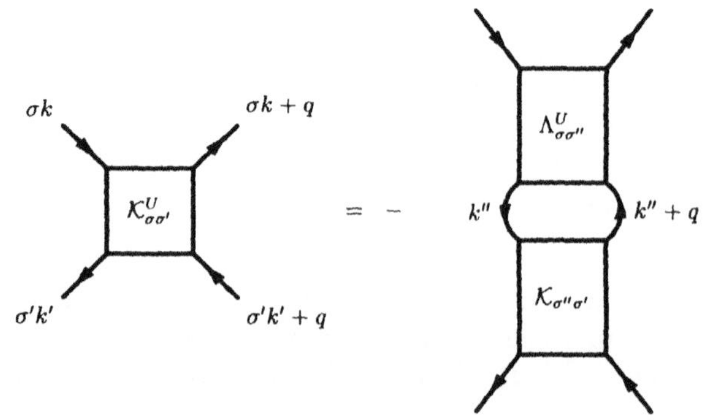


(4c)

Equations (4) are the Bethe-Salpeter equations and represent one set of the full parquet approximation. To complete it we must add relations connecting the two-particle reducible, $\mathcal{K}^\alpha$, and irreducible, $\Lambda^\alpha$, functions. For this purpose we introduced two-particle bubbles for each channel

$$\begin{aligned} G^{(2)eh}_{\sigma\sigma'}(k,k';q) &= G_\sigma(k+q)G_{\sigma'}(k'+q)\ ,\ G^{(2)ee}_{\sigma\sigma'}(k,k';q) = G_\sigma(k+q)G_{\sigma'}(k')\ , && (5a)\\ G^{(2)U}_{\sigma\sigma'}(k,k';q) &= G_{\sigma'}(k')G_{\sigma'}(k'+q)\ . && (5b)\end{aligned}$$

The functions $\Lambda^\alpha$ are then defined as

$$\begin{aligned}\Lambda^\alpha_{\sigma\sigma'}(k,k';q) &= I^\alpha_{\sigma\sigma'}(k,k';q)G^{(2)\alpha}_{\sigma\sigma'}(k,k';q)\\ &= \left[U\delta_{\sigma',-\sigma} + \Delta I_{\sigma\sigma'}(k,k';q) + \sum_{\alpha''\neq\alpha}\mathcal{K}^{\alpha''}_{\sigma\sigma'}(k,k';q)\right]G^{(2)\alpha}_{\sigma\sigma'}(k,k';q) \quad (6)\end{aligned}$$

where $\Delta I_{\sigma\sigma'}$ is a sum of all Feynman diagrams simultaneously irreducible in each two-particle channel. The sum contains only higher-order diagrams where three and more particles are multiply interconnected. In the usual treatment with only two-particle multiple scatterings this irreducible part is neglected. We hence put $\Delta I_{\sigma\sigma'} = 0$.

Parquet equations (4) define the two-particle functions as a functional of the renormalized one-particle propagators containing the self-energy. The self-energy is a functional of the full two-particle function $\mathcal{K}$ from the parquet equations. Hence to complete the parquet approximation we have to add formulas determining the dependence of the one-particle propagator on the self-energy (Dyson equation) and the self-energy on the full two-particle function. The consistency between one and two-particle quantities demands

$$\begin{aligned} G_\sigma(\mathbf{k},i\omega_n) &= \left[i\omega_n + \mu_\sigma - \epsilon(\mathbf{k}) - Un_{-\sigma} - \Sigma_\sigma(\mathbf{k},i\omega_n)\right]^{-1}\ , && (7)\\ \Sigma_\sigma(k) &= -\frac{U}{\beta^2\mathcal{N}^2}\sum_{\sigma'}\sum_{k'q}G_\sigma(k+q)\mathcal{K}_{\sigma\sigma'}(k+q,k';-q)G_{\sigma'}(k')G_{\sigma'}(k'+q) && (8)\end{aligned}$$

where $n_\sigma$ is the particle density, $\mu_\sigma = \mu + \sigma B$, and $\mu$ is the chemical potential and $B$ an external magnetic field. [1]

Equations (3)-(8) constitute the parquet approximation. In this approximation only fully renormalized one and two-particle functions appear. It means that not only all one-particle propagators are fully renormalized with the self-energy insertions but also the bare electron-electron interaction in the two-particle scatterings is replaced by corresponding irreducible two-particle (vertex) functions. What is missing in the parquet diagrams are three-particle and higher-order cumulant functions. They are, however, not expected to be relevant for the static Coulomb interaction. They contribute via dipole and higher-order multi-pole interactions and are weak in the Hubbard model.

## SIMPLIFIED PARQUET ALGEBRAS

Parquet diagrams differ significantly from simpler, single-channel approximations using multiple two-particle scatterings such as *RPA*, *GWA* or *T*-matrix (TMA). The difference manifest itself in the effective interaction in the two-particle scattering processes. The bare interaction in the two-particle scattering events remains unrenormalized in the single-channel approximations. In the parquet equations it is replaced by renormalized vertex functions representing a dynamically screened interparticle potential. A qualitative difference between a solution to the full parquet equations and to the

[1]The self-energy $\Sigma$ in our notation measures corrections to the static Hartree approximation contained in the particle densities $n_\sigma$ determined from a sum rule $n_\sigma = (\beta\mathcal{N})^{-1}\sum_k G_\sigma(k)$.

single-channel approximations becomes evident at the two-particle criticality. The vertex functions from one or more channels get critical and sharply peaked around the Fermi energy. The exchange between the scattered particles becomes strongly energy dependent and cannot be approximated by a static effective interaction. Moreover, the singularity at the critical point in the parquet approximation must be integrable, while the single-channel approximations usually display logarithmic divergences in the integrated quantities (e. g. self-energy).

On the other hand the nonlinear integral equations represented by the parquet diagrams are extremely complicated with an intricate analytic structure of the solution. Each two-particle function entering the parquet equations contains three independent (complex) variables. Neither of them can simply be neglected, since the variables are mutually interconnected due to the scatterings in different inequivalent channels. Consequently, a nonperturbative solution to the parquet equations has not yet been found. That is also why the parquet diagrams are relatively little used in condensed matter although being in existence more than thirty years.

The only hope that the parquet approximation may become useful for quantitative calculations of the effects of the electron correlations is that one succeeds in simplifying the full parquet algebra to a manageable form. To this purpose we have to decide where we want the parquet approximation to be reliable. As discussed in the Introduction it is the intermediate coupling with critical two-particle functions where we need to improve upon easy-minded approximations. We hence demand asymptotic reliability of the parquet diagrams at the two-particle critical points and in particular at the metal-insulator transition. We retain only the diagrams making the effective mass of the electrons divergent at the metal-insulator transition.

## Two-Variable Two-Particle Functions

It was recently argued that at least at half filling only two of the three inequivalent channels are relevant in the critical region of the metal-insulator transition [5]. Only multiple scatterings in the electron-hole and interaction channels contribute to divergent diagrams at the critical point. This simplification, called dipole approximation, however, does not reduce the number of the variables in the two-particle functions. It only simplifies the algebra of the parquet approximation. To lower the number of the relevant variables we utilize the fact that the singularities in the two-particle functions in the parquet approximation must be integrable. It means that when integrating over the variable in which the singularity arises we obtain a finite result. Our approximation consists in neglecting all finite (nondiverging) contributions at the critical point and keeping only the divergent ones. This must be done at the level of the diverging two-particle functions.

There is always one relevant bosonic variable in each two-particle channel. It is the variable remaining untouched in the scattering processes. It can easily be seen from multiplication rule (2) that it is $k - k'$, $k + k' + q$, and $q$ for the electron-hole, electron-electron, and interaction channels, respectively. Divergences in the reducible functions appear in the respective conserved variables. For the metal-insulator transition only $k - k'$ and $q$ are relevant.

In the first step we reduce the number of the relevant variables in the two-particle functions to two. This is achieved when we neglect mixing of the reducible functions from different channels. We replace the full two-particle function $\mathcal{K}$ on the right-hand side of (4) by $U + \mathcal{K}^{\alpha}$, i. e. by the sum of the completely irreducible vertex and the reducible function in the $\alpha$-channel. Such a replacement does not change the (universal) critical behavior of the resultant reducible function $\mathcal{K}^{\alpha}$. The bare interaction is a constant and consequently the solution does not depend on the outgoing variables. Further on, the triplet functions can be eliminated from this approximation, since $I_{\sigma,\sigma}$ gets irrelevant. If we denote $I^v_{\sigma,-\sigma} = U + \mathcal{K}^{eh}_{\sigma,-\sigma}$ and $I^h_{\sigma,-\sigma} = U + \mathcal{K}^{U}_{\sigma,-\sigma}$, the parquet equations reduce to

$$
\begin{aligned}
I^v_{\sigma,-\sigma}(k;q) &= U \\
&\quad - \frac{1}{\beta\mathcal{N}} \sum_{q'} I^h_{\sigma,-\sigma}(k;q') G_{\sigma}(k+q') G_{-\sigma}(k+q'+q) I^v_{\sigma,-\sigma}(k+q';q) \,, \qquad (9)
\end{aligned}
$$

$$
\begin{aligned}
I^h_{\sigma,-\sigma}(k;q) &= U + \frac{1}{\beta^2\mathcal{N}^2} \sum_{q',q''} I^v_{\sigma,-\sigma}(k;q') G_{-\sigma}(k+q') G_{-\sigma}(k+q'+q) \\
&\quad \times I^v_{\sigma,-\sigma}(k+q';q''-q') G_{\sigma}(k+q'') G_{\sigma}(k+q''+q) I^h_{\sigma,-\sigma}(k+q'';q) \,. \qquad (10)
\end{aligned}
$$

The bosonic variable $q$ has different meaning in the functions $I^h$ and $I^v$. It is the horizontal, vertical transfer momentum for the former and the latter, respectively.

Because of breaking of the symmetry between the incoming and outgoing variables in the two-particle functions we must be careful in deriving the equation for the self-energy. To be consistent we use a symmetrized formula with the bare interaction at the incoming and outgoing momentum. We obtain

$$
\begin{aligned}
\Sigma_{\sigma}(k) &= -\frac{U}{2\beta^2\mathcal{N}^2} \sum_{k',q} G_{\sigma}(k+q) G_{-\sigma}(k') G_{-\sigma}(k'+q) \Big[ I^h_{\sigma,-\sigma}(k+q;-q) \\
&\quad + I^h_{-\sigma,\sigma}(k';q) + I^v_{\sigma,-\sigma}(k+q';k'-k) + I^v_{-\sigma,\sigma}(k';k-k') - 2U \Big] \,. \qquad (11)
\end{aligned}
$$

The reduced two-channel parquet approximation with two-variable two-particle functions has the critical behavior of the full parquet approximation. The universal quantities derived from the divergent functions of the parquet equations (4) are reproduced in equations (9)-(11). Only nonuniversal quantities such as the critical interaction does not come out in this simplification as in the full parquet approximation.

We achieved a simplification of the full parquet algebra without loosing leading critical behavior of the solution, but the resulting equations have nonlinear convolutive character. We have to find a nonperturbative solution to these equations when addressing the metal-insulator transition. This is, however, still beyond the reach of our analytic-numerical tools. We must find a further reduction of complexity of the parquet equations to make them useful for a quantitative analysis.

## One-Variable Two-Particle Functions

It is the bosonic variable $q$ in the two-particle functions $I^h, I^v$ that is relevant at the critical point and in which the singularity arises. The fermionic variable $k$ in both functions labels the eigenvalues of the integral kernel in equations (9), (10). The maximal eigenvalue governs the critical behavior. If we are interested only in the critical asymptotic behavior we can approximate the actual maximal eigenvalue by a value at the lowest-lying fermionic four-momentum. By such an approximation we neglect mixing of the fermionic four-momenta in the equations and replace $k$-dependence in the two-particle functions by a constant, their value at $k = 0$. This approach corresponds to a low-energy expansion used by Hamann [6] by assessing the strong-coupling limit of the renormalized RPA of Suhl in the single-impurity problem.

The suggested simplification is possible only at zero temperature, where the difference between the fermionic and bosonic Matsubara frequencies vanishes. We hence put

$T = 0$, which is the most interesting case for the metal-insulator transition. Using the above ansatz, neglecting $k$-dependence in $I^h, I^v$, and putting $k = 0$ in the one-electron propagators we reduce the parquet equations to a manageable form. We introduce new functions

$$\Gamma_\sigma(q) = \frac{1}{\beta\mathcal{N}} \sum_{q'} I^h_{\sigma,-\sigma}(q') G_\sigma(q') G_{-\sigma}(q'+q) \tag{12}$$

$$\mathcal{K}_\sigma(q) = \frac{1}{\beta\mathcal{N}} \sum_{q'} I^v_{-\sigma,\sigma}(q') G_\sigma(q') G_\sigma(q'+q) \ , \tag{13}$$

with the aid of which we obtain for the two-particle functions

$$I^v_{\sigma,-\sigma}(q) = \frac{U}{1+\Gamma_\sigma(q)} \ , \tag{14}$$

$$I^h_{\sigma,-\sigma}(q) = \frac{U}{1-\mathcal{K}_{-\sigma}(q)\mathcal{K}_\sigma(q)} \ . \tag{15}$$

The above equations have the complexity of the single-channel approximations, but contain the full vertex renormalization in the two-particle scatterings. It is always the renormalized vertex function that enters the denominator of equations (14), (15) via the functions $\Gamma$ and $\mathcal{K}$. The single channel approximations replace the two-particle vertex functions $I^h, I^v$ in (12), (13) by the bare interaction $U$.

It is easy to derive the corresponding equation for the self-energy. It reads

$$\begin{aligned}\Sigma_\sigma(k) = & -\frac{U}{2\beta\mathcal{N}} \sum_q \Big[ G_\sigma(k+q) X_{-\sigma,-\sigma}(q) \left( I^h_{-\sigma,\sigma}(q) + I^h_{\sigma,-\sigma}(-q) - U \right) \\ & + G_{-\sigma}(k+q) X_{\sigma,-\sigma}(q) \left( I^v_{\sigma,-\sigma}(q) + I^v_{-\sigma,\sigma}(-q) - U \right) \Big]\end{aligned} \tag{16}$$

where we introduced a single bubble

$$X_{\sigma,\sigma'}(q) = \frac{1}{\beta\mathcal{N}} \sum_{k'} G_\sigma(k') G_{\sigma'}(k'+q) \ . \tag{17}$$

Equations (12)-(17) replace the original parquet approximation.[2] They fully determine the generating two- and one-particle functions. All thermodynamic and spectral quantities can be calculated from them. Since we used a number of approximate steps in the derivation of the reduced parquet algebra, any new quantity must first be formulated within the full parquet approximation. The simplifications are then introduced via the generating one- and two-particle functions. In this way a consistent theory is built up, since the parquet approximation can be viewed upon as a conserving approximation derived from a generating functional [7].

## Electron-Hole Symmetry

An exact solution as well as the full parquet approximation fulfill a number of identities and symmetries. One of the most important features of the exact solution is the electron-hole symmetry. It means, that the physics must not depend on whether we use the particle or the hole picture. The electron-hole symmetry is obeyed by the complete

[2]There is a different way how to reduce the parquet approximation to single-variable equations [2]. The other reduction is based on the so-called leading logarithmic divergences in single-loop diagrams. This approach is inappropriate in the strong-coupling limit, where such divergences cannot survive because of the integrability of the singularities in the parquet equations.

parquet approximation. It is not clear whether we have not lost some of the required symmetries in course of simplifying the parquet algebra.

We lost the symmetry between the incoming and outgoing variables in our approximate equations. Further on we also lost the electron-hole symmetry in only one particle of the pair in the two-particle scatterings. The former was lost during the reduction to two-variable two-particle functions and the latter due to the suppression of the electron-electron channel. These losts are not so serious if we are in the critical region of the metal-insulator transition. The one-sided electron-hole transformation changes the sign of the effective interaction and hence breaks the bound electron-hole pair.

The full electron-hole symmetry in all variables must hold for any approximate theory, otherwise we were unable to guarantee the Fermi liquid properties at weak coupling. The electron-hole transformation may be defined in the parameter space of the Hubbard model as

$$\mu \to \bar{\mu} = U - \mu \ , \quad B \to \bar{B} = -B \ , \quad \epsilon_{\mathbf{k}} \to \bar{\epsilon}_{\mathbf{k}} = -\epsilon_{-\mathbf{k}} \tag{18}$$

where the bar denotes a hole representation. It is easy to check by inspection that a solution to the reduced parquet equations (12)-(17) fulfill the following symmetry relations

$$G_\sigma(k) = -\bar{G}_\sigma(-k) \ , \quad \Sigma_\sigma(k) = -\bar{\Sigma}_\sigma(-k) \ , \quad n_\sigma = 1 - \bar{n}_\sigma \tag{19}$$

$$\Gamma_\sigma(q) = \bar{\Gamma}_\sigma(-q) \ , \quad \mathcal{K}_\sigma(q) = \bar{\mathcal{K}}_\sigma(-q) \ . \tag{20}$$

This electron-hole symmetry together with the analyticity of the one- and two-electron functions is used to prove Fermi-liquid properties of solutions to (12)-(17) for sufficiently weak interaction, i. e. before poles in the vertex functions $I^v, I^h$ appear.

## EFFECTIVE IMPURITY MODEL

Up to know we have worked with a very general formulation of the parquet approximation applicable in any spatial dimension. However, recent studies indicate that most of the effects of strong electron correlations can qualitatively be understood already within a dynamical mean-field theory or impurity models [8]. It is sufficient in many situations to reduce the effects of strong correlations to a dynamical mean-field where the self-energy and vertex functions are local, but frequency-dependent. Such an approach significantly reduces the numerical complexity but keeps much of the interesting physics of strongly correlated electrons.

We apply the simplified parquet approximation to a single impurity or a dynamical mean field model. The fluctuations in space do not contribute to the dynamics and the four-momenta from the general formulation of the parquet approximation collapse to frequencies. This simplification enables one to perform explicitly analytic continuation from the Matsubara to the real frequencies using contour integrals. Realizing that the reduced equations hold for $T = 0$ and introducing $\zeta, z$ for the bosonic, fermionic complex frequencies, respectively, we can write for the two-particle functions

$$\mathcal{K}_\sigma(\zeta) = -U \int_{-\infty}^{0} \frac{d\omega}{\pi} \left[ G_\sigma(\omega + \zeta) \mathrm{Im} \frac{G_\sigma(\omega_+)}{1 + \Gamma_{-\sigma}(\omega_+)} + \frac{G_\sigma(\omega - \zeta)}{1 + \Gamma_{-\sigma}(\omega - \zeta)} \mathrm{Im} G_\sigma(\omega_+) \right] , \tag{21}$$

$$\begin{aligned} \Gamma_\sigma(\zeta) = & -U \int_{-\infty}^{0} \frac{d\omega}{\pi} \left[ G_{-\sigma}(\omega + \zeta) \mathrm{Im} \frac{G_\sigma(\omega_+)}{1 - \mathcal{K}_\sigma(\omega_+)\mathcal{K}_{-\sigma}(\omega_+)} \right. \\ & \left. + \frac{G_\sigma(\omega - \zeta)}{1 - \mathcal{K}_\sigma(\omega - \zeta)\mathcal{K}_{-\sigma}(\omega - \zeta)} \mathrm{Im} G_{-\sigma}(\omega_+) \right] \end{aligned} \tag{22}$$

and analogously for the one-particle self-energy

$$\Sigma_\sigma(z) = U\int_{-\infty}^{0}\frac{d\omega}{\pi}\Big\{G_\sigma(\omega+z)\mathrm{Im}\left[X_{-\sigma,-\sigma}(\omega_+)\left(I^h_{-\sigma,\sigma}(\omega_+)-U\right)\right]$$
$$+X_{-\sigma,-\sigma}(\omega-z)\left(I^h_{-\sigma,\sigma}(\omega_+)-U\right)\mathrm{Im}G_\sigma(\omega_+)+G_{-\sigma}(\omega+z)\mathrm{Im}\left[X_{\sigma,-\sigma}(\omega_+)I^v_{\sigma,-\sigma}(\omega_+)\right]$$
$$+X_{\sigma,-\sigma}(\omega-z)I^v_{\sigma,-\sigma}(\omega-z)\mathrm{Im}G_{-\sigma}(\omega_+)\Big\} \tag{23}$$

where $\omega_+ = \omega + i0^+$. The local propagator is for the mean field

$$G_\sigma(z) = \int_{-\infty}^{\infty} d\epsilon\rho(\epsilon)\frac{1}{z+E_\sigma-\Sigma_\sigma(z)-\epsilon} \tag{24}$$

and for the impurity model

$$G_\sigma(z) = \left[z+E_\sigma-V^2g(z+\sigma B)-\Sigma_\sigma(z)\right]^{-1} \tag{25}$$

where $E_\sigma = \mu+\sigma B - Un_{-\sigma}$ for the mean-field and $E_\sigma = \mu + \sigma B - \varepsilon$ for the impurity model. In the former $\rho(\epsilon)$ is the density of states and in the latter, the parameter $\varepsilon$ is the energy of the impurity orbital, $V$ hybridization to the conduction electrons, and $g(z)$ the one-electron local Green function of the conduction electrons. Note that there is no difference at the two-particle level between the impurity and mean-field equations.

## Intermediate Coupling

We use the local formulation of the simplified parquet approximation to demonstrate that intermediate coupling is dominated by a two-particle criticality. We confine ourselves only to half filling and the spin symmetric case (paramagnet) where the singularity indicates a metal-insulator transition. We further assume a bipartite lattice, i. e. $\rho(\epsilon) = \rho(-\epsilon)$.

A solution to equations (21)-(23) has at weak coupling the required Fermi-liquid properties. The two-particle functions $X, \Gamma, \mathcal{K}$ are real at the Fermi energy. Iterations starting with $U = 0$ converge to a unique solution. Such a procedure becomes numerically unstable if we increase the interaction beyond $U_{c0}$ defined as

$$U_{c0} = -\left[X_0(0)\right]^{-1}, \quad X_0(0) = -\int_{-\infty}^{0}\frac{d\omega}{\pi}\,\mathrm{Im}\left[G^{(0)}(\omega_+)^2\right] \tag{26}$$

where $G^{(0)}$ is a noninteracting one-particle propagator. The value $U_{c0}$ corresponds to the critical interaction determined from the $RPA$ pole in the electron-hole susceptibility. Explicitly, $U_{c0} = V^2\pi^2\rho(0)$ for a mean-field ( $d=\infty$) model with a Lorentzian DOS (infinite bandwidth) and $U_{c0} = 3\pi w/4$ for a Bethe $d=\infty$ lattice (finite bandwidth). This interaction need not necessarily represent a critical point. It only indicates that we cannot start iterations from an unperturbed solution and that weak-coupling perturbation theories get numerically unstable. We can say that the weak-coupling regime goes over at $U_{c0}$ to an intermediate one. The transition is, however, smooth for effective impurity models.

At intermediate coupling the functions $I^v$ and $I^h$ approach a pole at the Fermi energy $\zeta = 0$. It means that $1+\Gamma(0)$ as well as $1-\mathcal{K}(0)^2$ approach zero. A new small (dimensionless) scale $\Delta = 1+\Gamma(0) \approx 1-\mathcal{K}(0)^2$ arises. This scale cannot be derived from the one-electron model parameters. It is a consequence of a two-particle criticality. In the single-impurity case it is related to the Kondo scale. The scale vanishes at the metal-insulator transition if it is of second order.

The vanishing scale $\Delta$ causes nonanalyticities in the two-particle functions $\Gamma(\omega)$ and $\mathcal{K}(\omega)$, and in the one-electron self-energy $\Sigma(\omega)$. Since the singularity in the vertex functions $I^v(\omega), I^h(\omega)$ must be integrable, the functions $\Gamma, \mathcal{K}, \Sigma$ remain bounded (continuous). Only their derivatives show divergences at the Fermi energy at the critical point. It is straightforward to show that leading-order divergences in (21)-(23) are

$$\text{Im } \mathcal{K}'(0_+) \doteq -\frac{U}{\pi} \frac{[\text{Im } G(0_+)]^2}{1+\Gamma(0)}, \quad \text{Im } \Gamma'(0_+) \doteq -\frac{U}{\pi} \frac{[\text{Im } G(0_+)]^2}{1-\mathcal{K}(0)^2}, \tag{27}$$

$$\text{Re } \Sigma'(0) \doteq -\frac{U}{\pi} UX(0) \text{ Im } G(0_+) \left[\frac{1}{1+\Gamma(0)} + \frac{1}{1-\mathcal{K}(0)^2}\right] \tag{28}$$

where the prime denotes the derivative w.r.t. the frequency variable. We utilized the symmetry of the half-filled case, Re $G(0) = 0$ in the derivation of (27), (28) .

The Mott-Hubbard metal-insulator transition is defined by divergence of the effective mass [9]. Since $m^*/m \propto |\Sigma'(0)|$ we have proved that the singularity in the two-particle vertex functions $I^v, I^h$ causes divergence of the effective electron mass and hence the Mott-Hubbard metal-insulator transition. The way the effective electron mass diverges is essentially determined by vanishing of the Kondo scale $\Delta$. It means that the Kondo scale sets an interval around the Fermi energy within which the Landau quasiparticle picture holds not only for the one-electron but also for the two-electron functions. The narrow quasiparticle Kondo-Suhl peak is then a consequence of sharply peaked two-particle vertex functions. No effective static interaction can describe the Kondo strong-coupling asymptotics and the Mott-Hubbard metal-insulator transition with (27), (28) reliably.

The simplified parquet approximation describes qualitatively well the weak-coupling Fermi-liquid regime. It also correctly binds divergence of the effective mass with the Kondo scale arising from a singularity in two-particle functions. The derived approximate theory does well on the metallic side of the metal-insulator transition. Whether the approximation can correctly capture the physics up to the transition point must be decided from the dependence of the Kondo scale $\Delta$ onto the interaction strength $U$ or better on the parameter $1 - U/U_c$. The correct behavior must reproduce the exponential law $\Delta \propto \exp\{-U\rho(0)\}$ for the single-impurity model. We need a deeper analysis of equations (21)-(23) to decide whether this really is the case. It will be done in a separate paper.

Next we have to address the strong-coupling regime for $U > U_c$. Fermi-liquid theory no longer holds and it is unclear how to iterate towards a stable solution. The problem is to choose the value of the self-energy at the Fermi level. It vanishes in the Fermi-liquid, metallic phase but diverges in the atomic, insulating solution. What value the self-energy acquires in a self-consistent insulating solution to (21)-(23) needs to be investigated in more details. One can, however, easily check that the simplified parquet approximation allows for a consistent $U \to \infty$ limit. After an appropriate scaling of the variables one recovers an atomic solution at zero temperature. Dynamical fluctuations then enter as perturbations in $w/U$, where $w$ is an effective bandwidth of the conduction electrons. A perturbation expansion in $w/U$ can be a good starting point for iterations to the full solution of equations (21)-(23) in the strong-coupling, insulating phase.

## CONCLUSIONS

We demonstrated in this paper that a successful description of intermediate and strong coupling regimes in correlated electron systems demands sophisticated approximations with a dynamical renormalization of two-particle vertex functions. A transition regime between the weak and strong coupling limits is governed by the formation and condensation of long-living electron-hole pairs resulting in poles in two-particle Green functions. These dynamical effects cannot properly be rendered by theories with only effective static interactions. The well-known single-channel two-particle approximations such as RPA, TMA, or GWA are inadequate at intermediate coupling, since the interaction in the multiple scatterings remains unrenormalized. Only the full parquet approximation with all two-particle irreducible diagrams can be expected to capture the relevant physics at the two-particle criticality of intermediate coupling.

The parquet diagrams lead to a qualitatively different critical behavior than that obtained from the simple-minded single-channel approximations. The only singularities that can appear in the parquet approximation are integrable ones, i. e. convolutions of singular and regular functions must be finite. This feature distinguishes the parquet theory from the other existing approximations in correlated electron systems.

The parquet algebra is too intricate to allow for a nonperturbative solution. Only an appropriate reduction in the complexity of the parquet approximation can lead to a successful application of the method. We proposed in this paper a scheme how to reduce the nonlinear convolutive integral parquet equations to a system of quasi-algebraic equations not more complicated than the single-channel approximations. The only difference is that the effective interaction in the scattering events in one channel is dynamically renormalized by scatterings from other channels.

The reduction of the parquet algebra is enabled by taking into account only leading-order divergences at the metal-insulator transition and a low-energy expansion around the maximal eigenvalue of the parquet equations. The resulting equations allow only for integrable singularities and lead to sharply peaked two-particle vertex functions in the vicinity of the critical metal-insulator transition. They correctly bind divergence of the effective one-electron mass with the Kondo scale stemming from a two-particle singularity. What still remains to do is to show how well the simplified parquet approximation reproduces the strong-coupling asymptotics of the Kondo scale in the single-impurity problem known from an exact solution. Only then we can be sure we have a reliable interpolation between the weak and strong-coupling regimes of Hubbard-like models.

### ACKNOWLEDGMENT

This work was supported in part by the Grant No. 202/98/1290 of the Grant Agency of the Czech Republic.

## References

[1] V. V. Sudakov, Dokl. Akad. Nauk SSSR **111**, 338 (1956), [English transl. Soviet Phys. - Doklady **1**, 662 (1957)].

[2] A. A. Abrikosov, Physics **2**, 5 (1964).

[3] B. Roulet, J. Gavoret, and P. Nozières, Phys. Rev. **178**, 1078 (1969).

[4] R. A. Weiner, Phys. Rev. Lett. **24**, 1071 (1970).

[5] V. Janiš, J. Phys.: Condens. Matter **10**, 2915 (1998).

[6] D. R. Hamann, Phys. Rev. **186**, 549 (1969).

[7] V. Janiš, Preprint *cond-mat/9806118*

[8] A. Georges, G. Kotliar, W. Krauth, and M. Rozenberg, Rev. Mod. Phys. **68**, 13 (1996).

[9] W. F. Brinkman and T. M. Rice, Phys. Rev. B**2**, 4302 (1970).

# CORRELATION EFFECTS ON STABILITY IN Pu METAL AND ITS ALLOYS

P. E. A. Turchi[1], A. Gonis[1], N. Kioussis[2], D. L. Price[3], and B. R. Cooper[4]

[1] Lawrence Livermore National Laboratory (L-353), P.O. Box 808, Livermore, CA 94551
[2] Dept. of Physics and Astronomy, California State University, Northridge, CA 91330
[3] Dept. of Physics, University of Memphis, Memphis, TN 38152
[4] Dept. of Physics, West Virginia University, Morgantown, WV 26506-6315

## INTRODUCTION

The existence of six crystallographic allotropes from room temperature up to the solid-liquid transition just above 913 K at atmospheric pressure makes solid Plutonium unique among the elements in the periodic table. Among these phases (labeled α, β, γ, δ, δ', and ε), the δ phase, stable between 593 K and 736 K, has commanded considerable interest in the metallurgical and solid state communities[1]. In contrast to the low-temperature monoclinic α phase, which is brittle, the face-centered cubic (fcc) δ phase is ductile, a property that makes it convenient for engineering applications. This phase can also be stabilized through alloying with a number of other elements such as Ga, Al, Sc, and Am.

The low-symmetry and small atomic volume of the low-temperature phases of U, Np, and Pu is attributed to the role played by the *f*–electrons in determining the bonding in these materials. The unusually large 24 % volume increase in going from the α to the δ phase of Pu[2,3] has been argued on phenomenological grounds to be the result of decreased *f* bonding. Many attempts have been made to study this transition from the point of view of first-principles electronic structure calculations[4-7].

These calculations have been carried out within the framework of density-functional theory (DFT) and its local-density approximation (LDA), as well as the generalized gradient approximation (GGA)[4] designed to account for non-local effects of the electron density. In spite of these attempts, a fundamental understanding of the electronic structure properties of Pu, and the mechanisms responsible for the δ to α volume collapse is still lacking. As a direct consequence of this lack of understanding, the phase stability properties of Pu alloys are still poorly described from a fundamental standpoint[10-12]. For example[13,14], the solubility limits of Ga in δ and ε Pu (respectively 12.5 and 20 at.%), and the definite tendency toward ordering and phase formation (with the existence of six intermediate phases stable at room temperature) still need to be explained.

This paper briefly reports on a study of the electronic structure of Pu and Pu-Ga alloys that is designed to examine directly the effect of the electron-electron interaction on the equilibrium properties and on stability of these systems. As part of this study, the electronic structure and ground-state properties of Pu and Pu-Ga alloys were calculated using the LDA, a constrained version of the LDA in which the *f*-electrons were confined to core states, and finally the so-called LDA+U method. Although essentially phenomenological, this latter treatment sheds some light on the role of the partially localized *f*-electrons on properties. It also emphasizes the need for an *ab initio* treatment that

*Electron Correlations and Materials Properties*, edited by Gonis *et al.*
Kluwer Academic/Plenum Publishers, 1999. 

accounts not only for spectral properties, but also for equilibrium properties, phase stability, and energetics of correlated metals and alloys.

## METHODOLOGY

### Electronic-structure method

The electronic structure calculations are based on the full-potential linear-muffin-tin-orbital (FLMTO) method[15], which includes non-spherical terms within the muffin-tin spheres. The calculations provided a fully relativistic treatment of electrons in core states, while valence electrons were treated semi-relativistically, but with spin-orbit coupling taken into account. The form proposed by Ceperley and Alder[16] and parametrized by Perdew and Zunger[17] was used for the exchange-correlation potential.

The basis sets include the 6s, 6p, 7s, 7p, 6d, and 5f partial waves for Pu and 3d, 4s, 4p, and 4d for Ga. Three energy panels were characterized by three different $\kappa$ values set to the average kinetic energy in the interstitial region. In the case of Pu-Ga alloys, two of these panels ($\kappa_1$ and $\kappa_2$) were associated with the semi-core states, 6s and 6p, and the third ($\kappa_3$) with the valence states, 7s, 7p, 6d, and 5f of Pu, whereas for Ga $\kappa_2$ and $\kappa_3$ were associated with 3d, and 4s, 4p, and 4d, respectively. The muffin-tin potential and the charge density were expanded in spherical harmonics up to an angular momentum $l$=8. A reciprocal-lattice vector cutoff of $6a_0^{-1}$ (where $a_0$ is the Bohr radius) was used, and the integrations over the Brillouin zone were carried out according to a special k-point method[18]. Specifically, in the irreducible wedge of the Brillouin zone, 56 k-points were used for fcc-based Pu and Ga, and 60 (75) k-points for the $L1_2$ ($L1_0$) ordered structures $Pu_3Ga$ and $PuGa_3$ (PuGa). In these calculations atomic relaxations were ignored and equilibrium properties were obtained by minimizing the total energy with respect to the atomic volume. The muffin-tin radii of Pu and Ga were chosen so that in the pure elemental systems as well as in the compounds the ratio of the interstitial volume over the total volume was kept constant.

### Constrained LDA

It is generally accepted that the unique properties of Pu and its alloys derive from the behavior of the 5*f*-electrons and their hybridization to the other valence states. For example, the more these "localized" electrons contribute to the bonding, the higher the density of the material. To examine this property, we considered the extreme case in which all 5*f*-electrons were constrained to occupy core states with no *f*-symmetry allowed for the valence states.

### The LDA+U method

The so-called LDA+U method is designed to correct in a phenomenological way for the treatment of the Coulomb interaction among the "localized" electrons that is accounted for rather poorly within the LDA. This methodology has been described in detail in the literature[19-21]. In its simplest form, one assumes that the LDA provides a correct treatment of the electron-electron interaction in terms of the total number (N) of electrons, here the 5*f*-electrons. Hence, one can write an expression for the total-energy functional through the addition of a Hubbard-like term with a renormalized Coulomb interaction parameter $U$ (here for simplicity the Coulomb exchange term is dropped along with non-spherical terms) as follows,

$$E = E_{LDA} - \frac{U}{2}N(N-1) + \frac{U}{2}\sum_{i \neq j} n_i n_j ,$$

where $E_{LDA}$ is the usual total electron energy of the ground state, and $n_i$ is the number of localized electrons in state $i$. The value of $U$ can be estimated from constrained LDA-type supercell calculations[22], according to

$$U = E(n_f + 1) + E(n_f - 1) - 2E(n_f).$$

The energy $E(n_f+1)$ $[E(n_f-1)]$ associated with a configuration containing one more (one less) *f* electron is obtained within LDA by considering a supercell based here on a fcc lattice, with one of the simple-cubic sublattices occupied by $n_f+1$ $(n_f-1)$ *f*-electrons (the other valence electrons being undisturbed).

## RESULTS

Figure 1 shows the total densities of states (DOSs) of fcc-based (δ) paramagnetic Pu obtained in the LDA, panel (a), in the constrained LDA, panel (b), and in the LDA+U scheme, panel (c), with a value of $U \approx 3.2$ eV, at the same lattice parameter, a=8.7 a.u. The DOS of the LDA is dominated by the 5*f* contribution. The 5*f* DOS is mainly composed of two peaks associated, respectively, with the $5f_{5/2}$ and $5f_{7/2}$ states, and with a Fermi energy lying in a valley between these two peaks, as has been found in previous calculations[5]. The constrained LDA DOS, with the five 5*f*-electrons treated as core states, consists primarily of contributions of *s*, *p*, *d* character, with the Fermi level located on a peak in the DOS. Finally, the LDA+U yields a DOS in which the *f*-electrons of 5/2 symmetry are partially localized as a result of the presence of *U*. It should be pointed out that experimentally the coefficient of the linear term in the specific heat suggests a high DOS at the Fermi level that is not found in these calculations.

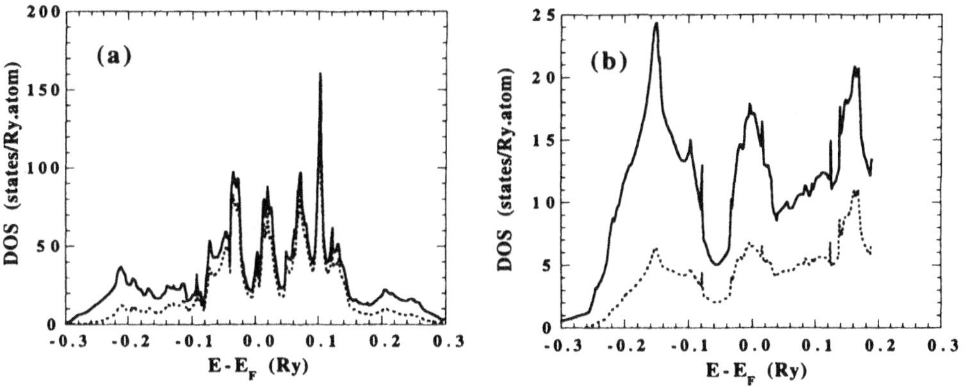


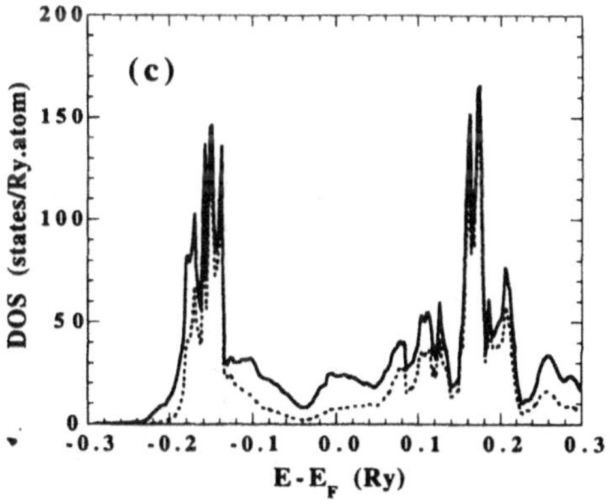


Figure 1. Total DOSs for fcc Pu as calculated within the LDA, panel (a), the constrained LDA described in the text, panel (b), and an implementation of the LDA+U method with $U \approx 3.2$ eV, panel (c). The partial *f*-, *d*-, and *f*-projected DOSs are shown with dotted lines in panels (a), (b), and (c), respectively.

The variation of *U* with the fcc lattice constant is presented in Fig. 2(a). This behavior is compatible with the fact that at small lattice constant the system is expected to behave as a Fermi gas, where the LDA provides an accurate description of the electronic behavior, whereas with increased expansion the *U* approaches a "screened atomic" value. Figure 2(b) shows the behavior of the equilibrium lattice constant as a function of *U*. This is consistent with the expectation that a large $U$ results in the localization of the $f$-electrons and hence a larger atomic volume.

The total energy versus volume for fcc-based Pu, calculated in the three different schemes outlined above, is displayed in Fig. 3. The volume shown is normalized to the LDA result. The usual LDA calculation yields an atomic volume in accordance with previous results[4] that is much smaller than what is observed experimentally. It also leads to a value of the bulk modulus of about 240 GPa, which is almost five times larger than the experimental value. On the other hand, the constrained LDA calculations provides a rather accurate value for the equilibrium volume and improves substantially the calculated bulk modulus, which now equals 96 GPa.

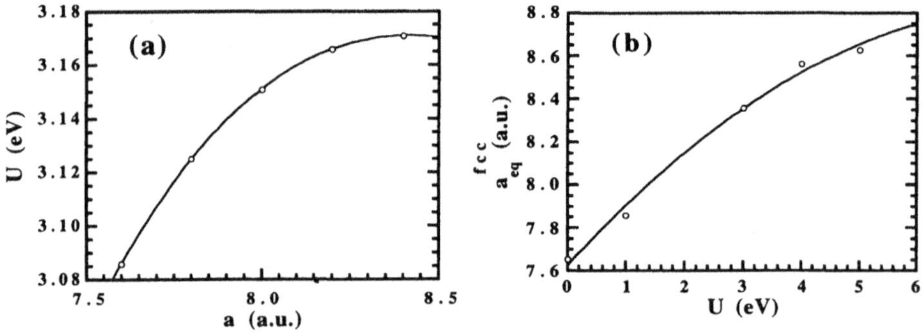


Figure 2. Variation of $U$ with lattice parameter, panel (a), and equilibrium lattice constant as a function of $U$ for fcc-based Pu.

Regarding the LDA+U results, the calculations give values for the equilibrium volume and the bulk modulus which are close to those of the constrained LDA. These calculations were performed by varying $U$ for each lattice parameter according to the behavior exhibited in Fig. 2(a). Although the constrained LDA and the LDA+U results display an evident similarity the former correspond to a higher ground-state energy. Moreover, we note that the LDA+U results derive from a self-consistent treatment of the on-site Coulomb interaction between the $f$-electrons and are not a simple translation downward by $U$ of the constrained LDA results.

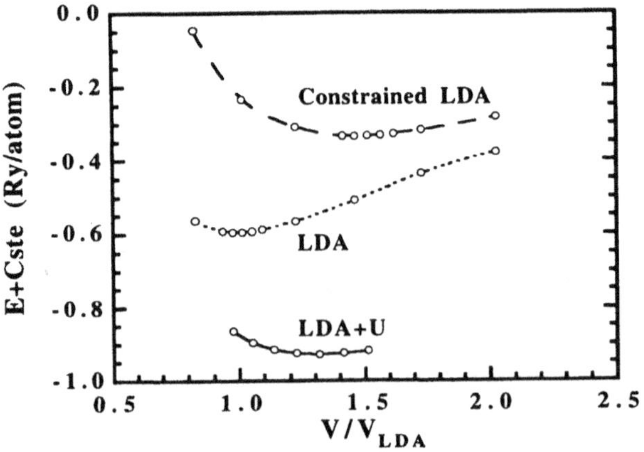


Figure 3. Total energy vs. normalized volume for fcc-based Pu calculated within the LDA (dotted line), the constrained LDA described in the text (dashed line), and the LDA+U scheme with varying $U$ as explained in the text (solid line).

Finally, we report on calculations of the energetics of stoichiometric Pu-Ga alloys. To estimate the alloy mixing energies, we consider a hypothetical fcc-based Ga solid. Here, LDA calculations lead to an equilibrium lattice constant of 7.831 a.u. that is consistent with the equivalent lattice constants of 8.086 and 7. 805 a.u. one derives, respectively, from the two observed complex structures oC8 and tI2 (in Pearson notation) of Ga. In addition, the calculated bulk modulus of 58.2 GPa is consistent with the experimental fact that Ga is extremely soft with a melting point close to room temperature.

Calculated energies of mixing of Pu-Ga alloys for the three ordered structures, namely $L1_2$ of $Cu_3Au$ type for $Pu_3Ga$ and $PuGa_3$ and $L1_0$ of CuAu-I type for PuGa, are shown in Fig. 4. Here, the dotted curve corresponds to an LDA calculation, the dashed curve to the constrained LDA, and the solid curve to LDA+U. We note that the LDA results fail to explain the definite tendency toward phase formation and chemical order exhibited by this alloy system. On the contrary this tendency is clearly reflected in both the constrained LDA and the LDA+U results. These results will be used as a basis for further analysis of the stability and chemical ordering trends in Pu-Ga alloys.

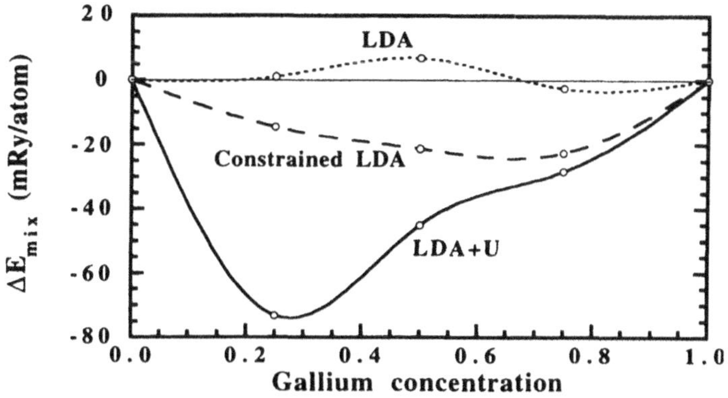


Figure 4. Energies of mixing of Pu-Ga alloys vs. Ga concentration at zero temperature within the LDA (dotted line), the constrained LDA described in the text (dashed line), and the LDA+U scheme with varying *U* as explained in the text (solid line).

## CONCLUSIONS

The work briefly reported here represents an attempt to understand at the fundamental level the electronic origins of the equilibrium and phase stability properties of Pu and Pu-Ga alloys. As expected, the results clearly indicate that for systems in which electron-electron correlations are known to be important, the LDA fails to provide an accurate description of the electronic structure and related properties. Some improvement can derive from constrained LDA calculations and from the application of the LDA+U method. However, the former of these leads to an unacceptably large ground-state energy, while the phenomenological nature of the second precludes a fundamental understanding at the *ab initio* level. In addition, the LDA+U also leads to values of the linear coefficient of the electronic specific heat that are far from experimental values (although the calculations neglect such physical effects as electron-phonon coupling which may be playing an important role in these systems). Also, effects associated with possible magnetic behavior are neglected[23]. At the same time, the LDA+U calculations lend support to the long-held view that the properties of Pu and its alloys are closely tied to the behavior of its *f*-electrons. As already alluded to above, they further increase the motivation to search for a predictive theory of correlated-electron systems that properly describes not only spectroscopic properties but also energetics at equilibrium and phase stability properties.

## ACKNOWLEDGMENTS

This work has been performed under the auspices of the U. S. Department of Energy by the Lawrence Livermore National Laboratory under contract No. W-7405-ENG-48. The work at California State University was supported through the National Science Foundation under Grant No. DMR-9531005, the Materials Research Institute at LLNL, and the Parsons Foundation. The work at West Virginia University was supported by the U. S. Department of Energy under Grant No. DE-FG07-97ER-45671.

## REFERENCES

1. F. L. Oetting, M. H. Rand, and R. J. Ackermann, *The Chemical Thermodynamics of Actinide Elements and Compounds, Part 1: The Actinide Elements*, F. L. Oetting, V. Medvedev, M. H. Rand, and E. F. Westrum, Jr., eds., International Atomic Energy Agency, Vienna (1976), and references therein.
2. F. H. Ellinger, Crystal structure of delta-prime and the thermal expansion characteristics of delta, delta-prime, and epsilon plutonium, *Trans. AIME* 206, 1256 (1956).
3. W. H. Zachariasen and F. H. Ellinger, The crystal structure of alpha plutonium metal, *Acta Cryst.* 16, 777 (1963).
4. Per Söderlind, O. Eriksson, B. Johansson, and J. M. Wills, Electronic properties of f-electron metals using the generalized gradient approximation, *Phys. Rev. B* 50, 7291 (1994), and references therein.
5. Olle Eriksson, L. E. Cox, B. R. Cooper, J. M. Wills, G. W. Fernando, Y. G. Hao, and A. M. Boring, Possibility of a δ-like surface for α-Pu: Theory, *Phys. Rev. B* 46, 13576 (1992).
6. M. Pénicaud, Electron localization in the series of actinide metals. The cases of δ-Pu and Es, *J. Phys. Condens. Matter* 9, 6341 (1997).
7. J. van Ek, P. A. Sterne, and A. Gonis, Phase Stability of Plutonium, *Phys. Rev. B* 48, 16280 (1993).
8. P. Weinberger, A. M. Boring, and J. L. Smith, Electronic structure of d-plutonium and of single Al, Ga, and Sc impurities in δ-plutonium, *Phys. Rev. B* 31, 1964 (1985).
9. P. Weinberger, A. Gonis, A. J. Freeman, and A. M. Boring, Electronic structure of Pu-rich $Pu_xAl_{1-x}$ alloys, *Phys. Rev. B* 31, 1971 (1985).
10. J. D. Becker, J. M. Wills, L. Cox, and B. R. Cooper, Electronic structure of Pu compounds with group-IIIB metals: Two regimes of behavior, *Phys. Rev. B* 54, R17265 (1996).
11. J. D. Becker, B. R. Cooper, J. M. Wills, and L. Cox, Calculated lattice relaxation in Pu-Ga alloys, *J. Alloys Comp.* 271-273, 367 (1998).
12. J. D. Becker, J. M. Wills, L. Cox, and B. R. Cooper, Calculated lattice relaxation in Pu-Ga, *Phys. Rev. B* 58, 5143 (1998).
13. F. H. Ellinger, C. C. Land, and V. O. Struebing, The plutonium-gallium system, *J. Nucl. Mat.* 12, 226 (1964).
14. *Phase Diagrams of Binary Actinide Alloys*, M. E. Kassner and D. E. Peterson, eds., ASM International, Materials Park, OH (1995).
15. D. L. Price and B. R. Cooper, Total energies and bonding for crystallographic structures in titanium-carbides and tungsten-carbides, *Phys. Rev. B* 39, 4945 (1989).
16. D. M. Ceperley and B. J. Alder, Ground-state of the electron gas by a stochastic method, *Phys. Rev. Lett.* 45, 566 (1980).
17. J. P. Perdew and A. Zunger, Self-interaction correction to density-functional approximations for many-electron systems, *Phys. Rev. B* 23, 5048 (1981).
18. D. J. Chadi and M. L. Cohen, Special points in the Brillouin zone, *Phys. Rev. B* 8, 5747 (1973).
19. I. V. Solovyev, P. H. Dederichs, and V. I. Anisimov, Corrected atomic limit in the local-density approximation and the electronic structure of *d* impurities in Rb, *Phys. Rev. B* 50, 16861 (1994), and references therein.

20. B. R. Cooper, S.-P. Lim, I. Avgin, Q. G. Sheng, and D. L. Price, Trade-off between increased size and increased broadening of magneto-optic effects in cerium and uranium systems, *J. Phys. Chem. Sol.* 56, 229 (1995).
21. D. L. Price, B. R. Cooper, S.-P. Lim, and I. Avgin, A variation of the LDA+U method appropriate to *f*-state localization: Application to magneto-optical properties, to be published in *Phys. Rev. B* (1999).
22. N. Kioussis, B. R. Cooper, and J. M. Wills, Magnetic instability with increasing hybridization in cerium compounds, *Phys. Rev. B* 44, 10003 (1991).
23. S. Méot-Reymond and J. M. Fournier, Localization of 5f electrons in δ-plutonium: Evidence for the Kondo effect, *J. Alloys Comp.* 232, 119 (1996).

# General discussion I

The moderator of the first part of the discussion was Borje Johansson.

**Johansson:** Let us start with you.

**Levy:** I will show some numbers for the He atom just to illustrate the discussion.These numbers come from Tulane, from the work that John Perdew and I did. The Hamiltonian is

$$H_\alpha = T + \alpha V_{ee} + \Sigma\, V_\alpha(\mathbf{r}_i).$$

The $\psi_\alpha$ yields the ground-state density $\rho_{GS}$ and minimizes $< T + \alpha V_{ee} >$. The table below shows the variation of $< \psi_\alpha \mid V_{ee} \mid \psi_\alpha >$ in Hartrees with $\alpha$.

| $\alpha$ | $< \psi_\alpha \mid V_{ee} \mid \psi_\alpha >$ |
|---|---|
| 0 | 1.0 |
| 1 | 0.9 |
| $\infty$ | 0.5 |

This is the Hamiltonian which gives the He density with $\alpha = 1$. At $\alpha = 0$, we have the KS Hamiltonian. At $\alpha = 1$ you have the true interacting potential. One changes $\alpha$ but adjsusting $V_\alpha$ so that the groun-state density remains the same. As one turns on $\alpha$, keeping the density fixed, one can examine the change of the electron-electron repulsion. The expectation value of the electron repulsion goes down as $\alpha$ increases. In order for the expectation value to reach a minimum, the repulsion term must decrease. The parameter $\alpha$ provides a measure of how strongly correlated the system is.

**Sawatzky:** It's not clear to me that this measures the degree of correlation, the correlation part of the energy, since the density is changing.

**Levy:** The density remains fixed. As I change the coupling constant from zero, the attractive potential changes so that the density remain fixed.

**Sawatzky**: My problem was that there are many ways for getting the density and the energy. And if you can get them with a Slater Determinant of restricted single-particle wave functions then you don't know if it's the right one.

**Levy**: This wavefunction has no correlations, which corresponds to the $\alpha = 0$ case.

**Sawatzky:** Can I do this for any system or for only a two-electron system?

**Levy:** For all systems.

**Sawatzky**: Is the $\psi_\alpha$ that you are using the true wave function?

**Levy:** It's the true many-body wave function.

**Sawatzky:** But I don't know the true wave function for the solid.

**Zieshe**: As $\alpha$ approaches infinity, it implies a Wigner-like state. Which quantity indicates this?

**Levy:** This is very interesting; we get it here with a $\varphi$ that minimizes $T + V_{ee}$. Is minimizing the repuslion while giving the right $\rho$ enouph? We model this to infinity but only integrate to one to get the $\varphi$. Consider the quantity $\langle 1/\alpha\ T + V_{ee} \rangle$ in terms of operators. The question is as $\alpha$ goes to infinity, is the minimum of this the same as the minimum of $V_{ee}$?

**Kohn**: We are off the subject. The only interest is $\alpha$ going from zero to one.

**Perdew**: There is the question, does DFT treat correlations? First of all, this $\varphi$ is single Slater determinant used to constuct the density as the expectation value of the density operator, but not the energy. These functional s are constructed nonempirically using what we know exactly of DFT and using the density functionals for exchange and for correlation. There is definitely correlation input in DFT. How about the output? Here are some typical results for atoms. The exchange energies are given within LSD with about a 10% error, GGA typically brings this down to one percent, correlation energies for atoms are overestimated by about a factor of two in the LDA, and the GGA brings this down to a level of about 5 %.

We have a number of model systems that are exactly soluble, such as the Hook's atom. This consists of two electrons with a Coulomb repulsion bound together by a spring constant force. This is a problem that can be solved analytically for many different values of the spring constant. You can actually write down analytically the GS wave function and some of the excited state wave functions. And this model can be used to test DFT functionals and give the right exchange and correlation energies for a wide range of spring constants. Even for very small spring constants where the density is very low but the correlation is very strong. I don't think I'll use the terms strong and weak correlation because they get me into trouble, so Ill use heavy correlation and light correlation. Qualitatively, I would say that we can compare variations of $dE_c/dE_x$; small means moderate or light, large means heavy correlation. I suggest we calculate these exactly and not within DFT because in the case of strong correlation DFT may not give a reliable estimate of the correlation energy. In the case of moderate or heavy correlation where this ratio becomes comparable to one, LSDA and GGA may give some errors including the possibility of false spin moments. I gave you before the example of the stretched $H_2$ molecule. The same effect occurs with the $Cr_2$ molecule at equilibrium bond lengths. The Cr molecule is very hard to get right, you can work very hard to get the wrong answer for the ionization energy. You can get the right atomization energy and the correct bond length with DFT, but what DFT can give wrong is the spin moment, it breaks the symmetry to put a localized up spin density on one atom and a down spin density on the other. I think the same effect occurs in solid transition metal oxides where the spin moments are not accurately predicetd. These criteria are not within DFT but to find the limits of these approximations we have to go outside them.

**Fulde:** There is no dought that DFT and LDA in contain correlations. My understanding is that DFT does not want to calculate the many-body wave function but properties. Slater determinants are not many-body wave functions. If we make statements what percentage of correlations are contained within DFT and LDA one has to be careful. We have to distinguish between interatomic correlations and intra-atomic correlations. The strong correlated systems are those with interatomic correlations that tend to reduce charge fluctuations between different atoms. Even though the majority of correlations come from the intra-atomic region, it is important to get correctly the bit between atoms in order to get properties right. So one may ask the question how can one use the nice features of DFT and LDA in order to take advantage of them in dealing with strong correlation. In my opinion, there are two possibilities. Speaking of low-energy excitations where you can formulate the problem in terms of phase shifts except for the strongly correlated channels where you use the LDA itself. Within this scheme you can calculate Fermi surfaces and anisotropies which are huge. For high-energy excitations one way to improve the LDA is to use the LDA+U within which one shifts the orrelated orbitals below the Fermi energy by

$U_{eff}/2$ down and those above the Fermi energy by $U_{eff}/2$ up. With this model we cannot hope to get the low-energy excitations right but we could get those of high energy. One can do a simple model in which one considers two orbitals, an f orital and a ligant orbital, where two electrons interact with a correlation energy U on the f orbital but do not interact in the ligant orbital because they hardly see each other. You put two electrons into those orbitals. Then assume a small hopping matrix element between the two orbitals. The first electron goes to the f orbital but the second cannot go there because of the U so it goes to the ligand orbital. This is a simple model, one you can solve by hand. And you can also do it within LSDA. This is very instructive. You can see what some of the problems are with the LDA.

**Sawatzky:** I also agree that the DFT has correlations in it in the ground-state energy and also in the ground-state density. Your definition of strong and weak correlation for the molecule, I agree with what Fulde said. In the solid, we are worried about the interatomic

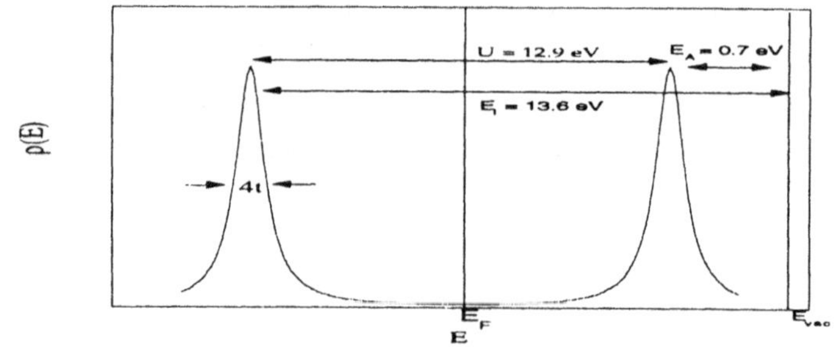


kinetic energy and the enrgy it costs to take an electron from here and put it in another place, that energy is U and that determines whether you have a strong or weak correlation. So it depends how large is this ionization potential minus the electron affinity. Strong is the picture shown below. Weak is when these peaks almost overlap. So, I think it is also a matter of terminology. You define strong and weak by comparing exchange and correlation. But if we want to know about properties, we have to know a bit more than the density, we have to know how the density has been composed. So we need to know more about the wave function so that we can see what the low-energy excitations might possibly be.

Can we divide materials into strongly, moderately and weakly correlated systems? The real questions one wants to answer at least in solids relates to the physical properties which are determined by the low energy scale excitations (i.e., of order kT). In normalization group-like methods we start from high energy scale interactions like U which may be about 10 eV in rare earths or 5 eV in 3d transition metals, and then try to find the influence of these on the low energy scale. For example, go from an Anderson impurity Hamiltonian with energy scales up to 10 eV, via a Schrieffer-Wolff transformation to a Kondo Hamiltonian, with an energy scale of 0.1 eV, and then from this to the low energy scale properties which may even in the end be describable in terms of a single free particle model, all be it that the particle may have a very large mass (heavy Fermions), or perhaps even exotic as in for example the High Tc's or 1D Hubbard systems where some talk in terms of new particles involving spin -charge separation. In the end we always try to get to a description of the low energy scale in terms of some kind of effective single particle theory and the effects of electron correlation are then incorporated in the nature of these new quasi particles. Of course these new particles need no longer be Fermions with a mass of me, a charge e, and a spin 1/2 as one uses in DFT-like calculations of solids in which one is working mostly with a Hamiltonian describing electrons. So in the use of DFT and band structure methods, the definition of correlation is more in terms of the bare interactions rather than the dressed interactions in which case it may be possible to define strong moderate and weak correlation. However, then I would like to ask the question how one can within DFT tell? After all the only exact results are the ground state energy and density and both of these can and usually are determined using a single Slater determinant of symmetry restricted one electron wave functions which in them selves contain no correlation what so ever. So in DFT theory, it seems to me that one cannot really tell whether the found energy and density is strongly determined by correlation since I can construct these with a wave function containing no correlation. Of course in correlated systems, it is obvious that this wave function is not the correct one but it does yield in principle the correct energy and charge density (but as discussed above, not the correct spin density).

**Kohn**: Is there a well-defined procedure which in principle could determine that U?

**Fulde:** I see the problem from a rather pragmatic standpoint. The correlations that can be treated in LDA are the RPA type of correlations. To treat strong correlations we can try the LDA +U. The U can suppress charge fluctuations one needs, and the U has to be put in by hand so to speak. So the scheme will be particlarly useful if the properties one calculates are not very sensitive to the size of U. So it is a pragmatic way to include correlations that have been missed in the LDA.

**Sawatzky**: The LDA+U is put in for specific orbital. This allows you to open up a gap where it is needed but it is a tricky business.

**Papaconstantopoulos:** I want to comment on what Fulde said on the use of phase shifts to parametrtize the band structure. This was done several years ago where experimental Fermi surface results were used to get phase shifts that were then used to construct the band structure. You were suggesting something like that but adjusting only the phase shifts that have to do with correlations.

**Fulde**: What I meant to say is to calculate the phase shifts from the LDA except for this one orbital, the 4f orbital, where correlations are strong and something is missing from the LDA. And for that orbital one adjusts the slope of the phase shift, which is what enters the determination of effective masses, to yield the gamma values.

**Perdew**: I just want to say that from my point of view the Hubbard U or SIC or the corrections the KS band gap are all the same thing at an underlying level. The derivative of the discontinuity of the band gap is quite large in insulators. In a system like solid neon it is very clear that what we are looking at is on-site interactions. In fact we were led to the discovery of the discontinuity by examination of the self-interaction correction. At least from my point of view one way of looking at complex systems, the physics is always the same.

**Kohn:** I just have to add that I have shown in a paper that any kind of gradient correction fails to correct this discontinuity. The difficulty is that any exchange correlation functional is continuous as you go from just below a full band to just above. The density is also continuous and you need a discontinuity to fix this problem.

**Gross**: I would like to ask the people who have discussed this band gap problem. Do you think it is impossible to get the right gap with one and the same local potential applied to both the highest orbital valence band and the lowest conduction band? I remember a statement in a book by Mott where he says that if one succeds in getting the right gap from the same local potential this must be an accident.

**Levy:** First, that work did not have the exact correlation potential. So there is something missing there. You said if it happens in one case it must be an accident. What if it happens in a thousand cases?

**Sawatzky**: There is a paper in which the authors solved the Hubbard model exactly for a finite system and then developed the exact DFT functionals and the LDA ones. And they found that indeed there is a discontinuity in the exact functional that is in fact approximately equal to U, or the difference in treating the ionization potential and the electron affinity as one suspects it should be.

# General discussion II

This part of the discussion was chaired by Angus Lawson.

**Lawson**: I think a good way to begin is to consider some of the questions that were raised yesterday afternoon. How many would say that the two camps (if we can think of them that way) are moving closer (the two camps referred to here are that of band theory and of many-body theory).

A show of hands indicates consensous that the camps are moving closer together.

**Kohn**: DFT came out of many body theory. There are maybe many-body theorists who do not think about DFT but those DFT theorists always think about many-body theory.

**Lawson:** Can we say that there is a convergence between the two camps?

**Sawatzky**: I don't know about a convergence, but there are many-body people who are starting to take a closer look at real systems. And in applying their models they take parameters that come out of DFT and LDA calculations.

**Lawson:** Anybody else wants to comment?

**Louie:** I would like to add that KS wave functions and eigenvalues are a good starting point for doing the next level many-body theory. So there is a merging of the two fields.

**Lawson:** As an experimentalist, I would like to know what people think are the prospects for doing realistic calculations for real materials with all the difficulties that implies.

**Cooper**: I would say that in regard to spectral properties we have a long way to go. Regarding integrated spectral properties we are there. There are some properties that are between the two.

**Sawatzky**: With respect to certain properties such as lattice defects, calculations give reasonable answers. But we can take some other systems such as the high-T superconductors, where at low T there is a transition from an antiferromagnetic insulating phase to a spin-glass phase to a superconducting phase as you change the oxygen dopping of the system. To go from the antiferromagnetic insulator to the superconducting state takes only a dopping of few percent. To understand this is the real challenge.

**Cooper**: I would disagree with you very strongly. That kind of problem we have addressed in the past quite successfully. We have spent a great deal of effort in understanding the physics of strongly correlated systems. I view as a challenge the calculation of excitation spectra such as gaps or optical spectra.

**Papaconstantopoulos**: I suppose the question is whether we have made any progress in making a contact with experiment in the investigation of real systems. Regarding this point I would say that the LDA has made better contact with experiment (than many-body theory). I would also say that TB is the common denominator between LDA and many-body theory. I also think that TB has a better chance to go towards real systems, large systems with hundreds of atoms than does LDA.

**Lawson:** I would now like to ask a different question. Are there any questions of principle that have not been properly addressed experimentally?

**Johansson:** What I would like to see addressed experimentally is possibly high-pressure properties. For example the behavior of Americium. Also the lighter rare earths could be put under high pressure and see if they have similar behavior.

**DeLong:** I would like to add to the last remark. The inclusion of local orbital degeneracy in many body phenomenological models is needed in order to make comparison with experiment. Some of these models can be checked with experiments on the rare earths, such as Samarium.

**Lawson**: We can now ask where we are going from here, what are the new directions, and what we will be able to calculate about ten years from now?

**Kohn:** The many-body wavefunction is extremely difficult to calculate. I'd like to express my view for two methods for practical use. Interest has developed in order N methods. These methods rely on the point that to understand what happens in a small region you need often know what goes on in a small neighborhood of that region.

**Fulde:** First I would like to reply to what Walter said about the wavefunction. Why are some people interested in the wavefunctions. For example if we ask what is the wavefunction for a solid in the independent electron approximation, we don't need to write it down for many years. For Si one can write the wavefunction very simply. Why are some people interested in the wavefunction? Precisely because we want to know what the effects of correlations are. In chemistry, for example, we have Mullikan on the one side and Pauli on the other. Mulliken started from the completely independent electron picture, Pauli from the valence-bond picture, and for some people they are interested to know how close to either picture a particular situation is. This is something that you can learn if you write down an improved expression for the wavefunction.

The second point is about order N methods. For the systems one has to develop these methods. But also it is not hopeless for the many-body theorist particularly if one starts to work with embedding schemes.

**Lawson:** Would you please tell us what you will be doing in the near future?

**Ziesche:** I enjoyed the remarks on the wave function. I would like to say that the pair distribution function is also important because it includes fluctuations. I think that this should be pursued in the future.

**Lawson:** Would you comment on what you expect to be doing in the near future?

**Perdew:** There are several interesting directions in which progress may be expected. One is the approach that Paul (Ziesche) and Tony (Gonis) are suggesting, to calculate the correlation function directly.

## List of Participants

Dr. Aloysius J. Arko
Los Alamos National Laboratory
MS K764
Los Alamos, NM 87545

Dr. Nick Bernhoeft
Institute Langevin
ILL4
B.P. 156 X
F-38042 Grenoble
France

Prof. David Ceperley
Department of Physics
University of Illinois Urbana
1110 W. Green Street
Urbana, IL 61801-3080

Prof. B. Coqblin
Laboratoire de Physique des Solides
Universites Paris-Sud
Batiment 510
91405 Orsay
France

Prof. Bernard. R. Cooper
Department of Physics
West Virginia University
Morgantown, WV 26506

Prof. L. D. De Long
Department of Phyaics and
Astronomy
University of Kentucky
Lexington, KY 40506-0055

Dr. Martin Divis
Department of Metal Physics
Charles University
Ke Karlovu 5
121 16 Praha 2
Czech Republic

Dr. Vaklav Drchal
Institute of Physics
Czech Academy of Sciences
Na Slovance 2
CZ- 180 40 Prague 8
Czech Republic

Prof. Adolfo Eguiluz
Department of Physics
University of Tennessee
Knoxville, TN 37996-1200

Prof. C. Y. Fong
Department of Physics
University of California
Davis, CA 95616-8677

Dr. Peter Fulde
Max-Planck-Intitute fuer Physik
Komplexer Systeme
Bayreuther Strasse 40
D-01187 Dresden
Federal Republic of Germany

Dr. Antonios Gonis
Chemistry and Materials
Science Division
P.O. Box 808, L-353
Lawrence Livermore
National Laboratory
Livermore, CA 94550

Dr. Andrew Görling
Lehrstuhl für Theoretische Chemie
Technische Universität München
D-8547 Garling, Germany

Prof. E. K. U. Gross
Institute of Theoretical Physics
Universisty Wirzburg
Am Hubland
Wirzburg D-97074
Germany

Dr. Ladislav Havela
Department of Metal Physics
Charles University
Ke Karlovu 5
121 16 Praha 2
Czech Republic

Dr. Vaclav Janiš
Institute of Physics
Czech Academy of Sciences
Na Slovance 2
CZ- 180 40 Prague 8
Czech Republic

Prof. Borje Johansson
Department of Physics
University of Uppsala
Box 530, S-751 21 Uppsala
Sweden

Prof. Nicholas Kioussis
Department of Physics
California State University Northridge
18111 Nordhoff Street
Northridge, CA 91330-8268

Prof. Walter Kohn
Department of Physics
University of California Santa
Barbara
Santa Barbara, CA 93106-9530
USA

Prof. Joseph Krieger
Physics Department
Brooklyn College
3900 Bedford Ave.
Brooklyn, NY 11210

Dr. Stefan Kurth
Department of Physics
Tulane University
New Orleans, LA 70118

Dr. Gerald H. Lander
EITU Tech Univ-EURATOM
Postfach 2340
D-76125
Karlsruhe, Germany

Dr. Michael Lang
Institut für Festkörperphysik
Technical University of Darmstadt
Technische Hochschule Darmstadt
Hochschulstrasse 8
D-64289, Darmstadt
Germany

Dr. Andrew C. Lawson
MLNSC H805
Los Alamos National Laboratory
PO Box 1663
Los Alamos NM, 87545

Prof. M. Levy
Department of Chemistry
Tulane University
New Orleans, LA 70118

Prof. Steven Louie
Department of Physics
366 LeConte Hall #7300
University of California Berkeley
Berkeley, CA 94720
USA

Prof. Norman March
Inorganic Chemistry Department
South Parks Road
Oxford, OX1 3QR
England

Prof. Ágnew Nagy
Institute of Theoretical Physics
Lajos Kassuth University
Delveceh, 4010
Hungary

Dr. J. Ortega
Depto de Fisica
UniversidadAutonoma
28049 Madrid
Spain

Prof. Hans R. Ott
Laboratorium fuer Festkoerperphysik
Eidgenossische Technische
Hochschule (ETH) Zurich
CH-8093, Zurich
Switzerland

Dr. Dimitri A. Papaconstantopoulos
Metal Physics - Code 6690
Naval Research Laboratory
Washington, DC 20375
USA

Prof. John P. Perdew
Department of Physics
Tulane University
New Orleans, LA 70118

Dr. G. A. Sawatzky
Laboratory of Solid State Physics,
Materials Science Center
University of Groningen
Nijenborgh 16, 9747 AG Groningen
Netherlands

Professor R. T. Scalettar
Department of Physics
University of California Davis
Davis, CA 95616-8677

Dr. Vladimir Sechovský
Department of Metal Physics
Charles University
Ke Karlovu 5
12116 Prague 2
Czech Republic

Prof. Jorge Seminario
Department of Chemistry and
Biochemistry
University of South Carolina
Columbia, South Caroline 29208

Dr. L. Soderholm
Division of Chemistry
Argonne National Laboratory
9700 S. Cass Avenue
Argonne, IL 60439

Dr. P. E. A. Turchi
Chemistry and Materials Science
PO Box 808, L-268
Lawrence Livermore National
Laboratory
Livermore, CA 94550

Prof. P. Ziesche
Max Planck Institute for
Physics of Complex Systems
Bayreuther Str. 40, Haus 16
D-01187 Dresden
Germany

Dr. Aleš Zupan
Institut Josef Stefan
Jamova 39
Ljubljana, Slovenia

Prof. A. Zdetsis
Physics Department
Faculty of Science
University of Patras
GR 261 10 Patra
Greece

## INDEX

GPSR Compliance

*The European Union's (EU) General Product Safety Regulation (GPSR) is a set of rules that requires consumer products to be safe and our obligations to ensure this.*

*If you have any concerns about our products, you can contact us on ProductSafety@springernature.com*

In case Publisher is established outside the EU, the EU authorized representative is:

Springer Nature Customer Service Center GmbH
Europaplatz 3
69115 Heidelberg, Germany

**Batch number: 10755801**

Printed by Printforce, the Netherlands